GAMMA-RAY BURSTS
IN THE SWIFT ERA

Previous Proceedings in the Series of Annual October Astrophysics Conferences in Maryland

	Year	Title	Publisher	ISBN
16th	2005	Gamma-Ray Bursts in the Swift Era	AIP Conference Proceedings 836	0-7354-0326-0
15th	2004	New Windows on Star Formation in the Cosmos	*"unpublished"*	
14th	2003	The Search for Other Worlds	AIP Conference Proceedings 713	0-7354-0190-X
13th	2002	The Emergence of Cosmic Structure	AIP Conference Proceedings 666	0-7354-0128-4
12th	2001	Two Years of Science with Chandra	*"unpublished"*	
11th	2000	Young Supernova Remnants	AIP Conference Proceedings 565	0-7354-0001-6
10th	1999	Cosmic Explosions	AIP Conference Proceedings 522	1-56396-943-2
9th	1998	After the Dark Ages: When Galaxies were Young	AIP Conference Proceedings 470	1-56396-855-X
8th	1997	Accretion Processes in Astrophysical Systems: Some Like It Hot	AIP Conference Proceedings 431	1-56396-767-7
7th	1996	Star Formations, Near and Far	AIP Conference Proceedings 393	1-56396-678-6
6th	1995	Cosmic Abundances	Astronomical Society of the Pacific	1-886733-20-1
5th	1994	Dark Matter	AIP Conference Proceedings 336	1-56396-438-4
4th	1993	The Evolution of X-Ray Binaries	AIP Conference Proceedings 308	1-56396-329-9
3rd	1992	Back to the Galaxy	AIP Conference Proceedings 278	1-56396-227-6
2nd	1991	Testing the AGN Paradigm	AIP Conference Proceedings 254	1-56396-009-5
1st	1990	After the First Three Minutes	AIP Conference Proceedings 222	0-88318-828-7

To learn more about these titles, or the AIP Conference Proceedings Series, please visit the webpage
http://proceedings/aip.org/proceedings

GAMMA-RAY BURSTS IN THE SWIFT ERA

Sixteenth Maryland Astrophysics Conference

Washington, DC   29 November – 2 December 2005

EDITORS

Stephen S. Holt
Franklin W. Olin College of Engineering
Needham, Massachusetts

Neil Gehrels
NASA/Goddard Space Flight Center
Greenbelt, Maryland

John A. Nousek
Pennsylvania State University
University Park, Pennsylvania

SPONSORING ORGANIZATIONS
NASA/Goddard Space Flight Center
University of Maryland
General Dynamics

Melville, New York, 2006
AIP CONFERENCE PROCEEDINGS ■ VOLUME 836

Editors:
Stephen S. Holt
Franklin W. Olin College of Engineering
Olin Way
Needham, MA 02492
U.S.A.

E-mail: steve.holt@olin.edu

Neil Gehrels
Code 661
NASA/Goddard Space Flight Center
Greenbelt, MD 20771
U.S.A.

E-mail: gehrels@milkyway.gsfc.nasa.gov

John A. Nousek
525 Davey Lab
Pennsylvania State University
University Park, PA 16802
U.S.A.

E-mail: nousek@astro.psu.edu

The articles on pp. 14 – 22, 97 – 102, 309 – 312, and 660 – 663, were authored by U.S. Government employees and are not covered by the below mentioned copyright.

Authorization to photocopy items for internal or personal use, beyond the free copying permitted under the 1978 U.S. Copyright Law (see statement below), is granted by the American Institute of Physics for users registered with the Copyright Clearance Center (CCC) Transactional Reporting Service, provided that the base fee of $23.00 per copy is paid directly to CCC, 222 Rosewood Drive, Danvers, MA 01923, USA. For those organizations that have been granted a photocopy license by CCC, a separate system of payment has been arranged. The fee code for users of the Transactional Reporting Services is: 0-7354-0326-0/06/$23.00

© 2006 American Institute of Physics

Permission is granted to quote from the AIP Conference Proceedings with the customary acknowledgment of the source. Republication of an article or portions thereof (e.g., extensive excerpts, figures, tables, etc.) in original form or in translation, as well as other types of reuse (e.g., in course packs) require formal permission from AIP and may be subject to fees. As a courtesy, the author of the original proceedings article should be informed of any request for republication/reuse. Permission may be obtained online using Rightslink. Locate the article online at http://proceedings.aip.org, then simply click on the Rightslink icon/"Permission for Reuse" link found in the article abstract. You may also address requests to: AIP Office of Rights and Permissions, Suite 1NO1, 2 Huntington Quadrangle, Melville, NY 11747-4502, USA; Fax: 516-576-2450; Tel.: 516-576-2268; E-mail: rights@aip.org.

L.C. Catalog Card No. 2006924949
ISBN 0-7354-0326-0
ISSN 0094-243X

Printed in the United States of America

CONTENTS

Preface xvii

HISTORY AND INTRODUCTION

A History of Gamma Ray Bursts and Other Astronomical
Conundrums 3
    V. Trimble
The Swift Gamma-Ray Burst Mission: First Results 14
    N. Gehrels *(On behalf of the Swift Team)*

SHORT GRBs AND SGRs

Observations of the Short-Duration Gamma-Ray Bursts: Prompt
Emission 25
    K. Hurley
The Afterglows and Host Galaxies of Short GRBs: An Overview 33
    E. Berger
Are Short GRBs Really Hard? 43
    T. Sakamoto, L. Barbier, S. Barthelmy, J. Cummings, E. Fenimore,
    N. Gehrels, D. Hullinger, H. Krimm, C. Markwardt, D. Palmer, A. Parsons,
    G. Sato, J. Tueller, R. Aptekar, T. Cline, S. Golenetskii, E. Mazets,
    V. Pal'shin, G. Ricker, D. Lamb, J.-L. Atteia, N. Kawai, and the Swift-BAT,
    Konus-Wind, and HETE-2 Teams
Late Flares from GRBs—Clues about the Central Engine 48
    A. I. MacFadyen, E. Ramirez-Ruiz, and W. Zhang
The Short/Hard GRB 050709 and Its Star-Forming Host Galaxy 54
    S. Covino, D. Malesani, G. L. Israel, P. D'Avanzo, L. A. Antonelli,
    G. Chincarini, D. Fugazza, M. L. Conciatore, M. Della Valle, F. Fiore,
    D. Guetta, K. Hurley, D. Lazzati, L. Stella, G. Tagliaferri, M. Vietri, and
    S. Campana
The Rate and Luminosity Function of Short GRBs 58
    T. Piran and D. Guetta
A Magnetar Flare in the *BATSE* Catalog? 64
    A. Crider
Short GRB Progenitors: Population Synthesis Predictions vs.
Observations 68
    T. Bulik, K. Belczyński, and B. Rudak
External Shock Model for the Radio Afterglows of Giant Flares from
Soft γ–Ray Repeaters 72
    X. Y. Wang, X. F. Wu, Y. Z. Fan, Z. G. Dai, and B. Zhang

v

PROMPT EMISSION

The Latest Two GRB Detected by *Hete-2*: GRB 051022 and GRB 051028..79
    A. J. Castro-Tirado, S. McBreen, M. Jelínek, S. B. Pandey, M. Bremer,
    A. de Ugarte Postigo, J. Gorosabel, S. Guziy, G. Bihain, J. A. Caballero,
    P. Ferrero, J. de Jong, K. Misra, and D. K. Sahu

The Swift Prompt Sample..85
    P. T. O'Brien, R. Willingale, J. Osborne, M. R. Goad, K. L. Page,
    A. P. Beardmore, O. Godet, D. N. Burrows, and N. Gehrels

Stochastic Wakefield Acceleration in Gamma-Ray Bursts..................91
    G. Barbiellini, F. Longo, N. Omodei, A. Celotti, and M. Tavani

External Shock Interactions of GRB Blast Waves and the Swift Observations..97
    C. D. Dermer

GRB 050315: A Step in the Proof of the Uniqueness of the Overall GRB Structure..103
    R. Ruffini, M. G. Bernardini, C. L. Bianco, P. Chardonnet, F. Fraschetti,
    R. Guida, and S.-S. Xue

Prompt γ-ray Emission Spectroscopy with *RHESSI*..................109
    M. E. Bandstra, S. E. Boggs, and E. Bellm

Constraints on TeV Emission from GRBs from the GeV Extragalactic Diffuse Gamma-Ray Flux..113
    S. Casanova and B. Dingus

Gamma-Ray Burst Jet Profiles and Their Signatures..................117
    C. Graziani, D. Q. Lamb, and T. Q. Donaghy

Are Long-Lag GRBs Different Than Other Long GRBs?..................121
    J. Hakkila, T. W. Giblin, C. D. Peters, S. M. Sonnett, and C. M. Nolan

Properties of the Intermediate Type of Gamma-Ray Bursts..................125
    I. Horváth, F. Ryde, L. G. Balázs, Z. Bagoly, and A. Mészáros

Statistical Analyses of RHESSI GRB Database..................129
    J. Řípa, R. Hudec, A. Mészáros, W. Hajdas, and C. Wigger

The Complete Spectral Catalog of Bright BATSE Gamma-Ray Bursts..................133
    Y. Kaneko, R. D. Preece, M. S. Briggs, W. S. Paciesas, C. A. Meegan, and
    D. L. Band

The "Supercritical Pile" Model of GRB: Spectra and Their Time Development..137
    D. Kazanas, A. Mastichiadis, and M. Georganopoulos

GRB 050717: A Long, Short-Lag Burst Observed by Swift and Konus..................141
    H. A. Krimm, C. Hurkett, V. Pal'shin, J. P. Norris, B. Zhang,
    S. D. Barthelmy, D. N. Burrows, N. Gehrels, S. Golenetskii, J. P. Osborne,
    A. M. Parsons, and M. Perri

Correlative Analysis of GRBs Detected by Swift, Konus and HETE..................145
    H. A. Krimm, S. D. Barthelmy, N. Gehrels, D. Hullinger, T. Sakamoto,
    T. Donaghy, D. Q. Lamb, V. Pal'shin, S. Golenetskii, G. R. Ricker,
    J.-L. Atteia, N. Kawai, and G. Sato

An Improved Redshift Indicator for Gamma-Ray Bursts, Based on the Prompt Emission 149
    A. Pélangeon, J.-L. Atteia, D. Q. Lamb, G. R. Ricker, and the HETE-2 Science Team

Numerical Simulation of 2-D Relativistic Hydrodynamics Using Adaptive Mesh Refinement Technique 153
    K. Kwak and F. D. Swesty

PIC Simulations of Prompt GRB Emissions 157
    E. Liang, K. Noguchi, and S. Sugiyama

Gamma-Ray Bursts and Giant Lighting Discharges in Protoplanetary Systems 161
    E. Winston, S. McBreen, B. McBreen, and L. Hanlon

Observations of Gamma-Ray Bursts with *INTEGRAL* 165
    S. McGlynn, S. McBreen, L. Hanlon, B. McBreen, S. Foley, J. French, G. Melady, A. von Kienlin, and R. Preece

A Key to the Spectral Variability of Prompt GRBs 169
    M. V. Medvedev

Temporal and Spectral Analyses of SGRs Observed by HETE-2 173
    Y. E. Nakagawa, A. Yoshida, M. Maetou, M. Suzuki, T. Tamagawa, T. Sakamoto, N. Kawai, Y. Shirasaki, K. Tanaka, M. Matsuoka, E. E. Fenimore, M. Galassi, J.-L. Atteia, K. Hurley, and G. R. Ricker

Properties of Multi Pulsed GRBs seen by *HETE-2* 177
    Y. E. Nakagawa, A. Yoshida, N. Ishikawa, K. Tanaka, T. Tamagawa, M. Suzuki, N. Kawai, M. Matsuoka, Y. Shirasaki, R. Vanderspek, E. E. Fenimore, M. Galassi, J.-L. Atteia, and G. R. Ricker

The Observable Effect of a Photospheric Component on GRB's Prompt Emission Spectrum: Peak Energy Clustering and Flat Spectra above the Thermal Peak 181
    A. Pe'er, P. Mészáros, and M. J. Rees

INTEGRAL Observations of the Nearby, Under-Luminous Gamma-Ray Burst GRB 031203 185
    C. R. Shrader

Mechanical Model for Relativistic Blast Waves and Stratified Fireballs 189
    Z. Uhm and A. M. Beloborodov

Neutrino-Cooled Accretion Disks around Spinning Black Holes 193
    W. Chen and A. M. Beloborodov

Compton Spectrum from Poynting Flux Accelerated e+e- Plasma 197
    S. Sugiyama, E. Liang, K. Noguchi, and H. Takabe

Suzaku Wide-Band All-Sky Monitor Observations of GRB Prompt Emissions 201
    K. Yamaoka, S. Sugita, M. Ohno, T. Takahashi, Y. Fukazawa, Y. Terada, Y. Endo, S. Hong, K. Abe, K. Onda, M. Tashiro, T. Enoto, R. Miyawaki, M. Kokubun, K. Makishima, G. Sato, K. Nakazawa, T. Takahashi, and the Suzaku HXD-II Team

Tail Emission of Prompt Gamma-Ray Burst Jets 205
    R. Yamazaki, K. Toma, K. Ioka, and T. Nakamura

The GRB Luminosity Function in the Internal Shock Model 209
    F. Daigne, R. Mochkovitch, and H. Zitouni

EARLY AFTERGLOW

The Swift XRT: Observations of Early X-Ray Afterglows 215
    D. N. Burrows, J. A. Kennea, J. A. Nousek, J. P. Osborne, P. T. O'Brien,
    G. Chincarini, G. Tagliaferri, P. Giommi, B. Zhang, and the XRT Team

Observations of Early Optical Afterglow 224
    P. W. A. Roming and K. O. Mason

Theories of Early Afterglow 234
    P. Mészáros

The Prompt and Early Afterglow X-Ray Spectra of Swift GRBs 244
    M. R. Goad, J. P. Osborne, A. P. Beardmore, O. Godet, and K. Page *(On behalf of the Swift XRT Team)*

Are XRFs Really Off-Jet GRBs? 253
    T. Q. Donaghy

X-Ray Line Emission in Swift XRT Spectra? 259
    N. Butler

Simulation Studies of Early Afterglows Observed with SWIFT 265
    K.-I. Nishikawa, P. Hardee, C. Hededal, C. Kouveliotou, G. J. Fishman, and
    Y. Mizuno

Observation of the Prompt and Early Afterglow of GRB 050904 by TAROT 271
    M. Boër, J. L. Atteia, Y. Damerdji, B. Gendre, A. Klotz, and G. Stratta

Rapidly Detecting Extincted Bursts with KAIT and PAIRITEL 277
    N. Butler, J. Bloom, A. Filippenko, W. Li, R. Foley, K. Alatalo,
    D. Kocevski, D. Perley, and D. Pooley

GRB 050421: A Possible Naked Burst with X-Ray Flares 281
    O. Godet, K. L. Page, J. P. Osborne, P. T. O'Brien, A. P. Beardmore,
    M. R. Goad, and the *Swift* Team

Swift and XMM Observations of the Dark GRB 050326 285
    A. Moretti, A. De Luca, D. Malesani, S. Campana, A. Tiengo,
    J. N. Reeves, M. Capalbi, G. Chincarini, S. Covino, G. Cusumano,
    P. Giommi, V. La Parola, V. Mangano, T. Mineo, M. Perri, P. Romano, and
    G. Tagliaferri

GRB 050117: Simultaneous Gamma-Ray and X-Ray Observations with the *Swift* Satellite 289
    J. E. Hill, D. C. Morris, T. Sakamoto, G. Sato, D. N. Burrows, L. Angelini,
    C. Pagani, A. Moretti, A. F. Abbey, S. Barthelmy, A. P. Beardmore,
    V. V. Biryukov, S. Campana, M. Capalbi, G. Cusumano, P. Giommi,
    M. A. Ibrahimov, J. A. Kennea, S. Kobayashi, K. Ioka, C. Markwardt,
    P. Mészáros, P. T. O'Brien, J. P. Osborne, A. S. Pozanenko, M. Perri,
    V. V. Rumyantsev, P. Schady, D. A. Sharapov, G. Tagliaferri, B. Zhang,
    G. Chincarini, N. Gehrels, A. Wells, and J. A. Nousek

Rapid GRB Afterglow Response with SARA 293
    K. V. Garimella, A. L. Homewood, D. H. Hartmann, C. Riddle, S. Fuller,
    A. Manning, T. McIntyre, and G. Henson

**Rapid Centroids and the Refined Position Accuracy of the *Swift*
Gamma-Ray Burst Catalogue** 297
    J. E. Hill, L. Angelini, A. Moretti, D. C. Morris, J. Racusin,
    D. N. Burrows, A. P. Beardmore, S. Campana, M. Capalbi, J. A. Kennea,
    J. P. Osborne, C. Pagani, G. Tagliaferri, G. Chincarini, N. Gehrels,
    A. Wells, and J. A. Nousek

Early Afterglow and Variability 301
    K. Ioka, S. Kobayashi, B. Zhang, T. Kenji, R. Yamazaki, and T. Nakamura

**Searching for Early Optical Transients of Gamma-Ray Bursts with
TAROT—Technical Status** 305
    A. Klotz, M. Boer, G. Buchholtz, Y. Damerdji, J. Eysseric, C. Pollas, and
    Y. Richaud

Rapid Identification of GRB Afterglows with Swift/UVOT 309
    F. E. Marshall and the Swift Team

**Can GRB Early X-Ray Afterglows be Explained by a Contribution
from the Reverse Shock?** 313
    F. Genet, F. Daigne, and R. Mochkovitch

The *Swift* X-Ray Flaring Afterglow of GRB 050607 317
    C. Pagani, D. C. Morris, S. Kobayashi, T. Sakamoto, A. D. Falcone,
    A. Moretti, K. Page, D. N. Burrows, D. Grupe, J. Racusin, J. A. Kennea,
    S. Campana, P. Romano, G. Tagliaferri, J. E. Hill, L. Angelini,
    S. Barthelmy, G. Chincarini, J. A. Nousek, and N. Gehrels

A Tale of Two Faint Bursts: GRB 050223 and GRB 050911 321
    K. L. Page, S. D. Barthelmy, A. P. Beardmore, D. N. Burrows,
    S. Campana, G. Chincarini, J. R. Cummings, G. Cusmano, N. Gehrels,
    P. Giommi, M. R. Goad, O. Godet, J. Graham, Y. Kaneko, J. A. Kennea,
    V. Mangano, C. B. Markwardt, P. T. O'Brien, J. P. Osborne, D. E. Reichart,
    E. Rol, T. Sakamoto, G. Tagliaferri, N. R. Tanvir, A. A. Wells, and
    B. Zhang

**GRB Follow-Up Observations with the Whipple Telescope and
VERITAS** 325
    D. Petry and the VERITAS Collaboration

**Colors and Absolute Magnitude of the Optical Afterglow of the Short
GRB050709** 329
    V. Simon, R. Hudec, and G. Pizzichini

**Near Infrared Monitoring of the Afterglow of the Very Bright Swift
Burst GRB 050525** 333
    G. Stratta, A. Klotz, J. L. Atteia, and M. Boer

**Shallow Decay of Early X-Ray Afterglows from Inhomogeneous
Gamma-Ray Burst Jets** 337
    K. Toma, K. Ioka, R. Yamazaki, and T. Nakamura

**Slow and Fast Components in X-Ray Light Curves of GRBs from
BeppoSAX WFC Archive** 341
    L. Vetere, E. Massaro, E. Costa, and P. Soffitta

**Automated Detection of Optical Counterparts to GRBs with
RAPTOR** 345
    P. R. Wozniak, W. T. Vestrand, S. Evans, R. White, and J. Wren

ROTSE-III Performance in the Swift Era 349
    S. A. Yost, E. S. Rykoff, F. Aharonian, C. W. Akerlof, M. C. B. Ashley,
    S. Barthelmy, N. Gehrels, E. Göğüs, T. Güver, D. Horns, Ü. Kiziloğlu,
    H. A. Krimm, T. A. McKay, N. Mirabal, M. Özell, A. Phillips,
    R. M. Quimby, G. Rowell, W. Rujopakarn, B. E. Schaefer, D. A. Smith,
    H. F. Swan, W. T. Vestrand, J. C. Wheeler, J. Wren, and F. Yuan

Bumps in the Optical Afterglow of GRB 030329: Refreshed Shocks and Evidence for Internal Shocks in GRBs? 353
    F. Gebet, F. Daigne, and R. Mochkovich

The Broadband Afterglow of GRB 030328 357
    E. Maiorano, N. Masetti, E. Palazzi, S. Savaglio, E. Rol, P. M. Vreeswijk,
    E. Pian, P. A. Price, B. A. Peterson, M. Jelínek, S. B. Pandey,
    M. I. Anderson, and A. A. Henden

Colors of Optical Afterglows of GRBs in Analysis of the Dust in their Environemnt 361
    V. Šimon, R. Hudec, and G. Pizzichini

LATE AFTERGLOWS & GRB/SN CONNECTION

Supernova and GRB Connection: Observations and Questions 367
    M. Della Valle

A Broadband Perspective on the GRB-SN Connection 380
    A. M. Soderberg

Late-Time X-Ray Flares during GRB Afterglows: Extended Internal Engine Activity 386
    A. D. Falcone, D. N. Burrows, P. Romano, S. Kobayashi, D. Lazzati,
    B. Zhang, S. Campana, G. Chincarini, G. Cusumano, N. Gehrels,
    P. Giommi, M. R. Goad, O. Godet, J. E. Hill, J. A. Kennea, P. Mészáros,
    D. Morris, J. A. Nousek, P. T. O'Brien, J. P. Osborne, C. Pagani, K. Page,
    G. Tagliaferri, and the Swift XRT Team

Physical Origin of X-Ray Flares Following GRBs 392
    B. Zhang

The Supernova Gamma-Ray Burst Connection 398
    S. E. Woosley and A. Heger

Evidence for Intrinsic Absorption in the Swift X-Ray Afterglows 408
    S. Campana, P. Romano, S. Covino, D. Lazzati, A. De Luca, G. Chincarini,
    A. Moretti, G. Tagliaferri, G. Cusumano, V. Mangano, V. La Parola,
    T. Mineo, P. Giommi, M. Perri, M. Capalbi, L. A. Antonelli,
    D. N. Burrows, J. E. Hill, J. L. Racusin, J. A. Kennea, D. C. Morris,
    C. Pagani, J. A. Nousek, J. P. Osborne, M. R. Goad, K. L. Page,
    A. P. Beardmore, O. Godet, P. T. O'Brien, A. A. Wells, L. Angelini, and
    N. Gehrels

Constraining the GRB Collimation with a Survey for Orphan Afterglows 414
    A. Rau, J. Greiner, and R. Schwarz

Long-Term Optical Monitoring of GRB 030329 420
    D. Bersier, K. Z. Stanek, P. M. Garnavich, T. Matheson, and P. Mazzali

XRF 011030: The Case of a Late X-Ray Flare Observed by
BeppoSAX 424
    A. Galli and L. Piro
A Catalog of X-Ray Afterglows Observed by BeppoSAX, XMM-
Newton and Chandra 428
    B. Gendre, A. Corsi, L. Piro, and M. De Pasquale
Swift Observations of GRB 050603 432
    D. Grupe, P. J. Brown, A. Retter, D. N. Burrows, J. A. Nousek,
    P. Mészáros, B. Zhang, and J. Cummings
Swift Panchromatic Observations of the Bright Gamma-Ray Burst
GRB 050525A 436
    S. T. Holland, A. J. Blustin, and the Swift Science Team
Correlation of Color Behaviour of Blazars and Optical Afterglows of
GRBs 440
    R. Hudec, R. Filgas, V. Simon, M. Basta, and M. Topinka
Identification of Two Categories of Optically Bright Gamma-Ray
Bursts and a Model-Independent Luminosity Indicator 444
    E. Liang and B. Zhang
Optically Selected GRB Afterglows 448
    F. Malacrino and J.-L. Atteia
Afterglow Calculation in the Blandford-Lyutikov Electromagnetic
Model of GRBs 452
    F. Genet, F. Daigne, and R. Mochkovitch
Interaction of Electromagnetic Dominated Outflow with
Inhomogeneous Ambient Medium 456
    K. Noguchi and E. Liang
Search for Correlations between BATSE Gamma-Ray Bursts and
Supernovae 460
    J. Polcar, M. Topinka, D. Nečas, F. Hroch, R. Hudec, V. Hudcová,
    G. Pizzichini, N. Masetti, and E. Palazzi
GRB Supernova Luminosities—Correcting for the Host Extinction 464
    A. Zeh, C. Riddle, S. Klose, D. A. Kann, and D. H. Hartmann
The Frontier of Darkness: The Cases of GRB 040223, GRB 040422,
GRB 040624 467
    P. D'Avanzo, P. Filliatre, P. Goldoni, L. A. Antonelli, S. Campana,
    G. Chincarini, S. Covino, A. Cucchiara, M. Della Valle, A. De Luca,
    S. Foley, D. Fugazza, N. Gehrels, D. Götz, L. Hanlon, G. L. Israel,
    D. Malesani, B. McBreen, S. McBreen, S. McGlynn, S. Mereghetti,
    L. Moran, J. A. Nousek, R. Perna, L. Stella, and G. Tagliaferri

PROGENITORS, COSMOLOGY, AND X-RAY FLASHES

Constraints on the Diverse Progenitors of GRBs from the Large-Scale
Environments 473
    J. S. Bloom and J. X. Prochaska
The Electromagnetic Model of Gamma Ray Bursts 483
    M. Lyutikov

The Progenitors of Short Gamma-Ray Bursts 493
    E. Ramirez-Ruiz
GRB Cosmology and the First Stars 503
    V. Bromm and A. Loeb
Long Gamma-Ray Bursts as Standard Candles 513
    D. Lazzati, G. Ghirlanda, G. Ghisellini, L. Nava, C. Firmani, B. Morsony,
    and M. C. Begelman
Variability in GRB Host Galaxy Line Fluxes 522
    A. K. Yoldaş, J. Greiner, and A. Rau
Missing GRB Host Galaxies in Deep Mid-Infrared Observations:
Implications on the Use of GRBs as Star Formation Tracers 528
    E. Le Floc'h, V. Charmandaris, B. Forrest, F. Mirabel, L. Armus, and
    D. Devost
GRB Afterglows as a New ISM and IGM Probe 534
    H.-W. Chen, J. X. Prochaska, and J. S. Bloom
GRB Host Studies (GHostS) 540
    S. Savaglio, K. Glazebrook, and D. Le Borgne
The Redshift Distribution of Long Gamma-Ray Bursts 546
    F. Daigne, E. Rossi, and R. Mochkovitch
GRB 050814 at z = 5.3 and the Redshift Distribution of *Swift* GRBs 552
    P. Jakobsson, A. Levan, J. P. U. Fynbo, R. Priddey, J. Hjorth, N. Tanvir,
    D. Watson, B. L. Jensen, J. Sollerman, P. Natarajan, J. Gorosabel,
    J. M. Castro Cerón, and K. Pedersen
The True Redshift Distribution of Pre-SWIFT Gamma-Ray Bursts 558
    B. Gendre and M. Boër
GRB 050904: The Oldest Cosmic Explosion Ever Observed in the
Universe 564
    G. Cusumano, V. Mangano, G. Chincarini, A. Panaitescu, D. N. Burrows,
    V. La Parola, T. Sakamoto, S. Campana, T. Mineo, G. Tagliaferri,
    L. Angelini, S. D. Barthelemy, A. P. Beardmore, P. T. Boyd,
    L. R. Cominsky, C. Gronwall, E. E. Fenimore, N. Gehrels, P. Giommi,
    M. Goad, K. Hurley, J. A. Kennea, K. O. Mason, F. Marshall, P. Mészáros,
    J. A. Nousek, J. P. Osborne, D. M. Palmer, P. W. A. Roming, A. Wells,
    N. E. White, and B. Zhang
X-Ray Flare in XRF 050406: Evidence for Prolonged Engine Activity 570
    P. Romano, A. Moretti, P. L. Banat, D. N. Burrows, S. Campana,
    G. Chincarini, S. Covino, D. Malesani, G. Tagliaferri, A. D. Falcone,
    M. Capalbi, G. Cusumano, P. Giommi, V. La Parola, V. Mangano, M. Perri,
    and C. Pagani
The Very Long X-Ray Afterglow of XRF 050416A 574
    V. Mangano, G. Cusumano, V. La Parola, T. Mineo, S. Campana,
    M. Capalbi, G. Chincarini, P. Giommi, A. Moretti, M. Perri, P. Romano,
    G. Tagliaferri, D. N. Burrows, O. Godet, J. A Kennea, K. Page, and
    J. L. Racusin
Confirmation of the $E^{src}_{peak}-E_{iso}$ (Amati) Relation from the X-Ray
Flash XRF 050416A Observed by Swift/BAT 578
    T. Sakamoto, L. Barbier, S. Barthelmy, J. Cummings, E. Fenimore,
    N. Gehrels, D. Hullinger, H. Krimm, C. Markwardt, D. Palmer, A. Parsons,
    G. Sato, and J. Tueller

The X-Ray Spectrum and Lightcurve of the Redshift 6.29 γ-Ray Burst GRB 050904 582
    D. Watson, J. N. Reeves, J. Hjorth, J. P. U. Fynbo, P. Jakobsson,
    K. Pedersen, J. Sollerman, J. M. Castro Cerón, S. McBreen, and S. Foley

NON-PHOTONIC AND HIGH ENERGY EMISSION FROM GRBs

Non-Photonic Emissions from γ-Ray Bursts 589
    E. Waxman
A New Search Paradigm for Correlated Neutrino Emission from Discrete GRBs Using Antarctic Cherenkov Telescopes in the Swift Era 599
    M. Stamatikos *(On behalf of the IceCube Collaboration)* and D. L. Band
Searching for Cataclysmic Cosmic Events with a Coincident Gamma-Ray Burst and Gravitational Wave Signature 605
    S. Márka and L. Matone
Spectroscopy of the Brightest Bursts up to Energies of 200MeV 612
    M. M. González, M. Carrillo-Barragán, B. L. Dingus, Y. Kaneko, and R. D. Preece
The Search for Neutrinos from Gamma Ray Bursts with AMANDA 616
    K. Kuehn *(On behalf of the IceCube Collaboration and the IPN Collaboration)*
A Search for Short Duration Very High Energy Emission from Gamma-Ray Bursts 620
    D. Noyes *(On behalf of the Milagro Collaboration)*
Milagro Search for Very High Energy Emission from Gamma-Ray Bursts in the *Swift* Era 624
    P. M. Saz Parkinson *(On behalf of the Milagro Collaboration)*

FUTURE MISSIONS AND INSTRUMENTATION

Post-*Swift* Gamma-Ray Burst Science and Capabilities Needed to EXIST 631
    J. E. Grindlay and the *EXIST* Team
GLAST, LAT and GRBs 642
    N. Omodei and the GLAST/LAT GRB Science Group
AGILE and Gamma-Ray Bursts 648
    F. Longo, M. Tavani, G. Barbiellini, A. Argan, M. Basset, F. Boffelli,
    A. Bulgarelli, P. Caraveo, P. Cattaneo, A. Chen, E. Costa, E. Del Monte,
    G. Di Cocco, G. Di Persio, I. Donnarumma, M. Feroci, M. Fiorini,
    L. Foggetta, T. Froysland, M. Frutti, F. Fuschino, M. Galli, F. Gianotti,
    A. Giuliani, C. Labanti, I. Lapshov, F. Lazzarotto, F. Liello, P. Lipari,
    M. Marisaldi, M. Mastropietro, E. Mattaini, F. Mauri, S. Mereghetti,
    E. Morelli, A. Morselli, L. Pacciani, A. Pellizzoni, F. Perotti, P. Picozza,
    C. Pittori, C. Pontoni, G. Porrovecchio, M. Prest, M. Rapisarda, E. Rossi,
    A. Rubini, P. Soffitta, A. Traci, M. Trifoglio, A. Trois, E. Vallazza,
    S. Vercellone, and D. Zanello

Prospects for GRB Polarimetry with GRAPE 654
    M. L. McConnell, P. F. Bloser, J. Legere, J. R. Macri, T. Narita, and
    J. M. Ryan

The Gamma-Ray Large Area Space Telescope and Gamma-Ray Bursts 660
    J. McEnery and S. Ritz *(On behalf of the GLAST Mission Team)*

In-Flight Calibration of the Swift XRT Effective Area 664
    G. Cusumano, S. Campana, P. Romano, V. Mangano, A. Moretti,
    A. F. Abbey, L. Angelini, A. P. Beardmore, D. N. Burrows, M. Capalbi,
    G. Chincarini, O. Citterio, P. Giommi, M. R. Goad, O. Godet,
    G. D. Hartner, J. E. Hill, J. A. Kennea, V. La Parola, T. Mineo, D. Morris,
    J. A. Nousek, J. P. Osborne, K. Page, C. Pagani, M. Perri, G. Tagliaferri,
    F. Tamburelli, and A. Wells

BOOTES-IR: The Extension of BOOTES towards the Near-IR 668
    A. de Ugarte Postigo, A. J. Castro-Tirado, M. Jelínek, P. Kubánek,
    R. Cunnife, J. Gorosabel, S. Vitek, S. Castillo-Carrión, S. Guziy,
    S. B. Pandey, T. de J. Mateo Sanguino, J. M. Castro Cerón, F. M. Zerbi,
    P. Conconi, S. Covino, A. Riva, V. de Caprio, P. Amado, A. Claret,
    C. Cardenas, S. Martín, J. M. Trigo-Rodriguez, C. Sánchez-Fernández,
    M. D. Sabau-Graziati, J. Díaz-Verdejo, and F. Vitali

Real-Time Optical Monitoring of GRBs 672
    R. Hudec and M. Křížek

In-Flight Calibration of the Swift XRT Point Spread Function 676
    A. Moretti, S. Campana, M. Capalbi, G. Chincarini, S. Covino,
    G. Cusumano, P. Giommi, V. La Parola, V. Mangano, T. Mineo, M. Perri,
    P. Romano, and G. Tagliaferri

GRB Astrophysics with Digitised Astronomical Archival Plates 680
    R. Hudec and L. Hudec

BART: Real Time Follow-Up of GRBs Since 2001 684
    P. Kubánek, M. Jelínek, R. Hudec, M. Nekola, and J. Štrobl

GRB Follow-Up with BOOTES Optical Chapter 5: The *Swift* Era 688
    M. Jelínek, A. J. Castro-Tirado, P. Kubánek, S. Vítek, A. De Ugarte
    Postigo, and R. Hudec

GLAST and GRBs: Probing Photon Propagation over Cosmological Distances 692
    F. Longo, N. Omodei, J. Cohen-Tanugi, J. D. Scargle, and F. Piron

The CASTER Black Hole Finder Probe 696
    M. L. McConnell, P. F. Bloser, G. L. Case, M. L. Cherry, J. Cravens,
    T. G. Guzik, K. Hurely, R. M. Kippen, J. R. Macri, R. S. Miller,
    W. Paciesas, J. M. Ryan, B. Schaefer, J. G. Stacy, W. T. Vestrand, and
    J. P. Wefel

SuperAGILE and Gamma Ray Bursts 700
    L. Pacciani, E. Costa, G. Barbiellini, E. del Monte, I. Donnarumma,
    Y. Evangelista, M. Feroca, M. Frutti, F. Lazzarotto, I. Lapshov,
    M. Mastropietro, E. Morelli, M. Rapisarda, A. Rubini, P. Soffitta, and
    M. Tavani

Burst Detector Sensitivity: Past, Present & Future 704
    D. L. Band

The In-Flight Spectroscopic Calibration of the *Swift* XRT CCD Camera 708
    A. P. Beardmore, O. Godet, A. F. Abbey, J. P. Osborne,
    K. L. Page, A. A. Wells, and M. R. Goad *(On behalf of the Swift XRT Team)*
GRB Astrophysics with LOBSTER 712
    R. Hudec, L. Pína, L. Švéda, and A. Inneman
High Redshift GRBs Observed by GLAST............................... 716
    N. Omodei *(On behalf of the GLAST/LAT GRB Science Group)*
List of Attendees 721
Author Index 727

Preface

This is the sixteenth installment of the current series of annual October Astrophysics Conferences in Maryland. These conferences are organized by astrophysicists at the Goddard Space Flight Center and the University of Maryland, with support from the Universe Division of the Science Mission Directorate at NASA Headquarters.

The topic for each conference is selected with the help of an International Advisory Committee, the current membership of which is:

Roger Blandford, Palo Alto
Claude Canizares, Cambridge US
Francoise Combes, Paris
Andy Fabian, Cambridge UK
Alex Filippenko, Berkeley
Andrea Ghez, Los Angeles
Guenther Hasinger, Munich
Steve Holt, Needham

Bob Kirshner, Cambridge US
Chris McKee, Berkeley
Lee Mundy, College Park
Anneila Sargent, Pasadena
Joseph Silk, Oxford
David Spergel, Princeton
Rashid Sunyaev, Moscow
Nicholas White, Greenbelt
Simon White, Munich

In the spirit of this series, where we attempt to identify "hot" topics with broad appeal to both observers and theoreticians, "*Gamma-Ray Bursts in the Swift Era*" was clearly an appropriate conference theme. The programme for this year's conference was developed by a Scientific Organizing Committee comprised of:

Guido Chincarini
Brenda Dingus
Neil Gehrels
Paolo Giomi
Steve Holt
Chryssa Kouveliotou
Shri Kulkarni
Don Lamb

Keith Mason
Lee Mundy
John Nousek
Julian Osborne
Luigi Piro
Alan Wells
Nicholas White
Ralph Wijers

After a brief welcoming address, the conference opened with what is now a traditional arrangement of introductory invited presentations. The first of these was a review of the history of astronomical burst phenomena by *Virginia Trimble*. The second, delivered by *Neil Gehrels*, introduced *Swift* and briefly reviewed some highlights of its first year of operation. The third, delivered by *Luigi Piro*, defined the outstanding issues to which the conference attendees would address their attention in the succeeding sessions. The programme of non-paralleled sessions then proceeded through the next three days. Each session was devoted to a specific topic with two or three invited talks and a similar number of contributed talks, with poster opportunities for additional contributions to the subject area of each session.

Almost all of our past conferences have been held at the Inn and Conference Center on the campus of the University of Maryland in College Park. In contrast, this year's was held at the l'Enfant Plaza Hotel in Washington, where the proximity to NASA Headquarters allowed a larger-than-usual participation from Headquarters scientists. Another logistical difference between this conference and others in the series was the banquet that is traditionally held at the conclusion of the second day. This year the location was Italian Embassy, in recognition of the contribution to the *Swift* mission of Italian scientists.

Thanks to *John Trasco* and *Susan Lehr* of the Astronomy Department of the University of Maryland, who oversaw the overall conference arrangements. *Sandy Barnes, Cathy Dicks* and *Ana Wilson* made sure that all the logistics were handled flawlessly. Thanks also to *Neil Gehrels* and *John Nousek*, the co-editors of these Proceedings, *Cathy Dicks, Sharissa Feasler, J.D. Myers and Kate Smerekar* for manuscript editing and to all of the attendees who contributed to the success of the meeting.

Some of the figures were originally presented by the authors in color, and although they are reproduced in the proceedings in black and white, many of the legends for these figures still cite these different colors. The reader may refer to the online version of these proceedings at the AIP Conference Proceedings Website, http://proceedings.aip.org/proceedings/confproceed/836.jsp for the original full-color graphics.

Steve Holt
January 2006

HISTORY AND INTRODUCTION

A History of Gamma Ray Bursts and Other Astronomical Conundrums

Virginia Trimble

Department of Physics & Astronomy
University of California, Irvine, CA 92697-4575
and
Las Cumbres Observatory, Santa Barbara, California

Abstract. The 24 years between the announcement of "gamma-ray bursts of cosmic origin" (1973) and the unambiguously cosmological 970228 seemed very long to the generation that lived through them. For a good many astronomical phenomena, however, the interval between the recognition of a puzzle and convergence of the community on a solution was even longer. Examples include periodic variable stars and coronal lines from the sun. In other cases, the right idea has appeared very soon after discovery, pulsars and quasars, for instance. Sometimes, on the other hand, the theorists are out in front with an explanation in advance of the discovery. This is called a prediction, and some have come very far in advance (heliocentric parallax surely holds the record), others only a few years (superluminal motion in quasars, for example). The talk and this paper explore a few examples of each class as a framework for the much-told GRB story.

Keywords: Gamma Ray Bursts, Pulsating Variable Stars, Astrophysics.
PACS: 92.60.

INTRODUCTION

Science can be described as the on-going comparison between data and ideas, with data sometimes laboriously sought, sometimes dumped on our plates, and ideas sometimes put forward sui generis, sometimes dragged up to deal with unexpected data, and, if all goes well, sometimes discarded when conflicts persist. I have pontificated on the history of GRBs on a number of previous occasions (Trimble 1992a, 1992b, 1993, 1994a, 1994b, 2004, 2005) and so here focus on how astrophysical consensus has been established in various cases.

There are two typical patterns, with variations. Sometimes the data lead, and an observation is perceived as puzzling, perhaps immediately, as in, obviously, the case of GRBs, perhaps not until many years later (1845 to 1896 for spiral arms). Multiple explanations are then put forward, and convergence occurs, again rapidly (pulsars) or very slowly (nature of dark matter). The convergence may hold down to the present, in which case we say we have a theory of the phenomenon (stellar structure and evolution may be the cleanest). Or it may be disrupted by additional data, to be followed by another consensus model (the GRB case). Naturally there are phenomena for which we are currently somewhere in the middle of these processes.

Alternatively, ideas may lead and a prediction be made. The phenomenon is then sought, sometimes immediately and repeatedly (heliocentric parallax and a neutron star in the remnant of SN1987A), sometimes not until a good deal later (gravitational radiation), and sometimes not at all, so that discovery surprises nearly everyone (the CMB).

The search can be long (parallax, from the Greeks to Henderson, Bessel, and Struve surely holds the record) or short (21 cm radiation). The first positive report may be confirmed (21 cm radiation within weeks; parallax in a couple of years) or falsified (the planet that does not orbit pulsar psr 1829-10). And if the search is long and unsuccessful, predictions will be forced downward (solar neutrino flux, fluctuations in the CMB) until, eventually, the twain meet and we have again consensus. Sometimes this sort of agreement can also fall apart (many observers reported Vulcan, the dragger on the orbit of Mercury, and, of course, cyclotron features in GRB spectra). But theories that predict phenomena before they are seen, like deflection of light by the sun, are generally held in high repute by the scientific community. Once again, we can find ourselves today at any stage of this process. Predictions for which direct evidence is still being sought include ultra-high-energy neutrinos (from GRBs or anything else), a pulsar with a black hole companion, proton decay, that neutron star (or even black hole) left from SN 1987A, Hawking radiation, gravitational radiation, and dark matter particles. What we all, of course, wish we could recognize are the cases where the current convergence is around the wrong idea, so that we could be in the vanguard of the revolution.

The following sections are brief "case studies" illustrating these patterns. GRBs come at the end.

STELLAR PULSATION

The novae stellae of Tycho and Kepler were followed by quite a number of reports of other new stars, including "a nova in the whale" spotted by Fabricius in 1596 and Howards in 1638. Helvelius recognized that the two were the same star (1662), and Boulliau showed that the variation was periodic (1667). He also put forward the first explanation, rotation of a star with much darker spots than those on the sun. This was the only idea in the bank for more than a century.

Edward Pigott and John Goodricke recognized the periodicity of Algol (reported as variable by Montanari in 1669) in 1782, when Goodricke was still a teenager (he was dead at 22), and proposed that they were seeing an eclipse by a large, dark companion. They applied the same idea to δ Cephei, whose periodicity was Goodricke's discovery, in 1784. Eclipsing binaries were then the consensus model for all stellar variability for more than a century. Curiously, Goodricke and Pigott themselves realized that the eclipse hypothesis made definite predictions – precise periodicity and symmetric light curves – not actually fulfilled by δ Cephei and η Aquilae, and seemingly not in their data by Algol, and abandoned the eclipse model.

More serious, eventually astrophysics developed to the point where one could estimate sizes of stars from their colors and luminosities, and it became clear that the orbital diameter of an eclipsing δ Cephei would be smaller than the sum of the radii. But theoretical understanding of stars had also advanced enough that, first, Plummer (1912), then Shapley (1914), and definitively Eddington (1926) recognized that a pulsating star, with simultaneous changes in radius and temperature could reproduce the observed light curves. Eddington identified the correct driving mechanism as well, a step in the ionization of an abundant element near the surface of the star. Hydrogen, we now know, but for Eddington H was 7% contaminant, not the solution.

This particular puzzle took more than 250 years for sorting out and arguably holds the long duration prize, although Kelvin could plausibly have demonstrated that pulsation yields at least the right time scales. Notice that orbital motion at contact and pulsation differ by only a factor of two or so, since both are manifestations of time $\propto (G\rho)^{-\frac{1}{2}}$.

THE SOURCE OF SOLAR AND STELLAR ENERGY, GIANTS, AND THE SOLAR NEUTRINO PROBLEM

That the heat and light of the sun would require maintenance became clear only with the 19th century establishment of universal conservation of energy. John Herschel suggested friction (presumably with some transparent substance) or electrical discharges. The story of how Julius Mayer and John Waterston recognized the problem in 1841 and 1843 respectively, proposed infall or contraction as the solution, and lived to see the credit go to Kelvin and Helmholtz has been told by Hufbauer (1991), and I will note here only the sad sideline that Mayer's paper was refused by the proceedings of the Paris Academy and Waterston's by the Royal Society. The 30,000,000 or so years the sun could live on contraction is invariably called the Kelvin-Helmholtz time scale. That it was not enough was gradually forced upon the physics and astronomy communities by uniformitarian geologists and Darwinian biologists. The laboratory discovery of radioactive decay led to the idea of subatomic processes as the dominant source of energy.

Key intermediate steps were (a) $E=mc^2$, (b) laboratory measurements of atomic masses in Cambridge (particularly hydrogen and helium) by Francis Aston starting in 1922, (c) the gradual recognition that stars contain very large amounts of hydrogen and helium, pioneered by the 1925 thesis of Cecilia Payne, and (d) the concept of barrier penetration, found in embryo form in Atkinson and Houtermans (1931) who suggested catalyzed cyclic reactions (details necessarily wrong because the neutron had not yet been discovered), but normally credited to Gamow and Condon and Gurney. And th0e problem was generally regarded as solved from the publication of Bethe (1939) onward. Notice that both special relativity and elementary quantum mechanics were needed. And we touch current events as well. Very many of us knew Bethe, and I was offered a job at Indiana by Atkinson in 1968.

Eddington (1926) had supposed that observed giants were actually contracting stars in the process of formation from diffuse material. A few of them of course are. The rest fall out of calculations of stellar evolution with little or no mixing. The role of composition (mean molecular weight) discontinuity was noted by Öpik (1938) and Hoyle and Lyttleton (1939). Schwarzschild's work (summarized in Schwarzschild 1958) made clear that red giants would indeed happen and that purely verbal descriptions of just why would never quite satisfy (which is why they continue to appear in the literature even now).

And then, after we had all been certain things were under control, came the discordance of prediction and observation of solar neutrino flux. See Bahcall (1989) for many details and references, but it is worth nothing here, first, that the range of explanations presented between about 1969 and 1989 surely rivaled the GRB inventory of hypotheses, ranging over chemistry, nuclear physics, astrophysics, and weak interactions and, second, that the right idea, neutrino oscillations, was in the inventory from the beginning, thanks to the inspired thoughts of Bruno Pontecorvo. I am not quite sure when this became the dominant idea (I switched when Bethe became firm about the issue, about 1994), but it became inescapable with the results from Super Kamiokanda and SNO, seeing the oscillation products first for cosmic-ray-induced neutrinos and then for the solar ones.

Durations? Well 106 years for the energy source if you count Herschel to Bethe; only a decade or two for red giants; and zero to 25 years for the solar neutrino puzzle, depending on when you personally came over to the oscillating side. Truthfully, a few living scientists still do not quite subscribe to a standard sun that began 4.56 Gyr ago as a homogeneous sphere of mostly hydrogen and helium, but none is young.

Well, while we are close to the sun, let's look at the coronal emission lines, first reported by Charles A. Young in 1869 and quite widely then attributed to an element to be called coronium by analogy with helium, responsible for some solar photospheric features, and nebulium, the source of emission lines from diffuse hot gas. Hufbauer (1991, p. 112) also tells the story. The culture heros are Walter Grotrian and Bengt Edlen, who engaged in some competition for first formal publication of the line identification with highly ionized iron and other elements in 1939. Duration a mere 70 years. Much new physics, from Bohr to Schroedinger, was required, and if you are absolutely sure you fully understand why stellar coronae are this much hotter than their photospheres (something times 10^6, vs. something times 10^3 K), and you can persuade most of us to agree with you, the duration of that puzzle can be truncated at 67 yr.

THE ADVANCE OF THE PERIHELION OF MERCURY AND SOME ALSO-RANS

Since the discovery was announced in 1859 (by U.J.J. Leverrier, see Hoskin 1999) and Einstein's explanation in 1915, we are down to a mere 56 year duration. But it was marked by another of those convergences around a wrong idea – the planet Vulcan, seen in transit some 20 times between 1859 and 1876, but apparently never again, though Leverrier had predicted additional transits in March 1877 and October 1882. About 1900, Simon Newcomb (who otherwise doesn't get much good publicity these days) was among the advocates of a deviation from pure $1/r^2$ gravity as the explanation. And, curiously, there was a very late attempt to break up the consensus, when Robert Dicke attempted to support a general relativistic variant called the scalar-tensor theory with measurements of the diameter of the sun implying an appreciable oblateness and quadruple moment which would, of course, give the effective gravity a $1/r^3$ component. In retrospect, Dicke and Goldenberg (1967) probably observed facular brightening of the solar equator, but if they had been right, the perihelion puzzle would have extended another 50 years. And the "what was needed" box would have held two new theories of gravity beyond Newton.

A few of my favorite also-rans include (a) stability of spiral arms in the presence of differential rotation, solved perhaps by C.C. Lin and F.H. Shu in 1964 (solitons) or perhaps not yet entirely, (b) the acceleration of cosmic rays, solved perhaps by E. Fermi in 1949 (38 years after the discovery that they are extraterrestrial), or perhaps not yet entirely, (c) the cause of nova explosions (discovery, arguably T Pyx in 1919, and attribution soon after by Milne to collapse of a normal star to a white dwarf). Eddington, however, recognized that novae must recur on statistical grounds, and the establishment of universal binarity and hydrogen fusion explosions on the surface of an accreting white dwarf stretched out from about 1946 to 1963.

A FEW PREDICTIONS AND THEIR FATE

Two of my favorites are polarization items, one with very rapid discovery of the predicted effect, one much more complicated. The quick case was optical polarization in the Crab Nebula, seen by Dombrowski (1953, who sadly spent most of the rest of his career looking unsuccessfully for polarization of other, thermal, nebulae), very soon after V.L. Ginzburg and I.S. Sklovsky said (separately!) it should be there. For details of this and of the prediction of circular polarization from the Crab, whose proposers honorably abandoned their hypothesis when it wasn't detected (Rees and Gunn 1974), see Trimble (2003). The Crab is also host to several other long-unconsensused phenomena and unfulfilled predictions.

The more convoluted case begins with a prediction from Chandrashkar (1946) that light coming to us from hot stars should be somewhat polarized by electron scattering in the stellar atmospheres. John Hall and William Hiltner, then both also at Yerkes, set out to look for the effect, though by the time they published, there were two separate papers, Hall (1949) and Hiltner (1949). They found polarization all right, but so systematically aligned with the galactic plane that anything happening in the stars themselves seemed most unlikely.

Aha! A "pattern I" of unexplained observation. But Davis and Greenstein (1951) and Gold (1952) soon came forward with (different) hypotheses. Davis and Greenstein invoked scattering by interstellar dust grains alighted by a galactic magnetic field, not yet then known to exist, while Gold proposed scattering by spinning grains whose alignment was due to large scale gas flows rather than magnetism. Incidentally, current understanding requires both magnetic field and spin.

Chandrasekhar's effect was finally reported by Kemp et al. (1982). That one took 36 years. James C. Kemp, a free spirited Oregonian much given to the wearing of Hawaiian shirts, belongs somewhere around here in another context, for the discovery of polarization of white dwarf light (Kemp et al. 1970), the first clear evidence for strong magnetic fields in these stellar cinders. It would be interesting to chase down the predictions and discoveries of magnetic fields in all the contexts where they are known or suspected and to sort out the stories into the various possible patterns. The fields one would have to worry about include that of earth (1600, Gilbert), the sun (1908, George Ellery Hale), the interstellar medium (briefly here, but also Fermi acceleration of cosmic rays), Ap stars (H.D. Babcock), white dwarfs, neutron stars and supernova remnants, intracluster and intergalactic media, and, of course, gamma ray bursts. I should think that the 80% polarization of one event, mentioned briefly at the present conference, would have counted as a detection if the polarization were a little more persuasive

The GRB field also has examples of both prompt and afterglow detections of predicted phenomena and of explanations that came along too early to have much influence. Meszaros and Rees (1997) forecast an optical afterglow very shortly before it was found. Rhoads and Paczynski (1993) forecast a radio one a little too early for it to have been in mental forefronts in 1997. They noted later that they "should have" realized that the radio tail would be part of more extensive emission, including visible light and X-rays.

GAMMA RAY BURSTS, PART I: WE HAVE TURNED EVERY ONE TO HIS OWN WAY

The discovery paper (Klebesadel et al. 1973) was squeezed into the length of an Astrophysical Journal Letter, and more of the process is described in Strong et al. (1975, the observational summary talk on GRBs at the Seventh Texas Symposium on Relativistic Astrophysics). The Los Alamos group knew about the Colgate (1968) prediction that supernova shock outbreak should radiate a burst of gamma rays and deliberately looked for signals in the direction of known supernovae during the period before 1973 without seeing any. They also knew that solar flares could be sources of very hard photons (Cline, Holt, and Hones 1968, whose third author was at Los Alamos, and whose first two were at the present meeting), looked for those, and saw them. They also decided to check for other (conceivably astronomical) events that could produce simultaneous triggers in both Vela 5 and Vela 6, without being supernovae, solar flares, or the bomb tests that were the original goal of that satellite series.

Thus were found the GRBs, and I would claim the discovery was not precisely serendipitous in the original sense of the word. The Russians had also worried about secret atmospheric or surface bomb tests, though at the time Mazets et al. (1974) reported that they had seen at least one of the Vela GRBs, I did not quite believe that the use of Cosmos 461 implied the existence of 460 previous launches. But the first was 16 March 1962, and the end of 1967 took the series to Cosmos 198, the majority of the payloads by then being "unannounced", not that the public description of, say, Vela 7 ("advanced nuclear detection satellite") was enormously more informative (TRW Space Log, Vol. 7, No. 4).

Truly spectacular were the thousand (well 118, Nemiroff 1994) flowers of theory that promptly blossomed. Among the more notable are the Colgate (1968) and Hawking (1974) predictions (the latter still in that category because the author does not give the impression of having heard of the observations); Jelley's (1974) proposal of a Leblanc-Wilson type collapse of a rapidly rotating, magnetized star, not so very different from the current understanding of long duration bursts; and Harris (1990) who looked for lines of GRBs across the sky that might be exhaust trails from interstellar space craft operating on electron-positron annihilation. This was in the era of general acceptance of spectral features, including a redshifted 511 keV line, that we face in the next section. And Fritz Zwicky's last published paper (Zwicky 1974) credited free explosions of chunks of escaping neutron star material, which he called Goblins.

Occasional models invoked entities that very probably exist in the universe but may not have much to do with gamma ray bursts, for instance Paczynski (1987 on gravitational lensing) and Paczynski (1988 on superconducting superstrings).

Less unconventional events in solar system, Milky Way, and cosmos also appeared. By the time Ruderman (1975) provided the theory review at the 7th Texas Symposium, he claimed that only anti-matter comets hitting white holes had gone unclaimed, though he himself was betting on "black hole ridden by accretion" to win and "glitch" to place. Very curiously, at that same meeting, Colgate disowned his prediction, saying that he thought that supernovae would indeed make hard photon bursts, they just hadn't been seen yet. Today we would probably say that he was right the first time. Ruderman also claimed that the only theorist who did not have a GRB model was J.P. Ostriker. Ostriker (personal communication 2003) reports that he still hasn't published one, but has given some thought to electromagnetic processes. The remark of Brecher and Morrison (1974) that the energy must be tightly beamed was also prescient, though they had in mind stellar flares as the energy source.

GAMMA RAY BURSTS PART II: ALL WE LIKE SHEEP HAVE GONE ASTRAY

A young participant suggested after my talk that so few astrophysicists now either read Isaiah or participate in Christmas Messiah sing-a-longs that the origin of this section title and the previous one might need their origins explained. There you have your pick of two.

Less than a decade after Klebesadel et al. (1973) saw light of print, the community had reached remarkably broad consensus on a model where GRBs were events on the surface of old, nearby, but still highly magnetized neutron stars. Lamb (1984, from the 1982 Texas Symposium) provides a good taste of the flavor of confidence of this conclusion. How did we manage to do this?

There were, I think, a handful of key events, beginning with the discovery of X-ray bursters (which is why the name is not available for gamma-poor GRBs!) in 1976-77 by Grindlay, Lewin, and their collaborators, followed quickly by a correct interpretation in terms of explosive helium burning on accreting neutron stars by Paul Joss. Woosley et al.'s (1976) idea of carbon detonation had originally been intended as a GRB model and so has the status of a prediction for recently-identified subclass of XRB where it is carbon that flashes! The X-ray bursters provided an honest, early hint of their brightnesses and nature by concentrating themselves in the general direction of the galactic center, while the GRBs were isotropic then and always.

In the same time frame came the discovery (from a balloon flight by the Truemper group in 1976) of cyclotron resonance features in the X-ray spectrum of Her X-1. Such have been found since in many XRBs and point us correctly toward strong magnetic fields. But similar features began popping up in GRBs: a redshifted positron annihilation line; 847 keV iron redshifted by the same amount; 20-70 keV cyclotron resonances, sometimes more than one per burst and in seemingly the right frequency ratio. Lamb (1984) provides extensive references, but let Mazets and Golonetskii (1981, a Venera result) stand for all the rest, including reports from SMM, HEAO-1, and especially Ginga. I was not in any way immune to this virus and in a 1990 review describe GRBs as "mergers of binary neutron stars with strong magnetic fields at cosmological distances."

Third, for a while it seemed that the events must be recurrent, because there were occasional optical flashes found on old photographic plates in the error boxes of recent GRBs, a 1978 burst on a 1928 Harvard plate, for instance. The implication was that the object involved must survive and be able to try again in a century or less. Next came the "no host" problem, with searches for host galaxies in GRB error boxes coming up empty right up to the time of 970228, when we all had to face the fact that at least some hosts were very distant, faint, blue galaxies.

Discovery that some neutron stars are very high velocity objects (the current record is something like 1000 km/sec) and so could populate an extended galactic halo contributed its mite of confusion as well, at the onset of the CGRO era. The reason this mattered was that the plots of numbers of GRBs vs peak flux or fluence (total energy received) finally began to turn over from the slope of 3/2 associated with homogeneous, isotropic populations. Such a turn-over had to occur at some flux for neutron stars in the galactic disk, and indeed there had been early reports, generally then attributed to assorted instrumental problems. I think the first that should have been credited was White et al. (1983) but nobody much asked me at that time.

Now, what does this have to do with runaway stars? Well, if we see the end of a distribution (the turn-over) but are at the center of it (the isotropy on the sky), only three possible sites remain: some part of the solar system much larger than 1 AU, some part of the Milky Way much larger than 10 kpc, or some part of the whole universe, larger than the distance at which redshift effects became important. Fishman (1995) concisely describes the situation at the point when CGRO had forced the simultaneous center/edge phenomena on us but before the first optical identifications, but the loudest voice crying in the wilderness that center + edge must mean cosmological distances was that of Paczynski (1991).

THE MISSED OPPORTUNITY AND THE BENEFICIARIES

Proposals for instruments to be carried on the Gamma Ray Observatory (later Compton Gamma Ray Observatory) exceeded the number of slots available, and so there was, of course, a peer review panel. It was my second (Space Telescope was first in 1977). Among the proposals were two (one from Walter Lewin and his colleagues, one from Paul Gorenstein and his) for X-ray imagers in the 20-50 keV range. These would have provided arc-minute positions from the long-wavelength tail of the brighter GRB events themselves. Neither was selected, primarily on the grounds that these were clearly X-ray instruments, and there were already several X-ray satellites with more in the pipe line. Bernie Burke's recollection (personal communication 2005) of how this view was expressed is harsher than mine. Since then I have merely said that the line between X-ray and gamma ray astronomy has nothing to do with photon energy or the emission process. It is merely that X-ray astronomy is done by X-ray astronomers and gamma-ray astronomy is done by gamma ray astronomers, and even now the twain don't seem to meet much oftener than they have to. I guess we should just all be glad the BATSE made the cut and that Fishman (these proceedings) was able to hold out for eight detectors on the eight corners!

If a CGRO X-ray imaging detector had flown, a good position leading to an optical counterpart might easily have been found by the fall of 1991 (BATSE started finding bursts in April.) Instead it was February 1997, and the requisitely accurate X-ray position came from the Beppo SAX, mostly Italian, satellite, launched in April 1996.

It has proven remarkably easy to get used to the extragalactic consensus, though as late as the 1995 diamond anniversary restaging of the Curtis-Shapley debate (Nemiroff 1995) a fairly sophisticated audience, asked to vote, were more or less equally divided between a Milky Way corona and the universe. It is also true that the fraction who voted "uncertain, or decline to state" was a good deal larger after the debate than before. And it remains true down to the present that GRBs seem to be the only class of astronomical source in which cosmological effects of redshift can be seen clearly over and above astrophysical ones of the evolution.

CONSENSUS AND AN EXERCISE FOR THE STUDENT

It is, I think, fair to say that the community now agrees that GRBs or at least the vast majority, are very high energy events that we can see from essentially the entire observable universe, though some details remain to be worked out, to put it tactfully. These include both the precise engines for the short duration events and many aspects of the processes by which the raw energy is transformed into the photons we see. Remember, however, that we felt the same way in 1985, except that the events were relatively low energy, repetitive and confined to some portion of the Milky Way, at least as far as the ones we could see.

The following is a partial list of astronomical phenomena that I think have interesting histories within the patterns of data leading vs ideas leading and fast vs slow progress to the next step. Some have been mentioned above, many not. Feel free both to decide which pattern you think each illustrates and to let me know of phenomena that should be added to the list. Blazhko effect. Second parameter in globular clusters. Oosterhoff types. Pulsars with BH companions. Ultra-high-energy neutrinos. Cosmological neutrinos. Proton decay. Extra-solar-system planets. Primary cosmic ray anti-protons. X-ray fluorescence from moon. Chandler wobble. Rings of Saturn. Lambda Boo stars. Extra-SS gamma rays. Radar echoes from cosmic ray showers. X-rays from cool, single magnetic white dwarfs. Cosmic rays above the GKZ limit. Intermediate mass black holes. Magnetic fields in white dwarfs. 3K microwave background. Solar and stellar flares.

ACKNOWLEDGMENTS

I am, as always, grateful to the SOC chairs, Stephen Holt and Neil Gehrels, for the opportunity to participate in the slightly-displaced Maryland October workshop; to Walter Lewin for sharing his September 7, 1978 letter to Albert Opp concerning the instrument package for CGRO; to the anonymous colleague who provided the pictures of Pigott, Goodricke, Leverrier, and all shown at the conference; to the UC Irvine Committee on Research for partial support of travel to the conference; to Alison Lara for bravely keyboarding the manuscript from my typed original, and to all the brave colleagues, living and dead, who have hallowed the ground of history of astronomy with their not-always-acknowledged contributions.

REFERENCES

1. R. d'E Atkinson and F.G. Houtermans,, *Zs. f. Ap* **54**, 652 (1929).
2. J.N. Bahcall, "Neutrino Astrophysics," Cambridge University Press (1989).
3. H.A, Bethe, *Phys. Rev.* **55**, 434 (1939).
4. K.R Brecher and P. Morrison, *ApJ* **187**, L97 (1974).
5. S. Chandrasekhar, *ApJ* **103**, 351, (1946).
6. T.E. Cline, S.S. Holt and E.W. Hones, *J. Geophys. Res.* **73**, 43 (1968).
7. S.A. Colgate, *Can. J. Phys.* **46**, S476 (1968).
8. L. Davis, and J.L. Greenstein, *ApJ* **114**, 206 (1951).
9. R.H. Dicke and J.M. Goldenberg, *Phys. Rev. Lett.* **18**, 313 (1967).
10. V.A. Dombrovski, *Dokl Akad. Nauk USSR* **94**, 1021 (1953).
11. A.S. Eddington, "Internal Constitution of the Stars," Cambridge Univ. Press, (1926, Dover Reprint 1959) pp. 183ff, 174ff
12. G.J. Fishman, *PASP* **107**, 1145 (1995).
13. T. Gold, *Nature* **169**, 322 (1952).
14. J.S. Hall, *Science* **109**, 166 (1949).
15. M.J. Harris, *J. Brit. Interplan. Assoc.* **43**, 551 (1990).
16. S.W. Hawking, *Nature* **248**, 30 (1974).
17. W.A. Hiltner, *Science* **109**, 165 (1949).
18. M.A. Hoskin, "The Cambridge Concise History of Astronomy," Cambridge Univ. Press, (1999), p. 165.
19. F. Hoyle, and R.A. Lyttleton, Proc. Cam. Phil Soc. 35, (1939), p. 592.
20. K. Hufbauer, "Exploring the Sun," Johns Hopkins Univ. Press, (1991), p. 55.
21. J.V. Jelley, *Nature* **249**, 747 (1974).
22. J.C. Kemp et al., *ApJ* **161**, L77 (1970).
23. J.C. Kemp et al., *ApJ* **273**, L85 (1982).
24. R.B. Klebesadel, I.B. Strong and R.A. Olson, *ApJ* **182**, L85 (1973).
25. D.Q. Lamb, in 11th Texas Symposium on Relativistic Astrophysics., Ed. D.S. Evans, Ann. NY Acad. Sci. 422, (1984), p. 237.
26. E.P. Mazets and S.V. Golonetskii, *Ap & SS* **75**, 41 (1981).
27. P. Meszaros and M.J. Rees, *ApJ* **476**, 232 and 482, L29 (1997).
28. E.P. Mazets et al., *JETP Lett.* **19**, 176 (1974).
29. R.J. Nemiroff, *Comm. on Astrophys.* **17**, 181 (1994).
30. R.F. Nemiroff, *PASP* **107**, 1131 (1995).
31. E. Öpik, *Publ. Obs. Astron. U. Tartu* **30**, No. 3 (1938).
32. B. Paczynski, *ApJ* **317**, L51 (1987).
33. B. Paczynski, *ApJ* **335**, 528 (1988).
34. B. Paczynski, *Acta Astron.* **41**, 257 (1991).
35. H.A. Plummer, *MNRAS* **73**, 665 (1911).
36. M.J. Rees and J.E. Gunn, *MNRAS* **167**, 1 (1974).
37. J.E. Rhoads and B. Paczynski, *Bull. Amer. Astron. Soc.* **25**, 1296 (1993).
38. M. Ruderman, *Ann. NY Acad. Sci.* **262**, p. 164 (1975, 7th Texas Symposium)
39. M. Schwarzschild, "Structure and Evolution of the Stars," Princeton University Press (1958, Dover 1965).

40. I.B. Strong. et al., *Ann. NY Acad. Sci* **262**, p. 145 (1975, 7th Texas Symposium).
41. V. Trimble, in "Gamma Ray Bursts," Ed. C. Ho et al., Cambridge Univ. Press, (1992a), p. 482.
42. V. Trimble, in "IAU Sym. 151 Evolutionary Processes, in Interacting Binary Stars," Ed. Y. Kondo et al., Kluwer, (1992b), p. 91.
43. V. Trimble, in "The Realm of Interacting Binary Stars," Ed. J. Sahade et al., Kluwer, (1993), p. 271.
44. V. Trimble, in the "Second Compton Symposium," Ed. C.E. Fichtel et al., *AIP Conf. Proc.* **304**, (1994a), p. 40.
45. V. Trimble, in the "Second Huntsville GGRB Workshop," Ed. G. Fishman et al., *AIP Conf. Proc.* **307**, (1994b), p. 717.
46. V. Trimble, in "Texas in Tuscany," Ed. R. Bandiera et al., *World Scientific*, (2003), p. 269.
47. V. Trimble, in "Gamma Ray Bursts: 30 Years of Discovery," Ed. E.E. Fenimore and M. Galassi, *AIP Conf. Proc.* **727**, (2004), p. 7.
48. V. Trimble, in "The New Astronomy," Ed. W Orchiston, Springer, (2005), p. 761.
49. S.E.Woosley et al., *Nature* **263**,101 (1976).
50. R.S. White et al., *Nature* **271**, 635 (1983).
51. F. Zwicky, *Ap & SS* **28**, 111 (1974).

The Swift Gamma-Ray Burst Mission: First Results

N. Gehrels[1], on behalf of the Swift Team

[1]*NASA/GSFC, Greenbelt, MD 20771, USA*

Abstract. Since its launch on 20 November 2004, the Swift mission is detecting ~100 new gamma-ray bursts (GRBs) each year, and immediately (within tens of seconds) starting simultaneous X-ray and UV/optical observations of the afterglow. It has already collected am impressive database of bursts, including prompt emission to higher sensitivity than BATSE, uniform monitoring of afterglows, and rapid follow-up by other observatories notified through the GCN.

Keywords: Gamma Ray Bursts, Astrophysics.
PACS: 98.70Rz.

INTRODUCTION

Despite impressive advances over the roughly three decades since GRBs were first discovered (Klebesadel et al. 1973), the study of bursts remains highly dependent on the capabilities of the observatories which carried out the measurements. The era of the Compton Gamma Ray Observatory (CGRO) led to the discovery of more than 2600 bursts in just 9 years. Analyses of these data led to the conclusion that GRBs are isotropic on the sky and occur at a frequency of roughly two per day all sky (Briggs 1996).

The BeppoSAX mission made the critical discovery of X-ray afterglows (Costa et al. 1997). With the accompanying discoveries by ground-based telescopes of optical (van Paradijs et al. 1997) and radio (Frail et al. 1997) afterglows, GRBs could start to be studied within the astrophysical context of identifiable objects in a range of wavelength regimes. Successful prediction of the light curves of these afterglows across the electromagnetic spectrum has given confidence that GRBs are the signal from extremely powerful explosions at cosmological distances, which have been produced by extremely relativistic expansion (Wijers, Rees & Mesaros 1997).

The Swift mission selected by NASA in 1997 combines the sensitivity to discover new GRBs with the ability to point high sensitivity X-ray and optical telescopes at the location of the new GRB as soon as possible. From this capability Swift has the goal to answer the following questions:

1. What causes GRBs?

2. What physics can be learned about black hole formation and ultra-relativistic outflows?

3. What is the nature of subclasses of GRBs?

4. What can GRBs tell us about the early Universe?

The general operations of the Swift observatory are as follows. The wide-field Burst Alert Telescope (BAT) detects the bursts in the 15-350 keV band and determines the position to a few arcminute accuracy. The position is provided to the spacecraft, built by Spectrum Astro General Dynamics, which repoints to it in less than 2 minutes. Two narrow-field instruments then observe the afterglow: the X-Ray Telescope (XRT) and UV-Optical Telescope (UVOT). Alert data from all three instruments is sent to the ground via NASA's TDRSS relay satellite. The full data set is stored and dumped to the Italian Space Agency's Malindi Ground Station.

Swift was built by an international team from the US, UK, and Italy. After five years of development it was launched form Kennedy Space Center on 20 November 2004. The spacecraft and instruments were carefully brought into operational status over an eight-week period, followed by a period of calibration and operation verification, which ended with the start of normal operations on 5 April 2004. A complete description of the Swift mission can be found in Gehrels et al. (2004).

As of 30 November 2005, the Swift achievements include: discovery of 92 new GRBs by the Swift Burst Alert Telescope (BAT) instrument (with a typical error region of less than 2 arcmin radius); observation of 72 X-ray afterglows by the Swift X-Ray Telescope (XRT) instrument (with a typical error region of less than 3 arcsec radius); and observations of 20 afterglows by the Swift UV-Optical Telescope (UVOT) instrument (with a typical error region of less than 1 arcsec). More than half of the afterglow observations start within two minutes of the BAT GRB trigger with a record of only 54 seconds. Afterglow data have been obtained from non-Swift discovered bursts with typical response times of 3 hours.

SWIFT HIGHLIGHTS

2.1 BAT Detected GRBs

The BAT (Barthelmy, et al. 2005a) on Swift has detected 77 GRBs between when it was turned on in mid-December 2004 and 30 November 2005. Thus in 345 days of operation, the BAT has detected GRBs at a rate of about 97 bursts per year. This value is close to the rate of 100 bursts estimated prior to launch.

Spectral analysis of the BAT bursts shows them to be consistent with the population of GRBs seen by the Compton Gamma-Ray Observatory BATSE experiment, both in the ratio of the fluxes in the 25-50 keV and 50-100 keV energy bands, and in flux and duration distributions.

2.2 XRT Detected GRBs

The XRT (Burrows, et al. 2005) is performing the first rapid-response observations of the X-ray afterglow of GRBs. In the first ~80 cases, all but 5 of the BAT GRB triggers resulted in detection of an X-ray counterpart for the BAT source. In 3 cases the XRT observations started while the BAT was still detecting hard X-ray prompt emission from the GRB.

The Swift afterglow observations are rapid, with more than half of the observations started in less than 300 seconds after the burst. When XRT arrives this quickly it is very common to see a fast X-ray decline within the first 300 seconds.

In addition to the BAT detected events, Swift can also observe GRBs discovered by other satellites. Swift has discovered X-ray afterglow emission in 4 cases for HETE-2 and 5 cases INTEGRAL. In a particularly impressive case, Swift was able to respond to the

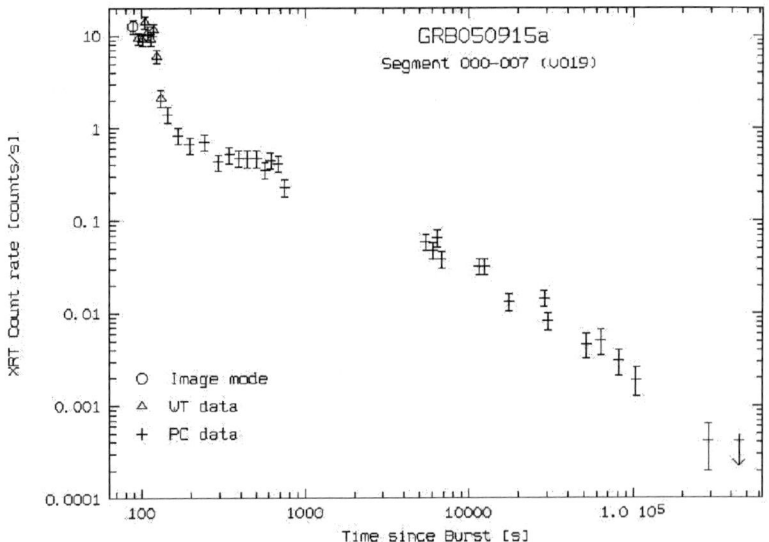

Fig. 1. Typical X-ray afterglow light curve.

ground control commands and start observations of the HETE-2 GRB050408 within 40 minutes of the GRB.

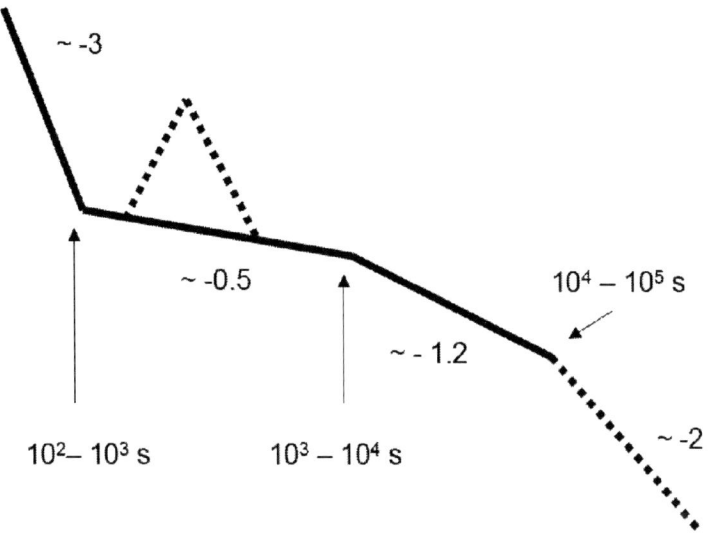

Fig 2. Generic lightcurve observed by XRT. Numbers shown next to the segment are the power-law index of the time decay. From Zhang et al. (2006).

2.3 UVOT Detected GRBs

The UVOT (Roming, et al. 2005a) is co-aligned with the XRT and so observes the GRB afterglows just as promptly as the XRT. Despite these prompt observations the UVOT has detected far fewer UV/optical counterparts than the XRT.

Of the first 68 GRBs observed by the UVOT, only 20 had detected emissions. The UVOT has generated important upper limits for these early times, which are lower than those for bursts studied by previous missions. Combining UVOT plus ground-based observations ~40 optical afterglows have been detected.

Reasons for this low number include the possibility that the Swift bursts are more distant than previous bursts; that a substantial number of GRBs have intrinsic dust extinction which suppresses the optical/UV emission compared to the I and R bands typically reported for earlier afterglows; or the possibility that some afterglows come from high magnetic field regions in the outflow which suppresses the optical and UV emission. These possibilities are discussed in Roming et al. (2005b).

Although not every GRB produces UV or optical flux, which can be detected by the UVOT, several bursts have produced early time light curves, including GRB050318 (Still et al. 2005) and GRB050319 (Mason et al. 2005).

2.4 XRT Early Light Curve Behavior

Swift has opened up a new regime for GRB afterglow studies. Never before has it been possible to study the X-ray behavior on timescales of minutes after the GRB happens. Swift has frequently started observations within a few minutes of the detection of GRBs by the BAT (with a record of only 52 seconds).

These extremely prompt observations have given rise to new findings. In roughly 40% of the cases, the X-ray afterglows can be characterized by a three-part light curve (see Figure 1 and 2). First comes an extremely rapid decay of a very bright source. At these early times the decay can be fit by a power law of index in the range of 2.5 or greater. After a few minutes the decay rate flattens, and we can fit it with an index approximately equal to 1 (plus or minus ~0.5). Finally after a delay ranging from hours to days, the decay rate steepens again, sometimes resulting in a behavior interpreted as a jet break (see Zhang et al. (2006) and Nousek et al. (2006) for summary papers). Tagliaferri et al (20005) and Barthelmy et al. (2005b) each consider two early XRT afterglows. They show that the X-ray emission during the prompt phase (estimated from extrapolation of the BAT spectrum) connects to the bright early XRT afterglow (see Figure 3). This suggests that the bright early afterglow is an extension of the prompt phase.

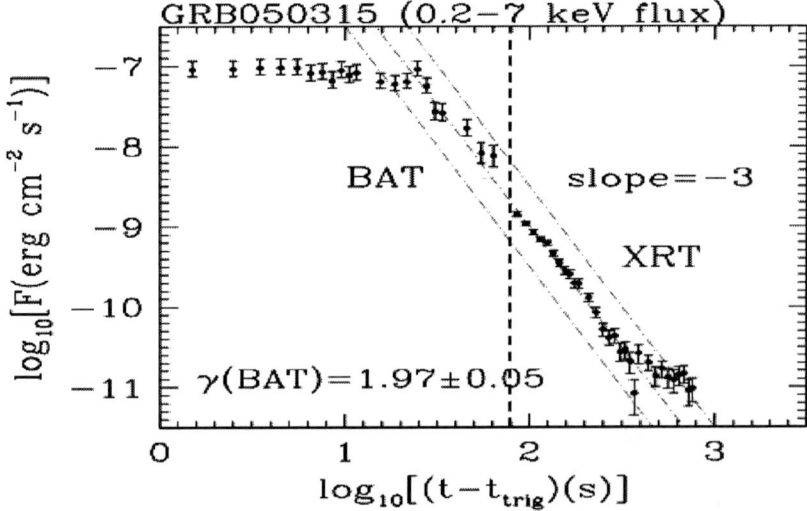

Fig. 3. The BAT spectrum is extrapolated to the 0.2 – 7 keV band. The early XRT lightcurve connects smoothly to the prompt emission. From Barthelmy et al. (2005b).

Swift also detects strong X-ray flares in afterglows at early times. In one case (GRB050502b) the X-ray flux increased by a factor of roughly 100. The dramatic flaring events seem to be superposed on a background, which follows the multipart behavior mentioned above. Burrows et al. (2005b) discuss the flaring behavior seen in GRB050502b and GRB050408. Flares are seen in 20%-30% of he observed afterglow.

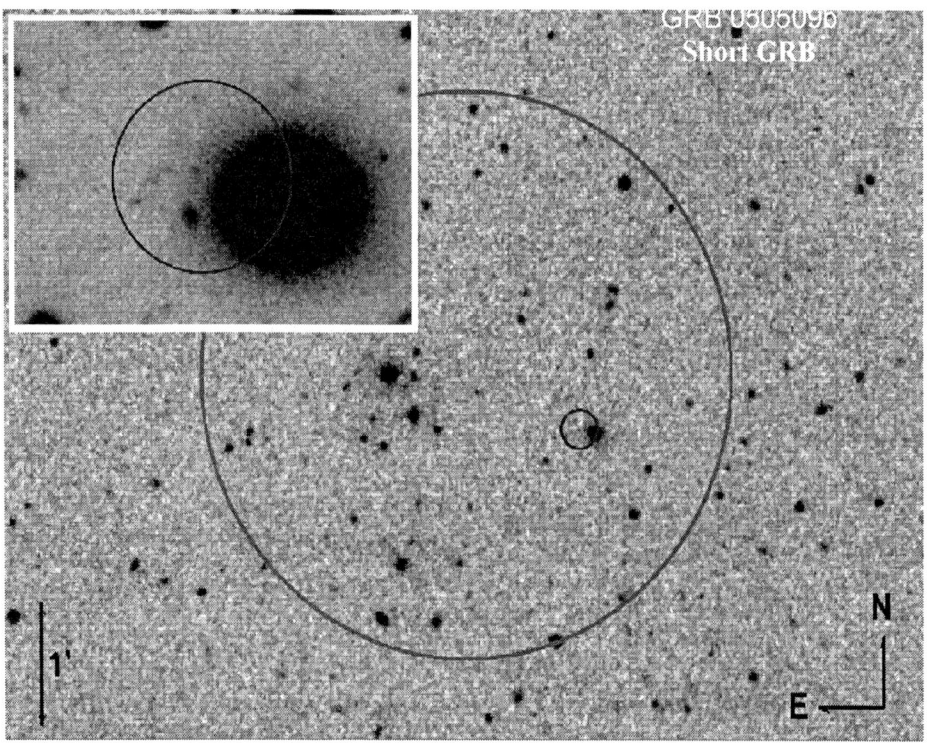

Fig. 4. Localization of short GRB 050509b. Large circle is BAT position; small circle is XRT position. The inset shows a bright elliptical galaxy in the XRT circle. From Gehrels et al. (2005).

2.5 Short GRBs

As of late November 2005, the BAT has detected 7 GRBs in the short-hard class. The first one of them (GRB 050202) had no prompt slew and no counterparts. From the next 2 events we have learned a great deal. GRB 050509b (Gehrels et al. 2005) had an X-ray afterglow that gave an error circle with a bright elliptical galaxy (cD galaxy in a cluster) in it (Figure 4). GRB 050724 (Barthelmy et al. 2005c) had an XRT afterglow, plus Chandra, optical and radio detections. The sub-arcsecond positions located it once again in the outer regions of a bright elliptical. The fact that these ellipticals have very low star formation rates argues strongly against a collapsar origin like that for long bursts. Also, the redshifts for the two are in the z = 0.2 to 0.3 range, a factor of ~3 closer than typical long GRBs. The evidence to date is consistent with an origin of short burst in merging binary neutron stars.

The picture for the last 4 short GRBs seen by Swift is still not well understood. GRB050813 appears to have a faint host at z=1.8. GRBs 050906 and 051105A did not have XRT detections despite rapid slews. GRB 050925 was in the galactic plane and had a soft spectrum; it may be a new galactic SGR.

2.6 GRB Redshifts

As of late November 2005, redshifts have been determined for 23 Swift GRBs. The average redshift (excluding short GRBs) is z=2.5. This is significantly higher than the pre-Swift average of z=1.2. The sensitivity of the Swift instruments is leading to a sampling of more distance GRBs.

On September 4, 2005, Swift detected a long, smooth GRB (Cusumane et al. 2005). The redshift was found to be the very large value of z =6.29, one of the highest redshift objects ever seen. The light curve for this GRB is shown in Figure 5.

2.7 Giant Flare from SGR 1806-20

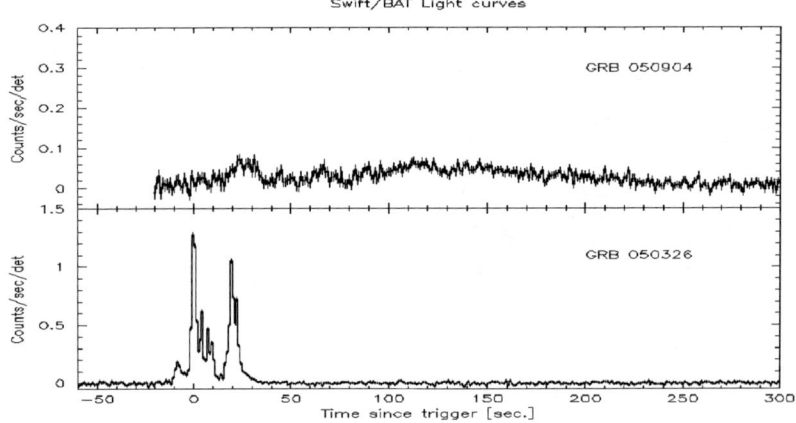

Fig. 5. Lightcurve of high-redshift GRB 050904 compared to a typical GRB. The long smooth nature of the lightcurve is due to cosmologic time dilation as the photons propagated to us from z=6.29.

On 27 December 2004, the Solar System was struck by the brightest gamma-ray transient ever observed. Every orbiting gamma-ray observatory detected the flash produced by the soft gamma-ray repeater SGR 1806-20. Although Swift was not pointed toward the target, the flux was so high that the BAT detector was swamped by more than a billion gamma-rays per cm2 passing through the structure of the spacecraft.

Palmer et al. (2005) present the Swift data on this dramatic event. Although the emitting system is located more than 10 kpc from the Earth, the energy flux was brighter than the full Moon for the 0.2 seconds. This giant flare was more than 100 times more luminous than the two previous flares seen in 1998 from SGR1900+14 and in 1979 from SGR0526-66.

Such events may be the cause of at least some short GRBs, in that the rapid, extremely bright flash of gamma rays had a similar duration and energy profile to a short GRB. Such an event in an external galaxy would be detectable out to 60 Mpc.

2.8 UV/Optical & X-ray Observations of SN2005am

Type Ia supernovae are critical to our understanding of the fundamental fabric of our Universe. They are the standard candles used to measure distances over the range in which cosmological effects become significant. Observations of nearby supernovae in the ultraviolet can be quite important for understanding and calibrating their light curves and luminosities.

Mission with UV capability such as the International Ultraviolet Explorer (IUE) and the Hubble Space Telescope began these studies, but they are limited in the intrinsically slower operational response time than offered by Swift. Thus Swift has been an ideal observatory for early observations of nearby bright supernovae, of which SN2005am is a prime example.

Brown et al. (2005) present ultraviolet and optical light curves for SN2005am, starting four days prior to maximum light, and extending to 69 days after peak. In addition, when the target was bright enough, Swift was able to carry out moderate-resolution grism UV/optical measurements. These data for SN2005am are the best sampled in time, and cover the widest range of any Type Ia supernova follow-up to date.

CONCLUSIONS

The Swift observatory is performing excellent scientific observations at high efficiency and with important progress toward its mission objectives.

The BAT is working flawlessly. The positional agreement with the XRT and ground-based detections suggests that the typical on-board positional accuracy for GRBs is roughly 65 arcsec, exceeding the pre-launch predictions.

The UVOT has demonstrated excellent UV and optical performance on GRBs and other sources. Source positions are acurate to 0.3 arcsec.

The XRT has demonstrated excellent X-ray sensitivity and rapid responsiveness. The average accuracy for the XRT positions confirmed with XMM or ground-based optical detection is 2.6 arcsec. XRT is observing afterglows at a level of 100 to 1000 times fainter than Beppo-SAX. This rapid acquisition with sensitive X-ray detection is revealing new lightcurve behaviors.

As Swift observations become more numerous we expect to build up a substantial database of prompt gamma-ray emission and early (to late) X-ray and UV/optical light curves. From these, new insights into GRB formation and GRB environments will be gleaned.

REFERENCES

Barthelmy, S., et al. 2005a, Sp Sci Rev. 120, 143.
Barthelmy, S., et al. 2005b, ApJ, (astro-ph 0511576).
Barthelmy, S., et al. 2005c, Nature, 438, 994.
Briggs, M. S. 1996, ApJ, 459, 40.
Brown, P. J., et al. 2005, ApJ, 635, 1192.
Burrows, D.N., et al. 2005a, Sp. Sci. Rev. 120, 165.
Burrows, D. N., et al. 2005b, Sci., 309, 1833.
Costa, E., et al. 1997, Nature, 387, 783.
Cusumano, G. et al. 2005, Nature accepted (astro-ph 0509737).
Frail, D. A., et al. 1997, Nature, 389, 261.
Gehrels, N., et al. 2004, ApJ, 661, 1005.
Gehrels, N., et al. 2005, Nature, 437, 851, 2005.
Klebesadel, R.W., Strong, I.B., & Olson, R.A. 1973, ApJ, 182, L85.
Mason, K., et al. 2005, ApJ, submitted (astro-ph 0511132).
Nousek, J., et al., 2005, ApJ, accepted (astro-ph 0508332).
Palmer, D., et al. 2005, Nature, 434, 1107.
Roming, P. et al., 2005a, Sp. Sci. Rev., 120, 95.
Roming, P., et al. 2005b, ApJ, (astro-ph 0511751).
Still, M., et al. 2005, ApJ, 635, 1187.
Tagliaferri, G., et al. 2005, Nature, 436, 985.
van Paradijs, J., et al. 1997, Nature, 386, 686.
Wijers, R. A. M. J., Rees, M. J., & Meszaros, P. 1997, MNRAS, 288, L51.
Zhang, B., et al., 2006, ApJ, (astro-ph 0508321).

SHORT GRBs AND SGRs

Observations of the short-duration gamma-ray bursts: prompt emission

K. Hurley

UC Berkeley, Space Sciences Laboratory, 7 Gauss Way, Berkeley, CA 94720-7450

Abstract. The observational properties of the short-duration gamma-ray bursts are reviewed. The time histories and energy spectra of these bursts, whether they are observed by the interplanetary network, HETE, Swift, or BATSE, appear to be drawn from the same sample. However, the fluences of the bursts differ considerably; in order of decreasing fluence, they are the bursts observed by the IPN, HETE, Swift, and BATSE. This suggests that, just as for the long duration bursts, Swift may be observing a more distant population. The question of whether some events could be extragalactic giant magnetar flares is examined. One recent event, GRB051103, may be in this category.

Keywords: gamma-ray bursts
PACS: 98.70.Rz

INTRODUCTION

The bimodal nature of the GRB duration distribution has been recognized for decades and studied by many people [1, 2, 3, 4, 5, 6]. Figure 1 shows the distribution of 1971 BATSE bursts, using data from the BATSE website (www.batse.msfc.nasa.gov/batse/). Both the long and the short duration distributions can be fit by lognormal probability functions [7, 8]. Figure 1 shows the fit to the BATSE distribution using a double lognormal function (i.e. the sum of two lognormal functions). From the fit to the long burst distribution, we can estimate that the probability that a burst of duration 1 s or less is actually a long burst in disguise is $< 10^{-6}$. That will be the approximate cutoff for the events to be discussed here: their durations must be less than one second.

Spectrally, the long and the short bursts tend to be different. First, the short bursts exhibit much shorter spectral lags than the long ones [9]. And second, there is a tendency for the short bursts to have harder spectra than the long ones, although it is just a tendency. Figure 2 shows the 1971 bursts on a hardness ratio – duration plot. The hardness ratio is defined here as the 100–300 keV fluence divided by the 25–100 keV fluence; earlier versions of this plot have been published [5, 10]. As figure 2 shows, there are short, soft bursts, and long, hard ones. In a recent study [11], the energy spectra of the long and short bursts were compared. It was found that the low energy spectral index α was larger for the short bursts than the long ones (consistent with the larger hardness ratios in figure 2), but that the peak of the spectrum, E_{peak} was, on average, slightly greater for the *long* bursts. This result was obtained when the time-integrated spectra of the long bursts were compared to the short bursts. But if the energy spectra of only the first two seconds of the long bursts were compared to the energy spectra of the short bursts, the α and E_{peak} distributions for the two samples were practically indistinguishable. Therefore we will make no cut on the energy spectra of the bursts

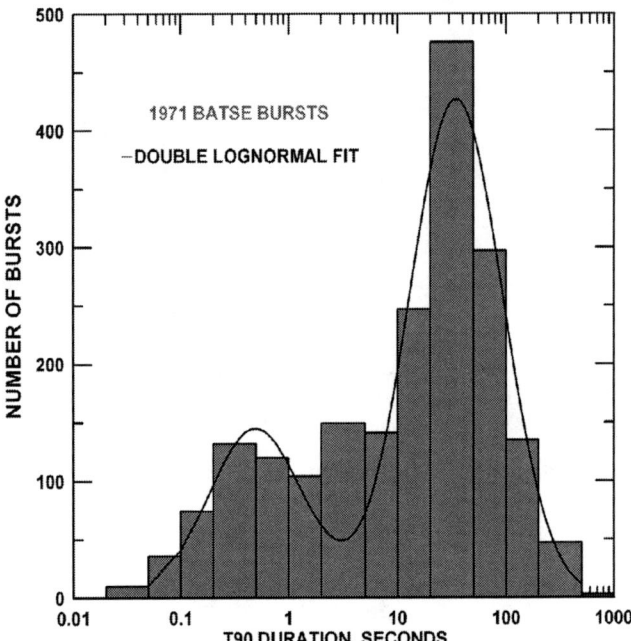

FIGURE 1. The duration distribution of 1971 bursts observed by BATSE. The T90 durations are taken from the BATSE website. Solid black line: a fit to the distribution using the sum of two lognormal probability distributions. The means, standard deviations, and modes for the two distributions are 0.31, 1.0, and 0.5 s for the short bursts, and 4.5, 1.0 and 33 s for the long bursts.

considered here.

A short burst, even with a hard energy spectrum, is not difficult to detect. However, it can be difficult to localize with an imaging system. In the <15 keV range, these bursts tend to be X-ray deficient, so although a coded mask will operate efficiently, there are often not enough photons to image. At higher energies, say >25 keV, coded masks tend to be partially transparent to the hard spectrum. The interplanetary network can localize short, hard bursts to very good accuracy ($\sim 1'$), but usually with delays of half a day or so. Since we now know that short bursts tend to have weak afterglows, this delay may preclude their detection. The relative numbers of short and long bursts that a given detection system will observe depend on the detector and its electronics, specifically efficiency as a function of energy and the trigger criteria. 25% short bursts and 75% long bursts (roughly the numbers observed by BATSE, figure 1) is probably a good estimate. But all detectors have a bias against the detection of very short events, due to detector saturation effects. As a burst becomes shorter, it must contain more photons per unit time to trigger a detector. Eventually, above a certain threshold, the detection system saturates, and either produces no more counts, or in some cases, fewer counts, as the number of photons increases. The point where saturation becomes important varies from detector to detector, and to date, the shortest burst detected had a duration ~ 8 ms

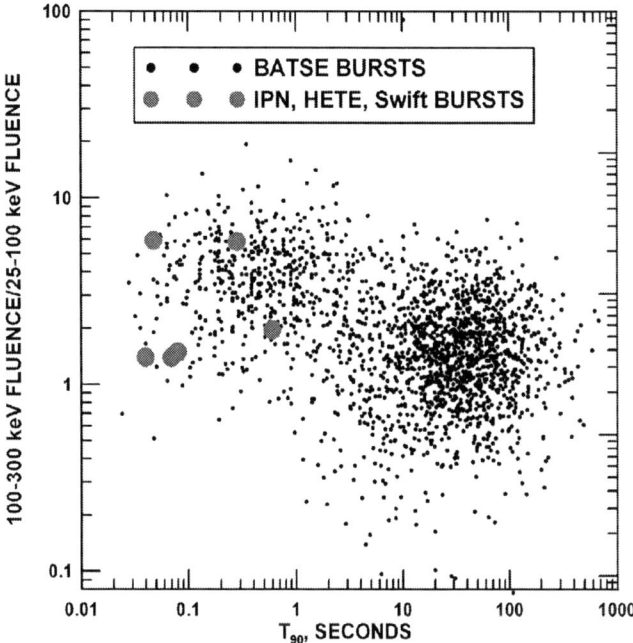

FIGURE 2. The hardness ratios and durations of 1971 BATSE bursts, using the data from the BATSE website, compared to those of several IPN, HETE, and Swift bursts.

[12].

TIME HISTORIES

Some structure was evident in the time histories of the short bursts even with the relatively small IPN detectors [13, 14]. Many short bursts revealed considerably more structure when observed with BATSE; these observations also made it possible to study time histories at times much longer than the burst durations, and evidence was found for long-lasting emission [15]. Other studies, using both BATSE and other data, have revealed similar behavior [16, 17, 18, 19]. This could either represent flaring, which is seen in the X-ray afterglows of long bursts [20], or it could be the X- and gamma-ray afterglow of the burst itself, or in some cases, part of the prompt emission. The HETE burst GRB050709B displayed the first clear evidence for a soft X-ray afterglow or flare following a short event [21]. Its time history is shown in figure 3.

Although the IPN, HETE, and Swift short bursts appear to be quite consistent with the BATSE sample as far as hardness ratios and durations are concerned, their fluences are very different. This is shown in figure 4. The IPN bursts tend to have the highest fluences, as would be expected from the relatively small sizes of the IPN detectors. HETE bursts go to fluences which are about an order of magnitude smaller, and Swift bursts go down

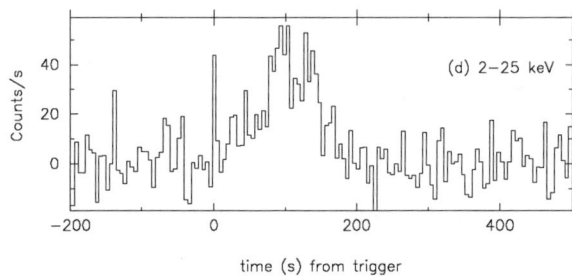

FIGURE 3. The HETE WXM time history of GRB050709B at low energies. This is the first short GRB to display a clear X-ray afterglow. Note that the GRB itself is only observed weakly in this energy range and with this binning. The low energy flaring or afterglow is clearly visible, lasting 200 s. This component of the burst has a soft spectrum.

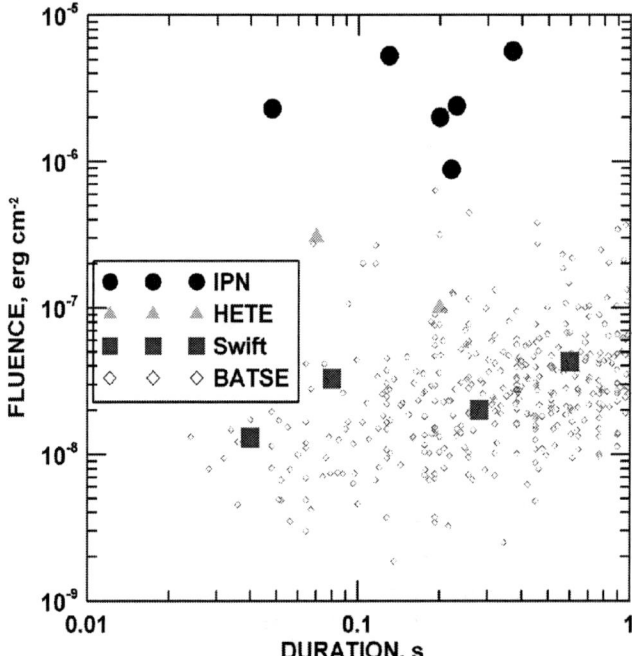

FIGURE 4. The fluences and durations of IPN, HETE, Swift, and BATSE short bursts.

still another order of magnitude. However, the BATSE burst fluences extend down yet one more order of magnitude. Since we know that the redshifts of the Swift long duration bursts tend to be higher than those of the pre-Swift bursts, one might guess from figure 5 that the same will be true for the short Swift bursts. Indeed, based on the statistics of just a few events, this appears to be the case (see table 1).

SIX WELL STUDIED SHORT GRBS

Studying the short bursts in detail is not easy. Since 1996, there have been ~350 GRBs (long and short) with relatively small, rapidly determined error boxes. Of course, over this period, our definitions of what constitutes a small box and what is rapid have evolved, but these are bursts for which counterpart searches could have been, and were, in some cases, carried out. Only 15 of the 350 had durations strictly < 1 s. Of the 15, redshifts have been measured for only 5. Long-wavelength afterglows have been observed for only 3 of the 5, and E_{peak} (the peak energy in a $\nu - F_\nu$ plot) and evidence for beaming have been found for only 1 of the 3. Note that, contrary to the case of the long bursts, it *is* possible to study the error boxes of the short bursts many years later, and attempt with some success to identify counterparts and measure redshifts [22].

ROTSE and TAROT have responded to 3 short-duration bursts within tens of seconds to minutes with varying coverage of the error box, but have detected no counterparts [23, 24]. There remains only one intriguing report of a bright optical afterglow: an $m_I \sim 9$ object 4 minutes after GRB000313 [25]. All the other optical afterglows of the short bursts are considerably fainter.

There are now actually 6 short bursts with reasonably well determined redshifts. They are summarized in table 1, which gives the date, the mission which localized it, the redshift, the approximate isotropic energy in the gamma-ray range, whether evidence was found for beaming (i.e. a break in the afterglow decay curve), and the peak energy. When the GRB error box is relatively large, and/or no afterglow is detected which can reduce its size, it contains more than one possible host galaxy, and one has been selected as the host using a probability argument. This is indicated with a question mark following the redshift; in these cases, the host galaxy identification should be considered plausible, but not proven. Note the tendency for the IPN, HETE, and Swift burst redshifts to increase in that order. In cases where no afterglow was detected, or where no break can be found in the decay light curve, no evidence can be found for or against beaming, and this is indicated with a question mark in this column. Finally, if a burst spectrum was not fit to a Band model, no E_{peak} was determined. In the case of the Swift bursts, this usually means that a simple power law fit is satisfactory over the measured energy range; E_{peak} could be outside that relatively restricted range, and again a question mark in this column indicates this. One event, 050724, has a 50–300 keV duration of 1.3 s, which is slightly beyond the cutoff. However, it posesses all the other properties of a short burst, and has been included in the table.

Evidence for beaming [26] and a peak energy [21] have been found for only one event (GRB050709B). Thus this is the only event that we can place on the Amati [27] and Ghirlanda [28] plots, which relate E_{peak} to the isotropic gamma-ray energy and the beaming-corrected energy, respectively, for long bursts. GRB050709B does not lie on the Amati relation, as noted by Villasenor [21], but it does come close to satisfying the Ghirlanda relation.

TABLE 1. Six well studied short GRBs

Date	Detected by	Redshift	$E_{\gamma,iso}$ erg	Evidence for beaming?	E_{peak}, keV
790613	IPN	0.09?	6×10^{49}	?	?
050509B	Swift	0.225?	2.7×10^{48}	?	?
050709B	HETE	0.16	3×10^{49}	Yes; 0.25 rad	84
050724	Swift	0.258	9.9×10^{49}	?	?
050813	Swift	0.722	1.7×10^{50}	?	?
050906	Swift	0.031?	1.2×10^{47}	?	?

SHORT BURSTS AND GIANT MAGNETAR FLARES

The observation of the giant magnetar flare from SGR1806-20 on December 27 2004 [29, 30, 31] has revived the question of whether some short duration bursts can be explained by extragalactic SGR flares. Figure 5 shows what the December 27 event would look like if it occurred at a distance 3 times greater and 300 times greater than its actual distance, as detected by RHESSI. If SGR1806-20 is at a distance of 15 kpc, these distances correspond roughly to the LMC and M81. At the LMC, the periodicity following the main peak is obvious to the eye (just as the periodicity in the March 5 1979 event was). But if the distance were 30 times greater than the December 27 event (roughly at Andromeda), the periodicity would probably not be detectable with RHESSI. At the distance of M81, the peak of a giant magnetar flare would be just detectable above the background. Of course, more sensitive detectors will respond more strongly.

The signatures of an extragalactic magnetar flare are first, a fast rise time and a short duration, and second, a hard energy spectrum. A third requirement is a total energy $< 10^{47}$ erg [29]. With luck, a relatively nearby flare would display a periodicity, probably in the 5–10 s range. And finally, repetition is expected, although the timescale is probably decades. The Swift burst GRB050906 [32] was initially proposed as a candidate [33], based on the time history and energetics (a bright galaxy at a distance of 130 Mpc was present in the 3 arcminute radius BAT error circle). However, later analysis has shown that the energy spectrum is too soft to be consistent with a giant flare [35]. GRB051103 [34] remains a plausible candidate. It has a rise time of 4 ms and a hard spectrum ($E_{peak} \sim 2$MeV). Its IPN position is consistent with the halo of M81, or possibly with M82. At the distance of M81, the total energy would be $\sim 4 \times 10^{46}$erg, although one question is why a magnetar, which is probably a young object, would be found in the halo of a galaxy.

CONCLUSIONS

HETE and Swift have begun to solve the long-standing mystery of the short bursts. Although Swift has begun to detect short events at a good rate, both HETE and the IPN are still needed. In addition to providing more short burst positions, they also complete the energy spectra of Swift bursts below 15 keV and above 350 keV. This means that

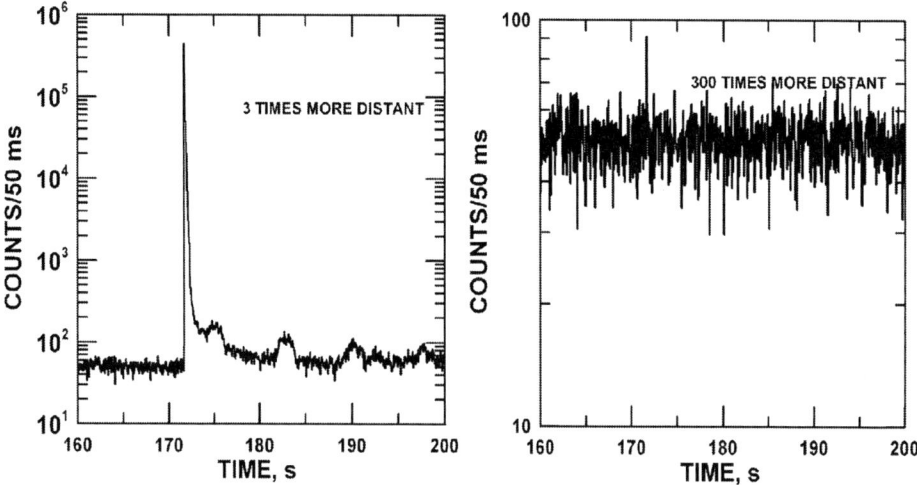

FIGURE 5. The giant flare from SGR1806-20, as it would be observed by RHESSI at distances 3 and 300 times greater.

early X-ray flares or afterglows can be studied in the first case, and peak energies above the Swift upper energy threshold can be measured, providing a better estimate of the total energy, and also allowing the short bursts to be placed on the Amati and Ghirlanda relations.

ACKNOWLEDGMENTS

I am grateful for support under NASA grants FDNAG5-9210 and NNG05GF72G.

REFERENCES

1. E. Mazets et al., *Astrophys. Space Sci.*, **84**, 173–179 (1982)
2. J.-P. Dezalay et al.,"Short Cosmic Events: A Subset of Classical GRBs?", in *Gamma-Ray Bursts*, edited by W. Paciesas and G. Fishman, AIP Conference Proceedings 265, American Institute of Physics, New York, 1992, pp. 304–309
3. J. Norris et al., *Nature*, **308**, 434–435 (1984)
4. K. Hurley, "Gamma-Ray Burst Observations: Past and Future", in *Gamma-Ray Bursts*, edited by W. Paciesas and G. Fishman, AIP Conference Proceedings 265, American Institute of Physics, New York, 1992, pp. 3–12
5. C. Kouveliotou et al., *Ap. J.*, **413**, L101–L104 (1993)
6. J. Norris et al., *B.A.A.S.*, **32(3)**, 1244–1244 (2000)
7. B. McBreen et al., *Mon. Not. R. Astron. Soc.*, **271**, 662–666 (1994)
8. S. McBreen et al., *Astron. Astrophys.*, **380**, L31–L34 (2001)
9. J. Norris, "Short Gamma-Ray Bursts are Different", in *Gamma-Ray Bursts in the Afterglow Era*, edited by E. Costa, F. Frontera, and J. Hjorth, ESO Astrophysics Symposia, Springer, Berlin, 2001, pp. 40–42
10. W. Paciesas et al., *Ap. J. Sup. Ser.*, **122**, 465–495 (1999)
11. G. Ghirlanda et al., *Astron. Astrophys.*, **422**, L55–L58 (2004a)

12. N. Bhat et al., *Nature*, **359**, 217–218 (1992)
13. J. Laros et al., *Ap. J.*, **245**, L63–L66 (1981)
14. C. Barat et al., *Ap. J.*, **280**, 150–153 (1984)
15. V. Connaughton, *Ap. J.*, **567**, 1028–1036 (2002)
16. R. Burenin, *Astron. Lett.*, **26(5)**, 269–276 (2000)
17. D. Lazzati et al., *Astron. Astrophys.*, **379**, L39–L43 (2001)
18. D. Frederiks et al., "Early Hard X-Ray Afterglows of Short GRBs with Konus Experiments", in *Proc. 3rd Rome Workshop on Gamma-Ray Bursts in the Afterglow Era*, edited by M. Feroci, F. Frontera, and N. Masetti, ASP Conference Proceedings 312, ASP, San Francisco, 2004, pp. 197–200
19. E. Montanari et al., *Ap. J.*, **625**, L17–L21 (2005)
20. D. Burrows et al., *Science*, **5742**, 1833–1835 (2005)
21. J. Villasenor et al., *Nature*, **437**, 855–858 (2005)
22. A. Gal-Yam et al., "The Progenitors of Short-Hard Gamma-Ray Bursts from an Extended Sample of Events", astro-ph/0509891 (2005)
23. R. Kehoe et al., *Ap. J.*, **554**, L159–L162 (2001)
24. A. Klotz et al., *Astron. Astrophys.*, **404**, 815–818 (2003)
25. A. Castro-Tirado et al., *Astron. Astrophys.*, **393**, L55–L59 (2002)
26. D. Fox et al., *Nature*, **437**, 845–850 (2005)
27. L. Amati et al., *Astron. Astrophys.*, **390**, 81–89 (2002)
28. G. Ghirlanda et al., *Ap. J.*, **616**, 331–337 (2004b)
29. K. Hurley et al., *Nature*, **434**, 1098–1103 (2005)
30. S. Mereghetti et al., *Ap. J.*, **624**, L105–L108 (2005)
31. D. Palmer et al., *Nature*, **434**, 1107–1109 (2005)
32. H. Krimm et al., *GCN Circ.* 3926 (2005)
33. A. Levan and N. Tanvir, *GCN Circ.* 3927 (2005)
34. S. Golenetskii et al., *GCN Circ.* 4197 (2005)
35. K. Hurley et al., in preparation (2006)

The Afterglows and Host Galaxies of Short GRBs: An Overview

Edo Berger

Carnegie Observatories, Pasadena, CA

Abstract. Despite a rich diversity in observational properties, gamma-ray bursts (GRBs) can be divided into two broad categories based on their duration and spectral hardness – the long-soft and the short-hard GRBs. The discovery of afterglows from long GRBs in 1997, and their localization to arcsecond accuracy, was a watershed event. The ensuing decade of intense study led to the realization that long-soft GRBs are located in star forming galaxies, produce about 10^{51} erg in collimated relativistic ejecta, are accompanied by supernovae, and result from the death of massive stars. While theoretical arguments suggest that short GRBs have a different physical origin, the lack of detectable afterglows prevented definitive conclusions. The situation changed dramatically starting in May 2005 with the discovery of the first afterglows from short GRBs localized by *Swift* and *HETE-2*. Here I summarize the discovery of these afterglows and the underlying host galaxies, and draw initial conclusions about the nature of the progenitors and the properties of the bursts.

Keywords: gamma-ray bursts; host galaxies
PACS: 98.70.Rz

HISTORY AND MODELS

The detection of short-duration gamma-ray bursts (GRBs) dates back to the *Vela* satellites [1]. However, only in 1993 the short bursts (with $T_{90} \lesssim 2$ s) were recognized as a separate sub-class from the long GRBs, and were furthermore shown to have on average a harder γ-ray spectrum [2]; hereafter I will refer to short-hard bursts as SHBs. The short durations of SHBs suggest that they are unlikely to result from the death of massive stars, for which the natural timescale (the free-fall time) is significantly longer, $t_{ff} \approx 30\,\text{s}\,(M/10M_\odot)^{-1/2}(R/10^{10}\,\text{cm})^{3/2}$.

Instead the main theoretical thrust has been focused on coalescing compact objects – neutron stars and/or black holes (DNS or NS-BH) – as the progenitors of SHBs [3, 4, 5, 6, 7]. In this context the duration, which is set by the viscous timescale of the gas accreting onto the newly-formed black hole, is short due to the small scale of the system. Other progenitors have been proposed in addition to the DNS and NS-BH systems, namely magnetars, thought to be the power source behind soft γ-ray repeaters [8], and accretion-induced collapse (AIC) of neutron stars [9, 10]. The magnetar model is unlikely to account for SHBs at cosmological distances, since even the 2004 Dec. 27 giant flare from SGR 1806-20 would only be detected by *BATSE* or *Swift* at $\lesssim 50$ Mpc; magnetars may contribute to a local population of SHBs. The AIC model has not been investigated in detail, but in the case of a white dwarf was shown to be too baryon rich to produce GRBs [11].

Naturally, without a distance and energy scale, or an understanding of the micro- and macro-environments of SHBs, it is nearly impossible to make any quantitative

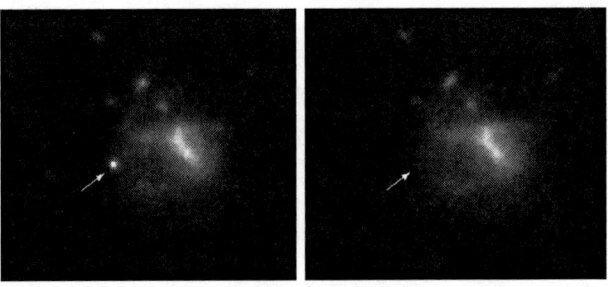

FIGURE 1. HST images of the optical afterglow and irregular star forming host of GRB 050709.

statements about their progenitors or the detailed underlying physics. As in the case of the long GRBs, this understanding relies on arcsecond positions, which in turn require the identification of afterglows. Several SHBs have been localized to sufficient accuracy (few arcmin^2) in the past to allow afterglow searches, but none have been detected at optical or radio wavelengths due to delayed and shallow searches [12].

THE DISCOVERY OF SHORT GRB AFTERGLOWS

The breakthrough in understanding the origin and properties of SHBs resulted from the localization of the first afterglows starting in May 2005. Here I provide a short account of these discoveries, highlighting the growing understanding achieved with each burst.

GRB 050509b was detected by *Swift* on 2005 May 9.167 UT with a duration of 40 ms and a fluence, $F_\gamma = 9.5 \times 10^{-9}$ erg cm^{-2} [13]. A fading X-ray afterglow was localized with the XRT onboard *Swift* to a positional accuracy of $9.3''$ radius [13]. Despite intense follow-up at optical and radio wavelengths no candidates were discovered to a limit of 24 mag at $t \approx 2.5$ hr [14, 15] and 0.1 mJy at $t \approx 2.2$ hr [16], respectively. Instead, a bright elliptical galaxy ($L \approx 3L^*$) at $z = 0.225$ was detected in coincidence with the X-ray error circle [14, 13]. The *a posteriori* probability of such a coincidence has been estimated at $0.01 - 1\%$ [14, 13, 17]. The implication of this possible association is that the progenitors of SHBs are related to an old stellar population. Unfortunately, the XRT error circle also contains over twenty fainter galaxies, which most likely reside at higher redshifts, thereby preventing a definitive association. The optical follow-up also placed a limit of $M_B > -13.3$ mag (5 mag fainter than SN 1998bw) on a supernova coincident with GRB 050509b [14, 15], providing additional support to a non-massive star origin *if the redshift of $z = 0.225$ is adopted*.

GRB 050709 was detected by *HETE-2* on May 9.942 UT with a duration of 70 ms and a fluence, 4.0×10^{-7} erg cm^{-2} $(2-400$ keV). The initial pulse was followed 25 s later by a softer X-ray bump with a duration of 130 s and a fluence of 1.1×10^{-6} erg cm^{-2} $(2-25$ keV) [19]. Observations with *Chandra* provided an arcsecond position [20], and led to the discovery of the optical afterglow [21, 20] (Fig. 1). No radio afterglow was detected [20]. Spectroscopy revealed that the host is a star forming galaxy at $z = 0.160$, but HST imaging showed that the burst did not coincide with a bright star forming region [20]. No

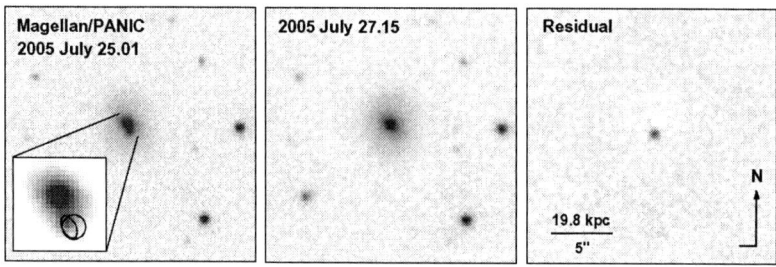

FIGURE 2. Magellan near-IR images of the afterglow and host galaxy of GRB 050724 on two separate occasions, and the residual image clearly showing the fading afterglow. The inset shows the radio (ellipse) and X-ray (circle) positions of the afterglow. From Ref. [18].

supernova component was detected [21]. These results suggest that while GRB 050709 occurred in a star forming galaxy, it was not related to a massive star.

GRB 050724 was localized by *Swift* on 2005 July 24.524 UT with a duration of 3 ± 1 s, dominated by an initial spike of 250 ms duration, and a fluence, $F_\gamma = 3.9 \times 10^{-7}$ erg cm^{-2} (15 – 150 keV) [22]. As in the case of GRB 050709, the initial pulse was followed 30 s later by a soft bump (15 – 25 keV) which lasted 120 s, but with a fluence of only 10% that of the main pulse. A bright X-ray afterglow detected by XRT [22] led to the discovery of the radio, optical and near-IR afterglow, which coincided with a bright elliptical galaxy at $z = 0.257$ (Fig. 2) [18]. This burst provided the first unambiguous association with a galaxy undergoing no current star formation; the limit at the position of the GRB is < 0.05 M$_\odot$ yr^{-1} [18]. The X-ray and optical light curves revealed a broad increase by nearly an order of magnitude at 0.65 d, suggestive of energy injection [22, 18]. The subsequent steep decline at $t \gtrsim 1$ d is reminiscent of the jet breaks detected in long GRBs, indicating a jet opening angle of about $8.5°$ [18].

GRB 050813 was detected by *Swift* on 2005 August 13.281 UT with a duration of 0.6 s and a fluence, 4.3×10^{-8} erg cm^{-2} (15 – 150 keV). The initial X-ray position included a pair of galaxies, of which one is at a redshift $z = 0.72$ [23, 24], and appears to be part of a galaxy cluster [25]. However, the revised XRT position excludes these galaxies, and instead contains a galaxy at $z \approx 1.8$, which appears to be part of a cluster at that redshift. No optical or radio afterglows were detected. Since this is the highest redshift SHB to date, it has significant implications for the progenitor lifetime (see below).

GRB 051221 is perhaps the best-studied SHB to date. It was discovered by *Swift* on 2005 December 21.077 with an initial hard pulse of 250 ms duration, followed by softer emission, which lasted about 1 s, and a total fluence, 3.2×10^{-6} erg cm^{-2} (20 – 2000 keV) [27, 28]. The X-ray afterglow was localized to $3.5''$ radius accuracy [29], leading to the identification of the optical, near-IR, and radio counterparts [26] (Fig. 3). A spectrum of the combined afterglow and host indicated a redshift of $z = 0.546$ [26]. The afterglow evolution follows a simple power law decay to $t \approx 13$ d, interrupted only by a period of flattening from 1.4 to 3.1 hr in the X-rays and reverse shock emission in the radio band (Fig. 4). The light curve evolution indicates an opening angle $> 13°$, an energy injection of about a factor of three, and a total energy of at least 1.0×10^{50} erg [26].

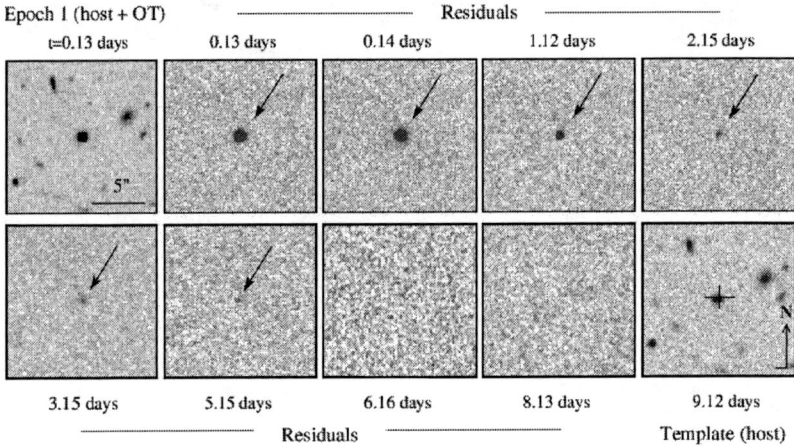

FIGURE 3. The fading optical afterglow and host galaxy of GRB 051221. From Ref. [26].

THE PROPERTIES AND PROGENITORS OF SHORT GRBS

The discovery of SHB afterglows allows us to address for the first time the basic questions regarding the nature and properties of these events:

- What is the energy release of SHBs (prompt emission and relativistic blast wave)?
- Is there evidence for prolonged engine activity?
- Are the ejecta collimated or spherical?
- Are SHBs accompanied by supernova-like events?
- What is the density and structure of the circumburst medium?
- In what type of host galaxies do SHBs occur?

Short GRB Energetics

The redshift distribution of the five SHBs with precise or putative redshifts ranges from about 0.16 to ~ 1.8, with three of the five bursts at $z \lesssim 0.3$ (Fig. 5); GRB 050813 stands out with a significantly higher redshift. At the same time, a statistical comparison of BATSE SHB positions to low-redshift galaxy catalogs suggests that $\sim 10-25\%$ of the BATSE SHBs may originate within 100 Mpc [30]. The isotropic γ-ray energies span a wide range, $E_{\gamma,\mathrm{iso}} \approx 1.1 \times 10^{48} - 3 \times 10^{51}$ erg, which is at the low end of the distribution for long GRBs (Fig. 5) [20]. The afterglow X-ray luminosities, which serve as a proxy for the blast-wave kinetic energy [31], span an equally wide range, $L_{X,\mathrm{iso}}(t=10\,\mathrm{hr}) \approx 7.0 \times 10^{41} - 6.4 \times 10^{44}$ erg s^{-1}. These values are again at the low end of the distribution for long GRBs [32].

As in the case of long GRBs, the true energy release is strongly dependent on collimation of the ejecta. Both GRBs 050709 and 050724 exhibit evidence for jets through

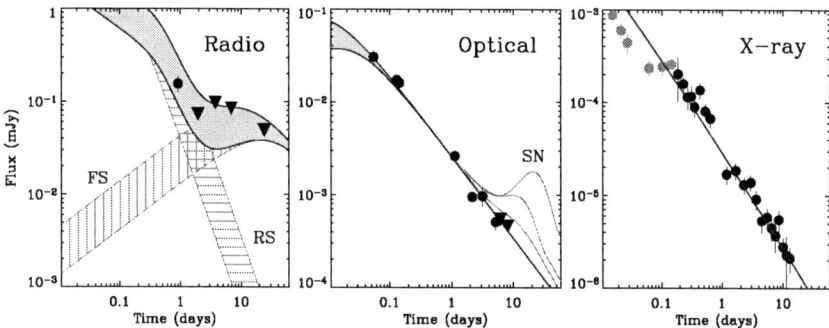

FIGURE 4. Broad-band light curves of the afterglow of GRB 051221. The black lines and shaded regions are model fits, which include forward shock emission and the reverse shock contribution in the radio, which is associated with the energy injection episode in the X-rays (at 0.05 d). From Ref. [26].

breaks and steep decays of their afterglow light curves [18, 20]. In the former case the opening angle is 14° while for the latter it is about 8.5°. GRB 051221, on the other hand, exhibits no clear break out to at least 13 days, or $\theta_j > 13°$ [26]. The inferred jet angles are consistently wider than the median value of $\langle \theta_j \rangle \approx 5°$ for long GRBs [26].

The broad-band light curves of GRBs 050709, 050724, and 051221 also allow a rough determination of the blast wave kinetic energies using the standard synchrotron model [33, 26] (Fig. 4). The parameters of interest are E_{KE}, the circumburst density (n), and the fractions of energy in relativistic electron (ε_e) and magnetic fields (ε_B). The values derived for the three SHBs are summarized in Tab. 1, and the derived beaming corrected energies are plotted in Fig. 5 [18, 20, 26]. The energies of GRBs 050709 and 050724 are about two orders of magnitude lower than those of long GRBs, $E \approx$ few $\times 10^{51}$ erg [34, 35, 36], but GRB 051221 has an energy that is at least an order of magnitude larger, and potentially as large as that of long GRBs.

The energy scale of GRB 051221 has important implications for the energy extraction mechanism, $\nu\bar{\nu}$ annihilation or MHD processes related to the black hole and/or the accretion disk [37, 7]. Numerical calculations reveal that $\nu\bar{\nu}$ annihilation is unlikely to power an SHB with a beaming-corrected energy in excess of few $\times 10^{48}$ erg, while MHD processes can produce $> 10^{52}$ erg s^{-1}. Thus, GRBs 050709 and 050724 could in principle be powered by $\nu\bar{\nu}$ annihilation, but it is unlikely that GRB 051221 was [26].

Prolonged Engine Activity

One of the intriguing results emerging from observations of the prompt emission and X-ray afterglows is evidence for prolonged engine activity and delayed energy injection. A hint of this result was already available from summed light curves of BATSE SHBs, which led to the possible detection of excess emission peaking ~ 30 s after the burst with a duration of ~ 100 s [38, 39]. This was originally proposed as evidence for afterglow emission, but there is now evidence from GRBs 050709 and 050724 that this is not the

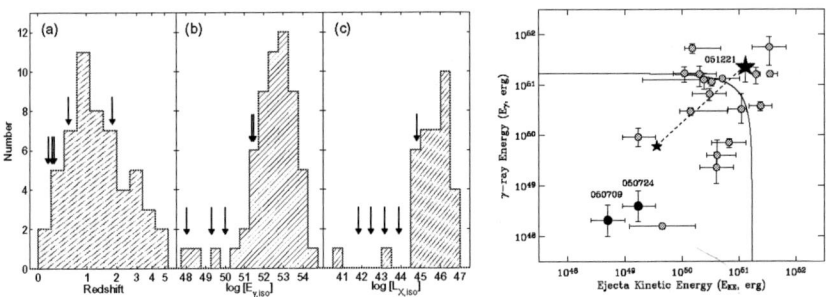

FIGURE 5. *Left:* Distributions of (a) redshift, (b) isotropic γ-ray energy, and (c) isotropic X-ray luminosity (a proxy for the blast wave kinetic energy). Adapted from Ref. [20]. *Right:* Beaming corrected γ-ray and afterglow kinetic energies for SHBs (black) and long GRBs (gray). From Ref. [26].

TABLE 1. Physical Properties of SHBs

	050709	050724	051221
Redshift	0.160	0.257	0.546
$E_{\gamma,\mathrm{iso}}$ (erg)	6.9×10^{49}	4.0×10^{50}	2.4×10^{51}
$E_{\mathrm{KE,iso}}$ (erg)	1.6×10^{48}	1.5×10^{51}	1.4×10^{51}
n (cm^{-3})	~ 0.01	~ 0.1	$\sim 10^{-3}$
ε_e	~ 0.3	~ 0.04	~ 0.3
ε_B	~ 0.3	~ 0.02	~ 0.1
θ_j (deg)	14	9	> 13
f_b	0.03	0.01	> 0.03
E_γ (erg)	2.1×10^{48}	4.5×10^{48}	$(6-240) \times 10^{49}$
E_{KE} (erg)	5.0×10^{48}	1.7×10^{49}	$(4-140) \times 10^{49}$
Reference	[20]	[18]	[26]

case (Fig. 6). In addition to the early flares, the light curves of GRBs 050724 and 051221 exhibit evidence for delayed energy injection [22, 26], which may arise from extended engine activity and/or a wide distribution of ejecta Lorentz factors.

Given the short lifetime and dynamical timescale in the case of DNS or NS-BH mergers, the theoretical expectation is for simple afterglow evolution and a short duration of the γ-ray emission. The X-ray flares do not fit naturally in this scenario, although delayed activity arising from fragmentation of the accretion disk has been proposed and argued to agree with the basic observational signatures [40]. An intriguing alternative, in the context of accretion-induced collapse, is that the flares result from the interaction of the emerging blast wave with the giant companion star, with the timescale and duration of the flare set by the binary separation and the radius of the companion [10]. Two observations can be used to test this scenario: (i) multiple flares, and (ii) if most SHBs are collimated, then only the small fraction whose jets intercept the companion should exhibit flares.

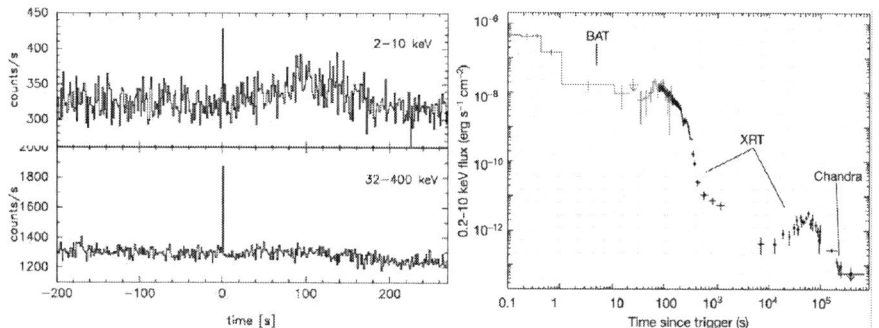

FIGURE 6. Flares in the prompt emission and X-ray afterglow of GRBs 050709 (left; Ref. [19]) and 050724 (right; Ref. [22]).

Host Galaxies and Offsets

One of the main clues that long GRBs are related to the death of massive stars came from their association with star forming galaxies. Long GRB hosts have star formation rates (SFR) typically in excess of 1 $M_\odot$ yr^{-1} (and sometimes > 100 $M_\odot$ yr^{-1} [41]), stellar populations younger than $\sim 10^8$ yr, and specific star formation rates that are higher than the general population of star forming galaxies [42]. In addition, the distribution of long GRBs relative to their hosts traces that of massive stars [43].

In a similar vein, we can use the hosts of SHBs to address the properties of the progenitors. The most striking difference compared to long GRBs is the existence of SHBs in elliptical galaxies. The limits on the SFR for the elliptical hosts of GRBs 050509b and 050724 are < 0.1 and < 0.05 $M_\odot$ yr^{-1}, respectively [14, 18]. Even the star forming hosts of GRBs 050709 and 051221, with ~ 0.5 and ~ 1.5 $M_\odot$ yr^{-1}, respectively, have lower SFRs than the median for hosts of long GRBs [20, 26]. The host of GRB 051221, moreover, exhibits evidence for an evolved stellar population and a near solar metallicity.

The mix of host types is reminiscent of type Ia supernovae (SNe Ia), which occur in both early- and late-type galaxies. It is interesting to note that the luminosity of SNe Ia is correlated with the host type, with intrinsically dimmer events occurring in early-type galaxies [44]. The current sample of SHBs is too small to address possible correlations of the burst and host properties, but this should become possible in the future.

It has also been noted that some SHBs occur in galaxy clusters. GRB 050509b is likely located in the cluster ZwCl 1234.0+02916, while GRB 050813 is most likely associated with a high redshift cluster. Several other cluster associations have been claimed for SHBs that lack arcsecond positions. At the present it is difficult to assess whether the latter associations are in fact correct, but some caution should be exercised given that the probability of finding a cluster in the NED archive within $10'$ from a random position on the sky is non-negligible, $\sim 5\%$ [45]. Moreover, since none of the SHBs with arcsecond positions are located in clusters, we consider the cluster associations claimed to date to be suggestive, though not secure.

Finally, as in the case of long GRBs the offset distribution has been proposed as

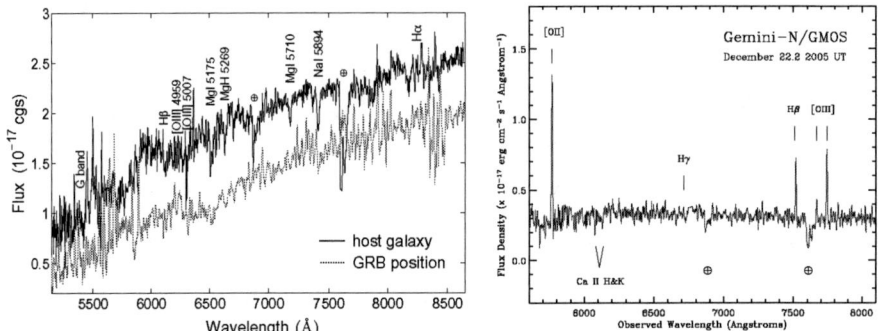

FIGURE 7. Spectra of the host galaxies of GRB 050724 (left) exhibiting typical features of an early-type galaxy, and GRB 051221 (right) exhibiting emission lines typical of star forming galaxies. From Refs. [18, 26].

a test of the progenitor population. The observed offsets in the coalescence model are a function of progenitor lifetimes, velocity kicks, and galaxy masses. Predictions from population synthesis models vary considerably, with possibly 50% of the bursts predicted to occur at offsets of > 10 kpc [11, 46]. Observationally, the offsets appear to be relatively small, $\lesssim$ few kpc [18, 20, 26], but we stress that given the uncertain input distributions, this test may not be as useful as originally envisioned.

Event Rates and Progenitor Lifetimes

Another set of useful constraints on SHB models may be obtained from the event rates and the lifetime of the progenitors. Several authors have attempted to constrain the progenitor lifetime distribution using the BATSE flux distribution and the observed redshifts, leading to conflicting results. The canonical time delay distribution, $P(\tau) \propto 1/\tau$ as inferred from Galactic DNS systems, may be disfavored by the relatively low redshifts of some SHBs [47, 48], but is not ruled out [49]. Other lifetime distributions (with a constant delay or constant rate) also provide adequate fits. The redshift distribution, particularly when augmented by claimed associations for IPN SHBs [47], has been used to argue for lifetimes of ~ 6 Gyr with a relatively small spread [48]. We caution here that the IPN associations may be spurious[1] and in particular bias the result to longer lifetimes. Moreover, with $z \approx 1.8$ for GRB 050813 the progenitor lifetime is constrained to be $\lesssim 3$ Gyr.

A more profitable approach may be to investigate the relative fraction of SHBs in early- and late-type galaxies. For SNe Ia, which are more prevalent in late-type galaxies, similar analyses have led to the conclusion that the progenitor lifetimes are $\tau \sim 1 - 3$ Gyr [50]. In the case of SHBs, several authors have argued that the fraction in early-type

---

[1] The "brightest galaxy" criterion clearly fails for some of the precisely localized SHBs.

galaxies is larger, and therefore the progenitors lifetimes are longer than for SNe Ia, i.e. $\gtrsim 3$ Gyr [47]. This analysis is highly uncertain at the present, however, since of the three precisely localized SHBs, two are in fact located in late-type galaxies.

Finally, the SHB local event rate is inferred to be at least $R \sim 10$ Gpc^{-3} yr^{-1} based on the BATSE rate and the current redshift distribution [48, 49]. The true rate is likely higher due to beaming, introducing a correction of $f_b^{-1} \sim 10-100$ for typical angles of $8-25°$. Another unknown correction is due to the low-luminosity cutoff, $R \propto L_{\min}^{-1}$.

SUMMARY AND FUTURE DIRECTIONS

The progress in our understanding of SHBs over the last several months has shifted the field from the realm of speculation to a quantitative study. The observations made to date have allowed us to determine some of the most important and basic properties:

- SHBs occur at cosmological distances with a relatively wide spread in redshift; a local population may contribute up to 20% of the BATSE sample.
- The energy scale is typically lower than that of long GRBs, ranging from few $\times 10^{48}$ erg to $\gtrsim 10^{50}$ erg.
- The ejecta appear to be collimated, but with opening angles that are larger than the median for long GRBs.
- SHBs tend to occur in lower density environments than long GRBs.
- SHBs occur in elliptical and star forming galaxies, with roughly equal numbers.

These properties are in broad agreement with the DNS or NS-BH coalescence models, but other possibilities remain equally viable. In particular the observed long-term flares and injection episodes, as well as the spread in energies and host types, may point to multiple progenitor systems and/or energy extraction mechanisms. Clearly, an increased sample of events will shed additional light on the progenitor population. Future tests may include studies of the correlation between burst and host properties, evolution of the burst properties as a function of redshift, and perhaps the detection of gravitational waves in coincidence with an SHB. An intriguing possibility if SHBs in fact tend to be associated with galaxy clusters is that they may provide a beacon to select clusters at high redshift, $z \gtrsim 1.5$, where blind searches are exceedingly difficult. This may already be the case with GRB 050813.

ACKNOWLEDGMENTS

It is a pleasure to thank my collaborators for their hard work in pursuit of an understanding of short GRBs, in particular Derek Fox, Dale Frail, Mike Gladders, Shri Kulkarni, and Alicia Soderberg. This work was supported by a Hubble Post-doctoral Fellowship grant, HST-HF-01171.01.

REFERENCES

1. Strong, I. B., et al. *ApJ* **188**, L1 (1974).
2. Kouveliotou, C., et al. *ApJ* **413**, L101 (1993).
3. Eichler, D., et al. *Nature* **340**, 126 (1989).
4. B. Paczynski, *AcA* **41**, 257 (1991).
5. Narayan, R., et al. *ApJ* **395**, L83 (1992).
6. L. M. Katz, J. I. & Canel, *ApJ* **471**, 915 (1996).
7. Rosswog, S., et al. *MNRAS* **345**, 1077 (2003).
8. R. C. Thompson, C. & Duncan, *MNRAS* **275**, 255 (1995).
9. Qin, B., et al. *ApJ* **494**, L57 (1998).
10. MacFadyen, A. I., et al. *astro-ph* **0510192** (2005).
11. Fryer, C., et al. *ApJ* **516**, 892 (1999).
12. Hurley, K., et al. *ApJ* **567**, 447 (2002).
13. Gehrels, N., et al. *Nature* **437**, 851 (2005).
14. Bloom, J. S., et al. *astro-ph* **0505480** (2005).
15. Hjorth, J., et al. *ApJL* **630**, L117 (2005).
16. D. A. Soderberg, A. M. & Frail, *GCN* **3387** (2005).
17. Pedersen, K., et al. *ApJ* **634**, L17 (2005).
18. Berger, E., et al. *Nature* **438**, 988 (2005).
19. Villasenor, J. S., et al. *Nature* **437**, 855 (2005).
20. Fox, D. B., et al. *Nature* **437**, 845 (2005).
21. Hjorth, J., et al. *Nature* **437**, 859 (2005).
22. Barthelmy, S., et al. *Nature* **438**, 994 (2005).
23. E. Berger, *GCN* **3801** (2005).
24. Foley, R. J., et al. *GCN* **3801** (2005).
25. Gladders, M., et al. *GCN* **3798** (2005).
26. Soderberg, A. M., et al. *astro-ph* **0601455** (2006).
27. Parsons, A., et al. *GCN* **4363** (2005).
28. Golenetskii, S., et al. *GCN* **4394** (2005).
29. Burrows, D. N., et al. *GCN* **4366** (2005).
30. Tanvir, N., et al. *Nature* **438**, 991 (2005).
31. E. Freedman, D. L. & Waxman, *ApJ* **547**, 922 (2001).
32. Berger, E., et al. *ApJ* **590**, 379 (2003).
33. Sari, R., et al. *ApJ* **497**, L17 (1998).
34. Frail, D. A., et al. *ApJ* **562**, L55 (2001).
35. Bloom, J. S., et al. *ApJ* **594**, 674 (2003).
36. Berger, E., et al. *Nature* **426**, 154 (2003).
37. E. Li, W. H. & Ramirez-Ruiz, *ApJ* **577**, 893 (2002).
38. Lazatti, D., et al. *A&A* **379**, L39 (2001).
39. Connaughton, V., et al. *ApJ* **567**, 1028 (2002).
40. Perna, R., et al. *ApJ* **636**, L29 (2006).
41. Berger, E., et al. *ApJ* **588**, 99 (2003).
42. Christensen, L., et al. *A&A* **425**, 913 (2004).
43. Bloom, J. S., et al. *AJ* **123**, 1111 (2002).
44. Hamuy, M., et al. *AJ* **120**, 1479 (2000).
45. E. O. Ofek, *GCN* **4553** (2006).
46. K. Perna, R. & Belczynski, *ApJ* **570**, 252 (2002).
47. Gal-Yam, A., et al. *astro-ph* **0509891** (2005).
48. Nakar, E., et al. *astro-ph* **0511254** (2005).
49. T. Guetta, D. & Piran, *A&A* **435**, 421 (2002).
50. Tonry, J. L., et al. *ApJ* **594**, 1 (2003).

Are Short GRBs Really Hard?

T. Sakamoto[*,†], L. Barbier[*], S. Barthelmy[*], J. Cummings[*,†], E. Fenimore[**], N. Gehrels[*], D. Hullinger[*,‡], H. Krimm[*,§], C. Markwardt[*,§], D. Palmer[**], A. Parsons[*], G. Sato[¶], J. Tueller[*], R. Aptekar[‖], T. Cline[*], S. Golenetskii[‖], E. Mazets[‖], V. Pal'shin[‖], G. Ricker[††], D. Lamb[‡‡], J.-L. Atteia[§§], N. Kawai[¶¶], Swift-BAT[***], Konus-Wind[***] and HETE-2 team[***]

[*]*NASA Goddard Space Flight Center*
[†]*National Research Council*
[**]*Los Alamos National Laboratory*
[‡]*University of Maryland*
[§]*Universities Space Research Association*
[¶]*Institute of Space and Astronautical Science*
[‖]*Ioffe Physico-Technical Institute*
[††]*Massachusetts Institute of Technology*
[‡‡]*University of Chicago*
[§§]*Observatoire Midi-Pyren'ees*
[¶¶]*Tokyo Institute of Technology*
[***]

Abstract. Thanks to the rapid position notice and response by HETE-2 and Swift, the X-ray afterglow emissions have been found for four recent short gamma-ray bursts (GRBs; GRB 050509b, GRB 050709, GRB 050724, and GRB 050813). The positions of three out of four short GRBs are coincident with galaxies with no current or recent star formation. This discovery tightens the case for a different origin for short and long GRBs. On the other hand, from the prompt emission point of view, a short GRB shows a harder spectrum comparing to that of the long duration GRBs according to the BATSE observations. We investigate the prompt emission properties of four short GRBs observed by Swift/BAT. We found that the hardness of all four BAT short GRBs is in between the BATSE range for short and long GRBs. We will discuss the spectral properties of short GRBs including the short GRB sample of Konus-Wind and HETE-2 to understand the hard nature of the BATSE short GRBs.

Keywords: Prompt gamma-ray emission, short GRBs
PACS: 43.35.Ei, 78.60.Mq

INTRODUCTION

In the year of 2005, there is a major progress in understanding of the nature of short GRBs. *Swift* X-Ray Telescope (XRT) found the X-ray afterglows from the short GRBs; GRB 050509B [1], GRB 050724 [2] and GRB 050813 [3, 4] detected by *Swift* Burst Alert Telescope (BAT). This discovery allows us to pinpoint the location of the short GRBs within $10''$ for the first time. And also the short GRB, GRB 050709, observed by *HETE*-2 allows us to determine the position less then $1''$ thanks to the follow-up observations by HST and *Chandra* [5]. Because of these accurate position measurements, we start to understand that the environment and/or the progenitor of short GRBs might be different from the long GRBs (e.g. [6]).

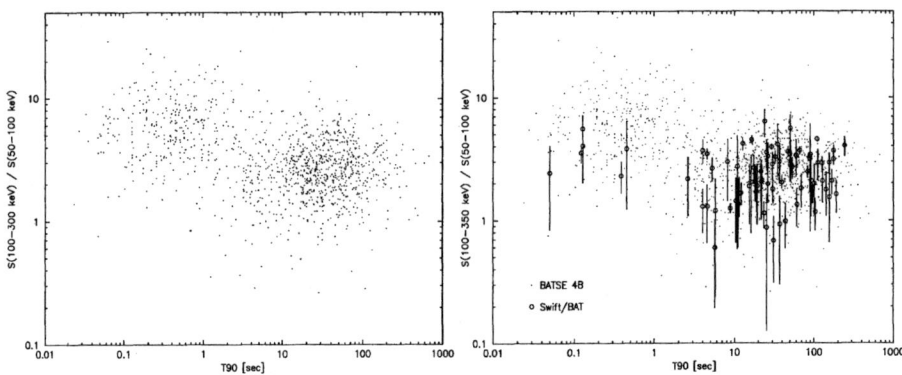

FIGURE 1. T_{90} - Hardness plot (left: BATSE and right: BAT). The error bar in the BAT sample is 90% confidence level.

From the prompt emission point of view, one of the most popular characteristic of short GRBs is the spectral hardness of these bursts. The left of figure 1 shows the T_{90} duration versus the fluence ratio between 100–300 (BATSE channel 3) and 50–100 (BATSE channel 2) keV band (hereafter HR32) for the BATSE GRBs [7]. As see in this figure, the short GRBs ($T_{90} < 2$ s) have a significantly larger hardness ratio comparing to that of the long GRBs. About 48% and 26% of the BATSE short GRBs having HR32 grater than 6 and 8 respectively.

In this paper, we will focus on the prompt emission spectral properties of short GRBs observed by *Swift*/BAT, *HETE*-2, and *Konus-Wind* to investigate the hardness of the short GRBs.

T_{90} - HARDNESS RATIO

Figure 1 and 2 show T_{90} vs. HR32 for the GRB sample of BATSE, BAT, *HETE*-2, and *Konus-Wind*. Although the HR32 for the long GRBs are consistent with the BATSE GRB sample, HR32 for the short GRBs observed by BAT, *HETE*-2 and *Konus-Wind* is not as hard as the BATSE short GRBs. There are about a quarter of the BATSE short GRBs having the HR32 grater than 8. However, none of the short GRBs observed by BAT, *HETE*-2 and *Konus-Wind* having the similar amount of HR32.

HR32 of the short GRBs of all four GRB instruments is summarized in figure 3. Although the number of the sample is limited for BAT and *HETE*-2, the distribution of HR32 for short GRBs observed by *Konus-Wind*, *HETE*-2, and BAT are all consistent. However, the BATSE HR32 distribution is spread over to much wider range, especially for the larger value of HR32.

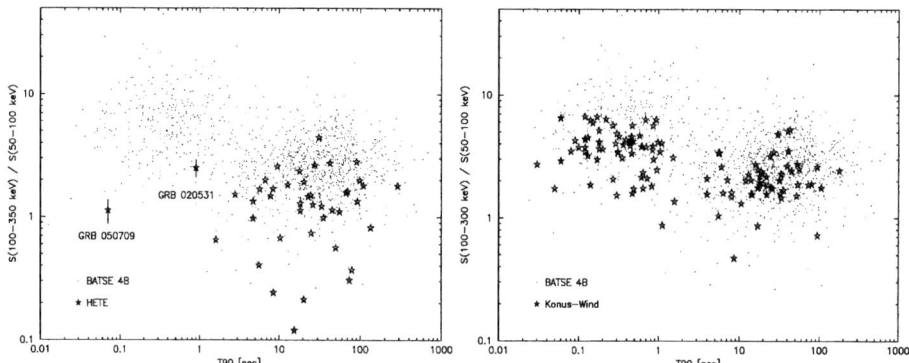

FIGURE 2. T_{90} - Hardness plot (left: HETE and right: Konus-Wind). Note that the reason for many long GRBs showing the smaller number of HR32 in the *HETE*-2 GRB sample is that *HETE*-2 is observing large number of "soft" GRBs, so called XRFs [8].

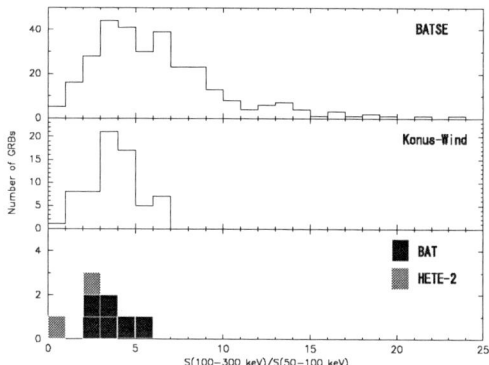

FIGURE 3. The histogram of HR32 for the short GRBs observed by BATSE, Konus-Wind, HETE-2, and Swift/BAT.

WHY BATSE SHORT GRBS LOOK HARDER?

To investigate the origin of the hardness seen in the BATSE short GRBs, we calculate HR32 with a power-law times exponential cutoff model[1] as a function of the power-law index, α, and the cutoff energy, E_0. The result is shown in the left panel of figure 4. To archive HR32 > 6 in a cutoff power-law model, α has to be grater than 0. If E_0 is less than 500 keV, α should increase rapidly when E_0 goes smaller. Thus, we may conclude that a large value of HR32 seen in the BATSE short GRBs is as a result of $\alpha > 0$, but not of a large E_0 energy.

To check our hypothesis, we made the comparison of α distribution between BATSE

---

[1] $dN/dE \sim E^{\alpha} \exp(-E/E_0)$

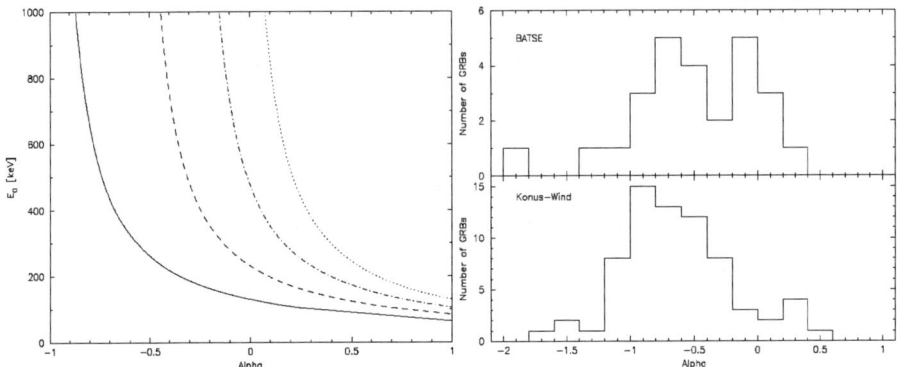

FIGURE 4. Left: The calculated α and E_0 as a function of HR32 (solid, dashed, dash-dotted, and dotted lines are HR32 of 4, 6, 8, and 10 respectively. Right: The photon index α distribution of the BATSE [9] and the Konus-Wind [10] short GRB sample.

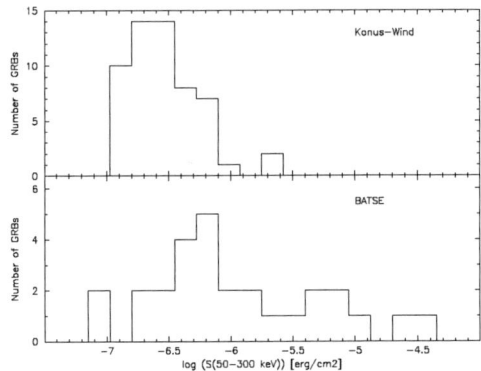

FIGURE 5. The 50-300 keV fluence distribution of the Konus-Wind and BATSE short GRB sample. Note that in the Konus-Wind short GRB sample of this figure, we only include the short GRBs with the time interval of the spectrum less than 256 ms. Thus, we are not including the large fluence bursts due to the difficulty in calculating the fluence in the 50-300 keV band.

[9] and Konus-Wind [10] sample. The right panel of figure 4 shows the comparison. The α distribution of the Konus-Wind short GRB sample has a tight distribution which is centroid around -0.8. On the other hand, as mentioned by Ghirlanda et al. [9], a large fraction of the BATSE short GRBs has $\alpha > 0$. This result tightens our conclusion that the hardness of the BATSE short GRBs are coming from the extremely flat photon index α.

As shown in figure 5, the BATSE short GRBs studied by Ghirlanda et al. [9] are the bright short GRBs. Since the Konus-Wind short GRBs studied by Mazets et al. [10] are well covering the fluence range of the BATSE short GRB sample, it is difficult to understand the systematic difference in α distribution between Konus-Wind and BATSE by the selection effect of the sample. However, we need the complete spectral information of the BATSE short GRBs to confirm our conclusion.

We study the hardness of the prompt emission of short GRBs observed by four different GRB instruments. We found that the hardness ratios of the short GRBs observed by BATSE have a systematically larger value than the short GRBs observed by other instruments. We also confirmed that the hardness of the BATSE short GRBs is as a result of the extremely flat power-law index α ($\alpha > 0$) which is not a dominant population in the *Konus-Wind* short GRB sample.

REFERENCES

1. Gehrels, N. et al. 2005, Nature, 437, 851
2. Barthelmy, S. et al. 2005, Nature, 438, 994
3. Retter, A. et al. 2005, GCN Circ. 3788
4. Morris, D. et al. 2005, GCN Circ. 3790
5. Fox, D. et al. 2005, Nature, 437, 845
6. Prochaska, J. et al. 2005, submitted to ApJL (astro-ph/0510022)
7. Paciesas, W. et al. 1999, ApJS, 122, 465
8. Sakamoto, T. et al. 2005, ApJ, 629, 311
9. Ghirlanda, G. et al. 2004, A&A, 422, L55
10. Mazets, E. et al. 2005, submitted to ApJS (astro-ph/0209219)

Late flares from GRBs - Clues about the Central Engine

Andrew I. MacFadyen*, Enrico Ramirez-Ruiz* and Weiqun Zhang[†]

*Institute for Advanced Study, School of Natural Sciences, Princeton, N.J. 08540
[†]Kavli Institute for Particle Astrophysics and Cosmology, Stanford University, Stanford, CA 94309

Abstract.
We discuss the implications of late flares following GRBs for determining the nature of the central engine. We argue that since the flare timescales are much longer than naturally expected for a stellar mass compact object, progenitor systems containing a non-compact star are favored for short bursts displaying minute-long flares. For the case of short bursts, we argue that a minute-long flare time scale may be naturally explained if the progenitor is a binary system consisting of a neutron star and a non-compact companion, e.g. a low mass X-ray binary. In this scenario, the neutron star accretes mass and angular momentum from its companion until it undergoes accretion-induced collapse to a black hole or quark star. Due to its rapid spin when it collapses, a small torus of neutron star material forms and accretes on a viscous time. The burst itself is expected to be short, ($\tau \lesssim 1$s), but the relativistic flow it accelerates hits the companion producing a minute-long flare. The delay between the GRB and beginning of the flare is given by the light travel time from the collapsed neutron star to the surface of the companion, tens of seconds, while the duration is given by the time for the relativistic shock to cross the star, a few minutes. No supernova-like emission is expected in this model. The progenitor systems are expected in both old and young stellar populations, consistent with the observed locations of short bursts in elliptical and star-forming galaxies. For long bursts, late flares are naturally explained by late accretion episodes if the massive stellar progenitor is not entirely disrupted during outburst.

Keywords: gamma rays: burst, shock waves, binaries: close, stars: neutron, X-rays: binaries, black hole physics
PACS: 98.70.Rz,97.80.Jp

SHORT GRB LOCATIONS

Short hard gamma-ray bursts (henceforth, SHBs) have recently been localized in host galaxies at cosmological distances [1, 2, 3, 4]. The properties of their host galaxies has been reviewed by Edo Berger at this meeting [5]. The presence of SHBs in elliptical galaxies rules out a progenitor system uniquely associated with recent star formation, e.g. collapsars or magnetars which form from collapse of young massive stars. The association of three SHBs with old stellar populations and of one with a star-forming galaxy is consistent with models involving old stars, e.g. binary systems of neutron stars or black holes with a long-lived companion. This is reminiscent of Type Ia supernovae which are also associated with both old and young stellar populations.

MINUTE-LONG SOFT X-RAY FLARES

Short, soft X-ray flares of ~ 100 s duration are observed to follow several SHBs by ~ 30 s. In addition, soft emission on these timescales is detected in stacked light curves of many bursts[6]. The source system responsible for SHBs must therefor be capable of producing emission on much longer timescales than expected for stellar mass compact objects. The short duration of the gamma-ray burst emission in SHBs ($0.001 < \tau < 3$s) implies a compact star, neutron star or stellar mass black hole, as the engine. Recent numerical work [7], has shown that the collapse of a supra-massive neutron star produces a rapidly rotating black hole with a compact accretion disk. The mass in these disks is found to range from $0.001 - 0.1 M_\odot$. The extraction of a fraction of the rest mass of these disks as they are accreted is sufficient to power a SHB. In addition, the neutron star material from which the disk forms is plausibly highly magnetized. Electro-magnetic extraction of black hole spin energy (up to $4.8 \times 10^{53} M_{hole}/M_\odot erg$) by magnetic field lines coupling the disk to the energy reservoir in the ergo sphere of the black hole is therefor possible[8].

Snapshots from numerical simulations of a GRB ejecta shell as it expands, collides with a red giant star, and sweeps around it, are shown in Figure 1. The interaction begins when a strong shock forms as the shell interacts with outer layers of the star, converting a large fraction of its kinetic energy into internal energy which is quickly generated and stays at a few 10^{49} erg for ~ 100 s (Figure 2). The initial increase in internal energy happens immediately after the GRB shell hits the red giant's surface, about 30 s after the neutron star collapses and an explosion is launched. A strong shock is formed, beginning where the two spheres first touch, then spreads as an increasingly larger portion of the stellar surface is shocked. The rate of energy dissipation as the shell sweeps across the star - if the neutron star was located one stellar radius from the red giant surface - is given by $\dot{E} \approx Ec/(4R_*) = 7.5 \times 10^{48}(E/10^{51} \text{ erg})(R_*/10^{12} \text{ cm})^{-1}$ erg/s, where E is the total energy in the ejected shell, c is the speed of light and R_* is the stellar radius. This estimate yields luminosities that are consistent with those found in numerical calculations shown in Figure 2. A few hundred seconds after the GRB (bottom panel of Figure 1), the shell has entirely wrapped around the star and fresh material is no longer being shocked. The internal energy declines rapidly after the GRB flare finishes shocking the star bringing the flare emission to an end. The decay slope of the corresponding light curve is expected to be modified by radiative effects. However, the dissipated energy has the correct magnitude, delay and duration to account for the observed properties of X-ray flares following short GRBs.

The soft X-ray flares observed following GRB 050724 and GRB 050709 correspond, for their redshifts of $z = 0.257$ and $z = 0.16$, to total energies of $E \sim 6 \times 10^{49}$ erg and $E \sim 3 \times 10^{49}$ erg respectively in the $2 - 25$ keV band. How can the energy dissipated during the GRB-star interaction (Figure 2) be transformed into soft X-ray radiation? The GRB ejecta shell can develop a stand-off shock before encountering the stellar envelope. A fraction of the energy behind the shock goes into generating turbulent magnetic fields and accelerating electrons to relativistic speeds. The relativistic electrons spiral in the magnetic field generating a synchrotron power-law radiation spectrum. The magnetic field strength is expected to be $\sim 10^4$ G at 10^{12} cm, strong enough to ensure that the shock-accelerated electrons cool quickly, yielding a power-law continuum extending

FIGURE 1. Collision of GRB ejecta with a companion red giant star showing the logarithm of pressure at $t = 10s$ (top), $t = 42s$ (middle) and $t = 184s$ after the GRB. The observer is located envisioned to be out of the binary plane toward the top of the figure. In the top panel, the small orange circle is the GRB ejecta and the larger purple sphere is a red giant star with radius 1.5×10^{12} cm. The ejecta form a spherical shell containing 2×10^{25} g ($10^{-8} M_\odot$) expanding at nearly the speed of light. 30 seconds after the GRB, the ejecta collide with the surface of the red giant and are rapidly decelerated in a strong shock. The internal energy produced during the collision is capable of explaining the energetics of the observed X-ray flares. The middle panel shows the GRB blast wave and shocked star 12 seconds after impact. The dark red region is the X-ray emitting region. In the bottom panel, the blast wave has wrapped around the star. At this time, the energy of the blast wave is mainly kinetic. The orange region in the middle panel and the green region in the bottom panel show the reflected ejecta shell traveling leftward toward to GRB source. The undisturbed shell expands close to the speed of light for at least two orders of magnitude in radius before producing the prompt GRB. We used the RAM code[9] to solve the conservative equations of special relativistic hydrodynamics. The computational domain for the simulations was $0 < r < 1.2 \times 10^{12}$ cm with a maximal resolution of $\Delta r = 4.7 \times 10^7$ cm for 1D and -1.2×10^{13} cm$< z < 1.2 \times 10^{13}$ cm, $0 < r < 1.2 \times 10^{13}$ cm, with $\Delta z = \Delta r = 7.32 \times 10^8$ cm for 2D. A gamma law equation of state with $\Gamma = 4/3$ was used and the ambient density was 100 baryons cm^{-3}.

into the X-ray band. Some of these X-rays will be deflected along the stellar surface before escaping, but about half (the exact proportion depending on the geometry and flow pattern) would irradiate the gas comprising the stellar envelope. For the large radiative efficiencies expected in relativistic shocks, radiative heating of the shocked material would be comparable to that of bulk heating[10]. However, radiative heating deposits energy near the stellar surface in layers with moderate scattering optical depth,

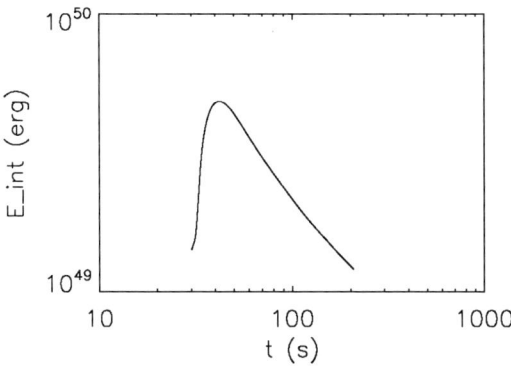

FIGURE 2. Total internal energy. As the cold GRB shell collides with the stellar surface a large fraction of the energy in the blast wave is converted to internal energy in a shock. Internal energy is produced with magnitude, duration and delay appropriate for being the source of X-ray flares observed to follow short GRBs. The sharp decline in the flare is easily explained by the finite time the companion's surface is actively shocked. In this figure, the remaining internal energy is being reconverted to kinetic energy and is not expected to contribute to the flare light curve.

the temperature is determined by photoionization equilibrium. This shallow radiatively heated layer is expected to be substantially hotter than a deeper bulk-heated region. For low radiative efficiencies, energy deposition by bulk heating would spread over a highly optically thick layer. In this case, the cooling rate, due principally to bremsstrahlung, recombination and Comptonization, would be high enough to reduce the temperature of the bulk-heated electrons to the equivalent black body temperatures of $T \sim$ a few tens of keV. This suggests that most of the flare luminosity in the X-ray band could be thermal, consistent with numerical results where a pressure of 10^{15} dyne cm^{-2} behind the shock (see the middle panel of Figure 1), corresponds to a black-body temperature of $\sim 2.5 \times 10^7$ K and a characteristic photon energy of $\sim$a few keV. We note that a radiatively-heated layer with a density up to $n_e \sim 10^{21}$ cm^{-3}, could produce strong Fe line emission provided that the ionization parameter $\xi = \beta L/(r^2 n_e)$ exceeds[11] 10^3, where L is the total luminosity of the GRB outflow and β is the fraction of the power that goes into X-ray continuum. This condition is indeed satisfied unless $\beta < 10^{-2}$. Thermal X-ray emission could also display line features, and such signatures should certainly be looked for. We note that the GRB shell must be optically thin when its radius is $\sim 10^{12}$ cm so that the shocked star is visible through the GRB shell. This requires that the GRB ejecta contain 10^{-8} M$_\odot$ or less and suggests that a magnetically dominated flow is preferred[12].

Shock heating of a non-compact binary stellar companion has other interesting consequences. During the interaction, $\sim 10^{-2}$ M$_\odot$ of the envelope of the companion star is compressed by the shock and heated. The deposited energy, a few 10^{48} erg, will cause the outer layers of the star to expand explosively. This will not, however, produce a supernova, for two reasons. First, the shock temperatures are too low for radioactive elements such as Ni^{56} to be produced. Second, the amount of ejected debris is small. Instead, after the expanded envelope becomes optically thin, a faint infrared/optical transient would appear a few weeks after the GRB.

Models in which the short GRB results from the collapse of a rapidly rotating neutron star in a close binary system provide are indicated by these considerations. The neutron star accretes matter and angular momentum from the binary stellar companion, eventually collapsing to form a black hole or quark star. The angular momentum in the equatorial region of the rapidly-spinning system is too large to be swallowed immediately when the neutron star collapses. The expected outcome, after a few milliseconds, in the black hole case, is therefore a spinning black hole orbited by a torus of neutron-density matter. The mass in these disks is thought to range from[7] $M_t \sim 10^{-3} - 10^{-2} M_\odot$. If magnetic fields anchored in the disk do not thread the black hole, then a relativistic outflow powered by torus accretion can at most carry the gravitational binding energy of the torus. The energy available for axtraction, in this case, is several times 10^{50} $\varepsilon(M_t/10^{-3} M_\odot)$ ergs, where ε is the efficiency in converting gravitational energy into relativistic outflow. If magnetic fields of comparable strength thread the black hole, its rotational energy offers an extra (and larger) source of energy that can in principle be extracted via the Blandford-Znajek mechanism[8].

LONG BURSTS

Finally, we offer some comments on long term emission observed following several long duration GRBs[14]. Figure 3 shows results from a numerical simulation of a relativistic jet in a collapsar with a mass-loaded biconical wind surrounding it. Even though the jet and wind propagate successfully through the polar regions of the star, the equatorial remains bound to the central black hole and is not ejected. MacFadyen et al (2001)[13] found that fall back of stellar material over timescales as long as days, can keep the central engine active if the star is not completely unbound by the jet and wind components of the explosion. This late time accretion can keep the central engine active for days and may be responsible for late time flaring activity observed following long duration GRBs. This work in preparation for ApJ.

ACKNOWLEDGMENTS

AIM acknowledges support from the Keck fellowship at the Institute for Advanced Study. ERR and WZ acknowledge support from NASA through the Chandra fellowship program. Part of this work was supported by NASA. Computations were performed with the Scheides Beowulf cluster at the Institute for Advanced Study. The software used in this work was in part developed by the DOE-supported ASC / Alliance Center for Astrophysical Thermonuclear Flashes at the University of Chicago. Specifically, the RAM code[9] utilizes the PARAMESH AMR and the IO tools from FLASH2.3.

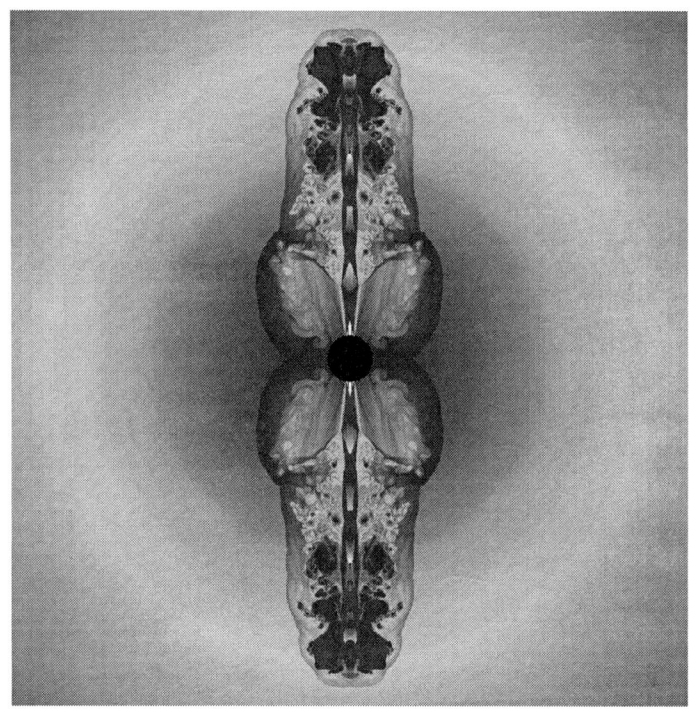

FIGURE 3. Collapsar jet and wind. The jet component has a maximum Lorentz factor of 50 and one Bethe (10^{51} erg) of energy. The wind component carries 10 Bethe with a maximum velocity of 40,000 km/s. Accretion is still possible at late times through the equatorial regions.

REFERENCES

1. J. S. Bloom et al *ArXiv Astrophysics e-prints* (2005), arXiv:astro-ph/0505480.
2. D. B. Fox, D. A. Frail, P. A. Price, S. R. Kulkarni, E. Berger, T. Piran, and A. M. Soderberg, *Nature* **437**, 845–850 (2005).
3. N. Gehrels, *ArXiv Astrophysics e-prints* (2005), arXiv:astro-ph/0505630.
4. J. X. Prochaska, J. S. Bloom, D. Pooley, C. W. Blake, R. J. Foley, S. Jha, E. Ramirez-Ruiz, and G. J., *Astrophys. J., Lett.* (2005), submitted.
5. E. Berger, *This Volume* (2006), arXiv:astro-ph/0602004.
6. D. Lazzati, E. Ramirez-Ruiz, and G. Ghisellini, *Astron. Astrophys.* **379**, L39–L43 (2001).
7. S. L. Shapiro, *Astrophys. J.* **610**, 913–919 (2004).
8. R. D. Blandford, and R. L. Znajek, *Mon. Not. R. Astron. Soc.* **179**, 433–456 (1977).
9. W. Zhang, and A. I. MacFadyen, *ArXiv Astrophysics e-prints* (2005), arXiv:astro-ph/0505481.
10. M. J. Rees, and P. Mészáros, *Astrophys. J., Lett.* **545**, L73–L75 (2000).
11. D. R. Ballantyne, and E. Ramirez-Ruiz, *Astrophys. J., Lett.* **559**, L83–L86 (2001).
12. V. V. Usov, *Nature* **357**, 472–474 (1992).
13. A. I. MacFadyen, S. E. Woosley, and A. Heger, *Astrophys. J.* **550**, 410–425 (2001).
14. Cusumano, G., et al. 2005, ArXiv Astrophysics e-prints, arXiv:astro-ph/0509737

The short/hard GRB 050709 and its star-forming host galaxy

S. Covino*, D. Malesani[†], G.L. Israel**, P. D'Avanzo*,[‡], L.A. Antonelli**,
G. Chincarini[§],*, D. Fugazza*, M.L. Conciatore**, M. Della Valle[¶], F.
Fiore**, D. Guetta**, K. Hurley[∥], D. Lazzati[††], L. Stella**, G. Tagliaferri*,
M. Vietri[‡‡] and S. Campana*

*INAF, Osservatorio Astronomico di Brera, via E. Bianchi 46, I-23807 Merate (LC), Italy
[†]International School for Advanced Studies (SISSA-ISAS), via Beirut 2-4, I-34014 Trieste, Italy
**INAF, Oss. Astronomico di Roma, via di Frascati 33, I-00040 Monteporzio Catone (Roma), Italy
[‡]Dip. di Fisica e Matematica, Università dell'Insubria, via Valleggio 11, I-22100 Como, Italy
[§]Univ. degli studi di Milano-Bicocca, Dip. di Fisica, piazza delle Scienze 3, I-20126 Milano, Italy
[¶]INAF, Osservatorio Astrofisico di Arcetri, largo E. Fermi 5, I-50125 Firenze, Italy
[∥]University of California, Berkeley, Space Sciences Laboratory, Berkeley, CA 94720-7450, USA
[††]JILA, University of Colorado, 440 UCB, Boulder CO 80309-0440, USA
[‡‡]Scuola Normale Superiore, piazza dei Cavalieri 7, I-56126 Pisa, Italy

Abstract. We present optical observations of the short/hard gamma-ray burst GRB 050709, the first such event with an identified optical counterpart. The object is located inside a galaxy at redshift $z = 0.1606 \pm 0.0002$ but slightly off-center. Late-time monitoring places strong limits on any supernova simultaneous with the GRB. The host galaxy is not of early type: it has blue colours and a star formation rate of $\sim$ 2–3.5 $M_\odot$ yr^{-1}.

Keywords: gamma-ray sources; gamma-ray burst
PACS: 98.70.Rz

INTRODUCTION

Two classes of GRBs are currently known, characterised by different durations and spectral properties [1]. Over the past years, great advances have been made in understanding the former class, thanks to the observation of their optical and radio counterparts. However, until recently, no optical emission from short GRBs had been identified [2], leaving fundamental questions about their nature, progenitors and distances unanswered.

Recently, thanks to the *Swift* and HETE-2 satellites, accurate and rapid localisations of short GRBs have become available, enabling deep, sensitive searches at long wavelengths. GRB 050709 was discovered by HETE-2 on 2005 Jul 9.94209 UT [3]. Followup observations with the *Chandra* X-ray observatory revealed a faint, uncatalogued X-ray source inside the HETE-2 error circle [4], coincident with a pointlike object embedded in a bright galaxy [5] at $z = 0.16$ [6]. The variability of this source led Price et al. [7] to propose it as the optical counterpart of GRB 050709.

We observed the field of GRB 050709 with the ESO Very Large Telescope (VLT), using the FORS 1 and FORS 2 instruments. In our first images the pointlike object reported by Jensen et al. [5] is clearly visible in the *V* and *R* bands (Fig. 1). The object lies inside a bright galaxy ($\approx 1''.2$ away from its nucleus).

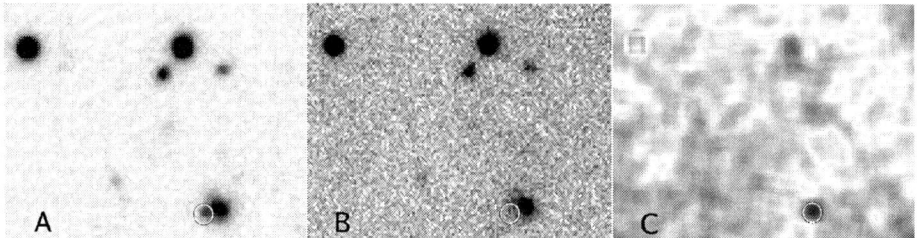

FIGURE 1. *R*-band image of the field of GRB 050709, on 2005 Jul 12.41 (A) and 20.42 (B). Panel C shows the result of the subtraction, evidencing a fading source coincident with the *Chandra* counterpart (circle). The boxes cover a region $30'' \times 20''$ wide.

On Jul 30 we took medium-resolution spectra of the host galaxy. Observations were carried out with the FORS 2 instrument at the VLT-UT1. From the detection of several emission lines, among them Hα, Hβ, and [O II], we derived a redshift $z = 0.1606 \pm 0.0002$. This is consistent with the results of Fox et al. [8]. Therefore, the rest-frame *B*-band luminosity of the host is[1] $L_B \sim 3.5 \times 10^{42}$ erg s^{-1} [$\sim 0.10 L_*$, assuming $M_B^* = -20.13$ as determined from the SDSS survey; 9].

DISCUSSION

Assuming a power-law flux decay ($F(t) \propto t^{-\alpha}$), our observations constrain α to be greater than 1.0 in the *V* band (3σ upper limit). This limit is consistent with the optical decay found by Fox et al. [8] using HST data between 5 and 10 d after the GRB [see also 10]. A similar limit was also put to the X-ray afterglow decay. Flaring activity was reported in X-ray light curve [8]. Given the limited available data, it is difficult to say whether a similar behaviour was present also in the optical band. For a deeper discussion about the results of our observations of GRB 050709 see Covino et al. [11].

The properties of the host galaxy are intriguing. Its colours are consistent with those of an irregular galaxy at $z \approx 0.2$ [12], and are much bluer than those of ellipticals (like those associated with other short GRBs). Moreover, hints of morphological structure are seen in our best-seeing images. A close inspection of the spectrum shows that both the Hα and Hβ lines have a narrow emission component (FWHM < 6 Å) partially filling a wider (~ 12 Å FWHM) absorption feature (Fig. 2). This classical signature [13] identifies a dominant stellar population ~ 1 Gyr old (mostly A-dwarf stars), together with a younger, hotter component. As indicated from the prominent nebular emission lines, star formation is still present. From the Hα and [O II] emission lines we infer a star formation rate of ~ 2–3.5 $M_\odot$ yr^{-1} once normalised to L_*. This is significantly less than that typically observed in long GRB host galaxies [14], but much larger than that in the hosts of GRB 050509B [15] and GRB 050724 [16], by factors of > 50 and > 150, respectively.

---

[1] Assuming a cosmology with $\Omega_m = 0.3$, $\Omega_\Lambda = 0.7$, and $h_0 = 0.71$.

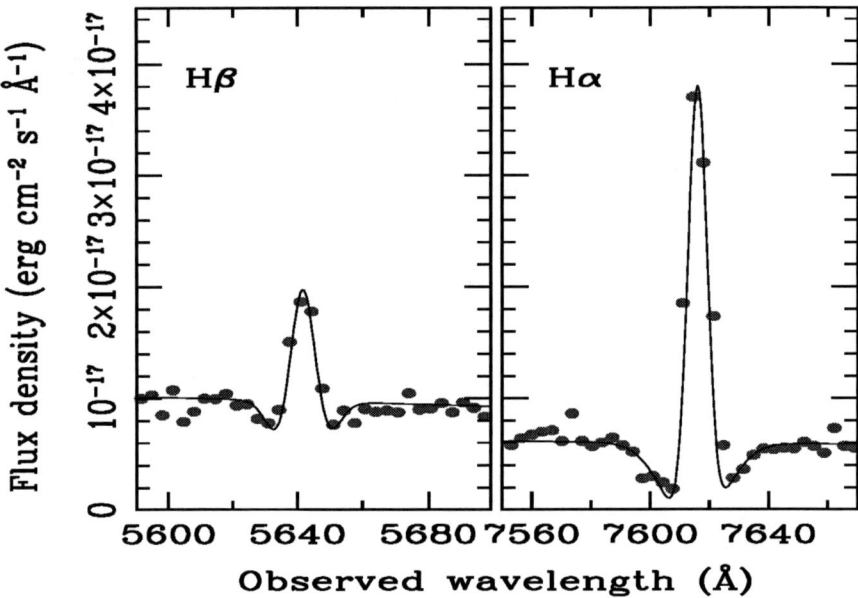

FIGURE 2. Details of the GRB 050709 host galaxy spectrum close to the Hα and Hβ lines, showing that the narrow emission emissions are superimposed on wider absorption features.

The most popular model for short-hard GRBs is the merger of a binary compact object system [e.g. 17]. Such events can occur in a late-type, star-forming galaxy [18], and give rise to short GRBs [19]. Since the merging timescales may be of the order of 10 Myr, small offsets between the explosion site and the galaxy core are possible. Therefore, GRB 050709 might have been produced in a tightly bound system, with a short merging time, similar to GRB 050724 [16]. However, we also note that according to the standard Faber-Jackson relation, the escape velocity from the GRB 050709 host is quite large (about 300 km s^{-1}), so that only a fraction of binary systems may be able to escape its potential well. In this case, a larger delay ($\sim$ 1 Gyr) would be consistent both with the observed offset and with the age of the older stellar population. A large instantaneous star formation rate would not be expected in this case, even if this does not pose any problem for the merger model.

Our photometry can put strong constraints on the presence of an unextinguished supernova (SN) exploded simultaneously with the GRB (see Fig. 3). Our limits impose a SN at least 100 times fainter than a typical type-Ia SN or a bright hypernova like SN 1998bw. Also fainter events like SN 1994I and even SN 1987A are incompatible with our data. An association with a SN seems therefore ruled out for GRB 050709 [see also 20, 10, 8]. The properties of the GRB 050709 host are however consistent with the model proposed by MacFadyen et al. [21], which advocates a collapsing neutron star accreting from a close non-compact companion. Such model would also explain the

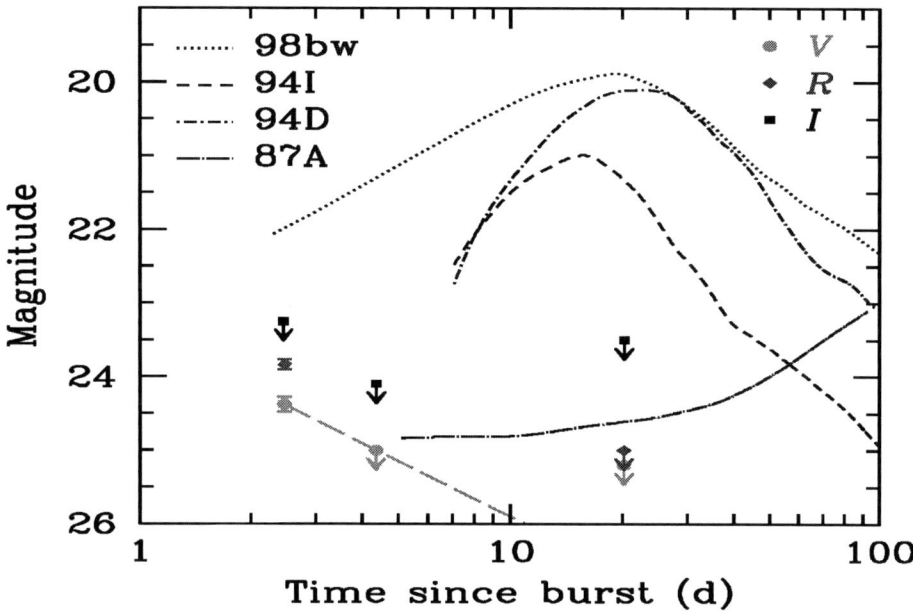

FIGURE 3. Light curve of the GRB 050709 afterglow (points), compared to those of several SNe (R band). Zero extinction is assumed at the GRB site.

flares observed in the X-ray light curve.

REFERENCES

1. Kouveliotou, C., Meegan, C.A., Fishman, G.J., et al. 1993, *ApJ*, **413**, 101
2. Hurley, K., Berger, E., Castro-Tirado, A.J., et al. 2002, *ApJ*, **567**, 447
3. Villasenor, J., Lamb, D.Q., Ricker, G.R., et al. 2005, *Nature*, **437**, 855
4. Fox, D.B., Frail, D.A., Cameron, P.B. et al., 2005a, *GCN* **3585**
5. Jensen, B.L., Jörgensen, U.G., Hjorth, J., et al. 2005, *GCN* **3589**
6. Price, P.A., Roth, K., & Fox, D.W. 2005a, *GCN* **3605**
7. Price, P.A., Jensen, B.L., Jörgensen, U.G., et al. 2005b, *GCN* **3612**
8. Fox, D.B., Frail D.A., Price, P.A., et al. 2005b, *Nature*, **437**, 845
9. Blanton, M.R., Hogg, D.W., Bahcall, N.A., et al. 2003, *ApJ*, **592**, 819
10. Hjorth, J., Watson, D., Fynbo, J.P.U., et al. 2005b, *Nature*, **437**, 859
11. Covino, S., Malesani, D., Israel, G.L. et al. 2005, *A&A*, in press (astro-ph/0509144)
12. Fukugita, M., Shimasaku, K., & Ichikawa, T. 1995, *PASP*, **107**, 945
13. Dressler, A., & Gunn, J.E. 1983, *ApJ*, **270**, 7
14. Christensen, L., Hjorth, J., & Gorosabel J. 2004, *A&A*, **425**, 913
15. Bloom, J.S., Prochaska, J.X., Pooley, D., et al. 2005, *ApJ*, in press (astro-ph/0505480)
16. Berger, E., Price, P.A., Cenko, S.B., et al. 2005a, *Nature*, **438**, 988
17. Eichler, D., Livio, M., Piran, T., & Schramm, D.N. 1989, *Nature*, **340**, 126
18. Belczyński, K., & Kalogera, V. 2001, *ApJ*, **550**, 183
19. Perna, R., & Belczynski, K. 2002, *ApJ*, **570**, 252
20. Hjorth, J., Sollerman, J., Gorosabel, J., et al. 2005a, *ApJ*, **630**, L117
21. MacFadyen, A.I., Ramirez-Ruiz E., & Zhang, W. 2005, astro-ph/0510192

The rate and luminosity function of Short GRBs

Tsvi Piran* and Dafne Guetta[†]

*Racah Institute for Physics, The Hebrew University, Jerusalem 91904, Israe
† Osservatorio astronomico di Rome v. Frascati 33 00040 Monte Porzio Catone, Italy

Abstract. We compare the luminosity function and rate inferred from the BATSE short hard bursts (SHBs) peak flux distribution with the redshift and luminosity distributions of SHBs observed by *Swift*/HETE II. The *Swift*/HETE II SHB sample is incompatible with SHB population that follows the star formation rate. However, it is compatible with a distribution of delay times after the SFR. This would be the case if SHBs are associated with binary neutron star mergers. The implied SHB rates that we find range from ~ 8 to $\sim 30 h_{70}^3 \mathrm{Gpc}^{-3}\mathrm{yr}^{-1}$. This rate is a much higher than what was previously estimated and it is comparable to the rate of neutron star mergers estimated from statistics of binary pulsars. If GRBs are produced in mergers the implied rate practically guarantees detection by LIGO II and possibly even by LIGO I, if we are lucky. Our analysis, which is based on *observed* short hard burst is limited to bursts with luminosities above 10^{49} erg/sec. Weaker bursts may exist but if so they are hardly detected by BATSE or *Swift* and hence their rate is very weakly constrained by current observations. Thus the rate of mergers that lead to a detection of a gravitational radiation signal might be even higher.

Keywords: cosmology:observations-gamma rays:bursts-gravitational radiation
PACS: 95.55.Ka,98.70.Rz

INTRODUCTION

The luminosity and rate of Gamma-Ray Bursts (GRBs) is one of the key issues in any astrophysical model. Shortly after the discovery that GRBs are cosmological Piran [1] and Mao and Paczynski [2] used the observed value of $\langle V/V_{max} \rangle = 0.32$ and estimated, assuming that the GRB rate is independent of redshift and that they are standard candles, the rate of GRBs. Already at that time Piran [1] suggested that if GRBs arise due to neutron star mergers [3] their rate will depend on redshift and it will follow the neutron star binary formation rate with a logarithmic distribution of time delays, reflecting the distribution of merger times.

In 1993 Koveliotou [4] noticed that the GRB distribution can be divided into two subsets of long and short bursts with a dividing duration of 2sec. As short bursts are also harder than long ones [5, 6], they are denoted as Short Hard Bursts (SHBs). Mao et al, [7] realized that BATSE is less sensitive to short GRBs than to long ones and pointed out that this should be considered when trying to fit the *observed* peak flux distribution to different models. Shortly afterwards Cohen and Piran carried out the first separate analysis of the luminosity and rates of long and short GRBs finding that the *observed* SHB population is located $z \leq 0.5$, namely they are much nearer than the *observed* distribution of long ones. Unfortunately a stubborn referee forced these authors

---

[1] talk given by Tsvi Piran

to take out the discussion of short GRBs from [8] and these results were reported only in a conference proceedings [9] and in Cohen's Phd thesis [10]. These finding were corroborated shortly later by Katz and Canel [11] and later by Tavani [12] who found that $\langle V/V_{max} \rangle$ of SHBs is significantly higher than the one measured for long bursts.

With the discovery in 1997 of GRB afterglow and the subsequent identification of host galaxies and redshift measurements direct estimates of the luminosity and rates were obtained for long bursts. It was discovered that long GRBs follow the SFR and that typical (isotropic equivalent) peak luminosities are about 10^{51} ergs/sec.

However, until recently no afterglow was detected from any short burst who remained as mysterious as ever. Last spring *Swift* and then HETE II detected X-ray afterglow from several short bursts. In some cases optical and radio afterglows were detected as well. This has lead to identification of host galaxies and to redshift measurements. While the current sample is very small several features emerge. First, unlike long GRBs that take place only in star forming spiral galaxies SHBs take place also in elliptical galaxies in which the stellar population is older. In this SHBs behave like type Ia Supernovae. Second, the redshift and peak (isotropic equivalent) luminosity distributions of the five short bursts (see Table 1) confirm earlier expectations [9, 10, 11, 12, 13] that the observed SHB population is significantly nearer than the observed long burst population.

We [13] (denoted hereafter GP05) have estimated, before the launch of *Swift* the luminosity function and formation rate of SHBs from the BATSE peak flux distribution. We have shown that the distribution is compatible with either a population of sources that follow the SFR (like long bursts) or with a population that lags after the SFR [1, 14]. If SHBs are linked to binary neutron star mergers [3] the SHB rate is given by the convolution of the star formation rate with the distribution $P_m(\tau)$ of the merging time delays τ of the binary system. These delays reflect the time it takes to the system to merge due to emission of gravitational radiation.

GRB	050509b	050709	050724	050813	051221
z	0.22	0.16	0.257	0.7 or 1.80	0.5465
$L_{\gamma,iso}/10^{51}$ erg/sec	0.14	1.1	0.17	1.9	3

Table 1: The *Swift*/HETE II current sample of SHBs with a known redshift.

THE LUMINOSITY FUNCTION OF THE BATSE SHB SAMPLE

As BATSE is less sensitive to short bursts than to long ones [7], even an *intrinsic* SHB distribution that follow the SFR gives rise to an *observed* distribution that is nearer to us. Still a delayed SFR distribution (that is *intrinsically* nearer) gives rise to even nearer *observed* distribution [13]. Therefore the recent *observed* redshift distribution of SHBs favors the delayed model and hence the merger scenario. Still the question was posed whether the predicted *observed* distribution is consistent with the current sample. Gal Yam et al., [15] suggested that the distributions are inconsistent and hence the delayed SFR model is ruled out. We re-examine the situation here and we show that while a delayed distribution with the best fit (maximal likelihood) parameters ia only marginally consistent ($p_{KS} = 0.05 - 0.06$) with the current sample, a delayed distribution with parameters within 1σ from the best fit parameters is compatible ($p_{KS} = 0.22 - 0.25$). We discuss the implications of this result to the binary Neutron star merger model and to the detection of gravitational radiation from such mergers. We also discuss the recent suggestion of Nakar et al. [16] that there rate of mergers producing very weak bursts is very high.

Our data set and methodology follow [17, 13]. We consider all the SHBs detected while the BATSE onboard trigger [18] was set for 5.5σ over background in at least two detectors in the energy range 50-300keV. These constitute a group of 194 bursts. We assume the functional form of the rate of bursts (but not the amplitude). We then search for a best fit luminosity function. Using this luminosity function we calculate the expected distribution of *observed* redshifts and we compare it with the present data. We consider the following cosmological rates:

- (i) A rate that follows the SFR (We do not expect that this reflects the rate of SHBs but we include this case for comparison.).
- (ii) A rate that follows the NS-NS merger rate. This rate depends on the formation rate of NS binaries, that one can safely assume follows the SFR, and on the distribution of merging time delays, $P_m(\tau)$. This, in turn, depends on the distribution of initial orbital separation a between the two stars ($\tau \propto a^4$) and on the distribution of initial eccentricities. Both are unknown. From the coalescence time distribution of six double neutron star binaries [19] it seems that $P_m(\log(\tau))d\log(\tau) \sim$ const, implying $P_m(\tau) \propto 1/\tau$,[1]. Therefore, our best guess scenario is a SBH rate that follows the SFR with a logarithmic time delay distribution.
- (iii) A rate that follows the SFR with a delay distribution $P_m(\tau)d\tau \sim$ const.
- (iv) A constant rate (which is independent of redshift.).

For the SFR we employ R_{SF2} of Porciani & Madau [20]: In models (ii) and (iii) the rate of SHBs is given by:

$$R_{\text{SHB}}(z) \propto \int_0^{t(z)} d\tau R_{SF2}(t-\tau)P_m(\tau). \tag{1}$$

	Rate(z=0) $Gpc^{-3}yr^{-1}$	L^* 10^{51} erg/sec	α	β	p_{KS} (z=0.7)	p_{KS} (z=1.8)
i	$0.11^{+0.07}_{-0.04}$	$4.6^{+2.2}_{-2.2}$	$0.5^{+0.4}_{-0.4}$	$1.5^{+0.7}_{-0.5}$	< 0.01	< 0.01
ii	$0.6^{+8.4}_{-0.3}$	$2^{+2}_{-1.9}$	$0.6^{+0.4}_{-0.4}$	2 ± 1	0.05	0.06
ii$_\sigma$	10^{+8}_{-5}	0.1	$0.6^{+0.2}_{-0.4}$	1 ± 0.5	0.22	0.25
iii	30^{+50}_{-20}	$0.2^{+0.5}_{-0.195}$	$0.6^{+0.3}_{-0.5}$	$1.5^{+2}_{-0.5}$	0.91	0.91
iv	8^{+40}_{-4}	$0.7^{+0.8}_{-0.6}$	$0.6^{+0.4}_{-0.5}$	$2^{+1}_{-0.7}$	0.41	0.41

Table 2: Best fit parameters Rate(z=0), L^*, α and β and their 1σ confidence levels for models (i)-(iv). Also shown are the KS probability (p_{ks}) that the five bursts with a known redshift arise from this distribution. We show two results for KS tests one with GRB 0508132 at $z = 0.7$ and the other at $z = 1.8$. Case ii$_\sigma$ corresponds to case ii with an L^* value lower by 1σ than the best fit one. Other parameters have been best fitted for this fixed number.

Following Schmidt [21] (see also [17, 13]) we consider a broken power law peak luminosity function with lower and upper limits, $1/\Delta_1$ and Δ_2, with power law indeces, α, β and luminosity break L^*.

We use $\Delta_{1,2} = (30, 100)$ [13]. In [22] (denoted GP06 hereafter) we show that both limits are chosen in such a way that a very small fraction (less than 1%) of the *observed* bursts are outside these range. Hence one cannot infer anything from the observations on the luminosity function in this range. Comparing the predicted distribution with the one observed by BATSE we obtain, the best fit parameters of each model and their standard deviation. The results are shown in table 2 and in figs 1 and 2 of GP06.

A COMPARISON WITH THE CURRENT *SWIFT*-HETE II SHB SAMPLE

We can derive now the expected redshift distribution of the observed bursts' population in the different models.

We assume that the minimal peak flux for detection for *Swift* is ~ 1 ph/cm^2/sec like BATSE (note the different spectral windows of both detectors which makes *Swift* relatively less sensitive to short bursts).

Fig. 1 depicts the expected *observed* integrated redshift distributions of SHBs in the different models. As expected, a distribution that follows the SFR, (i), is ruled out by a KS test with the current five bursts ($p_{KS} < 1\%$). This is not surprising as other indications, such as the association of some SHBs with elliptical galaxies suggest that SHBs are not associated with young stellar populations.

A distribution that follows the SFR with a constant logarithmic delay distribution, (ii), is marginally consistent with the data ($p_{KS} \sim 5-6\%$). The observed bursts are nearer (lower redshift) than expected from this distribution. If we use the Rowan-Robinson

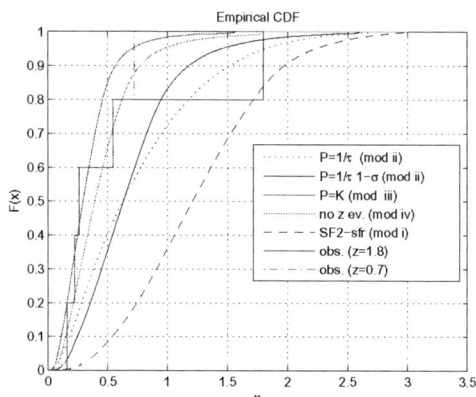

FIGURE 1. A comparison between the expected integrated *observed* redshift distributions of SHBs for models (i)-(iv) and (ii$_\sigma$) and the distribution of known redshifts of SHBs.

SFR [13], rather than SF2 of Porciani-Madau, the situation is even more promising $p_{KS} \sim 10\%$ (for either z=0.7 or z-1.8). When we move to a distribution that is 1σ away from the best fit distribution we find $p_{KS} \sim 22-25\%$ and even higher for the RR SFR. Thus the suggestion [15] that the observed sample rules out the NS merger model (with a logarithmic merger time distribution) was somewhat premature. Note however, that the local rate with this model (ii$_\sigma$) is sixteen times larger than the rate of the best fit model, (ii). This reflects the large flexibility in modeling the peak flux distribution.

To demonstrate the flexibility of the data we have considered two other time delay distributions. A uniform time delay distribution (iii) and an overall constant SHB rate (iv). Both models are compatible with the BATSE SHB distribution and with the sample of SHBs with a known redshift ($p_{KS} \sim 80\%$ and 40% respectively.). This result is not surprising. The BATSE peak flux distribution depends on two unknown functions, the rate and the luminosity function. There is enough freedom to chose one function (the rate) and fit for the other.

In all models compatible with the five bursts with a known redshift, the *intrinsic* SHB rate is pushed towards lower redshifts. The inferred present rates, ~ 30, ~ 8 and $\sim 10 h_{70}^3 \text{Gpc}^{-3} \text{yr}^{-1}$ for models (iii), (iv) and (ii$_\sigma$) respectively, are larger by a factor ten to fifty than those estimates earlier (assuming that SBHs follow the SFR with a logarithmic delay with the best fit parameters GP05). The corresponding "typical" luminosities, L^*, ranges from 0.1 to 0.7×10^{51} erg/sec.

CONCLUSIONS AND IMPLICATIONS

We have repeated the analysis of fitting the BATSE SHB data to a model of the luminosities and rates distributions. Our best fit logarithmic distribution model is similar to the best fit logarithmic model presented in GP05. A main new ingredient of this work is the fact that we consider several other models. We confirm our earlier finding that the BATSE data allows a lot of flexibility in the combination of the rates and luminosities.

A second new ingredient of this work is the comparison of the best fit models to the small sample of five *Swift*/HETE II SHBs. The *Swift*/HETE II data gives a new constraint. This constraint favors a population of SHBs with a lower intrinsic luminosity and hence a nearer *observed* redshift distribution. It implies a significantly higher local SHB rate - a factor of ten to fifty higher than earlier estimates. The new observations of Swift show that the SHBs are nearer than what was expected before and therefore, their luminosity is lower and their local rate is higher. We stress that this new result was within the 1σ error of the model presented in GP05, which had a very large range of allowed local rates and typical luminosities.

Provided that the basic model is correct and we are not mislead by statistical (small numbers), observational (selection effects and threshold estimates) of intrinsic (two SHB population) factors we can proceed and compare the inferred SHB rate with the observationally inferred rate of NS-NS mergers in our galaxy [23]. This rate was recently reevaluated with the discovery of PSR J1829+2456 to be rather large as 80^{+200}_{-66}/Myr. Although the estimate contains a fair amount of uncertainty [24]. If we assume that this rate is typical and that the number density of galaxies is $\sim 10^{-2}$/Mpc^3, we find a merger rate of 800^{+2000}_{-660}/Gpc^3/yr. Recently [25] have derived a beaming factor of 30-50 for short bursts. This rate implies a total merger rate of $\sim 240 - 1500/\text{Gpc}^3/\text{yr}$ for the three cases (iii), (iv) and (ii$_\sigma$). The agreement between the completely different estimates is surprising and could be completely coincidental as both estimates are based on very few events.

If correct these estimates are excellent news for gravitational radiation searches, for which neutron star mergers are prime targets. They imply that the recently updates high merger rate, that depends mostly on one object, PSR J0737+3039, is valid. These estimate implies one merger event within $\sim$ 70Mpc per year and one merger accompanied with a SHB within $\sim$ 230Mpc. These ranges are almost within the capability of LIGO I and certainly within the capability of LIGO II. If correct these estimates of the rate are excellent news for gravitational radiation searches, for which neutron star mergers are prime targets.

To conclude we stress that we have assumed that the luminosity function has a lower limit of $L^*/30$. This was just because even if such a limit does not exists weaker bursts would be barely detected. The current peak flux distribution of BATSE burst cannot confirm (or rule out) the existence of such population (note however, that Tanvir et al., [26] suggest that such a population exists on the basis of the angular distribution of BATSE SHBs). If such weak bursts exist then, of course, the overall merger rate will be much larger [16]. Such events will provide such a high rate that soon LIGO I will begin to constrain this possibility.

REFERENCES

1. Piran, T., 1992 ApJ,389, L83.
2. Mao, S., and Paczynski, B., 1992, ApJ, 388, L45.
3. Eichler, D., Livio, M., Piran, T., Schramm, D., 1989, Nature, 340, 126
4. Kouveliotou, C. et al., 1993, ApJ 413, L101.
5. Dezalay, J. P. et al., 1996, ApJ 471, L27.
6. Kouveliotou, C. et al., 1996, in *Gamma-ray bursts, Proceedings of the 3rd Huntsville Symposium. AIP Conference proceedings series, 1996, vol. 384, edited by Kouveliotou et al., ISBN: 156966859, p.42.*
7. Mao, S., Narayan, R. & Piran, T., 1994, ApJ 420, 171.
8. Cohen, E., and Piran, T., ApJ, 444, L25.
9. Piran, T. 1994, AIP Conf. Proc. 307: Gamma-Ray Bursts, 307, 543, (astro-ph/9412098).
10. Cohen, E., Phd Thesis, Hebrew University, 1996
11. Katz, J. I. & Canel, M., 1996, ApJ 471, 915.
12. Tavani, M., 1998, ApJ 497, L21.
13. Guetta, D., & Piran, T. 2005, A&A, 435, 421 (GP05)
14. Ando, S., 2004, JCAP, 06, 007.
15. Gal-Yam, A., et al.,2005, astro-ph/0509891.
16. Nakar, E., Gal Yam., A., and Fox, D., 2005, astro-ph/0511254,
17. Guetta, D., Piran, T. & Waxman E., 2005, ApJ, 619, 412.
18. Paciesas et al., 1999, ApJ. Supp. 122, 465.
19. Champion et al., 2004, MNRAS 350, L61.
20. Porciani, C., and Madau, P., 2001, ApJ 548, 522.
21. Schmidt, M., 2001, ApJ 559, L79.
22. Guetta, D. & Piran, T. 2006, submitted to A& A, astro-ph/0511239 (GP06)
23. Narayan, R., Piran, T., Shemi, A., 1991, ApJ 379, L17.
24. Kalogera, V. et al., 2004, ApJ 603, L41, erattum, 2004, ApJ. 614, L137.
25. Berger, E. *et al.* 2005, astro-ph/0508115.
26. Tanvir, N., et al., 2005, astro-ph/0509167

A Magnetar Flare in the *BATSE* Catalog?

A. Crider

Elon University, CB 2625, Elon, NC 27244

Abstract. To identify extragalactic magnetar flares, we have searched for their periodic tails by generating Lomb periodograms of the emission following short bursts detected by the *Burst and Transient Source Experiment* (*BATSE*). Out of 358 short bursts examined, one has a significant tail periodicity ($T = 13.8$ s, $P = 4 \times 10^{-5}$). The most probable host galaxy for this burst is "The Fireworks Galaxy" NGC 6946 ($d = 5.9$ Mpc). At this distance, the energy of the spike, $(2.7 \pm 0.3) \times 10^{44}$ ergs, is akin to those of the galactic magnetar giant flares, as are its duration (~ 0.4 s) and temperature (250 ± 60 keV). For the tail emission, however, our estimated temperature of 60 ± 5 keV is harder and the energy release of $(4.3 \pm 0.8) \times 10^{45}$ ergs is larger than those of the galactic magnetar flares. Regardless of the host, such a large ratio of tail-to-spike energy would imply that magnetar flare tails might be detectable out to further distances than previously thought.

Keywords: gamma ray bursts, magnetars
PACS: 98.70.Rz, 97.60.Jd

INTRODUCTION

Three times within the past thirty years, short (0.2-0.35 s), intense gamma-ray flares followed by softer ($kT \sim 25$ keV), periodic ($T = 5 - 8$ s) tails have erupted from the sources of the much fainter soft-gamma repeaters (SGR). The magnetar model proposed by Duncan and Thompson [1] has had considerable success in explaining SGR and the occasional giant flares. Assuming that our galaxy is not unique, then magnetar flares also occur in nearby galaxies. These would likely be labeled as short GRB, since the characteristic pulsating tail would near or below the background and the spectrum of the initial spike is similar to that of classical GRB [2]. To identify them, one can exploit three distinguishing attributes: (a) their locations relative to nearby galaxies, (b) their spectral temperatures, and (c) their faint oscillating tails. In this study, we searched for faint oscillations in the emission following short GRB.

PROCEDURES

The current *BATSE* catalog contains 2702 gamma-ray bursts; 2041 have calculated T_{90} durations. For the 358 short bursts ($T_{90} < 1.0$ s), we extracted the 30-50 keV, 64-ms lightcurve (as the galactic magnetar flare tails were soft) and refit polynomial backgrounds using data from 100 s before to 200 s after the trigger. We then generated a Lomb periodogram [3] for the 100-s intervals that immediately followed each short GRB. Of the 358 bursts analyzed, only one had a significance $P < 1/358$. The periodogram plotted in Figure 1 for GRB 970110 (*BATSE* #5770) reveals a highly significant (Lomb power=17.8, $P = 3 \times 10^{-5}$) peak periodicity of 13.8 s. We found the signal

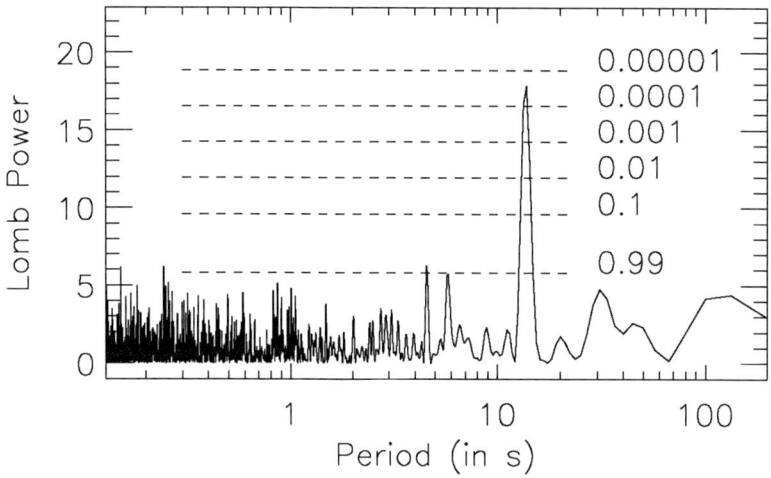

FIGURE 1. Lomb periodogram of the 100-s interval following GRB 970110. The dominant periodicity of 13.8 s has a Lomb power of 17.8. Significance levels are plotted as horizontal lines. The low chance probability of such an intense signal ($P = 3 \times 10^{-5}$) implies that this feature is real.

separately in both of the two *BATSE* Large Area Detectors (LADs) facing the event. Examining the 50-100 keV channel independently reveals the same strong periodicity (Lomb power=16.2, $P = 1 \times 10^{-4}$). No significant signal was found in the upper two channels (>100 keV), as might be expected for a soft magnetar flare tail. Figure 2 shows the countrate rebinned from 64-ms to 2-s bins and overlayed with a 13.8-s sinusoid to illustrate the periodicity found originally by the Lomb periodogram.

To test if this periodic signal was from a source unrelated to the burst, we examined periodograms of the pre-burst (-100-to-0 s) and subsequent (100-to-200 s) regions and found no significant periodicity, with maximum significances of $P = 0.55$ and 0.45 respectively. While the 1.024-s binning of the pre-burst emission limits the sensitivity of the Lomb periodogram to some extent, the period of interest (0-to-100 s) retains a marginally significant spike ($P = 0.01$) when resampled to this resolution. Curiously, there may be pulses in phase with the tail during the 50 s before the trigger, as can be seen in Figure 2. However, the lack of a sustained periodic signal before or 100 s after the trigger suggests that the pulsations are indeed a transient phenomena temporally coincident with the spike. Additionally, we found no hint of a 13.8-s periodicity in the DISCLA data of the other six LADs, suggesting the source is indeed from the direction of the burst. The Australia Telescope National Facility (ATNF) Pulsar Catalog [4] that includes the known gamma-ray pulsars and the anomalous X-ray pulsars contains no sources with a periodicity > 2 s inside the 99.7% confidence location of GRB 970110. While Cyg X-1, in its "soft/low" state on 1997 January 10, is just outside of the 99.7% *BATSE* confidence circle, there are no reports of it having a periodicity close to 13.8 s.

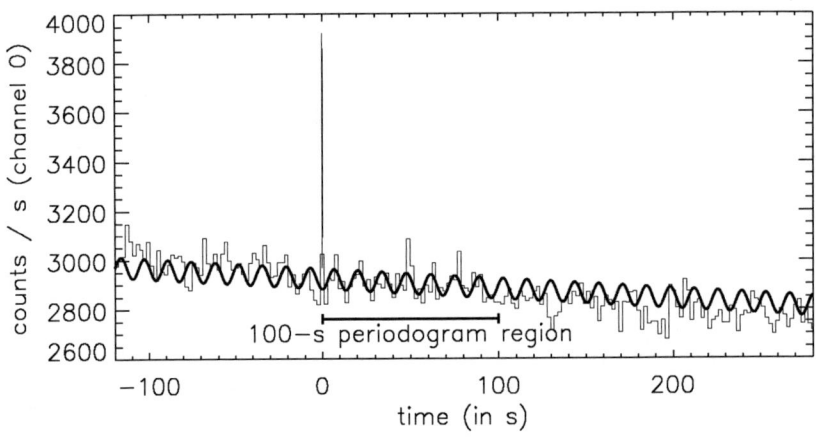

FIGURE 2. The *BATSE* 30-50 keV count rate around GRB 970110 at 2-s resolution. The best fit sinusoid with a 13.8-s period is plotted for reference. The burst itself is plotted at 64-ms resolution.

The duration of the spike (~ 0.4 s) is very similar to those of the magnetar flares. Using NASA-MSFC's *rmfit* analysis software and *BATSE* DISCSC data, we also found a similar spike spectrum. An optically-thin thermal bremsstrahlung (OTTB) spectrum gives an acceptable fit ($\chi^2 = 1.32, \nu = 2, P = 0.52$) with a temperature $kT_{\text{OTTB}} = 250 \pm 60$ keV. A blackbody spectrum fits more poorly ($\chi^2 = 8.2, \nu = 2, P = 0.017$) with $kT_{\text{BB}} = 30$ keV. Our result is consistent with the peak of the 1979 March 5 event ($kT_{\text{OTTB}} = 246$ keV; [2]), but is softer than the the 2004 December 27 flare ($kT_{\text{BB}} = 175 \pm 25$ keV; [5]). Using *rmfit*, we found the fluence of the spike to be $(6.5 \pm 0.5) \times 10^{-8}$ erg cm^{-2} in *BATSE*'s 25-2000 keV window.

Determining the tail's spectrum and fluence required calibrating the measured Lomb power and the total detector counts in each channel. We created several synthetic magnetar tails using the *Swift* BAT light curve for the 2004 December 27 magnetar flare ($t = 205$ to 505 s). These were scaled to represent what would be seen in each *BATSE* channel, with total counts ranging from 200 to 400,000. A periodogram for each allowed us to estimate the functional relationship (roughly quadratic in the region of interest) between the integrated detector counts in a channel and the Lomb power. We found our Lomb powers of 17.7 in channel 0 and 16.2 in channel 1 corresponded to 6000 counts and 5100 counts, respectively. We next convolved OTTB spectrum (kT_{OTTB} spanning 5 to 70 keV) with the detector response matrix for *BATSE* trigger #5770 to determine the number of counts expected in the four channels. From this, we estimated that the temperature of the tail is $kT_{\text{OTTB}} = 60 \pm 5$ keV, notably harder than other magnetar flares. With this spectrum, and the Lomb power for channel 0, we estimate a tail energy fluence of $(7.5 \pm 1.5) \times 10^{-7}$ erg cm^{-2}. The fraction of the total energy in the tail emission (94%) is much larger than the recent 2004 December 27 event (0.3%), but is comparable to the 1979 March 5 event (75%).

DISCUSSION

The *BATSE* localization of GRB 970110 includes few nearby galaxies. The dwarf spheroidal galaxy Draco ($d = 0.08$ Mpc) and the blue compact dwarf galaxy NGC 6789 ($d = 3.6$ Mpc) both fall just inside of the 95.4% confidence circle, but have low *a priori* probabilities of producing a magnetar flare based on their low star formation rates. Instead, we find a Bayesian probability of 87% that of the galaxies within 10 Mpc, the "Fireworks Galaxy" NGC 6946 ($d = 5.9$ Mpc) is the host. While just outside of the 95.4% confidence circle, its very high star formation rate of 3.12 $M_\odot$ yr^{-1} [6] makes it a likely source. In fact, we estimate a 38% *a priori* probability that a magnetar flare from NGC 6946 exists in the *BATSE* catalog. Assuming isotropic emission, the energy fluence in the spike corresponds to $(2.7 \pm 0.3) \times 10^{44}$ erg, comparable to 1979 March 5 flare, which had a spike energy of 1.2×10^{44} erg [7]. Its tail energy of $(4.3 \pm 0.8) \times 10^{45}$ ergs is larger than those of the three galactic magnetar flares, but only $\sim 10\times$ more than that of the 1979 March 5 event. The requisite dipole magnetic field strength of the magnetar that would constrain such a fireball would be $B_\star > 1.4 \times 10^{15}$ $(E_{\text{tail}}/4.3 \times 10^{45}$ erg$)^{1/2}$ $(\Delta R/10$ km$)^{-3/2}$ $(1/2 + \Delta R/2R_\star)^3$G, where $R_\star$ is the stellar radius and ΔR is the outer radius of the magnetic loop confining the plasma [8]. If instead GRB 970110 is from the Draco dwarf galaxy, then its energetics are more similar to the event that occurred two days after the 1997 August 27 giant flare [9]. In either case, the relatively large fraction of energy in the tail of this event suggests that the range to which *Swift* might detect similar periodicities should be extended. Hurley et al. [5] calculated that magnetar periods might be measured by *Swift* out to a distance of $\sim 2 - 8.5$ Mpc based on the fluence observed for the 2004 December 27 event. Approximately 15% of the tail energy in the 1997 January 10 event (6.5×10^{44} ergs) was released in the *Swift* XRT band (0.3-100 keV), suggesting that if this event is from NGC 6946, the Swift detection range for tails can be extended to $\sim 5 - 20$ Mpc.

ACKNOWLEDGMENTS

The author thanks Michael Briggs, Dana Hurley Crider, and his *PHY 251* students for helpful comments. This work was supported by a grant from Elon University.

REFERENCES

1. R. C. Duncan and C. Thompson, Astrophys. J. Lett. **392**, L9–L13 (1992).
2. E. E. Fenimore, R. W. Klebesadel, and J. G. Laros, Astrophys. J. **460**, 964 (1996).
3. W. H. Press, et al. *Numerical Recipes: The Art of Scientific Computing*, Cambridge University Press, Cambridge (UK) and New York, 1992, 2nd edn., ISBN 0-521-43064-X.
4. R. N. Manchester, G. B. Hobbs, A. Teoh, and M. Hobbs, Astron. J. **129**, 1993 2006 (2005).
5. K. Hurley et al., Nature **434**, 1098–1103 (2005).
6. I. D. Karachentsev, S. S. Kajsin, Z. Tsvetanov, and H. Ford, Astron. Astrophys. **434**, 935–938 (2005).
7. E. P. Mazets et al., Nature **282**, 587–589 (1979).
8. C. Thompson and R. C. Duncan, Mon. Not. R. Astron. Soc. **275**, 255–300 (1995).
9. A. I. Ibrahim et al., Astrophys. J. **558**, 237–252 (2001).

Short GRB progenitors: population synthesis predictions vs. observations

Tomasz Bulik*, Krzysztof Belczyński[†] and Bronisław Rudak**

*CAMK, Bartycka 18, 00716 Warszawa, Poland
Warsaw University, Astronomical Observatory, Al. Ujazdowskie 4, 00478 Warsaw, Poland
[†]New Mexico State University, Department of Astronomy, 1320 Frenger Mall, Las Cruces, MN 88003, USA
**CAMK, Rabianska 8, 87-100 Torun, Poland

Abstract. One of the leading models for the progenitors of of short GRBs are compact object mergers. The properties of coalescing compact object binaries have been obtained with the population synthesis method. The results show the differences between the NSNS and BHNS mergers. I review the status of these two models of short GRBs in the light of recently observed short bursts and their host galaxies.

Keywords: stars: binary, neutrons stars
PACS: 97.10.Cv, 97.80.Hn

INTRODUCTION

There was a number of presentations in this conferences in which the recent observations of short hard bursts (SHBs) were summarized. The HETE-2 and Swift observations have led to discoveries of afterglows and identification of host galaxies in which SHBs reside. These observations were analyzed with the aim of identifying the SHB progenitors. The leading model for the SHB progenitors emerging from these studies are compact object mergers. In this paper I wish to summarize the result of population synthesis analysis of the properties of compact object binaries and confront them with observations.

POPULATION SYNTHESIS RESULTS

In this paper I will concentrate on the results obtained with the use of the StarTrack population synthesis code. The code is described in detail in Belczynski et al. [1], and is based on several earlier papers [2, 3]. In the early papers [3, 4] we have mainly concentrated on the distribution of compact object mergers around galaxies. The understanding of such distributions has evolved along the understanding of the formation channels of such binaries. A very important point was the discovery of new evolutionary channels leading to formation of double neutron star binaries [5, 1]. These binaries have a very short lifetime because they are tightened by the additional common envelope phase with a helium star, provided that this star has small mass and can develop convective envelope. This new evolutionary channel has a significantly higher compact object binary production rate than the classical one. However, due to short lifetimes, tight orbits, and possibly weak recycling these binaries are hardly detectable as pulsars.

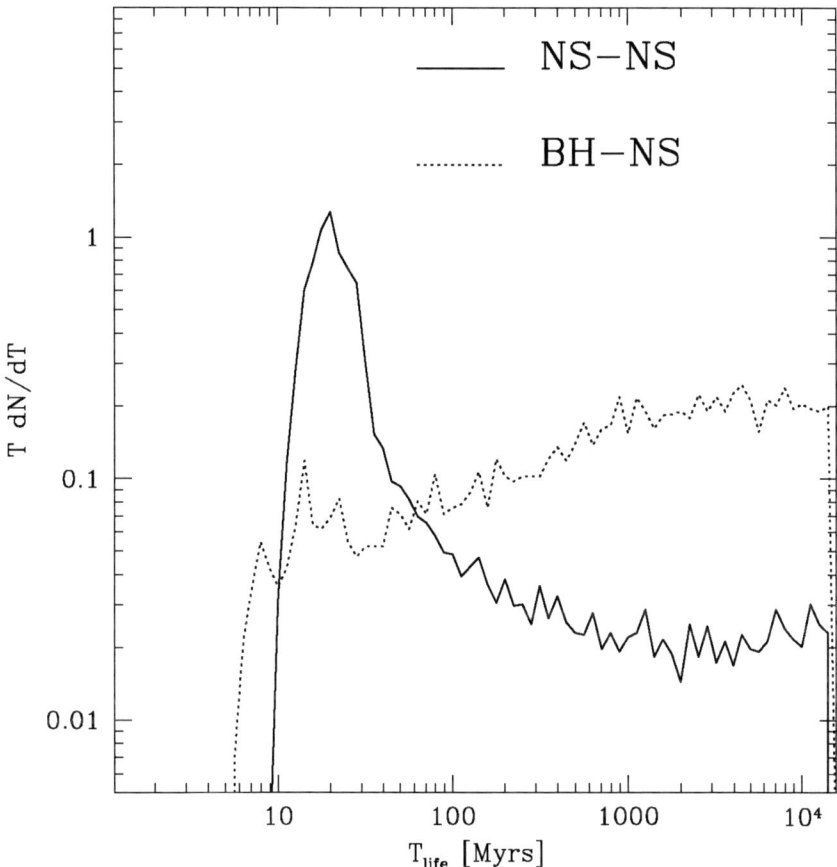

FIGURE 1. The distribution of lifetimes counted from zero age main sequence to the coalescence of possible binary progenitors of SHBs.

We present the results of the calculation of lifetimes of the double neutron star binaries (DNS), and back hole neutron star (BHNS) ones in Figure 1, see also [6]. There is a larger number DNS binaries with the short lifetimes in the range from 10 to 100Myrs, and above 100Myrs the distribution of lifetimes τ is approximately $\approx \tau^{-1}$. In the case of BHNS binaries the lowest lifetimes are of the same order as in the case of DNS, i.e. 10Myrs but the distribution of lifetimes falls approximately as $\approx \tau^{-0.6}$. The velocities of BHNS are probably smaller than those of DNS since the kicks in the explosions of supernovae probably decrease with increasing mass of the remnant. These two factors determine the distributions of merger sites around galaxies.

The distributions of merger sites around galaxies have been shown in several papers

[7, 8]. These have been calculated for the case of spiral galaxies of different masses with the assumption of continuous star formation. With such assumptions the DNS population mergers took place inside the host galaxies because of their short lifetimes, while the BHNS systems managed to escape especially from low mass hosts despite their low velocities because of the long merger times.

The difference in the distribution of lifetimes also means that there ought to be a different redshift distribution of DNS and BHNS merger sites. Because of short delays the DNS merger rate should trace the star formation rate, while the BHNS merger rate should be delayed by a few Gyrs. Thus the local SHB population up to $z \approx 1$ should consist of both DNS and BHNS mergers, while at higher redshifts there should be less BHNS mergers and more DNS ones. Therefore the SHB rate should drop above the redshift of unity. Additionally, if the total energetics of the bursts depends on the total mass of the system, or on the mass of the black hole that there should be a strong evolution of the SHB internal luminosity function with redshift. The local SHB internal luminosity function should be wide and its width should decrease with the redshift.

SUMMARY

We have presented the results of studies of SHB progenitors with the StarTrack population synthesis code. The possible binary progenitors of SHBs i.e. DNS and BHNS does not have simple uniform properties. The distribution of lifetimes stretches from 10 Myrs to the the Hubble time. The slope of the distribution differs between the DNS and BHNS population. The velocities of the binary SHB progenitors vary with their type and mass. It must be mentioned that the distribution of velocities is inferred from the studies of solitary radio pulsars and the distributions of velocities for binaries may be different.

Up to this point we have only studied the distributions of merger sites around galaxies with continuous star formation rate history. Currently we are working with the much improved population synthesis code and we include galaxies with different star formation histories (Belczynski etal. 2006, in preparation).

Observations show that SHBs indeed may come from a nonuniform population. There is a burst - GRB 050509B - at the outskirts of a large elliptical galaxy, and another one -GRB 050709- in a small starburst galaxy. While the redshifts of most bursts is small, there is also one outlier - GRB 050813 at the redshift of $z = 1.7$, which is hardly consistent with models where there is a significant delay between star formation and merger. The small number of SHBs and the difficulty in identifying breaks in their afterglow lightcurves still prevent studying in some detail their energetics. However, the continuing of the Swift mission will hopefully lead to progress in this field as well.

ACKNOWLEDGMENTS

This research was supported by the KBN grants number 2P03D00125 and 2P03D02228.

REFERENCES

1. K. Belczynski, V. Kalogera, and T. Bulik, *ApJ*, **572**, 407–431 (2002).
2. K. Belczyński, and T. Bulik, *A&A*, **346**, 91–100 (1999).
3. T. Bulik, K. Belczyński, and W. Zbijewski, *MNRAS*, **309**, 629–635 (1999).
4. K. Belczyński, T. Bulik, and W. Zbijewski, *A&A*, **355**, 479–484 (2000).
5. K. Belczyński, and V. Kalogera, *ApJ Lett*, **550**, L183–L187 (2001).
6. T. Bulik, and K. Belczynski, *Baltic Astronomy*, **13**, 280–283 (2004).
7. K. Belczynski, T. Bulik, and B. Rudak, *ApJ*, **571**, 394–412 (2002).
8. K. Belczynski, T. Bulik, and V. Kalogera, *ApJ Lett*, **571**, L147–L150 (2002).

External shock model for the radio afterglows of giant flares from soft γ-ray repeaters

X. Y. Wang[*,†], X. F. Wu[**], Y. Z. Fan[**], Z. G. Dai[*] and B. Zhang[‡]

[*]*Department of Astronomy, Nanjing University, Nanjing 210093, China*
[†]*Department of Astronomy and Astrophysics, Penn. State University, University Park, PA 16802, USA*
[**]*Purple Mountain Observatory, Chinese Academy of Sciences, Nanjing 210008, China*
[‡]*Department of Physics, University of Nevada, Las Vegas, NV 89154, USA*

Abstract. Radio afterglows have been detected following two giant flares from soft gamma repeaters (i.e. SGR1900+14, SGR 1806-20). Recent follow-up observations of the December 27 giant flare of SGR 1806-20 have detected a multi-frequency radio afterglow from 240 MHz to 8.46 GHz, extending in time from one week to about one month after the flare. The angular size of the source was also measured for the first time. Here we show that this radio afterglow gives the first piece of clear evidence that an energetic blast wave sweeps up its surrounding medium and produces a synchrotron afterglow, the same mechanism as established for GRB afterglows.

INTRODUCTION

Soft γ-ray repeaters (SGRs) distinguish themselves from classical gamma-ray bursts (GRBs) by their repetitive, soft bursts coming from nearby sources, likely strongly magnetized neutron stars dubbed "magnetars" [1]. Giant flares [2, 3] are rare events from SGRs, each characterized by an initial hard spike with more than a million Eddington luminosity and a pulsating tail persisting for hundreds of seconds. A radio afterglow was detected following the 1998 August 27 giant flare from the soft gamma repeater (SGR) 1900+14. This radio afterglow was explained as the blast wave mission as it interacts with the surrounding medium [4], the same model as invoked in the standard theory of GRB afterglows [5].

A super-giant flare was recorded on 2004 December 27 from SGR 1806-20 [6, 7]. In comparison with previous ones [2, 3], this flare is exceptionally strong with an isotropic energy of gamma-ray emission larger than 8×10^{45}ergs, given a source distance [10] $d = 15.1$kpc. Afterglow emission from giant flares was first detected[11] from the SGR 1900+14 at radio frequencies. The radio afterglow of the December 27 giant flare from SGR 1806-20 was detected about 7 days after the flare with a flux density at 8.46 GHz approximately 100 times brighter than the peak flux of the SGR 1900+14 radio afterglow.

Similar to classic GRBs, the extraordinarily high peak luminosity $L > 10^6 L_{\rm Edd}$ (where $L_{\rm Edd}$ is the Eddington luminosity), hard spectrum and rapid variability of the initial spike emission of giant flares imply that the spike emission must originate from a relativistically expanding fireball with the initial Lorentz factor of at least several [12] in order to avoid the pair production problem. Such a relativistic outflow, after emitting the hard spike, should retain some energy and then drives a blast wave expanding into the surrounding medium.

MODELLING THE RADIO AFTERGLOW OF THE GIANT FLARE FROM SGR1806-20

Now let's first consider the dynamics of a blast wave that is driven by the relativistic outflow and expands into the interstellar medium (ISM). The blast wave will be gradually decelerated by the swept-up ISM and enter the non-relativistic phase at $t_{nr} = (3E/4\pi nm_p c^5)^{1/3} = 4.5 E_{46}^{1/3} n_0^{-1/3}$ d, where E is the isotropic kinetic energy of the blast wave, c is the speed of light, n is the number density of ISM, m_p is the proton mass, and we have used the usual notation $a \equiv 10^n a_n$ in c.g.s. units except that the observer's time t is in unit of day. So, for $E_{46}/n_0 < 4$, the blast wave must have entered the non-relativistic phase during the observational time frame ($t > 7$d). During this non-relativistic phase, the speed (in units of c) and the radius of the blast wave are given by[4]

$$\beta = \left(\frac{12E}{125\pi nm_p c^5 t^3}\right)^{1/3} = 0.4 E_{46}^{1/5} t_1^{-3/5} n_0^{-1/5} \tag{1}$$

and $R = \frac{5}{2}\beta ct = 2.6 \times 10^{16} E_{46}^{1/5} t_1^{2/5} n_0^{-1/5}$ cm, respectively. The measured angular size at 8.46 GHz of the radio afterglow is ~ 80 mas around seven to ten days after the flare[8, 9] and has essentially no variation in diameter during this period. At a distance of $d = 15.1$kpc, this angular size implies a subluminal speed of ~ 0.35c for the averaged projected expansion and an apparent radius of $R_\perp = 9 \times 10^{15}$cm. So, if the outflow is spherical at this time (i.e. $R = R_\perp$), we obtain the constraint on the outflow energy, i.e. $E \simeq 10^{44} n_0$ ergs. If, instead, the outflow is mildly beamed in the non-relativistic phase, then the half opening angle θ satisfies $\sin\theta = R_\perp/R = 0.4(E_{46}/n_0)^{-1/5}$. In any case, the real kinetic energy (after beaming correction) of the outflow is about $10^{44} - 10^{45}$ergs for a typical ISM with $n = 1$cm^{-3}.

As the blast wave sweeps up the surrounding medium, the shock accelerates electrons and amplifies the magnetic field in the swept-up matter[13]. We assume that the shocked electrons and the magnetic field acquire constant fractions (ε_e and ε_B) of the total shock energy density, so that we obtain the minimum electron Lorentz factor $\gamma_m \sim \varepsilon_e \frac{m_p}{m_e}(\Gamma - 1)$ (where Γ is the Lorentz factor of the shock) and the magnetic field

$$B = (16\pi\varepsilon_B\beta^2 nm_p c^2)^{1/2} = 1.4 \times 10^{-2} \varepsilon_{B,-1}^{1/2} E_{44}^{1/5} t_1^{-3/5} n_0^{3/10} \text{G}. \tag{2}$$

The afterglow emission arises from synchrotron radiation of these shocked electrons in the magnetic field. The electron energy distribution in the downstream of the shock usually manifests as a single power law or broken power law. The spectrum[?] at $t \simeq 9$d after the giant flare appears to have a break, with a power law spectral index of $\beta_1 \simeq -0.7$ between 2.4 GHz and 4.8 GHz and $\beta_2 \simeq -1.0$ between 4.8 GHz and 8.6 GHz. The steepening of the light curve decay with frequency also favors that there is a spectral break between 1.43 and 8.46 GHz in the synchrotron blast wave model. This break is unlikely to be the cooling break because the cooling frequency is much higher than the radio band. As suggested by Li & Chevalier[14] to explain the afterglows of GRB991208 and GRB000301C, we consider that this break is caused by an intrinsic break in the energy distribution of the shock-accelerated electrons, which takes the form of $dN_e/d\gamma_e \propto \gamma_e^{-p_1}$ for $\gamma_{min} < \gamma_e < \gamma_b$ and $dN_e/d\gamma_e \propto \gamma_e^{-p_2}$ for $\gamma_e > \gamma_b$, where γ_{min} and γ_b are the minimum and break Lorentz factors respectively. From the values of β_1 and β_2, we infer $p_1 \simeq 2.4$ and $p_2 \simeq 3.0$. The detailed numerical fit to the early radio afterglow is plotted in Fig. 1 in [15].

Later time radio follow-up observations reveal bumps occurred at 20 days in the light curves of the afterglow. The late bump has been explained as an second component which is a quasi-spherical subrelativistic fireball [16]. The second component may be produced by the reconnection occurred near the surface, the resultant outflow of which should have high baryon contamination and thus subrelativistic. The overall evolution of the source size can be reproduced in this scenario as well.

IMPLICATIONS FOR FUTURE MULTI-WAVELENGTH OBSERVATIONS

SGR afterglows have been detected only in the radio band so far. Within the above picture, it is natural to expect optical and infrared afterglows as well produced by the same relativistic fireball accompanying the SGR giant flares. At early times $t < t_{nr}$, the blast wave is relativistic. The optical frequency is generally below the break frequency v_b for typical parameter values when $t < 0.1$d, and the flux at R-band (4.55×10^{14}Hz) is

$$F_R = 30 \left(\frac{\varepsilon_e}{0.3}\right)^{p_1-1} \varepsilon_{B,-1}^{\frac{p_1+1}{4}} E_{44}^{\frac{p_1+3}{4}} n_0^{1/2} t_{-1}^{-\frac{3(p_1-1)}{4}} \left(\frac{d}{15.1\text{kpc}}\right)^{-2} \text{mJy}. \quad (3)$$

For typical parameter values and the inferred value $\varepsilon_B = 0.037$ from the radio afterglow, the magnitude of the optical afterglow at $t = 0.1$d is $m_R \simeq 13.4$. Because the characteristic synchrotron frequency at early times is generally below the optical frequency for an initial Lorentz factor of $\Gamma_0 \sim 10$, the optical afterglow reaches its peak at the deceleration time of the relativistic outflow, which is $t_{dec} = 90 E_{44}^{1/3} n_0^{-1/3} \Gamma_{0,1}^{-8/3}$s. At this time, the optical afterglow has a flux as bright as $F_R \sim 5$Jy for typical parameter values. Such bright optical afterglows would be easily detected by rapid response detectors, such as Ultraviolet/Optical Telescope (UVOT) on *Swift* and Robotic Optical Transient Search Experiment (ROTSE), if the optical extinction to the source is not very large. However, SGR 1806-20 has a low Galactic latitude and has optical extinction[17] of $A_V \sim 30$ mag. This makes it impossible to observe the optical afterglow from this SGR. The difficulty also applies for the other two SGRs at low Galactic latitudes, SGR 1900+14 and SGR 1627-41, whose extinction is also large[18, 19]. The R-band extinction of SGR 1900+14 is moderate[20], i.e. $A_R \sim 7$ mag, leaving some hope for optical afterglow detection at its peak. The best target for optical follow-up observations among the four known SGRs is SGR 0526-66, which lies in the Large Magellanic Cloud and has $A_V \simeq 1$. At a distance of $d \simeq 50$ kpc, the optical afterglow of a similar giant flare from this SGR is expected to be bright enough for follow-ups. Even the less rare intermediate bursts[21] with the isotropic energies ranging from 10^{41} to 10^{43} ergs are expected to have optical afterglows brighter than $m_R \simeq 23$ at $t < 0.1$d for SGR 0526-66. This is well above the *Swift* UVOT sensitivity. Detecting such an optical afterglow would further test the blast wave model and provide more detailed information of the early evolution of a fireball as well as its composition.

Because of the high optical extinction of low-latitude SGRs, infrared observations to giant flares and intermediate bursts appear very useful. The K band flux would be, according to Eq. (3), $F_{2.2\mu m} \sim 2.8$mJy or $m_K \simeq 13.4$ at $t = 1$d for typical parameter values (i.e. $E = 10^{44}$ergs, $n = 1$cm^{-3}, $\varepsilon_e = 0.3$, and $\varepsilon_B = 0.037$) to model the radio afterglow of the December 27 giant flare. Such a bright infrared afterglow could be well detected by ground-based infrared telescopes. The flux of the X-ray afterglow from the blast wave can be estimated in the same way. For the above typical parameter values, we find the flux at $t = 0.1$d is about $F_X \simeq 2.6 \times 10^{-10}$ergcm^{-2}s^{-1}. This flux is, however, much dimmer than the observed X-ray flux of the SGR 1900+14 giant flare[22], which is thought to be originated from the neutron star surface immediately after the burst.

ACKNOWLEDGMENTS

This work was supported by the Special Funds for Major State Basic Research Projects, the National 973 Project (NKBRSF G19990754), the National Natural Science Foundation of China under grants 10403002, 10233010 and 10221001. B. Zhang is supported by NASA NNG04GD51G and a NASA Swift GI (Cycle 1) program.

REFERENCES

1. C. Thompson, and R. C. Duncan, *Mon. Not. R. Astron. Soc.* **275**, 255–300 (1995).
2. E. P. Mazets, et al., *Nature* **282**, 587-589 (1979).
3. K. Hurley,*et al. Nature* **397**, 41-43 (1999).
4. K. S. Cheng, & X. Y. Wang, *Astrophys. J.* **593**, L85-L88 (2003)
5. P. Mészáros, and M. J. Rees, *Astrophys. J.* **476**, 232-237 (1997)
6. K. Hurley, et al. *Nature* **434**, 1098-1102 (2005)
7. D. M. Palmer,et al., *Nature*, **434** 1107-1109 (2005)
8. P. B.Cameron , et al.*Nature*, **434**, 1112-1114 (2005)
9. B. M. Gaensler, *Nature*, **434**, 1104-1106 (2005)
10. S. Corbel, & S. S. Eikenberry, *Astron. Astrophys.* **419**, 191-201 (2004)
11. D. Frail, S. R. Kulkarni,, & J. Bloom, *Nature* **398**, 127-129 (1999)
12. C. Thompson,& R. C. Duncan, *Astrophys. J.* **561**, 980-1005 (2001)
13. R. Blandford, and D. Eichler, *Phys. Rep.* **154**, 1-75 (1987)
14. Z.-Y. Li, & R. A. Chevalier, *Astrophys. J.***551**, 940-945 (2001)
15. Wang, X. Y., et al. *Astrophys. J.*, **623**, L29-L32 (2005)
16. Dai, Z. G., et al. *Astrophys. J.*, **629**, L81-L84 (2005)
17. S. S. Eikenberry, *et al. Astrophys.* J.**563**, L133-L137 (2001)
18. F. J. Vrba, et al. *Astrophys. J.***533**, L17-L20 (2000)
19. S. Wachter,*et al. Astrophys. J.***615**, 887-896 (2004)
20. C. Akerlof, *et al. A*strophys. J.**542**, 251-256 (2000)
21. C. Guidorzi, *et al. A*stron. Astrophys. **416**, 297-310 (2004)
22. P. M. Woods, *Astrophys. J.***552**, 748-755 (2001)

PROMPT EMISSION

The latest two GRB detected by *Hete-2*: GRB 051022 and GRB 051028

A. J. Castro-Tirado*, S. McBreen[†], M. Jelínek*, S. B. Pandey*, M. Bremer**, A. de Ugarte Postigo*, J. Gorosabel*, S. Guziy*,[‡], G. Bihain[§], J. A. Caballero[§], P. Ferrero[¶], J. de Jong[‖], K. Misra[††] and D. K. Sahu[‡‡]

*Instituto de Astrofísica de Andalucía (IAA-CSIC), P.O. Box 3.004, E-18.080 Granada, Spain
[†]Astrophysics Missions Division, RSSD, ESA-ESTEC, Noordwijk, The Netherlands
**Institute de Radioastronomie Milimetrique (IRAM), 38406 Saint Martin d'Héres, France
[‡] Nikolaev State University, Nikolskaya 24, 54030 Nikolaev, Ukraine
[§]Instituto de Astrofísica de Canarias (IAC), 38205 La Laguna, Tenerife, Spain
[¶]Thüringer Landessternwarte Tautenburg, Sternwarte 5, D-07778 Tautenburg, Germany.
[‖]Max-Planck Institut für Astronomie, Koenigstuhl 17, D-69117 Heidelberg, Germany
[††]ARIES, Manora Peak, Nainital 263 129, India
[‡‡]Indian Institute of Astrophysics, 560034 Bangalore, India

Abstract. We present multiwavelength observations of the latest two GRB detected by *Hete-2* in 2005. For GRB 051022, no optical/nIR afterglow has been detected, in spite of the strong gamma-ray emission and the reported X–ray afterglow discovered by *Swift*. A mm afterglow was discovered at PdB confirming the association of this event with a luminous ($M_V = -21.5$) galaxy within the X-ray error box. Spectroscopy of this galaxy shows strong a strong [O II] emission line at z = 0.807, besides weaker [O III] emission. The X-ray spectrum showed evidence of considerable absorption by neutral gas with $N_{H,X-ray} = 4.5 \times 10^{22}$ cm^2 (at rest frame). ISM absorption by dust in the host galaxy at $z = 0.807$ cannot certainly account for the non-detection of the optical afterglow, unless the dust-to-gas ratio is quite different than that seen in our Galaxy. It is possible then that GRB 051022 was produced in an obscured, stellar forming region in its parent host galaxy.

For GRB 051028, the data can be interpreted by collimated emission (a jet model with $p = 2.4$) moving in an homogeneous ISM and with a cooling frequency v_c still above the X-rays at 0.5 days after the burst onset. GRB 051028 can be classified as a "gray" or "potentially dark" GRB. The Swift/XRT data are consistent with the interpretation that the reason for the optical dimness is not extra absorption in the host galaxy, but rather the GRB taking place at high-redshift.

Keywords: black hole formation, starburst galaxies, gamma-ray bursts
PACS: 97.60.-s, 98.54.Ep, 98.62.Py, 98.70.Rz

INTRODUCTION

Nowadays, dark GRBs seem to constitute a significant fraction of the GRB population. Although no optical/nIR afterglows have been detected for this class, transient X-ray and radio emission has allowed to pinpoint the parent galaxies where the bursts have occurred. About 48% (34/71) of well-localized GRBs by the *Swift*/XRT so far in the 1st yr of the *Swift* Era do not show an optical (or near-IR) afterglow, in spite of deep searches (down to 21-22 mag) being performed for nearly all in < 24 hr.

The main possible scenarios that have been proposed are the obscuration scenario, the low-density scenario and the high-redshift scenario (also discussed in [1,2]). The first one could be due to: i) a high column density of gas around the progenitor like a dusty, clumpy medium in giant molecular clouds (GMCs [3]) or ii) to dust in the host galaxy at larger distances. In the latter case, this will only account for no more than 10% of events (discussed in [4]). The occurrence of a burst in a low-density ambient medium [5] will result in a very dimmed afterglow as well. But this seems unlikely too due to the fact that it is now well accepted the relationship of long-duration GRBs with core-collapse supernova whose progenitors would have not been able to travel too far away from their parent star-forming regions where they were born. The high-z case, in which the Ly-α forest emission will affect the optical band, will not be expected for more than 10% of the events [6].

It is therefore, most essential to chase new potential dark GRBs and to study whether the reason for its *darkness* could due to the obscuration scenario (the a-priori most plausible scenario for the reasons given above).

GRB 051022

The bright GRB 051022 constituted a perfect study case. It was discovered by *Hete-2* on 22 Oct 2005 [7]. The burst started at 13:07: UT and lasted for ≈ 200 s, putting it in the "long-duration" class of GRBs [8]. It was also observed by *Mars Odyssey* and *Konus*/WIND [9], making of it the highest fluence event detected by *Hete-2* in its 5-yr lifetime, with the exception of the nearby GRB 030329. By the time when *Swift* slewed and started data acquisition (about 3.5 hr after the event onset), fading X-ray emission was detected by the *Swift*/XRT, what can be considered as the first clear detection of the GRB 051022 afterglow [10]. This triggered a multiwavelength campaign at many observatories aimed at detecting the afterglow at other wavelengths.

Multiwavelength observations

We triggered optical, near-IR and mm observations at different observatories. We also made use of the public X-ray data from *Swift*/XRT which were taken starting at 3.5 hr after the event. Further details are given in [11].

Following the reported discovery of the X-ray afterglow, the main observational result is the detection of the mm afterglow superimposed to a bright optical/nIR potential host galaxy, in spite of the afterglow remaining undetected at optical/nIR wavelengths.

The obscuring medium

According to its observed properties, this event is located undoubtely in the dark GRB locus of the $F_{opt} - F_{X-ray}$ diagramme of Jakobsson et al. [1]. It is obvious that the afterglow was not detected neither at optical nor at near-IR wavelengths due to obscuration in the line of sight. In order to estimate the amount of extinction by dust at optical wavelengths, the SED predicts a magnitude $R \sim 18.3$ at $T_0 + 33$ hr, from which we infer a considerable extinction toward GRB 051022, with a lower limit $A_R = 3.2$ mag in the observer frame, i.e. equivalent to $A_V = 1.7$ mag in the rest frame at $z =$

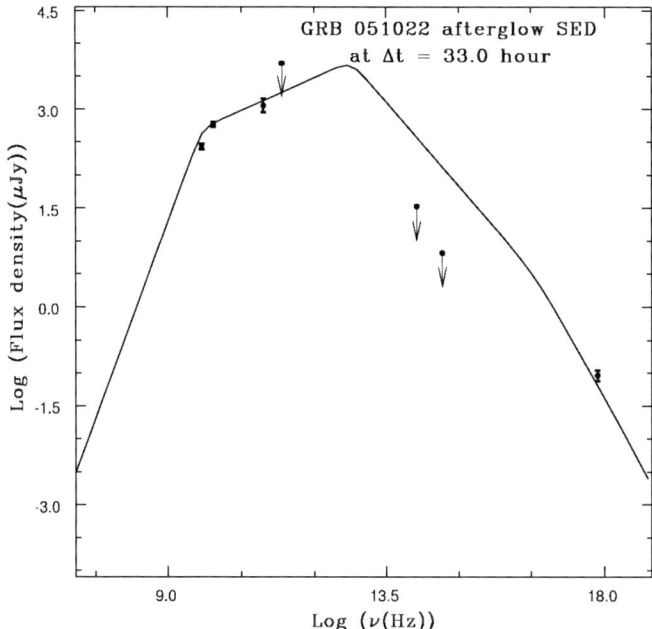

FIGURE 1. The SED of the GRB 051022 afterglow at $T_0 + 33$ hr (prior to the 2.9 day break time). We consider $p = 2.6$, assuming the slow-cooling regime. We have also included the radio detections at the WSRT [12] and VLA [13], as well as the 220 GHz and optical and near-IR upper limits reported in this paper. The initial parameters given at $T_0 + 0.06$ days are $v_a = 5.1 \times 10^9$ Hz, $v_m = 6.5 \times 10^{14}$ Hz, $v_c = 2.1 \times 10^{17}$ Hz and $F_{v,max} = 5.5$ mJy.

0.807 (see next subsection) from which a H equivalent column density at rest frame of $N_{H,opt} \geq 0.4 \times 10^{22}$ cm^2 is derived, assuming the relationship $A_V = 0.56 \times N_H$ (10^{21} cm^{-2}) + 0.23 [14].

From the X-ray spectral data, we have derived $N_{H,X-ray} = (0.9 \pm 0.2) \times 10^{22}$ cm^2 (observer frame) assuming the absorbing material being in a neutral cold state. This is equivalent to a $N_{H,X-ray}$ value at rest frame higher by a factor of $(1 + z)^{2.6}$ (= 5) at z = 0.807, i.e. $N_{H,X-ray} = (4.5 \pm 1.0) \times 10^{22}$ cm^2 (rest frame). Therefore, the $N_{H,X-ray}/N_{H,opt}$ ratio is ~ 9, as found in other GRBs [15,16] but not in two dark GRB host galaxies where $N_{H,X-ray}/N_{H,opt}$ was consistent with unity, or less: GRB 970828 [17] and GRB 000210 [4], where in both the dust-to-gas ratio is compatible with the Galactic one.

But could the absorption not take place in the circumburst environment of the GRB by in the ISM of the host galaxy itself? The value of $A_V = 1.0$ derived from the host galaxy SED (see next subsection) cannot account for the properties of this event considering a dust-to-gas ratio similar to the Galactic one.

Therefore we are left with the possibility of the burst arose from a giant Molecular cloud (GMC) were the $N_{H,X-ray}$ values are typical to the one found here and that the early gamma-ray radiation can destroy dust in the environment paving the way for the afterglow light going out through the molecular cloud.

The host galaxy

The GRB 051022 host galaxy is, unlike, most of the GRB hosts, brighter than L* ($M_V \geq -21.5$). We have also derived a SFR (UV) ≥ 20 $M_\odot$/yr and we foresee that this galaxy will be detected as sub-mm wavelengths as it was the case for the host galaxies of GRB 000210 and GRB 000418 [18].

GRB 051028

GRB 051028 was discovered by *Hete-2* on 28 Oct 2005 [19]. The burst started at T_0 = 13:36:01.47 UT and lasted for $\approx$ 16 s, putting it in the "long-duration" class of GRBs. *Swift* started to observe the field $\sim$ 7.1 hr after the event and detected the X-ray afterglow 5.2' away from the center of the initial 33' $\times$ 18' error box [20].

Optical and X-ray observations

Target of opportunity observations were triggered at several optical facilities in Asia and Europe. We also made use of the public X-ray data from *Swift*/XRT which consists of four observations starting $\sim$ 7.1, 120, 160 and 230 hr after the event respectively.

The X-ray data confirm the presence of a decaying X-ray source in the fraction (70 %) of the *Hete-2* error box covered by the *Swift*/XRT, as already pointed out by Racusin et al. [20]. The optical counterpart was discovered on our *R*-band images taken at the 4.2m WHT telescope starting 7.5 hr after the onset of the gamma-ray event. A faint R = 21.9 object was detected inside the *Swift*/XRT error circle. Further details are given in [21].

The optical data prior to 4 hr (i.e. in the range T_0 + 2.7 hr and T_0 + 4 hr) show a bumpy behaviour very similar to the one seen in other events like GRB 021004 [22], GRB 030329 (see [23] and references therein) and GRB 050730 [24]. In fact, the similarity with GRB 050730 is remarkable, if GRB 051028 is shifted up by 3 magnitudes (Fig. 5), and thus there is evidence for at least two of such bumps taken place, superimposed to the power-law decline. This could be explained in the context of the energy injection model [25]. Unfortunately there is no X-ray data available at this epoch.

High redshift and/or dust in the surrounding ISM ?

We have extrapolated the optical and X-ray fluxes of GRB 051028 afterglow to T_0 + 11 hr and derived a value of β_{opt-X} = 0.55 $\pm$ 0.05. Thus GRB 051028 is located in the "gray" or "potentially dark" GRB locus on the dark GRB diagram by Jakobsson et al. [1]. How can the optical faintness of GRB051028 be explained ?

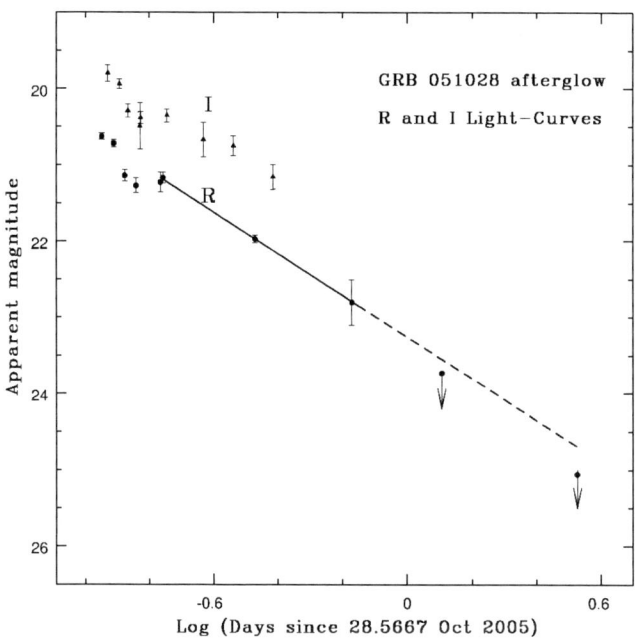

FIGURE 2. The GRB 051028 R and I-band lightcurves obtained at Hanle, Tautenburg and La Palma (WHT) starting 2.5 hr after the event onset and continuing up to 3.5 days later. The data after 4.0 hr are fit by a power-law decline exponent $\alpha_{opt} = 0.9 \pm 0.1$.

Although the redshift of this event could not be properly measured due to its faintness at the time of the discovery, we are able to constrain it on the basis of the VRI-band data presented in this paper. Using the magnitudes derived here and correcting them for the Galactic extinction in the line of sight, we determine a spectral optical index $\beta_{opt} = 2.1 \pm 0.4$. In the simplest fireball models [26], $F_\nu \propto \nu^{-\beta}$ with $\beta = p/2$ for $\nu > \nu_c$ and $\beta = (p-1)/2$ for $\nu < \nu_c$. Thus, for a typical range of p values in the range $1.5 < p < 3$ [27], β_{opt} should be in the range $0.25 < \beta_{opt} < 1.5$. In fact, the GRB 051028 data before $T_0 + 0.5$ day, are well fitted by a jet model with $p = 2.4$ in the slow cooling case, moving through the ISM (with ρ = constant) and with a cooling frequency ν_c still above the X-rays, so β_{opt} should be ~ 0.7. What is the reason for the discrepancy in the β values?

This is naturally explained by the fact that at $z \sim 3.2$ and ~ 4.0, the Lyman-α break begins affecting the V and R passbands respectively. Therefore, one obvious explanation for the β_{opt} value found for GRB 051028 is that it also arose at a $z \approx 3.6 \pm 0.4$. This value is in fact in agreement with the pseudo-z derived for this burst [28] and with the fact that no host galaxy is detected down to $R = 25.1$. This high-redshift is also supported by the late break time, as typical afterglows undergo a jet break episode before $T_0 + 1$ day in the rest frame [27].

The X-ray afterglow of GRB 051028 can be compared to other GRB afterglows in the sense that its flux at 11 hr is typical, i.e., one can assume that the burst has occurred on a classical n $\sim$ 1 cm^{-3} environment. The optical afterglow, on the other hand, is dim. This indicates that the faintness of the optical emission is not due to a low-density environment as in the case of some short GRBs, such as GRB 050509b [29]. Instead, we propose that GRB 051028 occurred in a galaxy at $z \approx 3.6 \pm 0.4$.

ACKNOWLEDGMENTS

We appreciate the *Swift*/XRT data being public as soon as a burst is observed. We also thanks F. J. Aceituno, G. C. Anupama, V. Béjar, R. González-Delgado, C. Gry, A. Henden, J. Licandro, D. A. Kann, S. Klose, R. Neri, S. Srividya, L. Valdivieso and S. Vanniarajan for fruitful conversations. This work is based partly on observations carried out with the IRAM Plateau de Bure Interferometer. IRAM is supported by INSU/CNRS (France), MPG (Germany) and IGN(Spain), and with the 1.5 OSN, 2.0 HCT, 3.5 TNG and 4.2 WHT on different astronomical observatories worldwide. This research has also been partially supported by the Ministerio de Ciencia y Tecnología under the programmes AYA2004-01515 and ESP2002-04124-C03-01 (including FEDER funds).

REFERENCES

. 1. Jakobsson, P., Hjorth, J., Fynbo, J. P. U. et al. 2004, ApJ 617, L21
. 2. Rol, E., Wijers, R. A. M. J., Kouveliotou, C., Kaper, L. & Kaneko, Y. 2005, ApJ 624, 868
. 3. Lamb, D. Q. 2001, in Gamma-ray Bursts in the Afterglow Era, ed. E. Costa, F. Frontera and J. Hjorth (Berlin, Springer), 297
. 4. Piro, L., Frail, D. A., Gorosabel, J., et al. 2002, ApJ 577, 680
. 5. Taylor, G. B., Frail, D. A., Kulkarni, S. R. et al. 2000, ApJ 502, L115
. 6. Gorosabel, J. et al. 2004, A&A 427, 87
. 7. Olive, J.-F., Ricker, G., Atteia, J.-L. et al. 2005, GCN Circ. 4131
. 8. Tanaka, K., Ricker, G., Atteia, J.-L. et al. 2005, GCN Circ. 4137
. 9. Hurley, K., Cline, T., Mitrofanov, I. et al. 2005a, GCN Circ. 4139
. 10. Racusin, J., Burrows, D., Gehrels, N. et al. 2005a, GCN Circ. 4141
. 11. Castro-Tirado, A. J.,Bremer, M., McBreen, S. et al. 2006a, A&A, submitted
. 12. van der Horst, A. J., Rol, E. and Wijers, R. A. M. J. 2005, GCN Circ. 4158
. 13. Cameron, P. and Frail, D. 2005, GCN Circ. 4154
. 14. Predehl, P. & Schmitt, J. H. M. M. 1995, A&A 293, 889
. 15. Galama, T. J. & Wijers, R. A. M. J. 2001, ApJ 549, L209
. 16. Campana, S. et al. 2005, A&A (TBC), submitted (astro-ph/0511750)
. 17. Djorgovski, S. G., Frail, D. A., Kulkarni, S. R. et al. 2001, ApJ 5 62, 654
. 18. Berger, E. et al. 2003, ApJ 588, 99
. 19. Hurley, K., Ricker, G., Atteia, J-L. et al. 2005b, GCN Circ. 4172
. 20. Racusin, J., Page, K., Kennea, J. et al. 2005b, GCN Circ. 4174
. 21. Castro-Tirado, Jelínik. M., Pandey, S. B. et al. 2006b, A&A, submitted
. 22. de Ugarte Postigo, A., Castro-Tirado, A. J., Gorosabel, J. et al. 2005, A&A 443, 841
. 23. Guziy, S. et al. 2006, A&A, submitted
. 24. Pandey, S. B. et al. 2006, A&A, submitted
. 25. Björnsson, G., Gudmundsson, E. H., & Jóhannesson, G. 2004, Ap J 715, L77
. 26. Sari, R., Piran, T. & Narayam R. 1998, ApJ 497, L17
. 27. Zeh, A., Klose, S. & Kann, D. A. 2006, ApJ, in press (astro-ph/050 9299)
. 28. Pelangeon, A., Atteia, J.-L., Lamb, D. Q., Ricker, G. E. et al. 2006, in: Gamma-ray bursts in the *Swift* Era, These Proceedings
. 29. Castro-Tirado, A. J., de Ugarte Postigo, A., Gorosabel, J. et al. 2005, A&A 439, L15

The Swift Prompt Sample

P.T. O'Brien*, R. Willingale*, J. Osborne*, M.R. Goad*, K.L. Page*, A.P. Beardmore*, O. Godet*, D.N. Burrows[†] and N. Gehrels**

Department of Physics and Astronomy, University of Leicester
[†]*Department of Astronomy & Astrophysics, 525 Davey Lab, Pennsylvania State University, University Park, PA 16802, USA*
**NASA/Goddard Space Flight Center, Greenbelt, MD 20771, USA*

Abstract.
We have analysed BAT and XRT data for 40 gamma–ray bursts (GRBs) observed using the *Swift* satellite. The earliest X-ray light curve can be well described by an exponential which relaxes into a power law, often with flares superimposed. The transition time between the exponential and the power law gives a physically defined timescale for the burst duration. In most GRBs the power law decay changes to a shallower decay within the first hour. The resultant "late emission hump" can last for several tens of ks. In other bursts the late hump is weak or absent. The observed variety in light curve shape can be explained as a combination of three components: prompt emission from the central engine; afterglow; and the late hump, which may also be due to the central engine. GRBs with stronger afterglow components are more likely to have brighter early optical emission. The late emission hump can have a total fluence equivalent to that of the prompt phase.

Keywords: gamma-ray bursts
PACS: 98.70.Rz

INTRODUCTION

Gamma-ray bursts (GRBs) are observed as a brief flash of gamma-rays seen at a random location on the sky. At that instant, the GRB becomes the intrinsically brightest single object in the Universe and can be seen at all redshifts. Understanding the observed emission is vital if we are to determine the nature of the progenitor. The *Swift* satellite [1], launched in November 2004, provides unique capability with which to study the earliest X-ray emission. In particular, by combining data from the Burst Alert Telescope and the X-ray Telescope, the temporal and spectral properties of a burst can be analysed from the first rise in flux out to days or weeks.

It is conventional to describe the duration of a GRB in terms of the timescale over which 90% of the gamma-rays were detected – the T_{90} parameter. The GRB X-ray flux can be represented as a function of time and frequency using a function $f_v \propto v^{-\beta} t^{-\alpha}$, where β is the spectral index and α is the temporal index. The photon index Γ is related to β by $\Gamma = \beta + 1$.

Initial *Swift* results show that some bursts, e.g. [2] [3], decline rapidly in the first hour, with $\alpha \geq 3$ and then break to a shallower decay, henceforth refered to as the "late emission hump". In other bursts the early X-ray flux declines more slowly with $\alpha \sim 1$, e.g. [4]. The observed differences in the rate of decay among bursts could be due to different relative contributions from emission components such as high-latitude (off-axis) emission [5] or afterglow produced by an external shock [6].

To disentangle the relative contribution of emission from the central engine and that due to the afterglow, we have systematically analysed the temporal and spectral properties of a large GRB sample combining data from the BAT and XRT. This analysis is described in detail in [7]. The sample comprises 40 GRBs detected by *Swift* prior to 2005 October 1 for which *Swift* slewed to point its narrow-field instruments within 10 minutes of the burst trigger time. Of the 40 GRBs, 38 are long bursts.

BAT AND XRT ANALYSIS

The BAT data were processed using the standard BAT analysis software (*Swift* software v. 2.0) as described in the BAT Ground Analysis Software Manual [8] and then light curves and spectra were extracted over 15–150 keV. Spectral indices ($f_\nu \propto \nu^{-\beta_b}$), were derived by fitting power laws over the T_{90} period. Power law fits were also used to parameterise the XRT spectra ($f_\nu \propto \nu^{-\beta_x}$), over 0.3–10 keV. Most of the GRBs show evidence for excess absorption above the Galactic column.

The spectra extracted from the XRT data are usually best fitted by a softer power law than the BAT data (i.e. $\beta_x > \beta_b$). Thus, we have formed unabsorbed, 0.3–10 keV flux light curves for each GRB by: (a) converting the XRT count rates into unabsorbed fluxes using the XRT power law model and (b) converting the BAT count rates into unabsorbed fluxes by extrapolating the BAT data to the XRT band using a power law model with an absorbing column derived from the XRT data and a spectral index which is the mean of the XRT and best-fit BAT spectral indices.

THE EARLY TEMPORAL AND SPECTRAL SHAPE

The GRB light curves often show considerable structure superimposed on the decaying light curve following the initial burst, often displaying flares which are usually small in amplitude but can be very large [9]. Some two-thirds of the light curves show an early, rapid phase of decline which then breaks to a late emission hump. This temporal break is usually observed within the first hour, and is clearly not a jet break. The remaining GRBs seem to have a continuous decline.

In order to compare light curves for GRBs with different power law decay indices, we have developed a procedure to fit light curves assuming there is a common intrinsic form to the early X-ray light curve. An average X-ray decay curve expressed by log(time) as a function of log(flux), $\tau(F)$, and log(flux) as a function of log(time), $F(\tau)$, was derived by taking the sum of scaled versions of each of the individual light curves, $f_i(t_i)$, where t_i is approximately the time since the largest/latest peak in the BAT light curve. The data points were transformed to normalised log(flux), $F_i = \log_{10}(f_i/f_d)$, and log(time) delay values, $\tau_i = \alpha_d \log_{10}(t_i - t_d) - \tau_d$. Four decay parameters (suffix d) specify the transformation for each GRB: f_d, the mean prompt flux; t_d, the start of the decay; τ_d, a time scaling; and α_d, a stretching or compression of time. The best fit f_d, t_d, α_d and τ_d for each GRB were found using a least squares iteration procedure, excluding bright flares. The resultant composite light curve for the entire sample is shown in Figure 1.

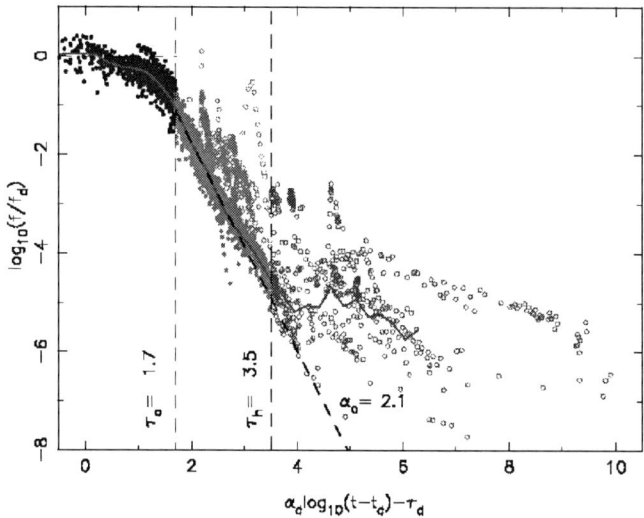

FIGURE 1. The composite X-ray light curve for 40 GRBs. The filled circles at $\tau < 1.7$ are within time T_p. This is followed by the power law decay. Flares or the late emission hump ($\tau > 3.5$) are shown as open circles. The average decay curve is shown as a solid curve while the mean power law curve is a dashed line.

Under the transformation all the light curves conform to an approximately universal behaviour with an initial exponential decline $\propto \exp(-t/t_c)$ followed by a power law decay $\propto t^{-\alpha_0}$. The transition between the two decay phases occurs when the exponential and power law functions and their first derivatives are equal, and is given for the average decay curve by $t_0 = t_c \alpha_0$ ($\tau_0 = 1.7$). Adopting this transition, for each GRB we define the division between the prompt and power law decay phases to be τ_0, corresponding to a prompt time $T_p = 10^{(\tau_0+\tau_d)/\alpha_d}$ seconds. This prompt time definition provides us with an alternative estimate of the duration of the prompt phase for each burst which depends on the physical shape of the light curve rather than the sensitivity of a particular instrument. T_p is comparable to T_{90} for many bursts, but it can be considerably shorter or longer.

The average decay curve relaxes into a power law with a decay index $\alpha_0 = 2.1$, found by linear regression on the average decay curve for $\tau_0 < \tau < 3.0$. This power law fit is shown as a dashed line in Figure 1. The fitting procedure results in those GRBs which follow a fairly continuous decay lying close to the power law. At $\tau \sim 3$ the average decay curve starts to rise above the power law decay in the majority of bursts. This is the start of the late emission hump, which we define to start at $\tau_h = 3.5$.

The initial temporal decay index for individual GRBs can be calculated by multiplying α_0 by the best fit α_d. GRBs with $\alpha_d > 1$ have decays steeper than average and those with

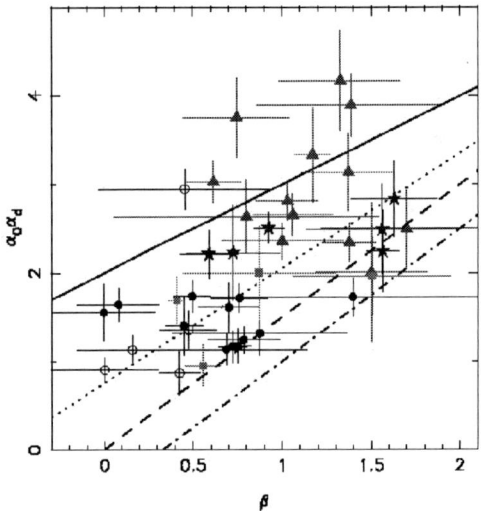

FIGURE 2. Correlation of $\alpha_0\alpha_d$ with β, where β is the average of the spectral indices from the BAT and XRT. The GRBs are plotted as: filled circles — no significant flares or hump; filled squares — flares but no hump; filled stars — flares and hump; filled triangles — hump but no flares; and open circles — no late data. The solid diagonal line shows the predictions of the high latitude model. The dashed and dot-dashed diagonal lines show the two afterglow models discussed in the text. The dotted diagonal line shows a fit to those bursts below the high latitude line.

$\alpha_d < 1$ shallower. The $\alpha_0\alpha_d$ are based on all the available data from both the BAT and XRT and are expected to be a more robust estimate of the global average than fitting a power law to a short section of light curve.

DISCUSSION

We can use the values of $\alpha_0\alpha_d$ to test the high latitude and afterglow models. The correlation of $\alpha_0\alpha_d$ with β, the average of the BAT and XRT spectral indices is shown in Figure 2. It is clear from Figure 2 that the decay indices, $\alpha_0\alpha_d$, correlate with the strength of flares and humps. The bursts with the most significant humps do not have large X-ray flares but they do have steep decays and straddle the high latitude line. The bursts with no significant flares or humps all lie below the high latitude line.

In principle, the relationship between the temporal decay index and spectral index has two components such that $\alpha = \alpha_v\beta + \alpha_f$. The coefficient α_v arises from the redshift of the peak of the spectral distribution of the synchrotron emission as a function of time and α_f arises from the temporal decay in the peak flux value of the same spectral

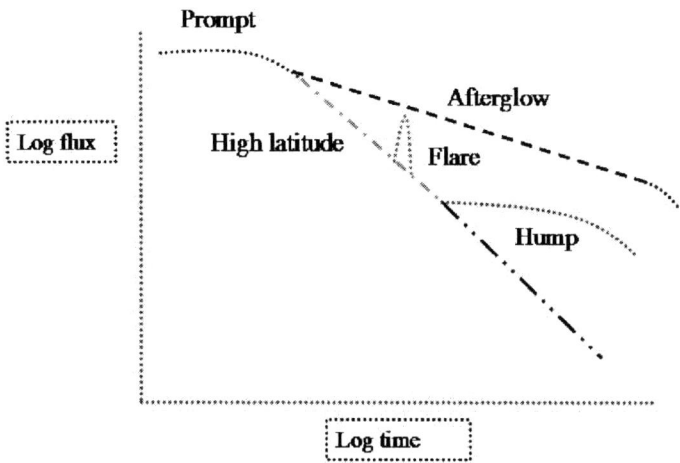

FIGURE 3. A schematic view of a GRB early X-ray light curve. Following the prompt emission the decay can either be gradual and fairly slow (afterglow dominated) or rapid (high latitude emission dominated) with a late emission hump. X-ray flares can occur in either case.

distribution. The solid line in Figure 2 shows the expected relationship for the high latitude model with $\alpha_v = 1$ and $\alpha_f = 2$. The dashed line shows the relationship expected from an afterglow expanding into a constant density ISM observed at a frequency below the cooling break ($v_x < v_c$) and before a jet break, with $\alpha_v = 3/2$ and $\alpha_f = 0$ [10]. If $v_x > v_c$ then α_v is unchanged and $\alpha_f = -0.5$. This is plotted as a dot-dashed line on Figure 2. All of the GRBs lie on or above these afterglow lines. The large majority of GRBs lie below the high latitude prediction. For these it is likely that we are seeing a combination of high latitude prompt emission and conventional, pre-jet-break afterglow.

To illustrate the multiple components causing the distribution seen in Figure 2, we show a schematic view of the X-ray light curve in Figure 3. Following the prompt emission bursts can either (a) decay rapidly and follow the high-latitude line, later displaying a late emission hump, or (b) decay more slowly and follow the afterglow line. In the latter case the late emission hump may be present but is masked by afterglow emission. Quantifying the strength of the late emission hump as that emission above the power law decay for $\tau > 3.5$, we find that the maximum late hump fluence is commensurate with the prompt fluence, suggestive of some kind of equipartition in energy between the emission phases.

We find a correlation such that those GRBs which decay more slowly are more likely to have an early optical detection. Using the initial UVOT V-band exposure to quantify the early optical brightness, for the 33 GRBs in our sample with UVOT observations in the first 10 minutes, those with $\alpha_0 \alpha_d < 2$ are four times more likely to have been optically detected.

CONCLUSIONS

We have analysed data for 40 GRBs observed by *Swift* prior to 2005 October 1 for which XRT observations began within 10 minutes of the BAT trigger. We have combined data from the BAT and the XRT to investigate the form of the X-ray emission (0.3–10 keV) during the first few hours following the burst.

The initial XRT spectral index is slightly steeper than that seen in the BAT, showing that spectral evolution occurs as the GRB ages. The light curve can be described by an exponential which relaxes into a power law whose decay rate varies considerably from burst to burst. The transition time between the exponential and power law provides a well determined measure of the burst duration. Some two-thirds of the GRBs display a light curve which shows a break to a shallower decay rate – the late emission hump – usually within an hour of the trigger. The remaining bursts decline fairly continuously.

Comparing the temporal and spectral indices of the power law decline, the distribution is consistent with a simple model in which the early emission is a combination of emission from the central engine and afterglow. Those GRBs in which the afterglow is weak early on decay fast during the power law phase and their X-ray light curves are consistent with the high latitude model. Some are dominated by afterglow, while the majority require a significant contribution from both components. The likelihood of an early optical detection strongly correlates with the strength of the X-ray afterglow component.

The late emission hump component may be present in all objects but can be masked by a strong afterglow component. The strongest humps seen have a total fluence which matches that of the prompt phase.

ACKNOWLEDGMENTS

The authors gratefully acknowledge funding for *Swift* at the University of Leicester by PPARC, in the USA by NASA and in Italy by ASI.

REFERENCES

1. N. Gehrels et al., 2004, ApJ, 611, 1005
2. G. Tagliaferri et al., 2005, Nature, 436, 985
3. S. Vaughan et al., 2005, ApJ, in press
4. S. Campana et al., 2005, ApJ, 625, L23
5. P. Kumar, A. Panaitescu, 2000, ApJ, 541, L51
6. P. Mészáros, M.J. Rees, 1997, ApJ, 476, 232
7. P.T. O'Brien et al., 2005, ApJ, submitted
8. H. Krimm, A. Parsons, C. Markwardt, 2004, "BAT Ground Analysis Software Manual" (http://heasarc.gsfc.nasa.gov/docs/swift/analysis)
9. D.N. Burrows et al., 2005, Science, 309, 1833
10. R. Sari, T. Piran, R. Narayan, 1998, ApJ, 497, L17

Stochastic wakefield acceleration in Gamma-ray Bursts

Guido Barbiellini, Francesco Longo*, Nicola Omodei[†], Annalisa Celotti** and Marco Tavani[‡]

Department of Physics, University of Trieste and INFN, section of Trieste
[†]*INFN, section of Pisa*
**SISSA, Trieste*
[‡]*IASF/INAF Rome and Department of Physics, University of Tor Vergata, Rome*

Abstract. Gamma-Ray Burst (GRB) prompt emission can, for specific conditions, be so powerful and short-pulsed to strongly influence any surrounding plasma. In this paper, we briefly discuss the possibility that a very intense initial burst of radiation produced by GRBs satisfy the intensity and temporal conditions to cause stochastic *wake-field particle acceleration* in a surrounding plasma of moderate density. We consider a simple but realistic GRB model for which particle wake-field acceleration can first be excited by a very strong low-energy precursor, and then be effective in producing the observed prompt X-ray and gamma-ray GRB emission.

Keywords: Waves, oscillations, and instabilities in plasmas and intense beams, particle acceleration, Gamma-ray Bursts
PACS: 52.35.-g, 94.20.wc, 98.70.Rz

INTRODUCTION

In this paper, we investigate a model of GRB production and evolution that assumes: (1) the existence of a thermal-like precursor of very large intensity following the energy release from relativistically moving shells or hydrodynamic fronts originating from the GRB compact object; (2) the radiative interaction of such a precursor (with a given power-spectrum as seen by an observer at rest) with a surrounding (initially not-ionized) gas shell at rest of density $n \sim 10^9 \mathrm{cm}^{-3}$ at the approximate distance $R \sim 10^{15}$ cm; (3) propagation of the radiative front within the shell, with possible activation of wake-plasma particle acceleration at the outwardly propagating radiative shock; (4) wake-field particle acceleration producing a generation of non-thermal electrons whose kinetic energy distribution function extends up to $> 100\,\mathrm{MeV}$; (5) betatron radiation of these energized electrons producing the observed non-thermal prompt GRB high-energy emission in the X-ray/gamma-ray energy ranges. The existence of an amount of material corresponding to a Thompson optical depth ~ 1 is derived by the observation made by BATSE of late bumps in the GRB light curve[1], interpreted as a Compton reprocessed emission from dense circumburst material at $R = 10^{15}$ cm[2]. The physical origin of this material is at present subject of very intense simulations of the WR wind, where the wind could be compressed by the preceding supernova explosion.

WAKEFIELD ACCELERATION MECHANISM

The very recent technical progress on high power laser beams has provided experimental evidence supporting the concept of wake-field particle acceleration in a plasma excited by an electromagnetic wave of duration comparable or shorter thanthe inverse of the plasma frequency[3]. A laser beam with a duration of 30 femtoseconds and a power of 50 TeraWatts, illuminating a gas jet 1 mm long with a density of 10^{19} particles per cm^3 is able to accelerate electrons to an energy of few tens of MeV. An X-ray beam of a few keV generated by the electron radiation loss in the plasma was observed by Ta Phuoc et al.[4] with a characteristic angular distribution resembling that of synchrotron radiation.

In this paper, following the general hypothesis already suggested by Tajima & Dawson[3], the processes of acceleration and radiative losses induced by ultra short laser pulses on the gas target, validated by recent experimental data, are suggested to be the source of the GRB prompt emission.

In our case, the laser role is played by GRB precursor photons emitted by the γ and $e^{\pm}$ shells collimated in an angle θ_j and moving with high relativistic velocity and the plasma density of the target gaseous nebula is assumed to be on the average 10^9 cm^{-3} over the region producing the prompt emission at a distance from the central source of about 10^{15} cm. The key point in the analogy GRB-Lab plasma acceleration is that the coherent radiation of laboratory lasers can be translated into non-coherent (stochastic) radiation from the precursor if a price is payed in terms of power density. The critical parameter in the coherent Lab plasma acceleration is the power surface density threshold. We show below that σ_{Wt} scales linearly with particle density. Its value is assumed from the experiment:

$$\sigma_{Wt} = 3 \times 10^{18} \left(\frac{n}{10^{19}\ cm^{-3}}\right)\ W\ cm^{-2} \qquad (1)$$

In the hypothesis that the GRB precursor carries an energy $E_{pr} = E_\gamma/\varepsilon$ and assuming $t_{pr} \sim t_{grb} \sim 30/(1+z)$ s the power density on the surface $R^2\theta^2 = 10^{27}$ cm^2 is:

$$\sigma_{Wgrb} = \frac{E_{pr}}{t_{grb}R^2\theta^2} \geq 2 \times 10^{16}\ W\ cm^{-2} \qquad (2)$$

which exceeds by several order of magnitude (around 10^8) the power density threshold (scaled at $n = 10^9$ cm^{-3}).

The threshold to initiate stochastically the wakefield mechanism is equal to the coherent power density multiplied by the square of the stochasticity parameter, proportional to $\sqrt{t_{pr}/\tau_p}$. This stochastic factor is the probability to produce by chance the coherent WFA, independently in time and space.

In other words, GRB not-coherent prompt emission is much more powerful than the coherent lab radiation: we assume that the loss of coherence is balanced by the enormous amount of power per square centimeter as well as the possibility of stochastic resonance: the wake field acceleration threshold is overcome.

Stochastic acceleration and deceleration are in principle related by plasma effects. However, since they are statistically independent, the two processes can be separated without loss of generality. In the following, the acceleration phase is not treated in detail. On the basis of the experimental results of Ta Phuoc et al.[4] it is assumed that a power law energy spectrum of electron and positron is produced. The accelerated beam enters in the plasma region with constant average gas density and radiates in the plasma, following again the results of Ta Phuoc et al.[4].

In the laser-driven betatron radiation the emission follows the scheme of Ta Phuoc et al.[4]. The angular distribution of the emitted radiation is dominated by the wiggling angle of the electron,

$$\theta = \frac{K}{\gamma}$$

where

$$K = \frac{\gamma \omega_b}{c} r_0$$

with

$$\omega_b = \frac{\omega_p}{\sqrt{2\gamma}}$$

the oscillating frequency, with $\omega_p = \sqrt{ne^2/(\varepsilon_0 m_e)}$ is the plasma fundamental frequency and r_0 is the impact parameter of the electron into the positive plasma column (e is the electron's charge).

GRB GAMMA EMISSION IN THE STOCHASTIC WAKE FIELD ACCELERATION REGIME

We address here the GRB prompt radiation produced by electrons radiating in a plasma excited by a very intense precursor playing the role of a "laser". The basic formulae of laser-driven betatron radiation are assumed taking into account the stochastic nature of the acceleration and the radiation phases. Let us now apply these results to the case of a plasma accelerated by the GRB precursor photons. We will treat the motion of the relativistic electrons radiating in the plasma. The fundamental difference with the above treatment resides in the stochastic nature of acceleration and radiation in the case of the GRB emission.

If the power surface density σ_W at any position of the electrons producing the GRB exceeds the value measured by the experiment[4] scaled at the GRB density and multiplied by the stochastic factor, the radiated photons of the GRB will be emitted from any accelerated electron provided that the stochastic angle of the electrons θ_{eff} remain below the jet opening angle θ_j. The elementary angle of the electron oscillation is:

$$\theta_i = \frac{r_0}{\lambda_b}$$

The stochastic motion of the electrons in the plasma results in an effective solid angle spanned by the electrons:

$$\theta_{\text{eff}}^2 = \left(\frac{R}{\lambda_b}\right) \theta_i^2$$

where we have introduced the adimensional quantity N:

$$N = \left(\sqrt{\frac{R}{\lambda_b}}\right) = \sqrt{\frac{R}{\sqrt{2\gamma}\lambda_p}}$$

where R is the distance travelled by the excited electron inside the plasma. This will be used to estimate the stochasticity of the electron geometrical path.

An electron with energy $E = \gamma mc^2$ wiggling inside the plasma with density n will contribute to the GRB detectable emission for a path R_θ if:

$$[\theta_{\text{eff}}(R_\theta, \gamma, n)]^2 \leq \theta_j^2 \tag{3}$$

Putting esplicitly the dependence on the path R_θ the previous equation could be written as:

$$R_\theta \frac{r_0^2}{\lambda_p^3}(2\gamma)^{-1.5} \leq \theta_j^2$$

This relation expressed in terms of the plasma density and using the scaling laws for r_0, λ_p and γ derived in Barbiellini et al.[5] could be written as:

$$R_\theta(\gamma) = \frac{2.8\,\theta_j^2 \gamma^{1.5} \lambda_p^3}{r_0^2} \sim 7\times 10^6 cm\, \frac{\gamma^{3/2}\theta_j^2}{n^{5/6}}$$

The equation 3 introduces a threshold for the electron emission over a distance R. The threshold Lorentz factor γ_t is obtained requiring that even in the case of a large energy loss, the electron trajectory still remains within the energized cone.

The effective energy loss in the stochastic acceleration regime is:

$$\frac{d\gamma}{dt} = -\frac{2}{3}r_e r_0^2 \gamma^4 \frac{\omega_b^4}{c^3}\left(\frac{R}{\lambda}\right)^{0.5}$$

where for simplicity we have introduced the classical electron radius r_e

$$r_e = \frac{e^2}{4\pi\varepsilon_0 m_e c^2}$$

and where we have multiplied the coherent energy loss by the stochastic factor $\sqrt{R/\lambda}$. This factor is our best estimation of the effect of the turbulence of the circumburst medium. Expressing the previous relation in terms of γ, and substituting $dt = c\,dR$, we obtain:

$$d\gamma \gamma^{-7/4} = 0.1 r_e r_0^2 \lambda_b^{-9/2} dR^{-3/2}$$

From this relation we estimate a typical "energy loss" distance R_e equivalent to the path where the electron looses half of its energy. Integrating the previous equation we get:

$$R_e \sim \frac{3.5 \lambda_p^3 \gamma^{-1/2}}{\left(r_e r_0^2\right)^{2/3}}$$

and
$$\frac{R_e}{\lambda_b} = \frac{2.5\lambda_p^2}{(r_e r_0^2)^{2/3}}\gamma^{-1}$$

The value of γ_t is then obtained by requiring
$$R_\theta(\gamma_t) = R_e(\gamma_t)$$

obtaining
$$\gamma_t = 1.1\left(\frac{r_0}{r_e}\right)^{1/3}\theta_j^{-1}$$

The opening angle of the jet θ_j imposes a threshold on the energy of the emitting electron wiggling with an effective angle so that the mean value of the peak energy has to be computed over the electron spectrum integrated only between γ_t and γ_{max}.

Taking into account the stochastic nature of the phenomenum due to the external density turbolence, we can calculate the predicted value for $\langle E_{peak}\rangle_{eff}$ in the stochastic regime, simply multiplying the coherent value of $\langle E_{peak}\rangle$ by the stochastic factor $\langle\sqrt{R_e/\lambda_b}\rangle$:

$$\langle E_{peak}\rangle_{eff} = \frac{3}{4}\hbar c\frac{r_0}{\lambda_p^2}\gamma_t\gamma_{max}\langle\sqrt{R_e/\lambda_b}\rangle$$

If we integrate in γ and in the Stochastic factor $\sqrt{R_e/\lambda_b}$ we obtain:

$$\langle E_{peak}\rangle_{eff} = 2.6\hbar c\left(\frac{r_0}{r_e}\right)^{1/3}\frac{\gamma_t\gamma_{max}^{1/2}}{\lambda_p}$$

$$\langle E_{peak}\rangle_{eff} = 1.4\times 10^{-9}\left(\frac{r_0}{r_e}\right)^{2/3}n^{1/3}\theta_j^{-1}\,\text{keV}$$

SPECTRAL - ENERGY CORRELATIONS

Noticing that the quantity γ_t introduces then a dependence of $\langle E_{peak}\rangle$ on θ_j, we use the above derived formulae for the peak energy of the emission and the effective angle of the radiation to derive an estimate for the recently obtained Amati[6] and Ghirlanda[7] relations.

Considering the electrons with an energy spectrum similar to that derived experimentally in the coherence regime, with Lorentz factor between γ_t and γ_{max}, the mean useful electron energy is:

$$E_e = N_0 mc^2\frac{\int\gamma\gamma^{-2}d\gamma}{\int\gamma^{-2}d\gamma}\sim N_0 mc^2\gamma_t\ln\frac{\gamma_{max}}{\gamma_t}$$

If each electron of energy greater than $mc^2\gamma_t$ emit n_γ of mean energy $\langle E_{peak}\rangle$ then the energy conservations requires:

$$mc^2\langle\gamma\rangle = n_\gamma \langle E_{\text{peak}}\rangle$$

$$mc^2\gamma_t \ln\frac{\gamma_{\max}}{\gamma_t} = n_\gamma \times \frac{3}{4}\hbar c \frac{r_0}{\lambda_p^2}\gamma_t\gamma_{\max}\left(\frac{R_e}{\lambda_b}\right)^{0.5}$$

We notice that the γ_t factor is present in both sides of the previous equation so that the θ_j dependence disappears.

The total energy E_γ is composed of the emission by the N_0 electrons.

$$N_0 mc^2\langle\gamma\rangle = E_\gamma = N_0 n_\gamma \langle E_{\text{peak}}\rangle$$

This permits us to derive the "Ghirlanda" relation:

$$E_\gamma = N_0 n_\gamma \langle E_{\text{peak}}\rangle$$

If the isotropic energy $E_{\text{iso}} = E_\gamma \times (\theta_j^{-2}/2)$ is introduced:

$$E_{\text{iso}} = \frac{2E_\gamma}{\theta_j^2} = \frac{2N_0 n_\gamma \langle E_{\text{peak}}\rangle}{\theta_j^2} \propto \langle E_{\text{peak}}\rangle^3$$

being $\theta_j^2 \propto \langle E_{\text{peak}}\rangle^{-2}$, obtaining a relation close to the "Amati" relation.

CONCLUSION

We studied the properties of a novel mechanism for prompt GRB radiation based on plasma laser-wake acceleration and radiation induced by an intense precursor irradiating a static gas nebula. The stochastic radiative emission of the energized plasma produces a photon emission with angular distribution dominated by the electron trajectories. Under general conditions applicable to our GRB model (precursor, static cloud, laser-wake acceleration conditions) we showed that there is a relation between the observed jet property (θ_j) and an electron energy threshold. In our mechanism the energy of the emitting wiggling electrons is the typical energy of the laser-wake betatron radiation. Furthermore, we derived a relation between source properties such as the collimation angle and the emitted energy.

REFERENCES

1. V. Connaughton, *ApJ*, **567**, 102 (2002).
2. G. Barbiellini et al., *MNRAS*, **350**, L5 (2004).
3. T.Tajima and J.M. Dawson, *Phys. Rev. Lett.*, **43**, 267 (1979).
4. K. Ta Phuoc, et al., *Physics of Plasmas*, **12**, 023101 (2005).
5. G. Barbiellini et al., *MNRAS*, submitted (2006)
6. L. Amati et al., *A&A*, **390**, 81 (2002).
7. G. Ghirlanda, G. Ghisellini, and D. Lazzati, *ApJ*, **616**, 331 (2004).

External Shock Interactions of GRB Blast Waves and the Swift Observations

Charles D. Dermer

E. O. Hulburt Center for Space Research, Code 7653, Naval Research Laboratory, Washington, DC 20375-5352

Abstract. Detailed calculations of radiation from a GRB blast wave that impacts a stationary cloud, including emission from both forward and reverse shocks, are presented. These results are used to determine whether the Swift observations are explained by a prompt explosion or an active central engine.

Keywords: Gamma-ray bursts; Jets
PACS: 98.58.Fd,98.70.Rz

INTRODUCTION

Observations with the BAT and XRT on Swift offer the first clear picture of the X-ray emission from gamma-ray bursts (GRBs) in the late prompt and early afterglow phases. In large numbers of GRBs, there are steep declines in the X-ray emission at the end of the prompt phase [1], an extended plateau phase [2], and X-ray flares [3] which, in some cases, have durations that are a small fraction of the time since the beginning of the GRB. These data have been used to argue [4] in favor of internal shocks and refreshed energization of the relativistic outflow.

The conclusions about refreshed shocks and active GRB central engines from the Swift data are made in the absence of a systematic study of a competing scenario to understand variability in the prompt and afterglow phases of a GRB, namely the external shock model [5, 6]. This is surprising because external shocks are acknowledged to operate during the afterglow phase, and external shock interactions can hardly be avoided in the prompt phase, given the prevalence of circumburst material and the small amount of matter that is needed to decelerate a relativistic blast wave.

Here we summarize a detailed analysis in preparation by the author of external shock emission, including both forward and reverse shocks, from a relativistic blast wave shell impacting a stationary cloud of material. These results will permit a critical comparison of the circumburst environment and physical properties of the GRB blast wave that are required to explain the Swift observations, with the goal of determining whether an impulsive explosion or an active central engine generates the emission from GRBs. Implications for the central engines of long duration GRBs are indicated.

RELATIVISTIC BLAST WAVE/CLOUD INTERACTIONS

A detailed study of the interaction between a relativistic blast wave shell and a stationary cloud with a spherical cap geometry is summarized here. The analysis is performed under the assumption that the cloud width $\Delta_{cl} \ll x$, where x is the distance of the cloud from the GRB explosion The interaction is divided into three phases: (1) a collision phase with both a forward and reverse shock; (2) a penetration phase where either the reverse shock has crossed the shell while the forward shock continues to cross the cloud, or vice versa; and (3) an expansion phase, where both shocks have crossed the cloud and shell, and the shocked fluid expands. Temporally evolving spectral energy distributions are calculated for the problem of the interactions of a blast-wave shell with clouds that subtend large and small angles compared with the Doppler beaming cone $\theta_0 \approx 1/\Gamma_0$, where Γ_0 is the coasting Lorentz factor of the blast wave.

For the elementary interaction, we consider a GRB taking place at redshift z and luminosity distance d_L that releases energy over a characteristic timescale Δ_0/c, with $\Delta_0 \ll 10^{10}$ cm. The apparent isotropic equivalent γ-ray energy released by a GRB explosion is E_0, and the explosion is assumed to form a fireball with initial Lorentz factor $\Gamma_0 = E_0/M_0 c^2 \gg 1$, where M_0 is the amount of baryonic matter mixed into the initial explosion. Due to shell spreading from internal motions in the blast wave at distance $x \gtrsim \Gamma_0^2 \Delta_0$, the shell width can be written as $\Delta(x) \cong \Delta_0 + \eta\, x/\Gamma_0^2$, with $\eta \ll 1$. The proper number density of the relativistic shell is given by $n(x) = E_0/4\pi x^2 \Gamma_0^2 m_p c^2 \Delta(x)$.

The relativistic shell collides with a cloud with uniform density n_{cl}, that is assumed to subtend a solid angle $\Delta\Omega_{cl} = \pi \theta_{cl}^2$ as measured from the explosion center. The angle between the axis through the cloud center and the line of sight to the observer is denoted θ_i. The coordinates $\vec{x} = (x, \theta, \phi)$ defining the location of the radiating material in the explosion frame are measured with respect to the line of sight to the observer, with the angle ϕ representing the projection of the azimuth on the plane normal to the direction to the observer. For simplicity, both the shell and cloud are assumed to be composed of electron-proton plasma.

The νF_ν flux, denoted by $f_\varepsilon(t)$, measured at observer time t and at dimensionless photon energy $\varepsilon = h\nu/m_e c^2$ is given by [7]

$$f_\varepsilon(t) = (2\pi d_L^2)^{-1} \int_{\theta_i-\theta_{cl}}^{\theta_i+\theta_{cl}} d\theta\, |\sin\theta|\, \phi_+ \int_0^\infty dx\, x^2 \varepsilon' j'(\varepsilon'; x, t')\, \delta_D^3, \quad (1)$$

for a spectral emissivity function $j'(\varepsilon'; x, t')$ describing isotropic emission in the comoving frame (primes refer to comoving quantities). The Doppler factor $\delta_D = [\Gamma(1 - \beta\cos\theta)]^{-1}$ is, in general, dependent on location and time. Here Γ is the emitting fluid's Lorentz factor and $\beta c = c\sqrt{1-\Gamma^{-2}}$ is its speed. The term $\phi_+ = \phi_+(\theta, \theta_{cl}, \theta_i)$ represents the maximal azimuthal angle subtended by the cloud for emission received from radiating plasma located at an angle θ with respect to the observer's line of sight. The integration is over volume in the stationary (explosion) frame.

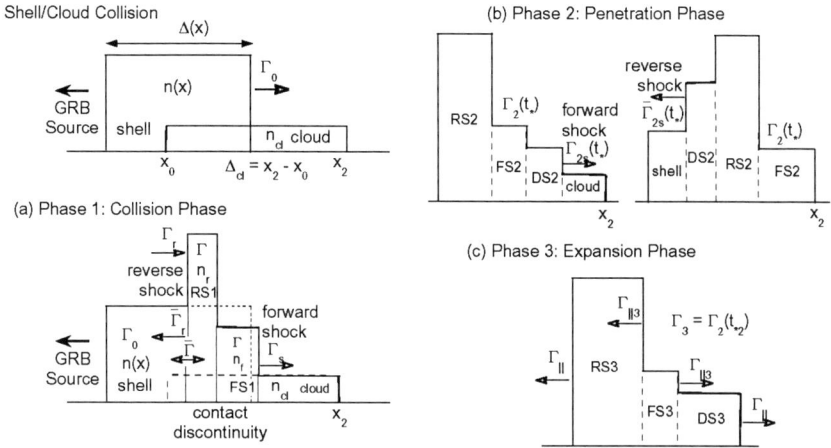

FIGURE 1. Phases of the blast-wave/cloud interaction. Upper left panel shows the planar geometry in the absence of a collision. The lower left panel illustrates Phase 1 and some of the derived hydrodynamical quantities (e.g., shocked fluid Lorentz factor Γ, reverse shock Lorentz factor $\bar{\Gamma}$). The upper right panel illustrates Phase 2, where either the RS has penetrated the shell before the FS has penetrated the cloud (left), or vice versa (right). The lower right panel shows the expansion phase for the first case of Phase 2.

PHASES OF THE INTERACTION EVENT

The hydrodynamical interaction is treated in a planar geometry, as illustrated in Fig. 1. Calculations using eq. (1) properly take into account radiative cooling at different comoving times as well as light-travel time effects for radiation emitted by different parts of the blast wave. The cloud is assumed to have uniform density n_{cl} between radii x_0 and x_2, and cloud width $\Delta_{cl} = x_2 - x_0$. A major simplifying assumption of the analysis is that $\Delta_{cl} \ll x_0$, so that $n(x) \approx n(x_0)$ and $\Delta(x) \cong \Delta(x_0) \equiv \Delta$, throughout the duration of the interaction. The interaction is divided into the following phases:

1. Collision Phase 1. Both a forward shock (FS) and reverse shock (RS) are found in this phase; the forward shock accelerates the cloud material, and the reverse shock decelerates the shell material.

2. Penetration Phase 2. This phase bifurcates into two cases, depending on whether the RS crosses the shell before the FS crosses the cloud, or whether the FS crosses the cloud before the RS crosses the shell. In the former case, the shocked fluid produced during the collision phase decelerates as more cloud material is swept up at the FS, during which time a newly accelerated particle population is introduced from the deceleration shock (DS). In the latter case, the shocked fluid produced during the collision phase is accelerated by the remaining shell material at the RS, and a new particle particle population is accelerated at the RS from an acceleration

shock. Only results for the former case (which is the most important case for GRB studies) are shown here.

3. Expansion Phase 3. After the FS crosses the remaining cloud material in the first case of Phase 2, or the RS crosses the remaining shell material in the second case of Phase 2, the shocked fluid, being composed of highly relativistic particles and magnetic fields, expands and adiabatically cools.

The hydrodynamical treatment of the FS and RS in Phase 1 follows Refs. [8] and [9], with the emissivity calculated according to the prescription of [10]. The shocked fluid is assumed to be hydrodynamically coupled in Phase 2, with an evolving Lorentz factor, $\Gamma_2(t_*)$, at stationary frame time t_* taken from the solution for a decelerating adiabatic relativistic blast wave [11]. In Phase 3, the expansion and cooling causes both the magnetic field and the electron Lorentz factor to decrease. The solution of Ref. [12] for electron energy evolution is used to describe this phase.

RESULTS

A numerical simulation code was written to calculate the full nonthermal synchrotron emission spectrum from the blast wave/cloud interaction. As an example of the results from this code, we consider a GRB that takes place at $z = 1$ and $d_L = 2.0 \times 10^{28}$ cm, with blast-wave properties given by $\Gamma_0 = 300, E_0 = 10^{53}$ ergs, $\Delta_0 = 10^7$ cm and $\eta = 0.01$. The blast wave is assumed to encounter a small cloud with density $n_{cl} = 10^6$ cm^{-3} located a distance $x_0 = 10^{15}$ cm from the explosion center[1] centered along the line of sight to the observer. For a quasi-spherical cloud of width $\Delta_{cl} = 2 \times 10^{13}$ cm, $\theta_{cl} = 0.01$.

Using the standard blast-wave physics formalism, we take $\varepsilon_e = 0.1$, $\varepsilon_B = 10^{-4}$, and a maximum electron Lorentz factor that is 10% of the radiation-reaction limit obtained by equating the Larmor timescale and the synchrotron energy loss time scale. A nonthermal power-law distribution of electrons are injected with index $p = 2.5$, and the ε_e and ε_B parameters of the FS and RS are assumed to be same.

Fig. 2 shows the νF_ν spectrum calculated near the peak of the $\varepsilon = 1$ ($h\nu \cong 500$ keV) light curve (see Fig. 3) at observer time $t = 0.46$ s (the emission is first received at time $(1+z)x_0/2\Gamma_0^2 c \cong 0.37$ s after the start of the GRB). The two panels have all standard parameters described above, except that $\varepsilon_B = 10^{-2}$ in the panel on the right. The various spectral components making up the spectrum are shown and, as can be seen, the DS fluid during Phase 2 makes the dominant contribution to the νF_ν spectrum at this time. The peak energy flux is at the level of 10^{-6} ergs cm^{-2} s^{-1}, the νF_ν peak energy $E_{pk} \sim 1$ MeV, and the spectrum is of the characteristic "Band"-shape.

The RS makes a lower energy emission component which, when $\varepsilon_B = 10^{-2}$, yields a strong optical flux. Even when $\varepsilon_B = 10^{-4}$, the RS emission produces concavity in the X-ray spectrum that should be detectable in spectral fitting of the Swift XRT data.

Fig. 3 shows light curves calculated for the standard parameter set with $\varepsilon_B = 10^{-4}$,

---

[1] The total mass of such clouds distributed in all directions around the GRB source is $\lesssim 2 \times 10^{-7} M_\odot$

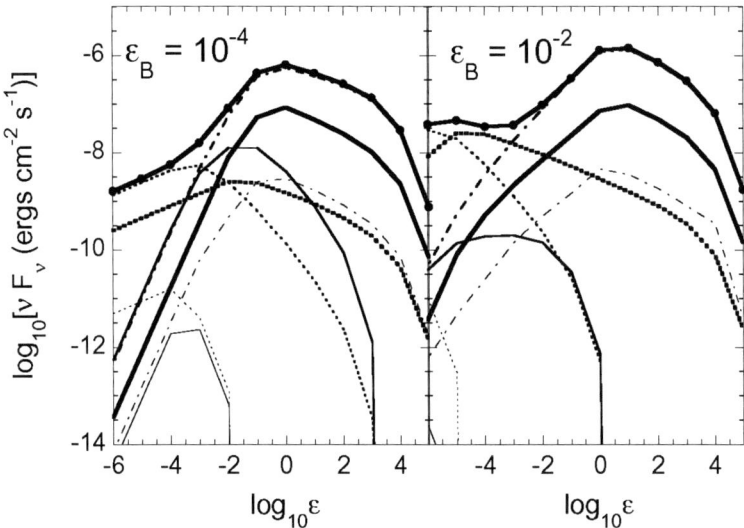

FIGURE 2. Separate spectral components and the total νF_ν flux observed at $t = 0.46$ s after the start of the GRB, due to the interaction of a relativistic blast wave with a small cloud for the parameters given in the text. The total spectrum is given by the heavy solid curve with data points, the FS components are given by the solid curves, the RS components by the dotted curves, and the DS components by the dot-dashed curves. The heavy, medium, and light curves correspond to Phases 1, 2 and 3, respectively.

while varying the cloud location by changing θ_i and cloud size by changing θ_{cl}, as labeled. In the collision phase, $\Gamma = 182$ and $\bar{\Gamma} = 1.13$, so this event involves a nonrelativistic RS. Due to strong relativistic beaming, emission from off-axis clouds reaches peak fluxes that are orders of magnitude less intense than on-axis events, and which are delayed by many seconds and stretched over a much longer time. Because of the small cloud size and shock compression of the radiating fluid into a narrow shell that has a small duration, the characteristic pulse profile is of the "curvature" form [7].

SUMMARY

We have developed a numerical simulation model of the interaction of a relativistic blast wave with a stationary external cloud described by a spherical cap geometry. As shown in our preliminary modeling, such events can produce flares with short durations and large energy fluxes. As shown in Ref. [6], flares with relative durations $\delta t/t \ll 1$ can be made in an external shock model provided that the cloud size is much smaller than $1/\Gamma_0$. Our more detailed modeling confirms these results, though it requires that the blast wave shell width $\Delta \ll x/\Gamma_0^2$ at large distances x from the GRB explosion. Thus it remains an open question whether observations of X-ray flares with Swift really require an active central engine, or whether a single impulsive explosion can produce the GRB variability. An impulsive central engine would conflict with the collapsar model, but would be in

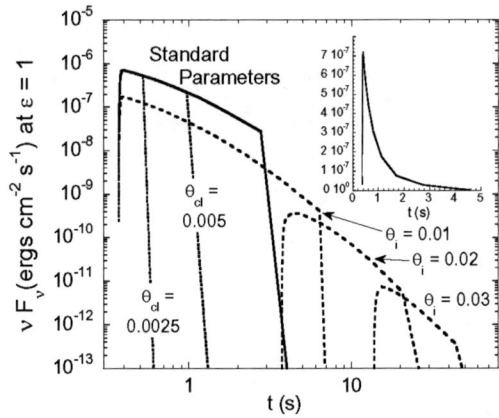

FIGURE 3. Light curves at $\varepsilon = 1$ ($h\nu = 511$ keV) from the interaction of a GRB blast wave with a cloud, with parameters as given in the text. Changes in the received light curves are shown for different cloud locations θ_i, with $\theta_{cl} = 0.01$, and cloud opening angles θ_{cl}, with $\theta_i = 0.0$. Inset shows the lightcurve for standard parameters on a linear scale.

accord with a two-step collapse mechanism for long duration GRBs.

ACKNOWLEDGMENTS

The work of CDD is supported by the Office of Naval Research, NASA *GLAST* Science Investigation No. DPR-S-1563-Y, and a NASA Swift Guest Investigator Grant.

REFERENCES

1. G. Tagliaferri, et al., *Nature*, **436**, 985–988 (2005).
2. S. D. Barthelmy, et al., *Astrophys. J. Lett.*, **635**, L133-L136 (2005).
3. D. N. Burrows, et al., *Science* **309**, 1833–1935 (2005).
4. B. Zhang, et al., *Astrophys. J.*, in press (2006), astro-ph/0508321.
5. P. Mészáros and M. J. Rees, *Astrophys. J.*, **405**, 278–284 (1993).
6. C. D. Dermer and K. E. Mitman, *Astrophys. J. Lett.*, **513**, L5–L8 (1999).
7. C. D. Dermer, *Astrophys. J.*, **614**, 284–292 (2004).
8. R. Sari and T. Piran, *Astrophys. J. Lett.*, **477**, L143–L146 (1995).
9. A. Panaitescu and P. Mészáros, *Astrophys. J.*, **526**, 707–715 (1999).
10. R. Sari, T. Piran, and R. Narayan, *Astrophys. J. Lett.*, **497**, L17–L20, (1998).
11. M. Böttcher and C. D. Dermer, *Astrophys. J.*, **532**, 281–2855 (2000).
12. S. Gupta, M. Böttcher, and C. D. Dermer, *Astrophys. J.*, submitted, (2005).

GRB 050315: A step in the proof of the uniqueness of the overall GRB structure

R. Ruffini*,†, M.G. Bernardini*,†, C.L. Bianco*,†, P. Chardonnet*,**, F. Fraschetti*,‡, R. Guida*,† and S.-S. Xue*,†

*ICRANet and ICRA, Piazzale della Repubblica 10, I-65100 Pescara, Italy.
†Dipartimento di Fisica, Università "La Sapienza", Piazzale Aldo Moro 5, I-00185 Roma, Italy.
**Université de Savoie, LAPTH - LAPP, BP 110, F-74941 Annecy-le-Vieux Cedex, France.
‡Osservatorio Astronomico di Brera, via Bianchi 46, I-23807 Merate (LC), Italy.

Abstract. Using the Swift data of GRB 050315, we progress in proving the uniqueness of our theoretically predicted Gamma-Ray Burst (GRB) structure as composed by a proper-GRB, emitted at the transparency of an electron-positron plasma with suitable baryon loading, and an afterglow comprising the "prompt radiation" as due to external shocks. Detailed light curves for selected energy bands are theoretically fitted in the entire temporal region of the Swift observations ranging over 10^6 seconds.

Keywords: gamma rays: bursts — radiation mechanisms: thermal — black hole physics
PACS: 98.70.Rz, 44.40.+a, 04.70.-s

INTRODUCTION

GRB 050315 [1] has been triggered and located by the BAT instrument [2, 3] on board of the *Swift* satellite [4] at 2005-March-15 20:59:42 UT [5]. The narrow field instrument XRT [6, 7] began observations ~ 80 s after the BAT trigger, one of the earliest XRT observations yet made, and continued to detect the source for ~ 10 days [1]. The spectroscopic redshift has been found to be $z = 1.949$ [8]. We present here the first results of the fit of this source in the framework of our theoretical model and point out the new step toward the uniqueness of the explanation of the overall GRB structure made possible by the Swift data of this source.

OUR THEORETICAL MODEL

GRB 050315 observations find a direct explanation in our theoretical model [see 9, 10, 11, 12, 13, 14, and references therein]. We determine the values of the two free parameters which characterize our model: the total energy stored in the Dyadosphere E_{dya} and the mass of the baryons left by the collapse $M_B c^2 \equiv BE_{dya}$. We follow the expansion of the pulse, composed by the electron-positron plasma initially created by the vacuum polarization process in the Dyadosphere. The plasma self-propels itself outward and engulfs the baryonic remnant left over by the collapse of the progenitor star. As such pulse reaches transparency, the Proper Gamma-Ray Burst (P-GRB) is emitted [15, 16, 10]. The remaining accelerated baryons, interacting with the interstellar medium (ISM), produce the afterglow emission. The ISM is described by the two additional parameters of the theory: the average particle number density $< n_{ISM} >$ and the ratio $< \mathscr{R} >$ between the effective emitting area and the total area of the pulse [17], which take into account the ISM filamentary structure [18].

The luminosity in fixed energy bands is evaluated integrating over the equitemporal surfaces [EQTSs, see 19, 13], computed using the exact solutions of the afterglow equations of motion [14], the energy density released due to the totally inelastic collisions of the accelerated baryons with the ISM measured in the co-moving frame, duly boosted in the observer frame. In the reference frame co-moving with the accelerated baryonic matter, the radiation produced by this interaction of the ISM with the front of the expanding baryonic shell is assumed to have a thermal spectrum [17].

We reproduce correctly in several GRBs and in this specific case (see e.g. Figs. 2–3) the observed time variability of the prompt emission as well as the remaining part of the afterglow [see e.g. 20, 11, 12, 21, and references therein]. The radiation produced by the interaction of the accelerated baryons with the ISM agrees with observations both for intensity and time structure.

As shown in previous cases (GRB 991216 [11, 22], GRB 980425 [23], GRB 030329 [24], GRB 031203 [21]), also for GRB 050315, using the correct equations of motion, there is no need to introduce a collimated emission to fit the afterglow observations.

The major difference between our theoretical model and the ones in the current literature [see e.g. 25, and references therein] is that what is usually called "prompt emission" in our case coincides with the peak of the afterglow emission and is not due to a different physical process [10]. The verification of this prediction has been up to now tested in a variety of sources like GRB 991216 [11], GRB 980425 [23], GRB 030329 [24], GRB 031203 [21]. However, in all such sources the observational data were available only during the prompt emission and the latest afterglow phases, leaving all the in-between evolution undetermined. Now, thanks to the superb data provided by the Swift satellite, we are finally able to confirm, by direct confrontation with the observational data, our theoretical predictions on the GRB structure [10] with a detailed fit of the complete afterglow light curve of GRB 050315, from the peak (i.e. from the so-called "prompt emission") all the way to the latest phases without any gap in the observational data.

GRB 991216

A basic feature of our model consists in a sharp distinction between two different components in the GRB structure: the proper GRB (P-GRB), emitted at the moment of transparency, followed by an afterglow completely described by external shocks and composed of three different regimes. The first afterglow regime corresponds to a bolometric luminosity monotonically increasing with the photon detector arrival time, corresponding to a substantially constant Lorentz gamma factor of the accelerated baryons. The second regime consists of the bolometric luminosity peak, corresponding to the "knee" in the decreasing phase of the baryonic Lorentz gamma factor. The third regime corresponds to a bolometric luminosity decreasing with arrival time, corresponding to the late deceleration of the Lorentz gamma factor.

In some sources the P-GRB is under the observability threshold. In Ruffini et al.

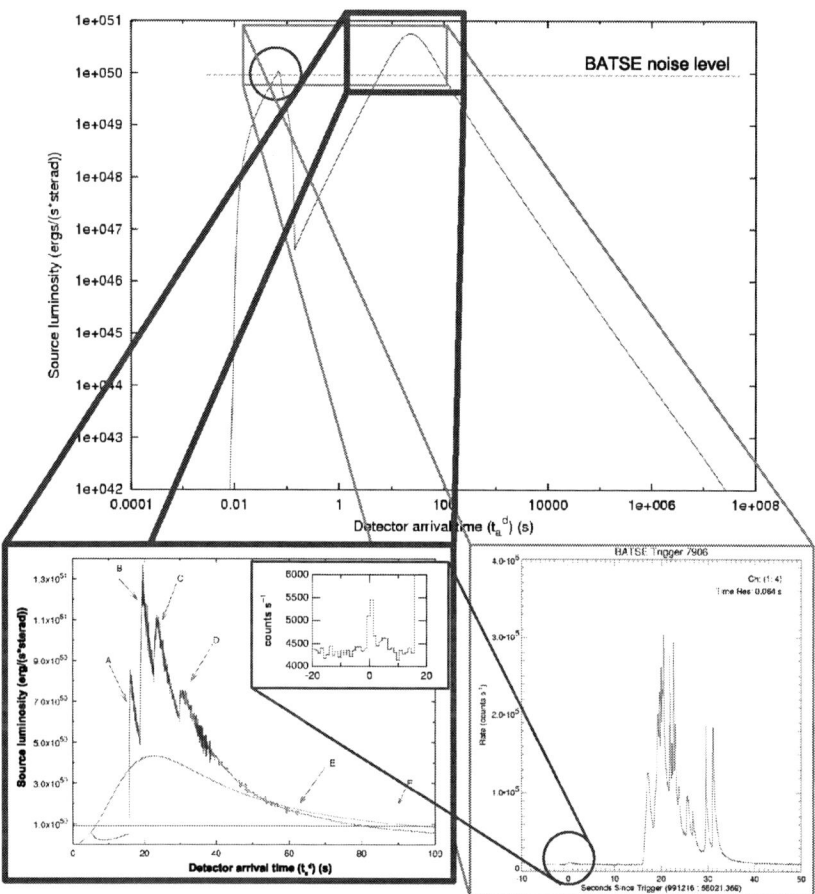

FIGURE 1. This picture shows our prediction on the GRB structure based on the analysis of GRB 991216 [10]. In the main panel there is the bolometric light curve computed using our model and composed by the P-GRB and the afterglow, together with the BATSE noise level. In the lower right panel there is represented the BATSE observation of the prompt emission [see 26, 27], with the clear identification of the observed "main burst" with the peak of the theoretical afterglow light curve and of the observed "precursor" with the theoretically predicted P-GRB (see enlargement). In the lower left panel is represented our theoretical fit of the BATSE observations of the afterglow peak using an inhomogeneous ISM. Details in Ruffini et al. [20, 11, 12].

[10] we have chosen as a prototype the source GRB 991216 which clearly shows the existence of this two components. Both the relative intensity of the P-GRB to the peak of the afterglow, as well as their corresponding temporal lag, have been theoretically predicted within a few percent (see Fig. 11 in Ruffini et al. [11]). The continuous line in the main panel of Fig. 1 corresponds to a constant ISM density averaged over the entire afterglow. The structured curve, shown in the bottom left panel, corresponds

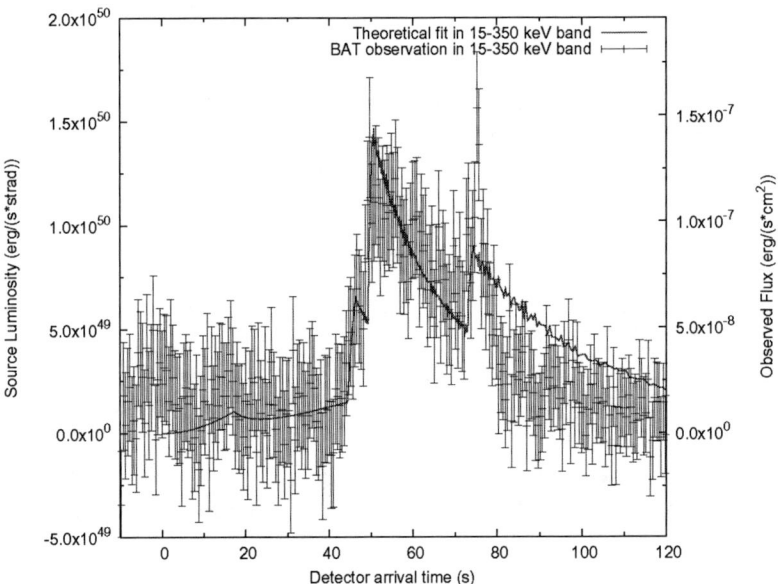

FIGURE 2. Theoretical fit (red line), computed using our model, of the BAT observations (blue points) of GRB 050315 in the 15–350 keV energy band [1].

to ISM density inhomogeneities which are assumed for simplicity to be spherically symmetric [20]. Clearly, a more precise description of the BATSE light curve (e.g. the two sharp spikes at ~ 30 s) will need a more refined 3-dimensional description of the ISM filamentary structure [18].

This same approximation of spherically symmetric description of the ISM inhomogeneities is in the following adopted for GRB 050315, and is sufficient to clearly outline the general behavior of the luminosity vs. photon detector arrival time in selected energy bands.

THE FIT OF THE OBSERVATIONS

The best fit of the observational data leads to a total energy of the Dyadosphere $E_{dya} = 1.47 \times 10^{53} erg$ [the observational Swift E_{iso} is $> 2.62 \times 10^{52}$ erg, see Ref. 1], so that the plasma is created between the radii $r_1 = 5.88 \times 10^6$ cm and $r_2 = 1.74 \times 10^8$ cm with an initial temperature $T = 2.05 MeV$ and a total number of pairs $N_{e^+e^-} = 7.93 \times 10^{57}$. The amount of baryonic matter in the remnant is assumed to be such that $B = 4.55 \times 10^{-3}$. The transparency point and the P-GRB emission occurs then with an initial Lorentz gamma factor of the accelerated baryons $\gamma_\circ = 217.81$ and at a distance $r = 1.32 \times 10^{14}$ cm from the Black Hole. The interstellar medium (ISM) parameters that we assume to best fit the observational data are: $< n_{ism} > = 0.121$ particles/cm^3 and $< \mathscr{R} > = 2.05 \times 10^{-6}$. The ISM density contrast is found to be $\Delta\rho/\rho \sim 10^2$ on a scale of 5.0×10^{16}

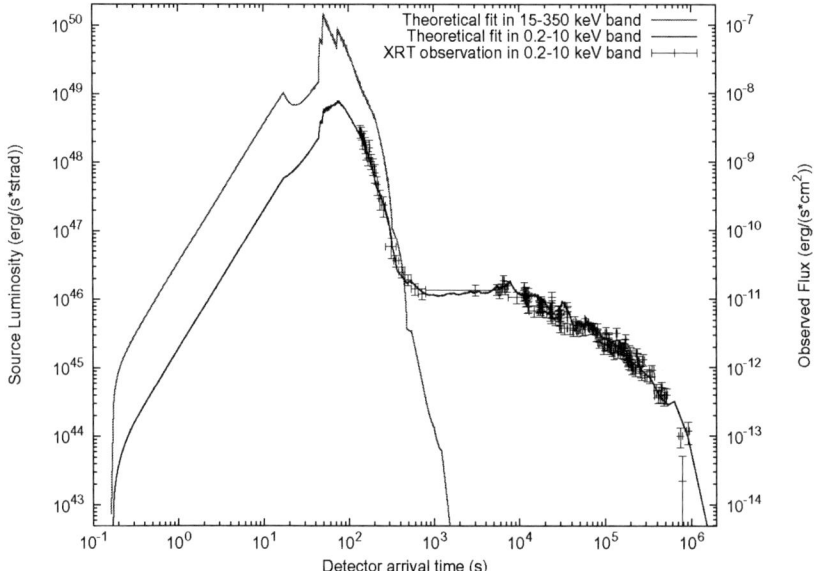

FIGURE 3. Theoretical fit (blue line), computed using our model, of the XRT observations (black points) of GRB 050315 in the 0.2–10 keV energy band [1]. The theoretical fit of the BAT observations (see Fig. 2) in the 15–350 keV energy band is also represented (red line).

cm.

In Figs. 2 and 3 we represent the theoretically computed GRB 050315 light curves, respectively in the 15–350 keV and in the 0.2–10 keV energy bands, which we obtained using our model, together with the corresponding data observed respectively by the BAT and the XRT instruments on board of the *Swift* satellite [1]. For completeness, in Fig. 3 is also represented the theoretically computed 15–350 keV light curve of Fig. 2, but not the BAT observational data to not overwhelm the picture too much.

The very good agreement between the theoretical curves and the observations is a most stringent proof of our predictions on the GRB structure [10].

It goes without saying that also in the case of GRB 050315 a more detailed correspondence between the theory and the temporal fine structure of the BAT observational light curve could be achieved with a full 3-dimensional description of the ISM filamentary structure [18].

CONCLUSIONS

In view of the above results, which clearly fits the overall luminosity in fixed energy bands, we will return in a forthcoming publication to the identification of the P-GRB using our theoretically predicted values for both its intensity and its time lag relative to the afterglow peak. We will also address the spectral analysis which is a most powerful theoretical prediction in order to evidence the continuity between the "prompt radiation" and the late phases of the afterglow and so to prove the uniqueness of the overall GRB structure.

ACKNOWLEDGMENTS

We thank P. Banat, G. Chincarini, A. Moretti and S. Vaughan for their help in the analysis of the observational data.

REFERENCES

1. S. Vaughan, et al., *ApJ*, in press (astro-ph/0510677).
2. S.D. Barthelmy, *SPIE*, **5165**, 175 (2004).
3. S.D. Barthelmy, et al., *Sp. Sc. Rev.*, in press (astro-ph/0507410).
4. N. Gehrels, et al., *ApJ*, 611, 1005 (2004).
5. A. Parsons, et al., *GCN 3094* (2005).
6. D.N. Burrows, et al., *SPIE*, **5165**, 201 (2004).
7. D.N. Burrows, et al., *Sp. Sc. Rev.*, in press (astro-ph/0508071).
8. D. Kelson, E. Berger, *GCN 3101* (2005).
9. R. Ruffini, C.L. Bianco, P. Chardonnet, F. Fraschetti, S.-S. Xue, *ApJL*, **555**, L107 (2001).
10. R. Ruffini, C.L. Bianco, P. Chardonnet, F. Fraschetti, S.-S. Xue, *ApJL*, **555**, L113 (2001).
11. R. Ruffini, C.L. Bianco, P. Chardonnet, F. Fraschetti, L. Vitagliano, S.-S. Xue, in *COSMOLOGY AND GRAVITATION: X^{th} Brazilian School of Cosmology and Gravitation; 25th Anniversary (1977-2002)*, edited by M. Novello and S.E. Perez Bergliaffa, AIP Conf. Proc. 668, American Institute of Physics, New York, 2003, p. 16.
12. R. Ruffini, M.G. Bernardini, C.L. Bianco, P. Chardonnet, F. Fraschetti, V. Gurzadyan, L. Vitagliano, S.-S. Xue, in *COSMOLOGY AND GRAVITATION: XIth Brazilian School of Cosmology and Gravitation*, edited by M. Novello and S.E. Perez Bergliaffa, AIP Conf. Proc. 782, American Institute of Physics, New York, 2005, p. 42.
13. C.L. Bianco, R. Ruffini, *ApJL*, **620**, L23 (2005).
14. C.L. Bianco, R. Ruffini, *ApJL*, **633**, L13 (2005).
15. R. Ruffini, J.D. Salmonson, J.R. Wilson, S.-S. Xue, *A&A*, **350**, 334 (1999).
16. R. Ruffini, J.D. Salmonson, J.R. Wilson, S.-S. Xue, *A&A*, **359**, 855 (2000).
17. R. Ruffini, C.L. Bianco, P. Chardonnet, F. Fraschetti, V. Gurzadyan, S.-S. Xue, *Int. J. Mod. Phys. D*, **13**, 843 (2004).
18. R. Ruffini, C.L. Bianco, P. Chardonnet, F. Fraschetti, V. Gurzadyan, S.-S. Xue, *Int. J. Mod. Phys. D*, **14**, 97 (2005).
19. C.L. Bianco, R. Ruffini, *ApJL*, **605**, L1 (2004).
20. R. Ruffini, C.L. Bianco, P. Chardonnet, F. Fraschetti, S.-S. Xue, *ApJL*, **581**, L19 (2002).
21. M.G. Bernardini, C.L. Bianco, P. Chardonnet, F. Fraschetti, R. Ruffini, S.-S. Xue, *ApJL*, **634**, L29 (2005).
22. R. Ruffini, M.G. Bernardini, C.L. Bianco, P. Chardonnet, F. Fraschetti, S.-S. Xue, *Adv. Sp. Res.*, in press, astro-ph/0503268.
23. R. Ruffini, C.L. Bianco, P. Chardonnet, F. Fraschetti, S.-S. Xue,*Adv. Sp. Res.*, **34**, 2715 (2004).
24. M.G. Bernardini, C.L. Bianco, P. Chardonnet, F. Fraschetti, R. Ruffini, S.-S. Xue, in *Proceedings of the tenth Marcel Grossmann Meeting*, edited by M. Novello and S.E. Perez-Bergliaffa, World Scientific, Singapore, in press.
25. T. Piran, *Rev. Mod. Phys.*, **76**, 1143 (2004).
26. BATSE GRB Light Curves, http://gammaray.msfc.nasa.gov/batse/grb/lightcurve/
27. BATSE Rapid Burst Response, http://gammaray.msfc.nasa.gov/~kippen/batserbr/

Prompt γ-Ray Emission Spectroscopy with *RHESSI*

M. E. Bandstra*, S. E. Boggs*,† and E. Bellm*

UC Berkeley Space Sciences Laboratory
†UC Berkeley Physics Department

Abstract. We describe ongoing efforts to use the nine Ge detectors on the Reuven Ramaty High Energy Solar Spectroscopic Imager (*RHESSI*) to perform spectroscopy on GRB prompt γ-ray emission. Since the detectors cover an energy range of 3 keV to 17 MeV, have excellent energy resolution (1-10 keV FWHM), and have a field of view of half the sky, *RHESSI* has the potential to be a powerful tool for performing high precision GRB spectroscopy. However, the situation is complicated by *RHESSI*'s high background variability, its relatively small collection area (150 cm^2), and its 4 s rotation period about its axis. Because *RHESSI* has limited calibrations for sources at angles outside of the primary field of view, we must rely heavily on simulations to determine the instrument response. We present our work so far on this problem, including a description of our simulations and preliminary GRB spectra with Band function fits.

Keywords: gamma-ray sources (astronomical); gamma-ray spectra; artificial satellites
PACS: 95.55.Ka, 98.70.Rz, 07.87.+v

THE *RHESSI* SPACECRAFT

The *RHESSI* instrument was designed to study solar hard X-ray and γ-ray emission in the range of 3 keV to 17 MeV with high spectral (∼ 1-10 keV FWHM) and temporal (1 μs) resolution. The heart of *RHESSI* is an array of nine large volume (∼300 cm^3) coaxial germanium detectors cooled to 75 K. For more information on the *RHESSI* instrument and spectrometer, see [1] and [2], respectively.

RHESSI has high angular resolution (2.3″) in its imaging field of view (∼1°); however there is a small chance that a GRB will ever be in this FoV. Nevertheless, the lack of shielding on the detectors makes *RHESSI* an effective all-sky GRB monitor [3]. Indeed, *RHESSI* has detected ∼20% of all Swift-era bursts to this date.

RHESSI has already been investigated as a GRB spectrometer [4, 5, 6]. It has been found that the detector fronts and cryostat walls block most photons below 100 keV [5], and we have likewise found that spectral fits below 100 keV suffer. A webpage showing lightcurves of all *RHESSI* GRBs is maintained by the *RHESSI* GRB team at the Paul Scherrer Institut (http://grb.web.psi.ch).

SIMULATIONS

Our main goals in performing this analysis are to exploit *RHESSI*'s high energy capabilities in order to constrain:

- GRB spectral parameters (α, β, E_{peak})

- GRB fluence (S_γ)

Unfortunately, *RHESSI* has limited calibrations for sources outside of its primary FoV, so we rely on MGEANT simulations. Because of the spacecraft's 4 s rotation period, we simulate annular sources centered around the rotation axis in order to generate instrument response to a GRB. We produced response matrices as functions of

- Off-axis angle of the source annulus (spaced by 15 °)
- Incident photon energy (64 logarithmic bins between 10 keV and 10 MeV)

To calculate spectral parameters, we assume a Band spectral model for the GRB and forward fit the data using these response matrices. The Band model is

$$N_E \propto \begin{cases} E^\alpha \exp(-E/E_0) & E < E_{break} \\ E^\beta & E > E_{break} \end{cases}$$

For some simulations in the GRB030329 case study, a very crude model of the atmosphere is used to simulate the effects of atmospheric scattering. This model consists of a dense cylinder of air centered about the spacecraft's rotation axis.

GRB030329 – A CASE STUDY

GRB030329 is one of the brightest bursts *RHESSI* has seen (see Fig. 1), with a HETE-detected fluence of $S_\gamma \approx 1.2 \cdot 10^{-4}$ erg s^{-1} (30-400 keV) and $T_{90} \approx 23$ s [7]. With an E_{peak} of 70.2 keV, this burst is too soft for *RHESSI* to measure α and E_{peak} well, so it will provide a "best marginal case." In addition, the *RHESSI* data for this burst have previously been investigated, which can provide a comparison [4, 5].

We ran an MGEANT simulation of the burst on a thin annulus 144° from *RHESSI*'s rotational axis. The simulation assumes the same spectrum and fluence in 30-400 keV as HETE. Figure 2 shows the actual measured spectrum after background subtraction and the simulated count spectrum. Table 1 compares the Band parameters in the various cases. The discrepancy between the actual data and the simulated data below ~25 keV is an MGEANT artifact. Note that the *RHESSI* data suggest a softer high-energy tail than the HETE measurements, demonstrating how *RHESSI*'s broad energy band can be used to improve higher-energy spectral fits.

GRB050904 – A HIGH REDSHIFT BURST

This long burst was identified by Swift (GCN 3910), and had a fluence of $S_\gamma \sim 5 \cdot 10^{-6}$ erg cm^{-2} and a $T_{90} \sim 225$ s [8]. Several hours after the burst, the redshift of the optical afterglow was measured to be $z \approx 6.29$, making it the highest redshift ever observed for a GRB [9]. During the burst, *RHESSI* was passing near the SAA, which causes the slow rise and fall in the background visible in the raw lightcurve (Fig. 3). Empirical polynomial fits are used in background subtraction. The main peak of the burst, from $T + 80$ s to $T + 220$ s, is mostly obscured by the background, but the two initial peaks at $T + 28$ s and $T + 56$ s are visible. Some emission is seen above 200 keV.

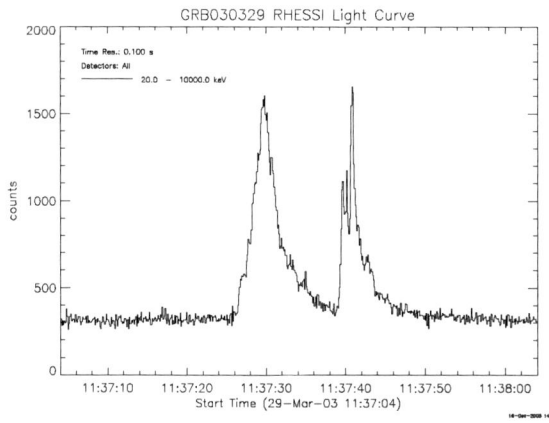

FIGURE 1. *RHESSI* raw light curve for GRB030329 (left)

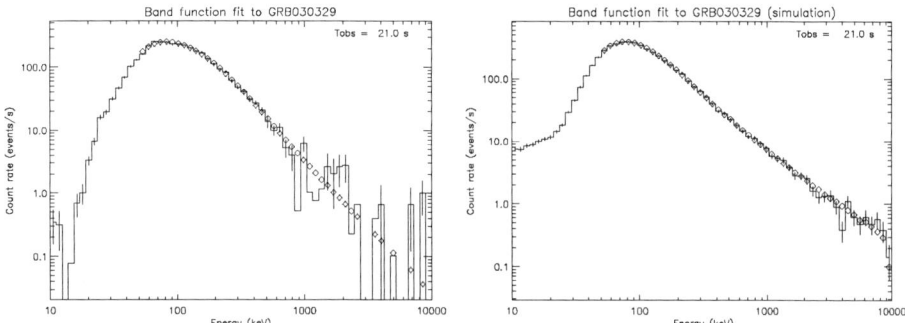

FIGURE 2. *RHESSI* actual measured spectrum for GRB030329 with Band fit (left), *RHESSI* response to simulated count spectrum for GRB030329 with Band fit (right).

TABLE 1. Comparison of spectral fits to GRB030329: HETE results; Measured (1) – *RHESSI* data; Measured (2) – *RHESSI* data holding α fixed and using the HETE energy band; Measured (3) – *RHESSI* data only fitting a high energy power law; Simulation (1) – simulated data assuming a Band model over entire *RHESSI* energy range; and Simulation (2) – simulated data including crude atmospheric backscattering (but using the same instrument response as the other fits). All fluences are calculated for the 30-400 keV bandpass for ease of comparison.

	Fit interval (keV)	α	β	E_{peak} (keV)	S_γ (erg cm^{-2})
HETE [7]	30 - 400	-1.32 ± 0.02	-2.44 ± 0.08	70.2 ± 2.3	$1.18 \cdot 10^{-4}$
Measured (1)	50 -10000	-1.69 ± 0.04	-3.13 ± 0.26	126.3 ± 15.4	$0.74 \cdot 10^{-4}$
Measured (2)	60 - 400	-1.32 (fixed)	-2.39 ± 0.04	114.4 ± 5.1	$0.71 \cdot 10^{-4}$
Measured (3)	200 -10000	n/a	-2.73 ± 0.05	n/a	n/a
Simulation (1)	50 -10000	-1.00 ± 0.20	-2.47 ± 0.01	79.4 ± 21.6	$1.05 \cdot 10^{-4}$
Simulation (2)	40 -10000	-1.50 ± 0.19	-2.56 ± 0.01	42.3 ± 12.3	$1.43 \cdot 10^{-4}$

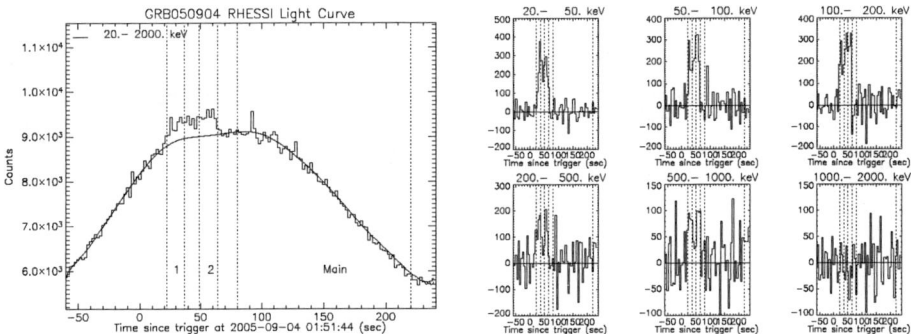

FIGURE 3. *RHESSI* light curve for GRB050904: raw counts with polynomial background fits in the 20-2000 keV band (left) and the resulting background-subtracted lightcurves in the bands 20-50, 50-100, 100-200, 200-500, 500-1000, and 1000-2000 keV (right). The two initial peaks (1 and 2) and the main peak (Main) are labeled.

Further analysis of this burst awaits a better understanding of the background and instrument response. Since we see emission above 200 keV, we can hope to one day constrain E_{peak} for this burst, which so far is only known to lie above 150 keV [10]. The possibility of constraining the spectra of this and other high-z bursts is exciting especially because of the cosmological implications.

FUTURE DIRECTIONS

This is a work in progress. Our goals for the near future include:

- Simulations of Earth backscattering
- Include azimuthal rotation dependence
- Investigation of correlation between Band function parameters
- Study systematics of *RHESSI* spectral fits

REFERENCES

1. Lin, R. P., et al. 2002, *Solar Phys.*, 210, 3
2. Smith, D. M., et al. 2002, *Solar Phys.*, 210, 33
3. Smith, D. M., et al. 2003, *Proc. SPIE*, 4851, 1163
4. Wigger, C., Hajdas, W., Smith, D. M., Güdel, M., Hurley, K., Mchedlishvili, A., & Zehnder, A. 2004, Nuclear Physics B Proceedings Supplements, 132, 331
5. Hajdas, W., et al. 2004, ESA SP-552: 5th INTEGRAL Workshop on the INTEGRAL Universe, 805
6. Wigger, C., Hajdas, W., Arzner, K., Güdel, M., & Zehnder, A. 2004, *ApJ*, 613, 1088
7. Vanderspek, R., et al. 2004, *ApJ*, 617, 1251
8. Sakamoto, T., et al. 2005, GRB Circular Network, 3938, 1
9. Kawai, N., Yamada, T., Kosugi, G., Hattori, T., & Aoki, K. 2005, GRB Circular Network, 3937, 1
10. Cusumano, G., et al. 2005, astro-ph/0509737

Constraints on TeV Emission from GRBs from the GeV Extragalactic Diffuse Gamma-Ray Flux

Sabrina Casanova and Brenda Dingus

Los Alamos National Laboratory, Los Alamos, New Mexico 87545, USA

Abstract. TeV gamma rays emitted by GRBs are converted into electron-positron pairs through absorption processes off the extragalactic infrared radiation fields. In turn the pairs produced, whose trajectories are randomized by magnetic fields, inverse Compton scatter off the cosmic microwave background photons. The beamed TeV gamma ray flux from GRBs is thus transformed into a GeV energy isotropic flux, which contributes to the total extragalactic gamma-ray background emission. Assuming a model for the extragalactic radiation fields, for the GRB redshift distribution and for the GRB luminosity function, we use the measured GRB flux in MeV range to set upper limits on the GRB emission in the TeV energy range that is predicted in several models.

Keywords: GRB high energy γ emission
PACS: 98.70.Rz

INTRODUCTION

The nature of the extragalactic gamma ray background emission has been a topic of great interest since EGRET collaboration evaluated its spectrum in the range from 30 MeV to 100 GeV. [1] Mukherjee and Chiang [2] suggested that blazars can explain up to 25 per cent of the extragalactic emission. This hypothesis seems to be reinforced by Strong, Moskalenko and Reimer's new evaluation of the extragalactic emission [3]. Kneiske and Mannheim [4] recently suggested that up to 85 per cent of the extragalactic emission could arise from blazars. The question about the origin of the extragalactic emission is though still open. In fact all unresolved discrete sources outside the Galaxy contribute to the extragalactic background emission. Prompt and delayed emissions from GRBs also contribute to the diffuse extragalactic emission, especially if some GRBs emit photons in the GeV-TeV energy. In fact, outside GRBs, due to interactions with cosmic infrared background photons, most of the high energy GRB photons produce high-energy electron-positron pairs. The pairs inverse Compton scatter off CMB photons and produce secondary photons, which in turn interact with IR photons and generate other pairs. Multiple inverse Compton scatterings occur until the energy of the secondary photons is no longer enough to trigger pair production with the IR photons. After multiple pair-production and inverse Compton processes, the initial energy of the TeV photons escaping the GRBs will have been shifted to MeV-GeV energies.[5, 6, 7] The data on the extragalactic diffuse emission in the MeV-GeV energy range can thus be used to put constraints on high energy emission from gamma ray bursts at GeV-TeV energy.

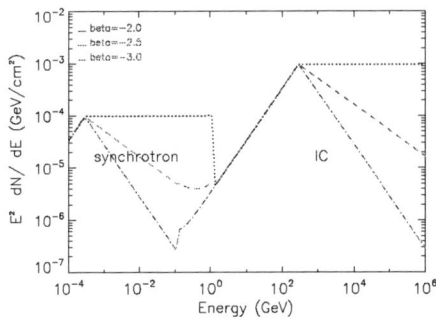 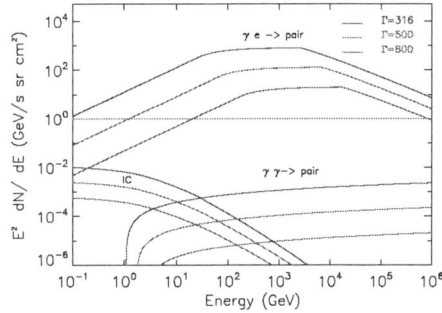

FIGURE 1. Fig.(1a) Band spectrum and Inverse Compton spectrum for a GRB having isotropic luminosity 10^{52} erg/s and located at $z = 1$. The slope $\alpha = -1$ and β varies bewteen -2 and -3 for both synchrotron and IC spectra. The ratio between the synchrotron peak flux and the TeV emission peak flux is assumed to be 10. Fig.(1b) Optical depths in the GRB fireball for different bulk Lorentz factors.

GRB MODEL

High energy protons and electrons are accelerated in GRBs in shocks by interaction with magnetic fields. The process of Fermi acceleration leads to electron and proton power-law spectra of index between -2 and -3. The observed GRB prompt spectrum arises from electron synchrotron cooling and is a Band function, proportional to E^{α} below the synchrotron peak energy, E_{pk}, with α about -1 and to E^{β} above the peak energy, with β usually between -2 and -3. [8] According to some authors [9, 10, 11] TeV photons are created during the prompt phase, either through proton synchrotron or inverse Compton. TeV photons are mostly absorbed through pair production with cosmic background radiation (CBR), but some indications of TeV gamma rays from GRBs at low redshift were provided by experiments like Milagro, Hegra and Tibet. We assume that the GRB prompt spectrum, plotted in Fig.(1a), is the sum of a synchrotron power law spectrum and of an additional broken power law at TeV energies with slopes $\alpha = -1$ and $\beta = -2$.

Photon-photon absorption through pair production inside GRBs controls the range in energy and the amount of radiation emitted. [7] In Fig.(1b) we plot the optical depths corresponding to the different processes taking place inside GRBs, $\gamma\gamma$ absorption, electron Compton scattering and $e\gamma \rightarrow e^{\pm}$. $\gamma\gamma$ pair production within the GRB attenuates the prompt spectrum, while electron Compton scattering and $e\gamma \rightarrow e^{\pm}$ have optical depths less than 1 for every set of values of the isotropic luminosity, the Lorentz factor and the time variability of the GRB. For high enough bulk Lorentz factors and for short enough time variabilities, the optical depth $\tau_{grb\gamma\gamma}$ is always less than 1 and GRBs are optically thin at all energies. TeV energy photons emitted by some process within the GRB will thus be able to leave the source and be observed by experiments like GLAST.

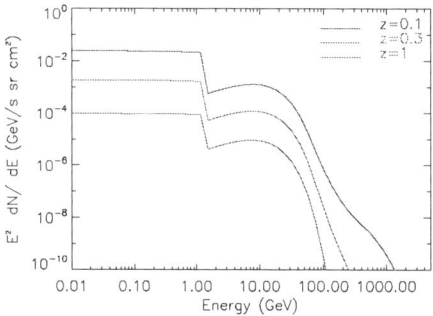 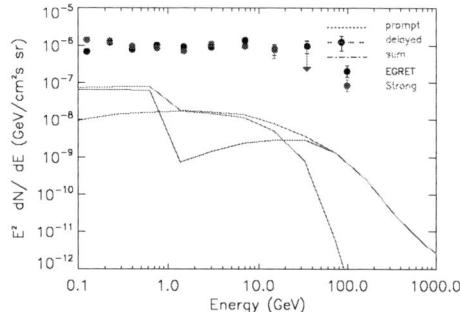

FIGURE 2. Fig (2a) Synchrotron and higher energy prompt flux from one GRB having isotropic luminosity 10^{52} erg/s, bulk Lorentz factor 316 and time variability 1 second. The slopes $\alpha = -1$ and $\beta = -2$ for both synchrotron and IC spectra. The ratio between the synchrotron peak flux and the TeV emission peak flux is assumed to be 10. Fig.(2b) GRB total emission from bursts having average bulk Lorentz factors 316, time variability 1 second and duration 20 seconds. The slopes $\alpha = -1$ and $\beta = -2$ for both synchrotron and IC spectra. The ratio between the synchrotron peak flux and the TeV emission peak flux is 1.

PROMPT AND DELAYED EMISSIONS

High energy gamma rays are also attenuated when travelling to us because they form pairs in collisions with low energy photons from the meta-galactic radiation field. We use the model for the extragalactic background fields by [12] to evaluate the attenuation rate $e^{-\tau_{bkg\gamma\gamma}}$.

The flux $(\frac{dN_{\gamma prompt}}{dE_\gamma dt dA})_{grb}$ which leaves each GRB and is attenuated in the extragalactic background fields is plotted in Fig.(2a). The total flux from prompt GRB emission is given by the following integral

$$Flux_{prompt} = \int dz \, \left(\frac{dN_{\gamma prompt}}{dE_\gamma dA}\right)_{grb} n_{GRB}(z) \, e^{-\tau_{bkg\gamma\gamma}} \qquad (1)$$

where $(\frac{dN_{\gamma prompt}}{dE_\gamma dA})_{grb}$ is the time integral of the differential gamma ray prompt flux from each GRB and $n_{GRB}(z)$ is the number of GRBs per unit time [13]. Outside the GRB, due to $\gamma\gamma$ interactions with cosmic infra-red background photons and CBR photons, most of the high energy photons produce high-energy $e^{\pm}$ pairs. The pairs will inverse Compton scatter off CMB photons and produce secondary photons, which will in turn interact with IR photons and generate other pairs. Multiple inverse Compton scatterings will happen until the energy of the secondary photons is no longer enough to trigger a pair production with the IR photons. When the energy of the scattered photons is not sufficient to produce a subsequent pair, the photons will travel to the detector without undergoing further absorption. When simulating the series of inverse Compton scatterings we assume that all interactions happen very close to the source, at the redshift of the source itself. We also assume that the magnetic field is stronger than 10^{-16} Gauss in order for the flux to be isotropically radiated. The number of secondary photons per unit volume per photon energy interval created in the vicinity of a GRB through one inverse Compton scattering of the pairs off cosmic background radiation photons is

$$(q_{IC}(E_\gamma,z,L_{iso}))_{grb} = \int dE_e \int d\varepsilon_{cmb} \frac{d\sigma_{IC}}{dE_\gamma}(E_\gamma,E_e,\varepsilon_{cmb}) u_{cmb}(\varepsilon_{cmb},z) \left(\frac{dN_e}{dA\,dE_e}\right)_{grb} \quad (2)$$

where $\frac{d\sigma_{IC}}{dE_\gamma}(E_\gamma,E_e,\varepsilon_{cmb})$ can be either the differential Thomson cross section (for low energy pairs) or the differential Klein-Nishina formula (for high energy pairs). $u_{cmb}(\varepsilon_{cmb},z)$ is the density of CBR photons per unit volume per photon energy interval at redshift z and $\left(\frac{dN_e}{dA\,dE_e}\right)_{grb}$ is the time integrated pair flux outside the GRB. We assume that the electron and the positron of the pair have energy E_e. The secondary source function in Eq.(2), iterated over multiple scatterings, is integrated over the comoving volume, divided by the area of the detector and weighted over the number of GRBs per unit time [13] to obtain the total counting rate

$$Flux_{delayed} = \int \frac{dV}{4\pi D_L^2} \int dz\, (q_{IC}(E_\gamma,z,L_{iso}))_{grb} n_{GRB}(z) e^{-\tau_{bkg\gamma\gamma}(E_\gamma,z)}. \quad (3)$$

In Fig.(2b) we plot the sum of all prompt and delayed GRB emissions. Considering that between 25 and 85 per cent of the extragalactic emission at MeV-GeV energies may be due to blazars, we can conclude that TeV energies fluxes from GRBs can be at least 10 times bigger than the synchrotron flux without violating the limit imposed by the extragalactic diffuse emission.

ACKNOWLEDGMENTS

We would like to thank Prof.Bing Zhang for valuable comments and discussions.

REFERENCES

1. P. Sreekumar et al. [EGRET Collaboration], Astrophys. J. **494**, 523 (1998) [arXiv:astro-ph/9709257].
2. R. Mukherjee and J. Chiang, Astropart. Phys. **11**, 213 (1999) [arXiv:astro-ph/9902003].
3. A. W. Strong, I. V. Moskalenko and O. Reimer, Astrophys. J. **613**, 956 (2004) [arXiv:astro-ph/0405441].
4. T. Kneiske and K. Mannheim, AIP Conf. Proc. **745**, 578 (2005) [arXiv:astro-ph/0411146].
5. R. Plaga, Nature **374**, 430 (1995).
6. Z. G. Dai and T. Lu, Astrophys. J. **580**, 1013 (2002) [arXiv:astro-ph/0203084].
7. S. Razzaque, P. Meszaros and B. Zhang, Astrophys. J. **613**, 1072 (2004) [arXiv:astro-ph/0404076].
8. D. Band et al., Astrophys. J. **413**, 281 (1993).
9. M. Gonzales et al., Nature **424**, 749 (2003).
10. K. Hurley et al., Nature **372**, 652 (1994).
11. R. W. Atkins et al., Astrophys. J. **583**, 824 (2003) [arXiv:astro-ph/0207149].
12. T. M. Kneiske, T. Bretz, K. Mannheim and D. H. Hartmann, Astron. Astrophys. **413**, 807 (2004) [arXiv:astro-ph/0309141].
13. D. Guetta and T. Piran, Astrophys. J. **619**, 412 (2005) [arXiv:astro-ph/0407429].

Gamma-Ray Burst Jet Profiles And Their Signatures

C. Graziani, D. Q. Lamb and T. Q. Donaghy

Department of Astronomy & Astrophysics, University of Chicago

Abstract. GRBs and XRFs with known redshifts and E_{pk} provide four empirical constraints on GRB jet structure: $\log(E_{iso})$ is approximately uniformly distributed over several orders of magnitude; $\log(E_\gamma)$ is narrowly distributed; the Amati relation holds between E_{iso} and E_{pk}; and the Ghirlanda relation holds between E_γ and E_{pk}. We infer the underlying angular profiles from the first two of the above constraints assuming a "universal jet" model, and show that such models cannot satisfy both constraints. We introduce a general and efficient method for calculating relativistic emission distributions and E_{pk} distributions from jets with arbitrary (smooth) angular jet profiles, and explore the behavior of universal jet models under a range of profile shapes and parameters, to map the extent to which these models can conform to the above four empirical constraints.

Keywords: Gamma Rays: Bursts — ISM: Jets and Outflows — Shock Waves
PACS: 98.70.Rz

INTRODUCTION

The debate over the meaning of the "GRBs are Standard Bombs" result of Frail et al. (2001) has given rise to two main alternative interpretations. In the "Variable Opening Angle" (VOA) interpretation, GRB sources all have a common amount of energy available for gamma-ray emission, and differ from one another by the opening angle of their conical jet (Frail et al. 2001, Lamb, Donaghy & Graziani 2004).

The alternative view is the "Universal Jet" model, according to which all GRB jets have a common angular energy emission profile that is richer than a simple uniformly-emitting cone. This picture attempts to explain the distribution of GRB energies using only the variation of observer lines-of-sight with respect to the jet axis (Rossi et al. 2002, Zhang & Mészáros 2002).

It is apparent that universal-jet models of GRBs are simpler than VOA models, in the sense that they require fewer parameters. Clearly, an important question that must be addressed in this connection is: do universal jet models have enough freedom to satisfy known observational constraints on GRBs?

EMPIRICAL CONSTRAINTS ON GRB JET MODELS

Certain constraints on universal jet shapes and distributions may be inferred from observations. The empirical constraints on jet models that we consider here are:

1. The distribution of GRB isotropic-equivalent energy E_{iso} appears to be quite broad, spanning at least four orders of magnitude, and appears approximately uniform in

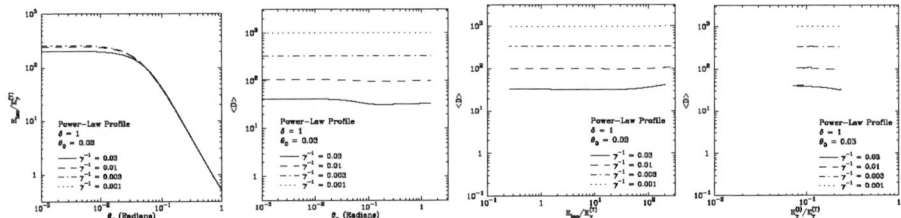

FIGURE 1. Emission characteristics for a $\delta = 1$ power-law profile universal jet model with profile width $\theta_0 = 0.03$ radians. From left to right: E_{iso} vs. θ_v; $\langle D \rangle$ vs. θ_v; "Amati" plot; "Ghirlanda" plot. The limited dynamic range of $\langle D \rangle$ for this profile prevents any realistic Amati or Ghirlanda correlation.

$\log E_{iso}$, that is,

$$dN/d\log E_{iso} \approx A, \qquad (1)$$

where A is a constant [1, 2].

2. The distribution of GRB total gamma-ray energy inferred using the Frail et al. [3] procedure — $E_\gamma^{(I)}$ — appears to be quite narrow, with most values clustered within a decade around a geometric mean whose (model-dependent) value is somewhere in the range 5×10^{50}–3×10^{51} erg [3].

3. Empirically, GRB isotropic energies E_{iso} and peak energies E_{peak} of GRB νF_ν spectra appear to be satisfy the "Amati Relation", a very tight correlation according to which $E_{peak} \sim E_{iso}^{1/2}$ [1, 2].

4. Empirically, GRB inferred total energies $E_\gamma^{(I)}$ and peak energies E_{peak} of GRB νF_ν spectra appear to be satisfy the "Ghirlanda Relation", a very tight correlation according to which $E_{peak} \sim E_\gamma^{(I)^{0.7}}$ [4].

EMISSION PROFILES

Neglecting the effect of relativistic kinematics, there is a unique emission profile $\varepsilon(\mu_v)$ (where $\mu_v \equiv \cos\theta_v$, with θ_v the angle between the line-of-sight and the center of the jet) that satisfies Eq. (1): The Fisher profile,

$$\varepsilon(\mu_v) \propto \exp(\theta_0^{-2}\mu_v). \qquad (2)$$

This functional form is called the "Fisher Distribution", and is well known from the theory of statistical distributions on spheres. In the small-angle approximation, when $\mu_v \approx (1 - \theta_v^2/2)$, it approaches a symmetric 2-dimensional Gaussian with standard deviation θ_0.

The universal "structured" jet model of Rossi et al. [5] and Zhang & Mészáros [6] does not attempt to satisfy constraint (1). Instead, it attempts to preserve the appearance of the Frail et al. [3] result that is, to satisfy constraint (2). It does this by imposing a power-law emission profile,

$$\varepsilon(\mu_v) \propto (1 - \mu_v)^{-\delta}, \qquad (3)$$

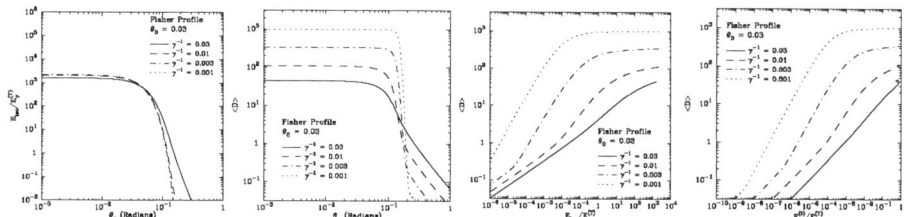

FIGURE 2. Emission characteristics for a Fisher profile universal jet model with profile width $\theta_0 = 0.03$ radians. From left to right: E_{iso} vs. θ_v; $\langle D \rangle$ vs. θ_v; "Amati" plot; "Ghirlanda" plot. Realistic Amati and Ghirlanda correlations are possible. The requirement that the flat-top part of the $\langle D \rangle$ profile should not subtend too much solid angle would likely result in strong fit constraints on γ and θ_0.

with $\delta = 1$.

RELATIVISTIC KINEMATICS

The observed emission profiles of GRB jets are related to the underlying intrinsic profiles by the action of relativistic kinematics — we observe the convolution of the jet profile with the Doppler factor. We carry out this convolution generally — for arbitrary smooth emission profiles — and efficiently, by means of the *Fast Discrete Legendre Transform* (FDLT) of Driscoll & Healy [7].

This method cannot be used for top-hat emission profiles, since these profiles are not smooth (the sharp edge of the profile contaminates the FDLT with unawanted aliased high-harmonic power). We therefore have also calculated exact analytical formulas for relativistic emission from top-hat profiles [8].

In order to furnish model predictions for Amati-like and Ghirlanda-like correlations, we require a quantity that can stand in for E_{peak}. Rather than attempting to superpose the appropriately Doppler-shifted and suitably Doppler-weighted spectra received from various parts of the jet, we have found it convenient to adopt $\langle D \rangle$, the *Average Doppler Shift*, where the averaging is carried out over the surface of the jet. Assuming a constant or slowly-varying $E_{peak}^{(rest)}$, we then have $E_{peak} \approx \langle D \rangle \times E_{peak}^{(rest)}$. $\langle D \rangle$ may be also computed by the FDLT technique for smooth profiles, and may be apressed analytically for top-hat profiles [8].

Figs. 1, 2, and 3 show the results of a selection of computed profiles and preditcted Amati/Ghirlanda correlations, for the $\delta = 1$ power-law, the Fisher profile, and the top-hat profile, respectively.

CONCLUSIONS

- It does not appear possible for a universal jet model to produce a narrow distribution of $E_\gamma^{(I)}$ simultaneously with a broad distribution of E_{iso}. This is a serious concern for universal models of GRBs.

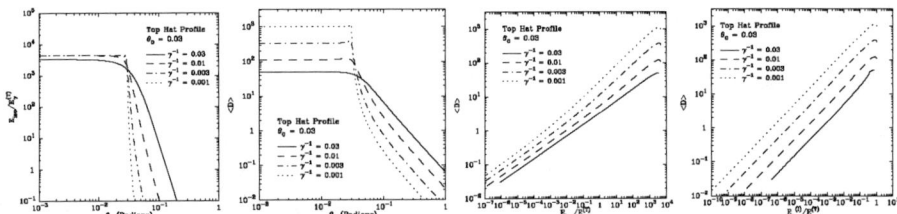

FIGURE 3. Emission characteristics for a top-hat profile universal jet model with profile width $\theta_0 = 0.03$ radians. From left to right: E_{iso} vs. θ_v; $\langle D \rangle$ vs. θ_v; "Amati" plot; "Ghirlanda" plot. Surprisingly, this profile family is also capable of convincing Amati and Ghirlanda slopes. However, the "GRBs are on-axis, XRFs are off-axis" picture of Yamazaki et al. [10] cannot be sustained in this context, since only XRFs would obey the correlations.

- Classic "structured" ($\delta = 1$ power-law) universal models cannot make Amati or Ghirlanda correlations with the correct slope. This is essentially because the dynamical range in $\langle D \rangle$ (and therefore in E_{peak}) is far too small.
- Universal models featuring steeper profiles (Fisher profiles, $\delta \approx 8$ power-law profiles, and top-hat profiles) *can* make respectable Amati/Ghirlanda correlations. The available regions of parameter space where this is possible without unbalancing the E_{iso} and $E_\gamma^{(I)}$ frequency distributions are constrained, meaning that real fits of these models to the data could produce significant constraints on the Lorentz factor γ and on the profile width θ_0.
- Interestingly, the Ghirlanda slopes are always steeper than the Amati slopes. This steepening is evidently a purely geometrical effect. It is clearly unrelated to the steepening effect noticed by Levinson & Eichler [9], since we have certainly not included any dependency of the break angle on E_{iso}, the source of the steepening noted in that paper. Apparently, with a break angle that is conceived purely in terms of geometry, it is still the case that Ghirlanda relations of universal models are expected to be steeper than their corresponding Amati relations.

REFERENCES

1. Amati, L., et al. 2002, A&A, 390, 81
2. Lamb, D. Q., et al. 2004, New Astronomy Review, 48, 423
3. Frail, D. A. et al 2001, ApJ, 562, L55
4. Ghirlanda, G., Ghisellini, G., & Lazzati, D. 2004, ApJ, 616, 331
5. Rossi, E., Lazzati, D., & Rees, M. J. 2002, MNRAS, 332, 945
6. Zhang, B., and Mészáros, P. 2002, ApJ 571, 876
7. Driscoll, J. R, and Healy, D. M. 1994, Adv. Appl. Math. 15, 202
8. Graziani, C., Lamb, D. Q., and Donaghy, T. Q. 2005, ApJ, submitted (astro-ph/0505623)
9. Levinson, A., and Eichler, D. 2005, astro-ph/0504125
10. Yamazaki, R., Ioka, K., & Nakamura, T. 2002, ApJ, 571, L31

Are Long-Lag GRBs Different than Other Long GRBs?

Jon Hakkila, Timothy W. Giblin,
Christopher D. Peters, Sarah M. Sonnett, and Christopher M. Nolan

Department of Physics and Astronomy, College of Charleston, Charleston, SC 29424-0001

Abstract. Long-lag GRBs have two different morphologies that can be delineated using the Internal Luminosity Function (ILF). Both burst types are typically dominated by a single broad pulse. Long-lag GRBs with large ILF indices are long FRED (Fast Rise Exponential Decay) bursts having well-defined peaks, while those with small ILF indices are *long smooth bursts* having no well-defined short timescale pulse structure. Long smooth bursts typically have have the longest lags of any type of GRB time history. We examine the hypothesis that long-lag GRBs are external shock signatures and contrast this with the hypothesis that they are merely broad internal shock pulses.

Keywords: gamma-ray sources; gamma-ray bursts
PACS: 98.70.Rz

INTRODUCTION

Long-lag gamma-ray bursts (GRBs) make up only a small fraction of all GRBs [11], yet they are interesting because they are *only* observed to contain long-lag pulses. This is in contrast to short-lag GRBs, which can contain long-lag pulses [5]. It is possible that long-lag pulses are indicative of external shock signatures [2, 5]. Long-lag bursts need to be studied in greater detail so that their role in the GRB puzzle can be clarified.

Two parameters have recently been shown [6] to describe the morphological characteristics of a gamma-ray burst's (GRB) time history. These two parameters are the *spectral lag* and the *internal luminosity function power law index*. These parameters (and others) describing GRBs in the BATSE Current Catalog www.batse.msfc.nasa.gov/batse/grb/catalog/current/ will be placed online alongside a suite of data mining tools for the purpose of GRB classification and pattern identification [4, 3].

ANALYSIS AND DISCUSSION

The spectral lag is the time delay between high energy emission and low energy emission; low energy emission typically lags behind high energy emission. Lags are obtained from the peak of the cross correlation function (CCF) [1] between two energy channels. BATSE's four broadband energy channels (Channel 1 : 25 – 50 keV; Channel 2 : 50 – 100 keV; Channel 3: 100 – 300 keV; Channel 4: 300 – 1000 kev) can produce six different lag measurements. A nonlinear fit to the CCF is used to obtain the lag, since the peak of the cross-correlation function can be difficult to identify due to background fluctuations. We use a GRB pulse function [9] to fit the CCF as an alternative to the simple asymmetric cubic [10] or symmetric quadratic [13] functions; we believe that the additional degrees of freedom provided by this model result in a more accurate CCF fit. However, the peak is difficult to measure to accuracies greater than 64 ms, since that is the minimum bin size available from this dataset. The channel 31 lag (lag_{31}) spans a large energy range (25 – 300 keV) and produces a long, easy-to-measure lag.

The Internal Luminosity Function (ILF), $\psi(L)$, uses 64-ms data from HEASARC's Compton Gamma-Ray Observatory Science Support Center to obtain the distribution of luminosity within a GRB; $\psi(L)dL$ represents the fraction of time during which a burst's luminosity lies between L and $L+dL$ [7]. A distribution function is constructed by binning count rates relative to a defined minimum (e.g. the 2σ above background is the value being referenced here), and expected Poisson background rates are subtracted from each bin so that only estimated source counts remain. This process allows the ILF to be measured closer to background than that obtained by Horack & Hakkila [7]. The number of bins used is reduced in an iterative fashion to optimize the signal in each bin, while the ILF is simultaneously normalized by the requirement that $\Sigma\psi(L)\Delta L = 1$. The ILF can generally be fit by a quasi power-law form such that $\psi(L) = AL^{\alpha}10^{\beta[log(L)]^2}$. We call α the power-law index and β the curvature index. A large α indicates that the burst has a large amount of high luminosity emission relative to low luminosity emission, while a small α indicates that the burst has a significant amount of emission at luminosities less than the peak luminosity. A large β indicates that there is no curvature such that the ILF is fit by a simple power law, whereas a small β indicates that the burst is also depleted in low luminosity emission relative to that expected from the power-law index α (e.g. such that the additional fit parameter β is required). Thus, α and β are highly correlated, such that only one of these fit parameters is needed to specify the fit.

The parameters α and lag_{31} are excellent at identifying GRBs with similar time history morphologies [6]. The relationship between general GRB time history characteristics are such that (1) longer lag bursts contain, on average, fewer pulses than short-lag bursts, and the pulses, in general, tend to be broader, and (2) a decrease in the fraction of high-luminosity emission to low-luminosity emission (small α contrasted with large α) is accompanied either by more pulses that are narrow (if the mean lag is small), or by pulses that have less pronounced peaks (if the mean lag is large). GRB luminosity anti-correlates with lag (e.g. [10]) and with complexity (e.g. [12]) such that luminous bursts have shorter lags and are more complex than low-luminosity bursts, yet cosmological time dilation should broaden pulses and increase lag times, but should not decrease the number of pulses. This argues that extrinsic cosmological effects are significantly less important than intrinsic ones.

Figure 1 is a plot of the ILF power-law index α vs. (lag_{31}) for 384 long-lag bursts ($lag_{31} \geq 0.3$). Figure 2 demonstrates sample time histories of long FRED bursts (left four panels) and long smooth bursts (right four panels). In contrast to FREDs, long smooth bursts have the longest lags of all GRBs but have a poorly-defined peak and few identifiable temporal structures.

It is possible that the broad, smooth pulses are signatures of external shocks, or conversely that they are signatures of long internal shocks.

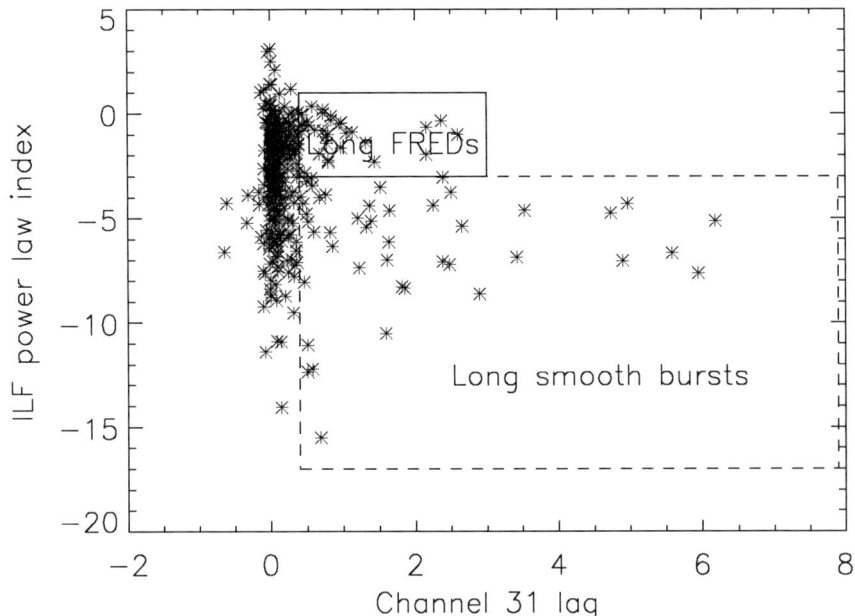

FIGURE 1. Long-lag (lag_{31}) GRBs delineated by ILF power-law index α.

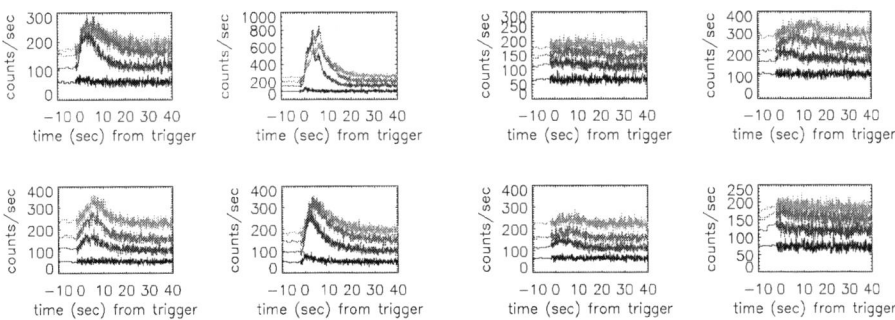

FIGURE 2. Sample time histories of long-lag GRBs (see text).

Evidence suggesting that broad pulses may be caused by external shocks:

- Long-lag pulses appear to be underluminous [5] and do not clearly follow the lag vs. luminosity relation of Norris, Marani, & Bonnell [10].
- Some GRBs characterized by narrow pulses also contain a broad pulse [5].

- GRBs with $\alpha \leq -1$ and lags shorter than $\text{lag}_{31} \leq 0.3$ have several moderately-broad pulses, while those with $\alpha \leq -1$ and $\text{lag}_{31} \geq 0.3$ are characterized by a single, broad pulse (Long smooth bursts). The abrupt separation of characteristics around $\text{lag}_{31} = 0.3$ seems to imply that long-lag bursts are produced by different physics or different physical conditions than short-lag bursts.
- Long smooth GRBs can have much longer lags than the longest FRED bursts, implying that physical differences my exist between these two long-lag GRB types.

Evidence suggesting that broad pulses may be caused by internal shocks:

- Emission tails, rather than specific pulse types, may provide evidence of external shocks (e.g. [2])
- Long flares found in GRB afterglows have been interpreted as refreshed internal shocks; it is not too dramatic an interpretation that long broad pulses in prompt emission (and therefore the long-lag GRBs) are also caused by internal shocks.
- Long smooth bursts in some ways appear related to long FREDs: a Long FRED can be a long smooth burst with an absent peak. GRBs having properties of both Long smooth bursts and long FREDs exist can be found with ILF power-law indices of $\alpha \approx -2$. It also appears that FREDs may be associated with a class of bursts having small afterglow luminosities as obtained from infrared observations [8]

We are currently working with BATSE and Swift data in an attempt to further study this problem.

ACKNOWLEDGMENTS

We would like to thank Jay P. Norris for many helpful discussions. This research is supported by NSF grant AST00-98499 and by NASA South Carolina Space Grant.

REFERENCES

1. Band, D. L., *Astrophys. J.*, **486**, 928 (1997).
2. Giblin, T. W., van Paradijs, J., Kouveliotou, C., Connaughton, V., Wijers, R. A. M. J., Briggs, M. S., Preece, R. D., & Fishman, G. J., *Astrophys. J. (Lett.)*, **524**, L47 (1999).
3. Giblin, T. W., Hakkila, J., Haglin, D. J., & Roiger, R. J., in *AIP Conf. Proc. 727: Gamma-Ray Bursts: 30 Years of Discovery*, 2004, p. 585.
4. Haglin, D. J., Roiger, R. J, Hakkila, J., & Giblin, T. W., *Data Science Journal*, **4**, 39 (2005).
5. Hakkila, J., & Giblin, T. W., *Astrophys. J.*, **610**, 361 (2004).
6. Hakkila, J., & Giblin, T. W., *Astrophys. J. (Lett.)*, submitted.
7. Horack, J. M., & Hakkila, J. 1997, *Astrophys. J.*, **479**, 371 (1997).
8. Liang, E. & Zhang, B., *Astrophys. J. (Lett.)*, submitted.
9. Norris, J. P., Nemiroff, R. J., Bonnell, J. T., Scargle, J. D., Kouveliotou, C., Paciesas, W. S., Meegan, C. A., & Fishman, G. J., *Astrophys. J.*, **459**, 393 (1996).
10. Norris, J. P., Marani, G. F., & Bonnell, J. T., *Astrophys. J.*, **534**, 248 (2000).
11. Norris, J. P., Bonnell, J. T., Kazanas, D., Scargle, J. D., Hakkila, J., & Giblin, T. W., *Astrophys. J.*, **627**, 324 (2005).
12. Reichart, D. E., Lamb, D. Q., Fenimore, E. E., Ramirez-Ruiz, E., Cline, T. L., & Hurley, K., *Astrophys. J.*, **552**, 57 (2001).
13. Wu, B. & Fenimore, E., *Astrophys. J. (Lett.)*, **535**, L29 (2000).

Properties of the intermediate type of gamma-ray bursts

I. Horváth*, F. Ryde [†], L.G. Balázs **, Z. Bagoly[‡] and A. Mészáros [§]

*Department of Physics, Bolyai Military University, Budapest, Box-12, H-1456, Hungary
[†]Stockholm Observatory, AlbaNova, SE-106 91 Stockholm, Sweden
**Konkoly Observatory, Budapest, Box-67, H-1525, Hungary
[‡]Laboratory for Information Technology, Eötvös University, Budapest, Pázmány P. s. 1/A,, H-1117, Hungary
[§]Astronomical Institute of the Charles University, V Holešovičkách 2, CZ 180 00 Prague 8, Czech Republic

Abstract. Gamma-ray bursts can be divided into three groups ("short", "intermediate", "long") with respect to their durations. The third type of gamma-ray bursts - as known - has the intermediate duration. We show that the intermediate group is the softest one. An anticorrelation between the hardness and the duration is found for this subclass in contrast to the short and long groups.

Keywords: gamma-rays: bursts – Cosmology: miscellaneous
PACS: 98.70 and 98.80

TWO-DIMENSIONAL GAUSSIAN FITS

Simultaneously Mukherjee et al. [1] and Horváth [2] found a third group of gamma-ray bursts (GRBs). Somewhat later several authors [3, 4, 5, 6] also suggested the existence of the third ("intermediate") group as well. The physical existence of the third group is, however, still not convincingly proven. For example, Hakkila et al. [3] believe that the third group is only a deviation caused by a complicated instrumental effect, which can reduce the durations of some faint long bursts. Later Hakkila et al. [7] published another paper which had different conclusions.

Using Principal Component Analysis (PCA), Bagoly et al. [8] have shown that there are only two major quantities necessary (called the Principal Components; PCs) to characterize most of the properties of the bursts in the BATSE Catalog. Consequently, the problem of the choice of the relevant parameters describing GRBs is basically a two-dimensional problem. For the statistical analysis the choice of two independent parameters is enough; they may be, but are not necessarily, the two principal components. This means that only two parameters, relevantly chosen, should be enough for the classification and determination of the groups. Here we have chosen **duration** T_{90} and **hardness** $H_{32} = F_3/F_2$ (F_3 and F_2 are the fluences) for these parameters.

We can assume that the observed probability distribution of GRBs in this plane is a superposition of the distributions characterizing the different types of bursts present in the sample. Introducing the notations $x = \log T_{90}$ and $y = \log H_{32}$ and using the law of

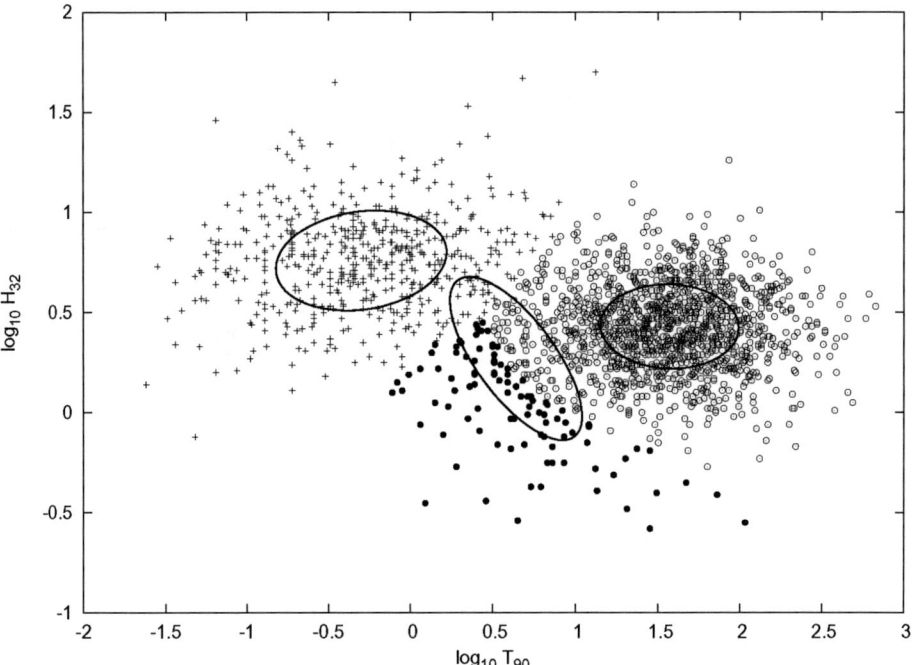

FIGURE 1. The 1956 GRBs in the $\{\log T_{90}; \log H_{32}\}$ plane. The 1σ ellipses of the three Gaussian distributions are also shown, which were obtained in the ML procedure. The different symbols (crosses, filled circles and open circles) mark bursts belonging to the short, intermediate and long classes, respectively.

full probabilities we can write

$$p(x,y) = \sum_{l=1}^{k} p(x,y|l) p_l. \qquad (1)$$

In this equation $p(x,y|l)$ is the conditional probability density assuming that a burst belongs to the l-th class. p_l is the probability for this class in the observed sample, where k is the number of classes. In order to decompose the observed probability distribution $p(x,y)$ into the superposition of different classes we need the functional form of $p(x,y|l)$. The probability distribution of the logarithm of durations can be well fitted by Gaussian distributions, if we restrict ourselves to the short and long GRBs, respectively [9]. We assume the same for the y coordinate as well. Therefore it holds

$$p(x,y|l) = \frac{(1-r^2)^{-\frac{1}{2}}}{2\pi\sigma_x\sigma_y} \exp\left[-\frac{1}{2(1-r^2)}\left(\frac{(x-a_x)^2}{\sigma_x^2} + \frac{(y-a_y)^2}{\sigma_y^2} - \frac{2r(x-a_x)(y-a_y)}{\sigma_x\sigma_y}\right)\right] \qquad (2)$$

The observational data from the Current BATSE GRB Catalog will be used. There are 2702 GRBs, for **1956** of which both the hardnesses and durations are measured.

TABLE 1. Parameters of the Gaussian fits ($k = 3$) in the $\{\log T_{90}; \log H_{32}\}$ plane.

classes	p_l	$a_x = \log T_{90}$	$a_y = \log H_{32}$	σ_x	σ_y	r (corr.coef.)
1	.245	-.301	.763	.525	.251	.163
2	.109	.637	.269	.474	.344	-.513
3	.646	1.565	.427	.416	.210	-.034

In order to find the unknown constants in Eq.(2) we use the Maximum Likelihood (**ML**) procedure of parameter estimation [10]. One can define the Likelihood Function in the usual way, after fixing the value of k, in the form $L = \sum \log p(x_i, y_i)$, where $p(x_i, y_i)$ has the form given by Eq.(1). Similarly, as it was done by Balázs et al. [10], the EM (Expectation and Maximization) algorithm is used to obtain the $a_x, a_y, \sigma_x, \sigma_y, r$ and p_l parameters at which L reaches its maximum value. We made the calculations for different values of k in order to see the improvement of L as we increase the number of parameters to be estimated.

The confidence interval of the parameters came from $2(L_{max} - L_0) = \chi_m^2$ equation. Moving from $k=2$ to 3 the number of parameters m increases by 6 (from 11 to 17), and L_{max} grows from 1193 to 1237, which means a very low probability of being a chance (10^{-10}).

Moving from $k=3$ to 4, the improvement in L_{max} is only 6 (from 1237 to 1243), which can happen by chance with a probability of 6.2%. Hence, the inclusion of the fourth class is not justified. Table 1. shows the parameters of the best fit with $k=3$.

RESULTS

The mathematical deconvolution of the $p(x,y)$ joint probability density of the observed quantities into Gaussian components does not necessarily mean that the physics behind the classes obtained mathematically is different. It could well be possible that the true functional form of the distributions is not exactly Gaussian and that the algorithm of deconvolution formally inserts a third one only in order to get a satisfactory fit. One needs detailed investigations based on the physical (e.g. spectral) properties of the individual bursts to prove its astrophysical validity.

Norris et al. [11] and Balázs et al. [10] found compelling evidence that there is a significant difference between the short and long GRBs. This might indicate that different types of engines are at work. The relationship of long GRBs to the massive collapsing objects is now also observationally well established, and the relation between the comoving and observed time scales is well understood [12, 13]. The short bursts can be identified as originating from neutron star (or black hole) mergers. Therefore the mathematical classification of GRBs into the short and long classes is physically justified.

An important question that must be answered in this context is whether the intermediate group of GRBs, obtained in the previous paragraph from the mathematical classification, really represents a third type of burst physically different from both the short and the long ones.

The classification into the short, intermediate and long classes is based mainly on the duration of the burst. From Table 1 one may infer that these three classes differ also in the hardnesses. The difference in the hardnesses between the short and long group is well known [14]. According to these data the **intermediate** GRBs are the **softest** among the three classes. This different small mean hardness and also the different average duration suggest that the intermediate group should also be a different phenomenon, that is, both in hardness and in duration the third group differs from the other two. On the other hand, no significant correlation exists between the hardness and the duration within the short and the long classes. Thus, these two quantities may be taken as two independent variables, and the short and long groups are different in both these independent variables.

In contrast, there is a strong **anticorrelation** between the hardness and the duration within the intermediate class. This is a surprising, new result, and because the hardness and the duration are not independent in the third group, one may simply say that only one significant physical quantity is responsible for the hardness and the duration within the intermediate group. Consequently, the situation is quite different here, because one needs two independent variables to describe the remaining two other groups. This is a strong constraint in modeling the third group. Hence, the question of the true nature of the physics in the intermediate group remains open, and needs further analysis.

In this paper we have shown that statistically a third group of GRBs exists. Also statistically no further groups are needed to describe the $\{\log T_{90}; \log H_{32}\}$ distribution of bursts. Finally, 11% of GRBs in the Current BATSE Catalog belong to the intermediate class. The memberships of the cataloged bursts are available on the internet [15].

ACKNOWLEDGMENTS

Thanks are due to D. L. Band, C.-I. Björnsson, J. T. Bonnell, L. Borgonovo, J. Hakkila, S. Larsson, P. Mészáros, J. P. Norris, H. Spruit and G. Tusnády for valuable discussions. This study was supported by OTKA, grant No. T048870.

REFERENCES

1. Mukherjee, S., et al. 1998, ApJ, 508, 314
2. Horváth, I. 1998, ApJ, 508, 757
3. Hakkila, J., et al. 2000, ApJ, 538, 165
4. Balastegui, A., Ruiz-Lapuente, P., & Canal, R. 2001, MNRAS, 328, 283
5. Rajaniemi, H.J., & Mähönen, P. 2002, ApJ, 566, 202
6. Horváth, I., et al. 2006, A&A, in print
7. Hakkila, J., et al. 2003, ApJ, 582, 320
8. Bagoly, Z., Mészáros, A., Horváth, I., Balázs, L.G., & Mészáros, P. 1998, ApJ, 498, 342
9. Horváth, I. 2002, A&A, 392, 791
10. Balázs, L.G., Bagoly, Z., Horváth, I., Mészáros, A., & Mészáros, P. 2003, A&A, 138, 417
11. Norris, J.P., Scargle, J.D., & Bonnell, J.T. 2001, Gamma-Ray Bursts in the Afterglow Era, Proc. Int. Workshop held in Rome, Italy, eds. E. Costa et al., ESO Astrophysics Symp. (Berlin: Springer), p. 40
12. Fox, D.B., et al. 2005, Nature, 437, 845
13. Ryde, F., & Petrosian, V. 2002, ApJ, 578, 290
14. Kouveliotou, C., et al. 1993, ApJ, 413, L101
15. Horváth, I., et al. 2005, class memberships http://itl7.elte.hu/~hoi/grb/class.html

Statistical Analyses of RHESSI GRB Database

Jakub Řípa[1,2], René Hudec[2], Attila Mészáros[1], Wojtek Hajdas[3] and Claudia Wigger[3]

[1] *Astronomical Institute, Faculty of Mathematics and Physics, Charles University, Prague, Czech Republic*
[2] *Astronomical Institute of the Academy of Sciences of the Czech Republic, CZ-251 65 Ondřejov, Czech Republic rhudec@asu.cas.cz*
[3] *Paul Scherrer Institut, Villigen, Switzerland*

Abstract. The Gamma-Ray Burst (GRB) database based on the data by the RHESSI satellite provides a unique and homogeneous database for the further scientific analyses. We present here the preliminary results of an analysis of this database.

Keywords: gamma-ray astrophysics, gamma-ray bursts
PACS: 95.55.Ka, 95.85.Pw

INTRODUCTION

The Ramaty High Energy Solar Spectroscopic Imager (RHESSI) is a NASA Small Explorer satellite designed to study hard X-rays and gamma-rays from solar flares [1]. It consists mainly of an imaging tube and a spectrometer. The spectrometer consists of nine germanium detectors (7.1 cm diameter and 8.5 cm height) [2]. They are only lightly shielded, thus making RHESSI also very useful to detect photons from any direction. The energy range sensitive for GRB detection extends from about 50 keV up to 20 MeV depending on the incoming direction. Energy and time resolutions are excellent for time resolved spectroscopy: $\Delta E = 3$ keV (at 1000 keV), $\Delta t = 1$ microsec. The effective area for near axis direction of incoming photons reaches up to 200 cm^2 at 200 keV. GRBs are observed with an effective solid angle of about half of the sky. With this wide field of view, RHESSI observes about one Gamma Ray Burst (GRB) per week (see also http://grb.web.psi.ch). Data are stored event-by-event in an onboard memory.

DISTRIBUTION OF THE DURATIONS PLOT

Totally 208 gamma-ray bursts from the years 2002 to 2005 observed by the RHESSI satellite were used. As one can see in Figure 1, we obtain a bimodal distribution of GRB durations. We fitted two Gaussian functions (5 parameters), as was done successfully for the BATSE Catalog (see [3] and the references therein). The maxima of our fit are at 0.2 s (short bursts) and 16.3 s (long bursts), respectively. A further result is that the short GRBs amount to about 11 % of all GRBs observed by RHESSI. We used the statistical χ^2 test to evaluate this fit. Strictly from the statistical point of view, the fit with the sum of two Gaussian distributions is a poor fit. The assumption that there are **only** two Gaussian subclasses is rejected on a less than 0.1 % significance level. Detailed exploration indicates that this feature is due to over-abundance of bursts in the time interval 1 – 4 seconds. Hence, RHESSI data point to a probably existence of a third subgroup of GRBs (intermediate bursts), like BATSE data analyses have shown [4].

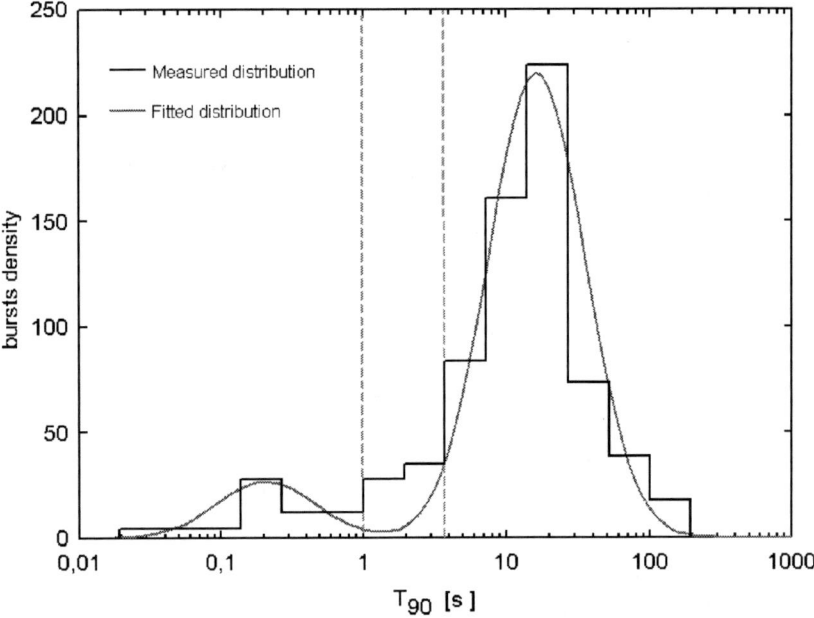

FIGURE 1. Distribution of durations of GRBs detected by RHESSI

The fitted two-lognormal function is

$$F(\log T_{90}) = N\left\{\frac{w_1}{\sigma_1\sqrt{2\pi}}\exp\left[-\frac{(\log T_{90} - \mu_1)^2}{2\sigma_1^2}\right] + \frac{w_2}{\sigma_2\sqrt{2\pi}}\exp\left[-\frac{(\log T_{90} - \mu_2)^2}{2\sigma_2^2}\right]\right\}$$

where: μ is average duration (logT$_{90}$), σ is width of Gaussian (logT$_{90}$), w is weight, f is degree of freedom, α is significance level, N = 208 is the number of used GRBs.

The second weight is $w_2 = 1 - w_1$ due to the normalization.

μ$_1$ (logT$_{90}$)	μ$_2$ (logT$_{90}$)	σ$_1$ (logT$_{90}$)	σ$_2$ (logT$_{90}$)	w$_1$	w$_2$	χ^2	f	α [%]
-0,69	1,21	0,35	0,33	0,11	0,89	86,8	6	>0,1

TABLE 1. Values of the parameters used in the two-lognormal fit

HARDNESS RATIO VS. DURATION PLOTS

The hardness ratio is defined as the proportion of two fluences F in two different energy bands integrated over the time interval T_{90}. The hardness ratio is a measure of how hard the gamma-rays in a burst are. Specifically we have three energy bands and its corresponding fluences: $F1_{T90}$ = total counts between 25 - 120 keV during T_{90}, $F2_{T90}$ = total counts between 120 - 400 keV during T_{90}, $F3_{T90}$ = total counts between 400 - 1500 keV during T_{90} and hardness ratios are: $H_{21} = F2_{T90} / F1_{T90}$ and $H_{32} = F3_{T90} / F2_{T90}$. In Figs. 2 and 3 we can clearly see two subgroups – short and long GRBs. We can also see that in general short bursts have harder spectra. Dashed lines bound the two bins where there is the possible over-abundance of GRBs in the plot 'Distribution of durations'.

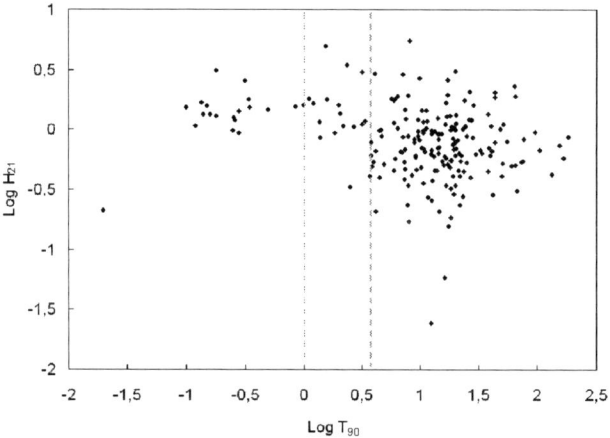

FIGURE 2. Hardness ratio H_{21} vs. duration T_{90}

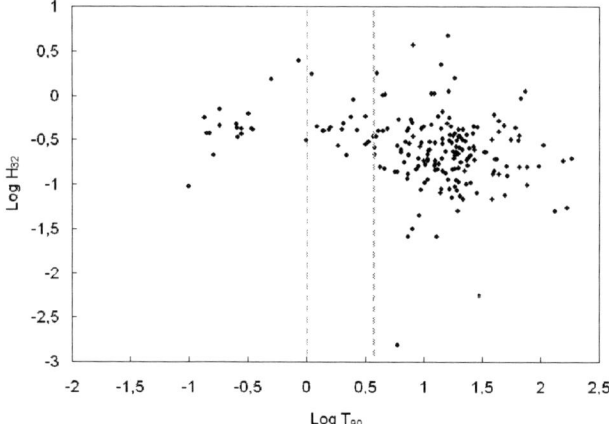

FIGURE 3. Hardness ratio H_{32} vs. duration T_{90}

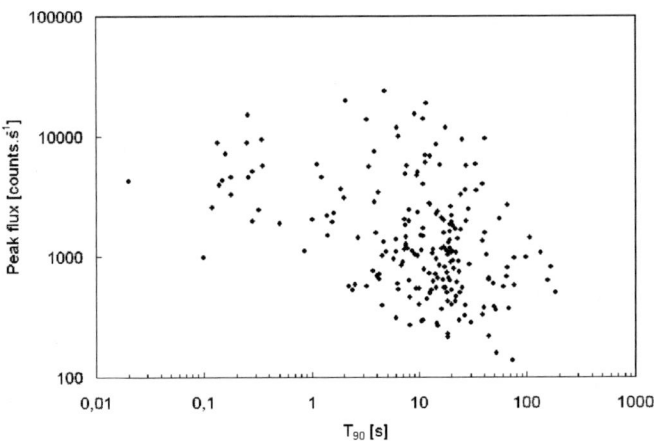

FIGURE 4. Peak flux vs. duration T90

In Figure 4 we obtained a similar dependence like in the plot 'Hardness ratio vs. duration'. As one can see there are noticeably separated two (short and long) subclasses of gamma-ray bursts.

CONCLUSION

The preliminary results (duration plot, hardness ratios) of analyses of RHESSI GRB database confirm the results and conclusions drawn from the CGRO BATSE database, including the possible existence of a third subclass of GRBs located between the short and long GRBs.

ACKNOWLEDGEMENTS

We acknowledge the support provided by the Grant Agency of the Academy of Sciences of the Czech Republic, grant A3003206 and by OTKA grant T048870.

REFERENCES

1. Lin, R. P. et al. 2002, *Solar Phys.*, 210, 3.
2. Smith, D. M., et al. 2002, *Solar Phys.*, 210, 33.
3. Horváth, I. 1998, *ApJ*, 508, 757.
4. Horváth, I., et al. 2005, astro-ph/0509909.

The Complete Spectral Catalog of Bright BATSE Gamma-Ray Bursts

Yuki Kaneko[*,†], Robert D. Preece[**,†], Michael S. Briggs[**,†], William S. Paciesas[**,†], Charles A. Meegan[‡,†] and David L. Band[§]

[*]*Universities Space Research Association*
[†]*NSSTC, 320 Sparkman Dr. Huntsville, AL 35805*
[**]*University of Alabama in Huntsville*
[‡]*NASA/Marshall Space Flight Center*
[§]*NASA/GSFC, Greenbelt, MD 20771 & JCA/UMBC, Baltimore, MD 21250*

Abstract. We present a systematic spectral analysis of 350 bright GRBs observed with BATSE, with high spectral and temporal resolution. Our sample was selected from the complete set of 2704 BATSE GRBs, and included 17 short GRBs. To obtain well-constrained spectral parameters, four different photon models were fitted and the spectral parameters that best represent each spectrum were statistically determined. A thorough analysis was performed on 350 time-integrated and 8459 time-resolved burst spectra. Using the results, we compared time-integrated and time-resolved spectral parameters, and also studied correlations among the parameters and their evolution within each burst. The resulting catalog is the most comprehensive study of spectral properties of GRB prompt emission to date, and provides constraints with exceptional statistics on particle acceleration and emission mechanisms in GRBs.

Keywords: Catalog – gamma-rays: bursts
PACS: 98.70.Rz

INTRODUCTION

Previous spectral studies of Gamma-Ray Burst (GRB) prompt emission have shown compelling evidence that a simple emission mechanism cannot entirely account for the observed spectra. However, spectral parameters that provide constraints on GRB emission mechanisms often depend on functional forms used to fit the data, as well as integration timescales of the spectra. A consistent, comprehensive spectral study of GRB prompt emission in large quantity is therefore crucial to unveiling the nature of GRBs.

A total of 2704 GRBs were observed with the Large Area Detectors (LADs) of the Burst and Transient Source Experiment (BATSE) on board the *Compton Gamma-Ray Observatory*, in the energy range of $\sim 30 - 1900$ keV. BATSE provided the largest database currently available of GRBs from a single experiment, with good spectral and temporal resolution. Many of the BATSE GRBs were bright enough for high-time resolution spectroscopy within the bursts as well as time-integrated spectroscopy. We present here a systematic spectral analysis of 350 bright BATSE GRBs. The sample size is more than twice that of the previous BATSE spectral catalog [1], which included only time-resolved spectroscopy. Our analysis included 350 time-integrated spectra and 8459 time-resolved spectra, and four different photon models were used to fit all the spectra. The resulting catalog is the largest and most comprehensive study to date of GRB prompt emission spectra.

THE ANALYSIS

To assure sufficiently good statistics, we selected the 350 brightest GRBs from the entire BATSE database, with either the peak photon flux (256 ms, 50 – 300 keV) > 10 photons s^{-1} cm^{-2} or the total energy fluence (30 – 2000 keV) > 2×10^{-5} ergs cm^{-2}.

One of eight BATSE modules consisted of a LAD and a Spectroscopy Detector (SD). In this work, we used only the LAD data, which provided higher sensitivity, to make the analysis more consistent throughout. In addition, a problem recently identified with the SD response matrices above ~ 3 MeV could render the SD data unreliable for spectral analysis at high energies [2]. The LAD data types used for the analysis are, in order of priority, High-Energy Resolution Burst (HERB), Medium Energy Resolution (MER), and Continuous (CONT) data [see 1, for BATSE data types]. All LADs were gain-stabilized with the usable energy range of $\sim 30 - 1900$ keV. We binned the spectra in time so that each resolved spectrum has signal > 45σ above background. An integrated spectrum is the sum of all the resolved spectra within the burst, and most integrated spectra cover the time range of T_{90}.

We used four photon models to fit each spectrum. Each photon model consists of two, three, four, and five free parameters, respectively in the order shown below.

1. **Power Law Model (PWRL)**
$$f_{\text{PWRL}}(E) = A \left(\frac{E}{100\text{keV}} \right)^{\lambda}$$

   Parameters: A = amplitude (photons s^{-1} cm^{-2} keV^{-1}) and λ = spectral index

2. **Comptonized Model (COMP)**
$$f_{\text{COMP}}(E) = A \left(\frac{E}{100\text{keV}} \right)^{\alpha} \exp\left(-\frac{E(2+\alpha)}{E_{\text{peak}}} \right)$$

   Parameters: A, α = low-energy spectral index, and $E_{\text{peak}} = \nu F_\nu$ peak energy (keV)

3. **GRB Model (BAND)** [3]
$$f_{\text{BAND}}(E) = \begin{cases} A \left(\frac{E}{100\text{keV}} \right)^{\alpha} \exp\left(-\frac{E(2+\alpha)}{E_{\text{peak}}} \right) & \text{if } E < (\alpha - \beta) \frac{E_{\text{peak}}}{2+\alpha} \\ A \left[\frac{(\alpha-\beta) E_{\text{peak}}}{(2+\alpha)100\text{keV}} \right]^{\alpha-\beta} \exp(\beta - \alpha) \left(\frac{E}{100\text{keV}} \right)^{\beta} & \text{if } E \geq (\alpha - \beta) \frac{E_{\text{peak}}}{2+\alpha} \end{cases}$$

   Parameters: A, α, β = high-energy spectral index, and E_{peak}

4. **Smoothly-Broken Power Law Model (SBPL)** [4, 5]
$$f_{\text{SBPL}}(E) = A \left(\frac{E}{100\text{keV}} \right)^{b} 10^{(a - a_{\text{piv}})}$$

   where $\quad a = m\Lambda \ln\left(\frac{e^q + e^{-q}}{2} \right) \quad a_{\text{piv}} = m\Lambda \ln\left(\frac{e^{q_{\text{piv}}} + e^{-q_{\text{piv}}}}{2} \right) \quad q = \frac{\log(E/E_b)}{\Lambda}$

   $q_{\text{piv}} = \frac{\log(100\text{keV}/E_b)}{\Lambda} \quad m = \frac{\lambda_2 - \lambda_1}{2} \quad b = \frac{\lambda_1 + \lambda_2}{2}$

   Parameters: A, λ_1, λ_2 = low- & high-energy spectral indices, E_b = spectral break energy (keV), and Λ = break scale (decades of energy)

THE RESULTS

We fitted the four photon models to each of the 350 integrated spectra and 8459 resolved spectra. From the overall performance of each model in fitting all spectra, determined by the χ^2 of the fits, we found that many spectra were fitted adequately by multiple photon models. We also confirmed that the low-energy index, α, of BAND and COMP models tends to be harder than the low-energy index λ_1 of SBPL.

Model Comparison within Each Spectrum. To find the model that best describes each spectrum (referred to as "BEST" here), we looked for significant improvements in χ^2 when fitted by more complicated models with more parameters, starting from the simplest PWRL model. When the significant improvements were found, we also required the additional parameters to be sufficiently constrained, to assure the parameters are meaningful and are indeed needed to represent the spectra. We found that the majority of the spectra required BAND or SBPL. COMP was considered BEST in much larger fraction of the resolved spectra than the integrated ones due to the existence of no-high-energy spectra within bursts as well as the lower signal-to-noise ratio. The distributions of the BEST spectral parameters for both integrated and resolved spectra are shown in Figure 1. To account for the difference between the low-energy indices of BAND, COMP, and SBPL, we use here the "effective α" [6, 7], the tangential slope in log scale at 25 keV, as the low-energy index where BEST is BAND or COMP. The effective α and λ_1 generally agree within 1σ. E_{peak} is the peak energy of the νF_ν spectrum, regardless of the photon model used.

Integrated vs. Resolved. For the comparison between the integrated and resolved parameter distributions (Figure 1), we calculated the Kolmogorov-Smirnov probabilities and statistics (P_{KS} and D_{KS}) for the null hypothesis of two distributions being consistent. We found a significant difference ($P_{\mathrm{KS}} \sim 10^{-16}$, $D_{\mathrm{KS}} = 0.23$) between the low-energy index distributions, and a moderate difference ($P_{\mathrm{KS}} \sim 10^{-2}$, $D_{\mathrm{KS}} = 0.10$) between their E_{peak} distributions. This is due to spectral evolution within bursts, commonly observed.

Spectral Parameter Evolution & Correlations. An example of the BEST spectral parameter evolution within a burst is shown in Figure 2. In this case, both the low-energy index and E_{peak} evolve from hard to soft, and the models with fewer parameters (PWRL,

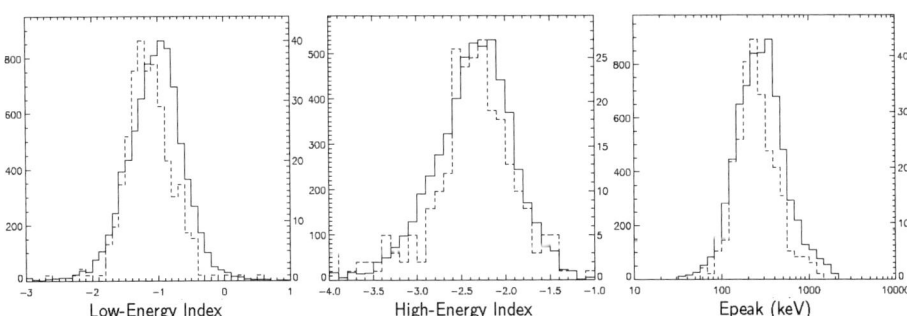

FIGURE 1. Comparison of BEST parameter distributions of integrated (dashed; right axis) and resolved spectra (solid; left axis). The last bins include values outside the edge values.

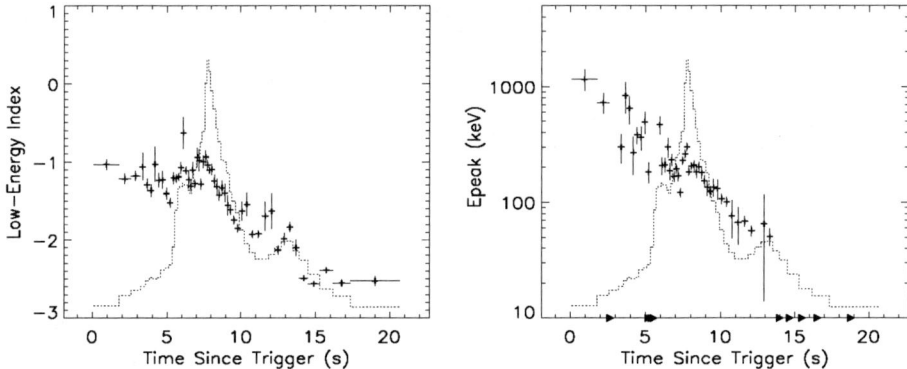

FIGURE 2. The BEST spectral parameter evolution of GRB 950403. Photon flux is overplotted as dotted lines. The arrowheads in E_{peak} plot indicate where the values cannot be determined.

COMP) are BEST at the tail as the burst spectra become softer. Within each burst, we also searched for the Spearman rank-order correlations among the BEST spectral parameters. The most significant ($> 3\sigma$) positive correlation was found between the E_{peak} and the low-energy index in the largest fraction of GRBs (26%).

Short GRBs. Our sample included 17 short GRBs ($T_{90} < 2$ s), three of which were bright enough for time-resolved spectral analysis. Within these three short GRBs, spectral evolution (either hard-to-soft or photon-flux tracking behavior) was clearly observed. However, we did not find any significant differences in the spectral parameters between the 17 short GRBs and long GRBs in our sample. A possible reason for this is the fact that only bright GRBs are considered here, which tend to be harder [8].

CONCLUSIONS

The GRB spectral database obtained in this work is derived from the most sensitive and largest database currently available. Therefore, these results set a standard for spectral properties of GRB prompt emission with exceptional statistics. Our results can provide reliable constraints for existing and future theoretical models of GRB emission and particle acceleration mechanisms.

REFERENCES

1. Preece, R.D., et al, *ApJS* **126**, 19 (2000).
2. Kaneko, Y., Ph.D. thesis, University of Alabama in Huntsville (2005).
3. Band, D.L., et al, *ApJ* **413**, 281 (1993).
4. Mallozzi, R.S., Preece, R.D., & Briggs, M.S., WINGSPAN, RMFIT (1994).
5. Ryde, F., *ApL&C* **39**, 281 (1999).
6. Preece, R.D., et al., *ApJ* **506**, L23 (1998).
7. Kaneko, Y. et al., *ApJS* (2006), submitted.
8. Mallozzi, R.S., et al., *ApJ* **454**, 597 (1995).

The "Supercritical Pile" Model of GRB: Spectra and Their Time Development

Demosthenes Kazanas*, Apostolos Mastichiadis[†] and Markos Georganopoulos**

*NASA/Goddard Space Flight Center, Greenbelt, MD 20771
[†]Dept. of Astronomy, University of Athens, Panepistimiopolis, Athens 15784, Greece
**Dept. of Physics, UMBC, 1000 Hilltop Circle, Baltimore MD 21250, also NASA/GSFC, Greenbelt, MD 20771

Abstract. The essence of the "Supercritical Pile" GRB model is a process for converting the energy stored in the relativistic protons of a Relativistic Blast Wave (RBW) of Lorentz factor Γ into electron – positron pairs of similar Lorentz factor. This is achieved by scattering the synchrotron radiation emitted by the RBW in an upstream located "mirror" and then re-intercepting it by the RBW. The repeated scatterings of radiation between the RBW and the "mirror", along with the threshold of the pair production reaction $p\gamma \rightarrow pe^-e^+$, lead to a maximum in the GRB spectral luminosity at an energy $E_p \simeq 1$ MeV, in general agreement with observation, independent of the value of Γ. Furthermore, the same threshold implies that the prompt γ–ray emission is only possible for Γ larger than a minimum value, thereby providing a "natural" account for the termination of this stage of the GRB as the RBW slows down. Within this model the γ–ray ($E \sim 100$ keV – 1 MeV) emission process is due to bulk Inverse Compton scattering and it is thus expected to be highly polarized if viewed at angles $\theta \simeq 1/\Gamma$ to the direction of the RBW. We have simulated the process of spectral formation by following the time development of the proton, electron and photon distribution functions. We present the results these calculations.

INTRODUCTION

The discovery of GRB afterglows by *BeppoSAX* and the ensuing determination of their redshifts [1], ushered a new era in GRB physics and in our understanding of their time development. While the issue of their distances and energetics was settled by these observations, a number of issues concerning the physics of the RBWs that give rise to the GRB phenomenon still remain open even though novel correlations obtained with the advent of observations [2] shifted the focus to other aspects of GRB physics. Chief among these still open issues is that of the dissipation of the RBW energy stored in relativistic protons (or magnetic fields). While the long energy-loss time scale of protons is useful for transporting the GRB energy to the distances required by observations, the same property makes the conversion of their energy into (the observed GRB) radiation exceedingly difficult. An altogether different, (and apparently) unrelated issue, is that of the narrow distribution of the GRB peak energy E_p [3], a rather serious problem in view of the strong dependence of this parameter on the RBW Lorentz factor ($E_p \propto \Gamma^4$) within the popular synchrotron model of the GRB emission in the γ–ray band.

The 'Supercritical Pile' model was conceived as a means for providing an answer to the first of the above questions, namely the internal energy dissipation in GRB, but it was found that the same physical considerations can provide also answer to the second open GRB issue, once one was willing to consider the implications of the model on the spectra of the resulting radiation.

"SUPERCRITICAL PILE": THE BASIC PHYSICS

The name of our model derives from the (at first sight incredulous) similarity between the RBW of a GRB and a nuclear pile. However, a closer inspection - albeit with our model in mind - reveals the following similarities: (a) They both contain a large amount of free energy, stored in relativistic protons in the GRB case and in nuclear binding energy in the case of the pile. (b) This energy can be released explosively - i.e. on the time scale of crossing the relevant size by photons or neutrons respectively, once certain criticality conditions are met. The process of energy transfer from protons to electrons is a *radiative instability*, studied first by [4], modified to include the effects of relativistic motion and the possibility of upstream reflection of the internally produced photons and their re-interception by the RBW as discussed in [5].

We assume the presence of a population of relativistic protons of form $n(\gamma_p) = n_0 \gamma_p^{-\beta}$ on the frame co-moving with the RBW, i.e. moving with Lorentz Factor (LF) Γ with respect to the observer (and also at zero angle to his/her line of sight; see [6] for details). We consider synchrotron photons from e^-e^+ pairs of energy γ_e, $\varepsilon_s = b\gamma_e^2$ (b is the value of the magnetic field in units of the critical value $B_c \simeq 4 \cdot 10^{13}$ G; all energies measured in units of the electron mass $m_e c^2$) reflecting off an upstream "mirror" and being re-intercepted by the RBW. Their energy will now be $\varepsilon'_s = \Gamma^2 b \gamma_e^2$. The threshold for e^-e^+ pair production of these photons with a proton of Lorentz factor equal to that of the electrons is $b\Gamma^2 \gamma_e^3 = 2$. To be conservative, we dispense with the notion of the presence of non-thermal particles and assume that the proton and electron distributions have a maximum LF equal to that of the RBW post-shock particle population, i.e. that $\gamma_e \simeq \Gamma$; this then leads to the following *kinematic threshold* condition for the instability

$$b\Gamma^5 \simeq 2 \quad \text{or} \quad \Gamma \gtrsim \left(\frac{2}{b}\right)^{1/5} \tag{1}$$

Assuming equipartition for the magnetic field, the above condition reads $\Gamma \gtrsim 250$, a value in agreement with that used in most models.

For the synchrotron-mirroring-pair production network to be self-sustained, at least one of the reflected synchrotron photons must pair-produce with the protons on the RBW (after its reflection by the "mirror") to replace the electron that produced it. Therefore the plasma optical depth τ to the pair producing reaction $p\gamma \to pe^-e^+$ must be at least as large as the inverse of the number of synchrotron photons produced by a given electron. An electron of energy γ produces $\mathcal{N}_\gamma \simeq \gamma/b\gamma^2 = 1/b\gamma$ photons, yielding $\tau = n_{\text{com}} \sigma_{p\gamma} \Delta_{\text{com}} \gtrsim 1/\mathcal{N}_\gamma$, where $n_{\text{com}}, \Delta_{\text{com}}$ are the comoving density and width of the RBW. Considering that the column density is a Lorentz invariant $n_{\text{com}} \Delta_{\text{com}} = nR$; then, taking into account the kinematic threshold relation (Eq. 1) the *dynamic threshold* condition reads

$$\sigma_{p\gamma}\Delta_{com} n_{com} = \sigma_{p\gamma} R n \gtrsim b\Gamma \quad \text{or} \quad \sigma_{p\gamma} R n \Gamma^4 \gtrsim 2 \ . \qquad (2)$$

This latter condition (and the physics behind it) are akin to those of those of a "supercritical" nuclear pile, hence the nomenclature of this model.

Assuming the width of the reflecting "mirror" to be thinner than the width of the RBW, blast waves with column densities higher than that implied by Eq. (2) will release the energy stored in relativistic protons explosively on times scales comparable to the RBW light crossing time scale; othewise, the duration of the burst will be comparable to the time it takes the RBW to cross the width of the "mirror". In this case, prominent emission is halted until the proton column has been built up significantly to conform to the dynamic threshold. For a RBW with Γ between those of Eqs. (2) and (1) and assuming the continuing presence of a "mirror", γ–ray emission will continue, as long as Γ is greater than that implied by the kinematic threshold (Eq. 1), in the *linear* regime, i.e. releasing its proton energy gradually over time as it accumulates. Eventually, when the value of the LF drops below this value, γ–ray emission stops and the RBW enters the stage of afterglow.

THE TIME-DEPENDENT GRB SPECTRA

Consider a RBW of Lorentz factor Γ. Because of the relativistic focusing of emitted radiation, we need only consider a section of the blast wave of opening half angle $\theta = 1/\Gamma$. The shocked electrons of the ambient medium (and pairs from the $p\gamma \to e^+e^-$ process) produce, as discussed above, synchrotron photons of energy $\varepsilon_s \simeq b\Gamma^2$. These, upon their scattering by the "mirror" and re-interception by the RBW, are boosted to energy $\varepsilon = \varepsilon_s \Gamma^2 = b\Gamma^4$ (in the RBW frame). These photons will then be scattered by the following electron populations: (a) By electrons of $\gamma \simeq 1$, originally contained in the RBW and/or cooled since the explosion. (b) By the hot ($\gamma \simeq \Gamma$), recently shocked electrons to produce inverse Compton (IC) radiation at energies correspondingly $\varepsilon_1 \simeq b\Gamma^4$ and $\varepsilon_2 \simeq b\Gamma^6$ at the RBW frame (the SSC process will also yield photons at $\varepsilon_{ssc} \simeq b\Gamma^4$, however it turns out that this is not as important and it is ignored here). At the lab frame, the energies of these three components, i.e. ε_s, ε_1, ε_2 will be higher by roughly a factor Γ, i.e. they will be respectively at energies $b\Gamma^3$, $b\Gamma^5$ and $b\Gamma^7$. Assuming that the process operates near its kinematic threshold, $b\Gamma^5 \simeq 2$, at the lab frame these components will occur at energies $\varepsilon_s \simeq \Gamma^{-2}$, $\varepsilon_1 \simeq 2 \simeq 1$ MeV and $\varepsilon_2 \simeq \Gamma^2 \simeq 10$ GeV $(\Gamma/100)^2$. This model therefore, produces "naturally" a component in the νF_ν spectral distribution which peaks in the correct energy range. It also predicts the existence of two additional components at an energies Γ^2 higher and $1/\Gamma^2$; such high energy emission has been observed from several GRBs [7].

We have modeled the process described above by following in detail the evolution of the proton, electron and photon distribution functions, taking into account for the latter the time-delay effects of scattering in the forward located mirror as well as the energy changes associated with this scattering. The interested reader can find the details in [8].

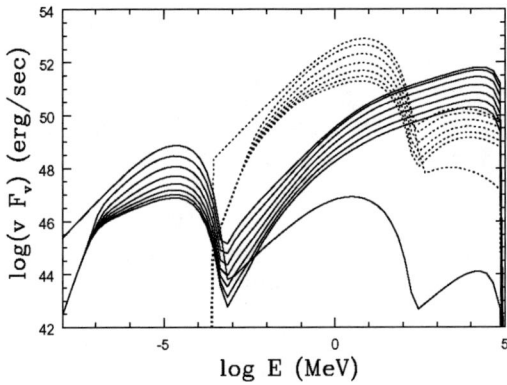

FIGURE 1. The evolution of the GRB spectra according to the model. The model parameters (on the RBW frame) are $B = 44$ G, $n_p = n_e = 10^3$ cm^{-3}, $\Gamma = 400$ and RBW radius $R = 3\,10^{16}$ cm.

These detailed calculations, performed assuming a constant value of Γ, by and large confirmed the qualitative arguments given above concerning the well defined components of the GRB spectra. Figure 1 presents an example of such a calculation which provides the prompt GRB spectra as a function of time as seen in the observer's frame. The three individual components attributed to synchrotron, inverse Compton and bulk-Compton components are clearly seen with each curve corresponding to a different time step, measured in units of $\Delta R/c$ where $\Delta R \simeq R/\Gamma^2$ is the width of the RBW and R is the size of the RBW. Since the calculations were performed with constant Γ and a given proton density the GRB emission continues until all the proton energy is radiated away.

The GRB emission detected by the detectors todate correcponds to the bulk-Comptonized component of the spectrum which occurs near 1 MeV which is appraent in the spectra. The synchrotron emission is much weaker and occurs at optical-UV frequencies. The optical prompt emission follows closely that of the γ−rays in disagreement with the observations of GRB 990123 but in agreement with those of GRB 041219a [9].

REFERENCES

1. Costa, E. et al. 1997, Nature, 387, 783
2. Amati, L. et al. 2002, A&A, 390, 81
3. Mallozzi, R. S. et al. 1995, ApJ, 454, 597
4. Kirk, J. G. & Mastichiadis, A., 1992, Nature, 360, 135
5. Kazanas, D. & Mastichiadis, A. 1999, ApJ, 518, L17
6. Kazanas, D., Georganopoulos, M. & Mastichiadis, A. 2002, ApJ, 578, L15
7. Dingus, B. L. 1995, Ap &Sp Sci, 231, 187
8. Mastichiadis, A. & Kazanas, D. 2006, ApJ (in press), also astro-ph/0512447
9. Vestrand, W. T. et al. 2005, Nature, 435, 178

GRB 050717: A Long, Short-Lag Burst Observed by Swift and Konus

H. A. Krimm[*,†], C. Hurkett[**], V. Pal'shin[‡], J. P. Norris[*], B. Zhang[§], S. D. Barthelmy[*], D. N. Burrows[¶], N. Gehrels[*], S. Golenetskii[‡], J. P. Osborne[**], A. M. Parsons[*] and M. Perri[‖]

[*]*NASA Goddard Space Flight Center, Greenbelt, Maryland, 20771, USA*
[†]*Universities Space Research Association, 10211 Wincopin Circle, Ste 500, Columbia, MD 21044*
[**]*Department of Physics and Astronomy, University of Leicester, Leicester, LE1 7RH, UK*
[‡]*Ioffe Physico-Technical Institute, Laboratory for Experimental Astrophysics, 26 Polytekhnicheskaya, St Petersburg 194021, Russian Federation*
[§]*Physics Department, University of Nevada, Las Vegas, NV 89154, USA*
[¶]*Department of Astronomy and Astrophysics, 525 Davey Lab., Pennsylvania State University, University Park, PA 16802, USA*
[‖]*ASI Science Data Center, Via Galileo Galilei, I-00044 Frascati, Italy*

Abstract. The long burst GRB 050717 was observed simultaneously by the Burst Alert Telescope (BAT) on Swift and the Konus instrument on Wind. Significant hard to soft spectral evolution was seen. Early gamma-ray and X-ray emission was detected by both BAT and the X-Ray Telescope (XRT) on Swift. The XRT continued to observe the burst for 7.1 days and detect it for 1.4 days. The X-ray light curve showed a classic decay pattern including evidence of the onset of the external shock emission at ~ 45 s after the trigger; the afterglow was too faint for a jet break to be detected. No optical, infrared or ultraviolet counterpart was discovered despite deep searches within 14 hours of the burst. The spectral lag for GRB 050717 was determined to be 2.5 ± 2.6 ms, consistent with zero and unusually short for a long burst. This lag measurement suggests that this burst has a high intrinsic luminosity and hence is at high redshift ($z > 2.7$). GRB 050717 provides a good example of classic prompt and afterglow behavior for a gamma-ray burst.

Keywords: gamma-ray bursts, gamma-ray telescopes
PACS: 98.70.Rz,95.55.Ka

INTRODUCTION

At 10:30:52.21 UT, 17 July, 2005, the Swift BAT triggered and located on-board GRB 050717 (BAT trigger 146372) [1]. The spacecraft began to slew to the source location 8.66 seconds after the trigger and was settled at the source location at T_0+63.46 s. This burst triggered Konus-Wind (K-W) at T_0(K-W) = 10:30:57.426 UT. The XRT began observing the burst at 10:32:11.49 UT. A fading source was discovered and the XRT continued to observe the source for 7.1 days. The Swift Ultra Violet/Optical Telescope (UVOT) observations [2] began at 10:32:10.7 UT, and no new source was detected within the XRT error circle in summed images in any of the six filters. GRB 050717 was not well positioned for follow-up observations. Consequently, no follow-up optical observations were made until more than 13 hours after the burst. In the several observations that were made after this time, no optical counterpart was detected.

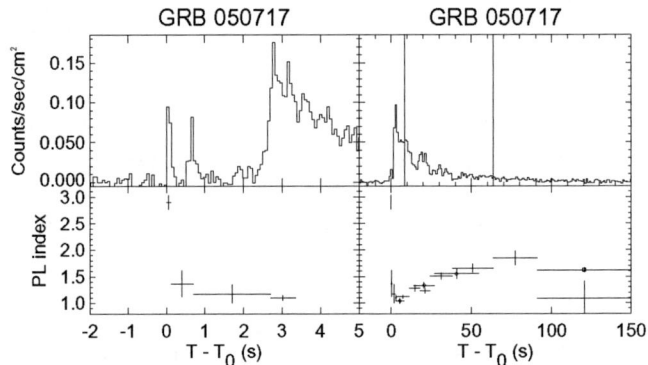

FIGURE 1. Background subtracted BAT light curves (top) and power-law fit indices (bottom). The panels on the right show the full duration of the prompt emission; those on the left show the precursor peaks more clearly. The start and end of the slew to the target are shown by vertical lines. Separate fits were made to each time interval indicated. Key: BAT data (plain symbols), joint BAT and Wind data (open diamonds), joint BAT and XRT data (open square). For the leftmost BAT/Wind point, the index α of the cut-off power law fit is shown. For the other fit points, the photon index from a power-law fit is shown.

LIGHT CURVES AND SPECTROSCOPY

Spectroscopy

The BAT triggered on the first of two short spikes that preceded the main emission (left-hand panels of Fig 1). These small precursors were followed by the main pulse, which displayed the common fast rise, exponential decay profile, rising within 450 ms then decaying with an average exponential decay constant $1.82^{+0.13}_{-0.11}$ (right-hand panels). On top of this slow decay, there were at least four other peaks. The duration T_{90} (15–350 keV) is 86 ± 2 s. The total fluence in this band is $(1.40 \pm 0.03) \times 10^{-5}$ erg cm^{-2}. The BAT data were binned into 11 time bins to track the spectral evolution of the prompt emission (lower panels of Fig 1). Starting with the main peak, there is clear evidence of spectral softening as the burst progresses until T_0+91 s, when the spectrum hardens.

Emission is seen in Konus-Wind up to $\sim$6 MeV. The spectrum of the main peak ($T_0+2.843 - 8.219$ s) is well fitted (20 keV–10 MeV) by a power law model with an exponential cutoff. In a joint fit with BAT, $\alpha = 1.04 \pm 0.05$ and $E_{peak} = 2401^{+781}_{-568}$ keV ($\chi^2 = 117$; 143 d.o.f.). The total fluence in (20 keV – 6 MeV) is $6.5^{+0.9}_{-2.2} \times 10^{-5}$ erg cm^{-2}.

The XRT spectrum from $T_0 + 91 - 310$ s has an average photon index of 1.65 ± 0.11, with an indication of an excess absorption (above galactic) of $2.75 \pm 0.57 \times 10^{21}$ cm^{-2}. The mean unabsorbed flux at 201 s is $5.76 \pm 0.31 \times 10^{-10}$ erg cm^{-2} s^{-1} (0.3-10.0 keV).

Post-Burst Emission

The γ-ray and X-ray light curve is shown in Fig 2. The light curve shows several features which can be interpreted based on models discussed in Zhang et al. [3]. First there is a

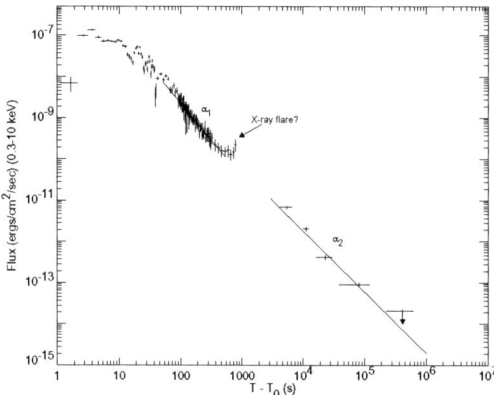

FIGURE 2. The combined BAT and XRT light curve. Points in the BAT light curve have been extrapolated from the BAT 15–150 keV energy band to the XRT 0.3–10 keV band and corrected for differences in the effective area. The broken power law fit to the X-ray light curve decay is also shown.

smooth transition from the prompt BAT emission into the early X-ray emission and a fairly steep decay (power law index α_1 below) until $T_0 > \sim 200$ s, followed by a possible superimposed X-ray flare. Observations resumed with a shallower power law index α_2.

Several models were fit to the data. First, was a broken power law. This gave a power law index $\alpha_1 = 2.10^{+0.17}_{-0.05}$, a break time of 203 ± 26 s and an index $\alpha_2 = 1.48 \pm 0.02$ (χ^2/dof = 159/111). If this can be interpreted as the curvature effect [3, 4, 5], the index should be $\alpha = 2 + \beta$, where β=0.62 is the *energy* index of the spectrum. Thus we should have α_1 = 2.62, as compared to the observed value of 2.10. Alternatively, the decay curve could be a superposition of two separate decay power laws, one steep due to the curvature component, and one shallow due to the forward shock component. A fit to a superposition model gave $\alpha_1 = 3.01^{+0.55}_{-0.23}$ and $\alpha_2 = 1.43 \pm 0.04$ (χ^2/dof = 161/110). Here the steep index is interpreted as the decay of the tail of the internal shock emission, which is superimposed on an underlying afterglow component.

An alternative model allows the time of the start of the afterglow to be a free parameter: $F(t) = A \times (t - t_0)^{-\alpha}$. This gave $t_0 = 28.8 \pm 2.4$ s and $\alpha = 1.44 \pm 0.02$ (χ^2/dof = 171.7/111) This decay index is consistent with the late time decay index in the other two fits and gives further credence to the idea that we are seeing the influence of the external shock throughout the afterglow light curve.

The shallow decay continues until the flux becomes unobservable to the XRT. Since there is no apparent break to a steeper decay in the light curve, the lower limit on a jet-break time is $t_b > 1.4$ days.

SPECTRAL LAG

It is possible to derive an estimate of the spectral lag of the BAT data between Ch 1 (15-25 keV) and Ch 4 (100–350 keV). The spectral lag was derived for the main peak of emission (from $T_0+2.26$ s $-$ 5.8 s). The lag was found to be $2.5^{+2.9}_{-2.4}$ ms. Hence the measured lag is statistically consistent with zero. Such a low value for lag is quite unique for a long burst since Norris [7] has shown that the dynamic range of lags for long bursts spans $\sim$ 25 ms to $\sim$ 300 ms. In fact out of the 90 brightest bursts studied by Norris [7], only 2% show a lag as small as that of GRB 050717.

One can use the lag, the measured peak flux ($1.69 \pm 0.16 \times 10^{-6}$ ergs cm^{-2} s^{-1} (15–350 keV; $T_0+2.752 - 3.008$ s), and E_{peak} to set lower limits on the distance to burst. Using the $+2\sigma$ limit on the lag (8.3 ms) and the methodology of Norris, Marani, & Bonnell [6] and Norris [7], one derives a redshift of 2.7 and a peak luminosity of 3.9×10^{53} ergs s^{-1} (15–350 keV). Since smaller values of spectral lag would lead to larger redshifts, this value, z = 2.7, can be considered the 2σ lower limit on the redshift. Such a large redshift is consistent with the non-detection of an optical or infrared counterpart to the afterglow and with the non-detection of a jet break.

A consistent interpretation of such a small lag is that the high energy emission from GRB 050717 has been redshifted downward more than usual into the BAT energy range. It has been shown [8, 9] that the high energy component of burst emission shows narrower peaks and more variation than is seen at lower energies. Shifting such spiky peaks into the BAT range would cause the measured lag to be smaller than what would be observed in long bursts at lower redshifts.

CONCLUSION

The long gamma-ray burst GRB 050717 shows a number of features which can be interpreted in light of the predominant models of bursts and their afterglows. The short spectral lag tells us that this burst is at a high redshift ($z > 2.7$) and hence has a large intrinsic luminosity ($L_{peak} > 8.7 \times 10^{53}$ erg s^{-1}). The features observed in the burst are likely representative of spiky high energy features red-shifted to the BAT energy range.

GRB 050717 also demonstrates many of the features of the unified picture of the late time evolution of GRB emission [3, 10]. During the early X-ray emission of GRB 050717, the decay index is somewhat less steep than would be expected if it were due solely to the tail emission of the prompt GRB. This can be interpreted as a superposition of tail and external shock emission.

REFERENCES

1. Hurkett, C. et al, GCN Circ 3633 (2005).
2. Blustin, A. et al, 2005, GCN Circ 3638.
3. Zhang, B. et al., astro-ph/0508321 (2005).
4. Dermer, C. D., *ApJ*, **614**, 284 (2004).
5. Kumar, P. & Panaitescu, A., *ApJ*, **541**, L51 (2000).
6. Norris, J.P., Marani, G.F., & Bonnell, J.T., *ApJ*, **534**, 248, (2000).
7. Norris, J.P., *ApJ*, **579**, 386, (2002).
8. Norris, J.P., et al , *ApJ*, **459**, 393, (1996).
9. Fenimore, E.E. & Bloom, J. S., *ApJ*, **453**, 25, (1995).
10. Nousek, J. A. et al.,*ApJ*, in press (astroph/0508332), (2005).

Correlative Analysis of GRBs detected by Swift, Konus and HETE

H. A. Krimm[*,†], S. D. Barthelmy[*], N. Gehrels[*], D. Hullinger[**], T. Sakamoto[‡], T. Donaghy[§], D.Q. Lamb[§], V. Pal'shin[¶], S. Golenetskii[¶], G. R. Ricker[‖], J-L Atteia[††], N. Kawai[‡‡] and G. Sato[§§,¶¶]

[*]*NASA Goddard Space Flight Center, Greenbelt, Maryland, 20771, USA*
[†]*Universities Space Research Association, 10211 Wincopin Circle, Ste 500, Columbia, MD 21044*
[**]*University of Maryland, College Park, MD, USA*
[‡]*National Research Council, Washington, DC, USA*
[§]*University of Chicago, Chicago, IL USA*
[¶]*Ioffe Physico-Technical Institute, Laboratory for Experimental Astrophysics, 26 Polytekhnicheskaya, St Petersburg 194021, Russian Federation*
[‖]*Massachusetts Institute of Technology, Cambridge, MA USA,*
[††]*Observatoire Midi-Pyrenees, Toulouse, France*
[‡‡]*Tokyo Institute of Technology, Tokyo, Japan*
[§§]*Institute of Space and Astronautical Science, JAXA, Kanagawa, Japan*
[¶¶]*University of Tokyo, Tokyo, Japan*

Abstract. Swift has now detected a large enough sample of gamma-ray bursts (GRBs) to allow correlation studies of burst parameters. Such studies of earlier data sets have yielded important results leading to further understanding of burst parameters and classifications. This work focusses on seventeen Swift bursts that have also been detected either by Konus-Wind or HETE-II, providing high energy spectra and fits to E_{peak}. Eight of these bursts have spectroscopic redshifts and for others we can estimate redshifts using the variability/luminosity relationship. We can also compare E_{peak} with E_{iso}, and for those bursts for which a jet break was observed in the afterglow we can derive Eg and test the relationship between E_{peak} and E_γ. For all bursts we can derive durations and hardness ratios from the prompt emission.

Keywords: gamma-ray bursts, gamma-ray telescopes
PACS: 98.70.Rz,95.55.Ka

METHODOLOGY

For the bursts used in this study (Table 1) we derived E_{peak}, E_{iso} and L_{iso} by the following method. First, for each burst we carried out a joint fit to the BAT and either the Konus or HETE time-averaged spectral data, depending which other instrument detected the burst. Fits were made to both the GRBM (Band) model and a power law model with an exponential cut-off. If the high energy photon index β was constrained in the Band model fit, then we used the Band model parameters (α, β, E_{peak}) in our further analysis; if there was no constraint on β, then we used the cut-off power law parameters (α, E_{peak}) and set β=-10. In either case, E_{peak} was converted to the source frame.

Next we derived a spectrum for the peak 1 s (L_{iso} calculation) and total burst (E_{iso} calculation) which we fit to the GRBM model. To allow direct comparison with the results of [1] for L_{iso}, the 1-s peak flux was defined in the energy range 30-10,000 keV.

TABLE 1. GRBs detected by Swift/BAT as well as either Konus-Wind (K) or HETE (H)

GRB	z	α	β	E_{peak} (keV)	$E_{iso}(10^{52}$ erg)	$L_{iso}(10^{52}$ erg/s)
041223 K	–	-0.83 ± 0.05	–	$337.6^{+27.7}_{-24.2}$	–	–
050215B H	–	-1.29 ± 0.75	–	28.6 ± 9.4	–	–
050219B K	–	-1.05 ± 0.07	–	162.8 ± 12	–	–
050326 K	–	-0.78 ± 0.16	$-2.48^{+0.24}_{-0.63}$	$206.4^{+32.8}_{-27.1}$	–	–
050401 K	2.9	-0.90 ± 0.30	$-2.55^{+0.22}_{-0.44}$	117.5 ± 18	41.2 ± 4.0	14.9 ± 1.1
050416B K	–	$-1.19^{+0.66}_{-0.40}$	–	$115.5^{+157}_{-43.4}$	–	–
050418 K	–	$-1.19^{+0.69}_{-0.37}$	–	55.2 ± 12.5	–	–
050525A K	0.606	-1.01 ± 0.11	$-3.26^{+0.23}_{-0.41}$	81.2 ± 2.3	2.3 ± 0.09	7.75 ± 0.75
050603 K	2.821	-1.03 ± 0.11	$-2.03^{+0.17}_{-0.29}$	343.7 ± 87	61.8 ± 3.1	67.8 ± 3.3
050713 K	–	-1.32 ± 0.06	–	445^{+129}_{-87}	–	–
050717 K	–	-1.04 ± 0.05	–	2400^{+780}_{-570}	–	–
050820A K	2.612	-1.25 ± 0.15	–	246.0^{+127}_{-66}	80	2.70 ± 0.16
050824 H	0.83	–	–	< 12.7	$0.179^{+6.8}_{-0.083}$	–
050904 K	6.29	-1.11 ± 0.10	–	436^{+335}_{-151}	140	10.6 ± 1.1
050922C H	2.198	-0.95 ± 0.11	–	196.8^{+64}_{-37}	4.64 ± 0.12	5.3 ± 0.37
051008 K	–	-1.07 ± 0.03	–	888 ± 669	–	–
051109A K	2.346	-1.38 ± 0.33	–	139.5^{+116}_{-45}	2.4	2.80 ± 0.40

For comparison with the results of [2] and [3], the total fluence was over 1-10,000 keV. We then used a simple cosmological model (H_0 = 70 km/s, Ω_m = 0.3 and Ω_L = 0.7) to derive E_{iso} and L_{iso} from the calculated flux.

COMPARISON TO PUBLISHED RELATIONS

Yonetoku Relation

For seven of the Swift/Konus/HETE bursts in the study set, we have a measurement of both E_{peak} and a spectroscopic redshift. For these bursts we can compare the parameters derived in this work to the results presented by Yonetoku et al. [1], who compare E_{peak} in the source frame to the peak luminosity of the burst. These data are shown in the left-hand plot of Figure 1.

We can see that while our results do not cluster as tightly as the data points Yonetoku et al. [1] use, all of the points save for GRB 050904 (see below) lie in or very near the error range of these results. Since three of the points are above the line and four below, there does not appear to be a systematic error in our calculations of either E_{peak} or L_{iso}. Note that GRB 050820A is plotted twice. Since the BAT data was cut off before the largest peak, when L_{iso} is derived from the BAT data alone (lower point), L_{iso} is underestimated. We used the Konus light curve which includes both peaks to derive a scale factor; this provides the upper point, which is closer to the relation.

The most extreme outlier is GRB 050904, the highest redshift burst seen by Swift (z=6.29). Though it had a high rest frame E_{peak}, it was also fairly low luminosity. Thus we urge caution in interpreting the Yonetoku et al. [1] relation for high redshift GRBs.

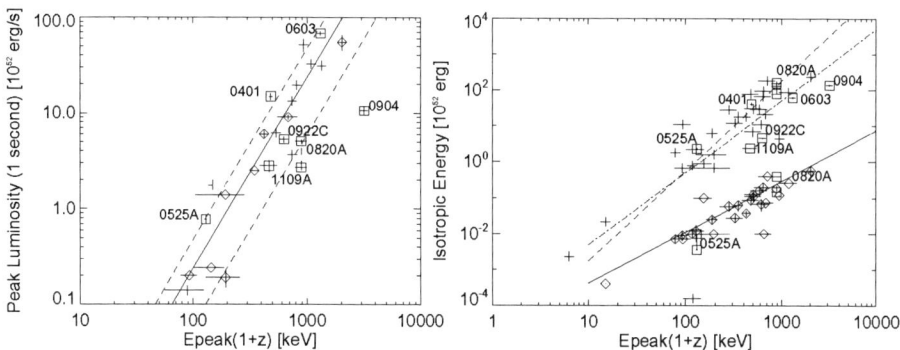

FIGURE 1. Comparison to the Yonetoku, Ghirlanda and Amati relations. **Left:** Yonetoku. Crosses: BATSE bursts; diamonds: BeppoSAX bursts; squares: Swift bursts. Swift bursts are identified by the month and day (all 2005) of the burst. The solid line shows the relationship derived by Yonetoku et al. [1]: $\log(L_{iso}) = 2.0\log(E_{peak}(src)) - 4.63$ and the dashed lines use the $\pm$ 1s uncertainties in the intercept, keeping the slope constant. **Right:** Ghirlanda/Amati. Comparison to the Ghirlanda and Amati relations. Data from [3]: crosses are E_{iso}; diamonds are E_γ The BAT/Konus/HETE points are shown as squares. The solid line shows the relationship derived by [3]: $\log(E_\gamma) = 1.42\log(E_{peak}) - 4.80$. The dashed line is the fit that [3] derived between E_{iso} and E_{peak} and the dot-dash line is the relationship derived by [2].

Amati and Ghirlanda Relations

For the same set of bursts discussed above we can compare the parameters derived in this work to the results presented in earlier analyses. Amati et al. [2] compare E_{peak} in the source frame to the total isotropic luminosity of the burst, assuming isotropic emission (E_{iso}). Ghirlanda, Ghisellini & Lazzati [3], confirm this relation and derive a separate, but related relationship which assumes beamed emission, $E_\gamma = (1 - \cos\theta)E_{iso}$, where θ is the jet collimation angle. The data derived from [3] are shown in the right-hand panel of Fig 1.

We are able to confirm the E_{peak}-E_{iso} ([2]) relation with our data. Again there is no evidence for a systematic error in our calculations. There are however, several outliers. As discussed above, correcting for the missing flux from GRB 050820A brings this data point very close to the parameterization in [3]. We note that GRB 050904 is also below the relations. Since we are integrating the entire burst, time dilation should not have an effect and we see that our point is closer to the relation than it is in the left-hand panel (Yonetoku). However, we still may not be properly accounting for all of the flux from this burst. There were several spectrally hard X-ray flares at later times that may be part of the prompt emission [5]. Thus E_{iso} for this burst should be treated as a lower limit. We have no explanation for the other two outliers, GRB 050922C and GRB 051109A.

Ghirlanda, Ghisellini & Lazzati [3] et al found a tighter relationship between E_{peak} and E_γ (E_{iso} corrected for the jet opening angle). Of the seven Swift/Konus/HETE bursts studied, only two, GRB 050525A and GRB 050820A, have a verified jet break. For GRB 050525A, Blustin et al. [6] reports two possible jet break times and corresponding opening angles: $t_j \sim 0.15$ d (3.2°) and $t_j \sim 0.6$ d (5.4°). These are both indicated in the

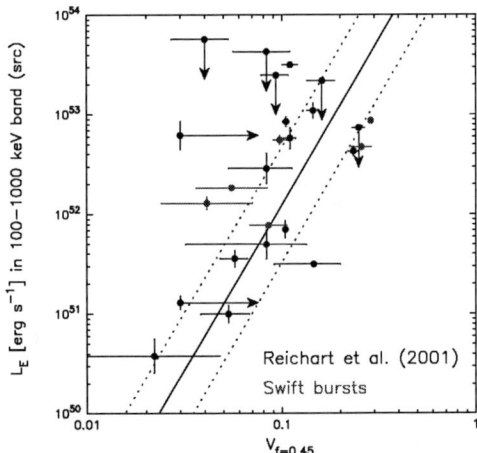

FIGURE 2. Comparison to the relation in [4]. Blue: Reichart's original points; red: Swift/BAT points.

figure, the higher point represents the larger opening angle and is more consistent with the points from [3]. The data for GRB 050820A is from Osborne et al. [7] and the two points are due to the scaling discussed above.

Reichart V-L Relation

Reichart et al. [4] derived a possible luminosity estimator for GRBs based on the variability of their light curves. We have repeated their analysis here for those GRBs in the study sample which had high enough SNR for the variability to be estimated. Of the seven sample bursts with known redshifts, we were able to derive variability for all save for GRB 050904. The six remaining points are plotted in Figure 2. The variability shown is the weighted average of BAT channels 2, 3, and 4 (whenever the SNR in channel 4 is large enough). These correspond to energy bands (25-50, 50-100 and 100-350 keV). This is close, although not perfect match to the energy ranges used by Reichart et al. [4]. We see in the figure that there is a reasonable match between the Swift data points and the original points. As in the E_{peak} relations discussed above, there is no sign of a systematic error in our calculations.

REFERENCES

1. Yonetoku, D., et al, *ApJ*, **609**, 935, (2004).
2. Amati, L., et al., *A & A*, **390**, 81, (2002).
3. Ghirlanda, G, Ghisellini, G., & Lazzati, D., *ApJ*, **616**, 331, (2004).
4. Reichart, D.E., et al, *ApJ*, **552**, 57, (2001).
5. Cummings, J. et al, these proceedings (2006).
6. Blustin, A. J., in press astro-ph/0507515 (2006).
7. Osborne, J. et al, these proceedings (2006).

An improved redshift indicator for Gamma-Ray Bursts, based on the prompt emission

A. Pélangeon*, J-L. Atteia*, D. Q. Lamb[†], G. R. Ricker** and the HETE-2 Science Team[‡]

*Laboratoire d'Astrophysique, Observatoire Midi-Pyrénées, 31400 Toulouse, France
[†]Department of Astronomy & Astrophysics, University of Chicago, Chicago, IL 60637, USA
**Center for Space Research, Massachussetts Institute of Technology, Cambridge, MA 02139, USA
[‡]An international collaboration of institutions in USA, France, Japan, Italy, Brazil and India

Abstract. We propose an improved version of the redshift indicator developed by Atteia [1], which gets rid of the dependence on the burst duration and provides better estimates for high-redshift GRBs. We present first this redshift indicator, then its calibration with HETE-GRBs with known redshifts. We also provide an estimation of the redshift for 59 bursts, and we finally discuss the redshift distribution of HETE-bursts and the possible other applications of this redshift indicator.

Keywords: gamma-rays:bursts
PACS: 95.85.Pw - 98.70.Rz

DESCRIPTION

In 2003, Atteia proposed $X_0 = N_\gamma/(E_{peak} \times \sqrt{T_{90}})$ as a possible redshift estimator [1], based on the $E_{peak} - E_{iso}$ correlation [2, 3] linking E_{peak}, the intrinsic peak energy of the νf_ν spectrum, and E_{iso}, the isotropic energy radiated by the source in its rest frame. In addition, Yonetoku et al. have shown that the $E_{peak} - L_{iso}$ correlation was less dispersed than the $E_{peak} - E_{iso}$ correlation (2004) [4].
The definition of our new redshift indicator is partly based on these two relations and is written as : $X = n_{15}/e_p$, where e_p is the observed peak energy and n_{15} the observed bolometric luminosity in units of photons and in the 15 sec. long interval containing the highest fluence. Thus, all the burst spectra are now done on the same duration in the observer frame.
To compute this estimator, the burst spectra are fit with a Band model [5] which gives us the spectral parameters : α, β, E_0 and the fluence in the energy range $[E_1 - E_2]$ of the detector. Then, as described in the paper of Atteia [1], the theoretical evolution of X with the redshift is computed for a "standard" GRB ($\alpha = -1$, $\beta = -2.3$, $E_0 = 250\ keV$), considering a "standard" cosmology ($\Omega_m = 0.3$, $H_0 = 65\ km.s^{-1}.Mpc^{-1}$, flat universe), and finally the estimation of the redshift is deduced by comparison between the value of X obtained for the GRB based on its spectral parameters, and the theoretical evolution of X. In the following we call this estimation *new pseudo-redshift* (hereafter *npz*).
In addition, errors on pseudo-redshifts are computed : considering first the errors on the spectral parameters obtained with the fit, 1000 values of X are simulated, then 1000 *npz* associated are also calculated, and errors on *npz* are so derived (the errors presented hereafter are at 90% confidence level).

CALIBRATION

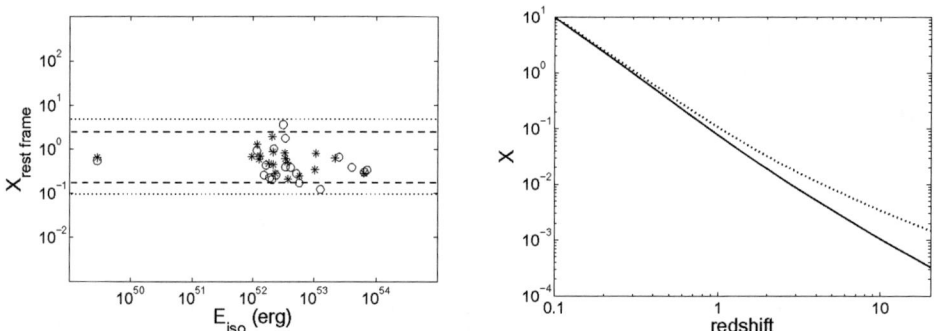

FIGURE 1. Left panel : intrinsic dispersion of the two redshift indicators : $X_0 = n_\gamma/(e_p \times \sqrt{t_{90}})$ *(circles)* and $X = n_{15}/e_p$ *(stars)* for 19 GRBs with spectroscopic redshift. Right panel : theoretical evolution of the two estimators *(dotted line for the previous indicator, solid line for the new one)* between $z = 0$ and $z = 20$.

A good redshift indicator must essentially satisfy two criteria : its independence on the bursts intrinsic characteristics, and its high dependence on the redshift.

Figure 1 *(left panel)* shows that whereas E_{iso} is extended on about 5 decades, the intrinsic dispersion of the new quantity X *(stars)* is only 1 decade against nearly 1.5 decade for the previous redshift indicator *(circles)*. In addition, the plot on the right panel shows the higher dependence of the new redshift indicator *(solid curve)* with the redshift, where the difference between the two estimators becomes significant for $z > 1$.

The redshift indicator is currently calibrated with 17 GRBs detected by *HETE-2* which have a spectroscopic redshift, and 2 additional GRBs (*050525* and *050603*) detected by *Konus-Wind* [6], [7]. The table on the figure 2 presents the results of the *npz* obtained with errors. We can notice *(right panel)* that the redshift estimate is always better than a factor 2 *(dashed lines)*, except for *GRB051022* which has a factor 2.15 and a small

GRB	npz	z	GRB	npz	z
010921	0.58 ± 0.35	0.45	030528	0.64 ± 0.1	0.78
020124	1.77 ± 1.35	3.2	040924	0.82 ± 0.7	0.86
020813	1.19 ± 0.1	1.25	041006	0.68 ± 0.7	0.72
020903	0.32 ± 0.3	0.25	050408	0.70 ± 0.6	1.23
021004	2.53 ± 1.45	2.33	050525	0.70 ± 0.1	0.61
021211	1.17 ± 0.9	1.01	050603	2.73 ± 0.3	2.81
030115	1.57 ± 1.2	2.2	050922C	2.63 ± 1.6	2.19
030226	2.9 ± 1.55	1.99	051022	1.72 ± 0.2	0.8
030323	3.15 ± 1.65	3.37	970508	1.11 ± 1	0.835
030328	1.75 ± 1.3	1.52	980326	1.19 ± 1.1	1
030329	0.22 ± 0.05	0.17	990712	0.46 ± 0.55	0.43
030429	2.34 ± 1.4	2.65	991216	0.65 ± 0.55	1.02

FIGURE 2. Left panel: redshift estimates for 24 bursts with known spectroscopic redshifts. Right panel: ratio npz/z for 20 bursts with known redshifts.

error *(see the arrow)*. We have yet to understand why this burst seems to be an outlier. Without this last GRB, the standard deviation is : $\sigma = 0.11\ dex$. We have also computed the pseudo-redshift for 4 other bursts referenced in the literature [3, 8] : *GRB970508, 980326, 990712* and *991216*, which have a duration similar to the one used ($\sim 15sec.$) in the derivation of the npz.

STUDY WITH THE NPZ

A sample of bursts without spectroscopic redshift

Taking into account all the long GRBs detected by the *FREGATE* instrument (6-400 keV) on-board of the satellite *HETE-2* which don't have spectroscopic redshift, we have computed a redshift-estimate for a sample of 34 bursts which had enough statistics to be correctly fit with *FREGATE* data. The results are given in the table presented in figure 3. The GRB redshift distribution is probably biased because of the small fraction of bursts which have a measure of their redshift. We tried to determine if this fact is confirmed for the *HETE-2* bursts, and what could be this distribution if we had a higher fraction of GRBs with known redshifts. Thus, we considered the redshift distribution of 3 groups of bursts.

The first group is composed of 19 *HETE*-bursts with redshift. This group has a cut in its redshift distribution at z = 3.3 *(figure 3, dotted line)*. Nevertheless, this cut seems to disappear and for the second group composed of 53 *HETE*-GRBs with redshift or pseudo-redshift *(solid line in figure 3)*, the cumulative distribution of this group is fully compatible with the group of 22 *SWIFT*-GRBs with measured redshift *(figure 3, dashed line)*. This result tends to show that the redshift-distribution of *HETE*-GRBs is biased at high-redshift.

Finally, we note that the sample studied contains few high-redshift GRBs : 4 GRBs only have a redshift higher than z = 4 (*GRB010612, 030913, 031026* and *051008*).

GRB	npz	GRB	npz
010612	5.25 ± 2.2	031109A	0.94 ± 0.2
010629	0.91 ± 0.9	031111A	2.14 ± 0.55
010928	3.64 ± 1.4	031203	2.17 ± 1.45
020127	2.21 ± 1.5	031220	1.53 ± 1.15
020305	1.98 ± 1.45	040319	1.79 ± 1.2
020331	2.21 ± 1.5	040423	1.26 ± 1
020418	1.4 ± 1	040425	2.23 ± 1.35
020801	1.21 ± 1	040511	1.83 ± 1.25
020812	3.48 ± 1.75	040709	1 ± 0.8
020819	1.21 ± 0.9	040912A	0.33 ± 0.35
021014	3.9 ± 1.9	040912B	2.94 ± 1.6
021016	2.8 ± 1.6	041016	3.49 ± 1.75
021104	1.22 ± 1.1	041211B	3.29 ± 1.6
030418	3.07 ± 1.7	050209	2.93 ± 1.6
030725	0.89 ± 0.3	051008	5.23 ± 2.2
030823	0.84 ± 0.7	051021	1.37 ± 1.2
030913	6.04 ± 2.7	051028	3.66 ± 1.8
031026	6.67 ± 2.9		

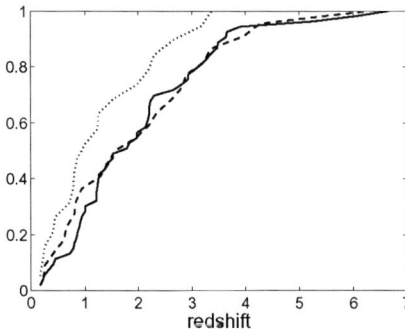

FIGURE 3. Left panel : redshift estimates for 35 bursts without redshift (sample of 34 *HETE*-bursts and *GRB051008* [9]). Right panel : cumulative distribution functions for *HETE*-bursts with redshift *(dotted line)*, *SWIFT*-bursts with redshift *(dashed line)* and a sample of 53 *HETE*-GRBs with redshift or pseudo-redshift *(solid line)*.

Comments on the *npz* and possible applications

Considering the most intense part of the GRBs seems a good way to improve the pseudo-redshift based on the prompt emission. Indeed, if we consider some examples such as *GRB010612* and *GRB031026*, they were previously found at $\hat{z} = 9.5$ and 14 [10], compared to the *npz* which now gives respectively 5.3 and 6.7, values probably closer to the real redshift.

Moreover, it has solved the problem of *multi-peaks* GRBs in which the background was taken into account in the determination of t_{90}, which had for consequence a biased value of X_0. For example, *GRB020305* had an estimation of 5.88 [10], and is now found at 1.98 ± 1.45, in agreement with the spectroscopic constraints ($z \leq 2.8$, [11]).

Finally, if we consider recent determinations of spectroscopic redshift for old bursts, we can notice that *GRB030528* ($z = 0.782$, [12]) has a close *npz* of 0.64. For *GRB020819*, we find a value of $npz = 1.21$, not close to the real redshift ($z = 0.41$, [13]), but the error (± 0.9) which is also large makes the estimation compatible with the true redshift.

The development of the pseudo-redshift finds several possible applications.

As the pseudo-redshift are rapidly computed, they can tell us very quickly whether the burst is at low or high-redshift, which permits to choose the appropriate way of observation.

Other applications of pseudo-redshift could be the verification of the validity of the $E_p - E_{iso}$ relation found by Amati [3] for a large sample of GRBs [14].

Finally, having a large sample of GRBs with redshift (or estimation) should let us study some of their cosmological aspects such as their luminosity function or the evolution of their rate with the redshift.

REFERENCES

1. J-L. Atteia, *A&A*, **407**, L1–L4 (2003).
2. N. M. Lloyd, V. Petrosian, and R. S. Mallozzi, *ApJ*, **534**, 227–238 (2000).
3. L. Amati, F. Frontera, M. Tavani, et al., *A&A*, **390**, 81–89 (2002).
4. D. Yonetoku, T. Murakami, T. Nakamura, et al., *ApJ*, **609**, 935–951 (2004).
5. D. Band, J. Matteson, L. Ford, et al., *ApJ*, **413**, 281–292 (1993).
6. S. Golenetskii, R. Aptekar, E. Mazets, et al., *GCNC* **3474 & 3660** (2005).
7. S. Golenetskii, R. Aptekar, E. Mazets, et al., *GCNC* **3518** (2005).
8. G. Ghirlanda, G. Ghisellini, and D. Lazzati, *ApJ*, **616**, 331–338 (2004).
9. S. Golenetskii, R. Aptekar, E. Mazets, et al., *GCNC* **4078** (2005).
10. J-L. Atteia, G. R. Ricker, D. Q. Lamb, et al., *in AIP Conf. Proc. 727: Gamma-Ray Bursts: 30 Years of Discovery*, pp. 37–41 (2004).
11. J. Gorosabel, J. P. U. Fynbo, A. Fruchter, et al., *A&A*, **437**, 411–418 (2005).
12. A. Rau, M. Salvato, and J. Greiner, *A&A*, **444**, 425–430 (2005).
13. P. Jakobsson, D. A. Frail, D. B. Fox, et al., *ApJ*, **629**, 45–51 (2005).
14. G. Pizzichini, P. Ferrero, M. Genghini, et al., *astro-ph/0503264* (2005).

Numerical Simulation of 2-D Relativistic Hydrodynamics Using Adaptive Mesh Refinement Technique

Kyujin Kwak and F. Douglas Swesty

Department of Physics and Astronomy, Stony Brook University, Stony Brook, NY 11790

Abstract. We have developed a code to numerically solve 2-D special relativistic hydrodynamics equations based on the LCPFCT algorithm. We implemented our code using PARAMESH, the Parallel AMR library. We present the details of our algorithm and exhibit its performance on a series of verification tests. Our code will be combined with 2-D transport code to build a code for 2-D special relativistic radiation hydrodynamics. The prompt and/or afterglow emission of Gamma-ray Bursts(GRBs) will be investigated in detail with the numerical simulation of relativistic radiation hydrodynamics.

INTRODUCTION

Gamma-ray Bursts (GRBs) are involved in highly relativistic fluid with Lorentz factors 100 – 1000 for the prompt emission and lower for the afterglows. Radiation from highly relativistic fluid and interaction between radiation and fluid has not been studied very well yet. As a part of IBEAM project [1], we have been developing a relativistic radiation hydrodynamics code in order to study the radiation process of GRBs in detail. In this paper, we present a relativistic hydrodynamics code we have developed recently. The future plans to build a radiation hydrodynamics code are given in the last section.

Our 2-D relativistic hydrodynamics code is based upon the Flux Corrected Transport (FCT) algorithm, which was first developed by Boris and Book [2, 3] for the Newtonian hydrodynamics. For the 2-D relativistic code, we modified LCPFCT [4], the latest, 1-D, version of FCT and used Zalesak's multi-dimensional limiter [5] to extend to 2-D. In order to use Adaptive Mesh Refinement (AMR) technique, we implemented PARAMESH, the Parallel AMR library, [6, 7] into our 2-D relativistic hydrodynamics code. In this paper, we present the basic equations for our code and two test problems. More details of finite differencing of the basic equations and implementation of PARAMESH will be given elsewhere [8].

2-D SPECIAL RELATIVISTIC HYDRODYNAMICS EQUATIONS

The 2-D special relativistic hydrodynamics equations are obtained by mass, momentum and energy conservation. In Cartesian coordinates, they are

$$\frac{\partial D}{\partial t} = -\frac{\partial (Dv_x)}{\partial x} - \frac{\partial (Dv_y)}{\partial y}, \quad (1)$$

$$\frac{\partial S_x}{\partial t} = -\frac{\partial(S_x v_x)}{\partial x} - \frac{\partial(S_x v_y)}{\partial y} - \frac{\partial p}{\partial x}, \quad (2)$$

$$\frac{\partial S_y}{\partial t} = -\frac{\partial(S_y v_x)}{\partial x} - \frac{\partial(S_y v_y)}{\partial y} - \frac{\partial p}{\partial y}, \quad (3)$$

$$\frac{\partial E}{\partial t} = -\frac{\partial(E v_x)}{\partial x} - \frac{\partial(p v_x)}{\partial x} - \frac{\partial(E v_y)}{\partial y} - \frac{\partial(p v_y)}{\partial y}, \quad (4)$$

where the variables are defined as $D \equiv \rho W$, $S_x \equiv W^2 \rho h v_x$, $S_y \equiv W^2 \rho h v_y$, $E \equiv W^2 \rho h - p$, $W \equiv \frac{1}{\sqrt{1-v_x^2-v_y^2}}$, and $\rho h \equiv \rho + \rho \varepsilon + p$. ε is the thermal energy per unit mass, time and velocity are measured in the lab frame and the speed of light is set to unity. The equation of state $p = p(\rho, \varepsilon)$ is combined with the above equations. For an ideal gas, $p = (\Gamma - 1) \rho \varepsilon$ with an adiabatic index of Γ.

Note that the relativistic equations (1)-(4) in their conservative form have exactly the same form as the Newtonian ones, but with different definition of the variables. In the Newtonian equations, the variables are defined as $D \equiv \rho$, $S_x \equiv \rho v_x$, $S_y \equiv \rho v_y$, and $E \equiv \rho \varepsilon + \frac{1}{2}\rho(v_x^2 + v_y^2)$.

The key difference between the relativistic and Newtonian equations lies in the fact that the conversion between the lab frame variables $\{D, S_x, S_y, E\}$ and the rest frame ones $\{\rho, v_x, v_x, p\}$ requires one to solve the nonlinear equations together with the equation of state.

TEST PROBLEMS

2-D Sedov-Taylor Point Explosion

Sedov [9] and Taylor [10] independently found the analytic solution of the Newtonian point explosion problem in 1-D Cartesian, 2-D cylindrical and 3-D spherical symmetry. The analytic solution is determined by three initial parameters, explosion energy E_0, ambient density ρ_1 and time after the explosion t. In our test run, $E_0 = 15$, $\rho_1 = 0.1$, and $t = 30$. Pressure of ambient matter is 10^{-10}. The adiabatic index Γ is $5/3$ with an ideal gas equation of state. The number of initial grids is 64×64 over the domain $x \in [0, 80]$ and $y \in [0, 80]$ and two more refinements are allowed according to density variation.

The result of our test run is shown in Figure 1 with analytic solution. The explosion is located at the center of domain and the x-axis is distance from the center. We can see that the numerical results are reasonably well matched with analytic solution. The location of shock wave is relatively accurate. The reason for the spread-out of the data points behind shock front is that the initial explosion is not a perfect point. We set up the initial explosion within four square boxes around the initial explosion point.

Relativistic Shock Tube

The analytic solution for the 1-D relativistic shock tube problem is well known and has often been used to test numerical relativistic hydrodynamics codes. We test our 2-D

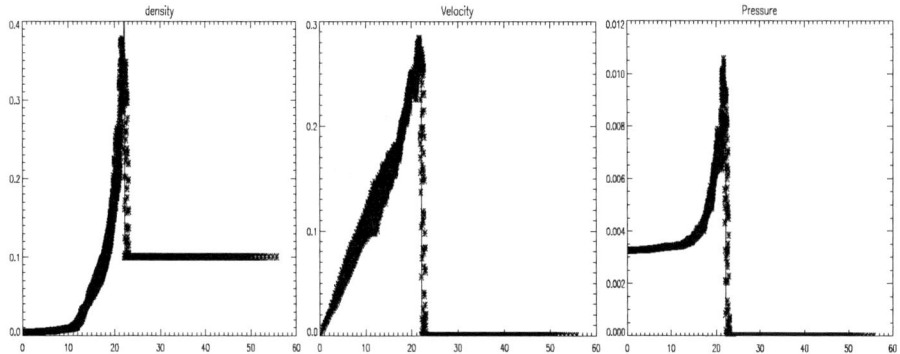

FIGURE 1. 2-D Sedov-Taylor Point Explosion

relativistic code against the 1-D analytic solution. In order to obtain the 1-D result out of the 2-D code, we set up the initial conditions in a way that they keep the symmetry of problem in one direction. The grids of the problem are spanned over $x \in [0,80]$ and $y \in [0,80]$ and the initial discontinuity is located at $x = 40$. At the left side $x < 40$, $\rho_L = 10, v_{x,L} = v_{y,L} = 0$, $p_L = 13\frac{1}{3}$ and at the right side $x > 40$, $\rho_R = 1.0$, $v_{x,R} = v_{y,R} = 0$, $p_R = \frac{2}{3} \times 10^{-6}$. The adiabatic index Γ is 5/3 and the running time is 36. This initial condition is the same as the second test problem of Schneider et al. [11]. Time step is controlled by $\Delta t = 0.4 \min[\Delta x, \Delta y]$. The initial number of grids is 32×32 and two more refinements are allowed according to density variation.

In Figure 2, we show the result of our test run. Each data point corresponds to the overlapped data along the y-direction. Overall, the numerical result is well matched to the analytic solution. The discrepancy of rarefaction region on the left side of each plot disappears with the increased number of initial grids. AMR puts more grids in the region of shock front and contact discontinuity thus giving better resolution. The discrepancy of the density profile behind shock front is also shown at the level of this spatial resolution in 1-D relativistic code of Schneider et al. [11]. The features of our 2-D result are similar to those of Schneider et al.'s 1-D code [11], such as the overshooting of pressure and density which corresponds to the undershooting of velocity at the rarefaction tail, the bump of the density profile behind the contact discontinuity and the undershooting of the density jump between shock front and contact discontinuity.

FUTURE PLANS

Swesty has already developed an implicit code to numerically solve the Boltzmann transport equation. The code is also implemented with PARAMESH in 2-D Cartesian coordinate. The movies of test problems are given in Swesty's website [12]. Our 2-D hydrodynamics code will be combined with transport code to build a relativistic radiation hydrodynamics code. At present, the implicit transport code solves only the lab frame transport equation. The comoving frame transport equation can also be solved by the current implicit transport code with some modification.

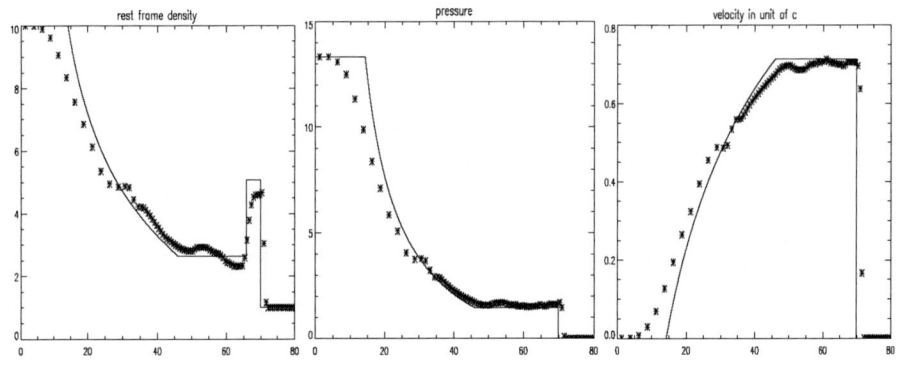

FIGURE 2. Relativistic Shock Tube

ACKNOWLEDGMENTS

This work was supported by NASA Cooperative Agreement No. NCC 5-615.

REFERENCES

1. http://www.ibeam.org
2. Boris, J. P., & Book, D. L. 1973, J. Comput. Phy., 11, 38
3. Boris, J. P., & Book, D. L. 1976, J. Comput. Phy., 20, 397
4. Boris, J. P., Landsberg, A. M., Oran, E. S., & Gardner, J. H. 1993, Naval Research Laboratory Report No. 6410-93-7192, also available from the website http://www.lcp.nrl.mil/lcpfct.
5. Zalesak, S. T. 1979, J. Comput. Phys., 31, 335
6. MacNeice, P., Olson, K. M., Mobarry, C., deFainchtein, R., & Packer, C. 2000, Comput. Phys. Commun., 126, 330
7. http://ct.gsfc.nasa.gov/paramesh/Users_manual/amr.html
8. Kwak, K., & Swesty, F. D. in preparation
9. Sedov, L. I. 1959, Similarity and Dimensional Methods in Mechanics (New York:Academic)
10. Taylor, G. 1950, in Proc. R. Soc. Lond. A., 201, 159
11. Schneider, V., Katscher, U., Rischke, D. H., Waldhauser, B., Maruhn, J. A., & Munz, C. -D. 1993, J. Comput. Phy., 105, 92
12. http://www.ess.sunysb.edu/dswesty/

PIC Simulations of Prompt GRB Emissions

E. Liang, K. Noguchi and S. Sugiyama

Rice University, Houston, TX 77005-1892

Abstract. We review PIC simulation results of GRB emissions for both the Poynting flux and shock scenarios.

INTRODUCTION

An outstanding problem in γ-ray bursts (GRBs) is the particle acceleration and radiation mechanisms, which are so efficient that they convert most of the primary energy from the central engine into γ-rays. While there is increasing evidence that at

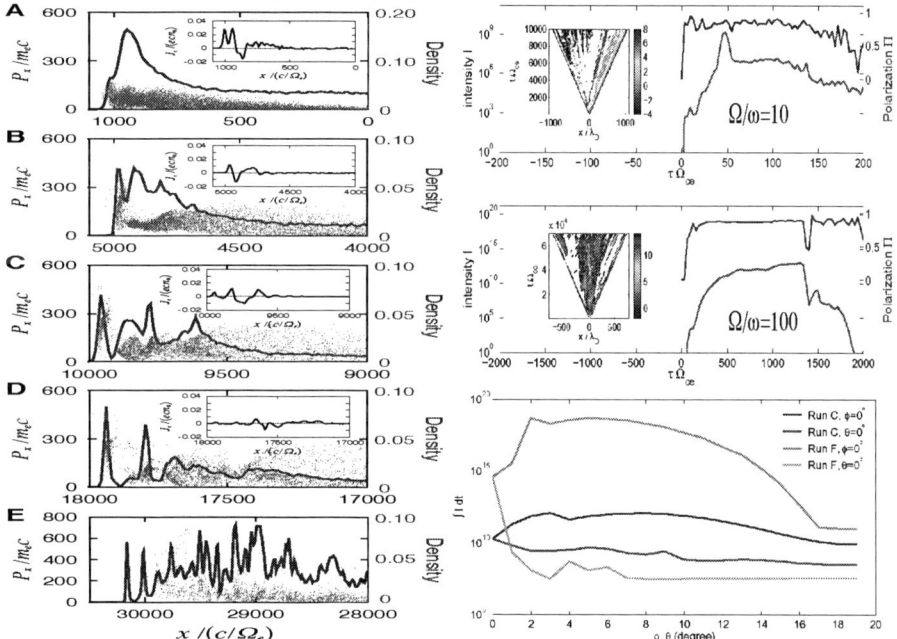

FIGURE 1 (left) PIC simulated PF accelerated e+e- slab showing repeated bifurcation of pulse profile (curves) and momentum profiles (dots) which resemble GRB light curves and hardness evolution [7].
FIGURE 2 (right) Using ray-tracing we compute the radiation intensity and polarization of a PF accelerated e+e- slab as functions of detector time (upper and middle panels) and angular fluence as function of view-angles (lower panel). Higher B field (middle panel) gives longer detected pulse [9].

least the classical long GRBs are related to the deaths of massive stars, there is yet little consensus on the primary energy source, or how it is converted into γ-rays. Two popular paradigms are relativistic hydrodynamic bulk flows (HBF) [1] or electromagnetic-dominated outflows (Poynting flux, PF) [2], driven by the formation of a new-born black hole or neutron star [3]. The former could be caused by disk accretion or a neutrino wind, while the latter could be caused by strongly magnetized accretion or ms magnetar wind. In both paradigms the difficulty is to find robust physical mechanisms which efficiently convert electromagnetic (EM) or bulk flow (BF) energy into the ultra-relativistic internal energy of a small number of nonthermal particles. A popular model of the HBF paradigm is the internal/external synchrotron shock model, where unsteady BF leads to dissipation via internal shocks in the prompt phase and snowplowing of the CSM/ISM leads to external shock emission in the afterglow phase. In shock models the popular mechanism is Fermi acceleration at the shock fronts followed by synchrotron or "jitter" radiation [4], though Comptonization [4] has also been invoked. Magnetic turbulence generated by Weibel instability is usually invoked in shock models [5]. In the PFA paradigm, particle acceleration and radiation are driven directly by large-scale ordered EM fields via collective plasma processes [2]. In both the HBF and PFA paradigms, the physics of relativistic particle acceleration and γ-radiation ultimately involve electromagnetic (EM) and plasma kinetic processes, which can only be correctly simulated using Particle-in-Cell (PIC) codes [6]. Such simulations have recently been pioneered by several groups [5,7]. Our group, working with LLNL and LANL, has developed and adapted the most advanced PIC codes for GRB simulation, including the first 3D PIC code with self-consistent radiation damping [8,9], and applied them to compute radiation outputs of both the HBF and PF scenarios [8,9]. Sample results are given in Fig.1-2.

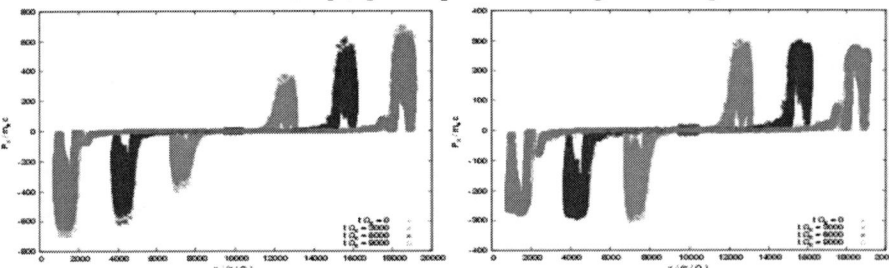

FIGURE 3 Evolution of $\Omega_e/\omega_{pe}=100$ PFA phase plots including Comptonization of external blackbody soft photons. Left panel is for low photon density and right panel is for high photon density, showing the rapid decline of the maximum Lorentz factors (Sugiyama these proceedings).

However, classical radiation damping using the Dirac term does not include Compton drag by external photons, which must be computed with Monte Carlo (MC) methods. We has recently implemented an approximate treatment of inverse Compton loss in the NN code by adding a time-averaged damping force $\mathbf{f}_D = -4\sigma_T g^2 \mathbf{v} U_s/3c$, where σ_T= Thomson cross-section and U_s=soft photon energy density, to represent the net effect of scattering by many isotropic soft photons [4]. Fig.3 gives sample results.

INTERACTION WITH AMBIENT PLASMA

Astrophysically it is important to study the interaction of PFAs with the ambient environment, both to see how the ambient plasmas damp the acceleration and absorb the Poynting flux, and how the heated ambient plasma radiate. Fig.4 shows sample

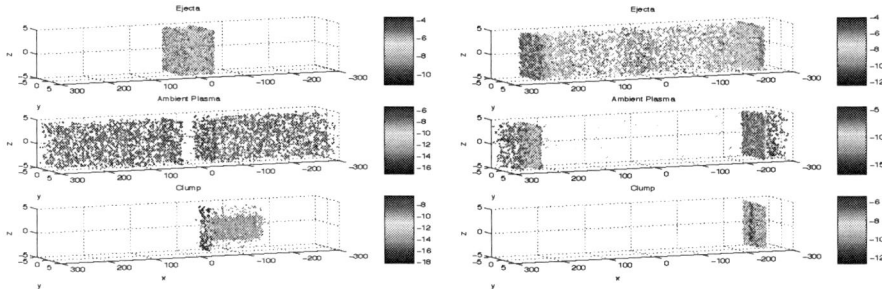

FIGURE 4 3D evolution of internal shocks when a warm e+e- ejecta slab with Lorentz factor of 100 at the front collides with e+e- plasmas upstream. The plasma on the left half of the grid is uniform but the plasma on the right has an additional high density "clump" to illustrate the effects of inhomogeneous shock. We plot 1/10000 of the particles and color code each particle with their instantaneous radiation power from shock acceleration. This example shows that the shock is not disrupted by inhomogeneity, and the transition layer is much thicker than gyroradii expected from MHD theory. We find that peak radiation power per particle is roughly the same for all three species: ejecta, upstream plasma and clump (from [8]).

results of 3D runs. Some of our major findings include: (a) the maximum Lorentz factors achieved even with ambient plasma are comparable to the vacuum case of Fig.1, though the number of high-energy particles decreases with increasing ambient density. (b) a complex multi-phase plasma exists in the contact region where high-γ ejecta plasma overruns low-γ "swept-up" plasma. The swept-up plasma consists of 2 phases: a cold unaccelerated phase coexisting with a PF-accelerated phase whose Lorentz factors < those of the ejecta. In the e-ion case, both the "swept-up" electrons and ions exist in 2 distinct phases. The transition region is very broad (>>> the ion gyroradii, Fig.5), contrary to MHD results. (c) the swept-up ion Lorentz factors are << the electron Lorentz factors because ions are pulled only by the charge separation electric fields, while leptons are accelerated by the EM pulse directly. This disproves the conventional MHD assumption that the electron and ion bulk Lorentz factors are the same and the transition layer thickness is ~ ion gyroradius. (d) there is no evidence of any disruptive plasma instability (Weibel, 2-stream, and lower-hybrid-drift instabilities) at the contact interface. Such instabilities are suppressed to first order by the strong transverse EM field. (e) the Poynting flux decays via ponderomotive acceleration of the ejecta and ambient electrons, plus mode conversion to longitudinal plasma waves which are absorbed by Landau damping. Fig.6 shows that PFA preferentially accelerates the e+e- component in a mixture of e-ion and e+e- plasma.

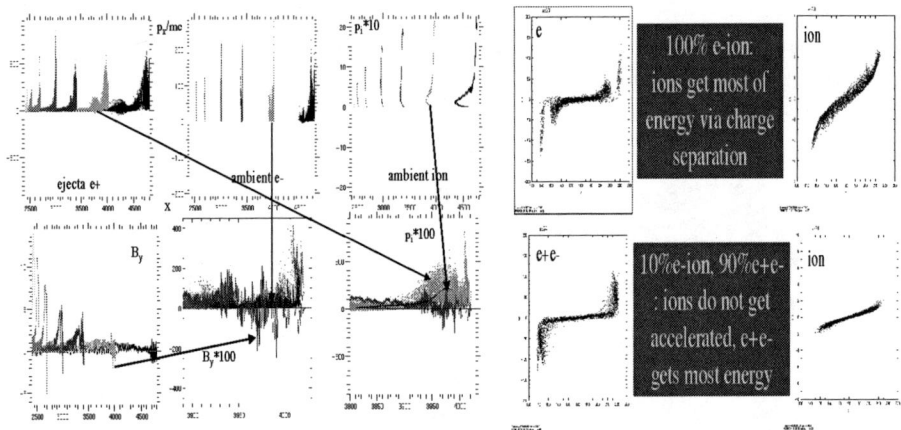

FIGURE 5 (left) Time-lapse phase plots and B-field profiles for a slab e+e- 2.5D PFA running into (0.001x ejecta density) cold e-ion ambient plasma. P_i is in units of $m_i c$. The acceleration of ejecta plasma stalls after it has swept up roughly equal mass of ambient plasma but the acceleration of ambient ions continues due to the pull by charge separation Ex. EM field decays by acceleration of swept-up plasma and mode conversion. Lower right panels show the blow-up details of the transition region [8].
FIGURE 6 (r) PF acceleration of pure e-ion plasma compared to mixture of e-ion and e+e- plasma [8].

ACKNOWLEDGMENTS

This work was partially supported by NASA NAG 5-9223 and NSF AST0406882

REFERENCES

1. T. Piran, Phys. Rep. **333**, 529 (2000); P. Meszaros, Ann. Rev. Ast. Astrophys. 40, 137 (2002); G. Fishman & C.A. Meegan, Ann. Rev. Ast. Astrophys. 33, 415 (1995); G. Fishman, in *Compton Gamma-Ray Observatory*, M. Friedlander, M. *et al.* Ed. (Conf. Proc. 280, Am. Inst. Phys., New York, 1993), p.669; R.D. Preece *et al.* Astrophys. J. Supp. 126, 19 (2000).
2. See for example M.V. Smolsky and V.V. Usov, Astrophys. J. 531, 764 (2000); M.H.P.M. Van Putten and A. Levinson, Astrophys. J. 584, 937 (2003); A. Levinson and M.H.P.M. Van Putten, Astrophys. J. 488, 69 (1997); M. Lyutikov and R. Blandford, Proc.1st N. Bohr Summer Inst., ed. R.Ouyed et al (2003, astroph020671); M. Lyutikov and E.G. Blackman, MNRAS 321, 177 (2002); E.G. Blackman, Proc.1st N. Bohr Summer Inst., ed. R.Ouyed et al (2003, astroph0211187).
3. S. Woosley et al., in AIP Conf. Proc. 727. 343 ed. E. Fenimore & M. Galassi (AIP 2004).
4. G. Rybicki & A.P. Lightman, *Radiative Processes in Astrophysics* (Wiley, New York, 1979); Medvedev, M. ApJ in press (2004); BAPS 50, 113 (2005).
5. Silva, L.O. et al. ApJ 596, L121 (2003); Nishikawa, K, et al. ApJ, 595, 555 (2003).
6. C. K. Birdsall and A. B. Langdon, *Plasma Physics via Computer Simulation*, (IOP, Bristol, UK 1991); K. Nishimura, S.P. Gary and H. Li, J. Geophys. Res. 107, 1375 (2002); A. B. Langdon, Comp. Phys. Comm. 70, 447 (1992); A.B. Langdon, and B. Lasinski, Meth. Comp. Phys.16, 327, ed. B. Alder et al. (1976)
7. E. Liang *et al.* Phys. Rev. Lett. 90, 085001 (2003); E. Liang and K. Nishimura, PRL 92, 175005 (2004); 28. K. Nishimura and E. Liang, Phys. Plasmas 11 (10) 4753 (2004); 29. K. Nishimura, E. Liang, and S.P. Gary, Phys. Plasmas 10(11) 4559 (2003).
8. E. Liang,, Astrophys. Sp. Sci. 298, 211 (2005); Rev. Mex. AA 23, 43 (2005).
9. K. Noguchi et al, Il Nuovo Cimento C28, 381 (2005); SLAC-R-752 eConf: C041213 (2005).
10. W. Coburn and S.E. Boggs, Nature 423, 415 (2003).

Gamma-ray bursts and giant lightning discharges in protoplanetary systems

E. Winston*, S. McBreen[†], B. McBreen* and L. Hanlon*

*School of Physics, University College, Dublin 4, Ireland
[†]Astrophysics Missions Division, Research Scientific Support Department of ESA, ESTEC, Noordwijk, The Netherlands

Abstract. Lightning in the solar nebula is considered to be one of the probable sources for producing the chondrules that are found in meteorites. Gamma-ray bursts (GRBs) provide a large flux of γ-rays that Compton scatter and create a charge separation in the gas because the electrons are displaced from the positive ions. The electric field easily exceeds the breakdown value of ≈ 1 V m^{-1} over distances of order 0.1 AU. The energy in a giant lightning discharge exceeds a terrestrial lightning flash by a factor of $\sim 10^{12}$. The predicted post-burst emission of γ-rays from a burst lasts for days and is more probable than the GRB because the radiation is beamed into a larger solid angle. The total amount of chondrules produced is in reasonable agreement with the observations of meteorites. Furthermore in the case of GRBs most chondrules were produced in a few major melting events by nearby GRBs and lightning occurred at effectively the same time over the whole nebula, and provide accurate time markers to the formation of chondrules and evolution of the solar nebula.

Keywords: <gamma-ray sources; gamma-ray bursts>
PACS: < 98.70.Rz>

INTRODUCTION

Chondrules are typically millimetre-sized stony spherules that constitute the major component of most chondritic meteorites. They appear to have crystallised rapidly from free floating molten or partially molten drops but the heat source responsible for melting chondrules remains uncertain. Proposed processes have been recently reviewed by [1] and [2]. One of the earliest proposals for the formation of chondrules is lightning caused by turbulence in the solar nebula [3, 4] but this model is not favoured for chondrule production for a number of reasons [eg 5]. If lightning were the mechanism responsible for chondrule formation, it would have to operate on a large scale comparable in size with the whole nebula [6].

Almost all proposed heat sources are local to the solar nebula. One exception is that the precursor grains near the surface of the nebula were melted when they efficiently absorbed X-rays and γ-rays from a nearby GRB [7]. This mechanism can produce a large amount of chondrules in the nebula (~ 30 Earth masses) but it has a low probability of occurrence ($< 0.1\%$). X-ray melting of material has recently been demonstrated in the laboratory for the first time using a powerful synchrotron [8, 9]. The model of McBreen et al. (1999) [7] did not include lightning caused by Compton scattering of γ-rays by the gas in the nebula or the γ-rays from post-burst emission. The effect of this new process is to induce giant lightning discharges over the whole nebula [10].

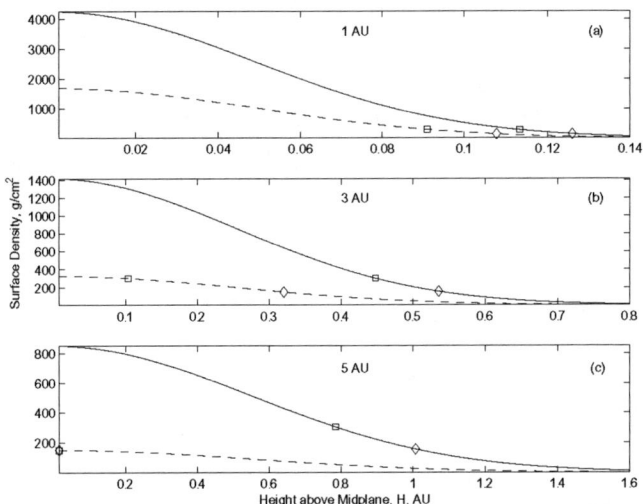

FIGURE 1. The surface density, Σ, as a function of height above the midplane, H, in astronomical units (AUs) for two models with normalising values, Σ_0 and exponent n, with $\Sigma_0 = 4250$ g cm^{-2} and n = -1.0 (solid line) and $\Sigma_0 = 1700$ g cm^{-2} and n = -1.5 (dashed line) and for three radial distances of (a) 1 AU, (b) 3 AU and (c) 5 AU. The diamonds and squares give path lengths of 150 and 300 g cm^{-2} measured from the top of the nebula.

ABSORPTION OF γ-RAYS IN THE NEBULA

The absorption of γ-rays in a gas with solar abundance is dominated by molecular hydrogen and it is sufficiently accurate here to consider only the absorption by H_2. Compton scattering is the dominant process in H_2 up to 70 MeV For γ-rays with energy above a few MeV, the kinematics of Compton scattering are such that most of the energy is taken by the electron, which is scattered in the forward direction. An incoming pulse of γ-rays in H_2 is gradually transformed into electrons that move further into the nebula leaving a cloud of positive charge in its wake.

We adopt the power law relationship for the surface density Σ of the solar nebula as a function of radial distance r from the Sun given by $\Sigma = \Sigma_0 [r/1AU]^n$ where Σ_0 has the normalising value of 4250 g cm^{-2} at 1.0 AU and exponent n is -1.0 [11]. The choice of Σ_0 and n varies between the models of the solar nebula. Hayashi et al. [12] proposed Σ_0 = 1700 and n = -1.5 in the minimum-mass model of the solar nebula and these values are consistent with minimum-mass models of extrasolar nebula [13]. In conventional models it is usually assumed that lightning occurs in the dusty midplane of the nebula. [eg 14]. However a GRB will preferentially interact with the outer region of the nebula that is in the direction of the GRB source. Therefore it is necessary to model the vertical structure of the nebula to obtain the surface density perpendicular to the midplane.

The vertical structure of the disk is given by $\rho = \rho_0 \exp(-z^2/H^2)$ where z is the vertical distance from the midplane where ρ_0 and H are the density at the midplane and scale height of the nebula. The surface density profile perpendicular to the midplane is

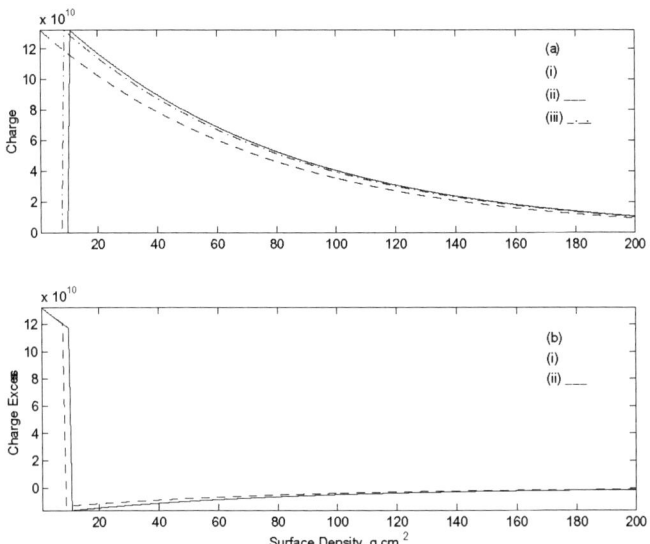

FIGURE 2. (a) The number of γ-rays that Compton scatter and the positive ions produced (i) as a function of path length in H_2 for an incident flux of 10^{13} photons m^{-2}. The number of electrons (ii) as a function of path length taking into account the ionization loss of 10 g cm^{-2} (solid line) and also including a retarding electric field of 1 V m^{-1} (dashed line). (b) The charge excess as a function of path length for the ionization loss (solid line) and including the retarding electric field (dashed line). The model used is for r = 3 AU and Σ_0 = 4250 g cm^{-2} with n = -1.0.

plotted in Fig. 1 for r = 1, 3 and 5 AU using the two models. Approaching the nebula from above the midplane, a path length of 300 g cm^{-2} will reach the midplane at 5 AU, and 0.1 AU above the midplane at 3 AU using the minimum mass model (Fig. 1c and Fig. 1b).

PROMPT GRB EMISSION AND CHARGE SEPARATION

To model the charge separation and electric field in the nebula, we adopt a GRB at 10 kpc that gives 10^6 erg cm^{-2} with 10% of the energy between 20 MeV and 100 MeV, which is assumed to be in 65 MeV γ-rays that yield Compton scattered electrons with energy of 50 MeV. The incident flux of γ-rays is $\approx 10^{13}$ photons m^{-2} and is attenuated exponentially in H_2 using a mass absorption co-efficient of $\mu = 1.33 \times 10^{-2}$ cm^2 g^{-1} (Fig. 2a (i)). The positive charge also declines exponentially in the same way because each Compton event creates a positive ion. The electrons travel on average an additional 10 g cm^{-2} into the nebula including only the ionization loss. The path length is further reduced by a retarding electric field of 1 V m^{-1} which is typical of the breakdown value (Fig. 2a (ii)). The charge excess is the difference between the two distributions (Fig. 2b). The net positive charge is confined to a layer of thickness 10 g cm^{-2} whereas the negative charge is distributed over a much wider range.

Electrical breakdown occurs when a normally insulating gas suddenly becomes conducting in a strong electric field. Breakdown occurs when the electric field is strong enough that a free electron accumulates ~ 1 eV of energy between successive collisions with gas molecules [6]. The value of the electric field scales with pressure and has a value of 20 V m^{-1} to 1 V m^{-1} at pressures typical of disks in planetary forming systems.

The total energy in γ-rays, over an area comparable in size to the charge separation of 0.1 AU, gives an upper limit of 10^{29} erg to the energy dissipated in the channel and this exceeds a large terrestrial lightning flash by $\sim 10^{12}$. The total amount of chondrules produced in the nebula within a radius of 5 AU is 5×10^{20} g assuming a GRB with 10^6 erg cm^{-2} has 10^5 erg cm^{-2} in 65 MeV γ-rays, 2×10^{10} erg g^{-1} to heat and melt the precursor dustballs and an efficiency of 10^{-2} to convert the γ-ray energy to chondrules [15]. The amount of chondrules produced is too small to account for the total mass of $\sim 3 \times 10^{24}$ g in the asteroid belt and the value of $\sim 10^{27}$ g when the asteroid belt was 300 times larger than at present. The mass of chondrules can be increased by 3×10^4 to 1.5×10^{25} g for a GRB at a distance of 300 pc ($\times 10^3$) with a higher isotropic luminosity of 10^{53} erg ($\times 10$) and an anomalous MeV component as observed in GRB 941017 ($\times 3$) [16].

DISCUSSION

GRBs and other sources are capable of producing significant quantities of chondrules in protoplanetary systems anywhere in the disk of a galaxy. Stars form in giant molecular clouds of size about 100 pc and hence there is a higher than average probability it will contain massive stars that produce GRBs and SGRs and irradiate protoplanetary systems in the same cloud to form large amounts of chondrules.

REFERENCES

1. Boss, A. 1996, in Chondrules and the Protoplanetary Disk, ed. R. H. Hewins, R. H. Jones, & E. R. D. Scott (Cambridge, Eds. R. H. Hewins et al.: Cambridge University Press), p257
2. Rubin, A. E. 2000, Earth-Science Rev. 50, 3
3. Whipple, F. L. 1966, Science, 153, 54
4. Cameron, A. G. W. 1966, Earth-Planet Sci. Lett., 1, 93
5. Desch, S. J. & Cuzzi, J. N. 2000, Icarus, 143, 87
6. Pilipp, W., Hartquist, T. W., & Morfill, G. E. 1992, ApJ, 387, 364
7. McBreen, B. & Hanlon, L. 1999, A&A, 351, 759
8. Duggan, P., McBreen, B., Hanlon, L., et al. 2002, in Gamma-ray Bursts in the Afterglow ERA, ed. E. Costa, J. Hjorth & F. Frontera (Springer Verlag), 294
9. Duggan, P., McBreen, B., Carr, A., et al. 2003, A&A, 409, L9
10. McBreen, B., Winston, E., McBreen, S., et al. 2005, A&A, 429, L41
11. Cameron, A. G. W. 1995, Meteoritics, 30, 133
12. Hayashi, C., Nakazawa, K., & Nakagawa, Y. 1985, in Protostars and Planets II, 1100
13. Kuchner, M. 2004, astro-ph/0405536
14. Horanyi, M., Morfill, G., Goertz, C. K., & Levy, E. H. 1995, Icarus, 114, 174
15. Jones, R. H., Lee, T., Connolly, H. G., Love, S. G., & Shang H. 2000, in Protostars and Planets IV, 927
16. González, M. M., Dingus, B. L., Kaneko, Y., et al. 2003, Nature, 424, 749

Observations of Gamma–Ray Bursts with *INTEGRAL*

S. McGlynn[*], S. McBreen[†], L. Hanlon[*], B. McBreen[*], S. Foley[*], J. French[*], G. Melady[*], A. von Kienlin[**] and R. Preece[‡]

[*]*School of Physics, University College Dublin, Belfield, Dublin 4, Ireland*
[†]*Astrophysics Missions Division, RSSD of ESA, ESTEC, Noordwijk, The Netherlands*
[**]*Max-Planck-Institut fur extraterrestrische Physik, 85748 Garching, Germany*
[‡]*Department of Physics, University of Alabama at Huntsville, USA*

Abstract. The *INTEGRAL* satellite has two coded–mask γ–ray instruments; the spectrometer (SPI) which is optimised for high resolution γ–ray line spectroscopy, and the imager (IBIS) which can localise GRBs to a precision of a few arcminutes. *INTEGRAL* was launched 3 years ago and the *INTEGRAL* Burst Alert System (IBAS) has detected 33 long duration GRBs, the most intense burst by far being GRB 041219 which also had prompt optical emission associated with it. The γ–ray properties of some of these bursts are presented with particular emphasis on spectral results. A subset of 6 GRBs were observed with *XMM-Newton* and a selection of these results is presented. New results from recent GRBs are also discussed.

INTRODUCTION

The European Space Agency's International Gamma-Ray Astrophysics Laboratory *INTEGRAL* [1] was launched in October 2002 from the Baikonur cosmodrome in Kazakhstan. *INTEGRAL* is composed of two coded-mask γ–ray telescopes, an X–ray imager and an optical monitor. Results from the γ–ray instruments IBIS [2] and SPI [3] are discussed. IBIS has a fully coded field of view (FCFOV) of $9° \times 9°$ and is composed of two solid state detector arrays: ISGRI, which is optimised at low energies (below 1 MeV) and is most sensitive in the range 15–300 keV and PIcSIT which has an energy range from $\sim$180 keV to 10 MeV. SPI is composed of 19 Germanium detectors surrounded by an active anti-coincidence shield of BGO and has excellent spectral resolution over an energy range of 18 keV–8 MeV with a FCFOV of 16°. The *INTEGRAL* Burst Alert System (IBAS) [4] has detected and localised 33 GRBs to date to an accuracy of a few arcminutes, with each localisation distributed to the GCN network [5] within seconds. The GRBs detected by *INTEGRAL* exhibit a range of spectral and timing characteristics. They are generally weak, and are particularly useful when combined with other observations such as *XMM-Newton* and ground based telescopes. An X–ray afterglow was detected for each of the 6 bursts followed up by *XMM-Newton*. A notable case was GRB 031203, where a dust ring was detected in the X–ray image [6]. A review paper by Mereghetti [7] presents some results up to mid-2005. Further information is available on the IBAS webpage (http://ibas.mi.iasf.cnr.it/IBAS_Results.html). Some highlight results from the mission to date, along with new data are presented here.

CP836, *Gamma-Ray Bursts in the Swift Era*,
edited by S. S. Holt, N. Gehrels, and J. A. Nousek
© 2006 American Institute of Physics 0-7354-0326-0/06/$23.00

GRB 040106

GRB 040106 was detected by IBAS and observed by *XMM–Newton* 5 hours after the burst alert [8]. It had a duration of ~52 s, with two bright pulses separated by a quiescent interval. A spectrum was extracted from the IBIS data (Fig 1a), giving a photon index $\Gamma = -1.72 \pm 0.15$, a peak flux of 6×10^{-8} erg cm^{-2} s^{-1} and a fluence of 6×10^{-6} erg cm^{-2} in the energy range 20–2000 keV. A fading X–ray source was detected by *XMM–Newton* within the error region. Its spectrum was fit by an absorbed power–law with spectral index $\beta_x = -0.47 \pm 0.01$ which is flatter than other X–ray afterglows seen with *XMM–Newton*. The possibility that the second pulse of the GRB was the onset of the afterglow was also examined by extrapolating the SPI and IBIS fluxes to X–ray energies to estimate the 2–10 keV flux during the GRB (Fig 1b). It was thus determined that the change in temporal decay index occurred between 50 and 5000 s after the onset of the GRB. This may be evidence of the passage of the cooling break through the X–ray band at that time. New observations of the initial X–ray afterglow phase by Swift indicate that the lightcurves are more complex than previously thought.

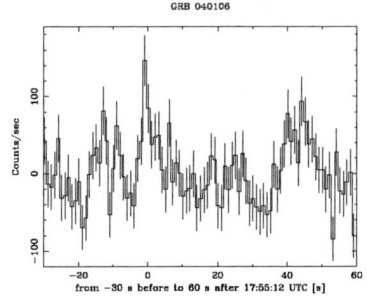
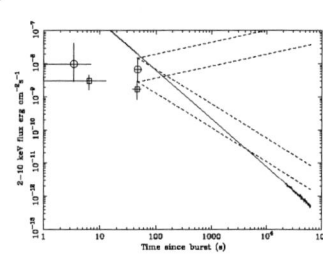

Fig. 1 a) *SPI lightcurve of GRB 040106 in the energy range 20 keV–8 MeV.*
b) *Extrapolated fluxes from IBIS (squares) and SPI (circles) for each pulse (2–10 keV). The solid line is the extrapolation of the X–ray data and the dashed lines represent the range of decay slopes.*

GRB 040223

GRB 040223 was detected by *INTEGRAL* in the direction of the Galactic plane [9] and was also followed up by *XMM-Newton*. A fading X–ray source was detected within the IBAS error region. This burst had an approximate duration of ~ 260 s with a multi-peaked lightcurve, making it one of the longest GRBs detected. The spectrum was best fit with a power–law with a photon index $\Gamma = -2.3 \pm 0.2$ in the energy range 20–200 keV(Fig 3a). The optical/near–IR afterglow was not detected due to the large absorption associated with the high column density. The X–ray spectrum was best fit by an absorbed power–law model giving a spectral index $\beta_x = -1.7 \pm 0.2$ and a large column density $N_H = 1.8 \times 10^{22}$ cm^{-2}. Further examples of *INTEGRAL* bursts that are X–ray rich include GRB 040624 [10], XRF 050522 and GRB 050626 (Table 1).

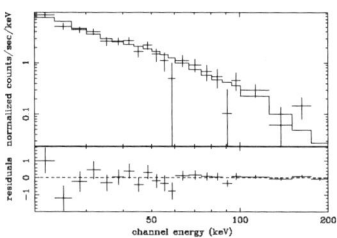 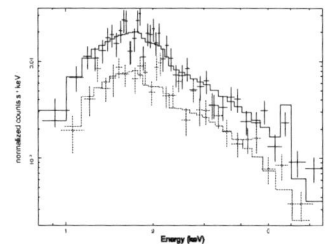

Fig. 2 a) *IBIS spectrum of GRB 040223 fit by a power–law model from 20–200 keV.* **b)** *EPIC-PN and MOS spectra of the GRB 040223 afterglow and the best fit absorbed power–law model.*

GRB 041219

GRB 041219 is the strongest and longest burst detected so far with *INTEGRAL*, with a peak flux higher than ~96% of the bursts observed by BATSE, and a T_{90} of 172 s. The lightcurve exhibits a multi-peaked structure, with a precursor of ~7 s duration, followed by a ~200 s period of quiescence before the main emission (Fig 4). The precursor was strong enough to trigger IBAS, so the localisation was available for multi-wavelength observations, and thus prompt optical [11] and NIR [12] emission was detected during the gamma ray emission phase The spectrum of the entire burst (~520 s) was well fit by the Band model [13] and a search for spectral lines during each period of emission of the burst is being carried out using SPI.

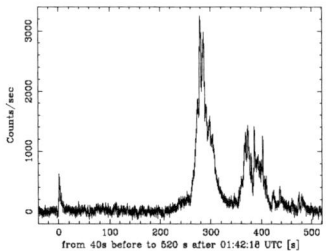 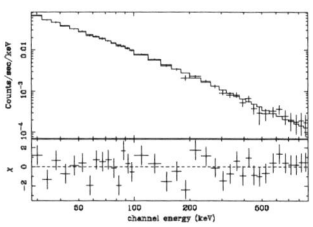

Fig. 3 a) *SPI lightcurve of GRB 041219 in the energy range 20 keV–8 MeV.* **b)** *SPI spectrum of GRB 041219 from ~20–900 keV fit by a Band model.*

RECENT RESULTS

Table 1 below shows spectral parameters of a selection of bursts recently detected by *INTEGRAL*. All spectra are extracted from IBIS in the energy range 20–200 keV and fit

TABLE 1. IBIS spectral results (20–200 keV) of a selection of recent GRBs.

	Photon Index	Peak Flux (erg $cm^{-2}\ s^{-1}$)
GRB 040422	1.26 ± 0.03	1.8×10^{-7}
GRB 050223	1.9 ± 0.3	3.5×10^{-8}
GRB 050504	1.3 ± 0.1	7.3×10^{-8}
GRB 050522	3.1 ± 0.5	2.0×10^{-8}
GRB 050626	2.2 ± 0.2	2.0×10^{-8}
GRB 050714A	2.2 ± 0.2	1.5×10^{-8}

with a power–law model. The peak flux is given for the brightest second of the burst. GRB 040422 was the only burst best fit by a Band model and so the low energy photon index (α) is given above [14]. Errors quoted are 1-σ for 1 parameter of interest.

SUMMARY

The GRBs detected by *INTEGRAL* are typically weak, with a peak flux in the range 0.5–3 photons $cm^{-2}\ s^{-1}$. A few exceptions have occurred, in particular GRB 041219, which has a peak flux of more than 30 photons $cm^{-2}\ s^{-1}$. The overall pattern however indicates a large number of faint and/or X–ray rich bursts with steep spectra that are relatively dim at optical wavelengths. Several bursts have occurred in highly obscured regions of the Galactic plane. Despite this, good localisations have been obtained and a number of multiwavelength observations have been carried out.

REFERENCES

1. Winkler, C., Courvoisier, T. J.-L., Di Cocco, G., et al., *A&A*, **411**, L1 (2003).
2. Ubertini, P., Lebrun, F., Di Cocco, G., et al., *A&A*, **411**, L131 (2003).
3. Vedrenne, G., Roques, J.-P., Schönfelder, V., et al., *A&A*, **411**, L63 (2003).
4. Mereghetti, S., Götz, D., Borkowski, J., et al., *A&A*, **411**, L291 (2003).
5. Barthelmy, S. D., *American Astronomical Society Meeting Abstracts*, **202** (2003).
6. Vaughan, S., Willingale, R., O'Brien, P. T., et al., *ApJL*, **603**, L5 (2004).
7. Mereghetti, D., S. Gotz, *astro-ph/0506198* (2005).
8. Moran, L., Mereghetti, S., Götz, D., et al., *A&A*, **432**, 467 (2005).
9. McGlynn, S., McBreen, S., Hanlon, L., et al., *astro-ph/0505349* (2005).
10. Filliatre, P., Covino, S., D'avanzo, P., et al., *astro-ph/0511722* (2005).
11. Vestrand, W. T., Wozniak, P. R., Wren, J. A., et al., *Nature*, **435**, 178 (2005).
12. Blake, C. H., Bloom, J. S., Starr, D. L., et al., *Nature*, **435**, 181 (2005).
13. Band, D., Matteson, J., Ford, L., et al., *ApJ*, **413**, 281 (1993).
14. Filliatre, P., D'Avanzo, P., Covino, S., et al., *A&A*, **438**, 793 (2005).

A key to the spectral variability of prompt GRBs

Mikhail V. Medvedev

Department of Physics and Astronomy, University of Kansas, Lawrence, KS 66045

Abstract. We demonstrate that the rapid spectral variability of prompt GRBs is an inherent property of radiation emitted from shock-generated, highly anisotropic small-scale magnetic fields. We interpret the hard-to-soft evolution and the correlation of the soft index α with the photon flux observed in GRBs as a combined effect of temporal variation of the shock viewing angle and relativistic aberration of an individual thin, instantaneously illuminated shell. The model predicts that about a quarter of time-resolved spectra should have hard spectra, violating the synchrotron $\alpha = -2/3$ limit. The model also naturally explains why the peak of the distribution of α is at $\alpha \sim -1$. The presence of a low-energy break in the jitter spectrum at oblique angles also explains the appearance of a soft X-ray component in some GRBs and their paucity. We emphasize that our theory is based solely on the first principles and contains no ad hoc (phenomenological) assumptions.

Keywords: gamma rays: bursts — radiation processes — shock waves — magnetic fields
PACS: 98.70.Rz; 95.30.Gv; 95.30.Qd

INTRODUCTION

Rapid spectral variability is a remarkable, yet unexplained feature of the prompt GRB emission. The variation of the hardness of the spectrum and the hard-to-soft evolution are the most acknowledged features [1, 2, 3]. A quite remarkable "tracking" behavior, when the low-energy spectral index α follows (or correlates with) the photon flux [2] is particularly intriguing. An example of such a trend is shown in Fig. 1 (we used data from [4]).

THEORY OF JITTER RADIATION

The angle-averaged spectral power emitted by a relativistic particle moving through small-scale Weibel-generated magnetic fields is given here without derivation (see [5, 6] for details):

$$\frac{dW}{d\omega} = \frac{e^2 \omega}{2\pi c^3} \int_{\omega/2\gamma^2}^{\infty} \frac{|\mathbf{w}_{\omega'}|^2}{\omega'^2} \left(1 - \frac{\omega}{\omega'\gamma^2} + \frac{\omega^2}{2\omega'^2 \gamma^4}\right) d\omega'. \tag{1}$$

Here γ is the Lorentz factor of a radiating particle and $\mathbf{w}_{\omega'} = \int \mathbf{w} e^{i\omega' t} dt$ is the Fourier component of the transverse particle's acceleration due to the Lorentz force. This temporal Fourier transform is taken along the particle trajectory, $\mathbf{r} = \mathbf{r}_0 + \mathbf{v}t$.

Because the Weibel-generated magnetic fields at a shock are highly anisotropic, as shown in Figure 2, the spectra of radiation emitted by an electron in such fields depend on the viewing angle Θ between the normal to the shock and the particle velocity in the

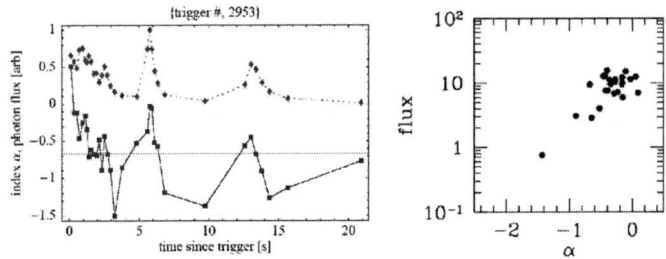

FIGURE 1. (a) — A tracking GRB: normalized flux (diamonds) and the soft spectral index α (squares) evolve similarly with time. (b) — Scatter plot of flux vs. α for GRB940429.

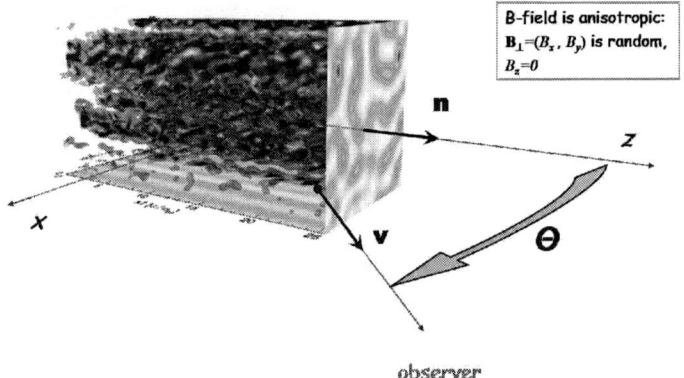

FIGURE 2. Magnetic filaments in a shock [7]. The radiation spectrum varies with the viewing angle Θ.

shock co-moving frame, which is approximately the direction toward an observer, for an ultra-relativistic particle. The acceleration spectrum is

$$\langle |\mathbf{w}_{\omega'}|^2 \rangle = C/2\pi \left(1 + \cos^2 \Theta\right) \int f_z(k_\parallel) f_{xy}(k_\perp) \delta(\omega' + \mathbf{k} \cdot \mathbf{v}) \, dk_\parallel d^2 k_\perp, \qquad (2)$$

where C is the mormalization of the magnetic feld spectrum over spatial scales to the total field energy density $B^2/8\pi$. Here f_{xy} and f_z determine spectra of the magnetic fields in the shock plane and in the direction of the shock motion. These spectra are independent of each other to a large degree, as indicated by numerical simulation (e.g., [8]). We assume f_{xy} and f_z are broken double-power-laws,

$$f_z(k_\parallel) = k_\parallel^{2\alpha_1} / (\kappa_\parallel^2 + k_\parallel^2)^{\beta_1}, \quad f_{xy}(k_\perp) = k_\perp^{2\alpha_2} / (\kappa_\perp^2 + k_\perp^2)^{\beta_2}, \qquad (3)$$

with the position of the break (peak) determined by the Weibel scale: the plasma skin depth, $\kappa_\perp \sim \kappa_\parallel \sim \omega_p/c$.

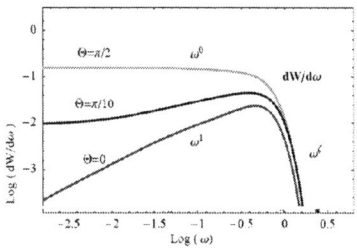
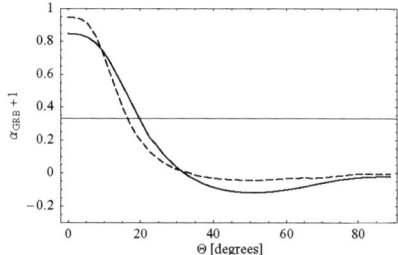

FIGURE 3. (a) — Typical spectra for three viewing angles, $\Theta = 0$, $\pi/10$, $\pi/2$. (b) — Soft sceltral index as a function of Θ at energies 10 (solid) and 30 (dashed) times below E_p.

INTERPRETATION OF PROMPT GRB SPECTRA

The typical jitter radiation spectra for three different angles are shown in Figure 3a. When a shock velocity is along the line of sight, the low-energy spectrum is hard $F_\nu \propto \nu^1$, harder than the "synchrotron line of death" ($F_\nu \propto \nu^{1/3}$). As the viewing angle increases, the spectrum softens, and when the shock velocity is orthogonal to the line of sight, it becomes $F_\nu \propto \nu^0$. Another interesting feature is that at oblique angles, the spectrum does not soften simultaneously at all frequencies. Instead, there appears a smooth spectral break, which position depends on Θ. The spectrum approaches $\sim \nu^0$ below the break and is harder above it. This softening of the spectrum at low ν's could be interpreted as the appearance of an additional soft X-ray component, similar to that found in some of GRBs [9].

Figure 3b shown the low-energy slope at a frequency 10 and 30 times lower than the spectral peak. These frequencies correspond to the edge of the *BATSE* window for bursts with the peak energy of about 200 keV and 600 keV, respectively. Hence, the spectral slope, α_{GRB}, will be close to those obtained from the data fits. Since $\Theta(t)$ increases with time during an individual emission episode, the curves roughly represent the temporal evolution of α_{GRB}. Assuming that time-resolved spectra are homogeneously distributed over Θ, one can estimate the relative fraction of the synchrotron-violating GRBs (i.e., those with $\alpha_{GRB} + 1 > 1/3$) as about 25%, which is very close to the 30% obtained from the data [4]. Most of the GRBs, $\sim 75\%$, should, by the same token, be distributed around $\alpha_{GRB} \sim -1$. Note also that time-integrated GRB spectra should have α_{GRB} around minus one, as well.

In the standard internal shock model, each emission episode is associated with illumination of a thin shell, — an internal shock and the hot and magnetized post-shock material. We assume that the shell is spherical (at least within a cone of opening angle of $\sim 1/\Gamma$ around the line of sight) and this shell is simultaneously illuminated for a short period of time. The observed photon pulse is broadened because the photons emitted from the patches of the shell located at larger angles, ϑ, from the line of sight arrive at progressively later times. The bolometric flux depends on ϑ, and hence on time as [3]:

$$F_{\text{bol}} = F_0 \mathscr{D}^2(\Theta)/\Gamma^2 \tag{4}$$

171

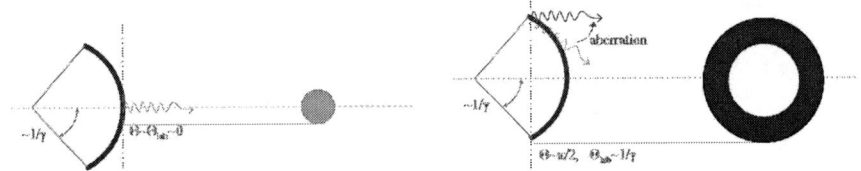

FIGURE 4. Cartoon explaining correlation of the spectral index and the flux as a combined effect of the anisotropy of jitter radiation and relativistic aberration.

with the Lorentz boost

$$\mathscr{D}(\Theta) = [\Gamma(1-\beta\cos\vartheta)]^{-1} = \Gamma(1+\beta\cos\Theta) = [\Gamma(1-\beta-\beta c\Delta t/R_0)]^{-1}. \quad (5)$$

Because of relativistic aberration, the comoving viewing angle, Θ, is greater than ϑ and approaches $\Theta \sim \pi/2$ (the shell is seen edge-on) when $\vartheta \sim 1/\Gamma$. Thus, there must be a tight correlation between the observed spectrum and the observed photon flux, because they are, in essence, different manifestations of the same relativistic kinematics effect. An observer first detects photons emitted close to the line of sight, for which $\vartheta \sim \Theta \sim 0$, see Figure 4a. The flux is the largest and α is the hardest (around 0) at this time. As time goes on, an observer sees photons emitted farer from the line of sight, as in Figure 4b. The flux decreases, whereas α becomes softer and approaches -1. Of course, we neglected cooling effects here, which can result in even softer spectra with $\alpha \sim -3/2$.

ACKNOWLEDGMENTS

This work has been supported by NASA grant NNG-04GM41G, DoE grant DE-FG02-04ER54790, and the KU GRF fund.

REFERENCES

1. P.N. Bhat, et al., *Astrophys. J.* **426**, 604 (1994).
2. A. Crider, et al., *Astrophys. J. Lett.* **479**, L39 (1997).
3. F. Ryde, and V. Petrosian, *Astrophys. J.* **578**, 290 (2002).
4. R.D. Preece, M.S. Briggs, R.S. Mallozzi, G.N. Pendleton, W.S. Paciesas, D.L. Band, *Astrophys. J. Suppl.* **126**, 19 (2000).
5. M.V. Medvedev, *Astrophys. J.* **540**, 704 (2000).
6. M.V. Medvedev, astro-ph/0510472, *Astrophys. J.* in press (2006).
7. M.V. Medvedev, L.O. Silva, M. Kamionkowski, astro-ph/0512709 (2005).
8. J.T. Frederiksen, C.B. Hededal, T. Haugbølle and Å. Nordlund, *Astrophys. J. Lett.* **608**, L13 (2004).
9. R.D. Preece, et al., *Astrophys. & Space Sci.*, **231**, 207 (1995).

Temporal and Spectral Analyses of SGRs Observed by HETE-2

Y. E. Nakagawa*, A. Yoshida*, M. Maetou*, M. Suzuki[†], T. Tamagawa[†], T. Sakamoto**, N. Kawai[‡], Y. Shirasaki[§], K. Tanaka*, M. Matsuoka[¶], E. E. Fenimore[||], M. Galassi[||], J. -L. Atteia[††], K. Hurley[‡‡] and G. R. Ricker[§§]

*Department of Mathematics and Physics, Aoyama Gakuin University, Sagamihara, Kanagawa 229-8558, Japan
[†]RIKEN (Institute of Physical and Chemical Research), Hirosawa, Wako, Saitama 351-0198, Japan
**NASA Goddard Space Flight Center, Greenbelt, Maryland 20771, USA
[‡]Department of Physics, Tokyo Institute of Technology, Meguro-ku, Tokyo 152-8551, Japan
[§]National Astronomical Observatory of Japan, Osawa, Mitaka, Tokyo 181-8588, Japan
[¶]Tsukuba Space Center, Japan Aerospace Exploration Agency, Tsukuba, Ibaraki 304-8505, Japan
[||]Los Alamos National Laboratory, Los Alamos, New Mexico 87545, USA
[††]Laboratoire d'Astrophysique, Observatoire Midi-Pyrénées, 14 Avenue E. Belin, 31400 Toulouse, France
[‡‡]University of California at Berkeley, Space Sciences Laboratory, Berkeley, California 94720-7450, USA
[§§]Center for Space Research, Massachusetts Institute of Technology, Cambridge, Massachusetts 02139, USA

Abstract. HETE-2 localized 63 flares from SGR1806−20 and 6 flares from SGR1900+14 in the summer periods from June 18 2001 through August 7 2005. We report on the temporal and spectral analyses of short flares from those SGRs. The estimation of T_{90} durations in 2-30 keV and 30-100 keV revealed that there is no difference between them with a few exceptions. For these exceptional short flares, there seems softening or possibly hardening during flares, but these spectral evolution are not common. We also found a good linear correlation between the rise and decay time, and they do not depend on a peak flux. We also found that the spectra are well reproduced by a sum of two blackbody models for all short flares. The temperatures are lying around 4 keV and 11 keV which are consistent with previous studies. They are not depend on either the magnitude of flare, the event morphology or the source.

Keywords: Neutron stars
PACS: 97.60.Jd

INTRODUCTION

Some authors reported that there is weak or no spectral evolution for SGR flares [1, 2, 3]. Up to now, the softening trends were reported for short flares from SGR1806−20 [4], a ~3.5 s long flare from SGR1900+14 [5] and a ~9 s long flare from SGR0526−66 [6].

The spectra of soft gamma-ray repeaters (SGRs) were well fitted by optically thin thermal bremsstrahlung (OTTB) in earlier days. It is suggested that the spectra of an intermidiate flare from SGR1900+14 detected by *High Energy Transient Explorer 2* (*HETE-2*) [7] were not described by OTTB below 15 keV, and were well reproduced by a sum of two blackbody models [8]. Similar results were reported for short flares from SGR1900+14 [9]. For SGR1806−20 flares, the specra detected by *International*

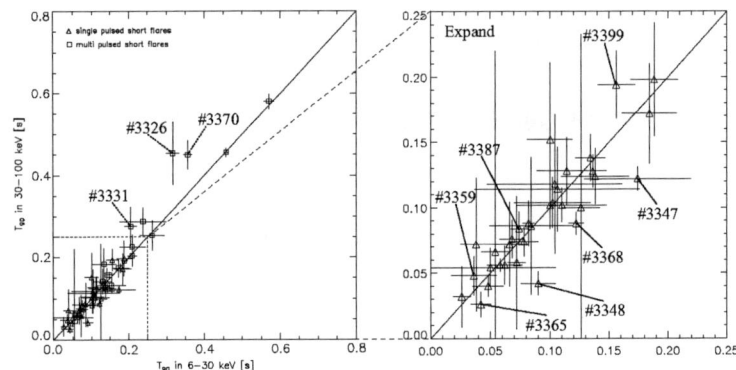

FIGURE 1. Distribution of T_{90}. The right panel shows the expanded view of the left panel.

Cometary Explorer were not reproduced by OTTB and other models [2, 10]. However, there is no report whether the spectra of flares from SGR1806−20 are well fitted by a sum of two blackbody models or not.

Here, we report the results of temporal and spectral analyses of short flares from SGR1806−20 and SGR1900+14 detected by *HETE-2*. In this study, we use the data of the Wide-Field X-Ray Monitor (WXM; 2-25 keV) [11] and French Gamma Telescope (FREGATE; 6-400 keV) [12] instruments on-board *HETE-2*.

OBSERVATION

HETE-2 triggered 181 flares from SGR1806−20 and SGR1900+14 in the summer periods from June 18 2001 through August 7 2005, of which 63 flares were localized to SGR1806−20 and 6 flares were localized to SGR1900+14, while 112 flares were out of the WXM field of view or too weak to localize. These localized flares include two intermidiate flares. Among these localized flares, we use 56 short flares for SGR1806−20 and 5 short flares for SGR1900+14 in our analyses because some flares do not have sufficient data for our analyses.

TEMPORAL ANALYSES

To investigate the spectral evolution, we computed the T_{90} durations for all short flares from SGR1806−20. Figure 1 shows the relationship between T_{90} duration in 6-30 keV and T_{90} duration in 30-100 keV, and the dotted lines show that they have same value. We found that 11 short flares (trigger numbers of *HETE-2* are #3326, #3331, #3347, #3348, #3351, #3359, #3365, #3368, #3370, #3387 and #3399) were out of the dotted lines, then we investigated the hardness ratios for them (Figure 2). We find that four short flares (#3348, #3351, #3365 and #3368) have clear softening trend, and two short flares (#3370 and #3399) might have a hard component after the onset. We also investigated a

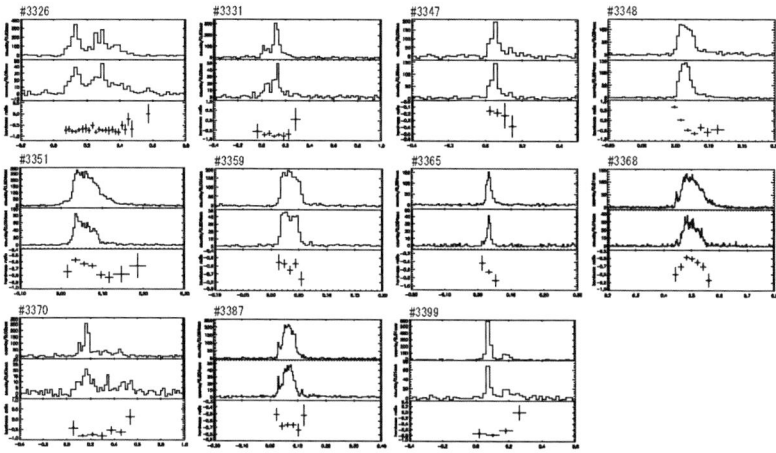

FIGURE 2. Summary of hardness ratio for 11 short flares.

relationship between the rise and decay time, and found a positive correlation between them. The rise and decay time do not depend on the peak flux of the flare.

SPECTRAL ANALYSES

As model functions, we use a power law model (PL), a cutoff power law model (CPL), a single blackbody model (BB), a sum of two blackbody models (two-BB), a disk blackbody (disk-BB) model and OTTB for all flares. If we adopt a photoelectric absorption model in case of CPL or two-BB, the values become consistent with zero for most cases. Therefore we do not adopt it.

Whereas the spectra of flares from SGR1806−20 and SGR1900+14 are well described by CPL or two-BB, we could reject PL, BB and OTTB. The resultant χ^2/d.o.f. values reject disk-BB for 4 short flares from SGR1806−20, while those of other short flares from SGR1806−20 and SGR1900+14 give significant results. The two-BB give acceptable fits for all short flares and best χ^2/d.o.f. values for almost all short flares, therefore we use two-BB for our spectral analyses.

The blackbody temperatures do not depend on the magnitude of flare (fluence), and are lying around 4 keV and 11 keV (left panel in Figure 3) which are consistent with previous studies [8, 9]. Furthermore, they does not depend on either the event morphology (i.e. single pulse or multi pulsed) or the source (i.e. SGR1806−20 or SGR1900+14).

The right panel in Figure 3 shows the positive correlation with $r = 0.544$ (where r denotes coefficient of correlation) between normalizations of low temperature (N_{LT}) and those of high temperature (N_{HT}). It implies that the spectral shape does not depend on the blackbody temperatures, but just normalizations vary.

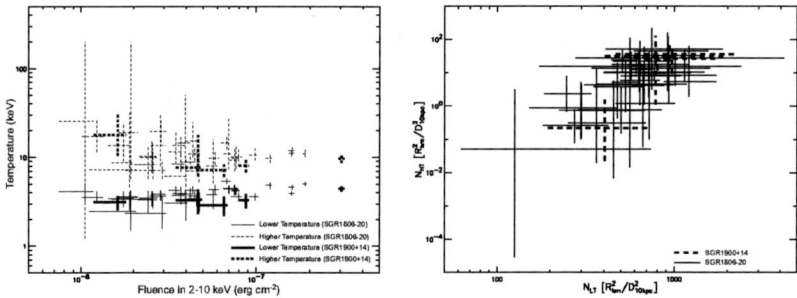

FIGURE 3. Relationship between fluences and temperatures (left panel), and relatshipship between low and high temperatures (right panel).

CONCLUSIONS

We performed temporal and spectral analyses of short flares from SGR1806−20 and SGR1900+14 observed by *HETE-2*. We found four short flares have clear softening trend and two short flares might have a hard component after the onset. There is a positive correlation between the rise time and the decay time. We also found that the spectra of short flares were well reproduced by two-BB with temperatures ∼4 keV and ∼11 keV which are consistent with previous studies. These temperatures were not depend on either the magnitude of flare (fluence), the event morphology (i.e. single pulse or multi pulsed) or the source (i.e. SGR1806−20 or SGR1900+14). It might implies that there is an unified view of flare mechanism and/or radiative transfer, between SGR1806−20 and SGR1900+14, and between single pulse events and multi pulsed events.

REFERENCES

1. C. Kouveliotou et al., *ApJ* **322**, 21–25 (1987).
2. E. E. Fenimore et al., *ApJ* **432**, 742–752 (1994).
3. D. Götz et al., *A&A* **417**, L45–L48 (2004).
4. E. E. Strohmayer, and A. Ibrahim, in *GAMMA-RAY BURSTS: 4th Huntsville Symposium*, edited by C A. Meegan, R. D. Preece, and T. M. Koshut, AIP Conference Proceedings 428, American Institute of Physics, Woodbury, 1998, pp. 947.
5. A. I. Ibrahim et al., *ApJ* **558**, 237–252 (2001).
6. S. V. Golenetskii et al., *Sov. Astron. Lett.* **13**, 166–168 (1987).
7. G. Ricker et al., *BAAS* **33**, 833 (2001).
8. J. -F. Olive et al., *ApJ* **616**, 1148–1158 (2004).
9. M. Feroci et al., *ApJ* **612**, 408–413 (2004).
10. J. G. Laros et al., *Nature* **322**, 152–153 (1986).
11. Y. Shirasaki et al., *PASJ* **55**, 1033–1049 (2003).
12. J. -L. Atteia et al., in *Gamma-Ray Bursts and Afterglow Astronomy*, edited by G. R. Ricker, and R. Vanderspek, AIP Conference Proceedings 662, American Institute of Physics, Melville, 2003, pp. 17–24.

Properties of Multi Pulsed GRBs seen by *HETE-2*

Y. E. Nakagawa*, A. Yoshida*, N. Ishikawa*, K. Tanaka*, T. Tamagawa[†],
M. Suzuki[†], N. Kawai**, M. Matsuoka[‡], Y. Shirasaki[§], R. Vanderspek[¶], E.
E. Fenimore[‖], M. Galassi[‖], J. -L. Atteia[††] and G. R. Ricker[¶]

Department of Mathematics and Physics, Aoyama Gakuin University, Sagamihara, Kanagawa 229-8558, Japan
[†]*RIKEN (Institute of Physical and Chemical Research), Hirosawa, Wako, Saitama 351-0198, Japan*
**Department of Physics, Tokyo Institute of Technology, Meguro-ku, Tokyo 152-8551, Japan*
[‡]*Tsukuba Space Center, Japan Aerospace Exploration Agency, Tsukuba, Ibaraki 304-8505, Japan*
[§]*National Astronomical Observatory of Japan, Osawa, Mitaka, Tokyo 181-8588, Japan*
[¶]*Center for Space Research, Massachusetts Institute of Technology, Cambridge, Massachusetts 02139, USA*
[‖]*Los Alamos National Laboratory, Los Alamos, New Mexico 87545, USA*
[††]*Laboratoire d'Astrophysique, Observatoire Midi-Pyrénées, 14 Avenue E. Belin, 31400 Toulouse, France*

Abstract. It is claimed by many authors that there are tight correlations between the spectral peak energy E_p, and the "brightness" such as isotropic energy $E_{\rm rad}$, collimation-corrected energy E_γ, or peak luminosity L_p. In many long GRBs, their light curves consist of several pulses or humps, and each pulse has a different spectrum. Using data of multi pulsed GRBs observed by *HETE-2*, we investigated these relations for each pulse and found it looked to have some correlations. This may imply that each pulse could behave as a single burst. Considering that the GRB rate follow the star formation rate of the Universe, we should have already detected a number of distant GRBs. Some methods are proposed to estimate a redshift by applying the above mentioned correlations. Extending the previous studies, we apply the redshift estimations using E_p-$E_{\rm rad}$ correlation for an individual pulse.

Keywords: gamma-ray sources; gamma-ray bursts
PACS: 98.70.Rz

INTRODUCTION

The recent studies suggested that the "Classical" hard gamma-ray bursts (GRBs), the X-ray rich GRBs (XRR) and the X-ray flashes (XRFs) are likely to belong to the same family of these phenomena. They are defined by the softness ratio $\log(S_X/S_\gamma)$ in the case of *HETE-2* [1]. Here, S_X and S_γ denote the energy fluences in 2-30 keV and 30-400 keV, respectively. There is a correlation between the peak energy of νF_ν spectrum E_p and the softness ratio $\log(S_X/S_\gamma)$ for the *HETE-2* GRBs [1].

Some authors claim that there are tight correlations between E_p, and the isotropic equivalent energy $E_{\rm rad}$ [2, 3, 4], the collimation corrected energy E_γ [3, 5] or the peak luminosity L_p [6]. Considering that the GRB rate follows the star formation rate of the Universe, we should have already detected a number of distant GRBs. Some methods are proposed to estimate a redshift by applying the above mentioned correlations to an observed GRB.

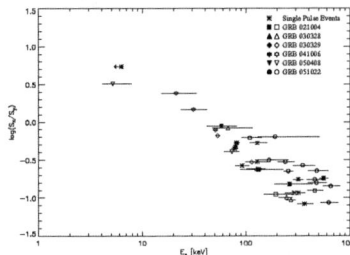

FIGURE 1. Correlation between E_p and $\log(S_X/S_\gamma)$ in source frame. The filled symbols denote the average spectra and the unfilled symbols denote the time-resolved spectra.

E_p-$\log(S_X/S_\gamma)$ CORRELATION IN SOURCE FRAME

In most long GRBs, their light curves consist of several pulses or humps, and each pulse has a different spectrum. We reported that E_p and $\log(S_X/S_\gamma)$ in each pulse in an event are lying on the E_p-$\log(S_X/S_\gamma)$ relation in observer frame [7]. We investigate the relationship between E_p and $\log(S_X/S_\gamma)$ in the source frame (Figure 1), and each pulse is also lying on this relation. The energy ranges of S_X and S_γ are $(2-30)/(1+z)$ keV and $(30-400)/(1+z)$ keV in the observer frame. These imply that each pulse has a spectrum similar to XRFs, XRRs or "Classical" hard GRBs, and therefore we can treat each pulse as a single event.

INTRINSIC ENERGY OF GRBS

Table 1 shows the list of multi pulsed GRBs with known redshift detected by *HETE-2*. We include these events in our analyses except for GRB 020813. We calculate $E_{\rm rad}$ and E_γ in each pulse in an event. The left panel in Figure 2 presents E_p-$E_{\rm rad}$ relation [2, 3, 4]. We find the pulses of GRB 050709 and of GRB 050408 are not lying on this relation. Furthermore, we investigate the relationship between E_p and luminosity $L_{\rm rad}$ in each pulse (the right panel in Figure 2), and find the pulses of GRB 050709 are lying on this relation, while those of GRB 050408 are not. The left panel in Figure 3 presents E_p-E_γ relation [3, 5]. We also find each pulse is lying on this relation except for GRB 050408. We show collimation-corrected luminosity L_γ and find correlation between E_p and L_γ except for GRB 050408 (the right panel in Figure 3).

DISCUSSIONS AND CONCLUSIONS

We found correlations of E_p-$E_{\rm rad}$, E_p-$L_{\rm rad}$, E_p-E_γ and E_p-L_γ for spectra of each pulse except for GRB 050408 and GRB 050709. In E_p-$E_{\rm rad}$ plane, spectra of GRB 050709 are not lying on the correlation, this is because GRB 050709 is a short burst followed by a soft long pulse [8]. However, they are lying on the correlation in E_p-$L_{\rm rad}$ plane. GRB 050408 is consists of two pulses and the second pulse is seen below 10 keV. The first pulse is lying on E_p-$E_{\rm rad}$ relation, whereas the second pulse is not lying on the relation.

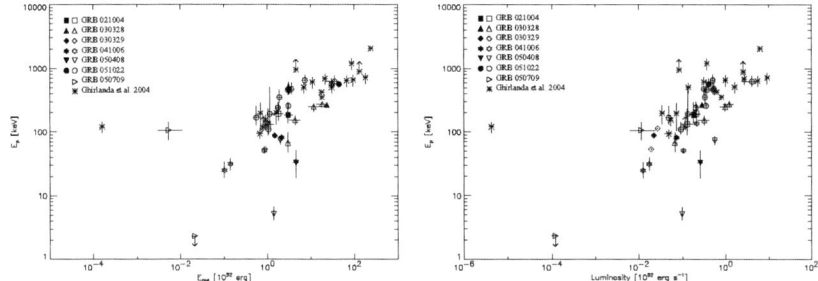

FIGURE 2. Relationship between E_p, and E_{rad} (left) or L_{rad} (right).

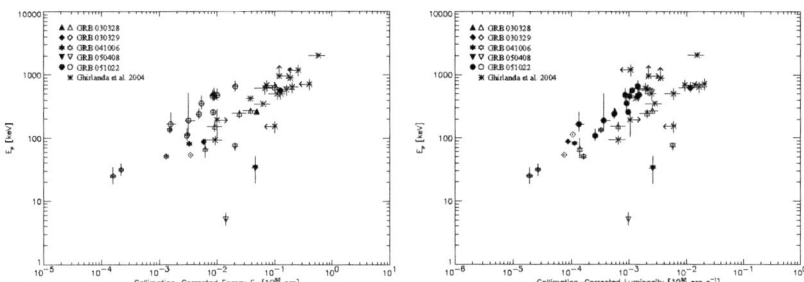

FIGURE 3. Relationship between E_p, and E_γ (left) or L_γ (right).

The second pulse is likely to be the beginning of afterglow. In addition, the pulses are not lying on other relations.

If we treat each pulse as an individual event and each pulse is lying on correlation between E_p and the intrinsic energies, the constraint of redshift estimations will

TABLE 1. List of multi pulsed GRBs with known redshift detected by *HETE-2*.

GRB	Date (UT)*	Redshift	θ_{jet}[†]	N_{pulse}**	T_{90}[‡]	T_{90}[§]	Ref.
020813	02:44:19	1.25	2.7	>7	-	-	[3, 10]
021004	12:06:14	2.3	n/a	2	40.2±2.3	54.7±4.0	[11, 12, 13, 14, 15]
030328	11:20:58	1.52	3.7	4	187±9	105±10	[16, 17, 18, 19]
030329	11:37:15	0.168	5.1	2	40.6±1.7	23.6±1.3	[20, 21, 22]
041006	12:18:09	0.716	3.2	4	36.8±0.5	23.6±2.1	[23, 24, 25]
050408	16:22:51	1.2357	8.2	2	34.0±0.6	5.70±0.34	[26, 27, 28]
050709	22:36:36	0.16	n/a	2	-	-	[29]
051022	13:07:58	0.8	4.3/4.4	12	178±8	-	[30, 31]

* Trigger time of GRB.
[†] θ_{jet} denotes the jet opening angle in units of degrees which is determined by the break time of the afterglow.
** N_{pulse} denotes the number of pulses in an event.
[‡] T_{90} durations in 2-25 keV. The quoted errors correspond to 68 % confidence region.
[§] T_{90} durations in 25-400 keV. The quoted errors correspond to 68 % confidence region.

TABLE 2. Result of our redshift estimations with the observed redshifts and the previous study.

GRB	Redshift by Observation	Our Estimation	Atteia et al. 2002 [9]
GRB 021004	2.3	4.7	n/a
GRB 030328	1.52	1.1	1.15
GRB 030329	0.168	0.2	0.17
GRB 041006	0.716	0.7	n/a
GRB 050408	1.2357	0.7	n/a

be improved. Extending the previous studies, we apply the redshift estimations using E_p-E_{rad} relation for an individual pulse of GRB 021004, GRB 030328, GRB 030329, GRB041006 and GRB 050408, estimate the most probable redshifts wihch satisfy spectra of all pulses. Table 2 shows the preliminary result of redshift estimations, and the empirical redshift [9]. The redshift estimations for GRB 020813 and GRB 051022 using E_p-E_{rad} correlation are underway.

REFERENCES

1. T., Sakamoto, et al., *ApJ*, 629, 311 (2005)
2. J. S. Bloom, et al., *ApJ*, 121, 2879 (2001)
3. J. S. Bloom, et al., *ApJ*, 594, 674 (2003)
4. L. Amati et al., *A&A*, 390, 81 (2001)
5. G., Ghirlanda, et al., *ApJ*, 616, 338 (2004)
6. D., Yonetoku, et al., *ApJ*, 609, 935 (2004)
7. A., Yoshida, et al., *Gamma-Ray Burst in the Afterglow Era: 4th Workshop* (2004)
8. J. S., Villasenor, et al., *Nature*, 437, 855 (2005)
9. J. -L. Atteia et al., *astro-ph*, 0304327, (2002)
10. P. A., Price, et al., *GCN Circ.*, 1475 (2002)
11. R., Chornock, et al., *GCN Circ.*, 1605 (2002)
12. P., Moller, et al., *A&A*, 396, L21 (2002)
13. S. B., Pandey, et al., *Bull. Astron. Soc. India*, 31, 19 (2003)
14. T., Matheson, et al., *ApJ*, 582, L5 (2003)
15. N., Mirabel, et al., *ApJ*, 595, 935 (2003)
16. E., Rol, et al., *GCN Circ.*, 1981 (2003)
17. M. I. Anderson et al., *GCN Circ.*, 1993 (2003)
18. B. A., Peterson, et al., *GCN Circ.*, 1974 (2003)
19. R. Burenin, et al., *GCN Circ.*, 1990 (2003)
20. J., Greiner, et al., *GCN Circ.*, 2020 (2003)
21. E. Berger et al, *Nature*, 426, 154 (2003)
22. A., Tiengo, et al., *A&A*, 409, 983 (2003)
23. D., Fugazza, et al., *GCN Circ.*, 2782 (2004)
24. P. A., Price, et al., *GCN Circ.*, 2791 (2004)
25. K. Z., Stanek, et al., *ApJ*, 626, 5 (2005)
26. E. Berger et al, *GCN Circ.*, 3201 (2005)
27. J. X., Prochaska, et al., *GCN Circ.*, 3204 (2005)
28. S., Covino, et al., *GCN Circ.*, 3508 (2005)
29. P. A., Price, et al., *GCN Circ.*, 3605 (2005)
30. A., Gal-Yam, et al., *GCN Circ.*, 4156 (2005)
31. J., Racusin, et al., *GCN Circ.*, 4169 (2005)

The observable effect of a photospheric component on GRB's prompt emission spectrum: peak energy clustering and flat spectra above the thermal peak

Asaf Pe'er[*], Peter Mészáros[†] and Martin J. Rees[**]

[*]*Astronomical Institute "Anton Pannekoek", Kruislaan 403, 1098SJ Amsterdam, the Netherlands*
[†]*Dpt. Astron. & Astrophysics, Dpt. Physics, Penn. State University, University Park, PA 16802*
[**]*Institute of Astronomy, University of Cambridge, Madingley Rd., Cambridge CB3 0HA, UK*

Abstract. A thermal radiative component is likely to accompany the first stages of the prompt emission of Gamma-ray bursts (GRB's) and X-ray flashes (XRF's). We study the properties of plasmas containing a low energy thermal photon component at comoving temperature $\theta \equiv kT'/m_e c^2 \sim 10^{-5} - 10^{-2}$ interacting with an energetic electron component. We show that, for scattering optical depths larger than a few, balance between Compton and inverse-Compton scattering leads to the accumulation of electrons at values of $\gamma\beta \sim 0.1 - 0.3$. For optical depths larger than ~ 100 and characteristic GRB bulk Lorentz factors of ~ 100 this leads to a peak in the observed photon spectrum at $0.1 - 1$ MeV, very weakly dependent on the values of the free parameters. For a wide range of the optical depths $0.03 \lesssim \tau_{\gamma e} \lesssim 100$ and comparable energy densities in the thermal and the leptonic component, a nearly flat energy spectrum ($\nu F_\nu \propto \nu^0$) above the thermal peak at $\approx 10 - 100$ keV and below $10 - 100$ MeV is obtained, regardless of the details of the dissipation mechanism or the strength of the magnetic field. In particular, these results are applicable to the internal shock model of GRB, as well as to slow dissipation models, e.g. as might be expected from reconnection, if the dissipation occurs at a sub-photospheric radii. We conclude that dissipation near the thermal photosphere can naturally explain (a) clustering of the peak energy at sub-MeV energies at early times; (b) steep slopes observed at low energies; and (c) a flat spectrum above 10 keV at late times. Our model thus provides an alternative scenario to the optically thin synchrotron - SSC model.

Keywords: gamma rays: bursts — gamma rays: theory — plasmas — radiation mechanisms: non-thermal
PACS: TBD

INTRODUCTION

The widely accepted interpretation of gamma-ray burst (GRB) phenomenology is that the observable radiation is due to the dissipation of the kinetic energy of a relativistic outflow, powered by a central compact object. The dissipated energy is converted to energetic electrons, which produce high energy photons by synchrotron radiation and inverse Compton (IC) scattering. While being consistent with a large number of GRB observations [1, 16, 12], two concerns are often raised: the γ-ray break energy of most GRB's observed by BATSE is in the range 100 keV - 300 keV [2, 13]. It is thought that clustering of the peak emission in this narrow energy range requires fine tuning of the fireball model parameters. In addition, there are increasing evidence for low energy spectral slopes steeper than the optically-thin synchrotron or synchrotron self Compton

(SSC) model predictions [3, 12, 4, 5, 15].

Motivated by this evidence, an additional thermal component was suggested which may contribute to the observed spectrum [7, 6, 14]. This radiation originates from the base of the relativistic flow, and is advected outward as long as the flow material remains opaque. As shown in [14], dissipation processes can occur at small enough radii, where the optical depth is high. As a result, the created spectrum is significantly different than the optically thin synchrotron-SSC model predictions. A detailed analysis of the resulting spectrum under these conditions was carried out by [8, 9]. Here, we give a short summary of the main calculations and results of these works.

ELECTRON ENERGY BALANCE

We consider a plasma cloud, containing a thermal component of low energy photons at a normalized comoving temperature $\theta \equiv kT'/m_e c^2$, with an external source (e.g., shocks or other dissipative processes) which continuously injects into the plasma isotropically distributed electrons at a characteristic Lorentz factor γ_m, at a constant rate during the comoving dynamical time t_{dyn}. We define the ratio of the energy densities in the photon component and the electron component to be $A \equiv u_{ph}/u_{el} \sim 1$ [8]. If the main electron energy loss is inverse-Compton scattering in the Thompson regime, the injected electrons cool down to $\gamma_f \simeq 1$ on a loss time

$$\frac{t_{loss}}{t_{dyn}} = \frac{3}{4A\gamma_m \tau_{\gamma e}}, \qquad (1)$$

where $\tau_{\gamma e}$ is the electron scattering optical depth. Thus, for $\tau_{\gamma e} > 1$, electrons accumulate at $\gamma_f \approx 1$ on a time shorter than the dynamical time.

While losing energy through inverse Compton scattering the low energy thermal photons, electrons are heated by down scattering energetic photons. The steady state electrons momenta $\gamma_f \beta_f$ is obtained by equating the electrons energy loss rate and the energy gain rate, and is given by [8]

$$(\gamma_f \beta_f)^2 e^{4/3(\gamma_f \beta_f)^2 n_{sc}} = \frac{3}{4A\tau_{\gamma e}}. \qquad (2)$$

For optical depths not much larger than a few, the exponent on the right hand side of equation 2 can be approximated as 1. In this approximation, the steady state electron momentum is given by $\gamma_f \beta_f (n_{sc} \lesssim 10) \approx (3/4A\tau_{\gamma e})^{1/2} \approx 0.3 A_0^{-1/2} \tau_{\gamma e,1}^{-1/2}$, where $\tau_{\gamma e} = 10^1 \tau_{\gamma e,1}$ and $A = 1A_0$ assumed. For optical depths higher than ~ 100, photons are upscattered to high energies above the Thompson regime. In this case, the energy is spread among the electrons and the photons, and the electrons momenta is $\gamma_f \beta_f = [3\theta(1+A^{-1})]^{1/2} = 0.08\,\theta_{-3}^{1/2}$, where $\theta = 10^{-3}\theta_{-3}$.

We calculated numerically the photon and particle energy distribution using the detailed time dependent numerical model described in Pe'er & Waxman [10, 11]. The results are presented in figure 1, which demonstrates the accumulation of electrons at $\gamma_f \beta_f \sim 0.1 - 0.3$ for various values of the optical depth, $\tau_{\gamma e} = 1 - 100$.

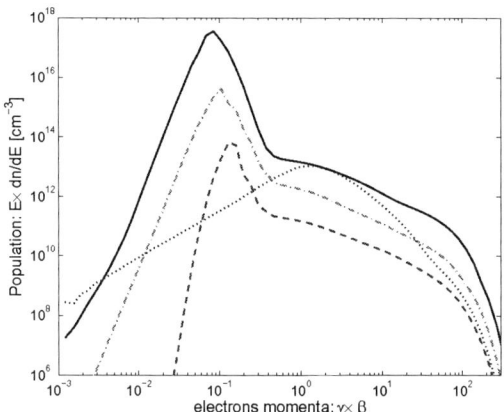

FIGURE 1. Detailed numerical model results show the electron momentum distribution at the end of the dissipation phase in GRB's. Optical depth are $\tau_{\gamma e} = 100$ (blue, solid), 10 (greed, dash-dotted), 1 (red, dash dash). The black dotted line shows the electrons momenta in the absence of Compton scattering (where synchrotron loses dominate the electrons cooling) for $\tau_{\gamma e} = 100$. $A = 0.44$ in all cases.

PHOTON SPECTRUM

For high value of the optical depth, $\tau_{\gamma e} \gtrsim 100$, the photons receive nearly all of the electrons energy, and a Wien peak is formed at comoving energy

$$\varepsilon_{WP} \approx 3\theta m_e c^2 \times (1 + A^{-1}) \simeq 10\,\text{keV}, \qquad (3)$$

irrespective of the value of the optical depth.

For intermediate values of the optical depth, $0.1 \lesssim \tau_{\gamma e} \lesssim 100$, scattering is in the Thompson regime, therefore the energy of a photon after n_{sc} scattering is $\varepsilon_{n_{sc}} \propto (\gamma_f \beta_f)^{2n_{sc}}$. The number density of photons that undergo n_{sc} scattering is $n_{ph,n_{sc}} \propto \tau_{\gamma e}^{n_{sc}}$. Since for A not much different than 1, $(\gamma_f \beta_f)^2 \approx \tau_{\gamma e}^{-1}$, we find that $\varepsilon dn/d\varepsilon \propto \varepsilon^{-1}$, or $\nu F_\nu \propto \nu^0$ above the thermal peak and below $\varepsilon_{max} = \gamma_f m_e c^2$ (in the plasma frame). The results of a detailed numerical model are presented in figure 2, for different values of the optical depth.

SUMMARY

We have considered the effect of a photospheric component on the observed spectrum after kinetic energy dissipation that occurs below or near the thermal photosphere of the outflows in GRB's or XRF's. For Thompson optical depths $\tau_{\gamma e} \gtrsim 1$ the electrons accumulate at $\gamma_f \beta_f \sim 0.1 - 0.3$, with only a weak dependence on the unknown parameter values. For $\tau_{\gamma e} \simeq 100$, the balance between Compton and inverse-Compton scattering by these electrons results in a spectral peak at $\sim 1 - 10\,\text{keV}$ in the plasma frame (sub MeV in the observed frame). For $\tau_{\gamma e}$ smaller than a few tens, an approximately flat energy per

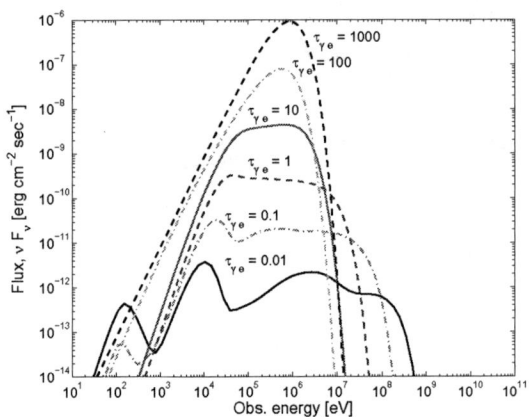

FIGURE 2. Time averaged spectra obtained for different values of the optical depth, $\tau_{\gamma e} = 10^{-2} - 10^3$ for the slow heating scenario. Similar results are obtained for dissipation by shock waves. For $\tau_{\gamma e} \gtrsim 100$, a Wien peak is formed at $\lesssim 1\,\mathrm{MeV}$ (the observed energy is lower by a factor of ~ 3 due to photon energy loss in the adiabatic expansion phase. This effect is not considered here). For lower values of $\tau_{\gamma e}$, the obtained spectrum is approximately flat, due to multiple Compton scattering.

decade spectrum is obtained above a break energy at few tens - hundred keV and up to sub GeV, while steep slopes are obtained at lower energy.

We find that the observed spectral slopes are significantly different than the synchrotron model prediction, due to the important role of inverse Compton scattering in this scenario. Further details as well as discussions about synchrotron emission, the role played by pairs and comparison of the resulting spectra under different dissipation mechanism are found in [8, 9].

REFERENCES

1. Band, D., *et al.* 1993, ApJ, 413, 281
2. Brainerd, J.J. 1998, in the 19th Texas Symposium on Relativistic Astrophysics and Cosmology, Eds.: J. Paul, T. Montmerle, and E. Aubourg (Paris: CEA Saclay)
3. Crider, A., *et al.* 1997, ApJ, 479, L39
4. Frontera, F., *et al.* 2000, ApJS, 127, 59
5. Ghirlanda, G., Celotti, A., & Ghisellini, G. 2003, A&A, 406, 879
6. Mészáros, P, Ramirez-Ruiz, E, Rees, MJ, & Zhang, B 2002, ApJ, 578, 812
7. Mészáros, P., & Rees, M.J. 2000, ApJ, 530, 292
8. Pe'er, A., Mészáros, P., & Rees, M.J. 2005, ApJ, 635, 476
9. Pe'er, A., Mészáros, P., & Rees, M.J. 2005, ApJ, submitted (astro-ph/0510114)
10. Pe'er, A., & Waxman, E. 2004, ApJ, 613, 448
11. Pe'er, A., & Waxman, E. 2005, ApJ, 628, 857
12. Preece, R., *et al.* 1998, ApJ, 506, L23
13. Preece, R., *et al.* 2000, ApJs.s., 126, 19
14. Rees, M.J. & Mészáros, P. 2005, ApJ, 628, 847
15. Ryde, F. 2005, ApJ, 625, L95
16. Tavani, M. 1996, ApJ, 466, 768

INTEGRAL Observations of the Nearby, Under-Luminous Gamma-Ray Burst GRB 031203

C.R. Shrader

Exploration of the Universe Division, NASA Goddard Space Flight Center
Code 661, Greenbelt MD 20771

Abstract. GRB 031203 was discovered by the INTEGRAL burst analysis system in near real time. It was about 10° off axis, thus well within the partially coded IBIS FoV and within the SPI fully coded FoV. This led to prompt follow up observations, reported on elsewhere, which resulted in red shift and luminosity information and estimates of the prompt soft X-ray intensity. Here we have revisited the INTEGRAL observations, notably including an analysis of the SPI germanium spectrograph data, which allows improved characterization of the spectral energy distribution to energies in excess of 200 keV. We explore the possibility of imposing new constraints on E_{peak} and on the hard-soft temporal lag determination.

Keywords: gamma-ray bursts - GRB 031203, lag-luminosity relation, E_{iso}-E_{peak} relation.
PACS: R98.70.Rz

INTRODUCTION

The gamma-ray burst GRB031203 was discovered by the INTEGRAL burst alert system in near real time on December 3, 2003. Its position was distributed quickly enough, 20 seconds after the burst trigger, for prompt after-glow study. Previously published studies of its after-glow emission indicated that it had one of the lowest recorded GRB redshifts, z=0.105 corresponding to a nominal distance of 460 Mpc. Furthermore, the prompt burst emission indicated a sub-300-keV isotropic-equivalent fluence of <10^{50} ergs, making its isotropic-equivalent luminosity one of the lowest among documented cases[5]. Prompt follow-up observation with XMM revealed a dust-scattering halo[9], which facilitated estimates of the soft X-ray fluence. This led to speculation that it could be categorized as an X-ray flash (XRF) or X-ray rich burst (XRR), and/or that it may be an outlier on the E_{peak} - E_{iso} "Amati" relation[1]. In the case of an XRF, E_{peak}<~50 keV, however, it has been argued that E_{peak} is apparently much higher[5], but this not well constrained by the IBIS spectrum, which does not exhibit any clear spectral inflection.

Similarities to GRB 980425, which is apparently associated with the z=0.009 SN 1998bw, have been noted[3,8]. Both are *(i)* Intrinsically under-luminous, *(ii)* Exhibit simple FRED light curves, *(iii)* Violate the E_{iso}-E_{peak} relation[5], *(iv)* Do not follow the luminosity-lag relation[4,6]. Additionally, radio afterglow observations of GRB 031203 have facilitated calorimetry, suggesting again a sub-luminous event[8]. However, the measurements on which some of these general conclusions are based are hampered by the limited sensitivity above 200 keV of the INTEGRAL/ISGRI instrument, and by the fact that the approximate 10-degree off-axis angle put the event outside of its fully coded field of view (FCFOV).

The contribution to the burst fluence above 200 keV, and the broad-band spectral energy distribution are critical issues in an accurate characterization of this apparently anomalous GRB. SPI offers superior spectral resolution and it has a larger FoV, although in practice the large and complex backgrounds limit the obtainable S/N. More significantly, it has greater sensitivity above 200 keV than IBIS. We were thus motivated to re-analyze the INTEGRAL data for GRB031203, with an emphasis on SPI spectral analysis.

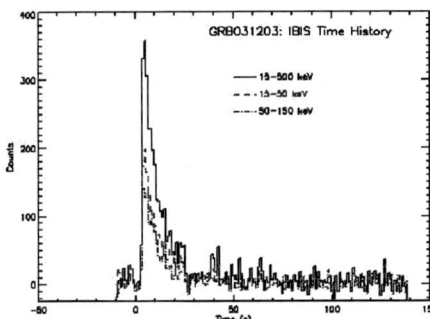

FIGURE 1. The IBIS/ISGRI light curve for GRB 031202 for the total (solid curve), and partial (15-50 keV and 50-150 keV) energy bands (dashed and dot-dashed curves). The counts were extracted, using the mask-shadow and time-window filtering as described in the text. The energy dependence of the profiles, i.e. the soft lag, is evident in a visual inspection of the light curves. For context, the Crab produces about 125 cts/s on this scale.

OBSERVATIONS AND DATA ANALYSIS

Light Curve Extraction

Both of INTEGRAL's main instruments, IBIS and SPI, have bandpass windows that sample the nominal peak spectral-emission region of typical GRBs, and both are wide FoV (9°×9° and 16° FCFoV) instruments. The IBIS upper-layer (lower-energy) detector array, ISGRI, has the largest effective collecting area below ~200 keV, thus we used data from that instrument to extract a light curve. It also has an energy resolution of ~10%. We first ran the standard analysis chain up to the image reconstruction using the single "science window" (*i.e.* segment of data) containing the GRB. We then constructed a pixel-illumination fraction (PIF) mask to quantify the shadow pattern for a source at the GRB position for the appropriate spacecraft orientation. Detector counts were then extracted for illuminated pixels, and from within a time window containing GRB 031203. The IBIS/ISGRI temporal resolution is about 200 μs.

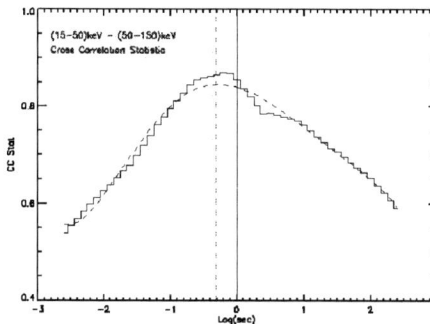

FIGURE 2. Cross-Correlation Function (CCF) computed for the 15-50-keV and 50-150-keV light curves. The smooth, dotted curve is a fit to the discrete points, used to determine the peak (overlaid by the vertical dotted line. The solid vertical line indicates the zero-lag point. It is evident that that an offset from zero, i.e. a negative lag, is present. We estimate $T_{lag} = -0.3 \pm 0.2$s, which would place GRB 031203 at least several orders of magnitude below the lag-luminosity curve of [4].

Spectral Lags

Time series derived from the 15-50-keV and 50-150-keV light curves were examined for the possible presence of spectral lags using a cross-correlation analysis. Specifically, we applied the discrete cross-correlation algorithm[2] to the two time series, using a lag range of ±3s with a sampling resolution of 0.1s. The resulting distribution is smoothly peaked near 0-s, but slightly skewed towards negative lag, *i.e.* hard leads soft (Figure 7). Formerly, we find $T_{lag} \approx -0.3 \pm 0.2$s, which we obtained by fitting a smooth curve to the CCF distribution, and identifying the peak. We then shifted the curve and computed delta-chi-square, degrading the confidence level to ~68%, to estimate the 1-σ uncertainty. It is evident that a lag is present, but that it is small. This is consistent with the notion that GRB 031203 deviates from the populated region of the lag-luminosity diagrams[2,9] thus furthering speculation that events such as this one and GRB 980425 may be representative of a separate subpopulation.

SPI Spectral Extraction

The SPI data from SCW 013900150010 were time filtered at the event level to derive pre- and post-burst segments, and a 40-sec segment spanning the duration of the GRB. The GRB start time and duration were based on the IBIS light curve of Figure 1. The standard SPI analysis was then done on each segment, leading to a spectral extraction of source plus background events. The data were binned into 55 logarithmically spaced channels from 30-500 keV. An exposure-time-weighted subtraction in detector count space was then performed to yield the net GRB source spectrum. The INTEGRAL/SPI specific capabilities of XSPEC *ver.12* were then used to analyze all 19 (i.e. shadowed plus un-shadowed spectra). A more complete description of this methodology is described elsewhere[7], but briefly, it allows apply the full (i.e. diagonal plus off-diagonal) instrumental response for *each detector* for *a given line of sight*. This is distinct from conventional coded-mask spectral analyses, which essentially entail an energy-dependent image reconstruction using a diagonal matrix. For a typical dithering observation, this entails modeling out all background and source content of the field using the mask pattern, but for a short, bright event such as the GRB, the background can be subtracted in detector count space using the pre- and post-burst data segments. The data is then fitted by simultaneously convolving a model to the response of each separate SPI detector. The key result

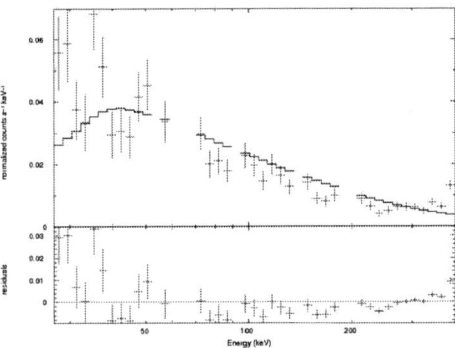

FIGURE 3. Time-averaged SPI spectrum for GRB 031203. The data were extracted as described in the text, and background subtracted in detector count space. The model fit depicted was for a simple power law with photon index $\Gamma=1.2$. Fits to the XSPEC GRBM spectral model were also attempted, but did not lead to an improved quality of fit. It seems likely that E_{peak} is higher than the ~10-keV value expected from the Amati relation.

is that no spectral break evident over the range of our detection. However, the spectral index derived is harder than that of [5]. This may have to do with the exact time window applied in addition to differences between the SPI and IBIS instrumental response functions. The hard spectrum may be consistent with the "alpha" portion of a GRB spectrum, in which case, E_{peak} is higher than ~200 keV.

DISCUSSION

We have revisited the INTEGRAL observation of nearby, under-luminous event GRB 031203. We included the SPI data in our analysis to attempt to characterize the high-energy extent of the emission. We find a harder spectrum than that previously reported based on IBIS. Our SPI spectral index was $\Gamma \approx 1.22 \pm 0.04$. The fluence and isotropic-equivalent energy output derived from our data are 1.1×10^{-7} erg·cm^{-2} (20-400 keV), or $E_{iso}=5.8 \pm 1.0 \times 10^{49}$ ergs for $z=0.105$, $(H_0, q_0, \Lambda_0)=(70,0,0.7)$. The emission is detected well above 300 keV, but becomes noise-limited by ~400-keV in the SPI data. The fact that GRB 031203 lies at least two orders of magnitude below published lag-luminosity curves, and that *either* the spectral energy distribution is very unusual, and consists of multiple components, *or* the event is an extreme outlier on the Amati relation, supports speculation of the existence of a distinct sub-population of under-luminous events.

REFERENCES

1. Amati, L., et al. 2002, A&A, 390, 81
2. Edelson, R. & Krolik, J., 1988, Ap.J., 333, 646
3. Galama, T.J., et al., 1998, Nature, 395, 670.
4. Norris, J.P., et al, 2000, 534, 248
5. Sazonov, S.Y., Lutovinov, A.A., & Sunyaev, R.A., 2005, Nature, 430,646
6. Schaefer, B.E., et al., 2001, Ap.J., 563, L123.
7. Shrader, C.R., 2004, proc. "5th INTEGRAL Workshop", ESA SP 552, pp. 901
8. Solderberg, A.M., et al, 2004, Nature, 430, 648
9. Watson, D., et al., 2004, Ap.J., 605, L101

Mechanical Model for Relativistic Blast Waves and Stratified Fireballs

Zuhngwhi Uhm and Andrei M. Beloborodov

Physics Department and Columbia Astrophysics Laboratory, Columbia University, 538 West 120th Street, New York, NY 10027. Email: zu@astro.columbia.edu, amb@phys.columbia.edu

Abstract. We propose a simple mechanical model for relativistic explosions with both forward and reverse shocks, which allows one to do fast calculations of GRB afterglow. The blast wave in the model is governed by pressures P_F and P_R at the forward and reverse shocks. We show that the simplest assumption $P_F = P_R$ is in general inconsistent with energy conservation law. The model is applied to GRBs with non-uniform ejecta. Such "stratified fireballs" are likely to emerge with a monotonic velocity profile after an internal-shock stage. We calculate the early afterglow emission expected from stratified fireballs.

Keywords: gamma-ray bursts, blast waves, afterglow, hydrodynamics
PACS: 98.70.Rz, *43.28.Mw, 82.33.Xj, 95.30.Lz

INTRODUCTION

Relativistic blast waves from GRBs are believed to produce the observed afterglow emission of the bursts. The blast-wave structure is schematically shown in Figure 1. To a reasonably good approximation, the whole blast [the region between FS and RS, including the contact discontinuity (CD)] may be assumed to have a common Lorentz factor γ_s [1]. The Lorentz factors of the forward shock (FS) and reverse shock (RS) are denoted by γ_F and γ_R, respectively. Pressures P_F and P_R behind the forward and reverse shocks are calculated from the jump conditions.

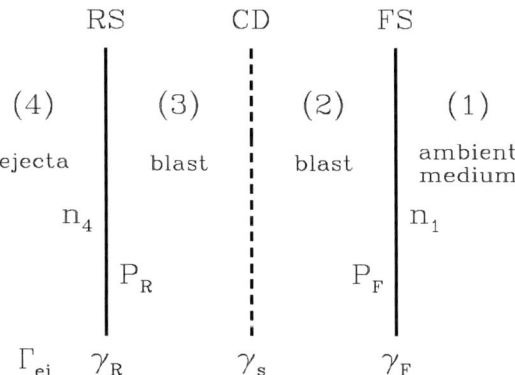

FIGURE 1. An illustrative diagram of 4 regions in a blast wave. Here n_1 and n_4 are the number densities of the ambient medium and the ejecta, respectively.

STRATIFIED FIREBALL

We focus here on the early stages of the GRB explosion when both forward and reverse shocks are expected to produce significant emission. Our calculations assume spherical symmetry. However, the results also apply to beamed explosions at early stages, when $\gamma_s > \theta_{jet}^{-1}$ where θ_{jet} is the opening angle of the ejecta.

The reverse shock (RS) is expected to propagate in the ejecta and produce significant emission at radii $r \sim 10^{16} - 10^{17}$ cm. The RS emission depends on the structure of the ejecta that forms at smaller radii $r < 10^{16}$ cm, when internal shocks take place [1].

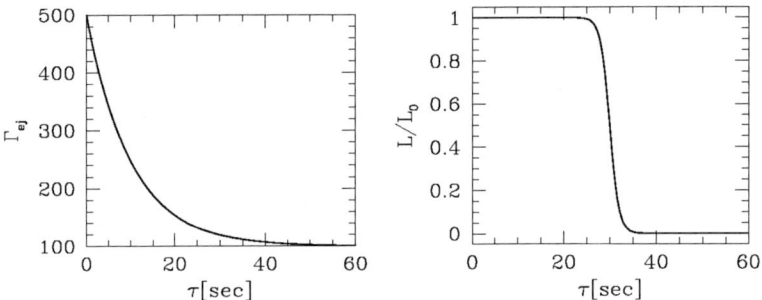

FIGURE 2. Lorentz factor (**Left**) and kinetic luminosity (**Right**) of ejecta

Theoretically, one expects the ejecta to emerge with a monotonic velocity profile (internal shocks tend to make the profile monotonic). Therefore we study here radially stratified ejecta and calculate the emission expected from such "stratified fireballs."

A simple example model considered here assumes ejecta with radial profile of Lorentz factor $\Gamma_{ej}(\tau) = (\gamma_1 - \gamma_2)e^{-a\tau/\tau_b} + \gamma_2$ shown in Figure 2 ($\gamma_1 = 500$, $\gamma_2 = 100$, $a = 3$, and $\tau_b = 30$ s is the burst duration). Here τ is the Lagrangian coordinate that labels the shells in the ejecta by their ejection times. The luminosity of the ejecta $L(\tau) = 4\pi r^2 v_{ej} \Gamma_{ej}^2 n_4 m_p c^2$ is also shown in Figure 2. Specifically, $L(\tau) = L_0/(1+\exp[(\tau - \tau_b)/\tau_d])$ is assumed here with $L_0 = 10^{52}$ erg/s and $\tau_d = 1$ s for the timescale of decay. The total energy of the burst is given by $E_{ej} = L_0 \tau_d \ln(1 + e^{\tau_b/\tau_d})$.

MODELING THE BLAST WAVE

1. Customary approximation: $P_F = P_R$

The customary approximate model blast wave assumes equal pressures at the forward and reverse shocks: $P_F = P_R$ [1]. This assumption, as we show in an accompanying paper, enables a simple analytical solution for γ_s and pressure $P = P_F = P_R$:

$$\gamma_s = \Gamma_{ej}\left[(1-2\sigma) + 2\left(\sigma^2 + (\Gamma_{ej}^2 - 1)\sigma\right)^{1/2}\right]^{-1/2} \quad \text{where} \quad \sigma = \frac{n_1}{n_4},$$

$$P = \frac{4}{3}(\gamma_s^2 - 1)n_1 m_p c^2,$$

It turns out however that the assumption of equal pressure leads to a contradiction with the law of energy conservation. The total energy of the blast wave and ejecta at a time t may be found by integrating the 00 component of the stress-energy tensor over volume, $E = \int T_{00} dV$. The result is shown in Figure 3. E decreases with time (dashed curve in the left panel), i.e. the energy is not conserved. We therefore propose another dynamical description.

2. Consistent mechanical model

Instead of finding γ_s from $P_F = P_R$ we solve the differential equation for γ_s:

$$\frac{d\gamma_s}{dt} = \frac{1}{\gamma_s H} v_s (P_R - P_F) - \frac{1}{H}(\gamma_s - 1/\gamma_s) \int_{RS}^{FS} \left(\frac{dP}{dt}\right) dr,$$

where $H = \int (U+P) dr$ is the integrated enthalpy of the blast. The first term on the right describes the external force applied to the blast due to the difference between P_F and P_R, and the second term describes the adiabatic acceleration of the blast. This equation is derived from the exact relativistic fluid equation. It is obtained by integrating over the blast and may be viewed as a mechanical equation. The mechanical model is consistent with the energy and momentum conservation (solid curve in the left panel of Fig. 3).

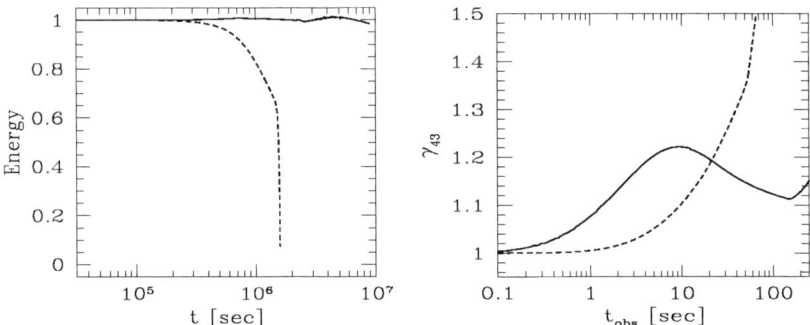

FIGURE 3. Left: Dashed curve shows the total energy of the blast wave calculated using $P_F = P_R$ assumption, solid curve – using our mechanical model. Right: Lorentz factor γ_{43} of region 4 relative to region 3 for the stratified (solid curve) and non-stratified (dashed curve) fireballs; our mechanical model is used in the calculations.

3. Light curves from stratified fireballs

We calculate the afterglow emission using the standard simple model of the postshock plasma. It assumes that a fraction $\varepsilon_e \sim 0.1$ of the shock energy goes to the electrons and accelerates them with a power-law energy spectrum $dN/dE_e \propto E_e^{-p}$. The postshock magnetic field is parameterized by the "equipartition" parameter $\varepsilon_B = (B^2/8\pi) U_{\text{th}}^{-1} < 1$ where U_{th} is the thermal energy density of the shocked medium.

We discretize the external medium and the ejecta into spherical mass shells m_i and use the Lagrangian description of the blast wave. Each m_i is impulsively heated at some point by a shock front (forward or reverse). We track the subsequent evolution of the shell and calculate its synchrotron emission in the shell comoving frame.

The shock-heated shell suffers radiative and adiabatic energy losses in the expanding blast wave, and its magnetic field decreases, so the shell emission weakens with time. We find its contribution to observed radiation at a given observed time t_{obs}. Integration over all m_i gives the exact total spectrum received from the blast wave at a given t_{obs}.

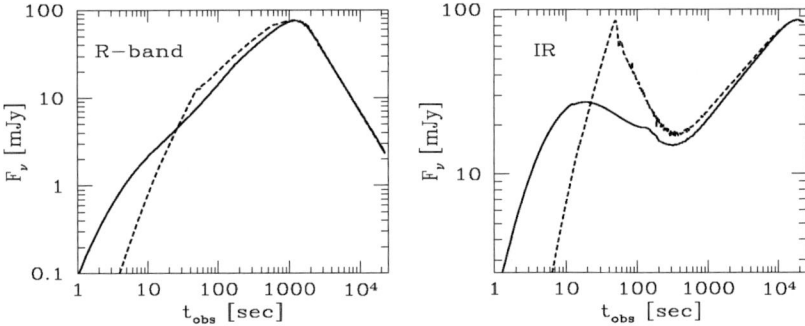

FIGURE 4. Left: Optical light curve (R-band) from stratified (solid) and non-stratified (dashed) fireballs. The explosion parameters are $\varepsilon_B = 10^{-3}$, $\varepsilon_e = 0.1$, $p = 2.5$, $n_1 = 10 \,\text{cm}^{-3}$, and redshift $z = 1$. **Right**: IR light curve ($\nu_{obs} = 10^{13}$ Hz) from the same models.

The optical and infrared light curves for two example models (stratified and non-stratified) are shown in Figure 4. We find weak optical emission from the reverse shock. It is weak because the RS Lorentz factor relative to the blast, γ_{43}, is modest and the emitted spectrum peaks in IR rather than in optical band. The peak frequency is

$$\nu_m \approx 10^6 B \gamma_m^2 \text{ Hz}, \qquad \gamma_m = 1 + \frac{p-2}{p-1}\frac{m_p}{m_e} \varepsilon_e (\gamma_{43} - 1).$$

The evolution of γ_{43} is shown in Figure 3 (right panel); its typical value is 1.2 and the factor $(\gamma_{43} - 1)^2$ that enters ν_m is small, $\sim 1/20$, shifting the peak frequency to IR.

The behavior of the RS Lorentz factor γ_{43} in the stratified fireball is significantly different from the non-stratified case: it reached maximum and then minimum (Fig. 3). Therefore, the RS emission in the stratified fireball is weaker and its peak in the light curve is less pronounced, with a smoother rise and slower decay.

ACKNOWLEDGMENTS

This work was supported by NASA grant NAG5-13382.

REFERENCES

1. T. Piran, *Rev. Mod. Phys.*, **76**, 1143 (2004).

Neutrino-cooled Accretion Disks around Spinning Black Holes

Wen-xin Chen and Andrei M. Beloborodov

Physics Department and Columbia Astrophysics Laboratory, Columbia University, 538 West 120th Street, New York, NY 10027

Abstract. We calculate the structure of accretion disk around a spinning black hole for accretion rates $\dot{M} = 0.01 - 10 M_\odot\ s^{-1}$. The model is fully relativistic and treats accurately the disk microphysics including neutrino emissivity, opacity, electron degeneracy, and nuclear composition. We find that the accretion flow always regulates itself to a mildly degenerate state with the proton-to-nucleon ratio $Y_e \lesssim 0.1$ and becomes very neutron-rich. The disk has a well defined "ignition" radius where neutrino flux raises dramatically, cooling becomes efficient, and Y_e suddenly drops. We also calculate other characteristic radii of the disk, including the ν-opaque and ν-trapping radii, and show their dependence on $\dot{M}$. Accretion disks around fast-rotating black holes produce intense neutrino fluxes which may deposit enough energy above the disk to generate a GRB jet.

Keywords: Gamma-ray bursts, accretion disks, black holes
PACS: 98.70.Rz, 97.10.Gz, 97.60.Lf

Most GRB models assume a short-lived accretion disk as a primary source of energy. Such disks have huge accretion rates $\dot{M} = 0.01 - 10 M_\odot\ s^{-1}$ and their cooling mechanism is neutrino emission [1, 2, 3]. Special features of neutrino-cooled disks are:

— The main cooling is due to reactions $p + e^- \to n + \nu$ and $n + e^+ \to p + \bar{\nu}$.
— Electron degeneracy affects the disk structure.
— The disk can be opaque to neutrinos which affects the cooling rate.
— Nuclear composition changes with radius, consuming energy.
— Neutrino cooling and electron degeneracy lead to very high neutron richness.

The disk model may be formulated following the classical vertically-integrated approach of Shakura & Sunyaev and approximating the azimuthal motion in the disk, u^ϕ, as Keplerian rotation. A small radial velocity is superimposed on this rotation. It is approximately given by $u^r = \alpha B (H/r)^2 r u^\phi$ where H is half-thickness of the disk (found from hydrostatic balance), B is a numerical factor determined by Kerr metric, and $\alpha = 0.1 - 0.01$.

The main equations of the accretion disk are as follows.

Energy conservation:
$$F^+ - F^- = c u^r \left[\frac{d(UH)}{dr} - \frac{U+P}{\rho} \frac{d(\rho H)}{dr} \right]. \quad (1)$$

Lepton conservation:
$$\frac{1}{H}(\dot{N}_{\bar{\nu}} - \dot{N}_\nu) = c u^r \left[\frac{\rho}{m_p} \frac{dY_e}{dr} + \frac{d}{dr}(n_\nu - n_{\bar{\nu}}) \right]. \quad (2)$$

Charge conservation:
$$n_{e^-} - n_{e^+} = Y_e \frac{\rho}{m_p}. \quad (3)$$

Chemical balance:
$$\mu_e - \mu_\nu = kT \ln\left(\frac{1-Y_e}{Y_e}\right) + (m_n - m_p)c^2. \quad (4)$$

Here F^+ is the viscously dissipated energy per unit area, F^- is the neutrino energy flux, $\dot{N}_\nu$ and $\dot{N}_{\bar{\nu}}$ are the number fluxes of neutrinos and antineutrinos, ρ is the mass density, T is temperature, Y_e is the proton-to-baryon ratio, μ_e and μ_ν are the chemical potentials of electrons and neutrinos, n_{e^-} and n_{e^+} are number densities of electrons and positrons. The energy density U and pressure P include contributions of baryons, electrons, positrons, radiation, and neutrinos. They are calculated by integrating the corresponding distribution functions (Fermi-Dirac distributions for $e^\pm$ and $\nu, \bar{\nu}$ if the disk is opaque to neutrinos).

Note that we include the advection terms in the laws of energy and lepton conservation (right-hand side of eqs. 1 and 2). They describe the most important effects of advection. Other effects of heat storage in the disk — in particular, the dynamical effects of the radial pressure gradient on u^r and u^ϕ — are relatively small and neglected here, i.e. the standard α-disk model is used to calculate u^ϕ and u^r.

The chemical balance and μ_ν are used only in the ν-opaque region; then the cooling rate is calculated as $F^- = cU_\nu/\tau_\nu + cU_{\bar{\nu}}/\tau_{\bar{\nu}}$. In the ν-transparent region, F^- equals $2H(\dot{U}_{e^-p} + \dot{U}_{e^+n})$ where $\dot{U}_{e^-p}$ and $\dot{U}_{e^+n}$ are the rates of energy release by reactions $e^- + p \to n + \nu$ and $e^+ + n \to p + \bar{\nu}$. The abundance of α-particles is determined by the nuclear statistical equilibrium.

We have solved the set of disk equations numerically starting from an outer region where no neutrino cooling takes place and assuming that the infalling matter is initially made of α particles (heavier elements are decomposed into α-particles as the matter approaches the black hole). A schematic picture of the accretion disk and its characteristic radii are shown in Figure 1.

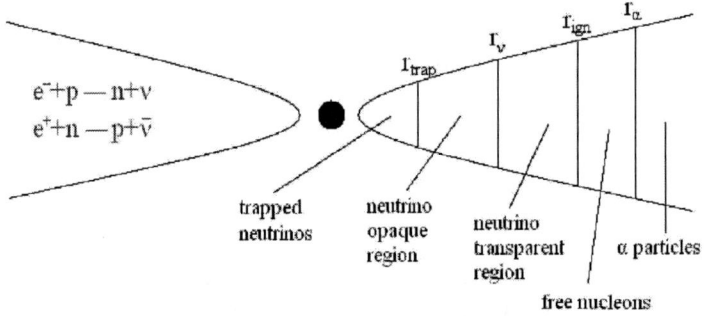

FIGURE 1. Schematic picture of the disk with characteristic radii indicated.

The outer region $r > 50r_g$ ($r_g = 2GM/c^2$ is the gravitational radius of the black hole) is dominated by α particles. Neutrino cooling is negligible in this region and the viscously dissipated heat is stored in the disk and advected inward. The destruction of α particles at $r_\alpha \sim 50r_g$ consumes ~ 7 MeV per baryon, which makes the disk somewhat thinner. At the "ignition" radius r_{ign} (slightly smaller than r_α), kT reaches $\sim$ MeV and neutrino cooling becomes significant, further reducing disk scale-height H/r. With decreasing radius the disk becomes opaque to neutrinos (first to ν and then to $\bar{\nu}$). In the innermost region of disks with $\dot{M} > 2M_\odot$ s^{-1} the neutrinos are trapped and advected toward the black hole. Figure 2 shows the found characteristic radii as functions of $\dot{M}$ for disks with $\alpha = 0.1$ around black holes with spin parameters $a = 0$ and 0.95.

We find two general properties of the disk:
1. — It regulates itself to a mildly degenerate state with $\mu_e = 1 \sim 3kT$. The reason of this regulation is the negative feedback of degeneracy and neutronization on the cooling rate: higher

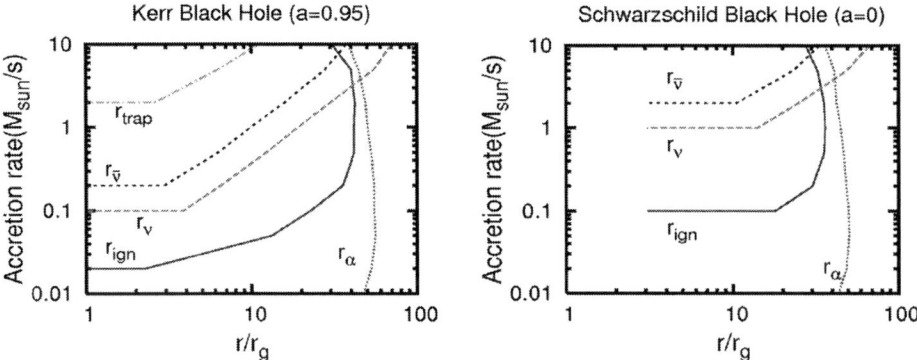

FIGURE 2. Characteristic radii in the accretion disk as a function of $\dot{M}$ for a black hole of mass $M = 3M_\odot$ and spin parameter $a = 0.95$ (left) and 0 (right). Viscosity parameter $\alpha = 0.1$ is assumed. Neutrino cooling does not ignite when $\dot{M}$ is below $0.02 M_\odot s^{-1}$ for Kerr and $0.1 M_\odot s^{-1}$ for Schwarzschild black holes. Neutrino trapping takes place for Kerr black holes when $\dot{M} > 2 M_\odot$ s^{-1}. The disk extends down to the marginally stable orbit, which is $3 r_g$ for $a = 0$ and $\approx r_g$ for $a = 0.95$.

degeneracy $\mu_e/kT \to$ fewer positrons ($n_+/n_- \sim e^{-\mu_e/kT}$) $\to$ lower equilibrium rate of neutrino emission $\to$ lower cooling rate $\to$ higher temperature $\to$ lower degeneracy.

2. — In a broad range of relevant accretion rates the disk is found to reach $Y_e \sim 0.1$. It corresponds to a very high neutron-to-proton ratio ~ 10.

Figure 3 shows examples of the disk structure around a Kerr black hole with $a = 0.95$ with accretion rate $0.2 M_\odot$ s^{-1}, for three different values of viscosity parameter α. The transitions at the characteristic radii manifest themselves in the non-monotonic behavior of the degeneracy parameter $\eta = \mu_e/kT, F^-/F^+, Y_e$, and H/r. In particular the sharp maximum of F^-/F^+ appears at the ignition radius. The disk is quite thin at $r < r_{ign}$ and cooled locally, $F^- \approx F^+$.

The high neutron richness has important implications for the global picture of GRB explosion. When the neutron-rich material is ejected in a relativistic jet and develops a large Lorentz factor, the neutrons decay at distance $\sim 10^{16}$ cm and affect the observed explosion.

Disks around rapidly rotating black holes produce a high neutrino flux. It deposits energy above the disk (via neutrino annihilation $\nu + \bar{\nu} \to e^+ + e^-$) and can drive an outflow with power $L > 10^{51}$ erg/s. A detailed study of $\nu\bar{\nu}$ annihilation around a Kerr black hole is in preparation.

ACKNOWLEDGMENTS

This work was supported by NASA grant NAG5-13382.

REFERENCES

1. R. Popham, S.E. Woosley, & C. Fryer, *Astrophys. J.*, **518**, 356 (1999).
2. A.M. Beloborodov, *Astrophys. J.*, **588**, 931 (2003).
3. K. Kohri, R. Narayan, & T. Piran, *Astrophys. J.*, **629**, 341 (2005).

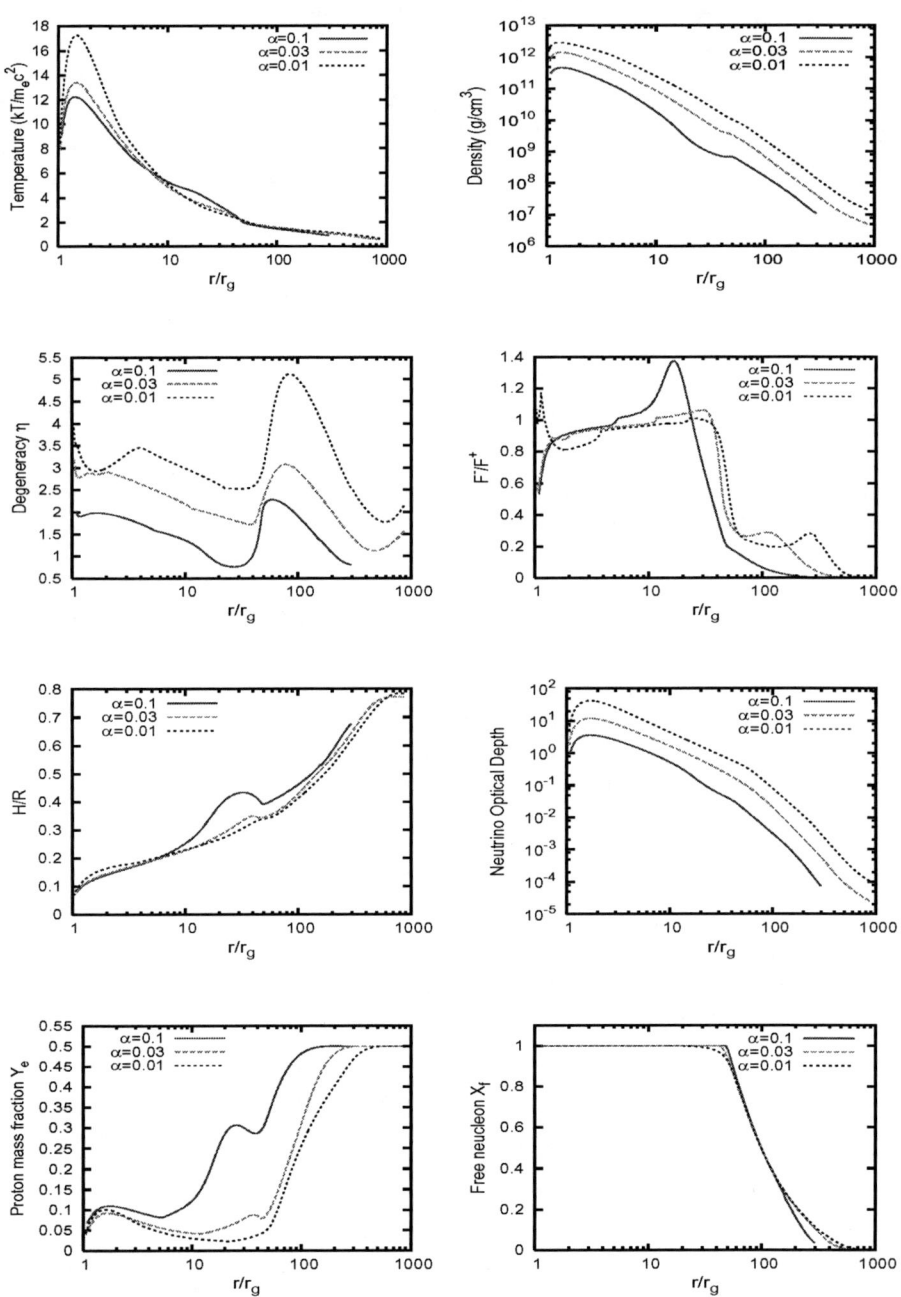

FIGURE 3. Numerical solution for accretion disk with $\dot{M} = 0.2 M_\odot s^{-1}$ around a black hole with mass $M = 3 M_\odot$ and spin parameter $a = 0.95$. Three models are shown with viscosity parameter $\alpha = 0.1, 0.03$ and 0.01. The degeneracy parameter η is defined as μ_e/kT, X_{free} is the mass fraction of free nucleons.

Compton Spectrum from Poynting Flux Accelerated e+e- Plasma.

Shinya Sugiyama*, Edison Liang*, Koichi Noguchi* and Hideaki Takabe[†]

*Rice University, 6100 Main St. Houston, TX 77005-1892, USA
[†]Institute of Laser Engineering Osaka University, 2-6, Yamada-oka, Suita, OSAKA 565-0871 JAPAN

Abstract.
We report the Compton scattering emission from the Poynting flux acceleration of electron-positron plasma simulated by the $2\frac{1}{2}$ dimensional particle-in-cell(PIC) code. We will show these and other remarkable properties of Poynting flux acceleration and Compton spectral output, and discuss the agreement with the observed properties of GRBs and XRFs.

Keywords: GRBs, Compton, Plasmas
PACS: 98.70.Rz

INTRODUCTION

We report the Compton scattering emission from the Poynting flux acceleration of electron-positron plasma simulated by the $2\frac{1}{2}$ dimensional particle-in-cell(PIC) code. We study the intense electromagnetic pulse loaded with e^+e^- plasma interacting with ambient blackbody soft photons. Driven by the strong electromagnetic field, the particles can be accelerated to ultrarelativistic energies with a power-law momentum distribution (We call this mechanism the diamagnetic relativistic pulse accelerator (DRPA) which is a special example of Poynting Flux acceleration by comoving transverse EM pulses [1]).

This mechanism results in the development of a power-law spectrum of photons via Compton scattering, (photon index is roughly $-1 \sim -3$). This is consistent with the Gamma-ray, X-ray power-law spectrum from the long duration GRBs. The emission has also strong angle dependence: the emission from 3-8 degree from the Poynting flux direction is harder than from the other angles. We will show these and other remarkable properties of Poynting flux acceleration and Compton spectral output, and discuss the agreement with the observed properties of GRBs and XRFs.

COMPTON DRAG FORCE

We use a $2\frac{1}{2}$ dimensional explicit simulation scheme based on the PIC method for time advancing of plasma particles and fields [1]. The dimensional system is double periodic in x and z directions. We solve the Maxwell equations and Lorentz equation of motion

TABLE 1. Plasma Parameters

	Frequency Ratio $[\Omega_{pe}/\Omega_e]$	X Length $[\lambda_d]$	Z Length $[\lambda_d]$	Plasma Temperature	Thermal Photon Temperature	Photon Density $[n_p/n_e]$
(A)	0.1	2000	4	1MeV	1eV	1.0
(B)	0.1	2000	4	1MeV	1eV	0.01
(C)	0.1	2000	4	1MeV	10eV	1.0
(D)	0.01	200	4	1MeV	1eV	1.0
(E)	0.1	2000	4	1MeV	-	-

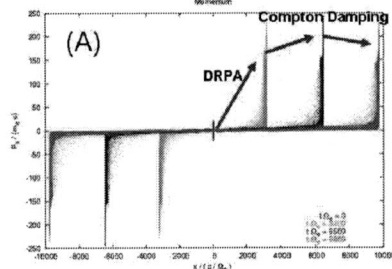 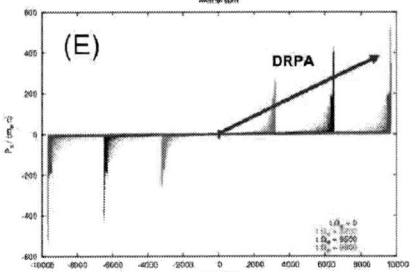

FIGURE 1. x-Px Evolution

of the plasma particles including the Compton drag force.

$$m\frac{d(\gamma\vec{v})}{dt} = q\left(\vec{E} + \frac{1}{c}\vec{v}\times\vec{B}\right) - F_\gamma\left(\frac{\vec{v}}{|v|}\right) \quad (1)$$

The last term of the equation ((1)) is the Compton drag force term. F_γ means the average photon drag force hitting each particle.

Initially, the electron and positron distributions are assumed to be Maxwellian with uniform temperature T_{e^-}, T_{e^+}. The spatial distributions of plasmas have a slab form located in the center of the grid. The initial background magnetic field $\vec{B}_0 = (0, B_0, 0)$ exists only inside the plasma.

CALCULATION RESULTS

The table(1) is the each initial plasma parameters we calculated. Ω_{pe} and Ω_e are the plasma and cyclotron frequency in the initial Maxwellian plasma respectively. λ_d is the electron Debye length. n_p/n_e is the number density ratio between photon and electron in a unit length λ_d.

Case[A] of the table(1) is a higher photon density case. Initially, the particles is accelerated by a DRPA, but, Compton scattering makes the strong drag force to the particle reaching the peak around $\gamma \sim 250$. Case[B] is a lower photon density case. This

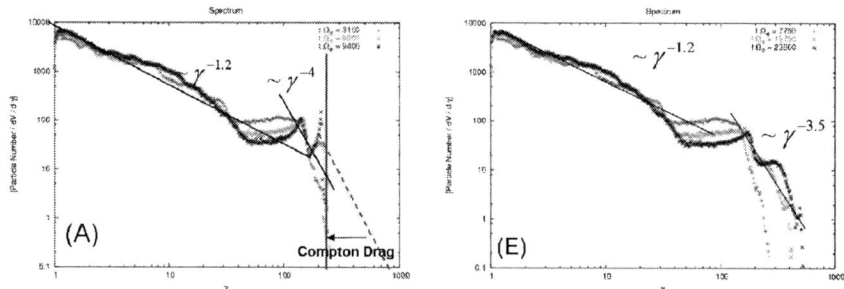

FIGURE 2. Gamma Distribution Evolution

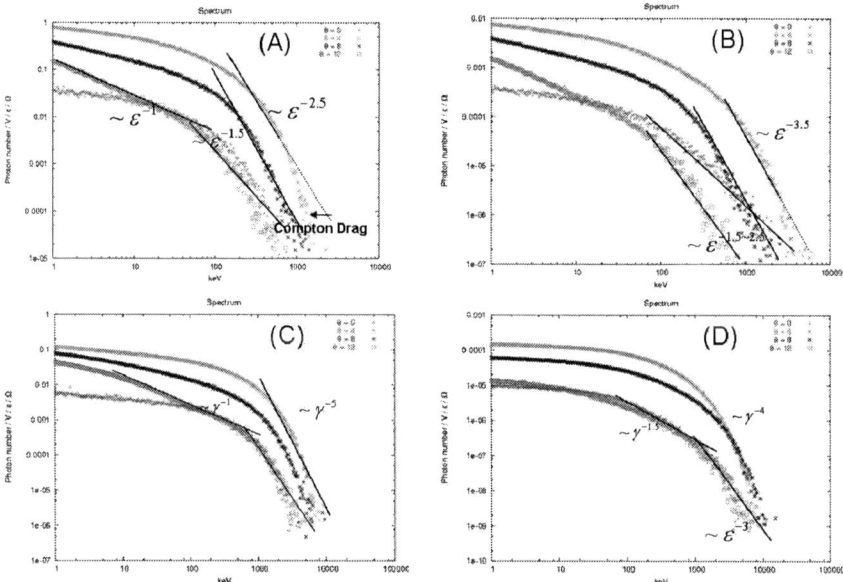

FIGURE 3. Angle Dependent Compton Spectrum

case is similar to the case[E] which is no photon drag. The case [C] and [D] are the case of higher temperature of the incident photon and strong magnetic field respectively.

Figure (1)shows the global evolution of a DRPA expanding into a vacuum and the Compton drag force effect.

Figure (2) shows the gamma distribution of the particles. The red and green dots on the figure is in the middle of the calculation time, and the blue dots is at the end of the calculation. Case [E] (no Compton effect) shows the strong acceleration, which increases the particle energy to $\gamma \sim 500$ or higher with the time glowing. However, Case [A] shows that the growth of the energy of the particles reach a peak around $\gamma \sim 250$. This means that Compton energy loss is strong and efficient in the case [A].

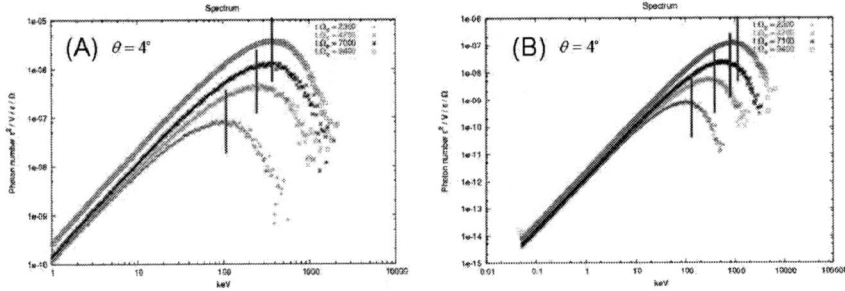

FIGURE 4. Spectrum Time Dependence

Figure (3) shows the angle dependent Compton spectrum. In each case, the spectrum around $4° \sim 6°$ has the peak. This is because the main part of DRPA is along the x direction, but particles are also accelerated to the vertical direction to the x and magnetic direction slightly. Moreover, it seems that there are two power-law indices in lower and higher energy region. The higher energy region of the case[A] has some similarity to the GRB spectrum. Case [A] and [C] have a cut off energy around few MeV which is from the strong Compton Drag Force. It is lower efficiency than the Case [B] and [D].

Figure (4) shows the spectrum time dependence along the angle $4°$. In case [A], E-peak is around a few hundred keV, this is consistent with the GRB spectrum.

CONCLUSION

The diamagnetic relativistic pulse accelerator (DRPA) can convert the magnetic energy into the kinetic energy of the surface particles effectively, resulting in ultrarelativistic particles. Most of particle acceleration is in x-direction, but slightly tilted to z-direction, $Pz/Px \sim 0.1 - 0.2$. This causes the Compton spectrum to have strong angle, time dependence.

The particle momentum distribution is roughly $\gamma^{-1.5 \sim 0}$ at low energy, $\gamma^{-3 \sim 4}$ at high energy. This distribution makes the broken power-law Compton spectrum. We can see the power-law Compton spectrum at high energy region of some cases. This power-law and E-peak might be consistent with GRBs or X-ray Flash. However, longer duration simulation will be needed to build up the correct power-law tail.

The Compton drag is effective in the case of high photon density and hot photon surrounding e+e- plasmas. We can see the high energy cut off by the Compton drag in the Compton efficient case because of the Compton drag $\propto \gamma^2$. This cutoff is testable with GRBs, however, The range of E-peak calculated agrees with observed classical GRBs.

REFERENCES

1. E. Liang and Kazumi Nishimura and Hui Li and S.Peter Gary, *Physical Review Letters* **90**, 085001 (2003).

Suzaku Wide-band All-sky Monitor observations of GRB prompt emissions

Kazutaka Yamaoka*, Satoshi Sugita*, Masanori Ohno†, Takuya Takahashi†, Yasushi Fukazawa†, Yukikatsu Terada**, Yasuhiko Endo‡, Soojing Hong‡,**, Keiichi Abe‡, Kaori Onda‡, Makoto Tashiro‡, Teruaki Enoto§, Ryohei Miyawaki§, Motohide Kokubun§, Kazuo Makishima§,**, Goro Sato¶, Kazuhiro Nakazawa¶, Tadayuki Takahashi¶ and the Suzaku HXD-II team

*Aoyama Gakuin University, 5-10-1 Fuchinobe, Sagamihara, Kanagawa 229-8558, Japan
†Hiroshima University, 1-3-1 Kagamiyama, Higashi-Hiroshima, Hiroshima 739-8526, Japan
**RIKEN, 2-1 Hirosawa, Wako, Saitama 351-0198, Japan
‡Saitama University, 255 Shimo-Ohkubo, Saitama-city, Saitama 338-8570, Japan
§University of Tokyo, 1-3-1 Hongo, Bunkyo-ku, Tokyo 113-0033, Japan
¶ISAS/JAXA, 3-1-1 Yoshinodai, Sagamihara, Kanagawa 229-8510, Japan

Abstract. The Suzaku Wide-band All-sky Monitor (WAM), realized by large thick anti-coincidence shields of the Hard X-ray Detector (HXD), can be powerful gamma-ray burst (GRB) detector which is sensitive to 50-5000 keV gamma-rays. The WAM is now in a full operational phase, and we have already detected some GRBs simultaneously with other satellites (Swift, Konus-Wind, HETE2 and INTEGRAL SPI/ACS). The most impressive event among detected GRBs is GRB051008, which was detected up to 2 MeV with the WAM. In this paper, we report on the WAM in-flight performance as a GRB monitor from initial three-months operations, focusing on the GRB trigger status and spectral analysis of GRB051008 and GRB051111 combined with Swift.

Keywords: X- and gamma-ray telescopes and instrumentation, gamma-ray bursts
PACS: 95.55.Ka, 98.70.Rz

SUZAKU WIDE-BAND ALL-SKY MONITOR

The Japanese 5th X-ray satellite Astro-E2 was successfully launched by M-V rocket from Uchinoura Space Center (USC) on July 10, and renamed as "Suzaku"[1]. It flows in near-circular orbit of 570 km with an inclination angle of 31 degree. Suzaku carries two kinds of instruments: four X-ray CCDs (XIS) on each focal plane of the X-ray Telescopes (XRT) and one hard X-ray Detector (HXD[2]) which covers a broadband energy range of 0.2–700 keV by a combination. These two instruments are now fully operated, and have observed more than 50 sources during performance verification phase so far with a narrow field of view. In addition, wide field of view observations in 50–5000 keV have been performed by HXD Wide-band All-sky Monitor (WAM[4][5]).

The WAM is a subsystem of the HXD, utilizing 20 BGO anti-coincidence detectors as a all-sky monitor with a half-sky coverage for bright celestial sources: GRBs, soft gamma-ray repeaters, solar flares, and etc. The WAM consists of four identical walls (WAM0 to 3) with an geometrical area of 800 cm^2. The excellent feature of the WAM is its on-axis large effective area of 400 cm^2 per wall even at 1 MeV. It will enable us

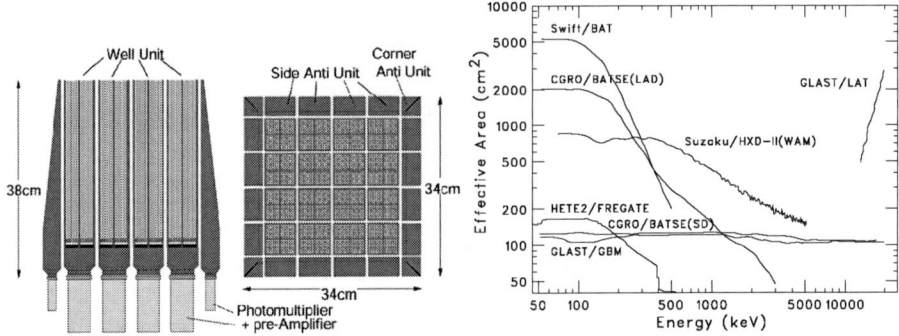

FIGURE 1. Left panel: Schematic view of the Suzaku HXD. The WAM consists of lateral 20 BGO anti-coincidence counters. Right panel: The WAM effective area in comparison with other GRB detectors.

to study high energy radiation of the GRBs at the MeV range and distribution of peak energy (Epeak) above 300 keV of the synchrotron emissions.

CURRENT WAM STATUS

The WAM also has been fully operated since August 19. All the 20 WAM sensors have been working well without significant problem. Automated GRB trigger system by onboard hardware is turned on from Aug. 22. Now we set the judgment level as above 5.7σ of background statistical fluctuation in the 100-240 keV band. The judgment time-scale is either 0.25 s or 1s. Table 1 shows the classified trigger list until the end of November. 109 events including 26 GRB candidates (13 GRBs confirmed by other satellites such as Swift, HETE2, Konus-Wind, and INTEGRAL SPI/ACS and 13 possible GRBs) triggered the WAM. For the confirmed GRBs, the WAM light curve is available at the web page [3]. We already submitted the GCN circulars concerning GRB051008 and GRB051111 detected simultaneously with Swift/BAT. Detailed spectral analyses are described in the next section. The estimated GRB detection rate will be about 100 per year, although the fine tuning to the trigger criteria is still going on.

TABLE 1. Summary of the classified events (2005 Aug.22 – Nov. 30)

Classified event	Number	Notes
Confirmed GRBs	13	e.x. GRB051008[7], GRB051111[8]
Possible GRBs	13	
Solar flares	15	e.x. Aug.22 GOES class M2.7
Particle events	36	
Statistical fluctuation	13	
Others (SAA entrance, mis-operation)	19	
Total	109	

The current response matrices are calculated using the Suzaku mass model[6] and the gain information. This mass model is based on the GEANT4 code and verified by pre-flight calibration with an accuracy of 20%. The energy-scale calibration is done by

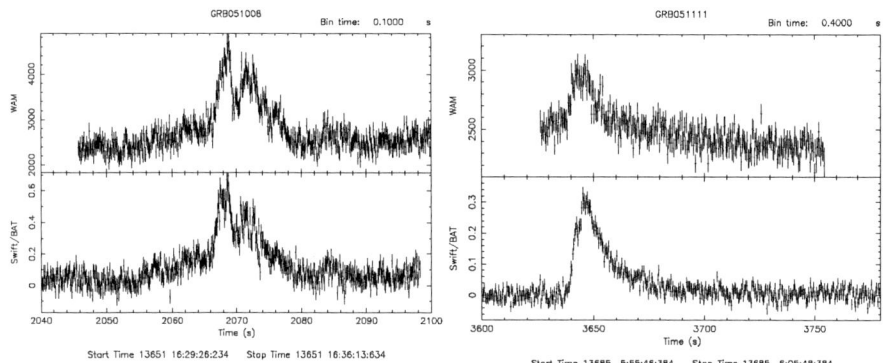

FIGURE 2. Swift/BAT and WAM light curve of GRB051008 and GRB05111.

monitoring 511 keV line seen in the WAM energy spectra. The current flux uncertainty above 100 keV is estimated at about 20% using the GRB events simultaneusly observed with Swift. We will also use solar-flare events and Crab Nebura in a near future. The absolute timing accuracy is estimated within 100 ms.

SPECTRAL ANALYSIS FOR SWIFT GRBS

In initial three-month operations, we have detected 5 Swift GRBs: GRB051008 and GRB051111 as triggered events, GRB050904, GRB050915b and GRB051006 as untriggered events. In order to investigate the combined Swift/BAT and WAM performance, we carry out the joint spectral analysis for the four events except for GRB050904. Fig 3 shows the energy spectra of GRB051008 and GRB051111. For GRB051008, the WAM could detect photon spectrum up to at least 2 MeV

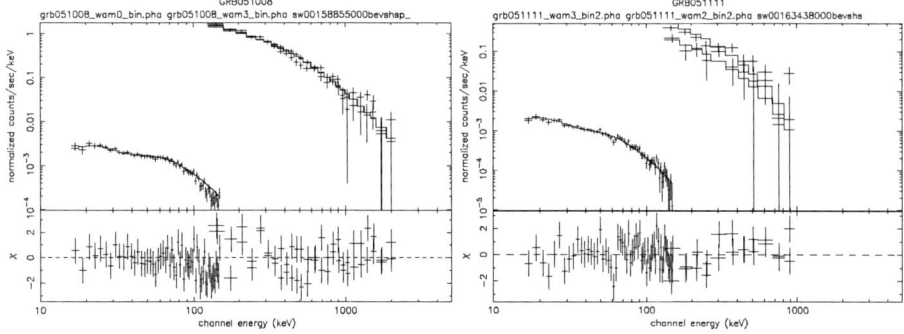

FIGURE 3. Combined Swift/BAT and WAM spectra of GRB051008 (Left) and GRB051111 (Right). Spectra were simultaneously fitted with a cutoff power law model.

We fitted the Swift/BAT and WAM spectra simultaneously with both cutoff power-law model and Band functions. The WAM normalization factor was fixed at the same as that of Swift/BAT. The best fitting parameter is shown in Table 2. We could determine the Epeak for three of the four GRBs with a higher accuracy than fitting with only the WAM data[7][8]. We also shows the unfolded νF_ν spectrum obtained by both detectors in using the cutoff power-law model. We found that the Epeak would be measurable in the range of 50–1000 keV. Furthermore, GRB051111 has known redshift (z=1.54 [9]). We calculated the isotropic energy (Erad=$(1.0\pm0.2)\times10^{52}$erg) and the peak energy (Epeak=354^{+125}_{-89} keV) in the source frame, which were found to be consistent with Amati relations [10]. Note that all these results are very preliminary, but combined Swift/BAT and WAM analysis will be very powerful to study broadband prompt emissions from 15 to 5000 keV or more.

TABLE 2. Best fitting parameters of the four Swift GRBs

	Cutoff powerlaw		Band function			
	α	Epeak	α	β	Epeak	
GRB050915b	$1.30^{+0.24}_{-0.26}$	57 ± 7	$-1.29^{+0.33}_{-0.23}$	<-2.43	57 ± 7	*
GRB051006	1.51 ± 0.10	—				
GRB051008	1.02 ± 0.05	956^{+173}_{-130}	-0.99 ± 0.06	$-1.77^{+0.18}_{-0.59}$	812^{+210}_{-156}	
GRB051111	1.18 ± 0.14	211^{+70}_{-42}	$-0.95^{+0.30}_{-0.20}$	$-1.96^{+0.18}_{-0.26}$	138^{+46}_{-37}	

* All the statistical errors. Not included in systematic errors.

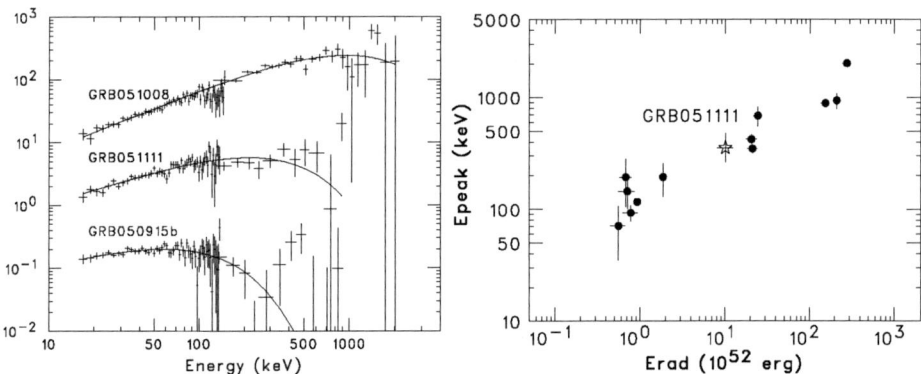

FIGURE 4. Left panel: Unfolded energy spectrum of the three Swift GRBs(GRB050915b, GRB051111, GRB051008). Right panel: GRB051111 parameters (open stars) on Amati relation.

REFERENCES

1. Suzaku web page http://www.astro.isas.jaxa.jp/suzaku/instrumentations
2. K. Makishima et al., New Century of X-ray Astronomy, ASP Conference Proceedings, 251, 564, 2001
3. WAM GRB page http://www.astro.isas.jaxa.jp/suzaku/HXD-WAM/WAM-GRB/grb/grb_table.html
4. K. Yamaoka et al., IEEE Trans. Nucl. Sci., 2005 in press
5. M. Ohno et al., IEEE Trans. Nucl. Sci., 2005 in press
6. Y. Terada et al., IEEE Trans. Nucl. Sci., vol. 52, 4, 909, 2005
7. M. Ohno et al., GCN circulars 4297, 2005
8. K. Yamaoka et al., GCN circulars 4299, 2005
9. J.X. Prochaska et al., GCN circulars 4271, 2005
10. L. Amati et al., A&A 390, 81, 2002

Tail Emission of Prompt Gamma-Ray Burst Jets

Ryo Yamazaki*, Kenji Toma†, Kunihito Ioka† and Takashi Nakamura†

*Department of Physics, Hiroshima University, Higashi-Hiroshima, Hiroshima 739-8526, Japan
†Department of Physics, Kyoto University, Kyoto 606-8502, Japan

Abstract. Tail emission of the prompt gamma-ray burst is discussed using a multiple emitting sub-shell (inhomogeneous jet, sub-jets) model. The tail is a superposition of a number of smooth, long-duration, dim, and soft pulses emitted by segments located far from the line of sight. We find that the behavior of the tail is not so much affected by the local inhomogeneity but affected by the global sub-jet distribution. Some observed tails may disfavor the power-law jets.

Keywords: gamma-ray bursts
PACS: 98.70.Rz

INTRODUCTION

The *Swift* satellite has revealed rich structures in early X-ray counterparts of gamma-ray bursts (GRBs) [1, 2, 3, 4, 5]. A distinct decaying component before the usual afterglow phase is identified. During this epoch, the decay is steeper and the spectral index is different compared with the subsequent phase [5], which suggests that this component is the tail emission of the prompt GRB [6, 7]. Most natural explanation for the tail is a high latitude emission from a relativistically moving shell [8, 9]. For a spherical uniform shell, the predicted decay index is around -3 [8], while events with more rapid decay exist [5]. Since the emission region sweeps the shell, the tail emission features, e.g., the decay index and smoothness, would diagnose the unknown GRB jet structure. We discuss the tail emission of the structured inhomogeneous GRB jets by using a multiple emitting sub-shell (inhomogeneous jet, sub-jets) model considered so far [10, 11, 12, 13, 14]. Our model well reproduces the observed features of the tail emission of the prompt GRB. Since the pulse duration is long for a large viewing angle, the local inhomogeneity is averaged and the smoothed tail is obtained. Then the global sub-jet distribution is important for the behavior of the tail. Therefore, the discrete multiple sub-jet model and the continuous surface model obtained by averaging sub-jets predict the same decay index. Some of the observed tails are not steep enough to be consistent with the power-law jet model. For details, see [15].

TAIL EMISSION OF PROMPT GRB

We consider the same model as discussed in our previous works. The whole GRB jet, whose opening half-angle is $\Delta\theta_{\rm tot}$, consists of $N_{\rm tot}$ emitting sub-shells (sub-jets). Each sub-jet departs at time $t_{\rm dep}^{(j)}$ ($0 < t_{\rm dep}^{(j)} < t_{\rm dur}$, where $j = 1, \cdots, N_{\rm tot}$, and $t_{\rm dur}$ is the active time of the central engine) from the central engine, and has an opening half-angle $\Delta\theta_{\rm sub}^{(j)}$, the

Lorentz factor $\gamma^{(j)}$, the emitting radius $r_0^{(j)}$, and the break frequency in the shell comoving frame $v_0'^{(j)}$. For each sub-jet, the emission model is the same as in the previous works [16, 17, 18, 19]. The whole light curve from the GRB jet is produced by the superposition of the sub-jet emission.

We present some of kinematical properties of prompt GRBs in the multiple sub-jet model. Let $\theta_v^{(j)}$ be the angle between the observer's line of sight and the axis of the jth sub-jet. The pulse starting and ending time at the observer are given by

$$T_{\text{start}}^{(j)} \sim t_{\text{dep}}^{(j)} + \frac{r_0^{(j)}}{2c\gamma_{(j)}^2}\left(1+\gamma_{(j)}^2 \theta_-^{(j)2}\right), \qquad (1)$$

$$T_{\text{end}}^{(j)} \sim t_{\text{dep}}^{(j)} + \frac{r_0^{(j)}}{2c\gamma_{(j)}^2}\left(1+\gamma_{(j)}^2 \theta_+^{(j)2}\right), \qquad (2)$$

where $\theta_+^{(j)} = \theta_v^{(j)} + \Delta\theta_{\text{sub}}^{(j)}$ and $\theta_-^{(j)} = \max\{0, \theta_v^{(j)} - \Delta\theta_{\text{sub}}^{(j)}\}$, and we use the formulae $\beta_{(j)} \sim 1 - 1/2\gamma_{(j)}^2$ and $\cos\theta \sim 1 - \theta^2/2$ for $\gamma^{(j)} \gg 1$ and $\theta \ll 1$, respectively. Then, the pulse duration, $\delta T^{(j)} = T_{\text{end}}^{(j)} - T_{\text{start}}^{(j)}$, is given by $\delta T^{(j)} \sim 1.5\, r_{14}(\theta_+^{(j)}/0.03)^2$ s for $\theta_v^{(j)} < \Delta\theta_{\text{sub}}^{(j)}$, and $\delta T^{(j)} \sim 26\, r_{14}(\Delta\theta_{\text{sub}}^{(j)}/0.02)(\theta_v^{(j)}/0.2)$ sec for $\theta_v^{(j)} > \Delta\theta_{\text{sub}}^{(j)}$, where $r_{14} = r_0^{(j)}/10^{14}$ cm. The peak energy $E_p^{(j)}$, that gives the peak of the vF_v spectrum, is approximated as $E_p^{(j)} \propto v_0'^{(j)} \delta^{(j)}$, where $\delta^{(j)} = [\gamma^{(j)}(1-\beta_{(j)}\cos\theta_-^{(j)})]^{-1}$. Using practical numerical calculations, we find $E_p^{(j)} \sim 6.5 \times 10^2 \gamma_2 v_5$ keV for $\theta_v^{(j)} \ll \Delta\theta_{\text{sub}}^{(j)}$, and $E_p^{(j)} \sim 2\,\gamma_2^{-1} v_5 (\theta_v^{(j)}/0.2)^{-2}$ keV for $\theta_v^{(j)} \gg \Delta\theta_{\text{sub}}^{(j)}$, where $v_5 = v_0'^{(j)}/5$ keV and $\gamma_2 = \gamma^{(j)}/10^2$. We note that the light curve from a single pulse becomes dim and smooth for large $\theta_v^{(j)}$.

For sub-jets with $\theta_v^{(j)} < \Delta\theta_{\text{sub}}^{(j)}$, we have $\delta T^{(j)} \ll t_{\text{dur}}$, because t_{dur} is about several tens of seconds. Then the bright pulses arising from these sub-jets are concentrated in the epoch $0<T<t_{\text{dur}}$. On the other hand, for sub-jets with $\theta_v^{(j)} > 0.1\, r_{14}^{-1/2}(t_{\text{dep}}^{(j)}/20\text{ sec})^{1/2}$, the second terms of the right hand sides of Eqs. (1) and (2) become larger than the first ones, so that $T_{\text{start}}^{(j)}$ and $T_{\text{end}}^{(j)}$ may be much larger than t_{dur} for large $\theta_v^{(j)}$. Such smooth, long-duration, dim, and soft pulses make the tail emission of the prompt GRB.

Now we concentrate on the temporal structure of the tail emission. We assume that each sub-jet is not drastically different. Since $T_{\text{start}}^{(j)}$ and $T_{\text{end}}^{(j)}$ are much larger than t_{dur}, they are not affected by $t_{\text{dep}}^{(j)}$. Thus we may consider that all the sub-jets contributing to the tail emission emit simultaneously at a mean radius r_0 in the central engine frame. The end of the tail emission T_{tail} is determined by the angular size of the whole jet: $T_{\text{tail}} \sim (r_0/2c)(\Delta\theta_{\text{tot}} + \vartheta_{\text{obs}})^2 \sim 2 \times 10^2 r_{14}[(\Delta\theta_{\text{tot}} + \vartheta_{\text{obs}})/0.3]^2$ sec. The tail flux at an observer time T is the superposition of the sub-jet emission with the pulse starting and ending time $T_{\text{start}}^{(j)} < T < T_{\text{end}}^{(j)}$. Sub-jets with viewing angles between $\theta_T - \Delta\theta_{\text{sub}}$ and $\theta_T + \Delta\theta_{\text{sub}}$ contribute to the tail flux at a time T, where $\theta_T = (2cT/r_0)^{1/2} \sim 0.2 r_{14}^{-1/2}(T/10^2\text{ s})^{1/2}$ rad. Then we may calculate the number of these contributing sub-jets $N_{\text{sub}}(T)$ and its variance $1/\sqrt{N_{\text{sub}}(T)}$ if the sub-jet distribution and observer's line of sight are given. While $N_{\text{sub}}(T)$ is sufficiently large, the tail light curve will be smooth. In the following, we actually draw the light curves of prompt GRB emission for various cases.

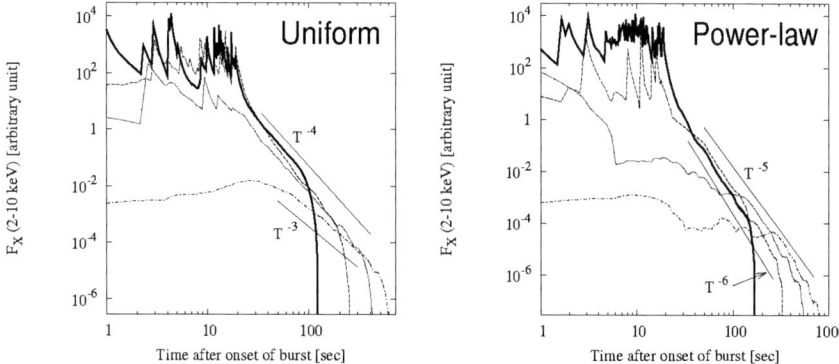

FIGURE 1. Examples of light curves of the prompt GRB emission calculated by the multiple emitting sub-shell model (multiple sub-jet model). All sub-jets have the same intrinsic properties. Thick-solid, dashed, dotted, and dot-dashed lines correspond the viewing angles of $\vartheta_{obs} = 0$, $\Delta\theta_{tot}/2$, $\Delta\theta_{tot}$, and $3\Delta\theta_{tot}/2$, respectively. The observer time is the time since the onset of the burst. *Left panel*: For the uniform sub-jet distribution. *Right panel*: For the power-law sub-jet distribution.

Example 1: Uniform jet. We first consider the uniformly distributed sub-jets. The number of sub-jets per unit solid angle is approximately given by $dN/d\Omega = N_{tot}/(\pi\Delta\theta_{tot}^2)$ for $\vartheta < \Delta\theta_{tot}$, where $\Delta\theta_{tot} = 0.25$ rad is adopted. The departure time of each sub-jet $t_{dep}^{(j)}$ is assumed to be homogeneously random between $t = 0$ and $t = t_{dur} = 20$ sec. The central engine is assumed to produce $N_{tot} = 1000$ sub-jets. We assume that all the sub-jets have the same values of the model parameters: $\Delta\theta_{sub} = 0.02$ rad, $\gamma = 100$, $r_0 = 1.0 \times 10^{14}$ cm, and $h\nu_0' = 5$ keV. The left panel of figure 1 shows the results, where $\vartheta_{obs} = 0$ (thick-solid line), $\Delta\theta_{tot}/2$ (dashed), $\Delta\theta_{tot}$ (dotted), and $3\Delta\theta_{tot}/2$ (dot-dashed) are considered. One can see the entire behavior of bursts. For the cases $\vartheta_{obs} < \Delta\theta_{tot}$, as expected, the bright pulses are observed in the period $0 < T < t_{dur} = 20$ sec, and subsequently the tail emission starts. However, when $\vartheta_{obs} = 3\Delta\theta_{tot}/2$, the whole jet is seen off-axis. Then the emission becomes dim and soft [17, 18, 19]. In any cases, the tail emission is smooth. For $\vartheta_{obs} = 0$, the number of contributing sub-jets $N_{sub}(T)$ is approximately given by $N_{sub}(T) \sim 4\pi\theta_T\Delta\theta_{sub}(dN/d\Omega)$. For our adopted parameters, we derive $N_{sub}(T) \sim 2.6 \times 10^2 \, (T/10^2 \, \text{s})^{1/2} \gg 1$, so that the tail light curve is smooth. Also in other cases ($\vartheta_{obs} \neq 0$), we obtain $N_{sub}(T) \gg 1$. The decay index of the tail emission is about -4 when it is determined by the whole light curve. When the time zero is shifted to the maximum of the last bright pulse, the decay index of the tail emission ranges between -3 and -2 [15].

Example 2: Power-law jet. Next we consider the power-law sub-jet distribution, i.e., $dN/d\Omega = C$ for $0 < \vartheta < \vartheta_c$ and $dN/d\Omega = C(\vartheta/\vartheta_c)^{-2}$ for $\vartheta_c < \vartheta < \Delta\theta_{tot}$, where $\vartheta_c = 0.03$ rad and $\Delta\theta_{tot} = 0.3$ rad, and $C = (N_{tot}/\pi\vartheta_c^2)[1 + 2\ln(\Delta\theta_{tot}/\vartheta_c)]^{-1}$ is the normalization constant [13, 14, 20, 21]. The departure time of each sub-jet $t_{dep}^{(j)}$ is assumed to be homogeneously random between $t = 0$ and $t = t_{dur} = 20$ sec and we adopt $N_{tot} = 350$. All the sub-jets have the same parameters as in the uniform jet case. The right panel of figure 1 shows the results. Compared with the uniform jet case, the decay index is steeper because the power-law jet is dimmer in the outer region, i.e., the sub-jets are sparsely distributed near the periphery of the whole jet. For example, when $\vartheta_{\text{obs}} = 0$, the number of contributing sub-jets $N_{\text{sub}}(T)$ is calculated as $N_{\text{sub}}(T) \sim 4\pi\theta_T \Delta\theta_{\text{sub}}(dN/d\Omega) \sim 28\ (T/100\ s)^{-1/2}$, which is a decreasing function of time contrary to the uniform jet case. We note that some of the observed tails may be too shallow to be consistent with the power-law jets.

Finally, a typical value of the viewing angle at an observer time T is $\theta_T \sim 0.2(T/10^2\ s)^{1/2}$ rad, and then the peak energy, E_p, is about a few keV. Therefore, the spectral index in the X-ray band is given by the high-energy photon index $\beta \sim -2.3$, which is consistent with the observed photon index (average value is 2.28) [5].

ACKNOWLEDGMENTS

This work was supported in part by Grants-in-Aid for Scientific Research of the Japanese Ministry of Education, Culture, Sports, Science, and Technology 09245 (R. Y.), 14047212 (T. N., K. I.), 14204024 (T. N.), and 17340075 (T. N.).

REFERENCES

1. Burrows, D. N. et al., Science, **309**, 1833 (2005).
2. Chincarini, G. et al., astro-ph/0506453 (2005).
3. Campana, S. et al., ApJ, **625**, L23 (2005).
4. Tagliaferri, G. et al., Nature, **436**, 985 (2005).
5. Nousek, J. A. et al., astro-ph/0508332 (2005).
6. Zhang, B. et al., astro-ph/0508321 (2005).
7. Panaitescu, A., et al., astro-ph/0508340 (2005).
8. Kumar, P., & Panaitescu, A., ApJ, **541**, L51 (2000).
9. Yamazaki, R., et al., PASJ, **57**, L11 (2005).
10. Nakamura, T., ApJ, **534**, L159 (2000).
11. Kumar, P., & Piran, T., ApJ, **535**, 152 (2000).
12. Yamazaki, R., Ioka, K., & Nakamura, T., ApJ, **607**, L103 (2004).
13. Toma, K., Yamazaki, R., & Nakamura, T., ApJ, **620**, 835 (2005).
14. Toma, K., Yamazaki, R., & Nakamura, T., ApJ, **635**, 481 (2005).
15. Yamazaki, R., et al., astro-ph/0509159 (2005).
16. Ioka, K., & Nakamura T., ApJ, **554**, L163 (2001).
17. Yamazaki, R., Ioka, K., & Nakamura, T., ApJ, **571**, L31 (2002).
18. Yamazaki, R., Ioka, K., & Nakamura, T., ApJ, **593**, 941 (2003).
19. Yamazaki, R., Ioka, K., & Nakamura, T., ApJ, **606**, L33 (2004).
20. Rossi, E., Lazzati, D., & Rees, M. J., MNRAS, **332**, 945 (2002).
21. Zhang, B., & Mészáros, P., ApJ, **571**, 876 (2002).

The GRB luminosity function in the internal shock model

Frédéric Daigne, Robert Mochkovitch * and Hannachi Zitouni[†]

Institut d'Astrophysique de Paris
UMR 7095 CNRS – Université Pierre et Marie Curie-Paris VI, 98 Bd Arago, 75014 Paris, France
[†]*Institut de Physique USTHB, Medea, Algeria*

Abstract. The intrinsic luminosity distribution of GRBs is currently poorly constrained by observations. In the framework of the internal shock model, it depends on the distribution of the kinetic energy injection rate in the relativistic outflow emerging from the central source and on the efficiency of internal shocks. We study the effect of this efficiency and show that it naturally leads a luminosity function with a high luminosity branch whose shape follows that of the kinetic injection rate, together with a power law tail at low luminosity.

Keywords: Gamma-ray sources; gamma-ray bursts, Luminosity and mass functions
PACS: 98.70.Rz, 97.10.Xq

INTRODUCTION

In the absence of an unbiased volume limited sample of GRBs with known distances, the GRB (isotropic) luminosity function (hereafter LF) remains uncertain. It should a priori extend from $L_{max} > 10^{54}$ erg.s^{-1} (GRB 990123) to $L_{min} < 10^{47}$ erg.s^{-1} (GRB 980425) but the true shape of the LF in this huge interval is not well known. Power laws or broken power laws are usually assumed and the BATSE Log N - Log P is then used to constrain the slopes (Daigne et al, this conference and [1, 2]). Here we show that the internal shock model predicts that a low luminosity power law branch of slope $-1/2$ can be expected in the LF. We obtain the condition for this branch to appear and discuss some consequences.

THE GRB LUMINOSITY FUNCTION

Analytical approach

In the context of the internal shock model the gamma-ray luminosity of a burst is given by

$$L = f(\kappa)\dot{E} \qquad (1)$$

where $\dot{E}$ is the rate of kinetic energy injection in the relativistic outflow produced by the central source and f the fraction of kinetic energy dissipated in internal shocks and eventually radiated. The value of f essentially depends on the contrast $\kappa = \Gamma_{max}/\Gamma_{min}$ of the Lorentz factor distribution in the relativistic wind.

The LF $\varphi(L)$ can then be written

$$\varphi(L) = \int_{\kappa_1}^{\kappa_2} d\kappa \frac{\psi(\kappa)\phi[L/f(\kappa)]}{f(\kappa)} \qquad (2)$$

where $\psi(\kappa)$ and $\phi(\dot{E})$ are the distribution functions of κ and $\dot{E}$ respectively. The LF is defined between $L_{\min} = f(\kappa_{\min})\dot{E}_{\min}$ and $L_{\max} = f(\kappa_{\max})\dot{E}_{\max}$. In the interval $L_* = f(\kappa_{\max})\dot{E}_{\min} < L < L_{\max}$, the limits of the integral in Eq.(2) are

$$\kappa_1 = f^{-1}\left(\frac{L}{\dot{E}_{\max}}\right) \quad \text{and} \quad \kappa_2 = \kappa_{\max} \qquad (3)$$

while for $L_{\min} < L < L_*$ we have

$$\kappa_1 = f^{-1}\left(\frac{L}{\dot{E}_{\max}}\right) \quad \text{and} \quad \kappa_2 = f^{-1}\left(\frac{L}{\dot{E}_{\min}}\right). \qquad (4)$$

An illustrative example

We suppose that $\phi(\dot{E})$ is a power law of index $-s$ between $\dot{E}_{\min}$ and $\dot{E}_{\max}$. If $\dot{E}_{\max}$ is sufficiently large we can write

$$\kappa_1 = f^{-1}\left(\frac{L}{\dot{E}_{\max}}\right) \sim \kappa_{\min} \qquad (5)$$

so that the GRB LF reads

$$\varphi(L) \propto L^{-s} \times \int_{\kappa_{\min}}^{\kappa_{\max}} d\kappa\, \psi(\kappa)[f(\kappa)]^{s-1} \propto L^{-s} \qquad (6)$$

for $L > L_*$ and

$$\varphi(L) \propto L^{-s} \times \int_{\kappa_{\min}}^{f^{-1}(L/\dot{E}_{\min})} d\kappa\, \psi(\kappa)[f(\kappa)]^{s-1} \qquad (7)$$

for $L < L_*$. The high luminosity branch of the LF is therefore also a power law of index $-s$. To obtain the low luminosity branch we first assume that $\kappa_{\min} = 1$. The minimum efficiency is zero (no internal shock) and for $\kappa \gtrsim 1$, we have

$$f(\kappa) \approx a(\kappa - 1)^\alpha \qquad (8)$$

where a and α are constant. At low luminosity the LF then behaves as

$$\varphi(L) \propto L^{-s} \times \int_{1}^{\kappa(L)} d\kappa\, \psi(1) a^{(s-1)} (\kappa - 1)^{\alpha(s-1)} \qquad (9)$$

where
$$\kappa(L) = 1 + \left(\frac{L}{a\dot{E}_{\min}}\right)^{1/\alpha} \tag{10}$$

which finally gives
$$\varphi(L) \propto L^{\frac{1}{\alpha}-1} \tag{11}$$

The slope of the low luminosity branch is fixed by the behavior of $f(\kappa)$ at $\kappa \gtrsim 1$ and does not depend on the slope of the high luminosity branch. In a toy model where IS are simply reduced to the collision of two shells of equal mass the efficiency can be obtained analytically
$$f(\kappa) = \varepsilon_e \frac{(\sqrt{\kappa}-1)^2}{\kappa+1} \tag{12}$$

where ε_e is the fraction of dissipated energy transferred to electrons and radiated [3]. Assuming that ε_e is a constant, $f(\kappa) \propto (\kappa-1)^2$ for $\kappa \gtrsim 1$ and therefore $\varphi(L) \propto L^{-1/2}$.

Numerical results

We have used a simple Monte-Carlo approach to generate synthetic GRBs for any given dirtibution $\psi(\kappa)$ and $\phi(\dot{E})$ and then obtain the resulting LF. We have first tested the case $\kappa_{\min} = 1$, $\kappa_{\max} = 10$ and $\phi(\dot{E}) \propto \dot{E}^{-1.5}$ with $\dot{E}_{\min} = 10^{51}$ erg.s^{-1} and $\dot{E}_{\max} = 10^{54}$ erg.s^{-1}. We compare in Fig.1a the LFs for $\psi(\kappa) \propto \kappa^{-\beta}$ and $\beta = 0$, 1 and 2. At the lowest luminosities the three LFs behave as $L^{-1/2}$ but they somewhat differ around $L = L_*$.

In Fig.1b we adopt $\psi(\kappa) \propto \kappa^{-1}$ ($\beta = 1$) but now $\kappa_{\min} \neq 1$, i.e. we suppose that the Lorentz factor in the outflow produced by the central source cannot be uniform. It appears that the low luminosity branch is now present only for $\kappa_{\min} \sim 1$. With $\kappa_{\min} \gtrsim 2$ it practically disappears. We also considered the case where $\phi(\dot{E})$ is lognormal rather than a power law. Again, a low luminosity power law branch of slope $-1/2$ is present if $\kappa_{\min} \sim 1$.

DISCUSSION AND CONCLUSION

The existence of a low luminosity branch of slope $-1/2$ is a robust prediction of the internal shock model as soon as the distribution of κ – the contrast between the maximum and minimum Lorentz factor in the relativistic flow emitted by the central source – extends down to $\kappa_{\min} \sim 1$. An evidence that this is the case may be the existence of XRFs if they indeed form in outflows with a small contrast as discussed in [3].

The high luminosity branch of the LF keeps the shape of the original distribution of injected kinetic power $\dot{E}$ and the transition between the two regimes occurs at a luminosity $L_* = f(\kappa_{\max})\dot{E}_*$ where $\dot{E}_*$ is the value of $\dot{E}$ where $\phi(\dot{E})$ has its (assumed to be single) maximum (in the case of a power law $\phi(\dot{E})$, $\dot{E}_* = \dot{E}_{\min}$). The low luminosity branch then has a practical interest only if L_* is large enough so that the bursts populating this branch can be observed at a reasonable rate. If L_* is too small the only accessible part of the LF will behave as $\phi(\dot{E})$, the distribution of injected power.

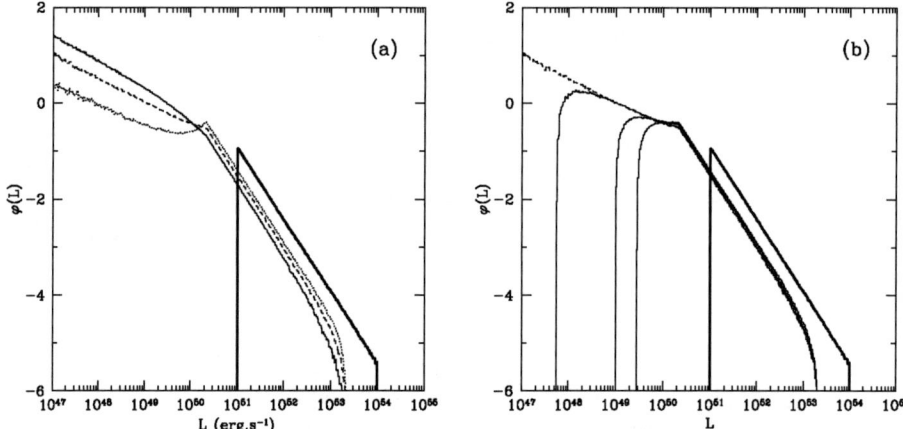

FIGURE 1. Left panel: GRB luminosity function for $\kappa_{min} = 1$ and three values of the index β for $\psi(\kappa)$. The full, dashed and dotted lines respectively correspond to $\beta = 0$, 1 and 2. The thick full line shows the distribution function for the injected kinetic power $\dot{E}$. Right panel: GRB luminosity function for $\beta = 1$. The dotted line shows the case $\kappa_{min} = 1$ while the other lines (from left to right) correspond to $\kappa_{min} = 1.1$, 1.5 and 2.

REFERENCES

1. Firmani, C., Avila-Reese, V., Ghisellini, G. and Tutukov, A.V., 2004, ApJ 611, 1033.
2. Guetta, D., Piran, T. and Waxman, E., 2005, ApJ 619, 412.
3. Barraud, C., Daigne, F., Mochkovitch, R. and Atteia, J.L., 2005, A&A 440, 809.

EARLY AFTERGLOW

The Swift XRT: Observations of Early X-ray Afterglows

David N. Burrows[*], J. A. Kennea[*], J. A. Nousek[*], J. P. Osborne[†], P. T. O'Brien[†], G. Chincarini[‡,#], G. Tagliaferri[‡], P. Giommi[&], B. Zhang[**] and the XRT team

[*]*Department of Astronomy and Astrophysics, The Pennsylvania State University, 525 Davey Lab, University Park, PA 16802, USA*
[†]*Department of Physics and Astronomy, University of Leicester, Leicester LE1 7RH, UK*
[‡]*INAF-Osservatorio Astronomico di Brera, Via Bianchi 46, 23807 Merate, Italy*
[#]*Università degli studi di Milano-Bicocca, Piazza delle Scienze 3, 20126 Milan, Italy*
[&]*ASI Science Data Center, via Galileo Galilei, 00044 Frascati, Italy*
[**]*Department of Physics, University of Nevada, Box 454002, Las Vegas, NV 89154-4002, USA*

Abstract. During the first year of operations of the *Swift* observatory, the X-ray Telescope has made a number of discoveries concerning the nature of X-ray afterglows of both long and short GRBs. We highlight the key findings, which include rapid declines at early times, a standard template of afterglow light curve shapes, common flaring, and the discovery of the first short GRB afterglow.

Keywords: GRB; Swift; XRT; afterglow.
PACS: 95.40.+s; 95.55.Ka; 95.85.-e; 95.85.Nv; 97.60.-s ; 97.60.Lf; 98.58.Fd; 98.62.-g; 98.62.Nx; 98.70.Rz

INTRODUCTION

The *Swift* observatory [1] was launched on November 20, 2004. It carries three instruments: the Burst Alert Telescope (BAT) [2], the X-ray Telescope (XRT) [3], and the UV-Optical Telescope (UVOT) [4]. The BAT, with a 2 sr field of view, discovers and localizes Gamma-Ray Bursts (GRBs). Once a new burst is discovered by the BAT, the *Swift* spacecraft autonomously performs a rapid slew to that position to point the XRT and UVOT telescopes at the burst. The XRT and UVOT are designed to observe GRB afterglows within 1-2 minutes of the burst.

The XRT is a Wolter I grazing incidence X-ray telescope with effective area of about 120 cm^2 at 1.5 keV [5]. The XRT obtained first light on December 11, 2004 (Figure 1a) and observed its first GRB afterglow on Dec. 23, 2004 [6]. Instrument testing and calibration continued until March 2005, but automated GRB observations began in mid-January. During most of the calibration period, GRB followup observations were limited to 20 ks, but by the commencement of normal operations in April 2005, automated GRB followups lasted for 60 ks of observing time, with most afterglows followed for days to weeks, and a few long-lasting afterglows followed for up to two months. The XRT is a low background instrument, making very sensitive observations routine, as indicated in Figure 1b.

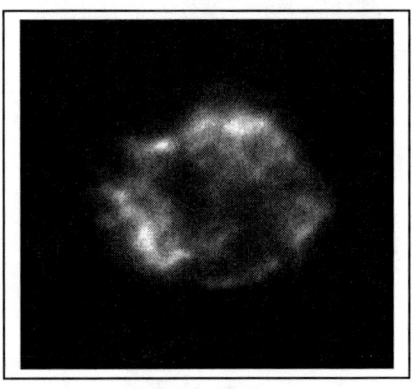 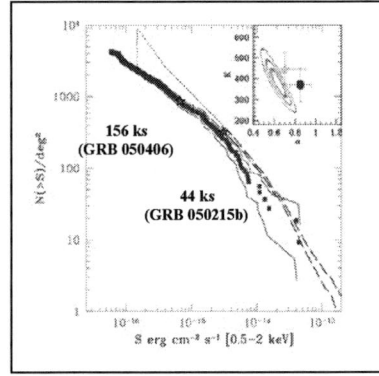

FIGURE 1. a) Left-hand panel: First-light observation of the Cas A supernova remnant. Photons are color coded according to their energy. b) Right-hand panel: The Chandra log N-log S curve (from [7]) for the 0.5-2.0 keV band, with two *Swift* data points from two deep XRT afterglow observations overplotted with blue crosses [8]. The XRT sensitivity limit is below 10^{-15} erg/cm²/s.

STATISTICAL PROPERTIES OF XRT AFTERGLOWS

The *Swift* BAT discovered 92 bursts over a period of ~47 weeks of live time between December 17, 2004 and November 24, 2005, for an average detection rate of 102 bursts per year (see Figure 2 for a graph of burst fluence vs. time).

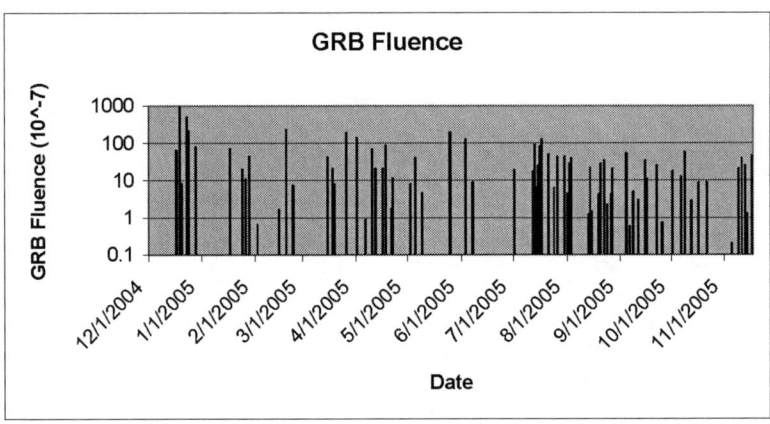

FIGURE 2. Fluence of BAT-discovered bursts (15-150 keV) vs time. Dates are given as MM/DD/YYYY (US convention).

The XRT observed 53 of these bursts within 350 s and a total of 77 within 200 ks of the burst. Afterglows were detected from 71 of these. The time since trigger for the first XRT observation is bimodal, primarily because some bursts detected by the BAT were too close to the Earth limb to allow an immediate slew, and the spacecraft had to wait roughly ½ orbit before observations by the XRT and UVOT could begin. There are also some bursts that were too close to the Sun or Moon to allow immediate observations. The distribution of XRT observation times is shown in Figure 3a. We note that in its first year of operations, the *Swift* XRT has more than doubled the total number of known GRB afterglows (from ~55 at the time of the *Swift* launch).

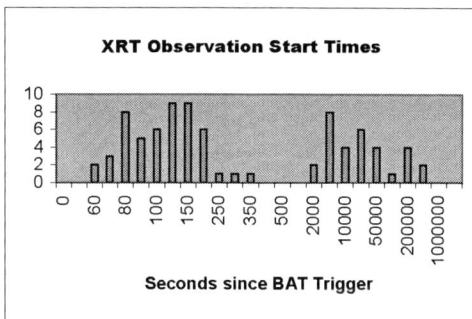
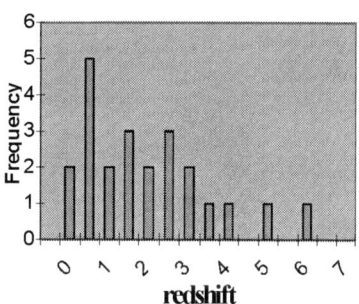

FIGURE 3. a) Left: Distribution of times for the beginning of XRT observations for BAT-discovered GRBs. The observations with times less than 500 s are visible nearly immediately. The observations that begin after a few thousand seconds were typically too close to the Earth limb when the burst occurred, and the spacecraft had to wait about ½ orbit before observations could commence. The higher times are due to Sun and Moon observing constraints. **b)** Right: Distribution of redshifts for *Swift* bursts (from GCN Circulars).

Averaged over the first year of operations, we slewed promptly to the burst ~80% of the time. This fraction has increased dramatically from ~50% early in the mission, due to improvements in our observing strategies. In 95% of the cases with prompt slews (defined as slews within 350 s of the burst), the XRT finds an afterglow. Most of the remaining cases are short bursts, which will be discussed further below. We have only one case (GRB 050911) in which a prompt slew occurred to a long burst and there was no detectable afterglow. This case could be due to a "naked" GRB: a GRB located in an extremely low density environment [9]. In about 80% of the cases, we find either a very rapid decline at early times [10], strong X-ray flares [11,12], or both. These behaviors will be discussed further below.

At the end of 2005, there are now 29 *Swift* GRBs with redshift measurements. The mean redshift for the 23 long GRBs is 2.5, while the mean redshift for the six short GRBs with measured redshift is 0.43. The distribution of redshifts found for *Swift* bursts is shown in Figure 3b. These are almost entirely based on ground-based spectroscopy, although a few bursts have photometrically-determined redshifts. The highest redshift burst found so far is GRB 050904, at z=6.29 [13,14].

The main purpose of the XRT is to obtain rapid, accurate positions for GRB afterglows. With recent corrections to the XRT boresight, XRT positions are now typically accurate to about 3.5 arcseconds [15]. Unfortunately, because the afterglows are much fainter than we anticipated when the instrument was designed, our on-board centroiding algorithm succeeds for only about 40% of GRBs. Changes to the flight software are underway that will allow us to determine rapid centroids for even faint afterglows.

EARLY X-RAY AFTERGLOWS

"Standard" X-ray Afterglow Template

Before launch, we expected that most X-ray afterglows would consist of a simple power-law decay, with a jet break around one day post-burst. Instead, we have found that the "standard" X-ray afterglow has a much more complex shape. The typical afterglow light curve can be described in terms of a template with 3 or 4 power law segments [16, 17, 18] with significant flaring superposed on all segments. Figure 4a, from [16], shows some typical afterglow light curves, while Fig. 4b shows the "Rosetta Stone" light curve of GRB 050315, which displays all four segments described in [17]: a steep decline, a very flat section attributed to energy injection,

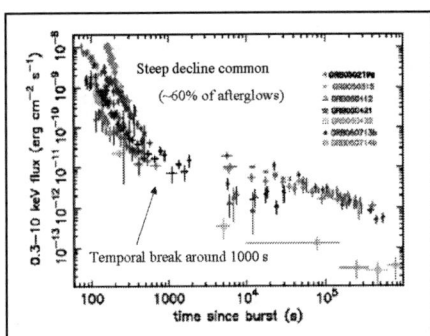
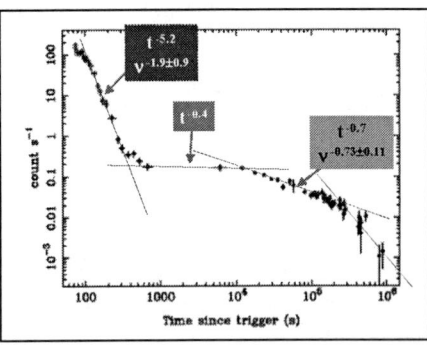

FIGURE 4. a) Left: Typical non-flaring X-ray afterglow light curves measured by the XRT. These show the typical steep, flat, steep sequence. **b)** Right: The afterglow of GRB 050315 [19], which shows all four segments identified by [17].

a more typical decay identified as the external shock of the afterglow, and finally, a steep decline associated with a jet break.

Early Rapid Decline

The identification of an early rapid decline in X-ray afterglows was a very early result from the XRT [10]. We now know that this segment is present in over half of the afterglows with prompt observations. The rapid decline is attributed to the "curvature effect", the result of light propagation delays from a curved wavefront. We have shown that the BAT light curves extrapolated into the XRT energy band join smoothly with the XRT light curves in nearly all cases [20], and that this steep component is indeed associated with the decaying tail of the prompt emission [20,21]. In fact, the spectral and temporal properties of this phase of the light curve are consistent with a combination of "curvature" radiation from the prompt emission and an underlying afterglow component, providing evidence that, in most cases, the afterglow is already established before the prompt emission decays away, and is revealed as a flatter segment of the light curve once the curvature radiation drops to a low enough flux [20].

X-ray Flares

XRT observations have shown that flares are very common in X-ray afterglows, particularly in the first hour after the burst [11,12]. We find evidence for X-ray flaring at some level in nearly half of all afterglows with prompt observations. In some cases the flares constitute a significant fraction of the total energy budget, exceeding the fluence of the prompt emission for GRB 050502B [22]. We believe that the flaring is dominated by late-time activity of the central engine. Several observational characteristics of the flares point in this direction, including the magnitude of the largest flares (20% - 100% of the prompt fluence), the rapid rise and decay of the flares ($\delta t/t \ll 1$), the fact that multiple flares are common, and the fact that the flares typically do not add energy into the underlying afterglow. Furthermore, spectral evolution during the flares is common, with the emission softening as the flare progresses [22,23,24], similar to what is seen in the prompt emission.

Short GRBs

At long last, 2005 has seen an explosion of information on afterglows of short GRBs. By the end of the year, the *Swift* BAT had discovered 11 short GRBs, and the XRT had found X-ray afterglows for six of them (GRB 050509B [25]; GRB 050724 [26]; GRB 050813 [27]; GRB 051210; GRB 051221A [28,29]; and GRB 051227 [30], although the classification of this burst is not entirely clear). In addition, the HETE-2 and Chandra satellites observed the X-ray afterglow of GRB 050709 [31,32].

The results of these observations are quite interesting. First, and perhaps most importantly, short GRBs are found to be located in regions of little or no star formation, consistent with the expectations of merger models of short GRBs, and inconsistent with collapsar models. At least two short GRBs have been identified with large elliptical galaxies, and most seem to be associated with galaxy clusters. This is in stark contrast to long GRBs, which are found in star-forming regions of isolated, irregular galaxies. The short GRBs are found at much lower redshifts than long GRBs, as noted above: the mean redshift for six short *Swift* GRBs is 0.43. The isotropic energy of short bursts is a factor of 100-1000 lower than that of long bursts [32], but in the absence of any measured jet breaks, it is not yet clear how the actual energetics compare. The X-ray afterglows tend to be much weaker than those of long bursts (undetectable in 3 cases in spite of prompt XRT observations, and barely detected in two others), but four of the afterglows are quite bright. Interestingly, all four short GRBs with bright X-ray afterglows have flares or significant late-time energy injection, an unexpected result for NS-NS merger scenarios, and an important feature that must be explained by theories of short GRBs. Finally, those cases with strong X-ray afterglows show no evidence for jet breaks out to very late times [33,34]; see Figure 5 for an example.

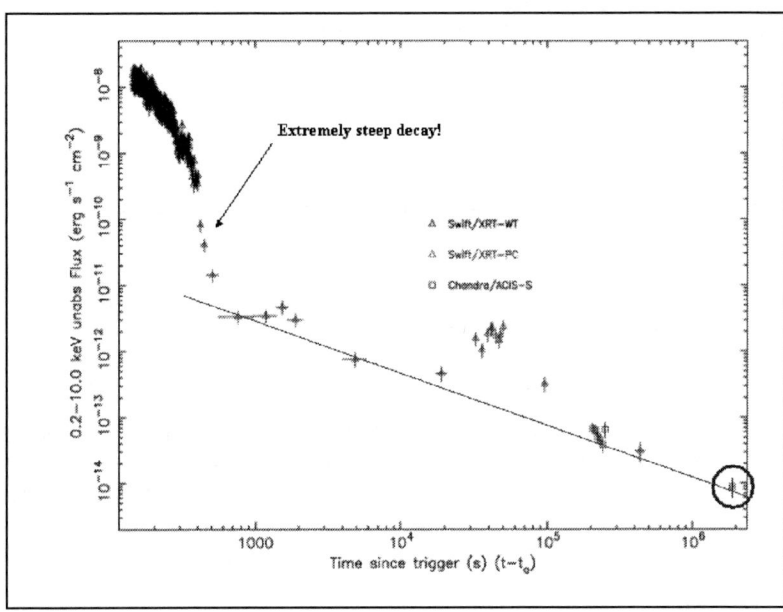

FIGURE 5. X-ray light curve for GRB 050724, combining BAT data (black, extrapolated into the XRT energy band, see [26,33] for details), XRT data (red) and Chandra data (blue). The underlying afterglow power law decay is shown by the green line. The late Chandra data point [34] is consistent with this power law decay, indicating that there is no jet break before 2×10^6 seconds.

High Redshift Bursts

One exciting prospect for GRB afterglows has been the localization of a substantial number of very high-redshift bursts that can be used to investigate the IGM, burst host galaxies, and cosmological effects out [37]. Indeed, in its first year of operations *Swift* has located several bursts at redshift ≥ 4, including GRB 050505 (z=4.27) [38], GRB 050730 (z=3.967) [39], GRB 050814 (z=5.3) [40]; and GRB 050904 (z=6.29) [13,14,35,36].

GRB 050904 is the highest redshift GRB ever found, as well as the highest redshift X-ray source ever found. In spite of this, it was also one of the brightest X-ray afterglows we have seen to date. The X-ray light curve is shown in Figure 6, plotted in the burst rest frame. Optically this burst was very bright for such a large redshift, with J=17.5 at 0.1 day after the burst; this brightness has enabled high resolution spectroscopy of the line of sight to the afterglow [14,41]. The burst happened only 900 Myr after the Big Bang, implying a collapsar origin from a massive star progenitor. With additional GRBs at similar redshifts, limits can be placed on star and galaxy formation in the early universe [35], and the reionization epoch can be probed directly [41]. Although the flaring in this afterglow was somewhat larger than is typical, it is not without precedent at this early rest frame epoch for flaring to be substantial in XRT afterglows. Energetically, this burst appears to be typical of long GRBs.

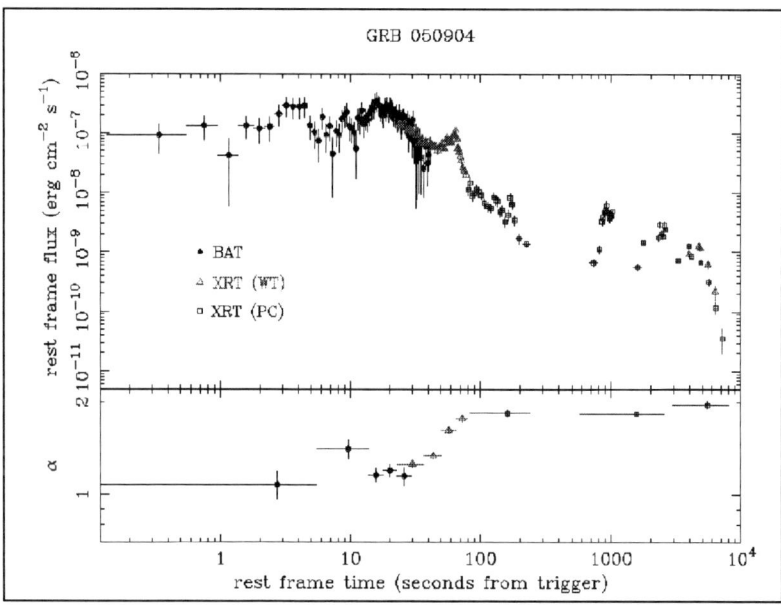

FIGURE 6. X-ray light curve of GRB 050904 at z=6.29 [35,36]. Times are in the burst rest frame. The BAT data have been extrapolated into the XRT energy range (black). XRT data in WT mode are in red, while XRT data in PC mode are in blue. The PC mode data are corrected for pile-up. Note the high level of flaring throughout the X-ray light curve, until the flux plummets at 7000 s in the rest frame. The lower panel shows the photon index, which increases from about 1 at early times to about 2 after the first 100 rest-frame seconds.

(We note that our light curve differs significantly from that presented by [42]. The light curve in [42] displays a sharp discontinuity by a factor of four in the flux at about 80 s in the rest frame, which occurs right at the transition between WT and PC mode data. This discontinuity is not present in our light curve or in the raw count rate, nor is there a discontinuity between PC and WT mode data in any XRT calibration source or in any other XRT afterglow.)

CONCLUSIONS

The *Swift* X-ray Telescope has had a remarkably successful first year, with progress made on many of our prime mission objectives, including the discovery of short GRB afterglows, important insights into the nature of both long and short GRBs, and the beginnings of a catalog of X-ray afterglow properties. Unexpected results included rapid early decays and flaring. The picture that is emerging is that X-ray afterglows are dominated at very early times by emission from the central engine, that the true afterglow (external shock) typically becomes dominant after a few thousand seconds, but that internal shocks can still produce significant energy inputs in the form of X-ray flares at very late times.

ACKNOWLEDGMENTS

This work is supported at Penn State by NASA contract NAS5-00136; at the University of Leicester by the Particle Physics and Astronomy Research Council on grant number PPA/Z/S/2003/00507; and at OAB by funding from ASI on grant number I/R/039/04. We gratefully acknowledge the contributions of dozens of members of the XRT team at PSU, UL, OAB, GSFC, ASDC, INAF-IASF-Pa and our subcontractors, who helped make this instrument possible. We thank the Swift Flight Operations team for their excellent work in operating the spacecraft, and the staff of the Swift Science Center and Swift Data Center for their contributions to the success of this mission.

REFERENCES

1. N. Gehrels et al., *The Astrophysical Journal* **611**, 1005-1020 (2004).
2. S. D. Barthelmy et al, *Space Science Reviews*, 120, 143-164 (2005).
3. D. N. Burrows et al., *Space Science Reviews*, 120, 165-195 (2005).
4. P. W. A. Roming et al., *Space Science Reviews*, 120, 95-142 (2005).
5. P. Romano et al., "In-flight calibration of the Swift XRT effective area", in *UV, X-Ray, and Gamma-Ray Space Instrumentation for Astronomy XIV*, edited by O. H. W. Siegmund, Proc. SPIE, 5898, 589819-1 – 589819-8.
6. D. N. Burrows et al., *The Astrophysical Journal (Letters)* **622**, L85-L88.
7. P. Tozzi et al., "AGNs and Clusters in Chandra Deep Fields", in *Where's the Matter?*, edited by L. Tresse and M. Treyer (2002), astro-ph/0111036
8. P. Giommi et al., in preparation (2006).
9. K. L. Page et al., *The Astrophysical Journal (Letters)*, in press (astro-ph/0512358)
10. G. Tagliaferri et al., *Nature* **436**, 985-989 (2005).
11. D. N. Burrows et al., *Science* **309**, 1833-1835 (2005).
12. D. N. Burrows et al., "Swift XRT Observations of X-ray Flares in GRB Afterglows", in *X-ray Universe 2005 Conference Proceedings*, in press (astro-ph/0511039)
13. G. Tagliaferri et al., *Astronomy and Astrophysics* **443**, L1-L5 (2005).
14. N. Kawai et al., *Nature*, submitted (astro-ph/0512052)
15. A. Moretti et al., *Astronomy and Astrophysics*, submitted.
16. J. Nousek et al., *The Astrophysical Journal*, in press (astro-ph/0508332).
17. B. Zhang, Y. Z. Fan, J. Dyks, S. Kobayashi, P. Mészáros, D. N. Burrows, J. A. Nousek, N. Gehrels, *The Astrophysical Journal*, in press (astro-ph/0508321).
18. A. Panaitescu, P. Mészáros, N. Gehrels, D. Burrows, J. Nousek, *MNRAS*, in press (astro-ph/0508340).
19. S. Vaughan et al., *The Astrophysical Journal*, in press (astro-ph/0510677).
20. P. T. O'Brien et al., *The Astrophysical Journal*, submitted.
21. E. W. Liang et al., in preparation.

22. A. Falcone et al., *The Astrophysical Journal*, in press (astro-ph/0512615).
23. P. Romano et al., *Astronomy and Astrophysics*, in press.
24. C. Pagani et al., *The Astrophysical Journal*, submitted.
25. N. Gehrels et al., *Nature* **437**, 851-854 (2005).
26. S. D. Barthelmy et al., *Nature*, in press (astro-ph/0511579).
27. D. Fox et al., in preparation.
28. A. Parsons et al., in preparation.
29. D. N. Burrows et al., in preparation.
30. L. Barbier et al., in preparation.
31. J. S. Villasenor et al., *Nature* **437**, 855-858 (2005).
32. D. Fox et al., *Nature* **437**, 845-850 (2005).
33. S. Campana et al., in preparation.
34. D. Grupe et al., in preparation.
35. G. Cusumano et al., *Nature*, in press (astro-ph/0509737).
36. G. Cusumano et al., *Astronomy and Astrophysics*, in preparation.
37. D. Q. Lamb and D. E. Reichart, *The Astrophysical Journal* **536**, 1-18 (2000).
38. C. Hurkett et al., *MNRAS*, submitted.
39. M. Perri et al., in preparation.
40. P. Jakobssen et al., *Astronomy and Astrophysics*, submitted (astro-ph/0509888).
41. T. Totani et al., *Publ. Astr. Soc. Japan*, submitted (astro-ph/0512154).
42. D. Watson et al., *The Astrophysical Journal*, in press (astro-ph/0509640).

Observations of Early Optical Afterglows

P. W. A. Roming[*] and K. O. Mason[†]

[*]*Penn State University, Department of Astronomy & Astrophysics,
525 Davey Lab, University Park, PA 16802, USA*
[†]*Particle Physics and Astronomy Research Council,
Polaris House, North Star Ave, Swindon, Wilts SN2 1SZ, UK*

Abstract. The Swift Ultra-Violet/Optical Telescope (UVOT) has performed extensive follow-up on 71 Swift Burst Alert Telescope triggered gamma-ray bursts (GRBs) in its first ten months of operations. In this paper, we discuss some of the UV and optical properties of UVOT detected afterglows such as XRF 050406, the bright GRB 050525A, the high redshift GRB 050730, the early flaring GRB 050801, and others. We also discuss some of the implications of why 75% of GRB afterglows observed by UVOT in less than one hour are "dark."

Keywords: gamma-ray bursts, Swift, UVOT, dark gamma-ray burst
PACS: 98.70.Rz, 95.55.Fw, 95.85.Mt, 95.85.Ls, 95.85.Kr

INTRODUCTION

The Swift [1] Ultra-Violet/Optical Telescope (UVOT) [2] began observing gamma-ray burst (GRB) afterglows on January 24, 2005, after a three week on-orbit checkout. Between January 24 and December 9, 2005, the UVOT observed 71 GRB afterglows discovered by the Swift Burst Alert Telescope (BAT) [3], comprising 85% of the BAT observed sample. In addition, 60 of the 71 and 49 of the 71 afterglows were observed within less than one hour and five minutes, respectively, of the BAT trigger.

UVOT's speed of response, sensitivity, multiwavelength nature, and length of semi-uninterrupted observations[1] offers a unique opportunity to study the properties of early optical and UV afterglows. In addition, the Swift X-Ray Telescope (XRT) [4] positional data within the optical field of these afterglows allows us to be confident that no bright afterglows were missed due to confusion with nearby stars.

Below, we review some of the UV and optical properties of a sample of UVOT detected afterglows. In addition, we investigate some possible reasons why a large fraction of GRB afterglows observed by UVOT in less than one hour are "dark."

DETECTED BURSTS

Between January 24 and December 9, 2005, the UVOT detected nineteen afterglows from GRBs discovered by the BAT. Table 1 identifies these UVOT detected afterglows as well as the broadband filters in which a detection was made. Redshifts, as determined by UVOT or ground-based data, are also provided. Below we

---

[1] Swift is in a low-Earth orbit and is therefore able to observe most objects continuously for no more than 40 minutes.

briefly describe the following GRBs: 050318, 050319, 050406, 050525A, 050730, 050801, 050802, and 051117A.

TABLE –1. UVOT Detected Afterglows

GRB	UVW2	UVM2	UVW1	U	B	V	White	z	Ref
050318				X	X	X		1.44	[5]
050319					X	X		3.24	[6,7]
050406				X	X	X		2.44	[8]
050416A	X			X	X	X		0.65	[9,10,11,12]
050525A	X	X	X	X	X	X		0.61	[13,14]
050603						X		2.82	[15,16]
050712				X		X			[17,18]
050726						X			[19,20]
050730					X	X		3.97	[21,22,23]
050801	X	X	X	X	X	X		<1.2	[24]
050802	X	X	X	X	X	X		<1.2	[25,26]
050815						X			[27,28]
050820A				X	X	X	X	2.61	[29,30]
050824		X		X	X	X	X	0.83	[31,32]
050922C				X	X	X	X	2.07	[33,34]
051016B	X			X	X	X		0.94	[35,36]
051109A				X	X	X	X	2.35	[37,38]
051111				X	X	X	X	1.55	[39,40]
051117A						X	X		[41]

GRB 050318

GRB 050318 is the first burst for which UVOT detected an afterglow. The first UVOT detection occurred in the V-filter observation, which started 3230 seconds after the BAT trigger. The decay profile reveals very little fluctuation in the lightcurves, although the number of data points is admittedly small. A weighted mean of the decay slopes produces _ = -0.94. A combined UVOT and XRT spectrum provides a redshift of z = 1.44 [5].

GRB 050319

Observations of GRB 050319 began much earlier, and were much more richly sampled, than GRB 050318. The first UVOT detection of GRB 050319 occurred in the V-filter observation, which started 90 seconds after the BAT trigger. The decay profile reveals potential fluctuations in both the V- and B-band lightcurves, particularly at early times. The best-fit power law is _ = -0.57 [6]. The XRT data reveals that a break in the fit to _ = -1.14 occurs at about 20,000 seconds after the trigger. No such break is evident in the late-time optical lightcurves. Based on an absorption line system in the spectrum of the GRB, a redshift of z = 3.24 is determined [7].

GRB 050406

GRB 050406 is the first burst classified as an X-Ray Flash (XRF) with a detected UVOT afterglow; it was also the earliest optical detection of an XRF afterglow at the time. The first UVOT detection occurred in the V-filter observation, which started 88 seconds after the BAT trigger. The XRT lightcurve manifests flaring several hundred seconds after the burst, but no correlated flaring is seen in the UVOT lightcurves. The UVOT decay profile also does not follow the XRT's [8]. The X-ray lightcurve is well fitted by a broken power law of $\alpha = -1.58$ at early times and $\alpha = -0.5$ at later times [42]. The best power-law fit to the combined B- and U-band lightcurves produces $\alpha = -0.75$. A UVOT and XRT broadband spectrum provides a redshift of $z = 2.44$ [8].

GRB 050525A

GRB 050525A is the brightest UVOT afterglow detection to date and was well sampled in all the broadband color filters. The first detection occurred in the V-filter observation, which started 65 seconds after the BAT trigger. The maximum observed flux occurs 68 seconds after the burst, with $m_V = 12.86$. The decay profile for this burst is much more complex than the XRT lightcurve or any previous UVOT detected burst. The best fit to the lightcurve is a double power law, $\alpha_1 = -1.56$ and $\alpha_2 = -1.14$, with a decay index after a jet break of $\alpha_j = -1.76$. The X-ray lightcurve is well fit by a broken power law of $\alpha = -1.2$ at early times and $\alpha = -1.62$ at later times [13]. Based on emission and absorption line systems in the spectrum of the GRB's host galaxy, a redshift of $z = 0.61$ is established [14].

GRB 050730

The first UVOT detection of GRB 050730 occurred in the V-filter observation, which started 119 seconds after the BAT trigger [21]. The UVOT lightcurve reveals flaring in the first 1000 seconds with the peaks of the flares corresponding to the XRT flare peaks (see Figure 1). The best-fit optical profile, after the flaring, is a power law of $\alpha = -0.41$, which is one of the shallowest decay profiles seen in UVOT lightcurves. The corresponding X-ray lightcurve is fit by a power law of $\alpha = -2.40$, but variability in the lightcurve is a dominant feature. Based on a broad Lyα absorption line, thought to originate in the host of the GRB, a redshift of $z = 3.97$ is determined [22].

GRB 050801

The first UVOT detection of GRB 050801 occurred in the V-filter, 52 seconds after the BAT trigger [24]. The X-ray lightcurve shows an initial decline and then a recovery. This is not unusual for the X-ray behaviour of bursts; however, what is unusual is that the optical curve follows it almost exactly (see Figure 2). The initial decline is even more significant in the UVOT data. After about 200 seconds both the X-ray and optical curves decline with $\alpha = -1.3$. The X-ray lightcurve has a pretty good fit out to 10^5 seconds, but there is excess optical emission with respect to the power law after about 10^4 seconds. Using a combined XRT and UVOT SED, the best-fit redshift was determined to be $z = 1.45$.

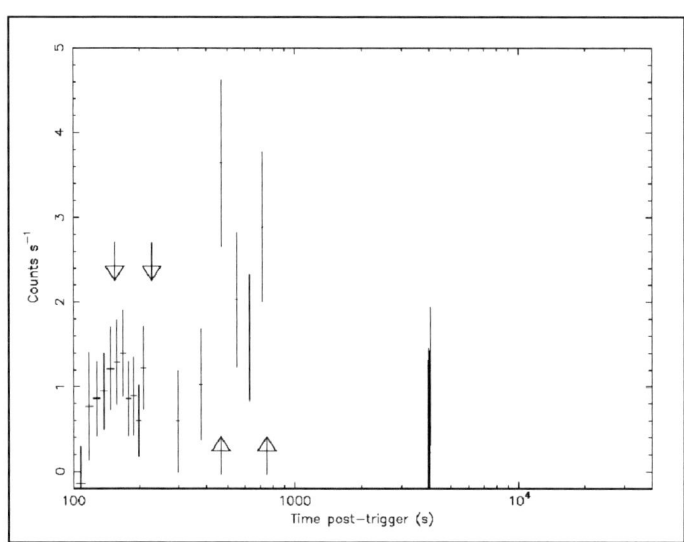

FIGURE 1. V-band lightcurve of GRB 050730. The arrows mark the times of the flares in the XRT lightcurve. Peaks in the V-band correspond with the locations of the X-ray flares.

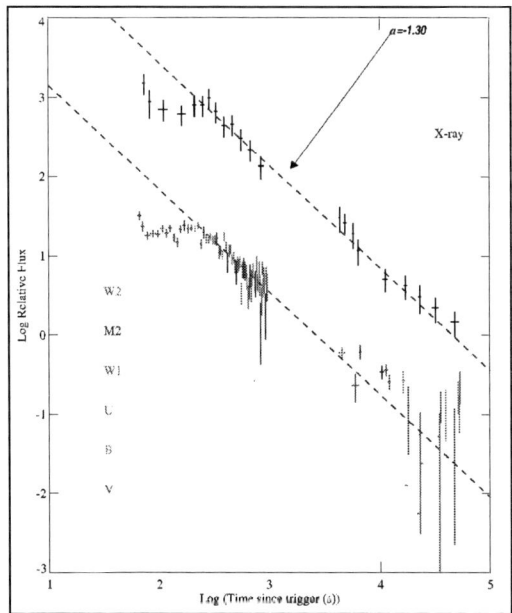

FIGURE 2. XRT and UVOT lightcurves of GRB 050801 plotted together in flux units on a log scale. The relative normalization between the X-ray and optical/UV curves is arbitrary for display purposes, but the same factor has been applied to the burst, so the separation does correspond to the relative X-ray to optical flux ratio. The UVOT data from all filters have been normalized to the V-band using data in the interval when the instrument was cycling rapidly between filters (10 second integration times) to determine the normalization factors. Filters distinguished by color.

GRB 050802

The first UVOT detection of GRB 050802 occurred in the V-filter, 286 seconds after the BAT trigger [25]. As with GRB 050801, there is some initial variability in the X-ray lightcurve which is rising followed by a small re-brightening at about 1000 seconds. There is no strong evidence for similar behaviour in the UVOT lightcurve, but the error bars are larger than GRB050801 (see Figure 3). The overall decline is much flatter than 050801, with _ = -0.9. There is a break in the X-ray band at about 10^4 seconds to _ = -1.6, but no corresponding break in the UV/optical. Based on several absorption features, a redshift of $z = 1.71$ was determined [43].

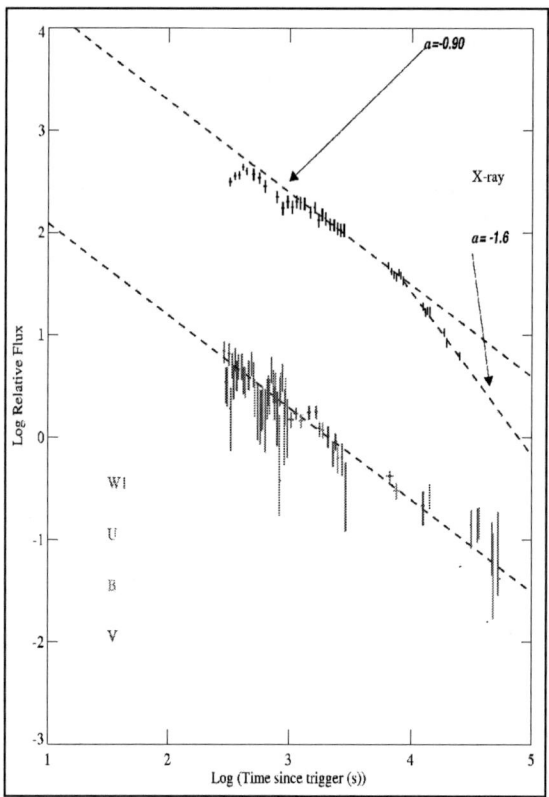

FIGURE 3. XRT and UVOT lightcurves of GRB 050802 plotted using the same technique as was used for GRB 050801 (see Figure 2).

GRB 051117A

The UVOT White-light filter was inserted into the automated observing sequence after the first nine months of operation. The White-light filter is effectively a clear filter, and the response curve is essentially the wavelength-dependant response curve of the detector. These observations are thus similar to the unfiltered observations made by ROTSE [44] and should be more sensitive than filtered observations for low-redshift bursts. The first UVOT detection of GRB 051117A occurred in the V-filter, 111 seconds after the BAT trigger [41]. However, later detections of the afterglow were made in the White-light filter revealing that the source was indeed fading (see Figure 4). Because the observations were made in the White-light filter, no redshift determination could be made.

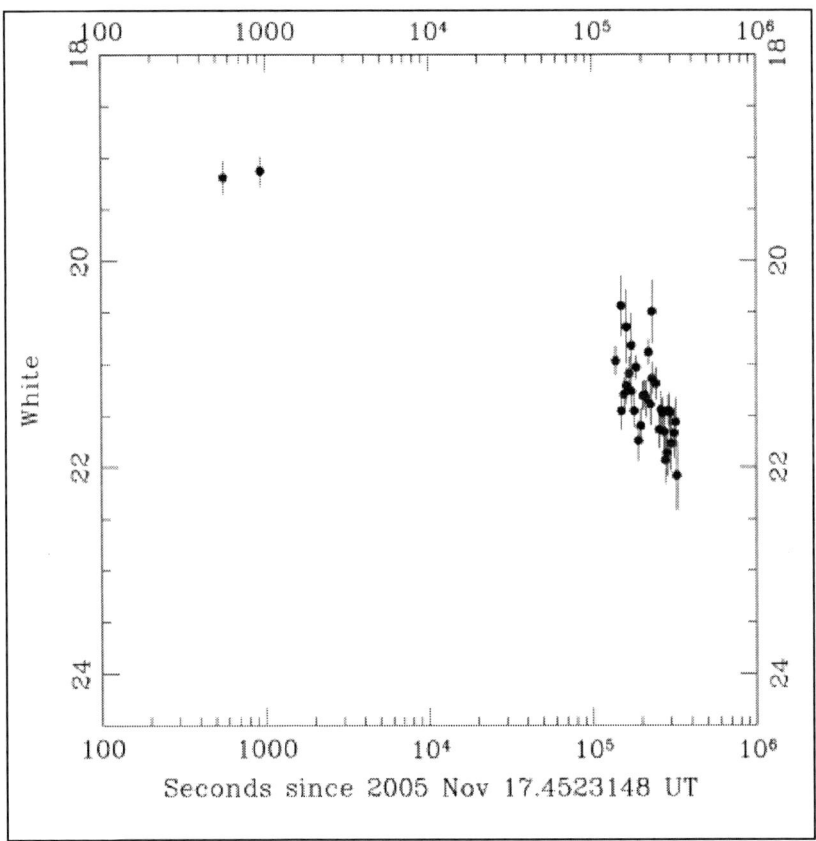

FIGURE 4. White-light observation of GRB 051117A. The source is clearly detected and is fading. The Figure shows all the White-light observations up to approximately 3.5 days after the BAT trigger that are above the 3-sigma detection limit.

General Properties of UVOT Detected Afterglows

Because the number of UVOT detected afterglows is still small, it is difficult to quantity the sample properties of these early afterglows at this time. However, we provide the distribution of the redshift (see Figure 5) as well as the temporal decay indices (see Figure 6) for the UVOT detected afterglows. From the figures, it appears that the normal distribution of UVOT bursts is in the $0 < z < 3$ range and the temporal decay indices are weighted to the 0.5-1.0 range. Qualitatively, we can describe a range of properties seen in UVOT detected afterglows, as follows:

- there is no UV/optical flaring although there is X-ray flaring,
- there is UV/optical flaring as well as X-ray flaring,
- fluctuations are seen in shallow decay UV/optical lightcurves,
- fluctuations are seen in steep decay UV/optical lightcurves,
- UV/optical lightcurves follow the X-ray lightcurves,
- UV/optical lightcurves do not follow the X-ray lightcurves,
- steep decay profiles are consistent with the standard fireball model,
- and, shallow decay profiles are difficult to reconcile with the standard fireball model.

From the description above, it can be seen that the optical lightcurves have large range of properties, pointing to different circumburst environments. Again, it is emphasized, that the sample size is still small. As the sample grows it may become apparent that a particular feature or set of features will dominate the properties of the early afterglows of GRBs.

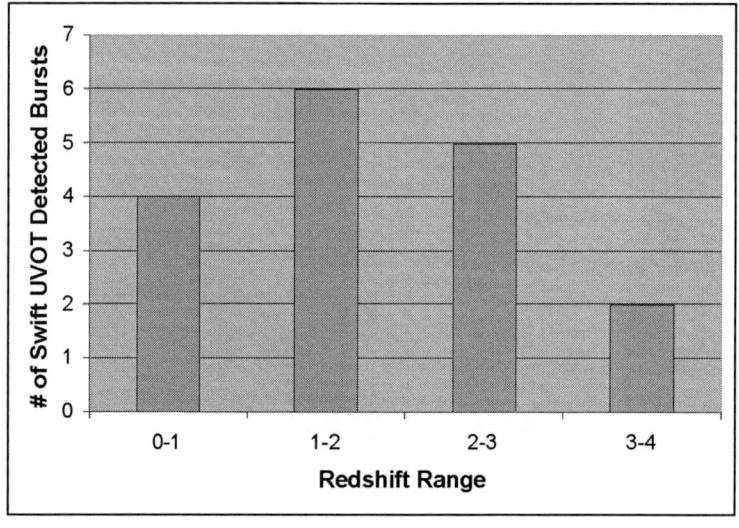

FIGURE 5. Distribution of the redshift range for UVOT detected afterglows.

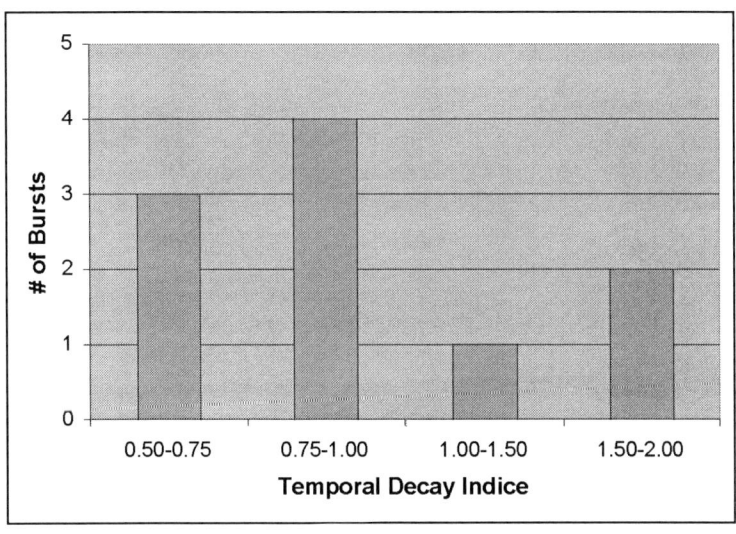

FIGURE 6. Distribution of the temporal decay indices for UVOT detected afterglows.

"DARK" BURSTS

UVOT observed 60 bursts in less than an hour (and in 49 cases in less than five minutes) after the BAT trigger with no detections being made in 45 instances. Of the 45 non-detections with UVOT, 13 *were* detections by ground-based observers via the GCN [45]. Most of these reported detections were in bands red-ward of the UVOT's, perhaps indicative of a high redshift such that the Lyman edge is beyond the UVOT range. It may be that some of the remaining 32 non-detections could be due to them being inaccessibility to ground-based R- and IR-band observations in that part of the sky. However, even if the number of ground detections is doubled, approximately 25% of bursts would still be classified as "dark."

Qualitative definitions have been used to describe dark bursts such as: no afterglow detection in the optical or NIR, and the optical counterpart is initially faint and declines too rapidly to be detected. More recent work has tried to quantify the definition of dark bursts in order to remove the dependence on instrument sensitivity. The optical–to–X-ray spectral index method (β_{OX}) [46] is a quick method to determine if Swift bursts are dark. If $\beta_{OX} < 0.50$, then the burst is classified as dark. If $0.50 < \beta_{OX} < 0.55$, then the burst is classified as potentially dark. A more rigorous treatment is the upper limit on the afterglow flux (ULAF) method [47]. The ULAF method uses the temporal (α) and spectral (β) indices to determine the minimum and maximum value for the electron index (p). It examines eight cases of the standard afterglow model and extrapolates the maximum & minimum X-ray flux to the optical epoch. Optical flux that is below the minimum extrapolated X-ray flux is considered dark.

We have selected 19 bursts – which include some that are detected and others that are not detected by the UVOT – as a test of the β_{OX} and ULAF methods. Using the ULAF method, only one of the 19, GRB 050219B, is classified as dark. Using the β_{OX}

method, GRBs 050315, 050319, 050326, 050412, and 050505 are classified as dark [48]. It is noted that GRB 050319 was detected by UVOT [6], and that GRBs 050315 [49] and 050505 [50] were detected by ground-based observations.

Although many of the bursts in our sample are not classified as dark in the ULAF or $_{OX}$ classification schemes, it is still unclear why the early optical afterglows are suppressed. Both the $_{OX}$ and ULAF methods assume that GRBs follow the standard fireball model. However, recent work [51,52] suggests that the standard fireball model is not valid in the first few hours after many bursts. What then are the explanations for these dark bursts at early times?

Conventional reasons put forth for dark bursts include: circumburst absorption [53,54], low-density environment [55,56,57], intrinsic faintness [53,54], rapid temporal decay [56], Ly-_ blanketing and absorption due to high redshift [54,58], and extinction by dust [59]. For UVOT observed afterglows in which there was no detection, the standard explanations for darkness can be invoked in many cases. However, for bursts such as GRBs 050223, 050421, & 050422, the conventional explanations can't explain the suppression of the afterglow.

Recent work suggests that the suppression of these early GRB afterglows can be explained by high gamma-ray efficiencies which are consistent with highly magnetized outflows [48]. By comparing the X-ray afterglow flux at one hour ($F_{x,1}$) to the prompt _-ray fluence (S_) a gamma-ray efficiency can be defined. As the sample of "dark" bursts increases, we expect that our understanding of the suppression mechanisms will also continue to grow.

CONCLUSIONS

In less than one year of operation, the UVOT, in conjunction with other ground based telescopes, has more fully mapped the early time optical parameter space of GRB afterglows than previous telescopes or missions. Currently, the UVOT optical lightcurves denote a large array of properties which indicates a range of circumburst environments. Additionally, Swift data on dark bursts points to a newly proposed mechanism than the standard explanations previously presented. As the sample continues to grow, we anticipate that a specific trend or trends will reveal themselves.

ACKNOWLEDGMENTS

This work is supported at Penn State by NASA contract NAS5-00136. We gratefully acknowledge the contributions from members of the Swift UVOT team at PSU, MSSL, NASA/GSFC, and our subcontractors.

REFERENCES

1. N. Gehrels, et al., *Astrophys. J.*, **611**, 1005-1020 (2005).
2. P. W. A. Roming, et al., *Space Sci. Rev.*, **120**, 95-142 (2005).
3. S. D. Barthelmy, et al., *Space Sci. Rev.*, **120**, 143-164 (2005).
4. D. N. Burrows, et al., *Space Sci. Rev.*, **120**, 165-195 (2005).
5. M. Still, et al., *Astrophys. J.*, **635**, 1187-1191 (2005).
6. K. O. Mason, et al., *Astrophys. J.* (accepted).

7. J. P. U. Fynbo, J. Hjorth, B. L. Jensen, P. Jakobsson, P. Moller, & J. Näränen, *GCN Circ.* 3136 (2005).
8. P. Schady, et al., *Astrophys. J.* (submitted).
9. P. Schady, T. Sakamoto, K. McGowan, P. Boyd, P. Roming, J. Nousek, & N. Gehrels, *GCN Circ.* 3276 (2005).
10. P. Schady, T. Sakamoto, K. McGowan, P. Boyd, P. Roming, J. Nousek, & N. Gehrels, *GCN Circ.* 3280 (2005).
11. D. B. Fox, *GCN Circ.* 3480 (2005).
12. S. B. Cenko, S. R. Kulkarni, A. Gal-Yam, & E. Berger, *GCN Circ.* 3542 (2005).
13. A. J. Blustin, et al., *Astrophys. J.* (accepted).
14. R. J. Foley, H. -W. Chen, J. Bloom, & J. X. Prochaska, *GCN Circ.* 3483 (2005).
15. D. Grupe, et al., *Astrophys. J.* (submitted).
16. E. Berger, and G. Becker, *GCN Circ.* 3520 (2005).
17. E. Rol, T. Poole, K. McGowan, D. Grupe, J. Nousek, W. Voges, & N. Gehrels, *GCN Circ.* 3575 (2005).
18. T. Poole, D. Grupe, A. Breeveld, L. Angelini, & J. Greiner, *GCN Circ.* 3596 (2005).
19. S. Barthelmy, et al., *GCN Circ.* 3682 (2005).
20. T. Poole, A. Moretti, S. T. Holland, M. Chester, L. Angelini, & N. Gehrels, *GCN Circ.* 3698 (2005).
21. S. T. Holland, et al., *GCN Circ.* 3704 (2005).
22. H. –W. Chen, I. Thompson, J. X. Prochaska, & J. Bloom, *GCN Circ.* 3709 (2005).
23. A. J. Blustin, S. T. Holland, A. Cucchiara, N. White, & D. Hinshaw, *GCN Circ.* 3717 (2005).
24. M. De Pasquale, et al., *Mon. Not. R. Astron. Soc.* (submitted).
25. K. McGowan, D. Band, P. Brown, C. Gronwall, H. Huckle, & B. Hancock, *GCN Circ.* 3739 (2005).
26. K. McGowan, A. Morgan, K. Mason, & T. Kennedy, *GCN Circ.* 3745 (2005).
27. D. Fox, et al., *GCN Circ.* 3811 (2005).
28. P. Schady, et al., *GCN Circ.* 3817 (2005).
29. J. X. Prochaska, J. S. Bloom, J. T. Wright, R. P. Butler, H. W. Chen, S. S. Vogt, & G. W. Marcy, *GCN Circ.* 3833 (2005).
30. M. Chester, M. Page, P. Roming, F. Marshall, P. Boyd, L. Angelini, J. Greiner, & N. Gehrels, *GCN Circ.* 3838 (2005).
31. P. Schady, R. Fink, M. Ivanushkina, S. Holland, K. McGowen, & N. Gehrels, *GCN Circ.* 3880 (2005).
32. J. P. U. Fynbo, et al., *GCN Circ.* 3874 (2005).
33. P. Jakobsson, J. P. U. Fynbo, D. Paraficz, J. Telting, B. L. Jensen, J. Hjorth, & J. M. C. Ceron, *GCN Circ.* 4017 (2005).
34. S. D. Hunsberger, F. Marshall, S. T. Holland, P. Brown, A. Morgan, P. Roming, & A. Cucchiara, *GCN Circ.* 4041 (2005).
35. A. J. Blustin, A. Parsons, S. T. Holland, P. Meszaros, M. Chester, & N. Gehrels, *GCN Circ.* 4107 (2005).
36. A. M. Soderberg, E. Berger, & E. Ofek, *GCN Circ.* 4186 (2005).
37. S. T. Holland, S. Campana, P. Smith, H. Huckle, & N. Gehrels, *GCN Circ.* 4235 (2005).
38. R. Quimby, D. Fox, P. Hoeflich, B. Roman, & J. C. Wheeler, *GCN Circ.* 4221 (2005).
39. G. Hill, J. X. Prochaska, D. Fox, B. Schaefer, & M. Reed, *GCN Circ.* 4255 (2005).
40. T. S. Poole, T. Sakamoto, A. J. Blustin, B. Hancock, & T. Kennedy, *GCN Circ.* 4263 (2005).
41. S. T. Holland, D. Band, K. Mason, F. Marshall, & N. Gehrels, *GCN Circ.* 4300 (2005).
42. P. Romano, et al., *Astron. Astrophys.* (accepted).
43. J. P. U. Fynbo, et al., *GCN Circ.* 3749 (2005).
44. E. S. Rykoff, et al., *Astrophys. J.*, **631**, L121-L124 (2005).
45. S. D. Barthelmy, P. Butterworth, T. L. Cline, N. Gehrels, G. J. Fishman, C. Kouveliotou, & C. A. Meegan, *Astrophys. J. Suppl. Ser.*, **231**, 235- (1995).
46. P. Jakobsson, et al., *Astrophys. J.*, **617**, L21-L24 (2004).
47. E. Rol, R. A. M. J. Wijers, C. Kouveliotou, L. Kaper, & Y. Kaneko, *Astrophys. J.*, **624**, 868-879 (2005).
48. P. W. A. Roming, et al., *Astrophys. J.* (submitted).
49. D. Kelson, & E. Berger, *GCN Circ.* 3100 (2005).
50. S. B. Cenko, C. Steidel, N. Reddy, & D. B. Fox, *GCN Circ.* 3366 (2005).
51. J. A. Nousek, et al., *Astrophys. J.* (submitted).
52. B. Zhang, Y. Z. Fan, J. Dyks, S. Kobayashi, & P. Mészáros, *Astrophys. J.* (submitted).
53. D. Lazzati, S. Covino, & G. Ghisellini, *Mon. Not. R. Astron. Soc.*, **330**, 583-590 (2002).
54. J. U. Fynbo, et al., *Astron. Astrophys.*, **369**, 373-379 (2001).
55. D. A. Frail, et al., *Astrophys. J.*, **525**, L81-L84 (1999).
56. P. J. Groot, et al., *Astrophys. J.*, **502**, L123-L127 (1998).
57. G. B. Taylor, et al., *Astrophys. J.*, **537**, L17-L21 (2000).
58. P. J. Groot, et al., *Astrophys. J.*, **493**, L27-L30 (1998).
59. G. B. Taylor, et al., *Astrophys. J.*, **502**, L115-L118 (1998).
60. A. Panaitescu, & P. Kumar, *Astrophys. J.*, **571**, 779-789 (2002).
61. P. Kumar, & A. Panaitescu, *Mon. Not. R. Astron. Soc.*, **346**, 905-914 (2003).
62. Y. Z. Fan, B. Zhang, & D. M. Wei, *Astrophys. J.*, **628**, L25-L28 (2005).

Theories of Early Afterglow

P. Mészáros

Dept. of Astronomy & Astrophysics and Dept. of Physics, Pennsylvania State University, University Park, PA 16802, USA

Abstract. The rapid follow-up of gamma-ray burst (GRB) afterglows made possible by the multi-wavelength satellite *Swift*, launched in November 2004, has put under a microscope the GRB early post-burst behavior, This is leading to a significant reappraisal and expansion of the standard view of the GRB early afterglow behavior, and its connection to the prompt gamma-ray emission. In addition to opening up the previously poorly known behavior on minutes to hours timescales, two other new pieces in the GRB puzzle being filled in are the the discovery and follow-up of short GRB afterglows, and the opening up of the $z \gtrsim 6$ redshift range. We review some of the current theoretical interpretations of these new phenomena.

CHALLENGES POSED BY NEW SWIFT OBSERVATIONS

Compared to previous satellites, Swift has made a large difference on two main accounts. First, the sensitivity of the Burst Alert Detector (BAT, in the range 20-150 keV) is a factor ~ 5 higher than for the corresponding instruments in the predecessor CGRO-BATSE, BeppoSAX and HETE-2. Second, Swift can slew in less then 100 seconds in the direction determined by the BAT instrument, positioning its much higher angular resolution X-ray (XRT) and UV-Optical (UVOT) detectors on the burst [1]

As of December 2005, at an average rate of 2 bursts detected per week, over 100 bursts had been detected by BAT, of which 90% were followed promptly with the XRT within 350 s from the trigger, and about half within 100 s [2], while $\sim 30\%$ were detected with the UVOT [3]. Of these, over 23 resulted in redshift determinations. Ten short GRB were detected, of which five had detected X-ray afterglows, three had optical, and one had a radio afterglow, and five had a redshift determination.

The new observations brings the total redshift determinations to over 50 since 1997 when BeppoSAX enabled the first one. The redshifts based on Swift have a median $z \gtrsim 2$, which is a factor ~ 2 higher than the median of those previously culled via BeppoSAX and HETE-2, [5]. This can be ascribed to the higher sensitivity of BAT and the prompt accurate positions from XRT and UVOT, making possible ground-based detection at a stage when the afterglow is much brighter. The highest Swift-enabled redshift so far is in GRB 050904, obtained with Subaru, $z = 6.29$ [6], the second highest being GRB 050814 at $z = 5.3$, whereas the previous Beppo-SAX era record was $z = 4.5$. The relative paucity of UVOT detections versus XRT detections may be ascribed in part to this higher median redshift, and in part to the higher dust extinction at the implied shorter rest-frame wavelenghts for a given observed frequency [3], although additional effects may be at work too.

The BAT light curves show that in some of the bursts which fall in the "long" category ($t_\gamma \gtrsim 2$ s) faint soft gamma-ray tails can be followed which extend the duration by a factor up to two beyond what BATSE could have detected [1]. A rich trove of information on the burst and afterglow physics has come from detailed XRT light curves, starting on average 100 seconds after the trigger, together with the corresponding BAT light curves and spectra. This suggests a canonical X-ray afterglow [18] with one or more of the following:
1) an initial steep decay $F_X \propto t^{-\alpha_1}$ with a temporal index $3 \lesssim \alpha_1 \lesssim 5$, and an energy spectrum $F_\nu \propto \nu^{-\beta_1}$ with energy spectral index $1 \lesssim \beta_1 \lesssim 2$ (or photon number index $2 \lesssim \Gamma = \alpha + 1 \lesssim 3$), extending up to a time $300\text{s} \lesssim t_1 \lesssim 500\text{s}$;
2) a flatter decay $F_X \propto t^{-\alpha_2}$ with $0.2 \lesssim \alpha_2 \lesssim 0.8$ and energy index $0.7 \lesssim \beta_2 \lesssim 1.2$, at times $10^3\text{s} \lesssim t_2 \lesssim 10^4\text{s}$;
3) a "normal" decay $F_X \propto t^{-\alpha_3}$ with $1.1 \lesssim \alpha_3 \lesssim 1.7$ and $0.7 \lesssim \beta_2 \lesssim 1.2$ (generally unchanged the previous stage), up to a time $t_3 \sim 10^5\text{s}$, or in some cases longer;
4) In some cases, a steeper decay $F_X \propto t^{-\alpha_4}$ with $2 \lesssim \alpha_4 \lesssim 3$, after $t_4 \sim 10^5\text{s}$;
5) In about half the afterglows, one or more X-ray flares are observed, sometimes starting as early as 100 s after trigger, and sometimes as late as 10^5s. The energy in these flares ranges from a percent up to a value comparable to the prompt emission (in GRB 050502b). The rise and decay times of these flares is unusually steep, depending on the reference time t_0, behaving as $(t-t_0)^{\pm \alpha_{fl}}$ with $3 \lesssim \alpha_{fl} \lesssim 6$, and energy indices which can be also steeper than during the smooth decay portions. The flux level after the flare usually decays to the value extrapolated from the value before the flare rise.

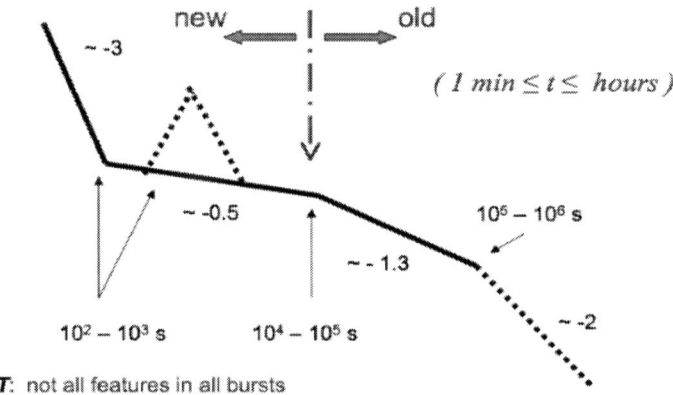

FIGURE 1. Schematic features seen by the XRT in bursts detected by Swift [16] (see text).

Another major advance achieved by Swift was the detection of the long burst GRB 050904, which broke through the astrophysically and psychologically important redshift barrier of $z \sim 6$. This burst was very bright, both in its prompt γ-ray emission ($E_{\gamma,iso} \sim 10^{54}$ erg) and in its X-ray afterglow. Prompt ground-based optical/IR upper limits and a J-band detection suggested a photometric redshift $z > 6$ [4]. Spectroscopic confirmation with the 8.2 m Subaru telescope gave a $z = 6.29$ [6]. There are several striking features to this burst. One is the enormous X-ray brightness, exceeding for a full day the X-ray brightness of the most distant X-ray quasar know to-date, SDSS J0130+0524, by up to a factor 10^5 in the first minutes [7]. The implications as a tool for probing the IGM are thought-provoking. Another feature is the extremely variable X-ray light curve, showing many large amplitude flares extending up to at least a day. A third exciting feature is the report of a brief, very bright IR flash [8], comparable in brightness to the famous $m_V \sim 9$ optical flash in GRB 990123.

The third major advance from Swift was the discovery and localization of short GRB afterglows. As of December 2005 nine short bursts had been localized by Swift, while in the same period HETE-2 discovered two, and one was identified with the IPN network. In five of the Swift short bursts an X-ray afterglow was measured and followed up, with GRB 050709, 050724 and 051221a showing an optical afterglow, and 050724 also a radio afterglow, while 040924 had an optical afterglow but not an X-ray one [46]. These are the first afterglows detected for short bursts. Also, for the first time, host galaxies were identified for these short bursts, which in four cases are early type (ellipticals) and in two cases are irregular galaxies. The redshifts of four of them are in the range $z \sim 0.15 - 0.5$, while another one was initially given as $z = 0.8$ but more recently has been reported as $z \simeq 1.8$ [49]. The median z is $\lesssim 1/3 - 1/2$ that of the long bursts. There is no evidence for significant star formation in any of these host environments, which corresponds to what one would expect for neutron star mergers or neutron star–black hole mergers, the most often discussed progenitor candidates (it would also be compatible with other progenitors involving old compact stars).

The first short burst seen by Swift, GRB 05059b, was a low luminosity ($E_{iso} \sim 2 \times 10^{48}$ erg) burst with a simple power-law X-ray afterglow which could only be followed for $\sim 10^4$ s [36]. The third one, GRB 050724, was brighter, $E_{iso} \sim 3 \times 10^{50}$ erg, and could be followed in X-rays for at least 10^5 s [37]. The remarkable thing about this burst's X-ray afterglow is that it resembles the typical X-ray light curves described above for long GRB – except for the lack of a slow-decay phase, and for the short prompt emission which places it the the category of short bursts, as well as the elliptical host galaxy candidate. It also has X-ray flares, at 100 s and another one at 3×10^4 s. The first flare has the same fluence as the prompt emission, while the late flare has $\sim 10\%$ of that. The interpretation of these pose interesting challenges, as discussed below.

MODELS OF EARLY AFTERGLOWS IN LIGHT OF SWIFT

The afterglow is expected to become important after a time

$$t_{ag} = \text{Max}[(3/4)(r_{dec}/2c\Gamma^2)(1+z), T] = \text{Max}[10^2(E_{52}/n_0)^{1/3}\Gamma_2^{-8/3}(1+z)|\text{s}, T], \tag{1}$$

where the deceleration time is $t_{dec} \sim (3/4)(r_{dec}/2c\Gamma^2)$ and T is the duration of the prompt outflow, t_{ag} marking the beginning of the self-similar blast wave regime.

Denoting the frequency and time dependence of the afterglow spectral energy flux as $F_\nu(t) \propto \nu^{-\beta} t^{-\alpha}$, the late X-ray afterglow phases (3) and (4) described above are similar to those known previously from Beppo-SAX. (For a review of this earlier behavior and its modeling see e.g. [9]). The "normal" decay phase (3), with temporal decay indices $\alpha \sim 1.1 - 1.5$ and spectral energy indices $\beta \sim 0.7 - 1.0$, is what is expected from the evolution of the forward shock in the Blandford-McKee self-similar late time regime, under the assumption of synchrotron emission.

The late steep decay decay phase (4) of §, occasionally seen in Swift bursts, is naturally explained as a jet break, when the decrease of the ejecta Lorentz factor leads to the light-cone angle becoming larger than the jet angular extent, $\Gamma_j(t) \gtrsim 1/\theta_j$ (e.g. [9]). It is noteworthy, however, that this final steepening has been seen in less than $\sim 10\%$ of the Swift afterglows, and then with reasonable confidence mainly in X-rays. The corresponding optical light curve breaks have been few, and not well constrained. This is unlike the case with the ~ 20 Beppo-SAX bursts, for which an achromatic break was reported in the optical [10], while in some of the rare cases where an X-ray or radio break was reported it occurred at a different time [11]. The relative paucity of optical breaks in Swift afterglows may be an observational selection effect due to the larger median redshift, and hence fainter and redder optical afterglow at the same observer epoch, as well as perhaps reluctance to commit large telescope time on more frequently reported bursts (an average, roughly, of 2/month with Beppo-SAX versus 2/week with Swift).

Steep decay

Among the new early afterglow features detected by Swift, the steep initial decay phase $F_\nu \propto t^{-3} - t^{-5}$ in X-rays of the long GRB afterglows is one of the most puzzling. There could be several possible reasons for this. The most immediate of these would be the cooling following cessation of the prompt emission (internal shocks or dissipation). If the comoving magnetic field in the emission region is random [or transverse], the flux per unit frequency along the line of sight in a given energy band, as a function of the electron energy index p, decays as $F_\nu \propto t^{-\alpha}$ with $\alpha = -2p\,[(1-3p)/2]$ in the slow cooling regime, where $\beta = (p-1)/2$, and it decays as $\alpha = -2(1+p)$, $[-(2-3p)/2]$ in the fast cooling regime where $\beta = p/2$, i.e. for the standard $p = 2.5$ this would be $\alpha = -5$, $[-3.25]$ in the slow cooling or $\alpha = -7$, $[-2.75]$ in the fast cooling regime, for random [transverse] fields [12]. In some bursts this may be the explanation, but in others the time and spectral indices do not correspond well.

Currently the most widely considered explanation for the fast decay, either in the initial phase (1) or in the steep flares, attributes it to the off-axis emission from regions at $\theta > \Gamma^{-1}$ (the curvature effect, or high latitude emission [13]. In this case, after the line of sight gamma-rays have ceased, the off-axis emission observed from $\theta > \Gamma^{-1}$ is $(\Gamma\theta)^{-6}$ smaller than that from the line of sight. Integrating over the equal arrival time region, this flux ratio becomes $\propto (\Gamma\theta)^{-4}$. Since the emission from θ arrives $(\Gamma\theta)^2$ later than from $\theta = 0$, the observer sees the flux falling as $F_\nu \propto t^{-2}$, if the flux were frequency independent. For a source-frame flux $\propto \nu'^{-\beta}$, the observed flux per unit frequency varies then as

$$F_\nu \propto (t-t_0)^{-2-\beta} \qquad (2)$$

i.e. $\alpha = 2 + \beta$. This "high latitude" radiation, which for observers outside the line cone at $\theta > \Gamma^{-1}$ would appear as prompt γ-ray emission from dissipation at radius r, appears to observers along the line of sight (inside the light cone) to arrive delayed by $t \sim r\theta^2/2c$) relative to the trigger time, and its spectrum is softened by the Doppler factor $\propto t^{-1}$

into the X-ray observer band. For the initial prompt decay, the onset of the afterglow (e.g. phases 2 or 3), which also come from the line of sight, may overlap in time with the delayed high latitude emission. In equation (2) t_0 can be taken as the trigger time, or some value comparable or less than by equation (1). This can be used to constrain the prompt emission radius [26]. When $t_{dec} < T$, the emission can have an admixture of high latitude and afterglow, and since the afterglow has a steeper spectrum than the high latitude (which has a prompt spectrum), one can have steeper decays [24]. Values of t_0 closer to the onset of the decay also lead to steeper slopes. Structured jets, when viewed on-beam produce essentially the same slopes as homogeneous jets, while off-beam observing can lead to shallower slopes [14]. For the flares, if their origin is assumed to be internal (e.g. some form of late internal shock or dissipation) the value of t_0 is just before the flare, e.g the observer time at which the internal dissipation starts to be observable [17]. This interpretation appears, so far, compatible with most of the Swift afterglows [16, 18, 19].

Alternatively, the initial fast decay could be due to the emission of a cocoon of exhaust gas [20], where the temporal and spectral index are explained through an approximately power-law behavior of escape times and spectral modification of multiply scattered photons. The fast decay may also be due to the reverse shock emission, if inverse Compton up-scatters primarily synchrotron optical photons into the X-ray range. The decay starts after the reverse shock has crossed the ejecta and electrons are no longer accelerated, and may have both a line of sight and an off-axis component [21]. This poses strong constraints on the Compton-y parameter, and cannot explain decays much steeper than $\alpha = -2$, or $-2-\beta$ if the off-axis contribution dominates. Models involving bullets, whose origin, acceleration and survivability is unexplained, could give a prompt decay index $\alpha = -3-5$ [15], but imply a bremsstrahlung energy index $\beta \sim 0$ which is not observed in the fast decay, and require fine-tuning. Finally, a patchy shell model, where the Lorentz factor is highly variable in angle, would produce emission with $\alpha \sim -2.5$. Thus, such mechanisms may explain the more gradual decays, but not the more extreme $\alpha = -5, -7$ values encountered in some cases.

Shallow decay

The slow decay portion of the X-ray light curves ($\alpha \sim -0.3-0.7$), ubiquitously detected by Swift, is not entirely new, having been detected in a few cases by BeppoSAX. This, as well as the appearance of wiggles and flares in the X-ray light curves after several hours were the motivation for the "refreshed shock" scenario [22, 23]. Refreshed shocks can flatten the afterglow light curve for hours or days, even if the ejecta is all emitted promptly at $t = T \lesssim t_\gamma$, but with a range of Lorentz factors, say $M(\Gamma) \propto \Gamma^{-s}$, where the lower Γ shells arrive much later to the foremost fast shells which have already been decelerated. Thus, for an external medium of density $\rho \propto r^{-g}$ and a prompt injection where the Lorentz factor spread relative to ejecta mass and energy is $M(\Gamma) \propto \Gamma^{-s}$, $E(\Gamma) \propto \Gamma^{-s+1}$, the forward shock flux temporal decay is given by [23]

$$\alpha = [(g-4)(1+s) + \beta(24-7g+sg)]/[2(7+s-2g)]. \qquad (3)$$

It needs to be emphasized that in this model all the ejection can be prompt (e.g. over the duration $\sim T$ of the gamma ray emission) but the low Γ portions arrive at (and refresh) the forward shock at late times, which can range from hours to days. I.e., it is not the central engine which is active late, but its effects are seen late. Fits of such refreshed shocks to observed shallow decay phases in Swift bursts [28] lead to a Γ distribution which is a broken power law, extending above and below a peak around ~ 45.

Another version of refreshed shocks, on the other hand, does envisage central engine activity extending for long periods of time, e.g. $\lesssim$ day (in contrast to the $\lesssim$ minutes engine activity in the model above). Such long-lived activity may be due to continued fall-back into the central black hole [32] or a magnetar wind [27]. One characteristic of both types of refreshed models is that after the refreshed shocks stop and the usual decay resumes, the flux level shows a step-up relative to the previous level, since new energy has been injected.

From current analyses, the refreshed shock model is generally able to explain the flatter temporal X-ray slopes seen by Swift, both when it is seen to join smoothly on the prompt emission (i.e. without an initial steep decay phase) or when seen after an initial steep decay. Questions remain concerning the interpretation of the fluence ratio in the shallow X-ray afterglow and the prompt gamma-ray emission, which can reach $\lesssim 1$ [24]. This requires a higher radiative efficiency in the prompt gamma-ray emission than in the X-ray afterglow. One might speculate that this might be achieved if the prompt outflow is Poynting-dominated. Alternatively, a more efficient afterglow might emit more of its energy in other bands, e.g. in GeV, or IR. Or [30] a previous mass ejection might have emptied a cavity into which the ejecta moves, leading to greater efficiency at later times, or otherwise the energy fraction going into the electrons increases $\propto t^{1/2}$.

X-ray flares

Refreshed shocks can also explain some of the X-ray flares whose rise and decay slopes are not too steep. However, this model encounters difficulties with the very steep flares with rise or decay indices $\alpha \sim \pm 5 - 7$, such as inferred from the giant flare of GRB 0500502b [29] around 300 s after the trigger. Also, the flux level increase in this flare is a factor ~ 500 above the smooth afterglow before and after it, implying a comparable energy excess in the low versus high Γ material. An explanation based on inverse Compton scattering in the reverse shock [21] can explain a single flare at the beginning of the afterglow, with not too steep decay. For multiple flares, models invoking encountering a lumpy external medium have generic difficulties explaining steep rises and decays [16], although extremely dense, sharp-edged lumps, if they exist, might satisfy the steepness [31].

Currently the more widely considered model for the flares ascribes them to late central engine activity [16, 18, 19]. The strongest argument in favor of this is that the energy budget is more easily satisfied, and the fast rise/decay is straightforward to explain. In such a model the flare energy can be comparable to the prompt emission, the fast rise comes naturally from the short time variability leading to internal shocks (or to rapid reconnection), while the rapid decay may be due to the high latitude emission following the flare, with t_0 reset to the beginning of each flare (see further discussion in [17]). However, some flares are well modeled by refreshed forward shocks, while in others this is clearly ruled out and a central engine origin is better suited [41]. Aside from the phenomenological desirability based on energetics and timescales, a central engine origin is conceivable, within certain time ranges, based on numerical models of the core collapse origin in long bursts. These are interpreted as being due to core collapse of a massive stellar progenitor, where continued infall into fast rotating cores can continue for a long time [32]. However, large flares with a fluence which is a sizable fraction of the prompt emission occurring hours later remain difficult to understand. It has been argued that gravitational instabilities in the infalling debris torus can lead to lumpy accretion [33]. Alternatively, if the accreting debris torus is dominated by MHD effects, magnetic instabilities can lead to extended, highly time variable accretion [34].

Short burst afterglows

Swift, and in smaller numbers HETE-2, have provided the first bona fide short burst X-ray afterglows followed up starting ~ 100 s after the trigger, leading to localizations and redshifts. In the first of these, GRB 050509b [36] the extrapolation of the prompt BAT emission into the X-ray range, and the XRT light curve from 100 s to about 1000 s (after which only upper limits exist, even with Chandra, due to the faintness of the burst) can be fitted with a single power law of $\alpha \sim 1.2$. The X-ray coverage was sparse due to orbital constraints, the number of X-ray photons being small. No optical transient was detected, but an E/S0 host galaxy was identified at $z = 0.225$ (e.g. [40]). In GRB050709 an optical transient was identified, as well as a host galaxy [39], an irregular galaxy at $z = 0.16$ (and the observations ruled out any supernova association). On the other hand, GRB 050724 was relatively bright, and besides X-rays, it also yielded both a decaying optical and a radio afterglow [40]. This burst, together with the greater part of other short bursts, is associated with an elliptical host galaxy. It also had a low-luminosity soft gamma-ray extension of the short hard gamma-ray component (which would have been missed by BATSE), and it had an interesting X-ray afterglow extending beyond 10^5 s [37]. The soft gamma-ray extension, lasting up to 200 s, when extrapolated to the X-ray range overlaps well with the beginning of the XRT afterglow, which between 100 and 300 s has $\alpha \sim -2$, followed by a much steeper drop $\alpha \sim -5-7$ out to $\sim 600s$, then a more moderate decay $\alpha \sim -1$. An unexpected feature is a strong flare peaking at 5×10^4 s, whose energy is 10% of the prompt emission, while its amplitude is a 10 times increase over the preceding slow decay. With about a half dozen of reasonably identified objects, the distribution of shorts bursts in redshift space and among host galaxy types, including fewer spiral/irregulars and more ellipticals, is typical of old population progenitors, such as neutron star binaries or black hole-neutron star binaries [38].

The main challenges posed by the short burst afterglows are the relatively long, soft tail of the prompt emission, and the strength and late occurrence of the flares. A possible explanation for the extended long soft tails ($\sim 100s$) may be that the compact binary progenitor is a black hole - neutron star system [37], for which analytical and numerical arguments ([44], and references therein) suggest that the disruption and swallowing by the black hole may lead to a complex and more extended accretion rate than for double neutron stars. The flares, for which the simplest interpretation might be as refreshed shocks (which would be compatible with a short engine duration $T \lesssim t_\gamma \sim 2$ s, for a Lorentz factor distribution), requires the energy in the slow material to be at least ten times as energetic as the fast material responsible for the prompt emission, for the GRB 050724 flare at 10^4 s. The rise and decay times are moderate enough for this interpretation. Another interpretation might be an accretion-induced collapse of a white dwarf in a binary, leading to a flare when the fireball created by the collapse hits the companion [45], which might explain moderate energy one-time flares. However, for repeated, energetic flares, as also in the long bursts, the total energetics are easier to satisfy if one postulates late central engine activity (lasting at least half a day), containing $\sim 10\%$ of the prompt fluence [37]. A possible way to produce this might be temporary choking up of an MHD outflow [34] (c.f. [43]), which might also imply a linear polarization of the X-ray flare [42]. Such MHD effects could plausibly also explain the initial ~ 100 s soft tail. However, a justification for substantial $\gtrsim 10^5$ s features remains so far on tentative grounds.

The similarity of the X-ray afterglow light curve with those of long bursts is, in itself, an argument in favor of the prevalent view that the afterglows of both long and short bursts can be described by the same paradigm, independently of any difference in the progenitors. This impression is reinforced by the fact that the X-ray light curve temporal slope is, on average, that expected from the usual forward shock afterglow model, and that in two short bursts (so far) there is evidence for what appears to be a jet break [40, 35]. However, while similar to zeroth order, the first order differences are revealing: the average isotropic energy is factor ~ 100 smaller, while the average jet opening angle (based on two breaks) is a factor ~ 2 larger [46, 35]. Using the standard afterglow theory, the bulk Lorentz factor decay can be expressed through $\Gamma(t_d) = 6.5(n_o/E_{50})^{1/8} t_d^{-3/8}$, where $t_d = (t/\text{day})$, n_o is the external density in units of cm^{-3}, and E_{50} is the isotropic equivalent energy in units of 10^{50} ergs. If the jet break occurs at $\Gamma(t_{br}) = \theta_j^{-1}$ the jet opening angle and the total jet energy E_j are

$$\theta_j = 9^\circ (n_o/E_{50})^{1/8} t_{d,br}^{3/8} \ , E_j = \pi \theta_j^2 E \sim 10^{49} n_o^{1/4} (E_{50} t_{d,br})^{3/4} \text{ erg} \ . \qquad (4)$$

For the first two well studied afterglows GRB 050709 and GRB 050724, together with the standard afterglow expressions for the flux level as a function of time before and after the break, this leads to fits [35] which are not completely determined, allowing for GRB050709 either a very low or a moderately low external density, and for GRB050724 a moderately low to large external density. The main uncertainty is in the jet break time, which is poorly sampled, and so far mainly in X-rays. As brighter short bursts are occasionally detected, e.g. GRB 051221A, the chances of tighter constraints on jet breaks should increase.

Prompt optical flashes and high redshift afterglows

Optical/UV afterglows have been detected with the Swift UVOT telescope in roughly half the bursts for which an X-ray afterglow was seen. For a more detailed discussion of the UVOT afterglow observations see [3]. Of particular interest is the ongoing discussion on whether "dark GRB" are really optically deficient, or the result of observational bias [5]. Another puzzle is the report of a bimodal intrinsic brightness distribution in the rest-frame R-band [47, 48]. This suggests possibly the existence of two different classes of long bursts, or at least two different types of environments.

Compared to a few years ago, a much larger role is being played by ground-based robotic optical follow-ups, due to the increased rate of several arc-second X-ray alerts from XRT, and the larger number of robotic telescopes brought on-line in the last years. For the most part, these detections have yielded optical decays in the $\gtrsim$ few 100 s range, initial brightness $m_V \sim 14-17$ and temporal decay slopes $\alpha \sim 1.1-1.7$ previously associated with the evolution of a forward shock [46, 49], while in a few cases a prompt optical detection was achieved in the first 12-25 s.

The most exciting prompt robotic IR detection (and optical non-detection) is that of GRB 050904 [8, 4]. This object, at the unprecedented high redshift of $z = 6.29$ [6], has an X-ray brightness exceeding for a day that of the brightest X-ray quasars [7], and its O/IR brightness in the first 500 s (observer time) was comparable to that of the extremely bright ($m_V \sim 9$) optical flash in GRB 990123, with a similarly steep time-decay slope $\alpha \sim 3$ [8]. Such prompt, bright and steeply decaying optical emission is expected from the reverse shock as it crosses the ejecta, marking the start of the afterglow [50]. However, aside from the two glaring examples of 990123 and 05094, in the last six years there have been less than a score of other prompt optical flashes, typically with more modest initial brightnesses $m_v \gtrsim 13$. There are a number of possible reasons for this paucity of optically bright flashes, if ascribed to reverse shock emission. One is the absence or weakness of a reverse shock, e.g. if the ejecta is highly magnetized [50]. A moderately magnetized ejecta is in fact favored for some prompt flashes [51]. Alternatively, the deceleration might occur in the thick-shell regime ($T \gg t_{dec}$. see eq. (1), which can result in the reverse shock being relativistic, boosting the optical reverse shock spectrum into the UV [52]. Another possibility, for a high comoving luminosity, is copious pair formation in the ejecta, causing the reverse shock spectrum to peak in the IR [53]. Since both GRB 990123 and GRB 050904 had $E_{iso} \sim 10^{54}$ erg, among the top few percent of all bursts, the latter is a distinct possibility, compatible with the fact that the prompt flash in GRB 050904 was bright in the IR I-band but not in the optical. On the other hand, the redshift $z = 6.29$ of this burst, and a Ly-α cutoff at ~ 800 nm would also ensure this (and GRB 990123, at $z = 1.6$, was detected in the V-band). However, the observations in these two objects but a suppression in lower E_{iso} objects appears compatible with having a relativistic (thick shell) reverse shock with pair formation.

ACKNOWLEDGMENTS

I am grateful to the Swift team for collaborations and to NASA NAG5 13286 for support.

REFERENCES

1. Gehrels, N, 2005, these procs.
2. Burrows, D, 2005, these proceedings; also astro-ph/0511039

3. Roming, P, 2005, these proceedings; also astro-ph/050927
4. Haislip, J, et al, 2005, Nature, subm. (astro-ph/0509660)
5. Berger, E, et al, 2005, ApJ 634:501
6. Kawai, N. et al, 2005, astro-ph/0512052
7. Watson, D. et al, 2005, ApJ(Lett) in press (astro-ph/0509640)
8. Boër, M, et al, 2005, Nature, subm (astro-ph/0510381)
9. Zhang, B and Mészáros, P, 2004, Internat.J.Mod.Phys. A, 19:2385
10. Frail, D et al., ApJ 562:L155
11. Berger, E, Kulkarni, SR and Frail, D, 2003, ApJ 590:379
12. Mészáros, P and Rees, MJ, 1999, MNRAS, 306:L39
13. Kumar, P and Panaitescu, A, 2000, ApJ 541:L51
14. Dyks, J, Zhang, B, Fan, YZ, 2005, ApJ subm (astro-ph/0511699)
15. Dado, S, et al, astro-ph/0512196
16. Zhang, B. et al, 2006, ApJ in press (astro-ph/0508321)
17. Zhang, B. et al, these proceedings
18. Nousek, J, et al, ApJ, in press (astro-ph/0508332)
19. Panaitescu, A, et al, 2005, MNRAS in press (astro-ph/0508340)
20. Pe'er, A, Mészáros, P and Rees, MJ, 2006, in prep.
21. Kobayashi, S, et al, 2005, ApJ subm (astro-ph/0506157)
22. Rees, MJ and Mészáros, P, 1998, ApJ, 496:L1
23. Sari, R and Mészáros, P, 2000, ApJ, 535:L33
24. O'Brien, P. et al, in preparation
25. Dyks, J, Zhang, B and Fan, YS, ApJ subm (astro-ph/0511699)
26. Lazzati, D. and Begelman, M, astro-ph/0511658
27. Zhang, B and Mészáros, P, 2001, ApJ 552:L35
28. Granot, J and Kumar, P, ApJ subm (astro-ph/0511049)
29. Burrows, D, et al, 2005a, Science, 309, 1833
30. Ioka, K, et al, astro-ph/0511749
31. Dermer, C, 2005, these proceedings
32. Woosley, S, 2005, these proceedings
33. Perna, R, Armitage, and Zhang, B, ApJ in press (astro-ph/0511506)
34. Proga, D and Begelman, M, 2003, ApJ, 592:767
35. Panaitescu, A, 2005, ApJ subm (astro-ph/0511588)
36. Gehrels, N. et al, 2005, Nature, 437:851
37. Barthelmy, S., 2005, Nature, in press (astro-ph/0511579)
38. Nakar, E, Gal-Yam, A and Fox, D, (astro-ph/0511254)
39. Fox, D. et al, 2005, Nature, 437:845
40. Berger, E, et al, 2005, ApJ, 634:501
41. Wu, X.F. et al, ApJ subm (astro-ph/0512555)
42. Fan, YZ, Proga, D and Zhang, B, 2005, astro-ph/0509019
43. Van Putten, M and Ostriker, A, 2001, ApJ 552:L31
44. Davies, M, Levan, A, King, A, 2005, MNRAS 365:54
45. MacFadyen, A, 2005, these proceedings
46. Fox, D, 2005, these proceedings.
47. Liang, EW and Zhang, B, 2005, astro-ph/0508510
48. Nardini, M, et al, D, 2005, astro-ph/0508447
49. Berger, E, 2005, these proceedings
50. Mészáros, P and Rees, MJ, 1997, ApJ 476:232
51. B. Zhang, S. Kobayashi and P. Mészáros, 2003, ApJ, 589:861
52. Kobayashi, S, 2000, ApJ, 545:807
53. Mészáros, P, Ramirez-Ruiz, E, Rees, MJ, Zhang, B, 2002, ApJ, 578:812

The prompt and early afterglow X-ray spectra of Swift GRBs

M.R. Goad, J.P. Osborne, A.P. Beardmore, O. Godet, K. Page - on behalf of the Swift XRT team

Department of Physics and Astronomy, University of Leicester, UK

Abstract. The Swift Gamma Ray Burst Explorer is providing exciting new insights into the nature of the prompt and early afterglow emission of GRBs, filling in the 8 hr observing gap immediately following the initial explosion. Here we report on the XRT 0.3-10 keV prompt and early afterglow spectra of those GRBs for which we have early (< few minutes) XRT data and which exhibit a common behaviour in their temporal decay. We also report on GRB 051117A, a source which is bright enough in XRT to allow the most detailed temporal spectral analysis to date.

Keywords: gamma-ray bursts
PACS: 98.70.Rz

INTRODUCTION

Analysis of ≈50 Swift detected GRBs with prompt (<few minutes) X-ray observations has revealed that many display a canonical X-ray decay light-curve (Nousek et al. 2006, O'Brien et al. 2006) with an initial steep decay breaking to a flatter slope on timescales of a few thousand seconds, followed by a steepening after several hours. In others, the extended "hump" in emission may be weak or entirely absent. Here we report on the spectral characteristics of those GRBs taken from the sample of O'Brien et al. (2006, and these proceedings), with a few later bursts, which display the canonical light-curve decay behaviour and for which sufficient counts are available for each of the light-curve decay segments (ie. see Figure 1) to perform a meaningful spectral analysis.

Of the 40 bursts listed by O'Brien et al. with the addition of those bursts observed by the end of October 2005, and with prompt X-ray observations, 30 have sufficient counts to perform a spectral analysis covering the first break in the light-curve, ie. segments 1 and 2. Excluding those bursts which do not display a clear break in their light-curves, leaves a sample of 16 bursts for which a statistical analysis of the spectral properties covering the 1st break has been performed.

Relationship between the BAT and early XRT spectra

Figure 2 left upper panel, shows the correlation between the BAT (segment 0) and XRT (segment 1, the steeply declining phase) photon indices. The majority of GRBs, are softer (ie. have steeper photon indices) in the XRT than in the BAT, consistent with their spectra becoming softer with time. The BAT/XRT photon indices are only weakly correlated, having a Pearson's r correlation r=0.51, with the significance level at which the null hypothesis of zero correlation is disproved, P=0.04. Excluding GRB 050714B, which has a very soft X-ray spectrum, consistent with it being an X-ray flash, the correlation is even weaker (r=0.35, P=0.2). Interestingly there is a much stronger correlation between the difference in the photon indices and the XRT pre-break photon index (Figure 2 lower panel) with r=0.92, P=4e-7, though again this weakens considerably if GRB 050714B is removed (R=0.65, P=0.009). A possible cause for such a strong correlation would be if the BAT photon indices were strongly clustered around a single value, though Figure 2 (left, upper panel) shows that this is not the case.

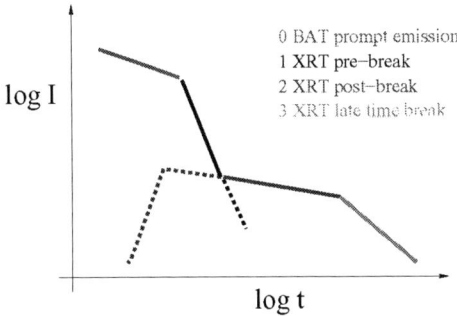

Figure 1 - Cartoon illustrating the canonical GRB decay light-curve (see also Nousek et al. 2006).

Relationship between the pre- and post-break and early XRT spectra

In Figure 2 (right, upper panel), we show the correlation between the XRT pre- (segment 1) and post-break (segment 2) photon indices. Interestingly in this case the majority of the post-break spectra are moderately harder than the pre-break spectra, possibly a result of late-time shock energisation giving rise to a shallower decay slope in the XRT light-curve, and their spectral indices are uncorrelated (r=0.19, P=0.49, excluding GRB 050714B). Similarly, there is a much stronger correlation between the difference in the photon indices of the pre- and post-break XRT spectra and the pre-break photon index (r=0.85, P=3e-5, Figure 2 (right, lower panel), which in this case is almost certainly due to the strong clustering of the post-break XRT photon indices.

Relationship between the BAT and post-break XRT spectra

Figure 3 (left, upper panel), shows the correlation between the post-break XRT photon index (segment 2), and the BAT photon index (segment 0). Once again, there is no correlation (r=0.28, P=0.3). However, in this case, the difference in the photon indices of the BAT and post-break XRT spectra are also uncorrelated (r=0.3, P=0.25), suggesting the two are unrelated.

The distribution in photon indices

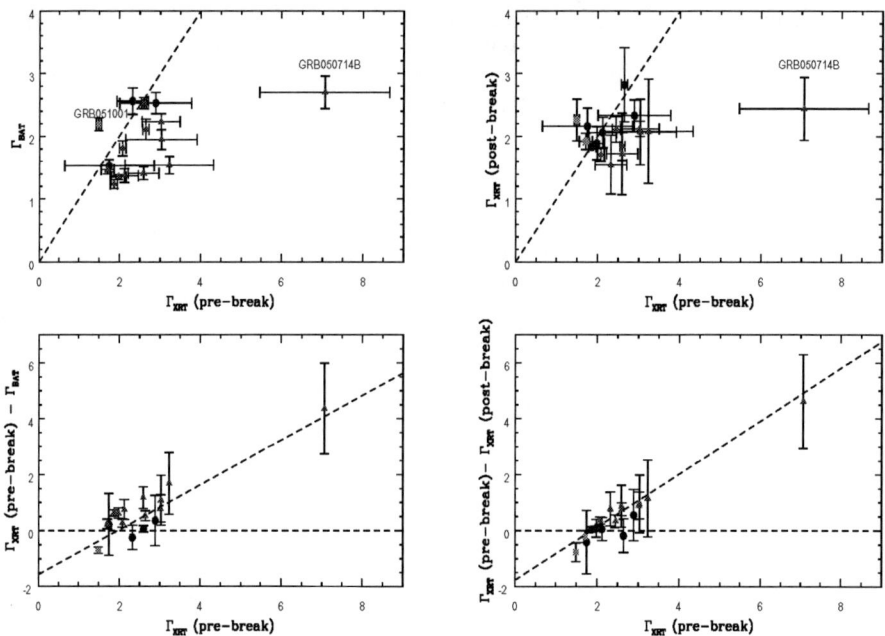

Figure 2 (*left, upper panel*) - The observed correlation between the pre-break XRT photon index (segment 1) and BAT photon index (segment 0). (*Left, lower panel*) - The observed correlation between the difference in XRT pre-break and BAT photon indices versus the XRT pre-break photon index. (*Right, upper panel*) - The observed correlation between the pre-break XRT photon index (segment 1) and the XRT post-break photon index (segment 2). (*Right, lower panel*) - The observed correlation between the difference in the XRT pre- and post break photon indices versus the pre-break photon index.

In Figure 3 (right) we show the distribution in spectral indices for segments 0 (BAT, upper panel), segment 1 (XRT pre-break, middle panel) and segment 2 (XRT post-break, lower panel). The BAT spectra have a mean photon index of 1.61 ± 0.02 and standard deviation (sd) $\sigma = 0.59$. The steep decay phase (segment 1) has a steeper mean photon index 2.12 ± 0.02 and $\sigma = 0.6$ (excludes GRB 050714B). By contrast, the post-break spectra have a harder mean photon index of 1.86 ± 0.03, and $\sigma = 0.36$. While there is significant overlap in the three distributions, the post-break XRT data exhibit a far narrower range in spectral indices. Our these three distributions significantly different? The results of a simple χ^2 test on the binned data are shown in Table 1.

Table 1 indicates a significant difference between the distributions of the BAT and XRT (post-break) photon indices. The XRT pre- and post-break photon indices are also significantly different, while the BAT and pre-break XRT photon indices show the smallest differences in their distributions. This confirms our earlier suggestion that while the XRT pre-break spectra are generally softer than the preceding BAT spectra and have a larger spread in values their two distributions are similar. By contrast, the spectra

TABLE 1. Significance test for differences between the distributions.

Data set	ndf	χ^2	Prob
BAT – XRT (pre-break)	5	5.94	0.32
XRT (pre-break) – XRT (post-break)	5	8.67	0.123
BAT – XRT (post-break)	3	8.47	0.037

taken during the shallow decline phase are much more narrowly distributed (small σ), are on average harder than the pre-break spectra and appear to be drawn from a different parent population, suggesting at least two different energy generation mechanisms are operating in these bursts.

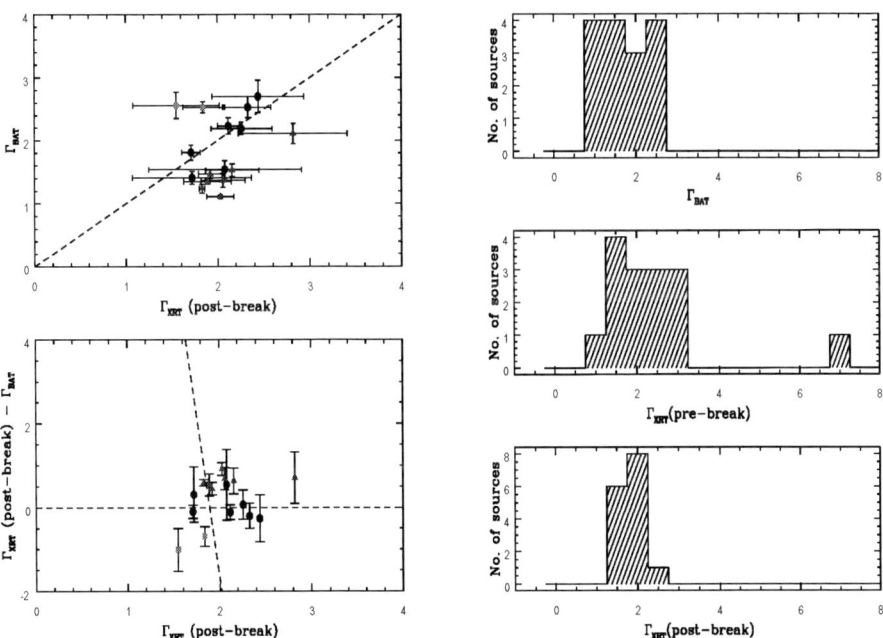

Figure 3 (*left, upper panel*) - The observed correlation between the post-break XRT photon index (segment 2) and BAT photon index (segment 0). (*Left, lower panel*) The observed correlation between the difference in XRT post-break and BAT photon indices and the XRT post-break photon index. (*Right*) - The observed distribution in photon indices of BAT (*segment 0, upper panel*), XRT pre-break (*segment 1, middle panel*), and XRT post-break (*segment 2, lower panel*). The post-break photon indices are strongly clustered.

The narrowness in distribution of photon index during the shallow phase of the XRT decay light-curve may indicate a universal energy generation mechanism for this segment of the light-curve. Proposed models include either : (i) late-time engine activity (Zhang et al. 2005), (ii) refreshed shocks due to collisions between shells ejected with different Lorentz factors (Granot and Kumar 2005), (iii) a two component jet model

(Peng et al. 2005), and (iv) time variable microphysics (Granot et al. 2006, submitted).

Spectral properties of post 2nd-break data

In our sample, we have 9 GRBs for which there are sufficient counts to derive a photon index after the 2nd break in the light-curve (segment 3). Of these, 3 show spectral steeping following the shallow decline phase, 5 shown no evidence for spectral variation, and 1 is inconclusive. Interestingly not one has a harder spectral index. The shallow temporal decay slope for the light-curve following the 2nd break and the absence of any significant spectral variation suggests that this segment is associated with the normal afterglow phase seen in pre-Swift bursts.

SUMMARY

In summary, for those GRBs which exhibit the canonical steep, shallow, steep, behaviour in their early XRT decay light-curves, the majority display a larger (or similar) photon index in the XRT band than in the BAT, with only one GRB showing a significantly harder spectrum in the XRT band. This is consistent with their being the tail of the prompt emission. Interestingly the segment 2 data (the shallow decline phase) present in approximately half of all bursts, display a narrow range of spectral slopes, which appear independent of the early decay slope. This most likely indicates separate origins for these two segments of the light-curve. Though the statistics for segment 3 data are generally poor, their spectral indices are on average higher than the segment 2 data.

A DETAILED SPECTRAL ANALYSIS OF THE X-RAY BRIGHT FLARING BURST GRB 051117A

GRB 051117A is one of the X-ray brightest GRBs observed to date, allowing an unprecedented temporal spectral analysis of the early X-ray light-curve of this source (Goad et al. 2006, in prep).

The combined BAT-XRT light-curve

Swift BAT detected and located on board GRB 051117A, at 10:51:20 UT Nov 17th 2005 (Band et al. 2005). The spacecraft slewed immediately to the source and began observations with XRT 107 s after the BAT trigger. The XRT found a very bright, fading uncatalogued source at RA=15h 13m 33.8s, Dec=+30d 52m 13.3s (J2000), with a positional uncertainty of 3.4 arcsec (90% containment). The BAT light-curve is FRED-like with a peak at $T_0 - 15$ s with the tail of emission lasting out to $T_0 + 190$ s, with a hint of further peaks at $T_0 + 225$ s and $T_0 + 350$ s. The XRT light-curve is one of the brightest and best-sampled XRT light-curves yet obtained, with an initial count rate >220 ct s^{-1} with data taken in Windowed Timing mode for the whole duration of the 1st orbit, and is still moderately piled up in the 2nd orbit (PC mode). Figure 4 (left panel) shows the combined BAT-XRT 0.6-10 keV light-curve. The long duration of this burst provides 190 s of overlap between BAT and XRT observations. The BAT light-curve was converted into the XRT 0.6-10 keV band using the combined BAT-XRT spectral fit to the data taken in the overlap region. The early XRT light-curve is highly structured with multiples flares of differing strength and duration, and joins smoothly with the BAT light-curve in the overlap region, suggesting that the X-ray emission is the tail of the prompt emission from the initial explosion. The initial decay slope for this burst is $\alpha = 0.77$ ($f(t) \propto t^{-\alpha}$) for $T < T_0 + 300$ s, breaks sharply for $T_0 + 7450 < t < T_0 + 16500$ s with a slope $\alpha > 5$, with a 2nd break to a shallower decay with $\alpha = 0.66$. At late times, $t = T_0 + 168$ ks, the light-curve steepens once more with a slope of $\alpha = 1.1$.

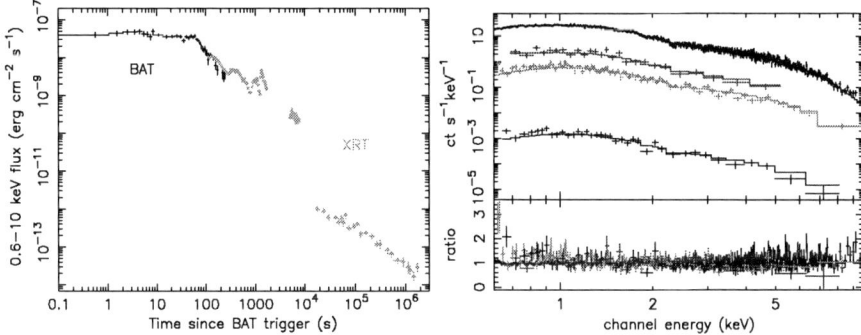

Figure 4 (*Left panel*) - The combined BAT and XRT light-curve in the 0.6-10 keV band. The BAT light-curve has been converted into the XRT band using the combined spectral fit to the BAT and XRT data in the overlap region of the light-curve. (*Right panel*) - Spectral fits to the 1st and 2nd orbit WT mode, and orbit 2 and later PC mode XRT data for GRB 051117A.

A detailed spectral analysis of the early X-ray light-curve

Global spectral properties

To investigate the long-term spectral evolution of GRB 05117A, we have formed average spectra covering the 1st orbit (WT mode 1605 s), 2nd orbit (WT mode 163 s, PC mode 2313 s) and orbit 3 onwards (PC mode 411 ks), see Figure 4. The data are well described by a single absorbed powerlaw with excess NH above the Galactic value of 1.82×10^{20} cm^{-2}, of NH$_{excess} = 2.1 \times 10^{21}$ cm^{-2} with $\Gamma = 2.24 \pm 0.02$, $\chi^2 = 708$ for 679 dof. Untying the spectral photon index between the spectra provides a significant improvement, $\Delta\chi^2 = 42$, F-statistic 14.3, null hypothesis 4.7×10^{-9} (Figure 3 - right panel). We note that tying Γ between the spectra and untying excess NH, provides an equally good description of the data, though excess NH is consistent with being constant within the errors. An absorbed broken powerlaw with excess NH and Γ pre- and post-break tied between the spectra, is only a slightly worse description of the data, $\chi^2 = 669$

for 674 dof.

We have searched for evidence of emission-lines in the early WT mode spectrum (1605 s exposure). At near the peak in the effective area (≈1 keV), we can rule out the presence of narrow spectral features ($\sigma = 100$ eV) with equivalent widths (EW) of less than 15 eV at greater than 90% confidence.

The superb quality (high observed count-rate) of the 1st orbit of WT mode data, (1605 s) allows us to perform a detailed temporal analysis of the X-ray spectral evolution of this source. For meaningful statistics we have divided the early X-ray light-curve into 37 time intervals each containing approximately 2000 ct/bin. The resultant 0.6-10 keV light-curve is shown in Figure 5 (left, upper panel). In detail the light-curve is well-described by a powerlaw decay with superposed FRED-like flares of varying strength and duration. A power spectrum of the light-curve is consistent with that expected from a superposition of random shots.

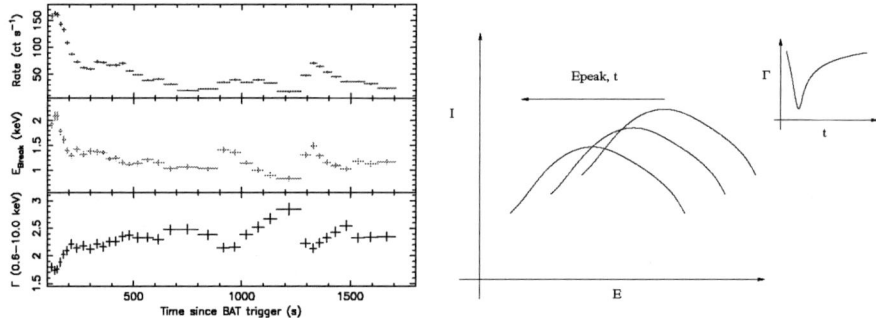

Figure 5 (*left, upper panel*) - The 1st orbit WT mode 0.6-10 keV light-curve binned to 2000 ct/bin. Middle panel - Temporal variation in the measured break energy for a broken powerlaw model fit to the individual spectra (see text for details. (*Left, lower panel*) - Variation in photon index Γ derived from a simultaneous fit to the 37 individual spectra. The adopted model is a simple absorbed powerlaw with excess NH tied between individual spectra. (*Right panel*) - A cartoon illustrating a simple mechanism for producing the observed variation in Photon index. As the source brightens, the break energy moves rapidly to higher energies, followed by a more gradual decline to lower energies as the source fades.

A simultaneous fit to the 37 spectra using a simple absorbed powerlaw, with NH fixed at the Galactic value and an excess NH of 2.1×10^{21} cm^{-2}, with Γ and the normalisation allowed to freely vary provides a good fit to the data, $\chi^2 = 2470$ for 2648 dof. Figure 5 (left, lower panel), shows the temporal variation in the derived photon index obtained from our model fits. Comparison of panels 1 and 3, indicates that the photon index and source intensity are strongly anti-correlated, such that the source is harder (smaller photon index) when brighter. A hardness ratio (HR) light-curve (not shown), where here we define HR as the ratio of the (2.0-10.0 keV)/(0.2-2.0 keV) bands, confirms this finding. A broken powerlaw fit to the spectra with NH fixed at the Galactic value, and the pre-and post-break Γ's tied together provides an acceptable fit to the data $\chi^2 = 2593$ for 2647 dof. Including excess NH only moderately improves this fit, $\Delta\chi^2 = 22$. Figure 5 (left, middle panel) shows the temporal variation in the break energy, E_{break}. The break energy is strongly correlated with source intensity such that the break energy is higher when the source is brighter. Analysis of the last flaring episode, $t > T_0 + 1300$ s, suggests that at the onset of the flare, the break energy moves rapidly to higher energies, before decreasing to lower energies as the intensity fades. For a single absorbed powerlaw fit, over the limited band-pass of the XRT, this will manifest as an anti-correlation between photon index and source intensity, as is observed (Figure 5, left lower panel). Figure 5 (right panel) shows a simple cartoon illustrating this effect.

DISCUSSION

The standard model for GRB afterglows, a spherical blastwave expanding into a uniform density ambient medium, predicts smooth afterglow light-curves. The early X-ray afterglow light-curve of GRB 051117A is far from smooth, displaying multiple FRED-like flares superposed on an underlying powerlaw decay. Similar flaring behaviour, though not at the level of detail presented here, has been reported in approximately half of all Swift detected GRBs with early (<few minutes) X-ray observations (e.g. Burrows et al. 2005, Falcone et al. 2005, King et al. 2005). Suggested models for the origin of flares, include (i) density fluctuations in the surrounding medium into which the blastwave expands, (ii) patchy shells, (iii) refreshed shocks, and (iv) late time engine activity (see Ioka, Kobayashi and Zhang 2005) for a thorough discussion of each of these competing scenarios. As noted by these authors, the relative fluctuation amplitudes and timescales can be used to rule out some scenarios for the origin of the flares. The early X-ray light-curve of GRB 051117A, is characterised by a broad range of flare rise-times (30 - 100 s) and decay-times (as measured by the e-folding time, 10-200 s).

Neither refreshed shocks, which require $\Delta t/t < 1/4$, or external density fluctuations, for which $\Delta t/t \ll 0.01$, can explain the observed flaring behaviour. The most plausible explanation is that we are seeing increased emission due to late-time internal engine activity. The observation that the BAT and XRT light-curves join up smoothly provides strong supporting evidence for this conclusion. Moreover, the timescale over which the flares are observed, is entirely consistent with known burst durations. The apparent steep break in the X-ray light-curve at $T > 7.4$ ks, has two possible origins. Either the flaring activity continues out to 16.5 ks, or the source activity ends abruptly, and we are simply observing emission arising from angles greater than Γ^{-1}, where in this case Γ is the Lorentz factor, with respect to our line-of-sight, the so-called 'curvature effect' (Kumar and Panaitescu 2000).

The following flattening of the light-curve is most-likely due to refreshed shocks, caused by collisions between shells emitted with a wide range in Lorentz factors. The slope of the late time data following the 2nd break in the light-curve at $T_0 + 168$ ks, and the lack of any detectable spectral variation suggests that this section of the light-curve is associated with the previously known afterglow phase of the GRB.

CONCLUSIONS

GRB 051117A displays the highest quality early X-ray light-curve to date. A detailed temporal spectral analysis shows an early X-ray light-curve composed of flares which are FRED-like in appearance and which span a broad range in amplitude and timescales, superposed on a powerlaw decay. Each flaring episode results initially in a rapid hardening of the spectrum followed by a more gradual softening as the flare fades. The simplest explanation for the observed spectral behaviour is the movement of the break energy to higher energies at the onset of the flare, which then falls to lower energies as the intensity decreases. The most likely origin for the flares is late time internal energy injection due to prolonged activity of the central engine.

REFERENCES

1. Band, D. et al. 2005, GCN 4280.
2. Burrows et al. 2005 Sp. Sc. Rev. 309, 1833.
3. Falcone et al. 2006, ApJ in press (astro-ph/0512615).
4. Granot, J. and Kumar, P., 2005, MNRAS in press (astro-ph/0511049).
5. Granot, J., Konigl, A. and Piran, T., MNRAS submitted (astro-ph/0601056).
6. Ioka, Kobayashi and Zhang 2005, ApJ 633, 1013.
7. King et al. 2005, ApJL 630 L113.
8. Kumar, P. and Panaitescu, A. 2000, ApJL 541, L51.
9. Nousek, J.A. et al. 2006, ApJ in press (astro-ph/0508332).
10. O'Brien, P.T. et al. 2006, ApJ submitted.
11. Peng, F., Konigl, A. and Granot, J. 2005, ApJ 626, 966.
12. Zhang, B. et al. 2005 ApJ submitted, astro-ph/0508321.

Are XRFs Really Off-Jet GRBs?

T. Q. Donaghy

Department of Physics, University of Chicago, 5640 S. Ellis Avenue, Chicago, IL 60637

Abstract.
Gamma-Ray Bursts (GRBs) are widely thought to originate from collimated jets of material moving at relativistic velocities. Emission from such a jet should be visible even when viewed from outside the angle of collimation. Such off-jet emission has been proposed as an explanation for X-Ray Flashes (XRFs). I summarize recent work on the special relativistic transformation of the burst quantities E_{iso} and E_{peak} as a function of viewing angle, in the context the top-hat shaped variable opening-angle jet model. The resulting formulae serve as input for a Monte Carlo population synthesis method. I find that the XRFs seen by *HETE-2* and *Beppo*SAX cannot be easily explained as classical GRBs viewed off-jet. For $\gamma < 300$, such models predict a large population of observable off-jet bursts that lie away from the $E_{peak} \propto E_{iso}^{1/2}$ relation. An inverse correlation between γ and Ω_0 has the effect of greatly reducing the visibility of off-jet events. Therefore, unless $\gamma > 300$ for all bursts or unless there is a strong inverse correlation between γ and Ω_0, top-hat variable opening-angle jet models produce a significant population of bursts away from the $E_{peak} \propto E_{iso}^{1/2}$ and $E_{peak} \propto E_\gamma^\beta$ relations, in contradiction of current observations.

Keywords: Gamma Rays: Bursts — ISM: Jets and Outflows — Shock Waves
PACS: 98.70.Rz

INTRODUCTION

Relativistic kinematics implies that even a "top-hat"-shaped jet will be visible when viewed outside its angle of collimation; i.e., off-jet. [1, 2] used this fact to construct a model where XRFs [3, 4] are simply classical GRBs viewed at an angle $\theta_v > \theta_0$, where θ_0 is the half-opening angle of the jet and θ_v is the angle between the axis of the jet and the line-of-sight. The authors showed that such a model could reproduce many of the observed characteristics of XRFs. Results from *HETE-2* observations [5] have shown that XRFs, X-ray-rich GRBs and GRBs lie along a continuum of properties and that XRFs with known redshift extend the $E_{peak} \propto E_{iso}^{1/2}$ relation [6] to over five orders of magnitude in E_{iso} [7]. [8] showed that for the off-jet model, the distribution of both on- and off-jet observed bursts was roughly consistent with the $E_{peak} \propto E_{iso}^{1/2}$ relation.

In this paper, I use the population synthesis method developed by [9] and incorporate the relativistic emission profiles calculated by [10], to predict the global properties of bursts localized by *HETE-2* and *Beppo*SAX. Following [8], I consider the possibility that XRFs are primarily classical GRBs observed off-jet and I show that it is difficult to account for the observed properties of XRFs in this manner. However, since the effect of special relativity on off-jet emission must exist, I seek to understand its relative importance in the context of current models of GRB jets. I revisit the top-hat variable opening-angle (THVOA) jet model put forward in [9], now including the effects of relativistic kinematics on off-jet emission. I describe my population synthesis method in §2, present results for six models which explore various regions of the parameter space in §3 and offer some conclusions in §4. For more details and discussion see [11].

METHOD

Relativistic kinematics causes frequencies in the rest frame of the jet to appear Doppler shifted by a factor $\delta = \gamma(1 - \beta \cos \theta)$, where β is the bulk velocity of the jet and θ is the angle between the direction of motion and the source frame observer. The simulations I describe chiefly deal with the kinematic transformation of two important burst quantities, E_{iso} and E_{peak}, as a function of viewing angle, θ_v.

The complete relativistic kinematic expressions involve convolution of the Doppler function and the intrinsic profile of the jet; for the case of the "top-hat" profile, an analytic expression can be given. The formulae are derived in [10], and I summarize them here. The current model differs slightly from that used by [8] in that we consider steady-state emission, rather than the evolution of burst properties due to time-of-flight effects. The model therefore applies to burst-averaged data products like E_{iso} and E_{peak}.

The observed E_{iso} of the jet as a function of θ_v is given by,

$$E_{\mathrm{iso}} = \frac{E_\gamma^{\mathrm{true}}}{2\beta\gamma^4(1-\cos\theta_0)} [f(\beta - \cos\theta_v) - f(\beta \cos\theta_0 - \cos\theta_v)], \quad (1)$$

where

$$f(z) = \frac{\gamma^2(2\gamma^2-1)z^3 + (3\gamma^2 \sin^2\theta_v)z + 2\cos\theta_v \sin^2\theta_v}{(z^2 + \gamma^{-2}\sin^2\theta_v)^{3/2}}, \quad (2)$$

and E_γ^{true} is the total energy emitted by the jet in gamma rays and serves as the energy scale for the emission profile.

The transformation of E_{peak} as a function of θ_v is slightly more complicated. The detailed physics underlying the prompt emission of gamma-ray bursts, including the origin of the Band spectrum and the value of E_{peak}, is not yet well understood. Instead we calculate the average Doppler shift across the jet as a proxy for E_{peak}. The average shift, $\langle D \rangle$, is given by,

$$\langle D \rangle = \gamma^{-1} \frac{f(\beta - \cos\theta_v) - f(\beta\cos\theta_0 - \cos\theta_v)}{g(\beta - \cos\theta_v) - g(\beta\cos\theta_0 - \cos\theta_v)}, \quad (3)$$

where

$$g(z) = \frac{2\gamma^2 z + 2\cos\theta_v}{(z^2 + \gamma^{-2}\sin^2\theta_v)^{1/2}}. \quad (4)$$

We require all bursts to obey the $E_{\mathrm{peak}} \propto E_{\mathrm{iso}}^{1/2}$ relation at the center of the jet, thereby fixing the normalization of $\langle D \rangle$ to that of E_{iso}. Thus,

$$E_{\mathrm{peak}} = \frac{\langle D(\theta_v) \rangle}{\langle D(0) \rangle} \cdot E_{\mathrm{peak}}^{(\mathrm{on})} = \frac{\langle D(\theta_v) \rangle}{\langle D(0) \rangle} \cdot C_A \cdot \left(E_{\mathrm{iso}}^{(\mathrm{on})} / E_A \right)^{0.5}, \quad (5)$$

where $E_{\rm iso}^{\rm (on)}$ is simply Equation 1 evaluated at $\theta_v = 0$. Figure 1 shows $E_{\rm iso}$ and $\langle D \rangle$ plotted as functions of θ_v for various values of γ and $\theta_0 = 0.01$ rad. Results from [9] correspond to the limit $\gamma \to \infty$.

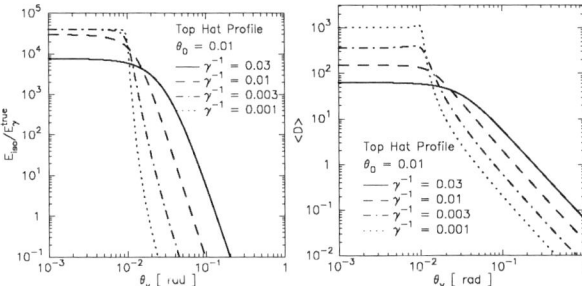

FIGURE 1. Emission profile (left) and average Doppler shift (right) as a function of viewing angle, for a range of γ and $\theta_0 = 0.01$ rad. From [10].

Beginning in the rest-frame of each burst I specify θ_0, $E_\gamma^{\rm true}$ and γ by drawing from various input distributions. I then calculate $E_{\rm iso}$ and $E_{\rm peak}$ from Equations 1 and 5. The rest of the simulations proceed according to the method presented in [9], employing the detector thresholds from the WXM on *HETE-2*.

In Figures 2, 3 and 4, I compare my results against the available data. For source-frame quantities, I compare with the $E_{\rm peak} \propto E_{\rm iso}^{1/2}$ relation [6], and with the $E_{\rm peak} \propto E_\gamma^\beta$ correlation recently discovered by [12]. In Figure 2, I also consider the larger sample of *HETE-2* localized bursts [13, 5], which has the advantage of having many more XRFs than the sample with known redshifts.

RESULTS

In the first model I consider (Y04), I adopt the parameters from [8]. This model attempts to explain classical GRBs in terms of the variation of jet opening-angles, while XRFs are interpreted as GRBs viewed off-jet. The Y04 model assumes $\gamma = 100$ and draws θ_0 values from the distribution $\theta_0^{-2} d\theta_0$, defined from 0.3 to 0.03 rad. $E_\gamma^{\rm true}$ is drawn from a narrow lognormal distribution centered on 1.2×10^{51} erg.

The left panel of Figure 2 shows that the standard E_γ value and a narrow range of opening angles is sufficient to explain the population of classical GRBs. By construction, the on-jet events follow the $E_{\rm peak} \propto E_{\rm iso}^{1/2}$ relation. Off-jet emission from similar bursts viewed at much larger θ_v accounts for the population of gray off-jet points that lie above the $E_{\rm peak} \propto E_{\rm iso}^{1/2}$ relation. For different values of θ_v, these bursts generally move along trajectories in the $[E_{\rm iso},E_{\rm peak}]$-plane that have a flatter slope than the $E_{\rm peak} \propto E_{\rm iso}^{1/2}$ relation. The observed off-jet bursts (gray points) are consistent with only a few observed bursts, and represent a population of events not seen by current instruments.

HETE-2 XRFs are not easily explained as classical GRBs viewed off-jet, as this model posits. The two XRFs with known redshifts lie along the $E_{\rm peak} \propto E_{\rm iso}^{1/2}$ relation, and furthermore, the larger sample of *HETE-2* XRFs without known redshifts do not fall in

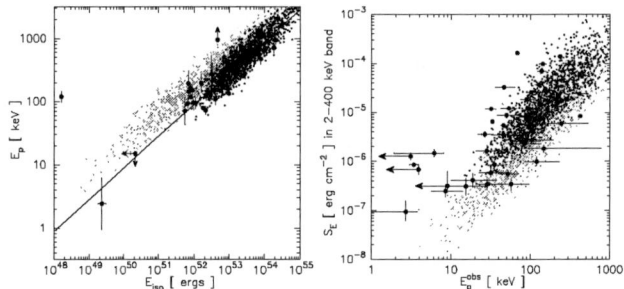

FIGURE 2. Model Y04, which uses the parameters of [8], where GRBs are seen on-jet and XRFs are explained by off-jet relativistic kinematics. Bursts detected on-jet (black) and off-jet (gray) are shown in the $[E_{\text{iso}}, E_{\text{peak}}]$-plane (left) and the $[E^{\text{obs}}_{\text{peak}}, S_E(2-400\text{ keV})]$-plane (right).

the region of the $[E^{\text{obs}}_{\text{peak}}, S_E]$-plane expected for this model (they lie at lower, rather than higher, $E^{\text{obs}}_{\text{peak}}$ values for a given S_E).

The next two models seek to explain both GRBs and XRFs by a wide distribution of jet opening-angles (see [9] for more details and discussion). These models generate XRFs that obey the $E_{\text{peak}} \propto E_{\text{iso}}^{1/2}$ relation by extending the range of possible jet opening solid-angles to cover five orders of magnitude. Hence, XRFs that obey the $E_{\text{peak}} \propto E_{\text{iso}}^{1/2}$ relation are bursts that are seen on-jet, but have larger jet opening-angles. Here I add the presence of off-jet relativistic kinematics to this picture.

I draw $\Omega_0 = 2\pi(1 - \cos\theta_0)$ values from the distribution $\Omega_0^{-2} d\Omega_0$, defined from 2π to $2\pi \times 10^{-5}$ sr. E_γ^{true} is drawn from a narrow lognormal distribution centered on 1.2×10^{49} erg. The lower central point for the E_γ^{true} distribution is a requirement for including in a unified model those events with measured E_{iso} values that are smaller than the usual standard energy of $\sim 10^{51}$ ergs. I consider $\gamma = 100$ (THVOA1) and $\gamma = 300$ (THVOA2).

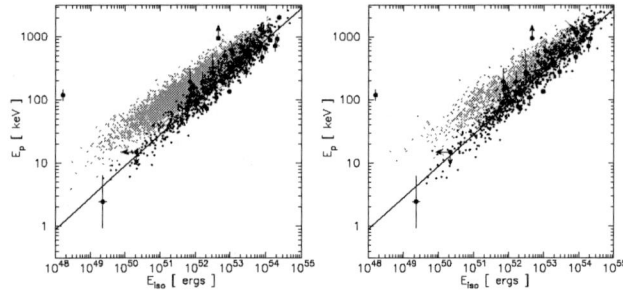

FIGURE 3. Two models which explain both GRBs and XRFs by a wide distribution of jet opening-angles, using $\gamma = 100$ (THVOA1, left) and $\gamma = 300$ (THVOA2, right). Bursts detected on-jet (black) and off-jet (gray) are shown in the $[E_{\text{iso}}, E_{\text{peak}}]$-plane.

For larger values of γ, the emission curves in Figure 1 drop off faster away from the edge of the jet. As can be seen in Figure 3, larger values of γ reduce the percentage of off-jet events observed by the WXM, and consequently the percentage of bursts seen away from the $E_{\text{peak}} \propto E_{\text{iso}}^{1/2}$ relation. This population of events is quite conspicuous in the $\gamma = 100$ case and less so for $\gamma = 300$. Furthermore, for any value of γ, XRFs are more easily explained by wide opening-angle jets than by off-jet emission.

The $E_{\text{peak}} \propto E_{\text{iso}}^{1/2}$ and $E_{\text{peak}} \propto E_\gamma^\beta$ relations can be mutually satisfied in these models for on-jet events by imposing an additional relation between Ω_0 and E_γ,

$$\Omega_0 = 2\pi(1 - \cos\theta_0) \propto E_\gamma^{(\alpha-\beta)/\alpha}, \quad (6)$$

where $\alpha = 0.5$ and $\beta = 0.7$ (for details see [14]). The natural minimum value for the distribution of E_γ^{true} is the point at which $E_\gamma = E_{\text{iso}}$, which is found to be $E^* \approx 3 \times 10^{44}$ ergs. Values of E_γ^{true} are generated by drawing from a power-law distribution that gives equal numbers of bursts per decade, defined from E^* through the largest observed E_γ value. Ω_0 and θ_0 are found via Equation 6, and the rest of the simulations proceed as above. Results for $\gamma = 100$ (GGL1) and $\gamma = 300$ (GGL2) are shown in Figure 4.

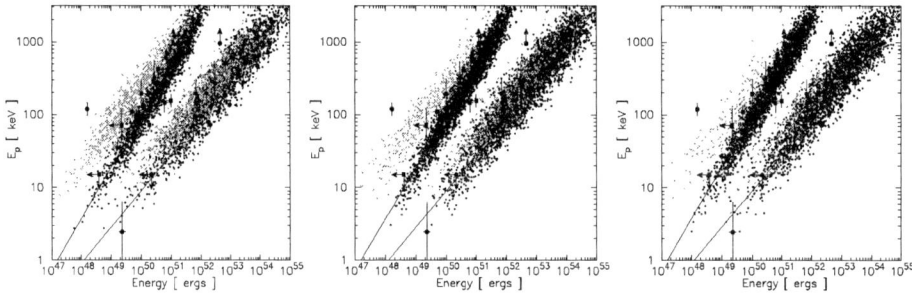

FIGURE 4. Three models which pick parameters to match both the $E_{\text{peak}} \propto E_{\text{iso}}^{1/2}$ and $E_{\text{peak}} \propto E_\gamma^\beta$ relations, using $\gamma = 100$ (GGL1, left) and $\gamma = 300$ (GGL2, center). The third model (GCOR, right) specifies $\gamma \propto \Omega_0^{-1}$ and $\gamma = 1000$ at the minimum value of Ω_0. Bursts detected on-jet (black) and off-jet (gray) are shown in the $[E, E_{\text{peak}}]$-plane.

One consequence of the Ω_0-E_γ correlation is a lack of very small jet opening-angles and a narrower range of θ_0. The high end of the E_{iso} distribution is now explained by jets with moderate opening angles but larger E_γ values. Larger θ_0 values produce trajectories in the $[E_{\text{iso}}, E_{\text{peak}}]$-plane that closer approximate a 0.5 power-law (compare Figures 3 and 4). However, the off-jet bursts deviate further from the $E_{\text{peak}} \propto E_\gamma^\beta$ relation than from the $E_{\text{peak}} \propto E_{\text{iso}}^{1/2}$ relation, leading to similar conclusions as above.

Finally, I impose an additional correlation, assigning narrower jets larger bulk γ values, that reduces the importance of off-jet emission even further. The exact relationship between the bulk γ of the material and the opening-angle of the jet is unknown, so I consider a simple model (GCOR) in which γ is given by $\gamma \propto \Omega_0^{-1}$. I fix the normalization by setting $\gamma = 1000$ for the bursts with the narrowest jets. Figure 4c shows that imposing this correlation greatly reduces the percentage of bursts seen off-jet ($< 7\%$ of detected bursts). Physically this is due to a combination of several effects. It is more probable to observe narrow jets at viewing angles slightly outside the jet than inside the jet, yet for such narrow jets a large value of γ ensures that the detectability of such slightly-off-jet bursts drops off very quickly with viewing angle. Broader jets may have smaller values of γ, but they are less likely to be observed off-jet.

CONCLUSIONS

The populations of off-jet bursts in these models are unable to account for the XRFs observed by *HETE-2* and *Beppo*SAX. The sample of *HETE-2* localized bursts contains XRFs which lie toward smaller $E_{\text{peak}}^{\text{obs}}$ values than is predicted by the off-jet emission model. XRF properties are more easily explained as the product of a wider jet opening-angle. This agrees with the evidence from X-ray afterglows of XRFs. [15] calculated the afterglow light curves predicted by various models of burst emission seen off-jet. They find that a general feature of off-jet afterglows is an initial rising light-curve that peaks at about the jet break time and then declines rapidly, similar to an on-jet event; such initially rising components have not been observed. In particular, XRFs with well-observed X-ray afterglows, for example XRF 020427 [16] and XRF 050215b [17], have afterglow light curves that join smoothly onto the end of the prompt emission and that show no evidence of a jet break for many days after the burst, implying large jet opening-angles.

Incorporating the $E_{\text{peak}} \propto E_\gamma^\beta$ relation in the THVOA model removes two of the main drawbacks of the THVOA model as presented in [9]. The requirement of very small ($\sim 2°$) jet opening angles to explain the largest E_{iso} values was criticized as being difficult to achieve in a hydrodynamic jet. The need to re-scale the central value of E_γ^{true} downward to $\sim 10^{49}$ ergs to incorporate the XRFs in a unified model was criticized as being difficult to reconcile with current afterglow models. The $E_{\text{peak}} \propto E_\gamma^\beta$ relation naturally produces a range of E_γ values that extends down into the XRF regime.

In conclusion, models GGL2 and GCOR are the most successful at matching the current data. The other models predict a large population of off-jet bursts that are observable and that lie away from the $E_{\text{peak}} \propto E_{\text{iso}}^{1/2}$ and $E_{\text{peak}} \propto E_\gamma^\beta$ relations. These discrepancies can be removed if $\gamma > 300$ for all bursts or if there is a strong inverse correlation between γ and Ω_0.

REFERENCES

1. Yamazaki, R., Ioka K., & Nakamura T. 2002, Ap.J., 571, L31
2. Yamazaki, R., Ioka K., & Nakamura T. 2003, Ap.J., 593, 941
3. Heise, J., in't Zand, J., Kippen, R. M., & Woods, P. M., in Proc. 2nd Rome Workshop: Gamma-Ray Bursts in the Afterglow Era, eds. E. Costa, F. Frontera, J. Hjorth (Berlin: Springer-Verlag), 16
4. Kippen, R. M., et al. 2003, in AIP Conf. Proc. 662, Gamma-Ray Burst and Afterglow Astronomy, ed. G. R. Ricker & R. K. Vanderspek (New York: AIP), 244
5. Sakamoto, T., et al. 2005a, Ap.J., 629, 311
6. Amati, L., et al. 2002, A&A, 390, 81
7. Lamb, D. Q., et al. 2005, Ap.J., submitted
8. Yamazaki, R., Ioka K., & Nakamura T. 2004, Ap.J., 606, L33
9. Lamb, D. Q., Donaghy, T. Q., & Graziani, C. 2004, Ap.J., 620, 355
10. Graziani, C., Donaghy, T. Q., & Lamb, D. Q. 2005, Ap.J., in press (astro-ph/0505623)
11. Donaghy, T. Q. 2005, Ap.J., submitted (astro-ph/0512577)
12. Ghirlanda, G., Ghisellini, G. & Lazzati, D. 2004, Ap.J., 616, 331
13. Barraud, C., et al. 2003, A&A, 400, 1021
14. Donaghy, T. Q., Lamb, D. Q., & Graziani, C. 2005c, Ap.J., submitted
15. Granot, J., Ramirez-Ruiz, E. & Perna, R. 2005, Ap.J., 630, 1003
16. Amati, L., et al. 2004, A&A, 426, 415
17. Sakamoto, T., et al. 2005b, Ap.J., submitted

X-ray Line Emission in Swift XRT Spectra?

Nat Butler

University of California at Berkeley, 445 Campbell Hall, Berkeley, CA 94720, USA

Abstract. We report initial results from an automated XRT spectral pipeline running at Berkeley. The data are fit in near real time and searched for emission lines. For the period between GRBs 050401 and 051117B, there are nearly 12 Msec of XRT data, which can be divided into 67 spectra each with > 500 cts. We discuss limits on Fe line emission for the bursts with measured redshifts. We find no highly significant emission lines as outliers in the overall population. However, there are interesting outliers in terms of the continuum fit parameters. From this we find evidence for thermal emission components in some of the bright XRT flares. Our automated software performs a blind search for emission lines and appears to also trigger on these regions of particularly soft spectra. This may indicate that some of the soft flare emission is in the form of lines.

Keywords: gamma rays: bursts — supernovae: general
PACS: 98.70.Rz

INTRODUCTION

One of the key open questions in the study of Gamma-ray bursts (GRBs) is that of the X-ray afterglow lines. Claims of low to moderate significance emission lines have been made based on data from several missions: Fe lines have been detected in afterglow data from *ASCA* [1], *Beppo-SAX* [2, 3], and *Chandra* [4]; lines from highly ionized light, multiple-α elements like Mg, Si, S, Ar, and Ca have been detected in afterglow data from *XMM* [5, 6] and *Chandra* [7]. The detections are challenging to theorists because they typically imply large, concentrated masses of metals in the circumburst material [see, e.g., 8] and a very efficient reprocessing of the non-thermal afterglow continuum into line radiation [see, 9, 10].

Sako, Harrison, & Rutledge [11] [see also, 12] argue that the claims made to date lack the necessary significance needed to prove that the X-ray lines are real. We address the line significance in the case of GRB 011211 in a recent paper [13] and find that the significance estimate is dominated by assumption made in the continuum modelling. For the analysis, we developed software to perform autonomous line detection. The software efficiently detects 1-5 emission lines (90% of the time for S/N=3 lines), and this facilitates straight forward Monte Carlo significance estimation as well as an unbiased emission line search through the mounting XRT data set.

DATA REDUCTION AND CONTINUUM FITS

Our automated pipeline at Berkeley downloads the XRT data in near real time and runs the detection software. The burst right ascension and declination are gleaned from the

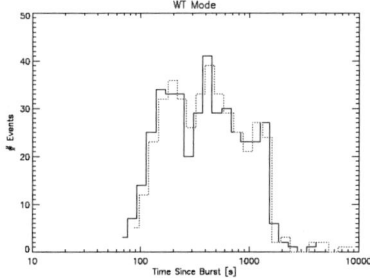 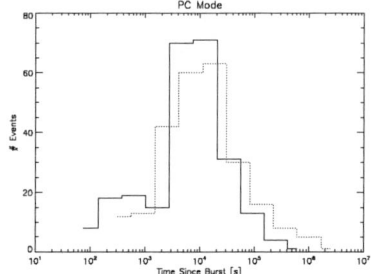

FIGURE 1. (A) Start (black) and stop (grey) times for the 25 WT mode spectra used in our analysis. (B) Start and stop times for the PC mode data.

GCN in order to run the xrtpipeline reduction task from the HEAsoft 6.0[1] software release. From there, we bin the data in time, exclude pileup regions, and produce spectra using custom IDL scripts. The spectra are fit in ISIS[2]. For each spectral bin, we require a S/N of 3.5.

We fit the integrated spectra for each burst with XRT data, and we also slice the data in time to search for transient emission features. For the period between GRBs 050401 and 051117B, we produce 42 PC-mode, 25 WT-mode spectra with $\gtrsim 500$ counts. The time coverage of the spectra are plotted in Figure 1. This is a number of counts for which we find line detection to be adequate. (We have also performed our full analysis using requiring 250 and 1000 counts per spectra for comparison purposes and to be sure that we are not missing credible detections). Overwhelmingly, the XRT data are well fit by absorbed power-law, with $\Gamma \sim 2$ on average (Figure 2). There are a handful of outliers to this trend, most noticeably the soft outliers occurring in the spectra of GRB 04015B and GRB 050822.

To search for line emission in each spectrum, we fit unresolved emission lines in addition to the power-law continua. We allow for the possibility of multiple emission lines [see, e.g., 5]. The lines are fitted successively, starting with one line and adding a total of five lines. The best-fit line location at each step is found by convolving the fit residuals through the line response function (LRF) [see, e.g., 12]. We model the LRF as a Gaussian with an energy dependent width, $\sigma(E) = 29.4(E/[1\text{keV}])^{0.355}$ eV. This functional form provides an excellent fit to the core of the XRT LRF.

FE LINE SEARCH AND UPPER LIMITS

In the sample discussed above, there are 15 XRT events with measured redshifts. Using the expected position of possible Fe lines, we search for emission lines within 20% of 6.7 keV. This allows for a possible blue- or redshift of the emitting material in the host

---

[1] http://heasarc.gsfc.nasa.gov/docs/software/lheasoft/
[2] http://space.mit.edu/CXC/ISIS/

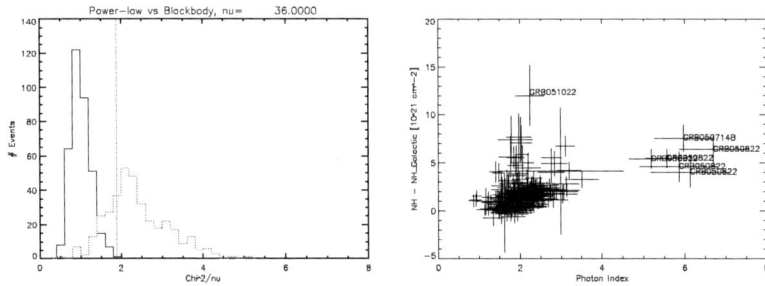

FIGURE 2. (A) Reduced χ^2 values for the continuum fits, shifted to a fiducial number of degrees of freedom $\nu = 36$. The dotted vertical line present the 99.9% confidence rejection threshold, to the right of which the fit should be rejected. A small fraction of the spectra are acceptable fit with a blackbody model (grey line), however most are well fit by absorbed power-laws (black line). (B) Continuum parameters N_H and photon index Γ for the absorbed power-law model fits. There are soft outliers (GRBs 050714B, 050822).

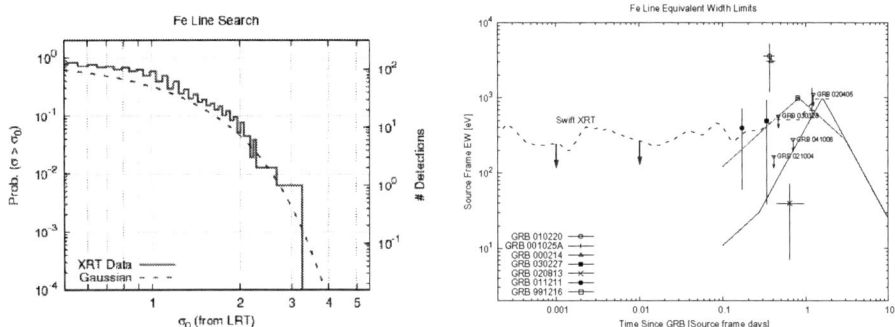

FIGURE 3. (A) The number of Fe-line detections versus the estimated significance from a χ^2 (LRT) test. The number of detection follows the expected curve (black dashed line) for Poisson noise. (B) The Fe line upper limits on the rest-frame equivalent width EW from the XRT data (dashed black line) in context. Also plotted are detection and upper limits from the literature [see, 14] and the EW [9] [see, also, 15].

frame. Our most significant detection is 3.4σ ($\Delta\chi^2 = 14.87$, for 2 additional degrees of freedom). For 158 spectra searched, this corresponds to a meager 90% confidence, and we do not consider it a credible detection. The distribution of detection significances is plotted in Figure 3A and is consistent with Poisson noise.

In Figure 3B we plot the average 90% confidence upper limit on the Fe-line rest-frame equivalent width versus detection epoch in the host frame. The limits for $t \lesssim 1$ day are ~ 300 eV, and the wiggle in the dashed line in Figure 3B gives a sense of the scatter. There are few spectra ($\sim 5\%$ of the sample; see Figure 1 at times $\gtrsim 1$ day, and the dashed line in Figure 3B rises there. The XRT spectra strongly constrain the existence of Fe lines prior to this time. They do not appear to be present.

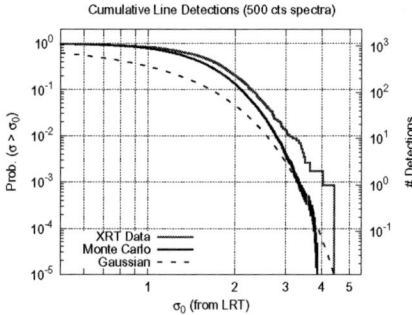

FIGURE 4. The number of detections (red line) of 1-5 emission lines in XRT spectra containing 500 or more counts. There is a modest excess in the number of detection over the model (dotted black line) and Monte Carlo curves.

MULTIPLE EMISSION LINE DETECTIONS IN BRIGHT FLARES

Figure 4 displays our detections of 1-5 emission lines from the blind search through the full data set. Searching ~ 460 spectra, a 4σ detection corresponds to 97% confidence. Because the most significant of our detection are near this value, none of our detection are credible in their own right. However, as we discuss below, most of the detection appear to coincide with XRT flaring. Arguably, this reduces the number of blind search trials and increases the line significance. Alternative to the possibility that the line emission during these periods is real, it is also possible that the line detection indicate discrepancies in the continuum modelling. We now discuss the individual detections.

GRB 050822 This event is a clear outlier in Figure 2. The continuum spectrum is well-fit by a blackbody during the flare at $t \sim 500$s. The spectrum shows a decreasing temperature, $kT = 0.25 - 0.17$ keV, during the flare. The blind emission line search finds 4.4σ evidence for lines ($\Delta\chi^2 = 40.56$, $v = 10$; Figure 5).

GRB 050714B Also a clear outlier in Figure 2, the WT mode spectrum for this event is well-fit by a blackbody. The spectrum during the flare (PC mode) requires a blackbody in addition to a power-law model or a power-law model plus lines. For 5 lines, the significance is 4.5σ ($\Delta\chi^2 = 42.02$, $v = 10$; Figure 6 shows the spectrum without lines).

GRB 051117A This event exhibited a flaring light curve. There is a modest significance line detection during a flare at 1.3 ksec (Figure 7). Three lines at $E = 0.85, 1.84, 3.6$ keV contain fluxes $2 - 3 \times 10^{-11}$ erg cm^{-2} s^{-1} and $EW = 25 - 250$ eV.

GRB 050502B? This event has a bright WT-mode flare at $t \sim 10^3$s. We find no significant evidence for lines in the time-sliced spectra, the full WT mode spectrum during the flare shows $\lesssim 4\sigma$ evidence for lines ($\Delta\chi^2 = 28.76$, $v = 10$) for lines (Figure 8).

CONCLUSIONS AND FUTURE WORK

There are two additional spectra (i.e., not discussed above) with $\gtrsim 4\sigma$ significant line emission, where the X-ray afterglow shows now evidence for flares (GRB 050505,

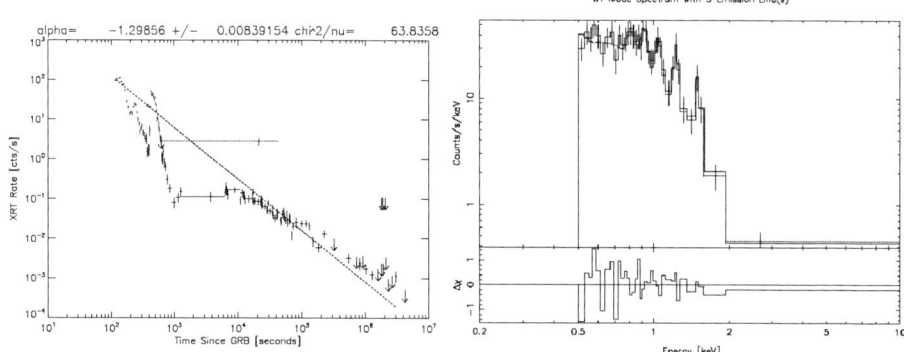

FIGURE 5. (A) The X-ray light curve for the GRB 050822 afterglow. There is a strong flare at $t \sim 500$s. (B) In one of the 500 counts spectra (a ~ 20s time interval) during the flare, emission lines improve the power-law model fit considerably.

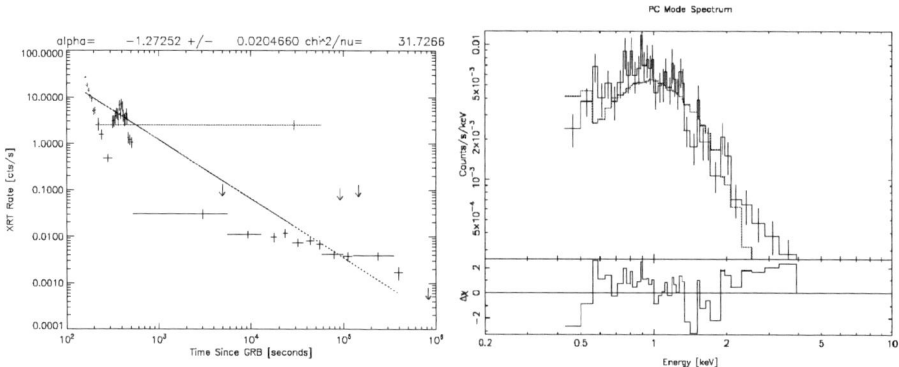

FIGURE 6. (A) The light curve for GRB 050714B. (B) The poor power-law fit to the spectrum for GRB 050714B during the flare at $t \sim 400$s.

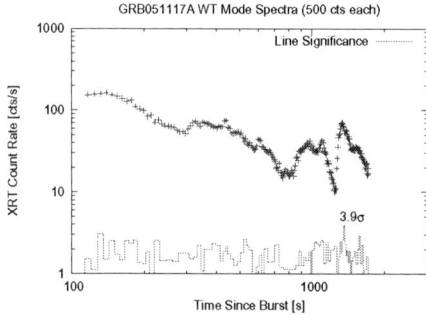

FIGURE 7. The line detection significance versus time for GRB 051117A. There is a modest significance detection near the peak time of the flare in a 15s duration WT mode spectrum.

263

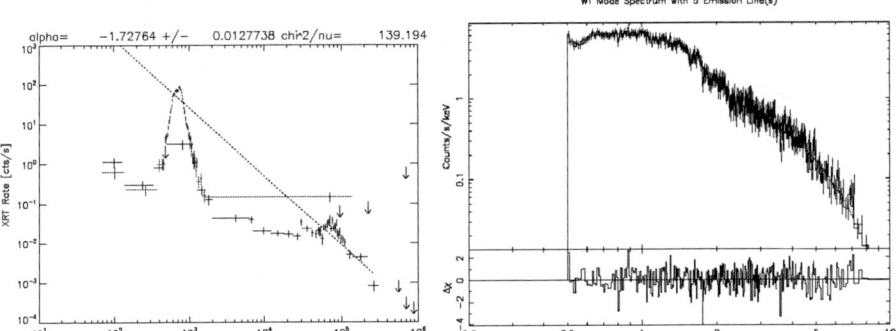

FIGURE 8. (A) Light curve and (B) spectrum for GRB 050502B. The WT mode spectrum plotted is for the entire flaring period at $t \sim 10^3$ s.

GRB 051022), but these appear to be due to systematics. Because all of our detection appear to occur during flares, we conclude that emission lines may be common there. This would shed light on previous detection by *XMM* of soft X-ray emission lines [5, 6] and thermal components [16]. It may be that line emission occurs during the flaring, or that the spectra during the flares are sufficiently different from the spectra outside of the flares as to regularly produce spurious line detection. We are in the process of investigating this further by examining the spectra during faint flares and also by investigating the line detections as a function of time during the bright flares. In any case, we find that some X-ray flares appear to have thermal emission components. We find that Fe lines are uncommon in spectra taken prior to 1 day in the host frame.

REFERENCES

1. Yoshida, A., et al. 1999, A&AS, 138 433
2. Piro, L., et al. 1999, A&AS, 138, 431
3. Antonelli, L. A., et al. 2000, ApJ, 545, L39
4. Piro, L., et al. 2000, Science, 290, 955
5. Reeves, J. N., et al. 2002, Nature, 415, 512
6. Watson, D., et al. 2003, ApJ, 595, L29
7. Butler, N., et al. 2003, ApJ, 597
8. Lazzati, D., Campana, S., & Ghisellini, G. 1999, MNRAS, 304, L31
9. Ballantyne, D. R., & Ramirez-Ruiz, E. 2001, ApJ, 559, L83
10. Lazzati, D., Ramirez-Ruiz, E., & Rees, M. J. 2002, ApJ, 572, L57
11. Sako, M., Harrison, F., & Rutledge, R. 2005, ApJ, 623, 973S
12. Rutledge, R., & Sako, M. 2003, MNRAS, 339, 600
13. Butler, N., et al. 2005a, ApJ, 627, L9
14. Butler, N., et al. 2005b, ApJ, 629, 908
15. Gou, L. J., Mészáros, P., & Kallman, T. R. 2005, ApJ, 624, 889
16. Watson, D., et al. 2002, A&A, 393, L1

Simulation Studies of Early Afterglows Observed with SWIFT

K.-I. Nishikawa*, P. Hardee†, C. Hededal**, C. Kouveliotou‡, G. J. Fishman‡ and Y. Mizuno‡

NSSTC, 320 Sparkman Drive, XD12 Huntsville, AL 35805
†*Department of Physics and Astronomy, The University of Alabama, Tuscaloosa, AL 35487*
**Dark Cosmology Center, Niels Bohr Institute, Juliane Maries Vej 30, 2100 Copenhagen Ø, Denmark*
‡*NASA-Marshall Space Flight Center, National Space Science and Technology Center, Huntsville, AL 35805*

Abstract. We have applied numerical simulations and modeling to the particle acceleration, magnetic field generation, and emission from relativistic shocks and plan to compare them with the observed gamma-ray burst emission. In collisionless shocks, plasma waves and their associated instabilities (e.g., the Weibel, Buneman and other two-stream instabilities) are responsible for particle (electron, positron, and ion) acceleration and magnetic field generation. A 3-D relativistic electromagnetic particle (REMP) code is used to study shock processes including spatial and temporal evolution of shocks in unmagnetized electron-positron plasmas with three different jet velocity distributions. The "jitter" radiation from the shocks is different from synchrotron radiation. The dynamics of shock microscopic process evolution may provide some insight into early afterglows. Our simulation studies provide insight into new GRB observations with Swift.

Keywords: Relativistic jets, Weibel instability, particle acceleration, emission
PACS: 98.70.Rz, 98.38.Fs, 94.20.wf, 94.05.Dd, 96.25.Tg

INTRODUCTION

This report presents the study of collisionless relativistic shocks associated with prompt gamma-ray bursts and their afterglows. Using a 3-D relativistic particle-in-cell code we have investigated the dynamics of relativistic shocks which play an essential role for early afterglows. There is now general agreement among theorists that the prompt emission from a gamma-ray burst (regardless of the central engine) results in the formation of a highly relativistic, highly collimated jet of out-flowing material that emits the observed prompt gamma-ray emission. Observational evidence suggests that GRBs are produced by Doppler beamed and boosted emission from shocks associated with a jet like flow. Within these general scenarios it is proposed that synchrotron radiation from shock accelerated particles in a shock-generated magnetic field produces the gamma-ray burst and produces the associated afterglows (e.g., [1]).

The Swift satellite is a multi-wavelength observatory designed to detect GRBs and their X-ray and UV/optical afterglows. Thanks to its fast pointing capabilities Swift is revealing the early afterglow phase. The Swift X-Ray Telescope (XRT) found that most X-ray afterglows fall off rapidly for the first few hundred seconds, followed by a less rapid decline [2]. In the early afterglows of GRB 050406 and GRB 050502b,

XRT detected strong X-ray flares: rapid brightening of the X-ray afterglow after a few hundred seconds post-burst [3]. These results suggest the existence of additional emission components in the early afterglow phase besides the conventional forward shock (blast wave) emission [4].

Zhang, Kobayashi, & Meszaros [5] have discussed a clean recipe for constraining the initial Lorentz factor γ_0 of GRB fireballs by making use of the early optical afterglow data alone. The input parameters are ratios of observed emission quantities, so that poorly known model parameters related to the shock microphysics (e.g. $\varepsilon_e, \varepsilon_B$, etc.) largely canceled out. This approach is readily applicable in the Swift era when many early optical afterglows are expected to be regularly caught. By combining this data with other information such as $\Re_B$ (which is defined as $(\varepsilon_B^r/\varepsilon_B^f)^{1/2}$; where the superscripts r and f represent the reverse shock region and forward shock region, respectively) and a parameter σ, the ratio between the electromagnetic energy flux and the particle energy flux, which is closely related to the initial magnetization of the outflow [6], this method has provided, for the first time, information about the magnetic content of the ejecta. Such information about the initial Lorentz factor of the fireball and whether the central engine is strongly magnetized are helpful for the identification of the GRB prompt emission site and mechanism, which are currently uncertain (e.g. [7]).

In collisionless shocks, plasma waves and their associated instabilities (e.g., the Weibel, Buneman and other two-stream instabilities) are responsible for particle (electron, positron, and ion) acceleration and magnetic field generation. Three-dimensional relativistic particle-in-cell (PIC) simulations have been used to study the microphysical processes in relativistic shocks. Recent PIC simulations using counter-streaming relativistic jets show that acceleration is provided in situ in the downstream jet, rather than by the scattering of particles back and forth across the shock as in Fermi acceleration [8, 9, 10, 11, 12, 13, 14, 15]. Three recent independent simulation studies have now confirmed that the relativistic counter-streaming jets excite the Weibel instability [16]. The Weibel instability generates current filaments and associated magnetic fields [17, 18, 20, 19], and accelerates electrons [8, 9, 10, 11, 12, 13, 14]. The current filaments and associated magnetic fields produced by the Weibel instability form the dominant structures in a relativistic collisionless shock. The growing current filaments generate highly nonuniform small-scale transverse magnetic fields around the current filaments. The "jitter" radiation to be expected from deflected electrons has different properties than synchrotron radiation [21, 22, 23], and may explain the complex time evolution and/or spectral structure in gamma-ray bursts [24, 25]. Particle acceleration perpendicular and parallel to the jet propagation direction accompanied by the non-linear development of the filamentary structures cannot be characterized as Fermi acceleration.

SIMULATIONS WITH 3-D REMP CODE

Three simulations were performed using an $85 \times 85 \times 640$ grid with a total of 380 million particles (27 particles/cell/species for the ambient plasma) and an electron skin depth, $\lambda_{ce} = c/\omega_{pe} = 9.6\Delta$, where $\omega_{pe} = (4\pi e^2 n_e/m_e)^{1/2}$ is the electron plasma frequency and Δ is the grid size. In all simulations, jets are injected at $z = 25\Delta$ in the positive z direction. Radiating boundary conditions were used on the planes at $z = 0, z_{max}$. Periodic boundary conditions were used on all other boundaries [26]. The ambient and jet electron-positron plasma has mass ratio $m_e/m_p \equiv m_{e^-}/m_{e^+} = 1$. The electron thermal velocity in the ambient plasma is $v_{th} = 0.1c$ where c is the speed of light.

The electron number density of the jet is $0.741n_b$ where n_b is the ambient electron number density. The jet makes contact with the ambient plasma at a 2D interface spanning the computational domain. Here the dynamics of the propagating jet head and shock region is studied. Effectively, we study a small portion of a much larger shock.

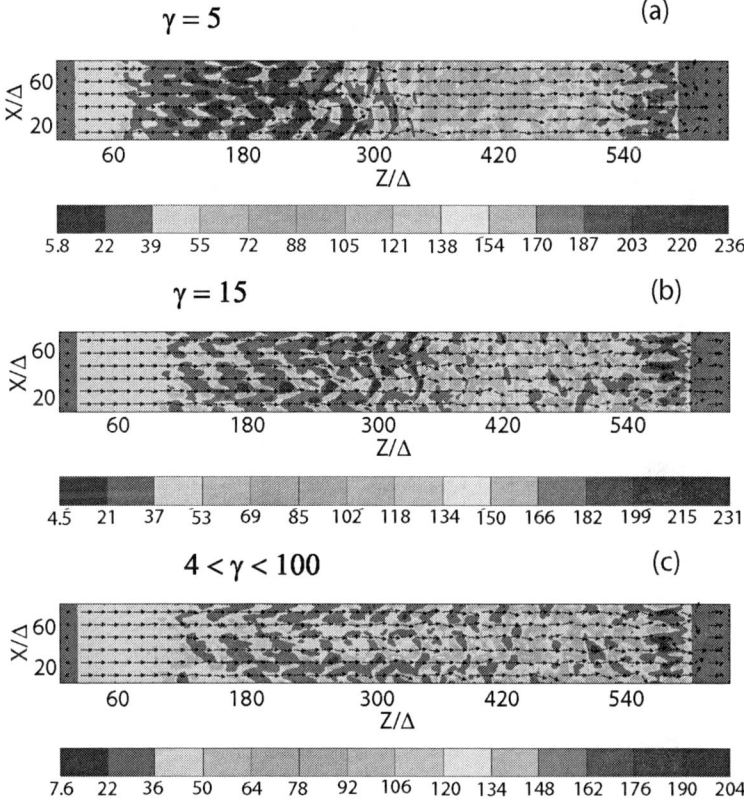

FIGURE 1. 2D images in the $x-z$ plane at $y = 43\Delta$ for the electron-positron jet injected into an unmagnetized ambient electron-positron plasma at $t = 59.8/\omega_{pe}$. The colors indicate the electron density with electron flux represented by arrows for $\gamma v_\parallel = 5$ (a) and 15 (b), and widely distributed pair injection $4 < \gamma v_\parallel < 100$ (c).

We have simulated three different initial distributions of jet electrons (positrons). Relativistic jets are injected into an unmagnetized ambient electron-positron plasma (initially $\sigma = 0$). Case A is the same as one simulation in previous papers [12, 13, 14] ($\gamma V_\parallel = 5$), but the system size is four (two) time longer. Case B uses a larger Lorentz factor ($\gamma V_\parallel = 15$). Case C mimics cold jet electrons and positrons created by photon annihilation ($4 < \gamma V_\parallel < 100$) [27]. For all three cases the temperature of jet particles is very cold ($0.01c$ in the rest frame).

Electron density and current filaments resulting from development of the Weibel instability behind the jet front are shown in Fig. 1 at time $t = 59.8/\omega_{pe}$. Electrons are accelerated by "radial" electric fields accompanying the current filaments. The electrons are deflected by transverse magnetic fields (B_x, B_y) via the Lorentz force: $-e(\mathbf{v} \times \mathbf{B})$, generated by current filaments (J_z), which in turn enhance the transverse magnetic fields [16, 17]. The complicated filamented structures resulting from the Weibel instability have diameters on the order of the electron skin depth ($\lambda_{ce} = 9.6\Delta$). This is in good agreement with the prediction of $\lambda \approx 2^{1/4} c \gamma_{th}^{1/2}/\omega_{pe} \approx 1.188 \lambda_{ce} = 10\Delta$ [17]. Here, $\gamma_{th} \sim 1$ is a thermal Lorentz factor.

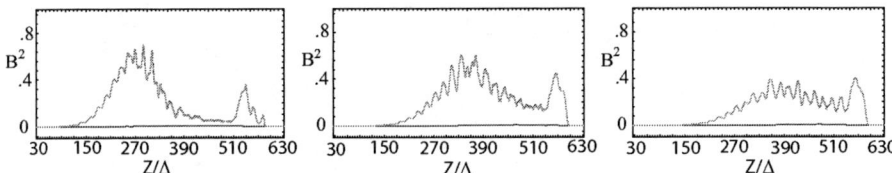

FIGURE 2. The magnetic field energy density averaged in the $x-y$ plane along the z-direction for the electron-positron jet injected into an unmagnetized ambient electron-positron plasma at $t = 59.8/\omega_{pe}$ $\gamma v_\| = 5$ (a) and 15 (b), and widely distributed pair injection $4 < \gamma v_\| < 100$ (c). The total energy density $(B_x^2 + B_y^2 + B_z^2)$ is plotted by the red curves, the perpendicular energy density $(B_x^2 + B_y^2)$, by green, and the parallel energy density (B_z^2), by blue curves.

The magnetic field energy averaged in the $x-y$ plane is plotted as a function of z (along the jets) for the three different cases in Fig. 2. The green curves show the perpendicular magnetic field energy $(B_x^2 + B_y^2)$. The blue curves present the parallel magnetic field energy (B_z^2). Since the parallel magnetic field energy is negligible, the total magnetic field energy (red curves) is overlapped by the perpendicular magnetic field energy (green curves). This indicates that the Weibel instability generates transverse magnetic fields.

All panels in Fig. 2 show similar structures: near the jet head a narrow region has a large magnetic field energy. Behind that region the magnetic fields become smaller where the current channels created by the Weibel instability merge nonlinearly. The particles are strongly accelerated in the perpendicular direction and particles are accelerated and decelerated in the flow direction. Where the magnetic field energy is largest and nonlinearly saturated at the some distance behind the jet front. In case A, the saturation level is the largest. However, the magnetic fields dissipate strongly in the nonlinear region. Since case C contains jet particles with higher Lorentz factors, the saturation level is higher in the nonlinear dissipation region behind the jet front of magnetic field energy (Fig. 2c). The spatial and temporal structure of magnetic field associated with accelerated electrons (positrons) produces different lightcurves and spectra; this will be investigated in the near future.

DISCUSSION

We have performed self-consistent, three-dimensional relativistic particle simulations of relativistic electron-positron jets propagating into unmagnetized electron-positron ambient plasmas (initially $\sigma = 0$) for a longer time and a larger simulation system than our previous simulations [12, 13, 14] in order to investigate the nonlinear stage of the Weibel instability. The main acceleration of electrons takes place in the downstream region. Processes in the relativistic collisionless shock are dominated by structures produced by the Weibel instability. This instability is excited in the downstream region behind the jet head, where electron density perturbations lead to the formation of current filaments. On average the nonuniform electric field and magnetic field structures associated with these current filaments decelerate the jet electrons and positrons, while accelerating the ambient electrons and positrons, and accelerate the jet and ambient electrons and positrons in the transverse direction. The nonlinear region current channels generated by the Weibel instability dissipate. Dissipation levels depend on the initial jet electron parallel velocity distributions.

The key issue is the microscopic dynamics of the reverse shocks which play a crucial role for the generation of early afterglows [5, 6, 4, 28]. We have examined the microscopic dynamics of relativistic shocks with three different velocity distributions in order to understand Swift observations and models of GRB emission. The important parameters are the initial Lorentz factors of relativistic jets (γ) and the ratio between the electromagnetic energy flux and the particle energy flux (σ). In this report we used $\sigma_0 = 0$ and have investigated the magnetization of the GRB outflows ($\Re_B \equiv (\varepsilon_B^r/\varepsilon_B^f)^{1/2}$). Figure 2 shows the total magnetic field energy along the jets. Depending on the initial jet Lorentz factors, We found that the values of the total magnetic field energy and the spatial and temporal evolution depend on the Lorentz factors. They will provide different lightcurves and spectra.

A self-consistent emission calculation based on the motion of electrons (positrons) in the simulation system has been developed. The calculation has reproduced the spectral structure of one GRB afterglow [29, 30]. In order to calculate the radiation (jitter-like) from the particles in the electromagnetic fields generated by the Weibel instability, the retarded electric field from a single particle is Fourier-transformed and gives the individual particle spectrum. The individual particle spectra are added together to produce a total spectrum over a particular simulation time span [29, 30]. In order to obtain lightcurves we can calculate a spectrum over short time spans, t_s, relative to the longer simulation time span. The change in power in an energy band can then be followed from one short time span to the next giving a light curve in the energy band. It should be noted that in this calculation very long (large) simulations are required using small time steps in order to increase the upper frequency limit to the spectrum. Frequency resolution is limited by the short time span ($\Delta\omega = 1/t_s$) [29]. We expect that lightcurves obtained in this way will provide reasonable evolution of the higher energy bands.

Future investigation with this newly-developed method will provide a spectrum of synchrotron/jitter emission, lightcurve, and allow the comparison of calculated lightcurves and spectra with early afterglows observed by Swift.

ACKNOWLEDGMENTS

K.I. Nishikawa was a NRC Senior Research Fellow at NASA Marshall Space Flight Center until June 2005. We thank Enrico Ramirez-Ruiz for suggesting a widely distributed velocity distribution for injected pair jets. This research (K.N.) is partially supported by the NSF awards ATM-0100997, INT-9981508, and AST-0506719. P. Hardee acknowledges partial support by a National Space Science & Technology (NSSTC/NASA) cooperative agreement NCC8-256 and NSF award AST-0506666. The simulations have been performed on IBM p690 at the National Center for Supercomputing Applications (NCSA) which is supported by the NSF.

REFERENCES

1. T. Piran, *Rev. Mod. Phys.* **76**, 1143–1210 (2005).
2. G. Tagliaferri, M. Goad, G. Chincarini, et al. *Nature* **436/18**, 985–988 (2005).
3. D. N. Burrows1, P. Romano, A. Falcone1, et al. *Science*, **309**, 1833–183X (2005).
4. S. Kobayashi, B. Zhang, P. Meszaros, and D. N. Burrows, *ApJL* submitted (2005) (astro-ph/0506157).
5. B. Zhang, S. Kobayashi, and P. Meszaros, *ApJ* **595**, 950–954 (2003).
6. B. Zhang, and S. Kobayashi, *ApJ* **628**, 315–334 (2005).
7. B. Zhang, and P. Meszaros, *ApJ* **572**, 876–879 (2002).
8. L. O. Silva, R. A. Fonseca, J. W. Tonge, J. M. Dawson, W. B. Mori, and M. V. Medvedev, *ApJ* **596**, L121–L124 (2003).
9. J. T. Frederiksen, C. B. Hededal, T. Haugbølle, and Å. Nordlund, *ApJ* **608**, L13–L16 (2004).
10. C. B. Hededal, T. Haugbølle, J. T. Frederiksen, and Å. Nordlund, *ApJ* **617**, L107–L110 (2004).
11. C. B. Hededal, and K.-I. Nishikawa, *ApJ* **623**, L89–L92 (2005).
12. K.-I. Nishikawa, P. Hardee, G. Richardson, R. Preece, H. Sol, and G. J. Fishman, *ApJ* **595**, 555–563 (2003).
13. K.-I. Nishikawa, P. Hardee, G. Richardson, R. Preece, H. Sol, and G. J. Fishman, *ApJ* **622**, 927–937 (2005).
14. K.-I. Nishikawa, P. Hardee, C. B. Hededal, and G. J. Fishman, *ApJ* submitted (2005) (astro-ph/0510590).
15. M. V. Medvedev, M. Fiore, R. A. Fonseca, L. O. Silva, and W. B. Mori, *ApJ* **618**, L75–L76 (2005).
16. E. S. Weibel, *Phys. Rev. Letters* **2**, 83-84 (1959).
17. M. V. Medvedev and A. Loeb, *ApJ* **526**, 697–706 (1999)
18. J. J. Brainerd, *ApJ* **538**, 628– (2000).
19. A. Gruzinov, *ApJ* **563**, L15–L18 (2001).
20. J. Pruet, K. Abazajian, and G. M. Fuller, *Phys. Rev. D.* **64**, 063002-1 (2001).
21. M. V. Medvedev, *ApJ* **540**, 704–714 (2000).
22. M. V. Medvedev, *ApJ* in press (2006) (astro-ph/0510472).
23. G. D. Fleishman, *MNRAS* in press (2006) (astro-ph/0511353).
24. R. D. Preece, M. S. Briggs, R. S. Mallozzi, G. N. Pendleton, W. S. Paciesas, and D. L. Band, *ApJ* **506** L23–L26 (1998).
25. R. D. Preece, M. S. Briggs, T. W. Giblin, R. S. Mallozzi, G. N. Pendleton, W. S. Paciesas, and D. L. Band, *ApJ* **581** 1248–1255 (2002).
26. O. Buneman, 1993, "Tristan", in *Computer Space Plasma Physics: Simulation Techniques and Software*, edited by H. Matsumoto Matsumoto & Y. Omura, 1993, Terra Scientific Publishing Company, Tokyo, 1993, pp. 76-84.
27. P. Meszaros, E. Ramirez-Ruiz, and M. J. Rees, *ApJ* **554** 660–666 (2001).
28. Y. Z. Fan, B. Zhang, and D. M. Wei *ApJ* **628** L25–L28 (2005).
29. C. B. Hededal, *Gamma-Ray Bursts, Collisionless Shocks and Synthetic Spectra*, PhD thesis, 2005 (astro-ph/0506559).
30. C. B. Hededal, and Å. Nordlund, *ApJ* submitted (2005) (astro-ph/0511662).

Observation of the prompt and early afterglow of GRB 050904 by TAROT[1]

M. Boër*, J.L. Atteia[†], Y. Damerdji*, B. Gendre**, A. Klotz*,[‡] and G. Stratta[†]

*Observatoire de Haute Provence, 04870 Saint Michel l'Observatoire, France
[†]LATT-OMP, 14, ave. E. Belin, 31400 Toulouse, France
**IASF/INAF, via fosso del cavaliere 100, 00133 Roma, Italy
[‡]CESR (CNRS/UPS), BP 4346, 31029 Toulouse, France

Abstract. We present the recent observation of the very high redshift burst source GRB 050904 made by the TAROT robotized telescope. We have compared our data with the SWIFT XRT light curve to analyze the broad ban spectrum. We show that the luminosity and the behavior of this event is comparable with that of GRB 990123, suggesting the existence of very bright events. They can be detected at very high redshifts, even with small or moderate aperture telescopes, and they may constitute a powerful means for the exploration of the young universe. An update of the last TAROT observations performed as a response from SWIFT alerts is made.

Keywords: Gamma-ray : bursts; Optical: Transients
PACS: 95.85.Jq : Near Infrared – 95.85.Nv : X-rays–98.70.Rz : gamma-rays sources, gamma-ray bursts

INTRODUCTION

Rapid observations of GRBs have been proven to be useful to probe GRB theory, specifically the internal and reverse shock models[1, 2]. They are useful also to understand the start of the afterglow (i.e. forward shock) emission. Prompt optical emission, when compared with the high energy part, gives access to the spectrum of the GRB in an extended range, both for the prompt and the early afterglow. Finally, given the transient nature and the rapidly declining brightness of the events, rapid observations, even with small aperture telescopes may prove to be competitive with observations acquired with large instruments (see e.g. [3]). Together with other experiments, the Télescope à Action Rapide pour les Objets Transitoires (TAROT, Rapid Action Telescope for Transients Objects [4]) has been designed to reach these goals. With the launch of SWIFT and the large number of alerts sent by this spacecraft, we could perform timely observations of GRB sources. We present here a short summary of the status of the TAROT experiment, and the result which could be obtained on the high redshift burst of September 4th, 2005.

TABLE 1. Summary of the current configuration of TAROT.

Aperture	25 cm
Focal ratio	f/3.4
Slewing time	1 - 2 s (up to 80 deg/s)
Acceleration	up to 120 deg/s^2
Limiting magnitude	R = 16.5 in 10s
CCD type	EEV 42-40 BI
CCD size	2048 x 2048, 3 x 3cm
Field of view	1.86 deg
Operating CCD temperature	-50^C
Readout speed	1MHz, i.e. 5s
Readout noise	8.5e$^-$ rms
GRB integration time	10 to 180s

SUMMARY OF TAROT OBSERVATIONS WITH SWIFT

TAROT is a fully autonomous 25 cm aperture telescope installed at the Calern observatory (Observatoire de la Cote d'Azur - France). Its 2 deg field of view ensures the total coverage of HETE error boxes. This telescope is devoted to very early observations of GRB optical counterparts. A technical description of TAROT can be read in [4] and [5]. The CCD camera has been recently replaced with an ANDOR 436BW camera based on the EEV 42-40, back illuminated chip. It is placed at the newtonian focus. The focal length is 0.85 meter and the pixel size is 13.5 microns. The spatial sampling is 3.3 arcsec/pixel. The readout noise is 8.5 electrons rms, for a readout time of 5s. For GRB observations the integration time varies from 10s to 180s in order to optimize the trade-off between sensitivity and time resolution. Table 1 presents the main characteristics of the telescope.

Since the launch of SWIFT, and until the end of 2005, i.e. in approximately one year, TAROT could perform observations of 9 GRBs (here, we take into account only alerts which where confirmed as true GRBs). The observations of GRB alerts received by TAROT are displayed on figure 1. The green area represents the minimum TAROT slewing time, the vertical lines are the T90 GRB duration, and the dots represent the start of the observations by TAROT. Some of the bursts could be observed during their prompt phase in gamma-ray or X-rays. Table 2 summarizes the results of the observations with the GRB name, its redshift when known, the start of observations by TAROT expressed in seconds after the event triggered BAT, and the peak magnitude or the upper limit, depending whether the source was detected or not. Five out of the nine events resulted in a positive detection, including distant events ($z \geq 3$). All but one events observed less than 1 hour after the trigger were detected by TAROT.

---

[1] Partly based on observations made at the Observatoire de Haute Provence (France) 80cm telescope

TABLE 2. Summary of SWIFT GRBs observed by TAROT.

Burst name	Redshift	Obs. start (s after trigger)	Magnitude
GRB 050306	?	90	R ≥ 17.5
GRB 050416	0.65	30300	R ≥ 19
GRB 050505	4.2	540	I = 18.2
GRB 050525A	0.6	360	R = 15.0
GRB 050730	3.97	66	I = 15.5
GRB 050803	0.42	3660	R ≥ 19
GRB 050824	0.83	600	R = 18
GRB 050904	6.29	86	I = 14.1
GRB 051221B	?	216	R ≥ 18.2

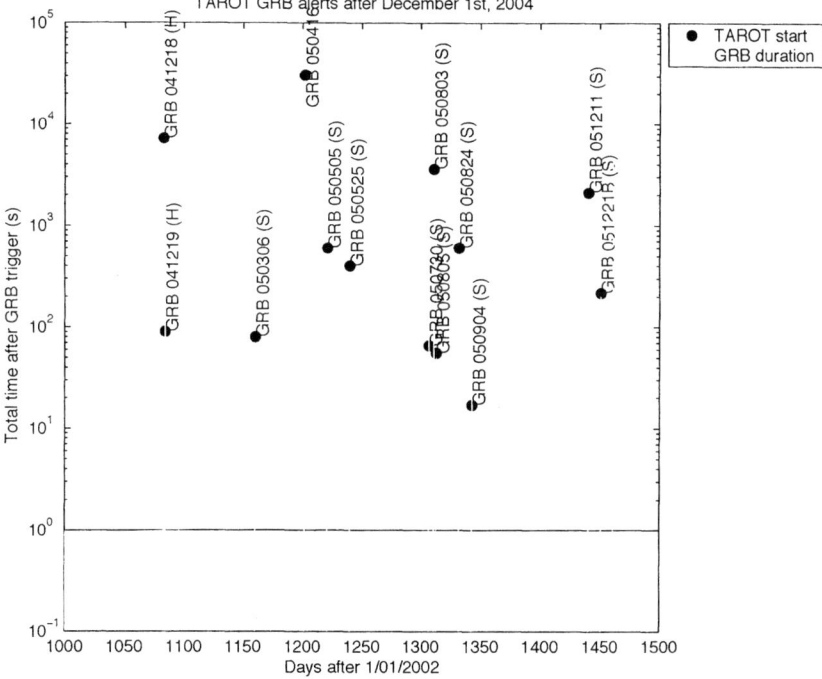

FIGURE 1. TAROT observations of GRB sources after Dec. 1st, 2004 to the end of 2005. The points represent the start of the observations relative to the trigger instant, the lines the GRB duration (T90), and the green area the minimal TAROT slewing time (1s). Note that several observations occurred during the prompt event.

OBSERVATIONS OF GRB 050904

GRB 050904 was detected by BAT at 1:51:44[6]. TAROT started to observe the source location 86 seconds later, including 5s of overhead time to process the alert and to slew

TABLE 3. Summary of the observations of GRB 050904 performed by TAROT

Interval #	Start time (s)	End time (s)	I band magnitude
1	86	144	>15.3
2	150	253	15.22± 0.26
3	312	443	15.26± 0.30
4	449	589	14.07± 0.24
5	595	978	>15.5
6	985	1666	>15.8

the telescope. The BOOTES experiment in Southern Spain observed simultaneously the field and failed to detect the counterpart: the use of a R filter prevented the detection of photons from this highly redshifted source[7]. SWIFT observations with the XRT and UVOT started about 160s after the trigger, allowing for simultaneous TAROT/XRT observations. Large telescopes started to image the field three hours later and discovered a bright, fading near-infrared source [8, 9, 10]. The Subaru telescope measured the spectroscopic redshift of the source [11] that was found[10] to be at a z = 6.29. To date GRB 050904 is the gamma-ray burst with the highest measured redshift and the farthest cosmological source ever observed with a small (25cm) telescope.

We acquired 22 unfiltered images with an exposure time varying from 15 to 90 seconds. As explained above, the exposure time increases to optimize the sensitivity. The long duration of this burst (thanks to its high redshift) allowed us to get images while it was still active at high energies. In order to increase the signal to noise ratio we co-added the images. A series of 6 images were computed, and the results of the processing is summarized in table 3.

Given the high redshift of the source we had to carefully recalibrate the frame relative to the Cousin I band. We acquired field VRI observations at the 80cm OHP telescope; we then modeled the flux ratio relative to three reference stars, assuming a spectral shape $f_v \propto v^{-0.7}$ for the GRB spectrum. Uncertainties on the order of ±0.5 on the spectral index induce errors lower than 20% on the flux estimated at 9500Å. The light curve obtained with TAROT is given on figure 2 together with the XRT light curve. We note that these observations are the only optical detections available for GRB 050904 during the first minutes after the trigger and that they display a continuous coverage of the prompt emission, from the reverse shock to the the early afterglow.

The optical light curve display a gradual increase of the emission during the first 150 seconds (interval 1), followed by a plateau lasting about 300s (intervals 2 and 3). A bright flare takes places approximately at the same time as the X-ray flare. Then we detect very marginally the transient after the end of the prompt event with a signal to noise ratio of 3.2. The broad-band spectrum has been reported elsewhere [12]. The results can be summarized as follows: the optical emission stands well above the extrapolation of the high energy spectrum, and represents another spectral component.

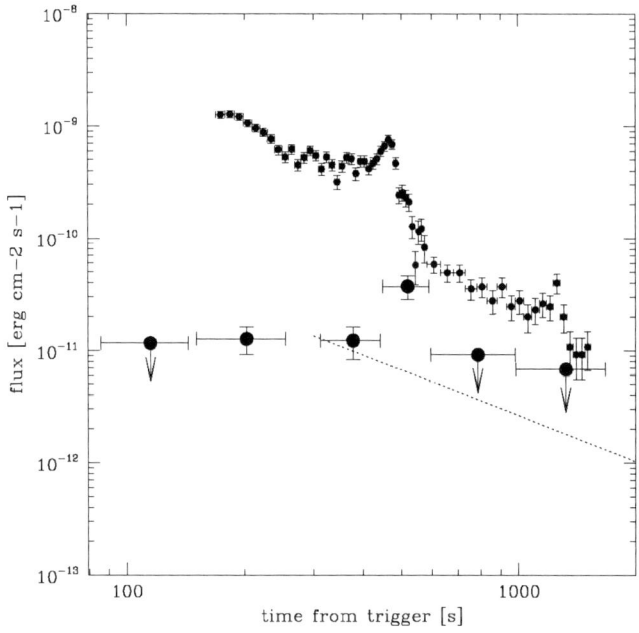

FIGURE 2. TAROT light curve of GRB 050904 (bottom) together with the XRT light curve

DISCUSSION AND CONCLUSIONS

The optical emission of GRB 990123[13] has been explained by reverse shock emission [14]. For GRB 050904 the optical emission can be explained with the reverse shock model[15]. In this case, the reverse shock would have started between intervals T1 and T2, about 150 sec after the trigger (20 seconds in the rest-frame). A long lasting optical emission could be produced when the reverse shock propagates into the various layers of the ejecta [16]. However, in that case we have to admit that the temporal coincidence of the optical flash with an X-ray peak is fortuitous [15]. Alternatively, it could find a natural explanation if the optical emission is attributed to the prompt emission [17].

If we compare GRB 990123 and GRB 050904 we can observe that both events where very bright. In both events the optical flare is detected near the end of the prompt emission (right after a high energy peak) and exceeds significantly the backward extrapolation of the late-time afterglow. Their rest-frame peak energy is comparable (2MeV for GRB 990123 and 1.5MeV for GRB 050904), and they have both a large isotropic energy release (respectively 2.8×10^{54} erg and $\sim 10^{54}$ erg). The plateau observed for GRB 990123 may be explained by the larger dead time of the ROTSE I experiment at the time of the observations. We conclude that both GRB 990123 and 050904 are very similar events. We note also that three out of the five detections made by TAROT (for SWIFT events) are highly redshifted sources (4.2, 3.97, and 6.29). This suggest a possible very bright end of the GRB luminosity function at visible wavelengths, or for the existence of a class of optically bright GRBs. The existence of GRBs with bright optical counterparts can have many important consequences: *i)* bright optical flares can be monitored with temporal resolutions of the order of the second, a sampling which is essential to understand the origin of the prompt optical emission; *ii)* the detection of OB-GRBs at high redshift may be the best way to hold the promises of GRBs for cosmology, since the measure of GRB redshifts beyond redshift 5 requires intrinsically bright optical and infrared emission; *iii)* finally, the importance of OB-GRBs is confirmed by the fact that GRB 990123 was the first GRB for which the prompt optical emission could be detected, while GRB 050904 was the first GRB at redshift >5 whose redshift could be measured and whose light curve could be studied in details. Rapid optical and infrared robotic dedicated telescopes, associated with high energy experiment able to deliver precise timely GRB source localizations may provoque breakthroughs in the GRB studies and in the exploration of the universes at redshifts in the range 5-15.

ACKNOWLEDGMENTS

BG and GS acknowledge the support by the EU Research and Training Network "Gamma-Ray Bursts, an enigma and a tool". TAROT has been funded by the Centre National de la Recherche Scientifique, Institut National des Sciences de l'Univers (INSU-CNRS), and built thanks to the support of the Technical Division of INSU-CNRS.

REFERENCES

1. Meszaros,P. & Rees, M. 1999, MNRAS, 306, L39
2. Katz, J.I. 1994, ApJ, 432, L107
3. Klotz, A., Boër, M., Atteia, J.L., 2003, A&A, 404, 815
4. Boër M. et al. 1999, A&AS, 138, 579
5. Bringer, M., Boër, M., Peignot, C., Fontan, G., Merce, C., 2001, Exper. Astrophys 12, 34
6. Cummings, J. et al. 2005, these proceedings
7. Jelinek, M. et al. 2005, GCN Circular 3929
8. Cusumano G. et al. 2005, these proceedings
9. Tagliaferri G. et al. 2005, A&A, 443, L1
10. Haislip J. et al. 2005, submitted to Nature, astro-ph/0509660
11. Kawai, N. et al., these proceedings
12. Boër, M., Atteia, J.L., Damerdji, Y., et al., 2005, ApJ, submitted
13. Akerlof C. et al. 1999, Nature, 398, 400
14. Sari, R. & Piran, T. 1999, ApJ, 517, 109
15. Wei, D.M. et al. 2005, astro-ph/0511154
16. Nakar E. & Piran T., 2005, Nuovo Cimento C, Proceedings of the 4th Workshop Gamma-Ray Bursts in the Afterglow Era, Rome, 18-22 October 2004, eds. L. Piro, L. Amati, S. Covino, B. Gendre, in press (astro-ph/0502475)
17. Vestrand et al. 2005, Nature, 435, 178

Rapidly Detecting Extincted Bursts with KAIT and PAIRITEL

N. Butler, J. Bloom, A. Filippenko, W. Li, R. Foley, K. Alatalo, D. Kocevski, D. Perley and D. Pooley

University of California at Berkeley, 445 Cambell Hall, Berkeley, CA 94720, USA

Abstract. Ground-based robotic telescopes have proven essential in the first 6 months of Swift operation. The robotic optical 0.7m Katzman Automatic Imaging Telescope (KAIT) and the robotic infra-red 1.3m Peters Automated Infrared Imaging Telescope (PAIRITEL) have contributed important observations during this period, and we discuss efforts to leverage the capabilities of each system to perform rapid, joint observations in the optical and infra-red. An autonomously paired KAIT/PAIRITEL system offers the potential to rapidly discover highly dust extincted bursts or bursts occurring at high redshifts ($z \gtrsim 5$) as near-IR dropouts.

Keywords: gamma rays: bursts — supernovae: general
PACS: 98.70.Rz

INTRODUCTION

The 0.76-m Katzman Automatic Imaging Telescope[1] (KAIT) is a fully robotic instrument operated by Alex Filippenko's group at Lick Observatory. The telescope control system checks the weather, opens the dome, points to the desired objects, acquires guide stars (in the case of long exposures), exposes the CCD ($6.7' \times 6.7'$ field of view), stores the data, and manipulates the data automatically, all without human intervention. KAIT reaches a limit of $R \sim 20$mag (4σ) in a 5-min exposure. KAIT is programmed to conduct multi-filter (VRIz) observations of GRB afterglows, with the V-band capability present as a consistency check with the *Swift* UVOT.

The Peters Automated Imaging Telescope[2] (PAIRITEL) project was designed by Josh Bloom expressly for the follow-up of *Swift* GRBs with the northern 2MASS 1.3m telescope, currently operated at Mt. Hopkins Arizona by the Harvard/Smithsonian Center for Astrophysics (CfA). In the Spring of 2003, the three-channel 2Mass IR camera was loaned from the University of Virginia on a long term basis to the CfA, which has designated routine rapid followup of all possible *Swift* GRBs as the top priority for the 1.3m.

---

[1] http://astro.berkeley.edu/~bait/kait.html
[2] http://cfa-www.harvard.edu/~jbloom/Autotel

TARGET SCIENCE

Detecting and Confirming $z > 5$ GRBs. Observations of VHR GRBs may address important cosmological questions, since GRBs are expected to occur out to redshifts as high as $z \approx 20$, and the bursts themselves and their *IR* afterglows are easily detectable out to such redshifts in the NIR or IR [1, 2, 3]. Such very high redshift bursts are expected to be "dark" at ultraviolet and optical wavelengths, due to absorption by hydrogen gas in the host galaxy and the IGM. A promising strategy for identifying the afterglows of such GRBs is to search for a transient source with PAIRITEL in the J, H, and K bands. Simultaneous observations with KAIT would allow us to check for strong R-band attenuation ("a dropout") and to establish $z \gtrsim 5$, as done for example for GRB 050904 [e.,g., 4]. The software we propose to develop would compare the data in near real-time, allowing us to rapidly publish our detection in order to maximize the potential for further observations of these exciting objects. Higher redshift events could be found from IR dropouts, and the consistency of the detection could be check by comparing against the KAIT data. Following Ciardi & Loeb [5], a burst at $z = 10$ with a typical forward-shock dominated afterglow will have H- and K- band magnitudes brighter than 17 (similar to the PAIRITEL sensitivity limits) for the first ≈ 1.5 hours. Rapid observations can confirm the heavily extincted J-band emission.

Probing Dust Extinction and Obscured Star Formation. A subset of the the "dark GRBs"–those with no detected optical afterglows despite early and deep searches [6, 7, 8]-which do not occur at high-redshift may occur in very dust regions. It is possible to detect these events in the IR and to use our broad V, R, I, z, J, H, and K band coverage separate out the host-frame reddening from the Galactic redenning. Thus we can hope to study the extinction law with the optical/IR spectral energy distribution. By folding in also the X-ray column density measurement from the XRT, it is possible to constrain the sizes of molecular clouds surrounding the progenitors [9]. Early and broad-band observations are crucial for separating out the dark GRBs which are not heavily extincted, because these are expected to be faint and to fade rapidly [see, e.g., 10].

STATUS, OBSERVATIONS OF GRB 051111

Both KAIT and PAIRITEL have demonstrated their feasibility and strong potential during the initial months of *Swift* operation. KAIT observed *Swift* GRBs 050412, 050416, 050522, and 050607 starting $\sim$ 1min after each GRB (see, GCN #s 3256, 3270, 3447, and 3533). PAIRITEL has observed several *Swift* afterglows. PAIRITEL discovered IR emission contemporaneous with GRB 041219a in observations beginning 7.2 minutes after the burst [11]. We are working to extend these capabilities by developing improved detection algorithms and visualization tools for the combined KAIT and PAIRITEL data pipelines. The rapid and autonomous observations by KAIT of GRB 051111 provide a benchmark for our nascent system. Observations with PAIRITEL began soon after the KAIT trigger. We summarize and compare the data below. Another burst—GRB 051109A—was recently also detected jointly by KAIT and PAIRITEL.

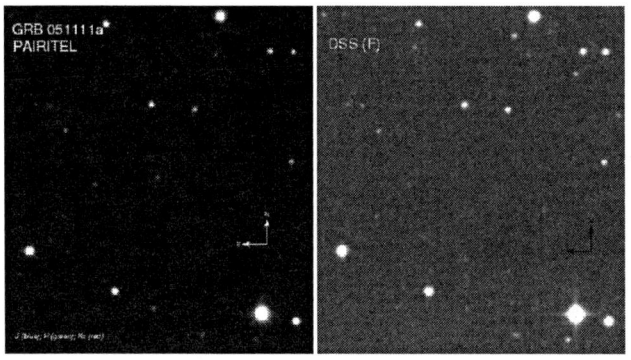

FIGURE 1. PAIRITEL JHK-band image of GRB 051111.

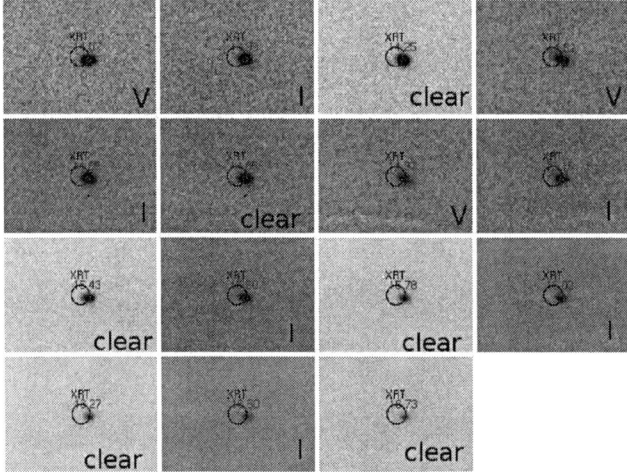

FIGURE 2. KAIT V, I Band and unfiltered images of GRB 051111.

KAIT detected Swift GRB 051111 (Trigger 163438; Sakamoto et al. [12]) in a series fifteen images automatically obtained starting at 06:00:25 UT (44s after the burst). The sequence includes a combination of images taken with the V and I filters, as well as some that are unfiltered (Figure 2. The optical transient originally detected by Rujopakarn et al. [13] is clearly present in our images. Although the transmission conditions were poor at Mt. Hopkins (i.e. clouds), we managed to obtain a total imaging exposure of 188s with PAIRITEL on the field of GRB 051111 before hitting a telescope limit (Figure 1). Our simultaneous JHK-band fluxes are shown in Figure 3 alongside the fluxes from KAIT.

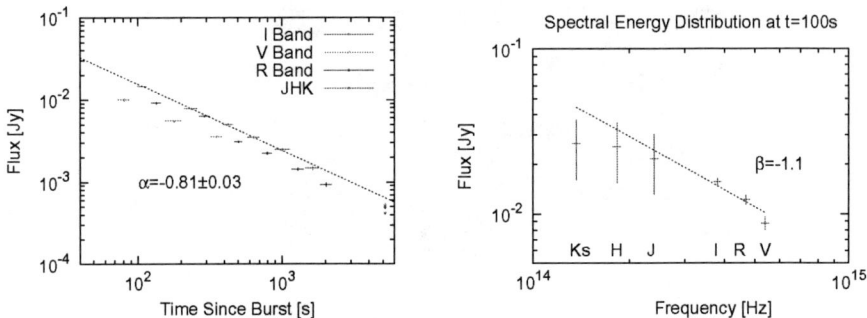

FIGURE 3. The KAIT/PAIRITEL lightcurve and spectral energy distribution at $t = 100s$ for GRB 051111.

ACKNOWLEDGMENTS

PAIRITEL is partially funded by a grant from the Swift Guest Investigator Program.

REFERENCES

1. Bromm, V., & Loeb, A., *ApJ*, **575**, 111, (2002).
2. Lamb, D. Q. & Reichart, D. E., *ApJ*, **535**, 1, (2000).
3. Reichart, D. E., et al., *ApJ*, **552**, 57, (2001).
4. Tagliaferri, G., et al., *A&A*, **443**, L1, (2005).
5. Ciardi, B., & Loeb, A., *ApJ*, **540**, 687, (2000).
6. Djorgovski, S. G., et al., *ApJ*, **562**, 654, (2001).
7. Piro, L., et al., *ApJ*, **577**, 680, (2002).
8. De Pasquale, M., et al., *ApJ*, **592**, 1018, (2003).
9. Galama, T. J., & Wijers, R. A. M. J., *ApJ*, **549**, L209, (2001).
10. Taylor, G. B., et al., *ApJ*, **537**, L17, (2000).
11. Blake, C. H., et al., *astro-ph/0503508*, (2005).
12. Sakamoto, T., et al., *GCN*, **4248**, (2005).
13. Rujopakarn, W., et al., *GCN*, **4247**, (2005).

GRB 050421: A possible naked burst with X-ray flares

O. Godet*, K. L. Page*, J. P. Osborne*, P. T. O'Brien*, A. P. Beardmore*, M. R. Goad* and the *Swift* team

X-ray and Observational Astronomy Group, Department of Physics & Astronomy, University of Leicester, LE1 7RH, UK

Abstract. We present *Swift* observations of the faint burst GRB 050421. The X-ray light-curve shows two flares: the first peaking at ~ 110 s after the BAT trigger (T_0) and the second one peaking at ~ 154 s. We argue that the mechanism producing these flares is probably late internal shocks. The X-ray light-curve is consistent with a power-law with a temporal index $\alpha_1 \sim 3.1$. The X-ray flux decays from $\sim 10^{-9}$ erg cm^{-2} s^{-1} at $T_0 + 100$ s to $< 7 \times 10^{-13}$ erg cm^{-2} s^{-1} at $T_0 + 1000$ s. An X-ray spectral softening is also observed with time from $\beta \sim 0.1$ to ~ 1.2, where β is the spectral index such as $F_\nu \propto \nu^{-\beta} t^{-\alpha}$. A good joint fit to the BAT and XRT spectra indicates that the early X-ray and Gamma-ray emissions are produced by the same mechanism. The X-ray spectral softening is likely due to a shift down to lower energies of the peak of the prompt emission, and the rapid decline of the X-ray emission is probably the tail of the prompt emission. This suggests that the X-ray emission is completely dominated by high latitude radiation and the external shock, if any, is extremely faint and below the detection threshold. GRB 050421 is likely the first "naked burst" detected by *Swift*.

Keywords: gamma-ray: bursts – Gamma-rays, X-rays: flares, individual(GRB 050421)
PACS: 98.70.Rz

INTRODUCTION

The *Swift* Gamma-Ray Burst (GRB) Explorer was launched on 20th November 2004 (Gehrels et al. 2004). It is a multi-wavelength observatory covering the Gamma-ray, X-ray and UV/optical bands. After the detection of a GRB by the Burst Alert Telescope (BAT, Barthelmy et al. 2005), the observatory slews automatically and rapidly to point the narrow field instruments: the X-ray Telescope (XRT, Burrows et al. 2005a), and the UV/Optical Telescope (UVOT, Roming et al. 2005). Due to its rapid pointing capability and high sensitivity, *Swift* is ideal for studying the properties of the early afterglow, in particular the transition between prompt and forward shock emissions.

We report here the *Swift* observations of the GRB discovered at 04:11:52 UT on 21st April 2005 by the BAT (Godet et al. 2005). The XRT started to observe 89 s after the BAT trigger (T_0) and observed an X-ray fading source at (J2000) RA=$20^{\rm h}29^{\rm m}02.44^{\rm s}$ and Dec= $+73°39'17.8''$ with a total uncertainty radius of 3.7 arc-seconds (90% containment). This XRT position is corrected for the effect of the boresight offset (Moretti et al. 2005). The UVOT started to observe 112 s after the BAT trigger, but detected no new source. No radio, infra-red or optical counterpart was detected by ground-based instruments (e.g. Bloom et al., Jelinek et al. 2005).

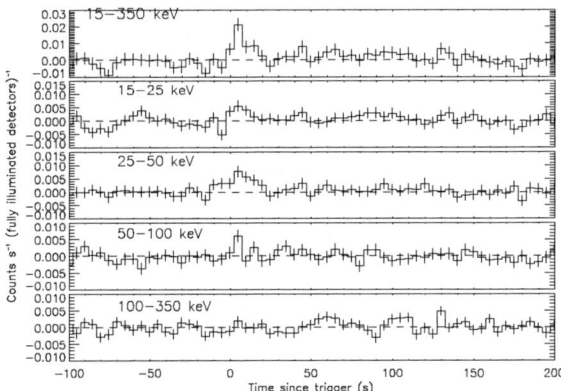

FIGURE 1. Background substracted BAT light-curve for 4 different energy bands from the top to the bottom: 15-350 keV, 15-25 keV, 25-50 keV, 50-100 keV and 100-350 keV. The count rate is given by units of counts s^{-1} (fully illuminated detector)$^{-1}$ and the time binning is 5 s.

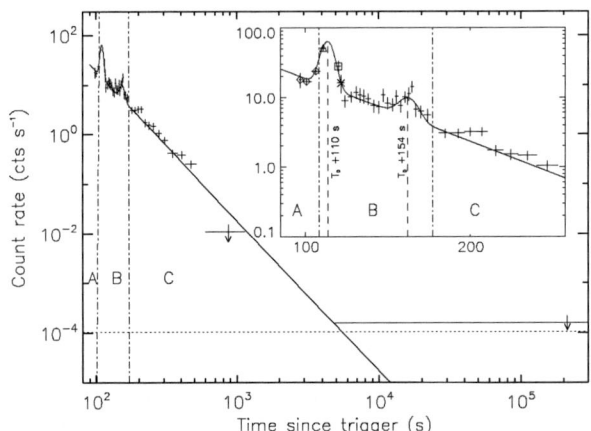

FIGURE 2. Background subtracted XRT light-curve of GRB 050421 over the energy band 0.3-10 keV: A) Low-rate Photo-Diode (LrPD) slew and settling data (diamonds); B) Image Mode (triangle), Piled-up Photo-Diode (PuPD) (square), LrPD (star) and Windowed Timing (WT) pointing data; C) Photon Counting (PC) pointing data. The best fit to the light-curve is plotted by the solid line. The dotted line corresponds to the background level in the PC data. The upper limits are 3 σ limits. The times of the two X-ray flares are also shown in the plot. For a description of the XRT readout modes, see Hill et al. 2004, 2005.

SUMMARY OF THE MAIN SPECTRAL AND TEMPORAL RESULTS

- The BAT light-curve showed a Fast-Rise and Exponential-Decay (FRED) peak with a duration of $T_{90} \sim 10.3$ seconds and a tail extending to $\sim T_0 + 70$ s (see Fig. 1).

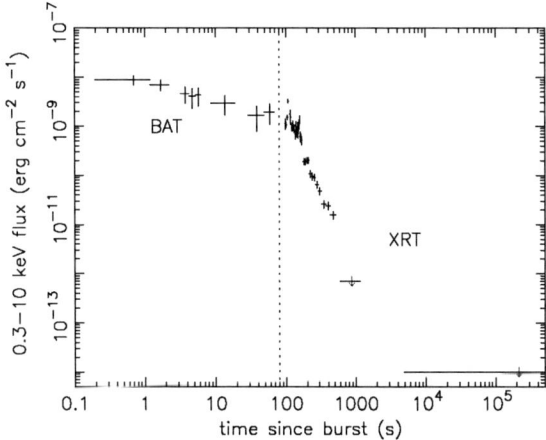

FIGURE 3. Combined light-curves of the BAT data (crosses) extrapolated in the 0.3-10 keV band and the XRT data (circles) in the 0.3-10 keV energy range. The unabsorbed fluxes are given in units of erg cm^{-2} s^{-1}. The BAT and XRT fluxes were derived from the spectral slopes given below in the text. For the first flare, peaking at 110 seconds, we assumed that its spectral slope is the same as that for the combined XRT-BAT spectrum.

The burst is faint, with a fluence of $1.1 \pm 0.7 \times 10^{-7}$ erg cm^{-2} in the 15-150 keV band. GRB 050421 is in the lowest 10% of GRBs detected by the BAT in term of fluence.

- The global XRT/(WT+PC) light-curve is well fit by a power-law with a temporal index $\alpha_g = 3.1$ with 2 flares superposed at $T_0 + 110 \pm 2$ s and $T_0 + 154 \pm 3$ s (see Fig. 2). However, note that if we only fit the XRT data points before $T_0 + 180$ s with the same model, we find a shallower temporal slope of $\alpha_p = 2.0 \pm 0.6$. Due to the rapid decline, the X-ray emission was no longer detected after $T_0 + 900$ s. The 3σ upper limit on the unabsorbed flux in the 0.3-10 keV band is 7×10^{-13} erg cm^{-2} s^{-1} at $\sim T_0 + 900$ s. The XRT light-curve also showed at least two flares peaking at $T_0 + 110 \pm 2$ s and $T_0 + 154 \pm 3$ s.

- There is a spectral softening from the early $T_0 + 115 - T_0 + 171$ s data with $\beta = 0.1 \pm 0.3$ to the later $\sim T_0 + 180 - T_0 + 771$ s data with $\beta = 1.2^{+0.3}_{-0.1}$. The uncertainties at a 99% confidence level for one parameter of interest ($\Delta\chi^2 = 6.63$) are $\beta_1 = 0.1^{+0.6}_{-0.5}$ and $\beta_2 = 1.2^{+0.5}_{-0.3}$ respectively for WT and PC spectra.

- The joint fit of the BAT and early XRT/WT data, which is consistent with an absorbed single power-law with a $\beta \sim 0.2$ slope, shows that they are likely to be produced by the same emission mechanism. Fig. 3 shows the BAT-XRT light-curve over the 0.3-10 keV energy band (in units of erg cm^{-2} s^{-1}). Note that the end of the BAT light-curve matches well with the beginning of the XRT light curve.

- Excess absorption in the observer frame, $\Delta N_H \sim 4.4 \times 10^{21}$ cm^{-2}, over that due to our Galaxy is seen in the XRT spectra. No variation in absorption with time is seen.

DISCUSSION & CONCLUSION

GRB 050421 was one of the faintest bursts observed by *Swift*. It showed a rapid temporal decline in X-rays on which were superposed at least two X-ray flares. The short rise and decay times and the amplitude of the flares indicate that the flares are probably produced by late internal shocks (e.g. Burrows et al. 2005b and Zhang et al. 2005). Simple kinematic arguments also support this interpretation (Ioka et al. 2005). The spectral softening seen in the XRT is likely due to the shift, in the XRT energy band, of the peak of the prompt spectrum with time (Ford et al. 1995). The temporal decline and the spectral results can not be explained by invoking the current afterglow models (e.g. Zhang et al. 2005). A natural explanation for the steep temporal decay and the spectral slopes seen in the previous section is then to invoke a "high latitude emission" model (e.g. Kumar & Panaitescu 2000). In this model, the tail of the prompt emission is produced at high angles relative to the observer line of sight (with angles $\theta > \Gamma^{-1}$ where Γ is the Lorentz factor), and α and β verify the following curvature relation $\alpha_c = 2 + \beta$. The relation is satisfied before $T_0 + 180$ s, since $\alpha_c = 2 + \beta_1 = 2.1 \pm 0.3$ is consistent with α_p. The curvature relation is also satisfied after $T_0 + 180$ s with $\alpha_c = 2 + \beta_2 = 3.2^{+0.3}_{-0.1}$, which is consistent with α_g.

A possible and simple explanation for the non detection of the forward shock emission would be that the very local surrounding density is so low that the blastwave did not sweep up sufficient matter to produce a sufficiently bright afterglow.

GRB 050421 would then be the first burst for which the XRT slewed promptly, for which no forward shock emission was observed. Therefore, GRB 050421 would be the first "naked burst" (Kumar & Panaitescu 2000) discovered by *Swift*.

REFERENCES

1. Barthelmy, S. et al. 2005, *Space Science Review*, **120**, 143, astro-ph/0507410
2. Bloom, J. S. et al. 2005, GCN Circular 3306
3. Burrows, D. N. et al. 2005, *Space Science Review*, **120**, 165, astro-ph/0508071
4. Burrows, D. N. et al. 2005, Science, astro-ph/0506130
5. Ford, L. A. et al. 1995, ApJ, **439**, 307
6. Gehrels, N. et al. 2004, ApJ, **611**, 1005
7. Godet, O. et al. 2005, accepted by *A&A*
9. Hill, J. E. et al. 2005, Proceedings of the SPIE, in press
9. Hill, J. E. et al. 2004, Proceedings of the SPIE, **5165**, 217
10. Ioka, K. et al. 2005, ApJ, **631**, 429
11. Jelinek, M. et al. 2005, GCN Circular 3298
12. Kumar, P. & Panaitescu, A. 2000, ApJ, **541**, 51
13. Moretti, A. et al. 2005, submitted to *A&A*
14. Roming, P. W. A. et al. 2005, *Space Science Review*, **120**, 95, astro-ph/0507413
15. Zhang, B. et al. 2005, submitted to ApJ, astro-ph/0508321

Swift and XMM observations of the dark GRB 050326

A. Moretti*, A. De Luca[†], D. Malesani**, S. Campana*, A. Tiengo[†], J.N. Reeves[‡], M. Capalbi[§], G. Chincarini*, S. Covino*, G. Cusumano[¶], P. Giommi[§], V. La Parola[¶], V. Mangano[¶], T. Mineo[¶], M. Perri[§], P. Romano* and G. Tagliaferri*

*INAF, Osservatorio Astronomico di Brera, via E. Bianchi 46, I-23807 Merate (LC), Italy
[†]INAF, Istituto di Astrofisica Spaziale e Fisica Cosmica di Milano, via E. Bassini 15, I-20133 Milano Italy
**International School for Advanced Studies (SISSA/ISAS), via Beirut 2-4, I-34014 Trieste, Italy
[‡]NASA/Goddard Space Flight Center, Greenbelt Road, Greenbelt, MD20771, USA
[§]ASI Science Data Center, via G. Galilei, I-00044 Frascati (Roma), Italy
[¶]INAF, Istituto di Astrofisica Spaziale e Fisica Cosmica di Palermo, via U. La Malfa 153, I-90146 Palermo, Italy

Abstract. We present a detailed analysis of the GRB 050326 prompt and afterglow emission. The combined capabilities of *Swift* (which sampled the light curve for a relatively long time span) and XMM (which ensured a large statistics) allowed to obtain a thorough characterization of the afterglow properties.

Keywords: gamma rays: bursts – X-rays: individual (GRB 050326)
PACS: 98.70.Rz

GRB 050326 was discovered by the Swift-BAT on 2005 Mar 26 at 9:53:55 UT. Its X–ray coordinates were $\alpha_{J2000} = 00^h 27^m 49.1^s$, $\delta_{J2000} = -71°22'16.3''$, with a 90% uncertainty radius of $1.5''$ containment. This burst was also detected by the *Wind*-Konus experiment, leading to the characterization of its broad-band gamma-ray spectrum (see [1] for the complete description of the observations and the analysis).

The prompt emission was relatively bright (with a 20–150 keV fluence of $\sim 8 \times 10^{-6}$ erg cm^{-2}). The spectrum was hard (photon index $\Gamma = 1.25 \pm 0.03$), suggesting a peak energy at the high end of the BAT energy range or beyond. Indeed, thanks to the simultaneous detection of this burst by the *Wind*-Konus experiment [2], the prompt spectrum could be fully characterized. The prompt bolometric fluence was $\mathscr{F} \sim 2.4 \times 10^{-5}$ erg cm^{-2} (1–10 000 keV), and the observed peak energy was $E_{p,obs} = 200 \pm 30$ keV.

Due to pointing constraints, Swift-XRT and Swift-UVOT observations could start only 54 min after the GRB. The X-ray afteglow was quite bright, with a flux of 7×10^{-11} erg cm^{-2} s^{-1} (0.3–8 keV) 1 hr after the GRB. However, no optical counterpart could be detected. The X-ray light curve showed a steady decline, with no breaks or flares. The best-fit power-law decay index was $\alpha = 1.70 \pm 0.05$ (Fig.1). Such regular behaviour is different from that usually observed by *Swift*, but this may be the result of the limited time coverage (observations could be carried out only between 54 min and 4.2 d after the burst). Indeed, extrapolation of the afterglow light curve to the time of the

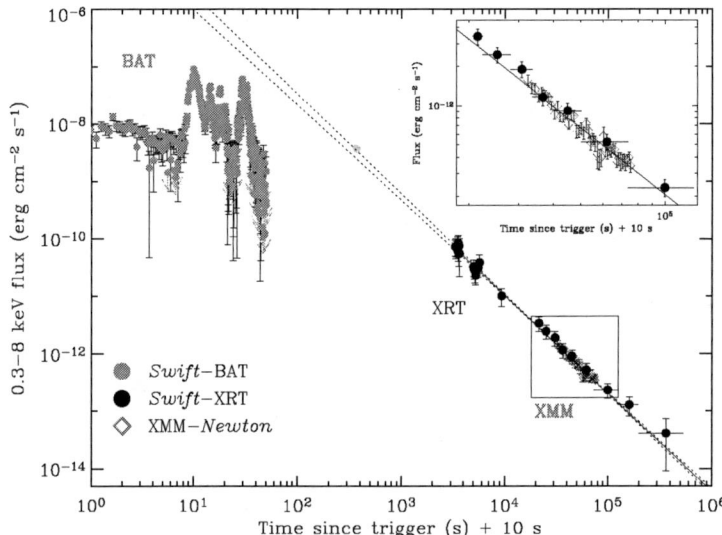

FIGURE 1. The light curve of GRB 050326 and of its afterglow in the 0.3–8 keV energy band. XRT (black circles) and XMM data (empty diamonds) show a very good agreement (see also the inset). The solid line shows the fit to the combined XRT/XMM afterglow light curve. The dotted lines indicate the 90% errors of the extrapolated X-ray light curve. Light filled circles indicate the extrapolation of the BAT data to the 0.3–8 keV energy range, assuming the Band model as the best-fit spectrum. In this figure, the time origin was set 10 s before the nominal trigger time, to show the weak, untriggered precursor. This has no effect on the determination of the afterglow decay slope, due to the late beginning of the XRT observation.

prompt emission overpredicts the burst flux, and may suggest a slower decay before the beginning of the XRT observation.

The analysis of the combined XRT and XMM data allowed to characterize in detail the afterglow spectrum. A fit with an absorbed power-law model provided a good description to the data, yielding a photon index $\Gamma = 2.09 \pm 0.08$ and a column density significantly in excess to the Galactic value. The best-fit model was thus computed adding an extra absorption component, leaving its redshift z free to vary. Although neither $N_{H,z}$ nor z could not be effectively constrained, a firm lower limit $N_{H,z} > 4 \times 10^{21}$ cm^{-2} could be set.

Therefore, GRB 050326 adds to the growing set of afterglows with large rest-frame column density [3, 4, 5]. The limits measured in the optical and ultraviolet region by UVOT lie well below the extrapolation of the X-ray spectrum (left panel of Fig 2). In particular, they violate the synchrotron limit that the optical-to-X-ray spectral index should be larger than 0.5. This implies a large extinction and/or a high redshift.

The X-ray spectral analysis also allowed us to set the lower limit $z > 1.5$ to the redshift of the absorbing component and, therefore, of the GRB (right panel of Fig. 2). The isotropic-equivalent gamma-ray energy was then $E_{\gamma,\text{iso}} > 1.4 \times 10^{53}$ erg. The temporal and spectral properties of the afterglow were nicely consistent with a spherical

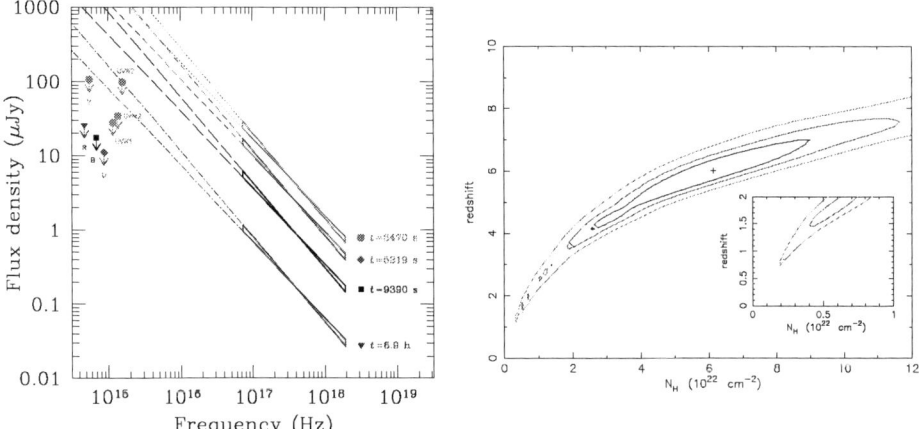

FIGURE 2. Left Panel. Broad-band spectral energy distribution of the afterglow of GRB 050326, computed at different times (which are identified by different symbols). The shape of the X-ray spectrum was assumed to be constant throughout the observation, and the decay law was adopted to report the X-ray flux at the time of the optical measurements. **Right Panel.** Confidence contours (68%, 90% and 99% levels for 2 parameters of interest) for the gas column density $N_{H,z}$ and the redshift z of the intrinsic absorber, as computed from the fit to the EPIC spectra. The Galactic column density was assumed to be $N_{H,MW} = 4.6 \times 10^{20}$ cm^{-2}. The inset shows a zoom-in of the low-redshift region.

fireball expanding in a uniform medium, with the cooling frequency above the X-ray range. We could therefore set a lower limit to the jet break time $t_b > 4$ d. The jet opening angle could be constrained to be $\vartheta_j > 7^o$, with only a weak dependence on the (unknown) fireball energy. The beaming-corrected gamma-ray energy was $E_{\gamma,j} = (3-8) \times 10^{51} (t_b/4 \text{ d})^{3/4}$ erg, independently from the redshift. GRB 050326 thus released a large amount of gamma rays (only GRB 990123 had a larger energy in the sample of [6]).

To be consistent with the Ghirlanda relation [6], two redshift ranges are allowed, either at low ($z < 0.8$) or high ($z > 4.5$) redshift (Fig. 3). However, to simultaneously satisfy the limits derived from the X-ray spectral analysis, only the high-redshift region is left. We note that the Ghirlanda relation is still based upon a small sample, so that any inference cannot yet be regarded as conclusive. However, the results from the X-ray spectra, the consistency of the GRB 050326 properties with the Ghirlanda relation, and the strong dearth of optical/ultraviolet afterglow flux, are overall consistent with a moderate/high redshift ($z > 4$). A search for the host galaxy through deep infrared and optical imaging may conclusively settle this issue.

REFERENCES

1. Moretti, A., De Luca, A., Malesani, D., et al. 2006, A&A in press, astro-ph/0512392
2. Golenteskii, S., Aptekar, R., Mazets, E., et al. 2005, GCN Circ. 3152

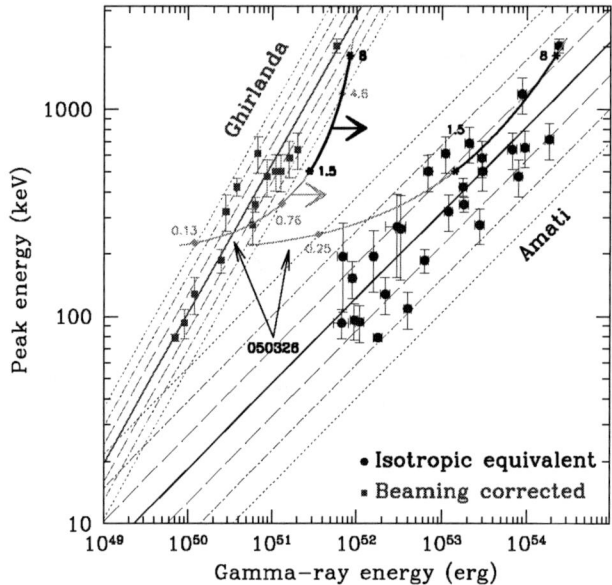

FIGURE 3. Comparison of GRB 050326 with the Amati (right) and Ghirlanda (left) relations [6, 7]. The thick solid curves (black and grey) show the position of GRB 050326 as its redshift varies in the interval $0.1 < z < 10$. The Ghirlanda track is actually a boundary (as the horizontal arrows indicate), since we can only infer a lower limit to the beaming-corrected energy at each redshift. Filled circles and squares indicate the GRBs which define the above two relations, plotted as straight solid lines (together with their 1-, 2- and 3-σ contours: long-dashed, dot-dashed and dotted lines, respectively). Data were taken from [6, 8]. Grey diamonds indicate the intersection of the GRB 050326 tracks with the 3-σ contours of the Amati and Ghirlanda relations. These points thus define the 3-σ redshift ranges for which GRB 050326 was consistent with the two relations. In the two GRB 050326 tracks, the region $1.5 < z < 8$ (indicated by the X-ray data) is shown in black, bound by asterisks.

3. Galama, T.J., & Wijers, R.A.M.J. 2001, Apj, 549, L209
4. Stratta, G., Fiore, F., Antonelli, L.A., Piro, L., & De Pasquale, M. 2004, Apj, 608, 846
5. Campana, S., Romano, P., Covino, S., et al. 2005b, A&A, in press (astro-ph/0511750)
6. Ghirlanda, G., Ghisellini, G., & Lazzati, D. 2004, Apj, 616, 331
7. Amati, L., Frontera, F., Tavani, M., et al. 2002, A&A, 390, 81
8. Ghirlanda, G., Ghisellini, G., Lazzati, D., & Firmani, C. 2005, Il nuovo cimento, in press (astro-ph/0504184)

GRB 050117: Simultaneous Gamma-ray and X-ray Observations with the *Swift* Satellite

J.E. Hill[1,2], D. C. Morris[3], T. Sakamoto[1], G. Sato[4], D.N. Burrows[3],
L. Angelini[1,5], C. Pagani[3,6], A. Moretti[6], A.F. Abbey[7], S. Barthelmy[1],
A.P. Beardmore[8], V.V. Biryukov[9], S. Campana[6], M. Capalbi[10],
G. Cusumano[11], P. Giommi[10], M.A. Ibrahimov[12], J.A. Kennea[3],
S. Kobayashi[3,13], K. Ioka[3,13], C. Markwardt[1], P. Meszaros[3], P.T. O'Brien[8],
J.P. Osborne[8], A.S. Pozanenko[14], M. Perri[10], V.V. Rumyantsev[15],
P. Schady[16], D.A. Sharapov[12], G. Tagliaferri[6], B. Zhang[17], G. Chincarini[6,18],
N. Gehrels[1], A. Wells[3,7], J.A. Nousek[3]

[1]*NASA/GSFC, Greenbelt, MD 20771, USA,*
[2]*USRA, Columbia, MD, 21044, USA,*
[3]*Department of Astronomy & Astrophysics, Penn State University, University Park, PA 16802, USA,*
[4]*ISAS/Japan Aerospace Exploration Agency, Kanagawa, 229-8510 Japan,*
[5]*The Johns Hopkins University, Baltimore, MD 21218, USA*
[6]*INAF -Osservatorio Astronomico di Brera, 23807 Merate, Italy,*
[7]*Space Research Centre, University of Leicester, Leicester LE1 7RH, UK,*
[8]*Department of Physics and Astronomy, University of Leicester, Leicester LE1 7RH, UK,*
[9]*Crimean Laboratory of Sternberg Astronomical Institute, Moscow University, Russia*
[10]*ASI Science Data Center, 00044 Frascati, Italy,*
[11]*INAF- Istituto di Astrofisica Spaziale e Fisica Cosmica Sezione di Palermo, 90146 Palermo, Italy,*
[12]*Ulugh Beg Astronomical Institute, Tashkent 700052, Uzbekistan,*
[13]*Center for Gravitational Wave Physics, Pennsylvania State University, University Park, PA 16802, USA,*
[14]*Space Research Institute, Moscow 117810, Russia,*
[15]*Crimean Astrophysical Observatory, Ukraine*
[16]*Mullard Space Science Laboratory, Holmbury St. Mary, Dorking, Surrey, UK,*
[17]*Department of Physics, University of Nevada, Las Vegas, NV 89154, USA*
[18]*Universit`a degli studi di Milano-Bicocca, Dipartimento di Fisica, I-20126 Milan, Italy*

Abstract. The *Swift* Gamma-Ray Burst Explorer performed its first autonomous, X-ray follow-up to a newly detected GRB on 2005 January 17, within 193 seconds of the burst trigger by the *Swift* Burst Alert Telescope. While the burst was still in progress, the X-ray Telescope obtained a position and an image for an un-catalogued X-ray source; simultaneous with the gamma-ray observation. The XRT observed flux during the prompt emission was 1.1×10^{-8} ergs cm^{-2} s^{-1} in the 0.5-10 keV energy band. The emission in the X-ray band decreased by three orders of magnitude within 700 seconds, following the prompt emission. This is found to be consistent with the gamma-ray decay when extrapolated into the XRT energy band. During the following 6.3 hours, the XRT observed the afterglow in an automated sequence for an additional 947 seconds, until the burst became fully obscured by the Earth limb. A faint, extremely slowly decaying afterglow, α=-0.21, was detected. Finally, a break in the lightcurve occurred and the flux decayed with α<-1.2. The X-ray position triggered many follow-up observations: no optical afterglow could be confirmed, although a candidate was identified 3 arcsecs from the XRT position.

Keywords: gamma rays: bursts - general: gamma-rays, X-rays - individual (GRB 050117)

INTRODUCTION

On 2005 January 17 at 12:52:36.037 UT, the *Swift* Burst Alert Telescope triggered and located GRB 050117 (Sakamoto et al. GCN 2952). For the first time, *Swift* responded autonomously to the BAT triggered burst, pointing the XRT at the GRB while the burst was still in progress, and allowing simultaneous gamma-ray and X-ray flux measurements of the prompt emission and follow up observations of the afterglow. The BAT lightcurve of the burst, which lasted 220 seconds, is multi-peaked. The XRT was on target and obtained a refined position and an image within 193 seconds of the BAT detection (Figure 1). The XRT detected the GRB at the end of the burst on-set, measuring a source position of RA(J2000)= 23^h 53^m 53.0^s Dec(J2000)=+65° 56' 19.8'' (Hill et al. GCN 2955) and an absorbed flux of $1.1 \pm 0.3 \times 10^{-8}$ ergs cm^{-2}s^{-1} in the 0.5-10 keV band. A faint afterglow was detected by the XRT during the subsequent orbits. The UVOT activation was not complete at the time of these observations and therefore it remained in a non-observing state throughout. No radio or optical afterglow was detected by the ground based follow-up observations.

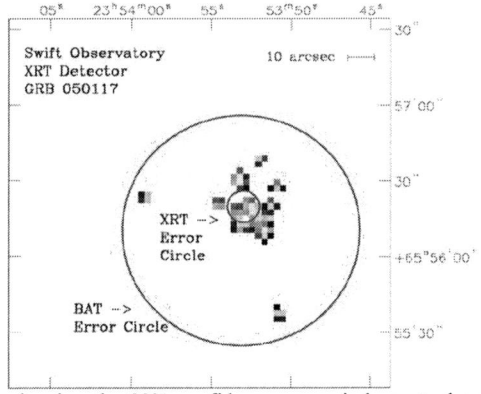

FIGURE 1. XRT Image showing the 90% confidence error circle centred on the onboard position, corrected for a non-nominal spacecraft configuration and the BAT 90% confidence error circle from ground processing.

Gamma-ray Analysis

The time-averaged spectral fit over the 15-150 keV energy band gave a photon index of 1.1 ± 0.2 and an E_{peak} of 123 ± 50 keV. The burst total energy fluence was $9.3 \pm 0.2 \times 10^{-6}$ ergs cm^{-2} in the 15-150 keV band in 220 seconds with more than one third of the fluence in the 100-150 keV band. The peak photon flux of 2.47 ± 0.17 ph cm^{-2}s^{-1} (integrated for one second from 15-150 keV using the best fit model of a simple power-law), occurred 87.22 seconds after the trigger.

XRT Analysis

The X-ray lightcurve is shown in Figure 2. The X-ray data indicate a decrease in flux of almost three orders of magnitude between the prompt emission at t=190 seconds and the afterglow at t=900 seconds. The decay becomes significantly flatter over the following ~6 hours and then becomes steeper again sometime later in order to be undetected 43 days after the burst.

A photo-absorbed power law spectral fit to the prompt emission (t=193 seconds) using a Galactic absorption column yields a photon index of 2.3±0.5. Fitting the same model to the summed data from the afterglow (t=900 seconds - 6.6 hours) yields a photon index of 2.0±1.1.

For the X-ray data we assume a spectrum of the form $F(t,\nu) \propto (t-t_0)^{\alpha} \nu^{\beta}$, where β is the spectral index and β =1-photon index, yielding a spectral index of -1.3 ± 0.5 and -1.0 ± 1.1 for the prompt and follow-up observations, respectively.

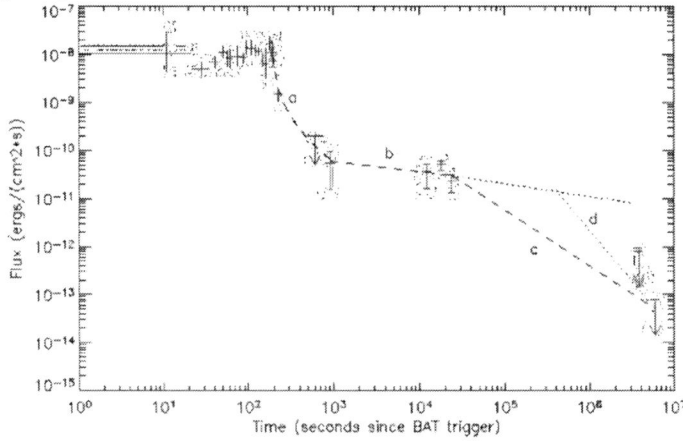

FIGURE 2. GRB 050117 lightcurve (absorbed fluxes): The BAT (blue) lightcurve using the spectral fits for each time bin and extrapolating the 15-100 keV flux into the 0.5-10 keV band and the XRT (red) lightcurve (0.5-10 keV), showing the upper limits for the observations at more than 43 days after the burst; a) A power law fit assuming high latitude emission from the internal shock where t_{shock}=187 seconds, α<-1.2; b) A power law fit to the afterglow decay with energy input from refreshed shocks assuming a t_0=trigger time, α=-0.2±0.2; c) Continuation of the afterglow decay assuming a break in the lightcurve at t=6.6 hours, t_0=trigger time, α=-1.2; d) Extrapolation of α=-2 from t=43 days, showing the latest expected time of the break in the lightcurve at ~4.5 days (t_0=trigger time).

The XRT and extrapolated BAT fluxes obtained from the simultaneous XRT and BAT observations (Figure 2) are well within the 90% confidence limits. If we consider that the multiple peaks in the BAT lightcurve are attributed to internal shocks from the collision of the faster expanding shells with slower shells in front, then it is reasonable to assume that the X-ray flux at this time was also produced by an internal shock collision.

The minimum expected emission following the peak from the internal shock is the high latitude emission from the curvature effect (Kumar & Panaitescu, 2000). The angular spreading time scale determines the decay timescale, and consequently the width of an internal shock peak, Δt. Therefore, t_0 of the internal shock is, t_{shock}=187 seconds. Due to the non-detection of gamma-ray flux between t=300 seconds and t=913 seconds we can assume that the X-ray flux after 900 seconds is dominated by the decaying afterglow and therefore we can use the XRT flux measurement at 900 seconds as an upper limit on the contribution from the decaying internal shock at this time.

This provides a constraint on the temporal decay of the internal shock of $\alpha<-1.1$. In the context of high latitude emission from the internal shock, where $\alpha=\beta-2$, and taking the photon index of 2.3 ± 0.5 into consideration, a decay of -3.3 ± 0.5 would be expected. This is within the constraints of the observation, where $\alpha<-1.1$.

Following the end of the prompt emission, the lightcurve enters a shallower decay phase where, for the following 6.3 hours, there is very little decay in flux. The 90% confidence upper limit for the decay index is -0.5, but the best fit to the data is shallower than this; $\alpha=-0.21\pm0.3$. A second break in the power law is implied by the steep decay between the data points at 23 ksec and the upper limit at 68 days. In order for the source to be undetected 68 days after the burst, the flux must decay with $\alpha<-1.2$ if the break occurred immediately after the last detection. If the break occurred later or the flux is significantly less than the upper limit, then the decay could be steeper.

The afterglow due to the collision with the ambient medium may have started while the internal shock emission was still in progress, although the sample rate is too low to confirm this. The emission from the afterglow appears to be enhanced by the additional input of energy from lagging shells of ejected material. This burst was long and multi-peaked, and therefore if the later shells were slow moving with a modest Lorentz factor, energy would be injected into the afterglow as each shell collides into the external medium. This would cause re-brightening super-imposed on the nominal afterglow decay and could explain the flatter than expected decay between 900 seconds and 6.6 hours. The lightcurve is not well sampled and so bumps, which may be expected from re-brightening, cannot be discerned from the lightcurve. The refreshed shock energy injection continues until at least five hours after the burst. Some time between five hours and 4.5 days after the burst trigger, the refreshed shocks ceased and the lightcurve turned over to a steeper decay rate of $\alpha<-1.2$ corresponding to the expected afterglow decay, thus the burst was below the XRT detection threshold at t=68 days.

To date, there have only been three other observations by *Swift* with simultaneous gamma-ray and X-ray detections. The observations of GRB 050117 demonstrate the unique capability of *Swift* to observe both the burst and the afterglow in the X-ray regime.

ACKNOWLEDGEMENTS

This work is supported at Penn State by NASA contract NAS5-00136; at the University of Leicester by the Particle Physics and Astronomy Research Council on grant number PPA/Z/S/2003/00507; and at OAB by funding from ASI on grant number I/R/039/04. We gratefully acknowledge the contributions of dozens of members of the *Swift* team at PSU, University of Leicester, OAB, GSFC, ASDC and our subcontractors, who helped make this Observatory possible and to the Flight Operations Team for their dedication and support.

REFERENCES

Cusumano, G. 2005, The Astrophysical Journal, in press
Tagliaferri, G., et al. 2005, Nature, Vol. 436, Issue 7053, 985
Kumar, P. and Panaitescu, A. 2000, The astrophysical Journal, 541, L51
Hill, J.E., et al. 2005, GCN 2955
Sakamoto, T., et al. 2005, GCN 2952

Rapid GRB Afterglow Response With SARA

K. V. Garimella[*], A. L. Homewood[*], D. H. Hartmann[*], C. Riddle[*],
S. Fuller[*], A. Manning[*], T. McIntyre[*], G. Henson[#]

[*]*Clemson University, Department of Physics & Astronomy, and SARA Observatory, Clemson, SC 29634*
[#]*Department of Physics and Astronomy, East Tennessee State University, Johnson City, TN 37614*

Abstract. The Clemson GRB Follow-Up program utilizes the SARA 0.9-m telescope to observe optical afterglows of Gamma Ray Bursts. SARA is not yet robotic; it operates under direct and Target-of-Opportunity (ToO) interrupt modes. To facilitate rapid response and timely reporting of data analysis results, we developed a software suite that operates in two phases: first, to notify observers of a burst and assist in data collection, and second, to quickly analyze the images.

Keywords: Rapid Response, Perl, website, automation, gamma ray bursts.
PACS: 95.75.-z, 95.75.Mn, 95.75.Rs

1. INTRODUCTION

Swift notifies other observatories of GRBs through email and socket messages called "GCN Notices". Socket communication is preferable to email alerts for automated systems, since information comes in smaller chunks with header information indicating the type of message. Messages arrive more quickly than emails, and a client-side script can more easily filter these socket messages into reasonable commands for a telescope. We utilize remote, human-controlled access to the SARA 0.9m telescope (not (yet) operational under a fully scripted schedule). Therefore the notices are more important for notifying the team rather than directly instructing the telescope to respond. For this purpose, the email method is sufficient. The format of the emails facilitates parsing through a regular expressions engine in any preferred language. We have written such a notification tool in Perl, a language well suited in an environment where string manipulations are numerous. This tool notifies group members by email and text-messages to phones. In addition, the tool also extracts key information from the Notices and assembles a page on a website with burst-specific data (coordinate history, finding charts, rise and set times, weather information, and in the case of a false particle trigger, retractions). Rapid data analysis becomes necessary to quickly return relevant information to the community. We have written scripts for standard image reduction sequences with minimum user-input.

2. TOOLS

2.1 The Notification and Website Generation Tool

The notification tool is a Perl script that continually monitors a POP3 email account for new notifications from the GCN system. The email account is, of course, subject to messages beyond the desired GCN notices (system maintenance alerts, spam). The notices are easily filtered from the irrelevant messages with a subject line regular expression filter. The notice is then distilled into key-value pairs for each piece of information (right ascension, declination, event time, etc.). Special attention must be paid to some fields. For instance, the GRB_RA and GRB_DEC fields are multi-line fields, and there are multiple COMMENTS fields. The parser must be aware of these departures from the traditional key-value list in order for the parse to be successful.

Some information in the notices may not be relevant (e.g., EXPOSURE_ID). Those that are important contain information like the GRB's RA and Dec, but this field is not necessarily provided by one field-name. The GCN notice system provides HETE and Integral alerts as well as Swift alerts. The script recognizes the differences between the notices to extract the required information.

A trigger number uniquely identifies each GRB event. Multiple notices on the same event may have revised information, but the trigger number will remain the same. The script keeps a list of all processed trigger numbers that is checked after the parsing is complete but before any notices are sent out. If the current trigger number is contained in the list, no notice is send to the team. This prevents a flood of emails and text messages to email accounts and phones.

At this stage, the script gets weather information for Kitt Peak and sends long-form and short-form messages to email accounts and phones (respectively) based on a prewritten recipient list. This information is then compiled into a trigger-specific page on a burst alert website (http://people.clemson.edu/~kgarime/burst). With the RA and Dec, the site can link to a DSS finding chart for that particular event. The website, written in PHP, monitors the directory in which the new trigger information is written and places a red banner at the top of the main page.

2.2 Image Analysis Tools

The tool described above ensures rapid response to a GRB event. The image analysis tools enable rapid photometry of the afterglow. Multiple images can be analyzed with these tools in order to produce a light curve. These tools implement a standard set of steps commonly adhered to in astronomical image analysis. These steps are dark-subtraction, flat-fielding, image registration, stacking in groups, and also calibration, using programs like IRAF, Source Extractor (Sextractor), and MIDAS. These programs serve as back-end tools for these automated analysis scripts.

2.2.1 Dark-Subtraction and Flat-Fielding

Dark-subtraction is performed by stacking a set of darks and subtracting them from the science images with a scripted call to IRAF. Flat-fielding is performed with calls to IRAF by stacking all of the flat-field images and normalization to the image maximum pixel value. This resultant image is divided out of the original data set. The script performs the subtraction and division with little to no user input. Pre-scripted regular expression file filters determine the appropriate dark and flat data sets.

2.2.2 Image Registration

Image registration is often the most complicated step. Tracking is less than perfect; average image shifts can be up to three pixels in either axis for back-to-back 300-second exposures. For large data sets, this can yield 100-pixel shifts from the first image in the set to the last. To perform the image registration, a script runs Source Extractor on every file in the data set. This produces a source list (a file containing five columns of data: an incrementing numerical id, x-pixel, y-pixel, magnitude, and magnitude error) for each file in the data set. Though it is expected that not every source will be in every file, it is expected that the brightest sources will be contained in every file. An image in the middle of the data set is chosen as reference. The brightest source is extracted and compared against every other file. In order to be deemed a matching source in another file, the selected source must fall within self-adjusting tolerances for the x,y-pixel coordinates and the magnitude. These tolerances are incrementally relaxed until a match is discovered or until, after a certain number of iterations, the script simply fails with an error message. This may result when the data set is exceptionally large, tracking is poor, or in crowded fields. While the algorithm is not robust enough to handle extreme situations, the script has yet to fail under typical conditions (fair tracking and weather, fields of low-to-moderate source density).

2.2.3 Stacking

For a particularly dim afterglow at the time of observation, it may be necessary to stack every file in the data set in order to detect the source. Other times, a relatively bright afterglow can be detected by stacking just a few images. To accomplish these tasks, and to produce light curves, we developed a user-input based stacking script.

2.2.4 Calibration

We compare instrumental magnitudes of several stars in our field to those in a standard catalog (such as USNO A2.0). Images from the telescope are not encoded with a World Coordinate System (WCS), so a direct comparison to the appropriate finding chart is necessary in order to obtain the RA and Dec of each individual source. Once the same source is identified in both files, a calibration file can be compiled with x,y location, apparent magnitudes, and magnitude errors from the source list and the actual magnitude from the catalog. With this data table, we generate a linear fit to obtain the calibrated magnitude, or an upper limit. To calibrate multiple files, only one calibration file is necessary. The software uses self-adjusting tolerances to match the calibration file from one image to the slightly offset sources in another file.

3. RESULTS

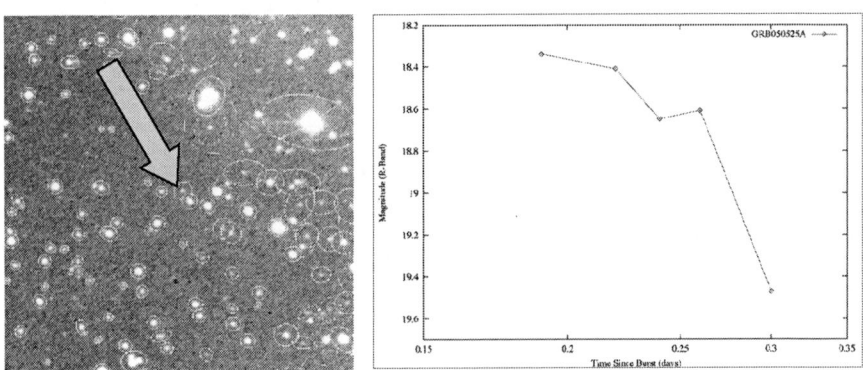

FIGURE 1. Source extractor result for GRB050525a (left), and associated light curve (right).

GRB050525A served as a software test. Figure 1 shows source detections and the light curve. SARA responded about four hours after the trigger, and we reduced and analyzed the data in about 20 minutes. GRB051227 proved more difficult. Imaging began 12 hours after the trigger. By this time, the optical transient had faded beyond the SARA threshold. Operating in "automatic" mode (in which the pixel coordinates of the afterglow need not be specified), the calibration software was able to determine an upper magnitude limit of R = 21.55 ± 0.20 mag, which was reported in GCN 4421.

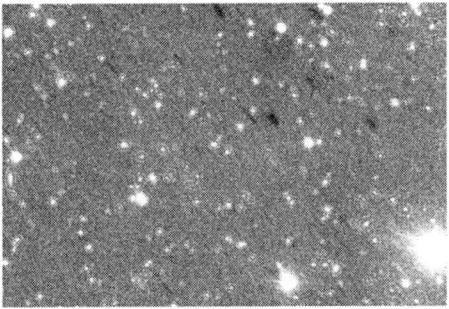

FIGURE 2. Source extractor result for GRB05227. No new source was discernable from DSS images.

Acknowledgments

We thank Jeremy McLaughlin (Radford University), and Adam Brimeyer (Iowa State University) for assistance with data collection and remote operations of SARA (see GCN 3491). We acknowledge partial support of the Clemson-SARA-GRB program from NASA's Swift team, and through NSF/DAAD grant INT-0128882.

Rapid Centroids and the Refined Position Accuracy of the *Swift* Gamma-ray Burst Catalogue

J.E. Hill[a,b], L. Angelini[a,c], A. Moretti[d], D.C. Morris[e], J. Racusin[e], D.N. Burrows[e], A.P. Beardmore[f], S. Campana[d], M. Capalbi[g], J.A. Kennea[e], J.P. Osborne[f], C. Pagani[d,e], G. Tagliaferri[d], G. Chincarini[d], N. Gehrels[a], A. Wells[e,f], J.A. Nousek[e]

[a]*NASA/Goddard Space Flight Center, Greenbelt, MD 20771, USA*
[b]*Universities Space Research Association, Columbia, MD, 21044, USA*
[c]*Department of Physics and Astronomy, The Johns Hopkins University, Baltimore, MD 21218, USA*
[d]*INAF - Osservatorio Astronomico di Brera, Merate, Italy*
[e]*Department of Astronomy & Astrophysics, Penn State University, USA*
[f]*Physics & Astronomy Department, University of Leicester, UK*
[g]*ASI Science Data Center, 00044 Frascati, Italy*

Abstract. The *Swift* X-ray Telescope autonomously refines the Burst Alert Telescope positions (~1-4' uncertainty) to better than 5 arcsec, within 5 seconds of target acquisition by the observatory for typical bursts. The results of the rapid positioning capability of the XRT are presented here for both known sources and newly discovered GRBs, demonstrating the ability to automatically utilise one of two integration times according to the burst brightness, and to correct the position for alignment offsets caused by the fast pointing performance and variable thermal environment of the satellite as measured by the Telescope Alignment Monitor. We present an evaluation of the position accuracy for both the onboard centroiding software and the ground software for the calibration targets and show that a significant improvement in position accuracy is obtained if the boresight detector position is optimised relative to the spacecraft pointing. Finally, we present an updated catalogue of *Swift* GRB X-ray positions obtained in Photon Counting Mode using the improved, calibrated boresight.

Keywords: Gamma-ray Bursts, X-rays, CCDs, Centroiding.

INTRODUCTION

The *Swift* X-ray position accuracy requirement is 5'' within 5 seconds of the observatory settling on the GRB. This requirement applies to the initial observation of the GRB by XRT in Image mode prior to switching into an observation mode according to the source countrate (Hill et al. 2005a). When the source has faded to less than ~1 mCrab, the XRT switches into Photon Counting (PC) mode. These data are analysed using the ground software to determine a refined position based on extensive follow-up and therefore better counting statistics. During the first 12 months on orbit, calibration and evaluation of the XRT position accuracy has provided improvements to the default parameters used for the position determination on-board and on the ground. This paper reports the position determination results to date.

Flight Software Position Accuracy

Observations of X-ray calibration targets were performed in Image mode to calibrate and optimise the XRT Flight Software (FSW) centroiding algorithm. Analysis of these data led to updates to the onboard centroiding parameters, which were then verified with observations of additional fainter calibration targets. Figure 1 (left) shows the accuracy of the XRT onboard centroiding algorithm for a variety of source positions and brightness's. Figure 1 (right) shows how the centroiding accuracy depends on the source counts and integration time, and the flux at which the source is bright enough for the flight software to use a shorter integration time of 0.1 s rather than 2.5 s. At the switch point between the long and short integration time, the total number of counts in the exposure is decreased and therefore the accuracy is decreased. As the flux is increased for the shorter exposure, the accuracy improves again.

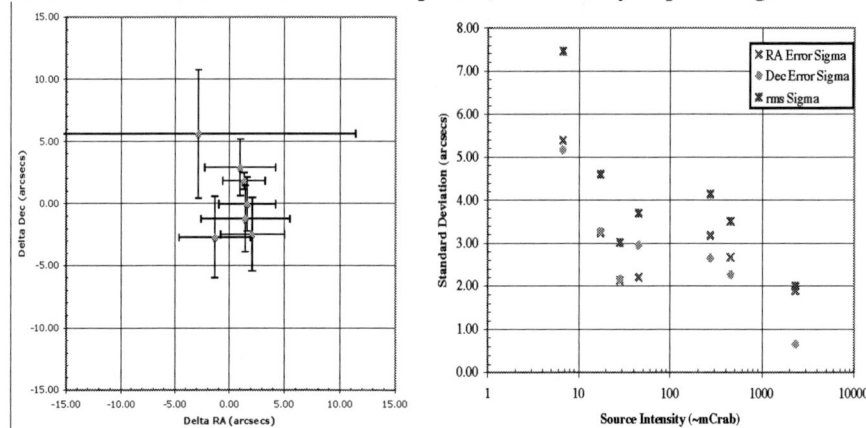

FIGURE 1. Left: XRT flight software position off-sets compared to the SIMBAD catalogue positions. Right: XRT flight software position off-sets versus source countrate, where ~0.7 cnts/sec/mCrab

Between launch (20 November 2004) and 17 October 2005, the Burst Alert Telescope (BAT) has detected 84 bursts. 70 of the bursts were observed by the XRT, 68 of which were detected at T<200 ks. Four of the bursts were observed by the XRT during the prompt phase while the burst was still in progress (e.g. Hill et al. 2005b). The narrow field instruments followed up 70% of the bursts in less than 350 seconds after the BAT trigger and the XRT provided prompt positions (<5 sec) in 41% of those cases. In the remaining cases the source was too faint or the XRT was in a calibration mode.

Following the initial boresight calibration between the XRT using Image mode and the Star Trackers, as an increasing number of roll angles were sampled, it became apparent that for some observations the positions obtained on the ground and onboard were less accurate and an update to the default boresight was necessary.

Image Mode Boresight Analysis

An *idl* script was written to recalculate source positions using different boresights, reproducing the FSW calculation. The script reads in the data from the header of the telemetered postage stamp message obtained in Image mode and regenerates the centroid position in RA and DEC with or without the Telescope Alignment Monitor (TAM) correction (Hill et al. 2005b). The script positions were verified against the onboard RA and DEC positions for the default boresight position (300 x 300) and found to be identical. The script was run for a matrix of boresight positions for each observation of 3 calibration sources. A position off-set was obtained for each observation by comparing the script position with the known optical position. An average off-set was calculated for each boresight position. From fitting a polynomial to the distribution of off-sets versus boresight position, a best-fit boresight position was obtained. This best-fit boresight was then verified against Image mode data of GRBs with optical counterparts. For the GRBs and the three calibration sources, the derived boresight was constant within the counting statistics. The largest variation from test-case to test-case was in the XRT y-axis. This analysis was performed for both TAM corrected and non-TAM corrected positions.

Photon Counting Mode Boresight Analysis

The best boresight position obtained for Image mode was verified against PC mode data where, due to the longer observing times, the counting statistics are, in general, much improved. 11 calibrations sources and 36 observations were selected with a distribution of roll angles. To limit the number of variables, this analysis was performed for data which were not corrected by the TAM.

Nine boresight positions were tested in a matrix around the best-fit boresight obtained from the Image mode data (e.g. Figure 2, left). As in the Image mode analysis the average off-sets for each boresight position were plotted against the boresight y- and boresight x-position. From fitting a curve to the data, the boresight was further refined to 298.2 x 299.3. This was confirmed as the best-fit position by reprocessing the observations with the refined boresight (Figure 2, right).

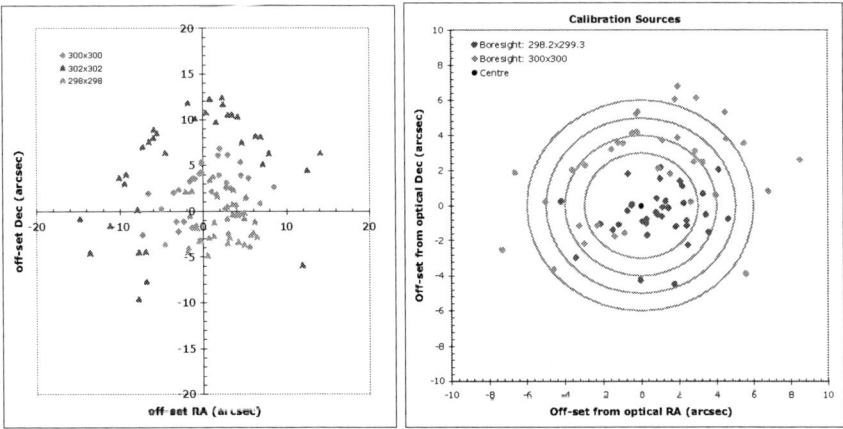

FIGURE 2. Left: Variation of position off-sets for 36 observations of 11 sources with three boresight positions. Default at 300x300, 298x298 and 302x302. Right: A comparison of calibration source position off-sets from optical positions for the default boresight and for the corrected boresight.

The mean off-set for the default boresight decreased from 4.5'' to 2.1'' for the new boresight. Additional verification was performed by plotting the off-set in RA and Dec against the roll angle of the observations and verifying that there was no accuracy dependence on roll angle. Following the boresight optimisation, an additional 12 sources were analysed over 35 observations with the new boresight. The average off-set from the optical positions was found to be 2.2'' as seen in the previous sample, with 97% of the positions less than 5'' from the optical position.

Gamma-ray Burst Refined Positions

A catalogue of 37 GRBs with optical counterparts observed between December 2004 and October 2005 were analysed with the default boresight (Figure 3, left) and the best-fit boresight position of 298.2 x 299.3 (Figure 3, right). The average off-set from the optical counterpart was 2.3'', the same as that obtained for the calibration sources. A significant improvement in accuracy can be seen in Figure 3. All 64 GRBs detected by XRT between December and October have been reprocessed with the new boresight and the updated positions are within the 90% error circle of the positions in Moretti et al. 2005.

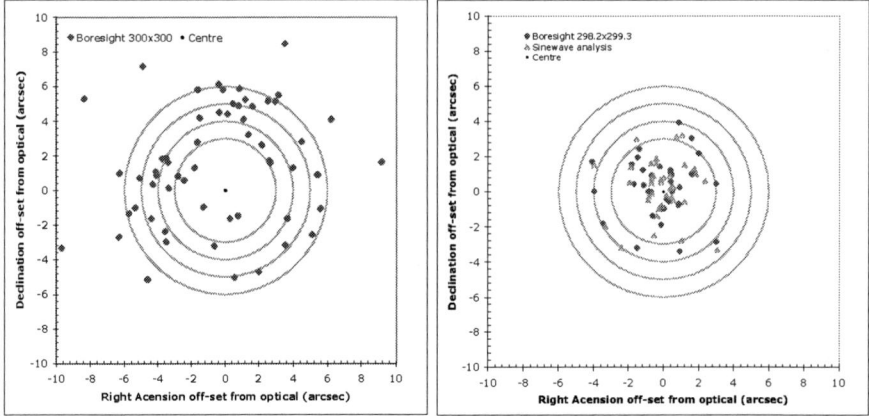

FIGURE 3. Left: XRT ground processed GRB positions for the default 300 x 300 boresight. Right: A GRB position off-sets from optical positions for a sinewave corrected fit and using a corrected boresight.

ACKNOWLEDGEMENTS

This work is supported at Penn State by NASA contract NAS5-00136; at the University of Leicester by the Particle Physics and Astronomy Research Council on grant number PPA/Z/S/2003/00507; and at OAB by funding from ASI on grant number I/R/039/04.

REFERENCES

Hill, J.E. et al., "The unique operating Modes of the Swift X-ray Telescope", 2005, Proc. SPIE, 5898, 589815-1
Hill, J.E, et al., The Astrophysical Journal, 2006, in press
Moretti et al, Astronomy & Astrophysics, 2005, submitted

Early Afterglow and Variability

Kunihito Ioka*, Shiho Kobayashi[†], Bing Zhang**, Toma Kenji*, Ryo Yamazaki[‡] and Takashi Nakamura*

*Department of Physics, Kyoto University, Kyoto 606-8502, Japan
[†]Astrophysics Research Institute, Liverpool John Moores University
**Department of Physics, University of Nevada, Las Vegas, NV 89154
[‡]Department of Physics, Hiroshima University, Higashi-Hiroshima 739-8526, Japan

Abstract. We show that simple kinematic arguments can give limits on variabilities in gamma-ray burst (GRB) afterglows. These limits are violated by X-ray flares in the early afterglows recently identified by the Swift satellite. We discuss that a probable solution is that the central engine continues to eject an intermittent outflow for a very long timescale up to 1 day. This long-lived engine model may also explain the flat decay of early X-ray afterglows, while the gamma-ray efficiency of GRBs should be incredibly high (>75–90%) in this model. We suggest new possible models to evade this efficiency crisis and discuss implications for future observations.

Keywords: gamma rays: bursts — gamma rays: theory — relativity
PACS: 07.85.Fv; 98.70.Rz; 47.75.+f; 52.27.Ny; 52.30.-q

X-RAY FLARES ARE NOT AFTERGLOW VARIABILITIES

Recently the *Swift* satellite has allowed us to observe early afterglows of gamma-ray bursts (GRBs) in the first few hours after the burst. Early X-ray afterglows observed by the *Swift* X-Ray Telescope have three kinds of canonical features that are not predicted by the standard model in the pre-*Swift* era [1, 2]. X-ray light curves show (i) an initial very steep decay ($\propto t^{-\alpha_1}$ with $3 \lesssim \alpha_1 \lesssim 5$) followed by (ii) a very shallow decay ($\propto t^{-\alpha_2}$ with $0.2 \lesssim \alpha_2 \lesssim 0.8$) that connects to the conventional late afterglow, while about half of the afterglows have (iii) strong, rapid X-ray flares minutes to days after the burst.

The steep decay component is most likely the tail emission of the prompt GRBs and/or of the X-ray flares [3, 4]. Even if the emitting surface stops shinning, we continue to see photons coming from the region at large angles relative to our line-of-sight because the emitting surface has a curvature. Most photons from the large angles are not emitted to our directions because of the relativistic beaming, so that the flux decays steeply. Since the emission region moves outward on the surface, the tail emission features, e.g., the decay index and smoothness, would diagnose the unknown GRB jet structure [4].

The X-ray flares are considered to be produced by the long activity of the central engine up to the time of the flares [5, 6]. This is because an afterglow cannot make a variability with a large amplitude and a short timescale by itself, i.e., such as by the ambient density fluctuations and the inhomogeneous emitting surface, as concluded by the kinematic arguments [6]. Under some standard assumptions we can derive the following limits for dips (bumps) that deviate below (above) the baseline of the afterglow

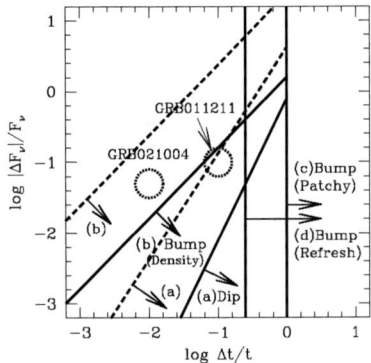

FIGURE 1. Kinematically allowed regions for afterglow variabilities are shown in the plane of the relative variability timescale $\Delta t/t$ and relative variability flux $\Delta F_\nu/F_\nu$. We have four limits: (a) for dips, (b) for bumps due to density fluctuations, (c) $\Delta t \geq t$ for bumps due to patchy shells and (d) $\Delta t \geq t/4$ for bumps due to refreshed shocks. When many regions fluctuate simultaneously, the limits (a) and (b) are replaced by dashed lines. We assume $F/\nu F_\nu \sim 1$ and $f_c \sim 1/2$. X-ray flares violate these limits.

flux F_ν with a timescale Δt and amplitude ΔF_ν:

(a) No dips in light curves can have a larger amplitude than the limit given by $|\Delta F_\nu|/F_\nu \leq (4/5)(\Delta t/t)^2$.
(b) Ambient density fluctuations cannot make a bump larger than $\Delta F_\nu/F_\nu \leq (4/5)f_c^{-1}(F/\nu F_\nu)(\Delta t/t)$ where the fraction of cooling energy is $f_c \sim (\nu_m/\nu_c)^{(p-2)/2} \sim 0.5$ and the ratio $F/\nu F_\nu \sim (\nu/\nu_c)^{(p-2)/2}$ is ~ 1 for typical standard afterglows.
(c) Patchy shells cannot make a bump with a timescale shorter than the observed time $\Delta t < t$, although the rising time $\Delta t_{\rm rise} < t$ is allowed.
(d) Refreshed shocks cannot make a bump with $\Delta t < t/4$.

These limits are summarized in Figure 1. When many regions fluctuate simultaneously, the limits (a) and (b) are modified and are given by $|\Delta F_\nu|/F_\nu \leq (6/\sqrt{2})(\Delta t/t)^{3/2}$ and $\Delta F_\nu/F_\nu \leq 12 f_c^{-1}(F/\nu F_\nu)(\Delta t/t)$, respectively. The observed variabilities of X-ray flares actually violate all the above limits (a)-(d). This suggests that the central engine is still active at the observed time ($\gtrsim 1$ day).

SHALLOW X-RAY AFTERGLOW AND EFFICIENCY CRISIS

So far two kinds of models are proposed for the shallow X-ray afterglows. One class of the models is the energy injection model [1, 2], in which continuous energy is injected into the afterglow so that the flux decay becomes slower than the usual $\propto t^{-1}$. The injection may be caused by (a) the long-lived central engine or (b) the short-lived central engine ejecting shells with some ranges of Lorentz factors. The other class is (c) the inhomogeneous jet model [8, 9]. In this model, early afterglows are not bright because the jet surface on the line-of-sight is dim and the surrounding off-axis region with ordinary brightness is observed later.

However, in all models, the shallow X-ray afterglows pose a serious problem, demanding an unreasonably high gamma-ray efficiency of the prompt GRBs, $\varepsilon_\gamma \equiv E_\gamma/(E_\gamma + E_k) \gtrsim 75\text{–}90\%$, where E_γ is the radiated prompt energy and E_k is the kinetic energy of the afterglow remained after the burst [7]. Even before the *Swift* era, one considers that the gamma-ray efficiency of the prompt GRBs is relatively high, i.e., $\varepsilon_\gamma \sim 50\%$ or more, and develops internal shock models that can manage to produce such a high efficiency. Since the required efficiency is further increased, we have a strong theoretical motivation to suspect the current models. In the following we suggest two more kinds of possible models for shallow X-ray afterglows without invoking an unreasonably high gamma-ray efficiency.

PRIOR ACTIVITY MODEL

One possible model is that a prior activity before the main burst changes both the density and velocity of the ambient matter. It is not unreasonable to consider such a prior activity because the X-ray flares suggest that the engine activity lasts very long after the burst, i.e., why not before the burst? Actually a sizable fraction of GRBs may have precursor activities. Note that a model that only changes the ambient density does not work. It is essential that the ambient density has a relativistic velocity.

Let us consider the following simple model. We assume that a explosion occurs at $t = -t_p \sim -10^4$ s (where we set $t = 0$ as the burst trigger) and mass

$$M(<\gamma_p) \propto \gamma_p^\alpha \tag{1}$$

with Lorentz factors less than γ_p is ejected, where we assume $\alpha > 0$ and $\gamma_p < \gamma_{max} \sim 30$. The energy associated with that mass is $E(<\gamma_p) = \gamma_p M c^2 \propto \gamma_p^{\alpha+1}$. Since $\alpha > 0$, almost all energy is concentrated near γ_{max}. We also assume a prior explosion is weaker than the main burst, $E(<\gamma_{max}) \equiv E_p \sim 10^{52}\text{erg} < E_\gamma \sim 10^{53}$ erg. The ejected mass sweeps the ambient density making an external shock. Since the explosion is weak, its afterglow is not so bright. Note that the Blandford & McKee [10] solution has the mass profile $M(<\gamma_p) \propto \gamma_p^{3/2}$ near the shock front and the index α is larger far from the shock.

We assume that the ejecta of the prompt burst at $t = 0$ is faster than the prior ejecta, i.e., $\gamma > \gamma_{max}$. Before catching up with the external shock, the burst ejecta will collide with the slower ejecta at a radius $R_p \sim c t_p \gamma_p^2$. Initially the reverse shock is Newtonian. Then the burst ejecta is not decelerated $\gamma \sim$ const. The internal energy is mainly released in the forward shock. We can show that the bolometric kinetic luminosity is given by [7]

$$L \propto t^{(\alpha-3)/2}. \tag{2}$$

Therefore, assuming that the X-ray luminosity is proportional to the bolometric kinetic one, we can explain the shallow decay if $\alpha \sim 1.5\text{–}2.5$. The shallow phase ends at around $t_s \sim 10^3\text{–}10^4$ s when the burst ejecta begins to be decelerated by the prior ejecta [7]. Since the prior explosion has less energy than the prompt burst, the final afterglow energy is comparable to the initial afterglow energy E_k and hence we have no efficiency crisis. This model predicts a precursor from the external shock due to the prior explosion. It is also predicted that the reverse shock emission from the burst ejecta is suppressed because it takes longer time for the reverse shock to become relativistic than usually considered.

TIME-DEPENDENT MICROPHYSICS MODEL

The other possibility to obtain the shallow X-ray afterglow without the efficiency crisis is to vary the microphysical constants, such as the energy fraction that goes into electrons ε_e and magnetic fields ε_B, during the observations [7, 11]. Even if the burst ejecta is decelerated and the internal energy is released, most internal energy is initially carried by protons. Without transferring the proton energy into electrons and magnetic fields, little radiation is emitted since protons are inefficient emitters.

If the index of the power-law electron distribution p is about $p \sim 2$ as usual and fast cooling electrons emit X-rays, the X-ray luminosity L_X is given by the bolometric kinetic luminosity L as $L_X \sim \varepsilon_e L$, and does not depend on the magnetic energy fraction ε_B so much ($L_X \propto \varepsilon_B^{(p-2)/4}$). Since $L \propto t^{-1}$, the shallow X-ray light curve $L_X \propto t^{-1/2}$ suggests that the electron energy fraction evolves as

$$\varepsilon_e \propto t^{1/2}, \tag{3}$$

which is saturated at the equipartition value $\varepsilon_e \sim 0.1$–1 when the shallow phase ends. Note that the initial value of ε_e at $t \sim 1$–100 s is still larger than the minimum energy fraction $\varepsilon_{e,\min} = m_e/m_p \sim 10^{-3}$. A multi-wavelength observations are useful to test this model. It is also interesting to study the time-dependent microphysics model in the reverse shock, which may explain the dim optical flashes from the reverse shock.

ACKNOWLEDGMENTS

This work is supported in part by Monbukagakusho Grant-in-Aid for Scientific Research 14047212 and 14204024.

REFERENCES

1. Nousek, J. A., et al. 2005, astro-ph/0508332
2. Zhang, B., et al. 2005, astro-ph/0508321
3. Kumar, P., & Panaitescu, A. 2000, ApJ, 541, L51
4. Yamazaki, R., Toma, K., Ioka, K., & Nakamura, T. 2005, astro-ph/0509159
5. Burrows, D. N., et al. 2005, Science, 309, 1833
6. Ioka, K., Kobayashi, S., & Zhang, B. 2005, ApJ, 631, 429
7. Ioka, K., Toma, K., Yamazaki, R., & Nakamura, T. 2005, astro-ph/0511749
8. Toma, K., Ioka, K., Yamazaki, R., & Nakamura, T. 2005, astro-ph/0511718
9. Eichler, D., & Granot, J. 2005, astro-ph/0509857
10. Blandford, R. D., & McKee, C. F. 1976, Phys. Fluids, 19, 1130
11. Ioka, K. 2005, Prog. Theor. Phys., 114, 1317

Searching for early optical transients of gamma-ray bursts with TAROT. Technical status

A. Klotz*,†, M. Boer†, G. Buchholtz**, Y. Damerdji*,†, J. Eysseric†, C. Pollas‡ and Y. Richaud†

*Centre d'Etude Spatiale des Rayonnements, Observatoire Midi-Pyrénées, Université Paul Sabatier, BP 4346, 31028, Toulouse Cedex 04, France
†Observatoire de Haute-Provence, 04870 Saint Michel l'Observatoire, France
**Division technique de l'INSU, 1 place Aristide Briand, 92195 Meudon cedex, France
‡Observatoire de la Calern, 2130 Route de l'Observatoire, Caussols, F-06460 St. Vallier de Thiey, France

Abstract. TAROT (Télescope á Action Rapide pour les Objets Transitoires = Rapid Action Telescope for Transient Objects) is a robotic observatory dedicated to observe the early optical transients of gamma-ray bursts. Since its first light in 1998, many improvements occurred. We describe the present technical status of TAROT that leads to the detection success of optical transients of gamma-ray bursts in 2005.

Keywords: Instrumentation, optical photometry, gamma-ray bursts
PACS: 98.70.Rz, 95.55.Qf

INTRODUCTION

TAROT is a fully robotized observatory linked to the GCN notices. Initially designed for the large error boxes provided by the BATSE instrument, its field of view is near $2°$. The 25 cm telescope $f/3.4$ is fit on a robust equatorial fork mount to slew up to $60°/s$. The telescope typically begins the first exposure less than 8 seconds after the GCN notice was send. TAROT observed 9 of the *Swift*'s GRBs in one year (see the Boer et al. paper in this issue).

Since 2003, we have used the time out GRB observations to many astrophysical programs: Blashko effects of RR Lyr stars [4], asteroid photometry [1] photometric follow up of cataclysmic stars [5], astrometry of satellites, etc. Moreover, every image is analyzed by Sextractor [2] and a huge catalogue of stars is updated each night. This photometric catalogue will be on-line on the next months in the context of virtual observatories.

RECENT TECHNICAL UPDATES

In 2005, we rewrote completely the observatory control software and the scheduler. Now the telescope status can be checked through a web page and the GRB observation strategy is optimized. We developed a software dedicated to manage the source catalogue

 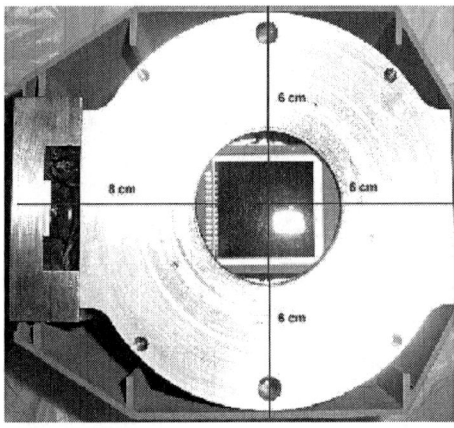

FIGURE 1. Left: The TAROT telescope located at the Calern Observatory. Right: The CCD camera *AndorMarconi*4240 has 2048 x 2048 pixels. The spatial sampling is 3.3 arcsec/pixel and the field of view 1.8°x1.8°. The readout time is 5 secondes and the readout noise is 8 electrons.

and to the variable sources identification. We currently work on the automatic GRB afterglow identification. A large part of the TAROT softwares are based on AudeLA [1], a suite of Tcl/Tk scripts extended by some C librairies.

FUTURE TECHNICAL UPDATES

In 2006, the scheduler will be upgraded to react in real time to new constraints. The goal is to optimize the exposure times from relevant informations provided by the processing software. For example, the strategy will not be the same when a bright afterglow is detected and when no new source is detected on the first exposures.

In late 2006, a clone of the TAROT Calern will be installed in Chile to cover a larger range of longitudes [3].

REFERENCES

1. R. Behrend, L. Bernasconi, R. Roy et al. 2006 A&A in press
2. E. Bertin, S. Arnouts, 1996, A&AS 117, 393–404
3. M. Boer, A. Klotz, J. L. Atteia et al. 2003 The Messenger 113, 45–48
4. J. F. Le Borgne, A. Klotz, M. Boer, 2005, Information Bulletin on Variable Stars, 5622, 1
5. H. Yamaoka, K. Itagaki, A. Klotz, C. Pollas, M. Boer 2004 IAU Circ. 8413, 2

---

[1] http://audela.ccdaude.com

FIGURE 2. Left: TAROT CALERN : long = 6.92389° E lat=43.75222° N alt=1270 m In use since 1998. Right: TAROT CHILE : long = 70.75° W lat=29.25° S alt=2347 m Will be installed in 2007.

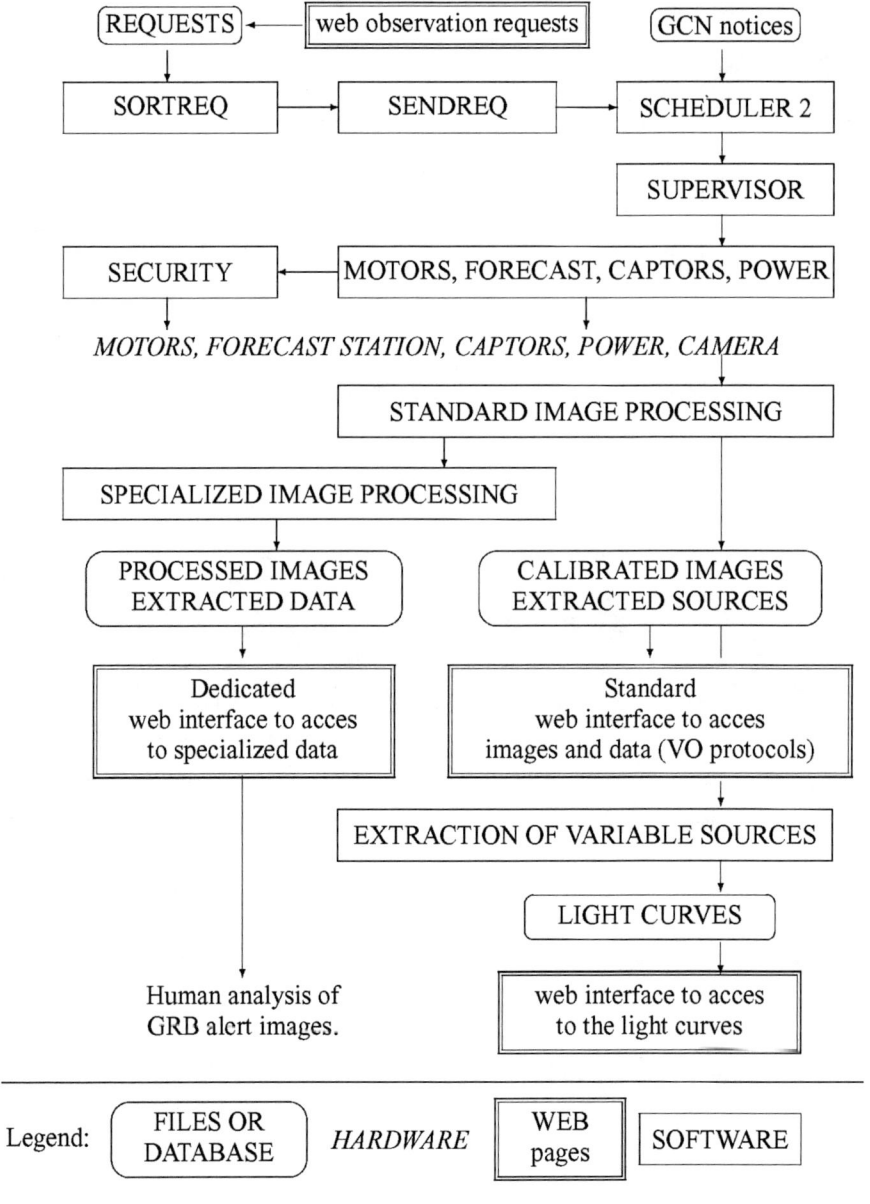

FIGURE 3. Flow diagram of TAROT components.

Rapid Identification of GRB Afterglows with Swift/UVOT

F. E. Marshall and the Swift Team

NASA Goddard Space Flight Center, Greenbelt, Maryland, 20771, USA

Abstract. As part of the automated response to a new GRB, the UVOT instrument on Swift starts a 200-second exposure with the V filter within ~ 100 seconds of the BAT burst trigger. The instrument searches for sources in a $8' \times 8'$ region, and sends the list of sources and a $160" \times 160"$ sub-image centered on the burst position to the ground via TDRSS. These raw products and additional products calculated on the ground are then distributed through the GCN within a few minutes of the trigger. We describe the sensitivity of these data for detecting afterglows, summarize current results, and outline plans for rapidly distributing future detections.

Keywords: gamma-ray bursts, gamma-ray telescopes
PACS: 98.70.Rz,95.55.Ka

INTRODUCTION

A major goal of the Swift mission is to enable ground-based observations of the afterglows of gamma-ray bursts (GRBs) by rapidly providing accurate positions to the community. In this paper we discuss the relevant capabilities of the Ultraviolet and Optical Telescope (UVOT), one of the three instruments of the Swift Observatory. UVOT is a 30-cm. telescope that is sensitive to photons in the wavelength range of 170 to 600 nm (Roming et al. [1]). Spectral information is available using a filter wheel which includes UV and optical grisms and 6 broad band filters.

Swift typically responds to a new GRB in the following fashion. After BAT determines the position of the GRB, the spacecraft maneuvers to point the co-aligned X-Ray Telescope (XRT) and UVOT at the GRB. After the maneuver ends (typically 1 to 2 minutes after the BAT trigger), UVOT takes a "finding chart" exposure lasting 200 seconds using the V filter. The response of UVOT can be configured. On October 3, 2005, the exposure was increased from 100 seconds to the current 200 seconds to increase the sensitivity. Early in 2006 the response will be changed again so that multiple finding chart exposures will be taken. The data from the finding chart exposure are then processed by the instrument and the results rapidly sent to the ground using NASA's Tracking and Data Relay Satellite System (TDRSS). The results and the additional processing on the ground are described in the following sections. The complete data from the initial exposure including event-by-event data from the full $17' \times 17'$ UVOT field-of-view are eventually sent to the ground using the Malindi Ground Station, but these data are not the subject of this paper.

UVOT DATA PRODUCTS

The on-board processing is designed to reduce the volume of telemetry so that it will fit into the limited available bandwidth while still providing the most important information about an afterglow. UVOT sends down two major products. The first is a list of detected sources with counts from a few surrounding pixels (the Source List data). The second is an image that is a subset of the full image.

The Source List data are produced by searching the $8' \times 8'$ image from the finding chart exposure for sources. For each of a maximum of 190 sources the location and counts of the central pixel and the counts in each of a small number of surrounding pixels is sent to the ground. The total number of pixels for a source can be either 1, 5, or 21. If a small number of sources is detected, then information is available for all the sources and for 21 pixels for each source. If a large number of sources is detected, not all of them will be sent to the ground and fewer pixels will be reported for some of the sources. The on-board algorithm is designed to send the brighter sources and to use more pixels for the brighter sources.

The Source List telemetry are processed automatically when received at GSFC to produce four files in standard formats. One (uvot_raw_srclist.fits) converts the data into FITS format with little additional processing. Another (uvot_sky_srclist.fits) maps the detector positions for the sources onto the sky to produce (a very sparse) sky image. A postscript file (uvot_field_image.ps) is also produced. It plots the source positions on a Digital Sky Survey (DSS) image. Fig. 1 shows an example of this file. The final file (uvot_catalog_srclist.fits) is of primary interest for this paper. An improved aspect solution is attempted by comparing the list of sources to the USNO-B.1 catalog of optical sources (Monet et al. [2]). Typically the improved aspect solution is offset by $\sim 2"$ from the original solution. If the attempt is not successful, then this file is not produced. When produced, the file contains the list of sources, their corrected sky positions, a flag indicating whether the source has been matched with a object in the ground catalog, the identification of the matched object, and an estimate of the V magnitude of the source based on the counts in the nearby Source List pixels. The list of sources is sorted by distance from the position of the afterglow determined with XRT if available; otherwise, it is sorted by distance from the GRB position determined with BAT. All these files are distributed to the community as soon as they are produced.

There are a few problems with the Source List data that need to be accommodated in their analysis. The most important is that spurious sources are found by the on-board source detection algorithm near bright ($V < 12.0$) sources. Tens of such spurious sources can be produced for very bright sources ($V < 8.0$). These sources are not flagged in the distributed data files.

The estimation of the magnitude for the sources is hampered by the limited number of pixels available. Even when 21 pixels are available, only part of the source image is available. Consequently it is necessary to correct for the "missing" counts before converting the source count rate to V magnitude. The correction factor was determined by comparing the counts reported in Source List data to the actual source counts determined after examining the complete finding chart image. Fig. 2 shows the results of this comparison for several finding chart images. The best-fit linear relationship between Source List counts and total counts has been used to compute the V magnitude. This slightly

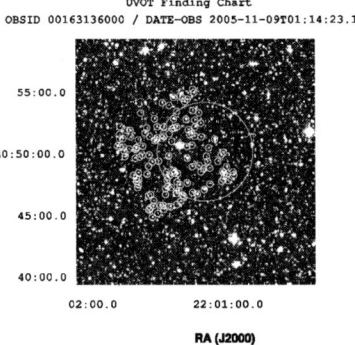

FIGURE 1. Source List positions (small circles) overlaid on the Digital Sky Survey image of the region near GRB 051109A. The distribution of the positions indicate the projection of the 8'x8' finding chart image on the sky. The large circle shows the initial BAT error circle.

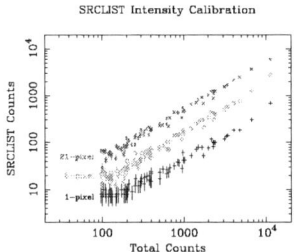

FIGURE 2. Source List counts vs. total counts for the 3 types of data.

overestimates the brightness when the total counts are less than ~ 150 (~ 18.1 V magnitude for a 200 second exposure).

The position given for an entry in the Source List is simply taken as the position of the pixel with the most counts. Consequently the accuracy of the Source List positions is significantly worse than the accuracy of positions obtained after the complete finding chart data are available on the ground. Fig. 3 shows the distribution of position errors for a Source List entries in a typical finding chart. 90% of the positions are within 1.3" of the correct position.

In addition to the Source List data, a small sub-image is created for each finding chart exposure and sent to the ground. The image is binned 2×2 (producing $1" \times 1"$ pixels) and covers $160" \times 160"$. Nominally the image is centered on the XRT position for the afterglow if it is available; otherwise the image is centered on the BAT position. The image is large enough to easily cover all of of a typical XRT error circle, but only covers $\sim 25\%$ of a typical BAT error circle.

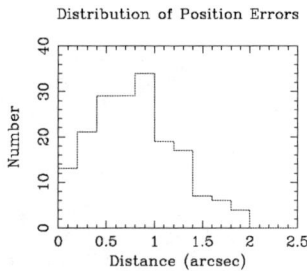

FIGURE 3. Observed distribution of position errors for Source List sources for a typical finding chart image.

Ground processing for the sub-image is similar to that for the Source List data except that a source detection algorithm is run on the ground instead of on the instrument. The same type of files are created with "image" replacing "srclist" in the names of the files. The sensitivity and positional accuracy that can be obtained is very similar to that for standard UVOT images.

RESULTS AND FUTURE PROSPECTS

Source List data products have been produced for most of the GRBs detected with Swift. Afterglows can be detected if they are brighter than ~ 18.0 magnitude in V. Magnitudes can be determined to ~ 0.5 (1σ), and positions determined to ~ 1.3". Through November, 2005, GRB afterglows are seen in Source List data for 7 GRBs (GRB050525A, GRB050712, GRB050726, GRB050730, GRB050802, GRB050922C, and GRB051109A). The optical afterglow for GRB050730 was discovered in the sub-image data (Holland et al. [3]).

The automated UVOT response will soon be changed to produce two finding charts using both the V and B filters. Changes are also being made to increase the fraction of the time that the sub-image covers the XRT error circle. Finally ground software is being improved so that afterglows can be detected automatically and reported to the community within ~ 10 minutes of the burst trigger.

REFERENCES

1. Roming, P. W. A. et al., *Space Sci. Rev.*, **120**, 95 (2005).
2. Monet, B. et al., *Astronomical Journal*, **125**, 984 (2003).
3. Holland, S. T. et al. GCN Circular 3704 (2005).

Can GRB early X-ray afterglows be explained by a contribution from the reverse shock?

F. Genet, F. Daigne and R. Mochkovitch

Institut d'Astrophysique de Paris
UMR 7095 CNRS – Université Pierre et Marie Curie-Paris VI, 98 Bd Arago, 75014 Paris, France

Abstract. We propose to explain the recent observations of early X-ray afterglows by SWIFT from the dissipation of energy in the reverse shock sweeping back into the ejecta. We compute the evolution of the dissipated power and discuss the possibility that it can be radiated in the X-ray range.

Keywords: Gamma-ray sources; gamma-ray bursts, Shock waves
PACS: 98.70.Rz, 96.50.Fm

INTRODUCTION

The X-Ray Telescope (XRT) on board the SWIFT satellite has for the first time allowed a follow-up of the X-ray afterglows of GRBs starting within one minute of the BAT trigger [?]. These early afterglow observations have revealed several surprising features which cannot be easily understood in terms of the usual interpretation where the afterglow comes from a forward shock in the Blandford-McKee regime. At very early times immediately after the prompt phase, the X-ray light curve exhibits a steep slope $\alpha \gtrsim 3$ (where $F(t) \propto t^{-\alpha}$) followed by a shallow part ($0.2 < \alpha < 0.8$) which can last for several hours until a more standard slope $1 < \alpha < 1.5$ is finally observed [?]. Moreover flares often appears superimposed to this typical light curve.

We propose an alternative to the standard view where the X-ray afterglow is explained by a contribution from the reverse shock. We develop a simplified model to follow simultaneously the internal, reverse and forward shocks which develop during the burst evolution. We compute the energy dissipated in the reverse shock and show that, for an initial distribution of the Lorentz factor in the flow which seems quite reasonable, its temporal evolution is very similar to the observed early X-ray afterglows. Since the reverse shock contribution is normally expected in the visible, we discuss under which conditions it can rather be radiated in X-rays.

INTERNAL, REVERSE AND FORWARD SHOCKS DURING BURST EVOLUTION

In the context of the internal shock model for the prompt emission, the pulses observed in GRB light curves are formed when fast moving material catches up with slower one which was previously emitted by the central source [?]. The shape of the pulses is initially dominated by hydrodynamical effects [?] while geometry (the curvature effect) becomes important at late times and leads to a temporal slope close to −3 in the absence of any other contribution (naked afterglow, [?]). But as the burst environment interacts with the ejected material, the resulting forward shock propagating through the circumstellar medium and the reverse shock sweeping back into the ejecta produce the long lasting afterglow emission.

We have obtained the energy dissipated in the reverse shock by generalizing the simple shell model we have developed for internal shocks [?]. This was done by considering the contact discontinuity which separates the ejecta and the shocked external medium. In our simple description it is represented by two shells moving at the same Lorentz factor. The first one corresponds to the ejecta already crossed by the reverse shock and the second to the shocked external medium. Two processes affect this two shell structure at the contact discontinuity : it collides either with shells of the external medium at rest, or with rapid shells from the relativistic ejecta. This represents both the forward and reverse shocks in our simplified scheme.

THE DISSIPATED POWER

The power dissipated in the reverse shock was computed for an initial distribution of the Lorentz factor of the form

$$\begin{aligned}\Gamma(t) &= 125 - 75\cos\left(\pi t/2\right) & \text{if } t < 2\,\text{s} \\ \Gamma(t) &= 200 & \text{if } 2 < t < 5\,\text{s} \\ \Gamma(t) &= 101 + 99\cos\left[\pi(t-5)/5\right] & \text{if } 5 < t < 10\,\text{s}\end{aligned} \quad (1)$$

corresponding to a relativistic flow emitted by the source for a total duration of 10 s. With such a profile a rapid part with $\Gamma = 200$ first catches up with a slower one ($\Gamma = 50$) producing a single pulse burst. During the late stages of source activity ($t > 5$ s) the Lorentz factor of the ejected material decreases to a very small value ($\Gamma = 2$). With this slow tail it takes several days for the reverse shock to cross the ejecta rather than a few tens of seconds when it is absent.

We have considered both a wind and uniform medium for the source environment. In the wind case, we tried three values of the parameter A_*: 1, 0.1 and 0.01 (such as $\rho(r) = 5\,10^{11} A_*/r^2$). In the constant density case, we also tried three values of n: 1000, 100 and 10 cm^{-3}. The resulting profiles for the dissipated energy are shown in Fig.1. They show a striking resemblance with the early X-ray afterglows observed by SWIFT. The reverse shock component smoothly connects to the tail of the burst profile. At late times the decline follows a constant slope $\alpha \sim 1$ - 1.5. The shallow intermediate region lasts longer and is less luminous in the wind case than in the constant density case. It is most sensitive to the density of the burst environment. The lower the density, the shallower is this part of the profile. At high density it essentially disappears, the constant slope $\alpha \sim 1$ - 1.5 following directly the initial steep decrease. Conversely at the lowest densities, this intermediate region can become completely flat and even fall to a temporary minimum. We checked the robustness of these results which do not depend on the details of the Lorentz factor distribution as long as Γ decreases to values of a few at late times.

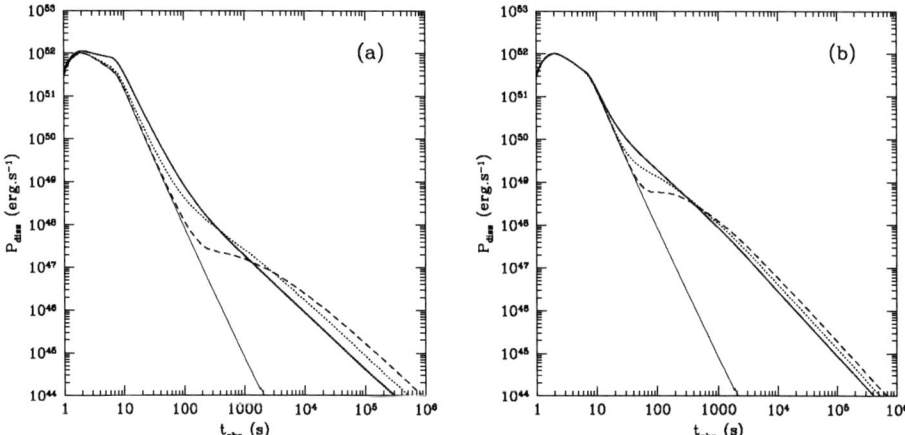

FIGURE 1. Dissipated power as a function of observer time during reverse shock propagation when the Lorentz factor is given by Eq.(1). (a): wind case with $A_* = 1$ (full line), $A_* = 0.1$ (dotted line) and $A_* = 0.01$ (dashed line); (b): uniform density case with $n = 1000$ (full line), 100 (dotted line) and 10 cm^{-3} (dashed line). In each panel the thin full line represents the naked burst.

CAN THE REVERSE SHOCK CONTRIBUTE IN X-RAYS?

Despite their similarity with the SWIFT observations, it must remain clear that the profiles shown in Fig.1 only trace the power dissipated in the reverse shock. With the assumptions ordinary made to compute the reverse shock contribution in GRBs it should manifest itself in the visible range [?]. Moreover most of the emission would take place in the slow cooling regime so that the observed light curve will not follow the instantaneous energy release.

One possibility to boost the reverse shock emission to X-rays and in the fast cooling regime would be to have only a small fraction ($\zeta \sim$ a few 10^{-3}) of the electrons accelerated in the shock. This was already considered by [?] in the context of GRB afterglows. They showed that ζ is not well constrained by the observations and, even if $\zeta \sim 1$ appears slightly favored, they included the whole interval $m_e/m_p < \zeta < 1$ in their discussion. In internal shocks, which are very similar to the reverse shock (both take place in the burst ejecta and are mildly relativistic) a small ζ is favored to insure that the emission takes place in the gamma-ray range as shown by [?] and more recently by [?] in the case of the short hard burst GRB 050509b.

CONCLUSION

We have developed a simplified model which enabled us to follow simultaneously the dynamics of the internal, external and reverse shocks in GRBs. We were mainly interested by dissipation in the reverse shock when the Lorentz factor in the material which is ejected in the final stages of source activity decreases to small values, $\Gamma_f < 10$.

The propagation of the reverse shock then extends over quite a long time needed to decelerate the fast moving part of the ejecta down to $\Gamma \sim \Gamma_f$. We have obtained the dissipated power as a function of observed time for different burst environments (wind or constant density). Its evolution shows a striking resemblance with the early afterglow light curves observed by SWIFT. However the reverse shock contribution is normally expected in the visible and to appear in X-rays it requires a transfer of the dissipated energy to only a small fraction ($\zeta \sim$ a few 10^{-3}) of the electron population. If this is possible, SWIFT XRT observations could be better explained by the reverse shock than by the standard afterglow produced by the forward shock. This would imply that the forward shock contribution generally lies below that of the reverse shock and could indicate an ineffective transfer of energy to electrons ($\varepsilon_e \lesssim 0.01$) at least at the beginning of the forward shock propagation.

Our proposal that a contribution from the reverse shock accounts for the early X-ray afterglows of GRBs relies on two well defined assumptions: the first one on the initial distribution of the Lorentz factor in the flow, the second on the post-shock acceleration of electrons. We believe that the former – a strongly decreasing Γ in material ejected during the late stages of source activity – is quite reasonable while the latter – the transfer of a large part of the shock dissipated energy to a small fraction of the electrons – is more uncertain. If both can be satisfied, the reverse shock rather than the forward shock could be responsible for the X-ray emission of GRBs from a few minutes to several days after the burst.

REFERENCES

- Burrows D.N., Hill, J.E., Nousek, J.A. et al. 2005, Space Sci.Rev. 120, 165.
- Nousek, J.A., Kouveliotou, C., Grupe, D. et al. 2005, astro-ph/0508332.
- Rees, M.J., & Meszaros, P. 1994, ApJ, 430, L93.
- Daigne, F., & Mochkovitch, R. 2003, MNRAS, 342, 587.
- Kumar, P. & Panaitescu, A. 2000, ApJ, 541, 51.
- Daigne, F., & Mochkovitch, R. 1998, MNRAS, 296, 275.
- Sari, R., & Piran, T. 1999, ApJ, 517, L109.
- Eichler, D., Waxman, E. 2005, ApJ, 627, 861.
- Lee, W.H., Ramirez-Ruiz, E. & Granot, J. 2005, ApJ, 630, L1.

The *Swift* X-ray flaring afterglow of GRB 050607

Claudio Pagani[*], David C. Morris[*], Shiho Kobayashi[†], Takanori Sakamoto[**], Abraham D. Falcone[*], Alberto Moretti[‡], Kim Page[§], David N. Burrows[*], Dirk Grupe[*], Judith Racusin[*], Jamie A. Kennea[*], Sergio Campana[‡], Patrizia Romano[‡], Gianpiero Tagliaferri[‡], Jamie A. Kennea[*], Joanne E. Hill[**], Lorella Angelini[**], Scott Barthelmy[**], Guido Chincarini[‡], John A. Nousek[*] and Neil Gehrels[**]

[*]*Department of Astronomy and Astrophysics, Pennsylvania State University, 525 Davey Lab, University Park, PA 16802, USA*
[†]*Astrophysics Research Institute, Liverpool John Moores University, Twelve Quays House, Birkenhead CH41 1LD, UK*
[**]*NASA Goddard Space Flight Center, Greenbelt, MD 20771, USA*
[‡]*INAF – Osservatorio Astronomico di Brera, Via Bianchi 46, 23807 Merate, Italy*
[§]*Department of Physics and Astronomy, University of Leicester, Leicester LE1 7RH, UK*

Abstract. The fast and autonomous response of the *Swift* satellite to newly detected Gamma-Ray Bursts (GRBs) has provided a new view on their X-ray and optical afterglows. We present here the X-ray Telescope observations of GRB 050607. The striking feature of the X-ray emission is the intense flaring activity during the early afterglow phase, indicative of central engine activity extended to several hundred seconds after the burst detection. The flares have very rapid timing variations and asymmetric shapes, similar to the FREDs that are frequently observed in the prompt GRB emission. After the flares, the X-ray lightcurve entered a phase of slow decay during which the forward shocks were probably being refreshed, followed by a late steepening consistent with the standard external shock model. Analysis of the X-ray emission showed spectral variations during the flares, with a harder spectrum at the flares onset and a softer component that lingered longer, as each flare decayed.

Keywords: GRB afterglow, individual (GRB 050607)
PACS: 98.70.Rz, 98.70.Qy

OBSERVATIONS

The Burst Alert Telescope triggered on the GRB 050607 at 09:11:22.81 UT on 2005 June 07. The BAT data showed a 3-peak lightcurve with a T_{90} of 26.5 s, a measured fluence of $5.8 \pm 0.5 \times 10^{-7}$ ergs cm^{-2} in the 15–150 keV band and a spectrum well fitted with a single power law model with index 1.83 ± 0.14.

The X-ray afterglow position, determined from ground-processed data, is α_{2000}=20h00m42.7s, δ_{2000}=+09d08m31.4s, with an uncertainty of 3.8 arcsec. The GRB afterglow was monitored by the XRT for 30 days after the burst for a total exposure time of 363 ks.

Optical detections of the GRB afterglow were obtained with ground-based telescopes in the *I* band with an initial magnitude of 21.46 ± 0.3 and in the *R* band ($22.5 < R < 23$ mag).

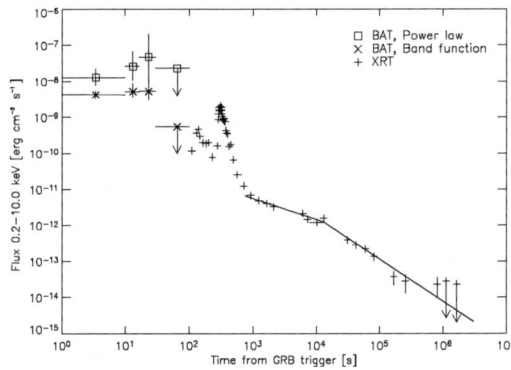

FIGURE 1. The combined BAT and XRT GRB 050607 lightcurve.

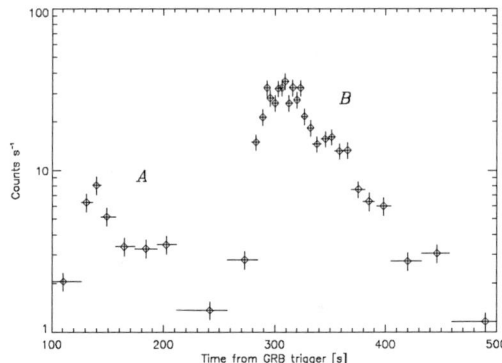

FIGURE 2. The early phase of the X-ray afterglow of GRB 050607 characterized by the flares' emission.

XRT TIMING ANALYSIS

The XRT lightcurve presented a complex behavior (Figure 1), with the early phase dominated by intense flaring emission. This was the fourth case of a GRB discoverd by *Swift* with X-ray flares superimposed on the overall fading afterglow. The second flare was the brighter ($\approx$ 23% of the BAT fluence) with a flux increase by a factor $\approx$ 25.

The flares were followed by a phase of slow decay that lasted approximately 12 ks and a late steepening to values typical of GRB afterglows. The best fit of the late lightcurve decay with a broken power law yields decay indices of 0.58 ± 0.07 and 1.17 ± 0.07.

The flares of GRB 050607 (Figure 2) were not symmetric as in previously reported cases [1, 2], showing a very steep rise and a shallower decay, similar to the FREDs that are frequently observed in the gamma-ray prompt emission. The choice of t_0 is critical in determining the flares' slopes indices. In fact, if t_0 is defined as the GRB trigger time

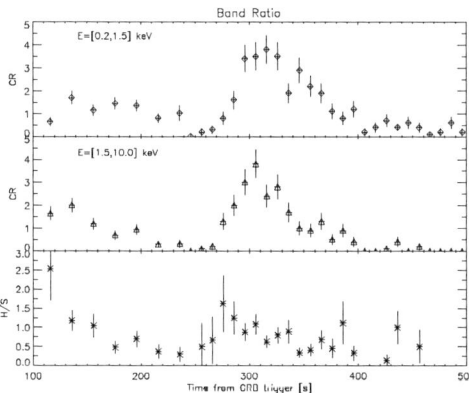

FIGURE 3. The hardness ratio clearly indicates spectral evolution during the flares with a harder component at the flares's onset, while the emission softens as the flares decay.

the slopes are steeper than with the choice of t_0 as the onset of the flares.

Spectral evolution during the flares is evident from the hardness ratio analysis (Figure 3). In particular, the harder component contributed to the overall emission mostly during the rising part of the flares, while the soft emission lingered longer as the flares decayed. The observed behavior indicates that the emission during the flares is harder than the underlying afterglow, and that the flares evolve spectrally, softening as they decay.

XRT SPECTRAL ANALYSIS

A simple absorbed power law model was used to fit the spectrum for the different afterglow phases. C-statistics were used in the fitting procedure during the first flare (flare A) and the late lightcurve, due to the low number of events, while χ^2 statistics were used for the brighter flare (flare B).

The best fit of flare A with an absorbed power law model yielded a photon index of $2.00^{+0.19}_{-0.18}$, with the N_H fixed to the Galactic value. The brighter flare B fit yielded a photon index of $2.27^{+0.13}_{-0.12}$,

The flares were also fitted with a cutoff power law model (equivalent to a Band function in the XRT energy band) to investigate a possible common origin of the X-ray flares and the prompt gamma-ray emission. The fit of flare B with a cutoff power law model is a better fit than an absorbed power law (χ^2_{red} improved from 1.54 to 1.14), yielding a photon index of $1.41^{+0.48}_{-0.53}$, harder than the simple power law, and a cutoff energy of $2.4^{+3.5}_{-0.9}$ keV. Our attempt to also fit the dimmer flare A with a cutoff power law did not properly constrain the fit parameters due to the low number of photons.

The fit to the late part of the X-ray afterglow yielded a photon index of $1.78^{+0.18}_{-0.13}$ for the shallow section of the lightcurve and of $1.97^{+0.36}_{-0.34}$ after the lightcurve break. The spectra of the different afterglow phases are shown in Figure 4.

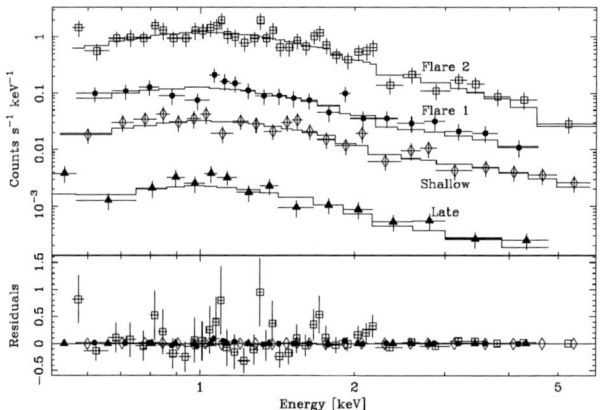

FIGURE 4. Spectral fit of the different XRT afterglow phases.

DISCUSSION

The XRT lightcurve of GRB 050607 presented most of the characteristics that have been observed in afterglows of GRB discovered by *Swift* [3, 4, 5]: early flares, a shallow phase and a late break to a steeper afterglow decay.

The flares of GRB 050607 had rapid time variations, with in particular a very steep rise, sharper than for previously reported cases, important in constraining the origin of the flares: an external shock emission can be excluded since inhomogeneities in the afterglow environment or in the outflow cannot generate the observed short timescale variations [6, 7]. The internal shocks origin of the flares is also supported by the flares asymmetric shape, similar to the FREDs frequently observed in the GRB prompt emission. In addition, the hardness ratio analysis showed spectral variations during the flares, with a harder component at the flares onset and a softer emission as the flares decayed.

Refreshed shocks during the shallow phase are the most likely explanation for the lightcurve behavior [5]. The steepening of the XRT lightcurve from the shallow to normal phase 12 ks after the GRB trigger was observed with no spectral evolution. Applying the standard blast wave model to explain the lightcurve after the late break, an electron index of 2.2 is obtained from the decay and spectral indices.

REFERENCES

1. Falcone, A. D., et al. 2005, ApJ, submitted
2. Romano, P., et al. 2005, A&A, submitted
3. Burrows, D. N., et al. 2005b, Science, 309, 1833
4. Nousek, J.A., et al. 2005, ApJ, accepted, astro-ph/0508332
5. Zhang, B., et al. 2005, ApJ, submitted
6. Lazzati, D., et al. 2002, A&A, 396, 5
7. Ioka, K., Kobayashi, S., & Zhang, B. 2005, ApJ, 631, 429

A Tale of Two Faint Bursts: GRB 050223 and GRB 050911

K.L. Page*, S.D. Barthelmy[†], A.P. Beardmore*, D.N. Burrows**, S. Campana[‡], G. Chincharini[‡], J.R. Cummings[†], G. Cusumano[§], N. Gehrels[†], P. Giommi[¶], M.R. Goad*, O. Godet*, J. Graham[‖], Y. Kaneko[††], J.A. Kennea**, V. Mangano[§], C.B. Markwardt[†], P.T. O'Brien*, J.P Osborne*, D.E. Reichart[‡‡], E. Rol*, T. Sakamoto[†], G. Tagliaferri[‡], N.R. Tanvir[§§], A.A. Wells* and B. Zhang[¶¶]

*Department of Physics and Astronomy, University of Leicester, UK
[†]NASA/GSFC, Greenbelt, MD 20771, USA
**Department of Astronomy & Astrophysics, PSU, University Park, PA 16802, USA
[‡]INAF - Osservatorio Astronomico di Brera, 23807 Merate (LC), Italy
[§]INAF - Istituto di Astrofisica Spaziale e Fisica Cosmica Sezione di Palermo, 90146 Palermo, Italy
[¶]ASI Science Data Center, 00044 Frascati, Italy
[‖]STScI, Baltimore, MD 21218, USA
[††]University of Alabama, NSSTC, Huntsville, AL 35805, USA
[‡‡]Department of Physics and Astronomy, University of North Carolina at Chapel Hill, NC 27599, USA
[§§]Centre for Astrophysics Research, University of Hertfordshire, Hatfield, AL10 9AB, UK
[¶¶]University of Nevada, Las Vegas, NV 89154-4002, USA

Abstract.
GRBs 050223 and 050911 were discovered by the *Swift* Burst Alert Telescope (BAT) on 23rd February and 11th September 2005 respectively. The observation of GRB 050223 showed a faint, fading X-ray source, which was identified as the afterglow; GRB 050911, however, was not detected, making any X-ray afterglow extremely faint. The faintness of the afterglow of GRB 050223 could be explained by a large opening or viewing angle, or by the burst being at high redshift. The non-detection of GRB 050911 may indicate the burst occurred in a low-density environment, or, alternatively, was due to a compact object merger, in spite of the apparent long duration of the burst.

Keywords: gamma-ray bursts
PACS: 98.70.Rz

INTRODUCTION

During its first year of operation *Swift* has triggered on 102 bursts, 87 of which were followed up by the X-ray Telescope (XRT). In almost all cases after a prompt slew, and often even after a substantial delay, an X-ray afterglow has been detected. Thus, *Swift* has significantly increased the number of few-arcsecond localisations of GRBs.

However, although many bursts are easily detectable by the XRT, some are very much fainter. Of the prompt slews (up until the start of December 2005), the only non-detections (besides GRB 050911) were GRB 050925 (although this trigger may have been due to a new SGR; Holland et al 2005; Beardmore et al. 2005), GRB 050906 (Fox et al. 2005a) and GRB 051105A (Mineo et al. 2005), which were both short bursts. [The

TABLE 1. γ- and X-ray parameters for GRBs 050223 and 050911. The times over which the X-ray fluxes were calculated are given in the last row of the table.

Burst	GRB 050223	GRB 050911	*Swift* mean
T_{90} (s)	23	16	46
15–150 keV T_{90} fluence (erg cm^{-2})	4.8×10^{-8}	3.0×10^{-7}	2.3×10^{-6}
0.3–10 keV unabs. flux (erg cm^{-2} s^{-1})	8.2×10^{-13}	UL: 1.7×10^{-14}	5.2×10^{-10}
time range post-burst (ks)	2.8–4.0	16–716	large range

afterglows of short bursts tend to be fainter and fall below the XRT detection threshold quite rapidly (e.g., Gehrels et al. 2005; Fox et al. 2005b).]

We present here the analysis of two faint bursts: GRBs 050223 and 050911. The X-ray afterglow of GRB 050223 was detected by the XRT after ~ 47 minutes, whereas GRB 050911 remained undetected in an observation starting ~ 4.6 hours after the burst.

DATA ANALYSIS

GRBs 050223 and 050911 were faint in both prompt and afterglow emission (see Table 1 and Figures 1 & 2). In the case of GRB 050223, the X-ray flux at 11 hours ($\sim 1 \times 10^{-13}$ erg cm^{-2} s^{-1} over 0.3–10 keV) was below all those detected by BeppoSAX (Piro 2004). The flux upper limit of 1.7×10^{-14} erg cm^{-2} s^{-1} for GRB 050911 shows that, at $\sim 10^4$ s, any X-ray afterglow emission was at least an order of magnitude fainter than all of the other long bursts detected by *Swift*, with the possible exception of GRB 050421 (Figure 1; Godet et al. 2005; Nousek et al. 2005).

GRB 050223 - a large opening/viewing angle or high redshift?

Using the standard GRB afterglow models (Zhang & Mészáros 2004), the data for this burst are inconsistent with post-jet-break evolution. A large opening angle could explain both a late jet-break and the faintness of the afterglow, as well as the BAT fluence being relatively low. Alternatively, the low afterglow flux and prompt fluence could be caused by the burst being at high redshift; *Swift* GRBs are at a mean redshift of ~ 2.1, while pre-*Swift*, the mean was ~ 1.2. More details on the analysis of GRB 050223 can be found in Page et al. (2005a).

GRB 050911 - a naked GRB or a short burst?

The complete non-detection of an X-ray afterglow is very unusual for *Swift* bursts, as mentioned above. Any afterglow corresponding to the burst GRB 050911 must have faded very rapidly or been extremely faint to be undetected at ~ 4.6 hours. One possible

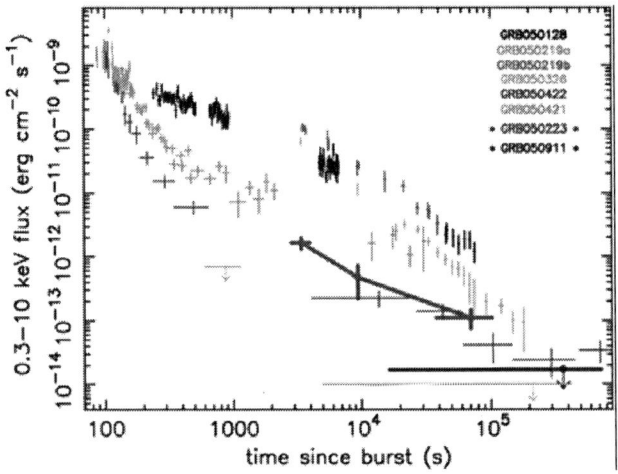

FIGURE 1. Flux light-curves for a selection of *Swift* GRBs, showing the faintness of the X-ray afterglows of GRB 050223 (thick red line) and GRB 050911 (blue 3σ upper limit). Adapted from Nousek et al. (2005).

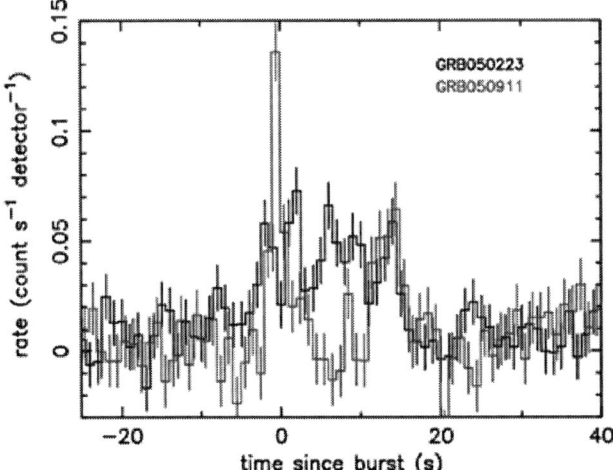

FIGURE 2. BAT light-curves (1-s bins) showing the count-rate per fully illuminated detector for each of the bursts.

explanation is the 'naked GRB' model, whereby the burst occurs in a low density environment, with the lack of surrounding material leading to a weak, or non-existent, forward shock. This may be the cause of the faintness of GRB 050421 (Godet et al. 2005).

Short bursts ($T_{90} < 2$ s; thought to be formed through compact object mergers) tend to show weak afterglows, fading below the XRT detection threshold within a few thousand seconds. Although $T_{90} > 2$ s for GRB 050911, there are two initial short (~ 0.5 s) spikes. Thus, GRB 050911 is like many short bursts in showing an initial short peak followed by longer, softer faint high energy emission (Lazzati, Ramirez-Ruiz & Ghisellini 2001; Connaughton 2002; Norris & Bonnell 2005). Simulations show that the later peak would not have been detected by BATSE at greater than the 1σ level; therefore, if BATSE had triggered on this weak burst at all, it is likely that it would have been classed as short.

GRB 050911 could, therefore, have been caused by a merger event: if one of the compact objects were a black hole, rather than a neutron star, the large mass ratio could lead to delayed accretion and, hence, high energy emission after 2 s (Davies, Levan & King 2005). See Page et al. (2005b) for more details.

CONCLUSIONS

The X-ray afterglows of both GRB 050223 and GRB 050911 are among the faintest observed at early times. Although *Swift* has the ability to measure faint X-ray emission out to many days after the burst, some afterglows are still too weak to be detected, indicating a difference in environment and/or formation mechanism.

REFERENCES

1. S. Barthelmy et al., 2005, Nature Letters, in press
2. A.P. Beardmore, K.L. Page., N. Gehrels, J. Greiner, J. Kennea, J. Nousek, J.P. Osborne, G. Tagliaferri, 2005, GCN4043
3. V. Connaughton, 2002, ApJ, 567, 1028
4. M.B. Davies, A.J. Levan & A.R. King, 2005, MNRAS, 356, 54
5. D.B. Fox, C. Pagani, L. Angelini, D.N. Burrows, J.P. Osborne & V. La Parola, 2005a, GCN 3956
6. D.B. Fox et al., 2005b, in prep.
7. N. Gehrels et al., 2005, ApJ, 611, 1005
8. O. Godet et al., 2005, A&A, in press
9. S.T. Holland et al., 2005, GCN 4034
10. D. Lazzati, E. Ramirez-Ruiz, G. Ghisellini, 2001, A&A, 379, L39
11. T. Mineo et al., 2005, GCN4188
12. J.P. Norris & J.T. Bonnell, 2005, ApJ, submitted
13. J.A. Nousek at al., 2005, ApJ, submitted (astro-ph/0508332)
14. C. Pagani, V. La Parola, & D.N. Burrows, 2005, GCN3934
15. K.L. Page et al., 2005a, MNRAS, 363, L76
16. K.L. Page et al., 2005b, ApJL, in press (astro-ph/0512358)
17. L. Piro, 2004, ASP Conference Series, Vol. 312, in press (astro-ph/0402638)
18. B. Zhang & P. Mészáros, 2004, Int.J.Mod.Phys, 19, 2385

GRB Follow-up Observations with the Whipple Telescope and VERITAS

Dirk Petry* and the VERITAS collaboration[†]

UMBC, Dept. of Physics, 1000 Hilltop Circle, Baltimore, MD 21250
[†]*see http://veritas.sao.arizona.edu*

Abstract. GRB observations above 100 GeV promise to constrain source models decisively and may also permit to study questions in other fields such as quantum gravity. The VERITAS collaboration has been working on GRB follow-up observations for several years using the Whipple 10 m telescope on Mt. Hopkins, AZ. We present a preliminary summary of the bursts for which interesting data is available. Even though the telescope mount was not optimized for this task, GRB follow-ups within less than 5 minutes have been possible. The new telescopes of the VERITAS array (under construction) will bring a further reduction in reaction time and an improvement in sensitivity.

Keywords: Gamma-ray bursts, Very High Energy gamma-ray observations
PACS: 95.85.Pw

INTRODUCTION

Gamma-Ray Bursts are a multi-wavelength phenomenon. So far, emission has been observed from radio wavelengths up to 18 GeV (a single photon detected by EGRET in GRB940217).

Many GRB spectra are hard and show no indication of a cut-off beyond several 100 MeV. Upper limits on the prompt emission at TeV energies obtained by air shower arrays such as AIROBICC and MILAGRO also do not constrain the emission in the several 100 GeV regime very much (in fact they report two marginal detections at higher energies [1, 2]), leaving it entirely possible that there is emission above 100 GeV in reach of the sensitivity of modern Imaging Atmospheric Cherenkov Telescopes (IACTs).

Many of the GRB models following the standard fireball shock scenario [3] predict significant emission at very high energies both during the prompt and afterglow phases. IACT observations can thus provide important tests for the present main-stream models.

Follow-up observations of GRBs with IACTs have been performed since the second half of the nineties (HEGRA, Whipple). The second generation telescopes were not optimized for fast follow-up. Their slow slew speed did not permit them to be on the burst position and take data within less than approx. 5 minutes. Also, the GRB positions given in the prompt notifications available from the Global Coordinate Network before the arrival of HETE2 and Swift were too uncertain to be covered by the small field-of-view of an IACT.

Now, with HETE2, Swift, and INTEGRAL operating and with the third-generation IACTs (VERITAS[4], MAGIC[5], HESS[6], CANGAROO III[7]) coming online, the possibility for the detection of very-high-energy (VHE) emission from GRBs opens up in earnest.

Here we present a summary of the GRB follow-up efforts by the VERITAS collaboration using the Whipple telescope and indicate the GRB-related plans for VERITAS. So far, no significant VHE emission from the direction of a GRB location was detected. A detailed analysis of the data is underway and will be published elsewhere [8, 9].

STATUS OF THE WHIPPLE TELESCOPE

The Whipple telescope on Mt. Hopkins (2300 m a.s.l.) has served for more than 20 years. It has been continuously upgraded and still takes valuable data, especially since it is presently the only well "broken-in" IACT on the Northern hemisphere (the first VERITAS telescopes and MAGIC have started observing only very recently).

The Whipple telescope has an energy threshold (peak photon detection rate for a Crab-Nebula-like spectrum) of 250 GeV at $0°$ zenith angle. The peak sensitivity is reached at 400 GeV. The telescope will operate until the full VERITAS-4 array goes online in 2007. In the remainder of its operation time, the observing schedule will focus on AGN monitoring and GRB follow-ups.

VERITAS STATUS

The first VERITAS telescope had "first light" in 2004. The second VERITAS telescope was completed in October 2005. Both telescopes are presently located at the Mt. Hopkins Smithsonian base camp. The new environmental assessment for the final Kitt Peak site will be completed early in 2006 and the collaboration hopes to get the go ahead from the NSF to proceed with the installation of VERITAS telescopes soon after that. The present plan is to be online with the four telescope array early in 2007.

FOLLOW-UP PROCEDURE

The Whipple telescope receives GCN notifications by socket connection. A dedicated software extracts the GRB location and time and sends them to the tracking computer. The software also sets off an acoustic alarm for the shift personnel in order to draw their attention to the burst. The personnel then verifies that it is safe and useful (in terms of weather conditions) to do a follow-up observation and starts the re-pointing of the telescope and the subsequent data taking.

It is not considered safe to make the GRB software repoint the telescope autonomously.

The Whipple telescope maximum slew speed is $3°/s$ which permits a response time of less than 5 minutes from the arrival of the notification to the time when the telescope begins acquiring data on the nominal GRB position assuming that the GRB position is within $60°$ of the zenith.

For the new VERITAS telescopes, the follow-up time will be further reduced to < 3 minutes by increasing the slew speed and optimizing the follow-up decision process.

FIGURE 1. The first two VERITAS telescopes at the Mt. Hopkins basecamp in November 2005. The mirror dish diameter is 12 m, f/d = 1.

REFERENCES

1. Padilla, L., et al., *A&A*, **337**, 43 (1998).
2. Atkins, R., et al., *ApJ*, **604**, L25 (2004).
3. Zhang, B. & Meszaros, P., *Int. J. of Mod. Phys. A*, **19**, 15 (2004).
4. Weekes, T.C., et al., *Astropart. Phys.*, **17**, 221 (2002).
5. Baixeras, B., et al., *NIM A*, **518**, 188 (2004).
6. Aharonian, F., et al., *Astropart. Phys.*, **22**, 109 (2004).
7. Enomoto, R. et al., *Astropart. Phys.*, **16**, 235 (2002).
8. Horan, D., et al., to be submitted to ApJ
9. Dowdall, C., et al., in preparation

TABLE 1. Successful GRB follow-ups with the Whipple telescope

GRB	Discovery Satellite	Redshift	Fluence* (ergs cm^{-2})	Duration (s)	Energy band (keV)	$T_{OBS}-T_{GRB}$ (minutes)	Exposure (minutes)
021112	HETE-2	0.8	—	—	—	221	193
021204	HETE-2	—	—	—	—	1014	55
021211	HETE-2	—	1.0×10^{-6}	> 5.7	8 - 40	1241	83
030329	HETE-2	0.17	1.0×10^{-4}	> 25	30 - 400	3873	334
030501	INTEGRAL	—	1.1×10^{-6}	~75	25 - 100	285	222
031026	HETE-2	—	2.3×10^{-6}	114	25 - 100	200	179
040422	INTEGRAL	—	—	8	—	101	249
050309	Swift BAT	—	—	26	—	7	101
050408	HETE-2	1.24	1.9×10^{-6}	15	30 - 400	719	28
050416	Swift BAT	0.65	2.8×10^{-7}	2.4	15 - 50	11	24
050502	INTEGRAL	3.79	1.4×10^{-6}	21	20 - 200	83	168
050504	INTEGRAL	—	1.5×10^{-6}	80	20 - 200	7	140
050509	Swift BAT	0.22	4.6×10^{-7}	12	15 - 350	4	155
050607	Swift BAT	< 5	8.9×10^{-7}	27	15 - 350	7	84

* The fluence, where available, is quoted for the energy range given in column 6 over the duration listed in column 5.

Colors and absolute magnitude of the optical afterglow of the short GRB050709

V. Šimon*, R. Hudec* and G. Pizzichini[†]

*Astronomical Institute, Academy of Sciences of the Czech Republic, 251 65 Ondřejov, Czech Republic
[†]INAF/IASF Bologna, via Gobetti 101, 40129 Bologna, Italy

Abstract. We find that the color indices of the optical afterglow (OA) of the short GRB050709 are consistent with those of the ensemble of 25 OAs of the long GRBs. This suggests that the spectral profiles of the OAs of both kinds of GRB are similar to each other and that the reddening inside their host galaxies is quite small ($E_{B-V} < 0.2$). The k-corrected R band absolute magnitude of the OA of GRB050709 in $(t - T_0)_{rest} = 2.13$ days (i.e. in the rest frame of the GRB) is $M_R \approx -16.7$, which is considerably fainter than that of the ensemble of the OAs of the long GRBs studied by [13]. This can suggest that the properties of a fireball do not depend very much on the phenomenon which originated it while the intensities differ considerably.

Keywords: radiation mechanisms: non-thermal; dust, extinction; galaxies, starburst; gamma-rays: bursts
PACS: 95.85.Kr; 95.85.Pw; 97.10.Zr; 98.54.Ep; 98.58.-w; 98.70.Rz

INTRODUCTION

GRBs can be divided into two groups according to their duration – the long ($\sim 10 - 100$ s), and short (< 2 s) ones [8]. The detection of the optical afterglows (OAs) of long GRBs (e.g. [16]) showed that they are mostly located in star-forming galaxies at cosmological redshifts. Nowadays, the long GRBs are interpreted as originating from hypernovae (e.g. [12]). The origin of short GRBs, also at cosmological redshifts, remains uncertain, although they are speculated to occur during a merger in a binary compact system, for example two neutron stars [3]. This is consistent with the location of their recently detected OAs in the outskirts of the host galaxy [1].

GRB050709 is a short GRB discovered by HETE-2 [17]. An X-ray afterglow and OA of this event were detected [4, 7, 5, 6], which enabled to determine the redshift $z = 0.16$ and locate this event in a distant galaxy [9, 2].

Here we present an analysis of the color indices and absolute magnitude of GRB050709 in comparison to those of the OAs of the long GRBs. Color indices are a powerful and innovative approach to the study of such events. They help us to investigate the common properties of the OAs, resolve among the individual radiation mechanisms, search for the relations among colors, luminosities and the decay rates of the OAs, and constrain the properties of the local interstellar medium of GRBs.

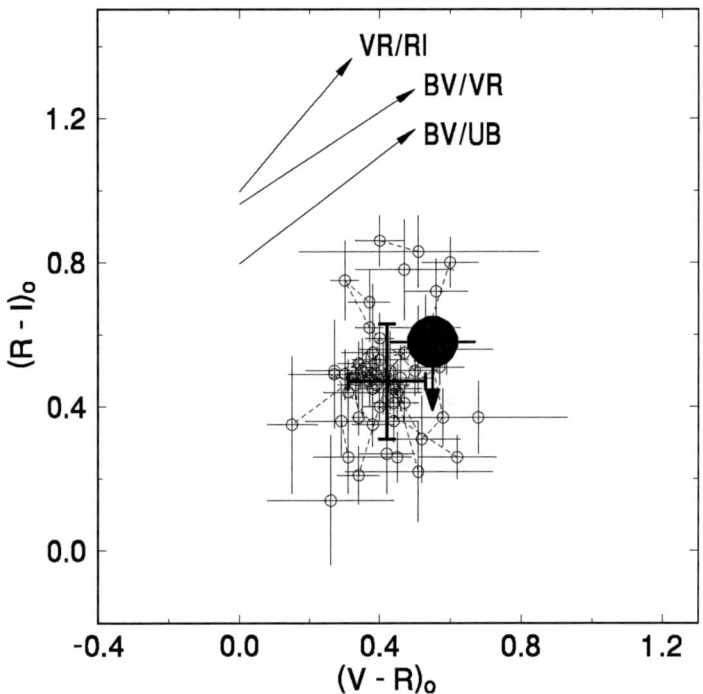

FIGURE 1. Color-color diagram which shows a comparison of the color indices of the OA of GRB050709 (big solid circle) with those of the OAs of the long GRBs (open circles) analyzed by [13]. The colors are in the observer frame. The indices of the OA of GRB050709 correspond to $t - T_0 = 2.473$ days in the observer frame while those of the remaining ones fall to 0.14 days$< t - T_0 < 10.2$ days. The colors of all OAs were corrected for the Galactic reddening according to [11]. The mean colors (centroid) of the OAs of the long GRBs, including their standard deviations, are marked by the large cross. The vectors denote the representative reddening paths for $E_{B-V} = 0.5$, using the Galactic extinction curve (see [13] for details).

COLLECTION AND ANALYSIS OF THE DATA

The magnitudes of the OA of GRB050709 were taken from [2] while the color indices and absolute magnitudes of an ensemble of the OAs of the long GRBs were taken from [13, 14]. Because of space limitations, the reader is referred to http://gcn.gsfc.nasa.gov/gcn/gcn3_archive.html (GCN Circulars, ed. S. Barthelmy) and J. Greiner's Web page http://www.mpe.mpg.de/~jcg/grbgen.html for full bibliographic references on each OA. The details of the data analysis and the determination of the color indices of the OAs can be found in [13].

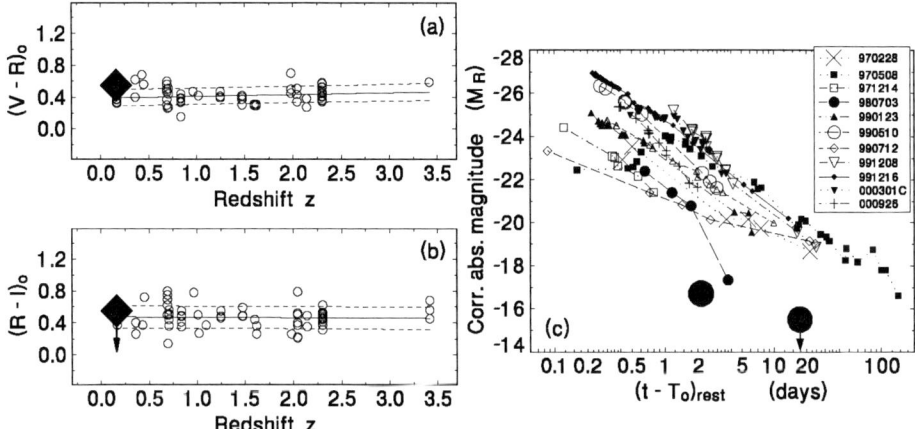

FIGURE 2. (a), (b) Color indices of the ensemble of OAs vs. their redshift z. The colors of the OAs of the long GRBs are marked by the open circles and correspond to $t - T_0 < 10.2$ days. Notice that the color index of the OA of GRB050709 in $t - T_0 = 2.473$ days (solid diamond) is consistent with that of the ensemble of the OAs of the long GRBs. (c) A comparison of the absolute R magnitudes of the ensemble of the OAs of the long GRBs from [13] and the OA of GRB050709. The k-corrections according to [13] were applied to all the measurements. Notice the very low brightness of the OA of GRB050709.

RESULTS

We show that the color indices of the OA of the short GRB050709 at $t - T_0 = 2.473$ days are consistent with those of the ensemble of 25 OAs of the long GRBs for $0.14 < t - T_0 < 10.2$ days (Fig. 1). The limit of $(R-I)_0$ of the OA of GRB050709 suggests that either its I mag is close to the detection limit or the slope of its spectrum is flatter than those of the spectra of the OAs of the long GRBs.

The mutual consistency of the color indices suggests that the spectral profiles of the OAs of both kinds of GRB are similar to each other. This speaks in favour of the common emission mechanism in the OA of GRB050709 and the OAs of the long GRBs. $(R-I)_0$ and $(V-R)_0$ of the OA of GRB050709 are consistent with a power-law spectral distribution, in accordance with the theoretical treatment of the GRB afterglow fireball emission model [10].

Our finding further implies that GRB050709 is not significantly reddened in its host galaxy. As shown in Fig. 2ab and in our companion paper [15], the very broad range of the redshifts ($0.17 < z < 3.5$) suggests that the long GRBs presumably occur in the different times of the evolution of their host galaxies, and hence in the regions with different metallicity and dust content. We conclude that the reddening of all the OAs in Fig. 1 is $E_{B-V} < 0.2$.

The k-corrected R band absolute magnitude of the OA of GRB050709 at $(t - T_0)_{rest} = 2.13$ days (i.e. time interval measured in the rest frame of the GRB) is $M_R \approx -16.7$, which is considerably fainter than that of the ensemble of the OAs of the long GRBs studied by [13] (Fig. 2c). Although the available data enable us to determine only the lower limit of the decay rate of the OA of GRB050709, it appears that this event extends the range of the absolute magnitudes of the OAs toward the fainter values. This probably implies that the OAs of short GRBs have similar colors but much fainter absolute magnitudes than those of the long ones.

ACKNOWLEDGMENTS

This study was supported by the grant A3003206 provided by the Grant Agency of the Academy of Sciences of the Czech Republic, the project ESA PRODEX INTEGRAL 90108, ESA PECS project 98023, and the CNR-AVČR collaborative project Investigation of GRBs (2004/2007).

REFERENCES

1. Barthelmy, S. D., Chincarini, G., Burrows, D. N., et al., *Nature*, **438**, 994 (2005).
2. Covino, S., Malesani, D., Israel, G. L., et al., *astro-ph/0509144v1* (2005).
3. Eichler, D., Livio, M., Piran, T., Schramm, D. N., *Nature*, **340**, 126 (1989).
4. Fox, D. B., Frail, D. A., Cameron, P. B., et al., *GCN 3585* (2005).
5. Fox, D. B., Frail, D. A., Price, P. A., et al., *Nature*, **437**, 845 (2005).
6. Hjorth, J., Watson, D., Fynbo, J. P. U., et al., *Nature*, **437**, 859 (2005).
7. Jensen, B. L., Jörgensen, U. G., Hjorth, J., et al., *GCN 3589* (2005).
8. Kouveliotou, C., Meegan, C. A., Fishman, G. J., et al., *ApJ*, **413**, L101 (1993).
9. Price, P. A., Roth, K., and Fox, D. W., *GCN 3605* (2005).
10. Sari, R., Piran, T., and Narayan, R., *ApJ*, **497**, 17 (1998).
11. Schlegel, D. J., Finkbeiner, D. P., and Davis, M., *ApJ*, **500**, 525 (1998).
12. Stanek, K. Z., Matheson, T., Garnavich, P. M., et al., *ApJ*, **591**, L17 (2003).
13. Šimon, V., Hudec, R., Pizzichini, G., and Masetti, N., *A&A*, **377**, 450 (2001).
14. Šimon, V., Hudec, R., Pizzichini, G., and Masetti, N., "Colors of Optical Afterglows of GRBs and Their Time Evolution," in **Gamma-Ray Bursts: 30 Years of Discovery: Gamma-Ray Burst Symposium**, edited by E. E. Fenimore and M. Galassi, AIP Conference Proceedings 727, American Institute of Physics, New York, 2004, pp. 487–490.
15. Šimon, V., Hudec, R., Pizzichini, G., "Colors of optical afterglows of GRBs in analysis of the dust in their environment," in **these proceedings**.
16. van Paradijs, J., Groot, P. J., Galama, T., et al., *Nature*, **386**, 686 (1997).
17. Villasenor, J. S., Lamb, D. Q., Ricker, G. R., et al., *Nature*, **437**, 855 (2005).

Near Infrared monitoring of the afterglow of the very bright Swift burst GRB 050525

G. Stratta*, A. Klotz[†], J.L. Atteia* and M. Boer**

*Observatoire Midi-Pyrenees, 14 Av. E. Belin, 31400, Toulouse Cedex 04, France
[†]Centre d'Etude Spatiale des Rayonnements, Observatoire Midi-Pyrenees, Universite' Paul Sabatier, 31028, Toulouse Cedex 04, France
**Observatoire de Haute Provence, St. Michel l'Observatoire, France

Abstract. We present the results from an extended monitoring of GRB 050525a taken minutes after the trigger time with the robotic telescope TAROT (Klotz et al. 2005) and more than one day later with the infrared camera WIRCam mounted on the 3.6-mt Canadian French Hawaii Telescope. A re-brightening feature was detected 33 minutes after the burst by TAROT. Together with GRB 021004 and GRB 050319, this is the third burst that shows an optical rebrightening in the rest-frame time interval of 0.01-0.02 days after the burst. We found no evidence of spectral variability before the re-brightening. The late times infrared data are nicely consistent with the expected power law decay with slope -1.6 derived from previous near infrared observations.

Keywords: 98.70.Rz
PACS: gamma-ray sources; gamma-ray bursts

INTRODUCTION

GRB 050525a was a very bright Gamma-Ray Burst (GRB) detected on May 25th, 2005 at 00:02:53 UT by the Swift/BAT instrument (Band et al. 2005). The afterglow was observed by both the X-Ray Telescope (XRT) and UV-Optical Telescope (UVOT) on-board Swift[1]. Minutes after the burst the afterglow was detected by the robotic telescope TAROT (Klotz et al. 2005). Several optical and infrared observations were also taken for this burst until days after trigger. A presumed host galaxy was measured with a redshift of z=0.606 (Foley et al. 2005). In this work we present the optical and near infrared (NIR) data of the early afterglow observed by the robotic telescope TAROT and of the late afterglow (more than one day post-burst) based on observations obtained with WIRCam, at the 3.6-mt Canada-France-Hawaii Telescope (CFHT) which is operated by the National Research Council (NRC) of Canada, the Institute National des Sciences de l'Univers (INSU) of the Centre National de la Recherche Scientifique (CNRS) of France, and the University of Hawaii.

---

[1] A comprehensive temporal and spectral analysis of the XRT and UVOT data has been published by Blustin et al. (2005).

OBSERVATIONS

TAROT data

The robotic telescope TAROT monitored the afterglow from 6 to 136 minutes after the BAT trigger. Evidence for rapid re-brightening of 0.65 mag in the R band was observed 33 minutes after the trigger (Fig. 1) (Klotz et al. 2005).

WIRCAM data

We observed the afterglow 1.6 and 2.6 days after the BAT trigger with the CFHT camera WIRCAM. The camera was still using engineering-grade chips instead of the final science grade arrays. We detected the afterglow only during the first night in the K_s filter while it was under the detection limit in the J filter. By computing the median of the two nights we found $K_s = 19.61 \pm 0.14$ mag and J> 20.93 mag (2σ upper limit) (Fig.1). To derive these magnitude values we took into account the color index inferred for this burst of J-K~ 1 (Cobb et al. 2005, Garnavich et al. 2005).

DISCUSSION

Early afterglow. For this burst, two rebrightenings features were detected in the X-rays (Blustin et al. 2005) and in the R band (Klotz et al. 2005) 300s and 33 minutes after the burst (observer frame), respectively. The temporal slopes in the X-ray and R-band energy ranges before the rebrightenins are consistent within the errors, being $\delta_X = -1.20 \pm 0.03$ in the X-rays and $\delta_R = -1.15 \pm 0.06$ in the R-band. Both the rebrightening features are characterized by similar morphology, with $\delta t/t \sim 0.1$. The lack of R-band temporal coverage during the X-ray rebrightening and of X-rays data during the R-band rebrightening prevented to check if this is the case of a multiple, achromatic rebrightening or rather is the propagation in time and in frequency of a single rebrightening. Together with GRB 021004 and GRB 050319, this is the third burst that shows an optical rebrightening in the rest-frame time interval of 0.01-0.02 days after the burst. Shao & Dai (2005) have interpreted the optical rebrightening as the emerging of the forward shock emission from the reverse shock component, without excluding, however, other possible explanations.

From the TAROT data, we computed the R-band fluxes at $t - t_0 = 250$ s and at $t - t_0 = 800$ s (where t_0 is the burst trigger time). We found that the optical/UV Spectral Energy Distribution (SED) extracted at the same two epochs from Swift/UVOT data (Blustin et al. 2005) smoothly reconnects to the R-band fluxes (Fig.2). We fitted a simple power law model to the NIR-optical-UV (0.8-0.2μ) SED corrected for the Galactic extinction towards this burst. We found no significant spectral variation, with spectral indexes $\alpha_{250} = -0.82 \pm 0.14$ (χ^2/d.o.f=9.3/5) and $\alpha_{800} = -1.10 \pm 0.20$ (χ^2/d.o.f=5.7/5) at 250s and at 800s respectively. Assuming that the optical range is below the cooling frequency (Shao & Dai 2005), we fixed the spectral index to the expected value (Sari et al. 1998) and we added to the power law model an absorption component in order

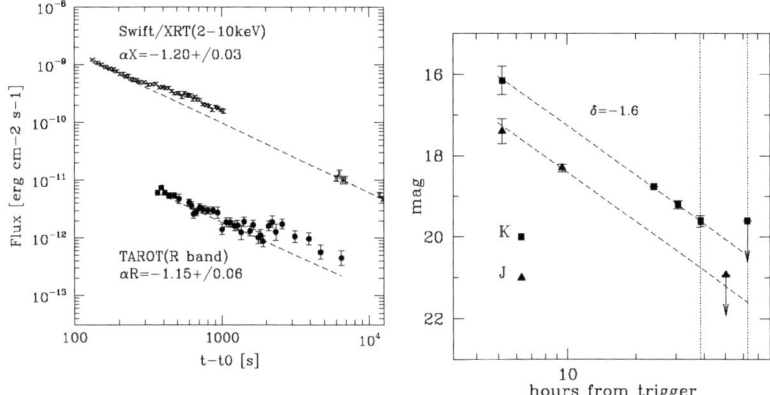

FIGURE 1. *Left panel:* X-ray (Swift/XRT) and R-band (TAROT) early afterglow light curve (Blustin et al. 2005, Klotz et al. 2005). *Right panel:* NIR late afterglow light curve (Cobb et al. 2005, Kaplan et al. 2005). The two vertical lines indicate the two WIRCAM observing nights. The J-band upper limit is obtained from the median of the two nights.

to reproduce a possible host galaxy extinction effect. We found a rest frame visual extinction of $A_V \sim 0.10$ assuming the Small Magellanic Cloud extinction curve, in agreement with previous analysis (Blustin et al. 2005).

Late afterglow. A temporal break was observed in the Swift XRT and UVOT data at 0.15 day after the burst and has been interpreted to be the evidence of a jet (Blustin et al. 2005). As noticed by Blustin et al. (2005), the post-break slope is shallower than expected from the jet model predictions possibly indicating a not significant sideway expansion of the jet or a smoothed temporal jet break with the time of the break shifted to 0.6 day after the burst. Our late time K_s photometry 1.6 days after the burst does not show any steepening of the light curve and is nicely consistent with the decay slope measured from previous NIR observations of $\delta = -1.6$ (Kaplan et al. 2005). Unfortunately, the upper limit we obtained during the second night could not provide any constraints on the decay slope and therefore on the two different jet scenarios. We note that from the NIR data, no evidence of a constant background from the host galaxy or of a slowly emerging of a supernova component (Soderberg et al. 2005) are detected at least up to 1.6 days after the burst.

SUMMARY

- An optical rebrightening of $\Delta R = 0.65$ mag was observed 33 minutes after the GRB. This is the third known afterglow that exhibits a rebrightening event beginning at 0.01-0.02 days in the rest frame, along with GRB 021004 and GRB 050319 (see Klotz et al. 2005).

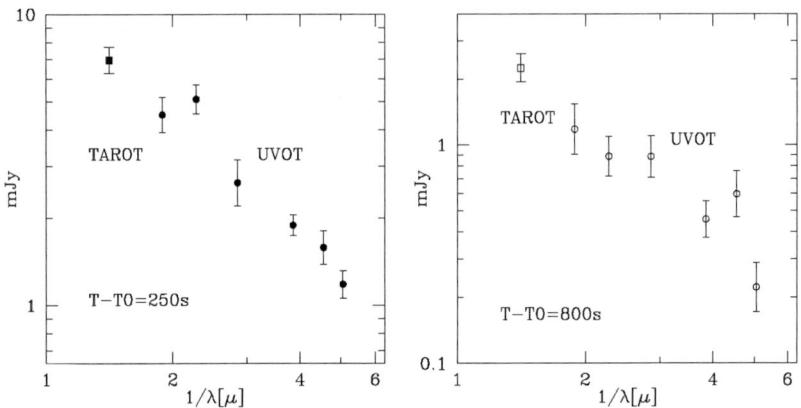

FIGURE 2. The 0.8÷0.2μ SED obtained from TAROT and Swift/UVOT data, not corrected for Galactic and extra galactic extinction at two different epochs before the optical re-brightening.

- With the advantage of the additional TAROT observations that extends the Swift/UVOT SED to the R-band, we confirm the absence of any spectral variation in the 0.8÷0.2μ wavelenghts range before the re-brightening and the presence of an additional absorption from the host galaxy of $A_V \sim 0.10$ assuming a SMC extinction curve.
- Our late time NIR observations from CFHT-WIRCam images 1.6 and 2.6 days after the burst, are consistent with the expected power law temporal behavior inferred from previous NIR band observations (Kaplan et al. 2005) with no evidence of steepening or shallowing at least up to 1.6 days after the burst.

ACKNOWLEDGMENTS

We thanks Y. Mellier and his collaborators for usefull refined WIRCam data analysis at the Terapix data center. GS acknowledge the support by the EU Research and Training Network "Gamma -Ray Bursts, an enigma and a tool".

REFERENCES

- A. Klotz, et al. , *A&A* **439**, 35 (2005).
- D. Band, et al. , *GCN* 3466 (2005).
- A. J. Blustin, et al. , *ApJ*, in press (2006)
- R. J. Foley et al., *GCN* 3483, (2005).
- B. E. Cobb and C. D. Bailyn, *GCN* 3506 (2005).
- P. Garnavich et al., *GCN* 6532 (2005)
- L. Shao & Z. Dai, *ApJ*, **633**, 1027 (2005)
- R. Sari, T. Piran and R. Narayan, *ApJ*, **497**, L17 (1998)
- D. Kaplan, et al., *GCN* 3507, (2005).
- A. Soderberg et al., *GNC* 3550 (2005)

Shallow Decay of Early X-Ray Afterglows from Inhomogeneous Gamma-Ray Burst Jets

K. Toma[*], K. Ioka[*], R. Yamazaki[†] and T. Nakamura[*]

[*]*Depatment of Physics, Kyoto University, Kyoto 606-8502, Japan*
[†]*Department of Physics, Hiroshima University, Higashi-Hiroshima 739-8526, Japan*

Abstract. Almost all the X-ray afterglows of gamma-ray bursts (GRBs) observed by the *Swift* satellite have a shallow decay phase in the first thousands of seconds. We show that in an inhomogeneous jet (multiple-subjet or patchy-shell) model the superposition of the afterglows of off-axis subjets (patchy shells) can have the shallow decay phase. The necessary condition for obtaining the shallow decay phase is that γ-ray bright subjets (patchy shells) should have γ-ray efficiency higher than previously estimated, and should be surrounded by γ-ray dim subjets (patchy shells) with low γ-ray efficiency. Our model predicts that events with dim prompt emission have the conventional afterglow light curve without the shallow decay phase like GRB 050416A.

Keywords: γ-ray sources; γ-ray bursts
PACS: 98.70.Rz, 01.30.Cc

INTRODUCTION

Before the *Swift* era, most of the X-ray and optical afterglows of gamma-ray bursts (GRBs) were detected only several hours after the burst trigger. *Swift* observations are unveiling the first several hours of the afterglows [e.g., 1, 2, 3]. Recently, Nousek et al. [3] analyzed the first 27 afterglows detected by *Swift* XRT, and reported that almost all the early X-ray afterglows of *Swift* GRBs do not show a simple power-law flux decline. They show a "canonical" behavior, where the light curve begins with a very steep decay, turns into a very shallow decay $\sim t^{-0.5}$, and finally connects to the conventional late-phase afterglow $\sim t^{-1}$ which is similar to what was observed in the pre-*Swift* era.

The shallow decay phase implies that more time-integrated radiation energy is observed at later time. This is unexpected in the standard model that can explain the late-phase afterglows, i.e., the synchrotron shock model of an impulsive homogeneous jet. There seems to be essentially no spectral variation at the transition from the shallow decay phase to the conventional decay phase. This suggests that the origin of the transition is either hydrodynamical or geometrical.

In the hydrodynamical model, the GRB jet is not impulsive but the energy is injected continuously into the blast wave for $\sim 10^4$ s [4, 3, 5, 6, and references therein]. Since the afterglow is dim in the shallow decay phase, the γ-ray efficiency for the front shells is much higher than previously estimated.

In the geometrical model, it is assumed that we observe more energetic regions of the GRB jet later as the afterglow shock decelerates and the visible region increases. The shallow decay phase of the "canonical" afterglow may be a combination of the tail part of the prompt emission and the delayed afterglow emission from an off-axis jet [7]. Since

Eichler & Granot [7] discussed a specific "ring-shaped" jet, more general studies for the jet angular structure are desirable to know the general characteristics of geometrical model [see also 5].

In this paper, we develop an inhomogeneous jet model to reproduce the "canonical" X-ray afterglows of GRBs in the framework of the geometrical model. In order to study the angular energy distribution in the jet, we consider an extremely inhomogeneous jet (a multiple-subjet model [8, 9]). Figure 1 (Left) illustrates the setup for our analysis of an inhomogeneous jet. We assume that the whole jet (*dashed circle*) consists of multiple subjets (*solid circles*), and the energy injected among subjets is negligible compared to the energy inside each subjet. Each subjet is assumed to make a prompt γ-ray radiation and a subsequent afterglow following the standard scenario. We may consider the initial opening half-angle of each subjet $\Delta\theta_0^i$ to be $\gtrsim \Gamma_0^{i\,-1}$, where $\Gamma_0^i \simeq 10^2 - 10^3$ is the initial Lorentz factor of each subjet. The superscript 'i' and 'w' denote each subjet and the whole jet, respectively, while the subscript '0' denotes the initial time when each subjet begins to decelerate. We calculate the early phase of the afterglow by superposing the contribution of each subjet, and study necessary conditions for reproducing the "canonical" afterglows of the *Swift* GRBs. See Toma et al. [10] for detailed discussion.

NECESSARY CONDITIONS FOR THE SHALLOW DECAY

The discussion is separated into two cases: Case (i) the line of sight is along a subjet. Case (ii) the line of sight is off-axis for any subjet. For both cases we will obtain the necessary conditions to reproduce the "canonical" afterglow.

In Fig 1 (Right), the dot-dashed line represents the X-ray afterglow light curve expected before the *Swift* era, i.e., the afterglow from a homogeneous jet with a typical afterglow energy of $E_{k,iso}^w = 10^{52}$ erg and an opening half-angle of $\Delta\theta_0^w = 0.1$ rad. The thin solid flat line around $L_X \sim 10^{50}$ erg s^{-1} and for $t < 10$ s represents a typical prompt burst with a duration of $\simeq 10$ s. Its isotropic γ-ray energy is $\gtrsim 10^{52}$ erg, which is comparable to or larger than $E_{k,iso}^w$ as in the actual observations [12]. The thin curved solid line for $t > 10$ s is the tail part of the prompt burst, which comes from the region with large viewing angles in the whole jet. The shaded line shows the "canonical" afterglow light curve obtained for both of the two cases in our model.

Let us consider the case (i). The $E_{k,iso}^i$ of the on-axis subjet afterglow (the dashed line (1)) should be at most $1/3$ of that of the dot-dashed line afterglow so that the tail part of the prompt emission for the steep decay phase is not hidden. The superposition of the off-axis subjet afterglows (2)-(5) with $E_{k,iso}^i = 3 \times 10^{52}$ erg produces the shallow decay phase. All the subjets expand sideways and finally merge into one shell producing the conventional afterglow emission. Thus the shallow decay phase would smoothly connect to the dot-dashed line and the final afterglow would be like the shaded line.

The prompt emission is dominated by that from the on-axis subjet because of the beaming effect. Thus the prompt burst energy $E_{\gamma,iso}^i$ of the on-axis subjet is $\gtrsim 10^{52}$ erg. Since $E_{k,iso}^i \sim 3 \times 10^{51}$ erg, this implies that the γ-ray efficiency for the on-axis subjet is $\varepsilon_\gamma \equiv E_{\gamma,iso}^i/(E_{\gamma,iso}^i + E_{k,iso}^i) \gtrsim 75\%$, which is larger than previously estimated.

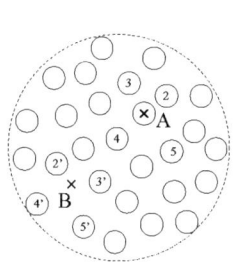
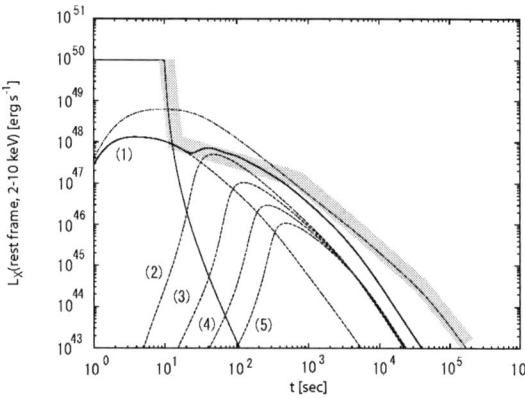

FIGURE 1. Left: Setup for our analysis of an inhomogeneous jet. A whole jet (*dashed circle*) consists of multiple subjets (*solid circles*). Points 'A' and 'B' describe the lines of sight for our calculations in case (*i*) and (*ii*), respectively. We take the initial opening half-angle of the subjets and the whole jet as $\Delta\theta_0^i = 0.01$ rad and $\Delta\theta_0^w = 0.1$ rad. Right: Example of the afterglow light curve (the isotropic-equivalent luminosity) in the range $2-10$ keV measured in the cosmological rest frame of the GRB. We calculate X-ray afterglow emission from the external shock of an impulsive homogeneous jet with sharp edge following Panaitescu & Kumar [11]. The jet dynamics is calculated by the mass and energy conservation equations with the effect of sideways expansion at the local sound speed and radiative energy losses. We fix the number density of the circumburst medium as $n = 1$ cm^{-3}, the ratio of the magnetic energy and the accelerated electron energy to the shocked thermal energy as $\varepsilon_B = 0.01$ and $\varepsilon_e = 0.1$, respectively, and the index of the energy distribution function of the accelerated electrons as $p = 2.3$. The dot-dashed line is the afterglow from a jet with $E_{k,iso}^w = 10^{52}$ erg, $\Delta\theta_0^w = 0.1$ rad, and $\theta_v = 0$. The thin solid flat line for $t < 10$ s represents a typical prompt burst which corresponds to the late-phase of the dot-dashed-line afterglow. The thin solid curved line $t > 10$ s represents the tail part of the prompt emission which we set as $\propto (t-9.0)^{-3.5}$. The dashed line (1) is the afterglow from a subjet with $E_{k,iso}^i = 3 \times 10^{51}$ erg, and $\theta_v^i = 0$. The dashed lines (2), (3), (4), and (5) are the afterglows from subjets with $E_{k,iso}^i = 3 \times 10^{52}$ erg, and $\theta_v^i = 0.025, 0.03, 0.035,$ and 0.045 rad, respectively. These subjets correspond to (2), (3), (4), and (5) for the line of sight 'A' (or (2'), (3'), (4'), and (5')) for the line of sight 'B' in the left panel. The thick solid line is the superposition of all the dashed lines (1) – (5). The shaded line is what we expect for the afterglows from inhomogeneous GRB jets.

Now what is observed when our line of sight is along the subjet with an energetic afterglow of $E_{k,iso}^i = 3 \times 10^{52}$ erg? Let us assume that the "canonical" afterglow is observed also in this case. Then the energy of the prompt emission should be $E_{\gamma,iso} \gtrsim 10^{53}$ erg in order for the tail emission to be larger than the afterglow emission from the on-axis subjet. Since the number of the energetic afterglow subjets should be larger than that of the high γ-ray efficiency subjets, this leads to larger event rate of more energetic prompt bursts, which is not consistent with current observations. Therefore the subjets with energetic afterglows should have low γ-ray efficiency and dim prompt emissions so that they are hard to be observed.

Next, we consider the case (*ii*). The "canonical" afterglow light curve is obtained by the same calculation as in case (*i*) removing the contribution from the on-axis subjet. In case (*ii*) also, the γ-ray efficiency ε_γ should be large. Since the beaming effect is strong for the prompt emission phase, the gamma-ray energy should be larger than the afterglow energy for the off-axis subjets. See Toma et al. [10] for detailed discussion.

DISCUSSION

We have investigated early X-ray afterglows of GRBs within inhomogeneous jet models by using a multiple-subjet model. We find that the shallow decay phase is produced by the superposition of the afterglows from off-axis subjets, and it connects to the conventional late-phase afterglow which is produced by the merged whole jet.

We found necessary conditions for obtaining the "canonical" afterglow by separating our discussions into two cases, i.e., whether the line of sight is along a subjet (case (i)) or not (case (ii)). In both cases (i) and (ii), subjets producing a bright prompt emission should have γ-ray efficiency larger than previously estimated. This requirement is similar to the hydrodynamical model [4, 3] and is problematic in the framework of the internal shock model.

In case (i), a subjet producing a bright prompt burst should have a dim afterglow emission and should be surrounded by several subjets producing a dim prompt and a bright afterglow emissions. When the line of sight is along the subjet with a dim prompt and a bright afterglow emissions, the conventional afterglow light curve without a shallow phase is observed. Therefore we predict that small $E_{\gamma,iso}$ events should have the conventional afterglow light curve. Among 10 *Swift* GRBs with known redshifts GRB 050416A has an extremely small $E_{\gamma,iso}$ of $\lesssim 10^{51}$ erg, and it does not have a shallow decay phase [3]. This event may support the case (i) of the inhomogeneous jet model, although more statistics is required to confirm the validity.

Case (ii) suggests that for most events both the prompt and the afterglow emissions arise from off-axis viewing angles, which is similar to the scenario of Eichler & Granot [7]. In this case, we found that most of the subjets should produce a bright prompt and a dim afterglow emissions. When the line of sight is along such a subjet, the conventional but dim afterglow is observed. Then we predict that there should be large $E_{\gamma,iso}$ events with a conventional afterglow in case (ii). We should observe such bright γ-ray events with a similar order of rate as the "canonical" events, which may be tested in future.

REFERENCES

1. Tagliaferri, G., et al. 2005, Nature, 436, 985
2. Chincarini, G., et al. 2005, preprint (astro-ph/0506453)
3. Nousek, J. A., et al. 2005, preprint (astro-ph/0508332)
4. Zhang, B., et al. 2005, preprint (astro-ph/0508321)
5. Panaitescu, A., et al. 2005, preprint (astro-ph/0508340)
6. Granot, J., & Kumar, P. 2005, preprint (astro-ph/0511049)
7. Eichler, D., & Granot, J. 2005, prepint (astro-ph/0509857)
8. Nakamura, T. 2000, ApJ, 534, L159
9. Kumar, P., & Piran, T. 2000, ApJ, 535, 152
10. Toma, K., et al. 2005, preprint (astro-ph/0511718)
11. Panaitescu, A., & Kumar, P. 2001, ApJ, 554, 667
12. Lloyd-Ronning, N. M. & Zhang, B. 2004, ApJ, 613, 477

Slow and Fast Components in X-Ray light curves of GRBs from BeppoSAX WFC archive

L. Vetere[*,†], E. Massaro[*], E. Costa[**] and P. Soffitta[**]

[*]*Dipartimento di Fisica, Università La Sapienza, Piazzale A. Moro 2, I-00185 Roma, Italy*
[†]*INAF-ASI Science Data Center, Via Galileo Galilei, I-00044 Roma, Italy*
[**]*INAF, IASF - Sezione di Roma, via del Fosso del Cavaliere, I-00133 Roma, Italy*

Abstract. Gamma Ray Bursts light curves show quite different patterns: from very simple to extremely complex. For some events the time profile in the 2-5 keV energy range is characterised by peaks superposed on a slowly evolving pedestal, which at high energy is less apparent. We describe this behaviour with the presence of two components (slow and fast) having different time scales. The slow component, SC, is usually less pronounced in the 5-10 keV energy range and almost disappears at higher energies. We present a time and spectral study of the SC of several GRBs detected by the WFC on board BeppoSAX. The origin of the SC is likely different from that of the fast component, FC, and can be related either to the presence of a hot photosphere or to the overlapping of the prompt emission with the initial phase of the afterglow.

Keywords: X rays: bursts - gamma rays: bursts
PACS: 95.85.Nv, 98.70.Rz

INTRODUCTION

GRBs light curves (LC) are very variable. They can be very simple or extremely complex. This particular behaviour has been studied many times looking for possible relations between different parameters of the lightcurve. Using the entire database of the Wide Filed Cameras on board BeppoSAX we realized that some bursts' time profiles are characterised by peaks superposed on a slowly evolving pedestal, which frequently becomes less apparent at higher energies while in some others remains stable or even grows. We describe this behaviour with the presence of two components (slow and fast) having different variability time scales. In this contribution we present the result of a temporal and spectral study of the light curves in three energy bands (2-5, 5-10, 10-26 keV) of ten GRBs detected by the Wide Field Cameras (for the complete analysis see [5]).

SLOW COMPONENT MODELLING

Using the WFC database, containing data on 56 GRBs, we derived the LCs in three energy bands and selected those showing a presence of a possible component evolving over the entire duration of the burst. In the following we will refer to it as Slow Component (shortly SC), while the residual prompt emission structure, often characterised by a series of short duration peaks will be indicated as Fast Component (FC).

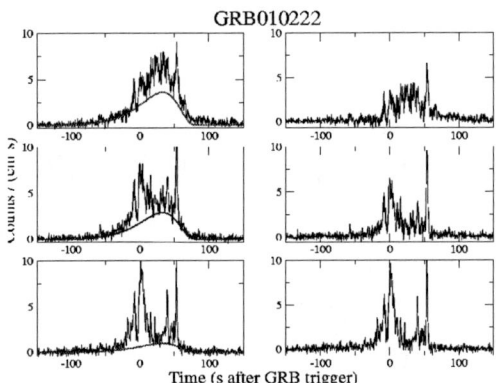

FIGURE 1. X-ray light curves of GRB010222 in the three considered energy ranges. Left panels show the total counts and the SC model; right panels show the corresponding FCs after the SC subtraction.

There is no direct (and unambiguous) way to separate SC and FC in a X-ray LC of a GRB. We therefore applied a simple heuristic approach and modeled the SC by means of the following analytical formula containing only a small number of parameters:

$$F_S(t,E_n) = A(E_n) \, (t-t_0)^b \, exp[-C \, (t-t_0)^s] \quad (1)$$

where $F_S(t,E_n)$ is the time dependent number of counts in the energy band n (hereafter we use $n=1, 2, 3$ for the lowest, mid and highest energy bands specified before), $A(E_n)$ is the amplitude of the SC in the considered energy band, and the values of remaining shape parameters b, C and s were chosen by interpolating the local minima of the LC in the band where the SC was more apparent.

Residual FCs were obtained by subtracting the SCs from the original data sets:

$$F_F(t,E_n) = F(t,E_n) - k_{ni} \, F_S(t-t_\star,E_i) \quad (2)$$

where $F(t,E_n)$ is the original LC, F_F the fast component, and $k_{ni} = A(E_n)/A(E_i)$ is the ratio between the amplitudes for the energy bands n and i (usually $i = 1$), where parameters were estimated; $t_\star$ is a possible time delay between the SCs at different energies. As examples we show the cases of GRB 010222 and GRB 990123 in Fig. 1, 2 where the X-ray light curves of these GRBs are plotted in the three considered energy ranges. The panels on the left show the total counts and the SC model; the panels on the right show the corresponding FCs.

SPECTRAL ANALYSIS

To obtain spectral informations we computed the Hardness Ratios (HR) between the counts in the three used energy bands both for the SC that for the FC comparing them with the corresponding values of the original one. We then estimated the corresponding spectral indices Γ by convolving a power law input spectrum with the WFC response

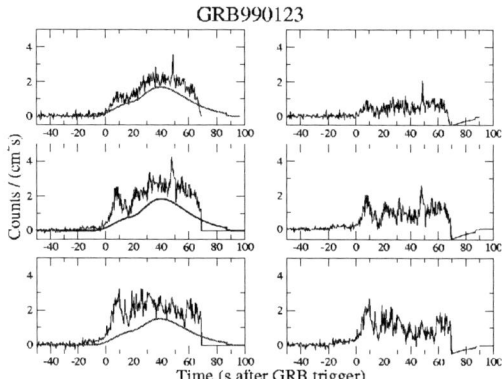

FIGURE 2. X-ray light curves of GRB 990123 in the three considered energy ranges. Left panels show the total counts and the SC model; right panels show the corresponding FCs after the SC subtraction.

matrix in the direction of the burst. We will indicate with HR_1 the HR between 2-5 keV and 5-10 keV energy range, HR_2 the HR between 5-10 keV and 10-26 keV. The calculated values of HR_1 for the SC, the FC are plotted in Fig. 3.

THE NATURE OF SC

Spectral differences between FCs and SCs suggest that their emission can be due to different emission mechanisms: one producing highly variable structures and another responsible of the SCs. We tried to investigate whether this second mechanism could be the initial phase of the afterglow but failed to find evidence that SC spectra are similar to those of the afterglows. The latters are usually characterised by photon indices $\Gamma \simeq 2$ (De Pasquale et al. 2005) while the mean Γ values of SCs are all smaller. Moreover, we also verified that SCs are not originated by blending of narrow peaks and of features closely spaced in time as a consequence of the low-energy broadening found by Fenimore et al. (1995). Furthermore, the subtraction of a SC model from original LCs gave residual fast components showing time structures very similar to those of LCs above 10 keV and having peak broadening in agreement with Fenimore's relationship.

Another possible interpretation is that SC is originated in an outflow photosphere as in the model developed by [2], [3], [4]. Usually the highly variable emission of GRBs is attributed to internal shocks occurring at a certain distance from the centre of the relativistic outflow, that is beyond the photosphere at which the flow become optically thin. At small enough radii the shock through dissipative effects can create a number of pairs sufficient to reestablish a second leptonic photosphere with a limiting radius beyond which the shock remain optically thin. Thus, above this radius, there will be a region favorable for producing the highly variable signals while the region below it will be favorable to produce a more stable emission. Moreover, sub photospheric dissipation leads to an increase of the radiative efficiency of the outflow, consequently boosting the quasi-thermal photospheric component so that it can dominate the synchrotron compo-

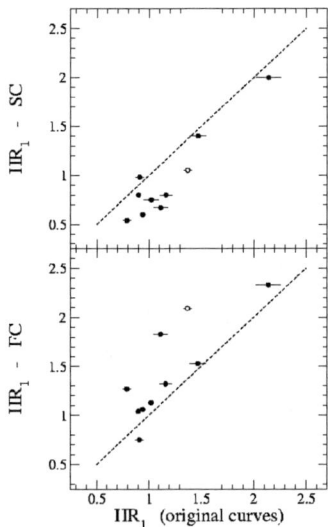

FIGURE 3. Hardness ratios between 2–5 keV and 5–10 keV of the SCs (upper panel) and FCs (lower panel) plotted against those of the original light curves of the eleven studied GRBs. Open circle indicates GRB 990123 which showed the largest change.

nent due to non-thermal shocks outside the photosphere. The expected spectrum of this thermal component is that of Black Body (BB), but it can also show a Comptonized component (Mészáros and Rees 2000). The kT of the BB component depends on the dimensionless entropy $\eta = L/\dot{M}c^2$ (here L is the fireball luminosity and $\dot{M}$ the mass rate of the outflow) and on the time scale variability $\xi = t_v/t_0$ with $t_0 = 3.25 \times 10^{-3}\mu_1$ s the Kepler rotation timescale (μ_1 is the black hole mass in units of 10 solar masses). In particular kT values ranging from a few keV to tens of keV are obtained for $\eta < 100$ and $\xi > 10^3$, the latter corresponding to long duration GRBs, like those in our sample.

In this scenario it is natural to associate the SC with the photospheric emission and FC with the more variable and harder non-thermal emission. It is important, therefore, to verify that the SC is or not a trace of a photospheric GRB emission, but this require the development of a complex physical model that is beyond the aim of this work.

REFERENCES

1. Fenimore, E.E., in 't Zand, J.J., Norris, J. P., Bonnell, J. T., Nemiroff, R. J. 1995, ApJ, 448, L101
2. Mészáros,& P., Rees, M.J., 2000, ApJ 530, 292
3. Mészáros, P., Ramirez-Ruiz, E. et al. 2002, ApJ 578, 812
4. Rees, M.J., & Meszaros, P., 2005, ApJ in press, astro-ph/0412702
5. Vetere, L., Massaro, E., Costa, E. et al. 2005, A&A in press.

Automated detection of optical counterparts to GRBs with RAPTOR

P. R. Wozniak, W. T. Vestrand, S. Evans, R. White and J. Wren

Los Alamos National Laboratory, MS-D466, Los Alamos, NM 87545

Abstract. The RAPTOR system (RApid Telescopes for Optical Response) is an array of several distributed robotic telescopes that automatically respond to GCN localization alerts. Raptor-S is a 0.4-m telescope with 24 arc min. field of view employing a 1k x 1k Marconi CCD detector, and has already detected prompt optical emission from several GRBs within the first minute of the explosion. We present a real-time data analysis and alert system for automated identification of optical transients in Raptor-S GRB response data down to the sensitivity limit of ~ 19 mag. Our custom data processing pipeline is designed to minimize the time required to reliably identify transients and extract actionable information. The system utilizes a networked PostgreSQL database server for catalog access and distributes email alerts with successful detections.

Keywords: Gamma-ray Bursts, Optical transients, Automated detection
PACS: 98.70.Rz, 98.62.Qz, 98.62.Tc, 89.20.Ff

1. THE RAPTOR PROJECT

RAPTOR is a long-term project at the Los Alamos National Laboratory with the goal of surveying the optical sky and autonomously detecting transient phenomena on timescales as short as ~ 30 seconds [1]. The search is utilizing an array of networked robotic telescopes with real-time data processing [2]. Each instrument can identify and report transients as well as follow up interesting triggers from alert feeds inside and outside the Raptor network. Here we describe a GRB followup system based on the fast response capability of Raptor with the Raptor-S telescope responding to GRB localization alerts from GCN. Fig. 1 presents the overall architecture of the system.

Raptor-S is a 0.4-m, f/5, fully autonomous telescope owned by Los Alamos National Laboratory and located at an altitude 2,500 m in the Jemez Mountains of New Mexico. It employs a 1k x 1k pix back-illuminated Marconi CCD47-10 chip with 13-micron pixels. The rapidly slewing mount allows re-pointing to a random direction in the sky in ~ 5 seconds from the time of receiving the trigger.

2. EVENT DETECTION

The response sequence for Raptor-S telescope consists of a series of ten 5-second duration exposures followed by 20 10-second exposures and finally by ~ 200 30-second exposures. A typical response data set covers the first 90 minutes after the GRB. All images are unfiltered. The magnitude scale of each image is adjusted to that of the first image using the median magnitude offset of a few tens to a few hundred objects in the field. The current event finding code uses statistics that can be computed incrementally,

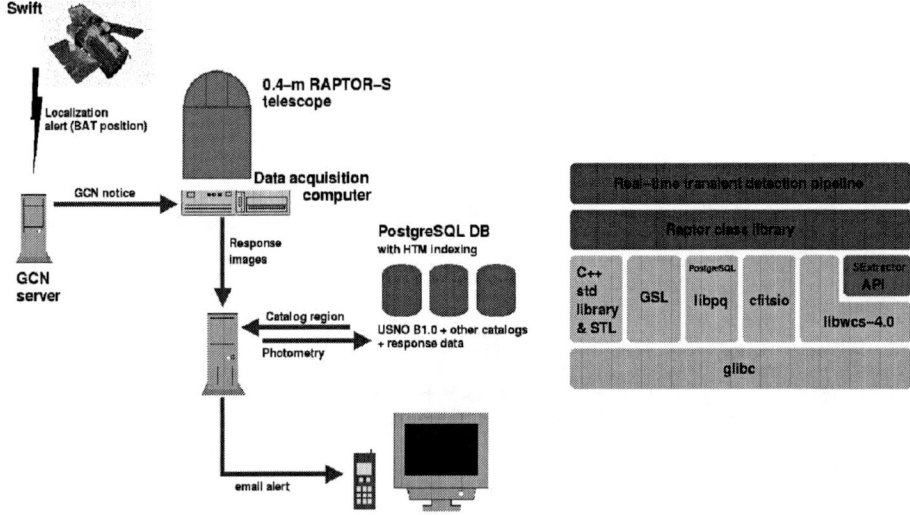

FIGURE 1. RAPTOR system for GRB followup and automated OT identification. Left: Schematic view of the system. Right: Architecture of the Raptor data processing software.

i.e. without the need to access all previous measurements in the response sequence. The selection criteria are as follows:

1. object is "in the error box"
   - within 2 nominal error radii of the trigger position and closer than 4 arc min.
   - or within 10 arc sec of the trigger position
2. object is variable
   - determined from the last series of consecutive 5-sigma detections
   - chi2/dof > 10.0 mag
   - scatter > 0.15 (r.m.s.)
3. object is well detected but not present in the catalog (currently USNO B1.0)
   - detected at > 5 sigma in at least 3 consecutive images
   - no catalog counterpart within 2.5 arc sec (unless it is at least 2.5 mag fainter)
4. various quality checks
   - reject problem flags (e.g. saturation and deblending problems)
   - reject poor quality frames and calibration problems
   - reject frames in which more than 5% of all objects are "variable"

3. DATA PROCESSING PIPELINE

Design considerations: The primary scientific goal of the RAPTOR project, i.e. the untriggered search for fast transients, places serious demands on software performance.

Ideally, one would like to meet the speed requirements without resorting to monolithic designs that are hard to maintain and evolve. The new generation Raptor pipeline is written in C/C++. We adopted an object oriented approach on top of a set of library APIs (Application Programming Interfaces) to promote good modularity. In particular, the SExtractor code [3] was modified to provide a convenient API (c.f. [4]). The use of a compiled language allows aggressive optimizations for speed critical processing.

Features:

- plugin architecture for algorithms
- SQL database connectivity with PostgreSQL
- flexible networked catalog querying with string interface
- extensive caching (e.g. calibration images and catalog regions)

4. PERFORMANCE

The automated pipeline delivers relative photometry accurate to about 1.5%, as measured by the r.m.s. magnitude scatter of bright unsaturated stars. Data processing speed was measured on a 64-bit machine with Intel Xeon CPU (3.6 GHz) and 2 GB of RAM. There is an overhead of about 2 seconds for the first image of the response sequence to send an SQL query to the database server and load the appropriate catalog region into memory. This overhead is subject to variability due to changes in network traffic. All remaining stages of the pipeline (basic corrections, source extraction, astrometric/photometric calibration, event identification and writing results to the database) are completed in less than 2 seconds for a typical 1k x 1k image taken in a medium density field. Presently, the largest overhead in response processing comes from image transfers between the DAQ computer and the pipeline computer. The latter delay will be eliminated when the rapid response pipeline is installed on a dedicated computer near the telescope.

5. FIRST RESULTS AND FUTURE DEVELOPMENTS

Raptor-S has been routinely responding to GCN alerts since the fall of 2004 and has detected optical couterparts to several GRBs including 041219a [5], 050319 [6] and 050820a [7]. The capability to automatically detect OTs in GRB response data and email notification were added to the system in October 2005. Over the time interval of ~ 2 months, the system responded to ~ 5 Swift/BAT localization alerts under conditions that could potentially result in OT detection. In that sample only GRB 051109a was detected by the automated pipeline (none of the remaining events showed detectable optical emission). The OT associated with GRB 051109a reached the peak brightness of ~ 15.1 mag [8] and triggered our event detection algoritm after the third 5-second image of the response sequence, thus validating our current event finding strategy. In Fig. 2 we present sample Raptor-S images from 051109a response data.

The present event detection criteria were designed towards maximizing the rate of successful detections rather than minimizing the number of false positives. This approach is

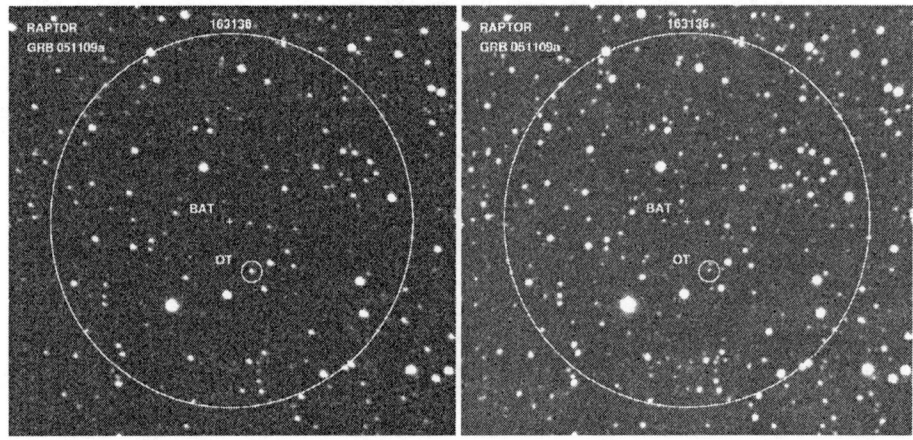

FIGURE 2. Images of the early afterglow of GRB 051109a taken by Raptor-S telescope during its automated response to Swift trigger 163136. Left: The first image in the response sequence, a 5-second exposure beginning 6.5 s after the GCN notice and 33.9 s after the trigger. Right: A 30-second exposure taken ~ 6 minutes later. The automated OT detection algorithm was triggered after the first three 5-second frames of the response sequence.

suitable for a triggered search down to magnitude 17.5–18.5 characteristic of Raptor-S telescope, given that initial positions from Swift/BAT are accurate to a few arc minutes and therefore the error box is not crowded. In the future we plan to implement "higher order" alerts that could be reported to transient alert networks (e.g. VOEvent initiative http://ivoa.net/twiki/bin/view/IVOA/VoeventWorkshop2) without human intervention.

ACKNOWLEDGMENTS

This work was performed as part of the Thinking Telescopes project supported by the Laboratory Directed Research and Development (LDRD) program at LANL. PRW was partially supported by the Oppenheimer Fellowship.

REFERENCES

1. Vestrand et al. 2002, Proceedings of the SPIE, 4845, 126
2. Vestrand et al. 2004, Astronomishe Nachrichten, 325, 549
3. E. Bertin & S. Arnouts 1996, Astronomy and Astrophysics Supplement, 117, 393
4. Galassi et al. 2003, ADASS XII, ASP Conference Series, 295, 225
5. Vestrand et al. 2005, Nature, 435, 178
6. Wozniak et al. 2005, Astrophysical Journal, 627, L13
7. Vestrand et al. 2006, AIP Conference Proceedings, this volume
8. Wozniak et al. 2005, GCN Circular, 4239

ROTSE-III Performance in the Swift Era

Yost, S. A.*, Rykoff, E. S.*, Aharonian, F.†, Akerlof, C. W.*,
Ashley, M. C. B.**, Barthelmy, S.‡, Gehrels, N.‡, Göğüş, E.§, Güver, T.§,
Horns, D.†, Kızıloğlu, Ü.¶, Krimm, H. A.‡, McKay, T. A.*, Mirabal, N.*,
Özel, M.‖, Phillips, A.**, Quimby, R. M.††, Rowell, G.†, Rujopakarn, W.*,
Schaefer, B. E.‡‡, Smith, D. A.§§, Swan, H. F.*, Vestrand, W. T.¶¶,
Wheeler, J. C.††, Wren, J.¶¶ and Yuan, F.*

*University of Michigan, 2477 Randall Laboratory, 450 Church St., Ann Arbor, MI, 48104
†Max-Planck-Institut für Kernphysik, Saupfercheckweg 1, 69117 Heidelberg, Germany
**School of Physics, University of New South Wales, Sydney, NSW 2052, Australia
‡NASA Goddard, Greenbelt, MD 20771
§Sabancı University, Orhanlı-Tuzla 34956 Istanbul, Turkey
¶Middle East Technical University, 06531 Ankara, Turkey
‖Çanakkale Onsekiz Mart Üniversitesi, Terzioğlu 17020, Çanakkale, Turkey
††Department of Astronomy, University of Texas, Austin, TX 78712
‡‡Department of Physics and Astronomy, Louisiana State University, Baton Rouge, LA 70803
§§Guilford College, Greensboro, NC 27410
¶¶Los Alamos National Laboratory, NIS-2 MS D436, Los Alamos, NM 87545

Abstract.
We report the successful performance of the Robotic Optical Transient Search Experiment (ROTSE) to promptly-disseminated Swift GRB triggers. ROTSE-III is a worldwide network of 4 unfiltered 0.45m optical telescopes. The telescopes operate robotically, automatically responding to GRB position triggers with a preset observations sequence. Including weather and other downtime, ROTSE-III can immediately respond to ∼40% of Swift trigger positions, with shutter open within approximately 6-8 seconds from the trigger dissemination. We discuss improvements made possible in the automated response the the small, accurate Swift error boxes. We report ROTSE-III's general results, including OTs discovered or confirmed, the distribution of imaging start times relative to the GRB duration, and an overview of OT lightcurves.

Keywords: gamma-rays: bursts, optical instruments
PACS: 98.70.Rz, 95.55.Cs

RESPONSE CAPABILITIES

ROTSE-III is a global network of four 0.45 m robotic, automated telescopes, designed for fast (∼6 sec) responses to GRB triggers. They have a wide (1.85×1.85 deg) field of view, and operate without filters. Under good conditions, they reach an R-equivalent limiting magnitude of ∼17 in a single 5 sec exposure, and ∼18.5 in a 60 sec image. The 4 ROTSE-III telescopes are located at Siding Springs Observatory, Australia (A), McDonald Observatory, Texas (B), the H.E.S.S. site in Namibia (C), and the Turkish National Observatory near Bakirlitepe, Turkey (D). At any time, at least one ROTSE is in astronomical night. The ROTSE-III telescopes are described in detail in [1].

ROTSE-III has responded promptly to 30 of the 85 rapid triggers (coordinate distribution delay < 1000 sec) since the *Swift* satellite went online in February 2005. Its response

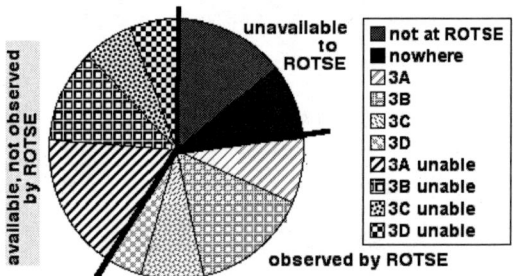

FIGURE 1. ROTSE-III's Responses by Telescope. Overall, ROTSE-III has responded promptly to 30 of the 85 rapidly-disseminated triggers since Swift went online (2005/02/15 – 2005/11/18), or 35%. The chart above shows the fraction of triggers promptly observed by each telescope, the location labelled by letters A-D (see text). Also indicated (darker slices) are the trigger fractions that could have been observed by the ROTSE sites if they had not been affected by weather or hardware downtime. When more than one site responded or was a potential responder, the trigger is subdivided to form the fractions given. Also indicated are the triggers that had a small ground window, not including the ROTSE sites, where the sun was down and the trigger was above the horizon at the burst time. The black slice indicates the triggers that could not be promptly observed from the ground (i.e., in the daytime sky).

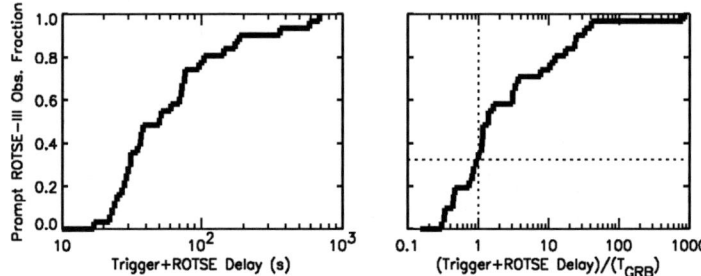

FIGURE 2. ROTSE-III Times From GRB Onset to Start of Imaging. ROTSE-III's 30 *Swift*-era rapid responses to non-delayed triggers are displayed as cumulative probability distributions. The delay from GRB onset to the distribution of a trigger is added to ROTSE's response time, and cases < 1000 sec define a rapid response. The top panel gives the response in sec, showing typical responses are ~ 30 sec, with a broad tail at 100 sec and above. The lower panel gives these rapid responses in units of the GRB duration. Approximately 1/3 of the responses begin during the GRB, and most have begun just after its end. The response at $\sim 800 \times$ the GRB duration is an ordinary response to a short GRB.

fraction of 35% includes weather and hardware downtime, and the distribution of burst trigger positions, as some are too close to the Sun for ground-based optical observing.

Fig. 1 charts the distribution of these responses among ROTSE sites. ROTSE-IIIB (Texas) has responded most frequently (12.5 times, note that responses by more than one telescope are divided between them), and ROTSE-IIID (Turkey) the fewest times (4). The observability distribution of rapid triggers has been uneven, with IIIC and IIID receiving fewer potential triggers than IIIA or IIIB. Approximately 10% of the rapid triggers have not been promptly observable from the ground, and a further 15% have had small prompt ground observability windows that did not include any ROTSE sites.

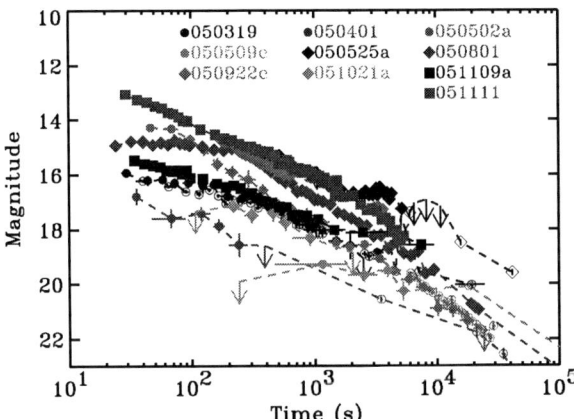

FIGURE 3. Early Lightcurves of OTs Detected by ROTSE-III in the *Swift* Era. ROTSE-III has detected 10 afterglows since February 2005. Their *R*-equivalent magnitudes are plotted here, with upper limits as arrows. ROTSE-III points are solid circles; in a few cases, additional data from the literature [3-11] are included as empty circles for comparison. The lightcurves are coded as indicated. Some show an underluminosity at the earliest times relative to the afterglow's subsequent evolution. There are no 990123-type cases with strong early overluminosity pointing to a reverse shock's emission.

RESPONSE AND REPORT SPEED

ROTSE-III's response speed, combined with promptly-disseminated GRB triggers (principally from *Swift*), facilitates early observations. As shown in Fig. 2 (top panel), a typical response time is 30-100 sec from γ-ray onset to the start of ROTSE's first 5 sec image. This total delay includes the computation and dissemination of the trigger position, as well as ROTSE's $\sim$6-8 sec slew & settle time. Only cases with a delay is $<$ 1000 sec are considered. A response time $>$ 1000 sec is delayed from ROTSE's perspective.

ROTSE-III's automated pipeline reduces images and flags potential optical counterparts. Collaborators then quickly evaluate candidates, allowing very rapid OT identifications. ROTSE-III presently holds the record for the fastest dissemination of an OT to the GCN Notices, just under 7 minutes from the GRB onset to the GCN submission (GRB 050801), as well as other cases with submission at $\sim$ 10 minutes. The next most-rapid cases are $\sim$ 1/2 hour from GRB to GCN.

ROTSE's typical response time is comparable to the duration of a long GRB. This capability yields cases of "prompt" optical observations, imaging of the GRB location while γ-ray photons are still being emitted. This is evident in the lower panel of Fig. 2, where the total delay as described above is given in units of the GRB duration (typically the reported T90). The response at 800× the GRB duration was for a normal $\sim$ 30 sec response to a short GRB.

ROTSE-III has detected optical photons contemporaneous with γ-ray emission for 3 GRBs of typical duration: 050401 [2], 051109A, and 051111 [Yost, Swan *et al.* in prep]. We have also observed 7 other GRB positions before the end of γ-ray emission, but no OTs were detected in those cases.

DETECTIONS

ROTSE-III has detected 10 OTs thus far in the Swift era. Their lightcurves in R-equivalent magnitudes are shown in Fig. 3. The earliest points are more densely sampled in the later events. *Swift*'s small, accurate error boxes have made an unjittered mode with only the central subframe read out possible, used for the first 10 (5-sec) exposures. This reduces the overhead from 9-10 sec to 2 sec per image.

There is significant lightcurve diversity, including a late rise (050509C), a long initial plateau (050801), and a steady decay connecting to observations by other instruments a day later (050401).

Some of these events are underluminous relative to back-extrapolations of later data. There is no strong evidence for early overluminosity like that expected from reverse shocks in these 10 cases. The bright reverse shock seen in 990123 remains unique among prompt GRB observations.

ACKNOWLEDGMENTS

The ROTSE project is made possible by grants from NASA, NSF, and the Australian Research Council, and through participation and support from the University of Michigan, University of Texas, Max-Planck-Institut fur Kernphysik, Los Alamos National Laboratory, and the University of New South Wales.

REFERENCES

1. C. W. Akerlof, R. L. Kehoe, T. A. McKay, E. S. Rykoff, D. A. Smith, D. E. Casperson, K. E. McGowan, W. T. Vestrand, P. R. Wozniak, J. A. Wren, M. C. B. Ashley, M. A. Phillips, S. L. Marshall, H. W. Epps, and J. A. Schier, *Pub. of the Astr. Soc. of the Pacific* **115**, 132–140 (2003).
2. E. S. Rykoff, S. A. Yost, H. A. Krimm, F. Aharonian, C. W. Akerlof, K. Alatalo, M. C. B. Ashley, S. D. Barthelmy, N. Gehrels, E. Göğüş, T. Güver, D. Horns, Ü. Kızıloğlu, T. A. McKay, M. Özel, A. Phillips, R. M. Quimby, W. Rujopakarn, B. E. Schaefer, D. A. Smith, H. F. Swan, W. T. Vestrand, J. C. Wheeler, and J. Wren, *Astrophys. Journal Letters* **631**, 121–124 (2005).
3. P. R. Woźniak, W. T. Vestrand, J. A. Wren, R. R. White, S. M. Evans, and D. Casperson, *Astrophys. Journal Letters* **627**, 13–16 (2005).
4. T. Yoshioka, C. W. Chen, S. Nishiura, M. Isogai, T. Soyano, W. P. Chen, Y. Urata, T. Tamagawa, K. Y. Huang, W. H. Ip, Y. Qiu and Y. Q. Lou, *GCN Notice* 3120 (2005)
5. R. McNaught and P.A. Price *GCN Notice* 3163 (2005)
6. P. D'Avanzo, D. Fugazza (INAF-OAB), N. Masetti (INAF-IASF, Bologna) and M. Pedani *GCN Notice* 3171 (2005)
7. B. Kahharov, M. Ibrahimov, D. Sharapov, A.Pozanenko (IKI), V.Rumyantsev, and G.Beskin *GCN Notice* 3174 (2005)
8. K. Misra, A. P. Kamble, and S. B. Pandey *GCN Notice* 3175 (2005)
9. J. Gorosabel, B. L. Jensen, G. Galaz, R. Salinas, J. Hjorth, J. P. U. Fynbo, H. Pedersen, D. Watson, P. Jakobsson, J. M. Castro Ceron *GCN Notice* 3425 (2005)
10. D. Malesani, S. Piranomonte, F. Fiore, G. Tagliaferri, D. Fugazza, and R. Cosentino *GCN Notice* 3469 (2005)
11. N. Mirabal, D. Bonfield, and K. Schawinski *GCN Notice* 3488 (2005)

Bumps in the optical afterglow of GRB 030329: refreshed shocks and evidence for internal shocks in GRBs?

F. Genet, F. Daigne and R. Mochkovitch

Institut d'Astrophysique de Paris
UMR 7095 CNRS – Université Pierre et Marie Curie-Paris VI, 98 Bd Arago, 75014 Paris, France

Abstract.
The afterglow of GRB 030329 presents a high variability, made of approximately constant duration bumps, which we explain by a series of late energy injections from slow shells (refreshed shocks). We model the GRB 030329 afterglow assuming that the central source first emits rapid material ($\Gamma \sim 100$) immediately followed by slower one ($\Gamma \sim 10$), which can catch up with the rapid material when it has been decelerated by the external medium. We fit the afterglow lightcurve, and extract from this fit some evidence that internal shocks must have occurred previously in the ejecta.

Keywords: Gamma-ray bursts, Shock waves
PACS: 98.70.Rz, 96.50.Fm

INTRODUCTION

The gamma ray burst of the 29th of March 2003 was one of the closest GRBs with known redshift and had the brightest afterglow to date. This afterglow showed several irregularities or bumps all observed after a break at about 0.5 day. Following Granot, Piran and Nakar [1] we discuss several possible origins for these bumps and concentrate on the "refreshed shock" hypothesis where slower shells progressively add their energy to the forward shock as it is decelerated by the external medium. In each of the successive bumps the increase in observed flux takes place on a short time scale, a condition not easy to satisfy. We concentrate on the first bump and present a detailed fit of its temporal evolution.

THE OPTICAL AFTERGLOW OF GRB 030329

The optical lightcurve of GRB 030329 first presents a power-law decay of slope $\alpha_1 \approx 0.9$ then, after a break at $t_{break} \approx 0.5$ day, a steeper decrease with $\alpha_2 \approx 1.9$ [2]. After the break the afterglow exhibits a high variability with many rebrightenings between 1.3 day and 6 days. The first and main bump begins at $t_{bump} \approx 1.3$ day, and has a rise time $\Delta t_{bump} \approx 0.3$ day. We then have $\Delta t_{bump} < t_{bump}$ which is a strong constraint on the possible mechanisms for the bump production. Before and after the bump the lightcurve approximately follows the same power-law, with a jump in observed flux of a factor ~ 2. The first bump is followed by at least two others well sampled ones, between 2.4 and 2.8 days and between 3.1 and 3.5 days.

ORIGIN OF THE BUMPS

Granot, Piran and Nakar [1] have considered several mechanisms that could give rise to a bump in the optical afterglow lightcurve, but all except one encounter problems that make them inappropriate in the case of GRB 030329. The first possibility discussed in [1] is that there is some variability in the external medium density. To explain the increase of observed flux after each bump, the external medium density would have to increase with radius in discret steps (rather than in a smooth manner). Such a density profile seems quite unlikely and moreover cannot easily explain the fact that $\Delta t_{bump} < t_{bump}$.

A second possibility would be to invoke the patchy shell model [3] where the energy per unit solid angle injected in the GRB outflow is not constant. However if we interpret the break in the lightcurve as jet break, the whole emitting surface is visible when the bumps are observed, so that the effect induced by a non-uniformity of energy injection should be negligible.

The last option considered in [1] is that energy is added to the forward shock at late times when slower material emitted by the central source can eventually catch up. This seems to be the most promising explanation, as we show below.

REFRESHED SHOCKS: LATE ENERGY INJECTION FROM A SLOW SHELL

We suppose that the central source has emitted some fast material with an average Lorentz factor $\Gamma_f \geq 100$, responsible for the prompt GRB emission, together with some slower material with an average Lorentz factor $\Gamma_s \sim 10$. We assume that these rapid and slower portions of the ejecta are produced simultaneously. The slow material initially lags behind before it becomes able to catch up with the rapid one which is progressively decelerated by the external medium.

To obtain an analytical estimate of the time of collision, we consider only two shells, representative of the fast and slow material. We make the simplifying assumption that the fast shell moves at a constant Lorentz factor $\Gamma = \Gamma_f$ until the deceleration radius R_D after which it follows the Blandford-McKee solution. Since the slow shell is "protected" by the preceding fast material, it moves at the constant Lorentz factor Γ_s. It can then be shown that the two shells collide at a time

$$t_{shock} \approx 1.66 E_{53}^{1/3} n^{-1/3} \Gamma_{s,10}^{-8/3} \quad \text{days} \tag{1}$$

for a uniform external medium of density n; E_{53} is the burst isotropic energy in unit of 10^{53} erg and $\Gamma_{s,10}$ the Lorentz factor of the slow shell in unit of 10. With $E_{53} = 1$ and $n = 1$, $t_{shock} = 1$ day for $\Gamma_s \approx 12$. At the time $t_{shock} \sim t_{bump}$ the slow material adds its energy to the forward shock, producing the bump. The time of the bump will however be spread over a duration Δt_{bump} for two different reasons:

- Angular spreading: Before the break the angular spreading time is simply given by $\Delta t_{ang} = \frac{R}{2c\Gamma^2} \gtrsim t_{obs}$ while after the break we have

$$\Delta t_{ang} = \frac{R\theta_s^2}{2c} = \Delta t_b \left(\frac{R}{R_b}\right) = \Delta t_b \left(\frac{t_{obs}}{t_b}\right)^a \quad (2)$$

where the index "b" refers to quantities at the moment of the break and θ_s is the opening angle for the slow material. To obtain Eq.(2) Granot, Nakar and Piran [1] supposed that the slow material, in the wake of the forward shell, stays cold and does not expand laterally so that $\theta_s = \theta_0$, the opening angle at the source. The exponent a depends on the dynamical behavior of the forward shell: $a \approx 0$ for lateral spreading at the local sound speed while $a = 1/4$ for negligible spreading and a uniform external medium. Since in GRB 030329 the bumps are observed after the break Eq.(2) applies and allows to have $\Delta t_{ang} < t_{obs}$.

- Dispersion in the Lorentz factor of the slow material: Equation (1) gives the moment of the shock for a fixed value Γ_s of the Lorentz factor of the slow material but the distribution of Γ_s has a non zero width $\Delta\Gamma_s$ leading to

$$\frac{\Delta t_{shock}}{t_{shock}} = \frac{8}{3}\frac{\Delta\Gamma_s}{\Gamma_s} \quad (3)$$

Since $\Delta t_{bump}/t_{bump} < 0.3$ for the bumps of GRB 030329, this limits the allowed dispersion in Γ_s to only $\Delta\Gamma_s/\Gamma_s < 0.1$

AFTERGLOW CALCULATION

In order to test the refreshed shock hypothesis we fitted the afterglow evolution between one and two days which corresponds to the moment of the first bump. We assumed that the central source emitted simultaneously fast material of Lorentz factor $\Gamma_f = 100$ and slower one with $\Gamma_s \approx 8$. The isotropic energy in the fast moving material is $E_f = 10^{53}$ erg and the density of the (uniform) external medium is $n = 20$ cm^{-3}. The initial opening angle of the jet is $\theta_0 = 3°$ but it has extended to $\theta_{shock} = 9°$ when the slow material catches up.

Following [1] we assumed that the slow material does not expand laterally. We moreover adjusted its Lorentz factor distribution between 7.5 to 8.5 to reproduce the shallow part of the light curve just before the bump rise. The resulting fit is shown in Fig.1.

DISCUSSION AND CONCLUSION

The good fit of the bump shown in Fig.1 critically depends on the assumption of a narrow distribution of the Lorentz factor $\Delta\Gamma_s/\Gamma_s < 0.1$ in the slow material. Such a condition may appear quite artificial. How can the central source generate a distribution

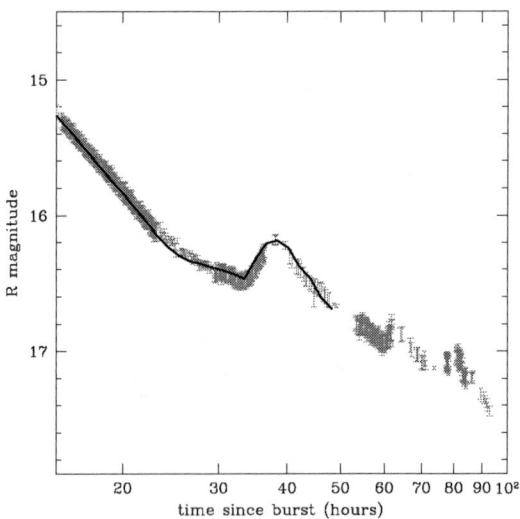

FIGURE 1. The first bump in the GRB 030329 afterglow light curve. The thick full line is our best fit.

of Γ_s with such a small dispersion while the internal shock model instead requires large fluctuations in the Lorentz factor? The answer may come from the internal shock model itself. During internal shocks fast moving material catches up with slower one and mass accumulates at the original location of the slow shells (in the flow rest frame). At the end of the internal shock phase the distribution of Lorentz factor steadily increases outwards in the ejecta with a succession of mass clumps sharing nearly the same Γ value. Such a behavior suggested by simplified models [4] has been confirmed by detailed hydrodynamical simulations [5].

The successive bumps in the GRB 030329 afterglow therefore probably results from the arrival at the forward shock of material having previously experienced internal shocks. Taking place at a Lorentz factor $\Gamma < 10$ these internal shocks will occur while the outflow still has a large optical depth. The radiation produced by the shocked material will be reprocessed and no observational signature will be detected.

REFERENCES

1. Granot, J., Nakar & E.,Piran, T., 2003, Nature 426, 138.
2. Lipkin, Y.M. et al, 2004, ApJ 606, 381.
3. Kumar, P. & Piran, T., 2000, ApJ 535, 152.
4. Daigne, F. & Mochkovitch, R., 1998, MNRAS 296, 275.
5. Daigne, F. & Mochkovitch, R., 2000, A&A 358, 1157.

The broadband afterglow of GRB 030328

E. Maiorano*, N. Masetti*, E. Palazzi*, S. Savaglio[†], E. Rol**, P.M. Vreeswijk[‡], E. Pian[§], P.A. Price[¶], B.A. Peterson[||], M. Jelínek[††], S.B. Pandey[††], M.I. Andersen[‡‡] and A.A. Henden[§§]

INAF-IASF, Bologna, Italy
[†]*The Johns Hopkins Univ., Baltimore, USA*
**Univ. of Leicester, UK*
[‡]*ESO-Santiago, Chile*
[§]*INAF-Oss. Astron. Trieste, Italy*
[¶]*Univ. of Hawaii, Honolulu, USA*
[||]*Australian National Univ., Weston, Australia*
[††]*IAA-CSIC, Granada, Spain*
[‡‡]*AIP, Potsdam, Germany*
[§§]*US Naval Observatory, Flagstaff, USA*

Abstract. We here report on the photometric, spectroscopic and polarimetric monitoring of the optical afterglow of the Gamma-Ray Burst (GRB) 030328 detected by *HETE-2*. We found that a smoothly broken power-law decay provides the best fit of the optical light curves, with indices $\alpha_1 = 0.76 \pm 0.03$, $\alpha_2 = 1.50 \pm 0.07$, and a break at $t_b = 0.48 \pm 0.03$ d after the GRB. Polarization is detected in the optical V-band, with $P = (2.4 \pm 0.6)$ % and $\theta = 170° \pm 7°$. Optical spectroscopy shows the presence of two absorption systems at $z = 1.5216 \pm 0.0006$ and at $z = 1.295 \pm 0.001$, the former likely associated with the GRB host galaxy. The X–ray-to-optical spectral flux distribution obtained 0.78 days after the GRB was best fitted using a broken power-law, with spectral slopes $\beta_{opt} = 0.47 \pm 0.15$ and $\beta_X = 1.0 \pm 0.2$. The discussion of these results in the context of the 'fireball model' shows that the preferred scenario is a fixed opening angle collimated expansion in a homogeneous medium.

Keywords: gamma-ray bursts; astronomical observations: visible; photometry; spectroscopy; polarimetry
PACS: 95.75.De; 95.75.Fg; 95.75.Hi; 95.85.Kr; 98.70.Rz

INTRODUCTION

GRB 030328 was a long, bright GRB detected on 2003 Mar 28.4729 UT, by the FREGATE, WXM, and SXC instruments onboard *HETE-2*, and rapidly localized with sub-arcminute accuracy (Villasenor et al. 2003). About ~ 1 hour after the GRB, its optical afterglow has been detected by the 40-inch Siding Spring Observatory (SSO) telescope (Peterson & Price 2003). A study of the X–ray afterglow of GRB 030328 was performed by Butler et al. (2005) using *Chandra* data.

We report here on the study of the optical afterglow emission of GRB 030328 made, within the GRACE[1] collaboration, performed with 7 different optical telescopes. A more detailed presentation of these data will appear in Maiorano et al. (2005).

---

[1] GRB Afterglow Collaboration at ESO; see http://www.gammaraybursts.org/grace/

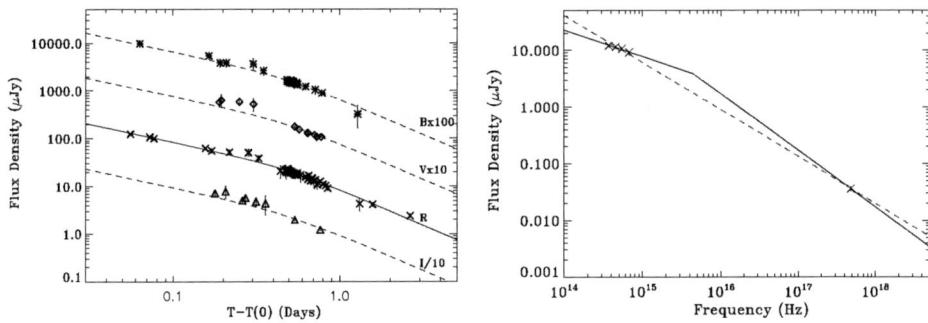

FIGURE 1. *Left panel*: BVRI light curves of the GRB 030328 afterglow. The solid and dashed lines represent the best-fit of the optical light curves. *Right panel*: broadband afterglow spectrum at 0.78 d after GRB 030328. The dashed line corresponds to a single optical-X–rays powerlaw spectral slope; the broken solid line indicates the description with v_c between the optical and the X–ray ranges.

OBSERVATIONS

Optical *UBVRI* data of the GRB 030328 Optical Transient (OT), for a total of 130 photometry points, were acquired at the 40-inch SSO (Australia), 1m ARIES (India), 2.5m NOT (Spain), 1.54m ESO Danish, 2.2m ESO/MPG, ESO VLT-*Antu* (Chile) and 1m USNO-FS (USA) telescopes.

A series of six 10-min optical spectra was obtained at ESO-Paranal with VLT-*Antu* starting 0.59 d after the GRB. The Grism 300V was used with a nominal spectral coverage of 3600–8000 Å and a spectral dispersion of 2.$''$7 Å/pixel.

Linear polarimetry *V*-band observations were acquired between 0.66 and 0.88 d after the GRB at VLT-*Antu*. Five complete imaging polarimetry cycles were performed.

RESULTS

Photometry

In Fig. 1 (left) we plot our photometric measurements together with those reported in the GCN circulars[2]. For the cases in which no error was reported, a 0.3 mag uncertainty was assumed. The *UBVRI* zero-point calibration was performed using the photometry by Henden (2003).

The optical data of Fig. 1 (left) were corrected for the Galactic foreground reddening assuming $E(B-V) = 0.047$ mag (Schlegel et al. 1998). The GRB 030328 host galaxy emission in the *BVRI* bands was computed from the data of Gorosabel et al. (2005) and subtracted from our optical data set.

---

[2] http://gcn.gsfc.nasa.gov/gcn/gcn3_archive.html

The best fit of the R-band data (in Fig. 1, left) is obtained using a smoothly broken powerlaw (Beuermann et al. 1999), with temporal indices $\alpha_1 = 0.76 \pm 0.03$ and $\alpha_2 = 1.50 \pm 0.07$ before and after a break occurring at $t_b = 0.48 \pm 0.03$ d from the GRB trigger, and with $s = 4.0 \pm 1.5$ the parameter modeling the slope change rapidity. This best-fit curve describes well the data in the other bands also (see Fig. 1, left). This means that the decay of the OT can be considered as achromatic.

From the measured jet break time we can compute the jet opening angle value for GRB 030328 which is, following Sari et al. (1999), $\theta_{jet} \sim 3°\!.2$.

Broadband analysis

By using the available information, we have constructed the optical-to-X–ray spectral flux distribution of the GRB 030328 afterglow at the epoch 0.78 days after the GRB, that is, the time with best broadband photometric coverage.

As Fig. 1 (right) shows, the best descriptions are a single powerlaw (dashed line) with a spectral index $\beta_{X-opt} = 0.83 \pm 0.01$, or a broken powerlaw with $\beta_{opt} = 0.47 \pm 0.15$ and assuming $\beta_X = 1.0 \pm 0.2$ from Butler et al. (2005).

However, the single powerlaw description of the broadband afterglow does not fit any of the synchrotron fireball scenarios (Sari et al. 1998, 1999). Instead, in the broken powerlaw description, which means that the synchrotron cooling frequency ν_c lies between the optical and X–ray bands, we obtain that the GRB 030328 afterglow broadband evolution is consistent with a jet-collimated expansion in a homogeneous medium with fixed opening angle (Mészáros & Rees 1999) and with an electron distribution index $p = 2$.

Assuming a negligible host absorption and using the optical and X–ray spectral slopes above, we obtain for ν_c the value 5.9×10^{15} Hz, which places this frequency in the ultraviolet band.

Spectroscopy

Figure 2 shows the spectrum of the GRB 030328 OT. Most of the significant features can be identified with Fe II, Mg II, Al II and C IV absorption lines in a system at a redshift of $z = 1.5216 \pm 0.0006$. These lines are associated with the circumburst gas or interstellar medium in the GRB host galaxy. A lower redshift absorption system at $z = 1.295 \pm 0.001$ is also found: for it, only two lines can be identified (Fe II $\lambda 2600$ and the unresolved Mg II $\lambda\lambda 2796, 2803$ doublet). Its detection indicates the presence of a foreground absorber.

Polarimetry

After correcting for spurious field polarization, we found $Q_{OT} = 0.029 \pm 0.008$ and $U_{OT} = -0.004 \pm 0.008$. The fit of the data with the relation of Di Serego Alighieri (1997)

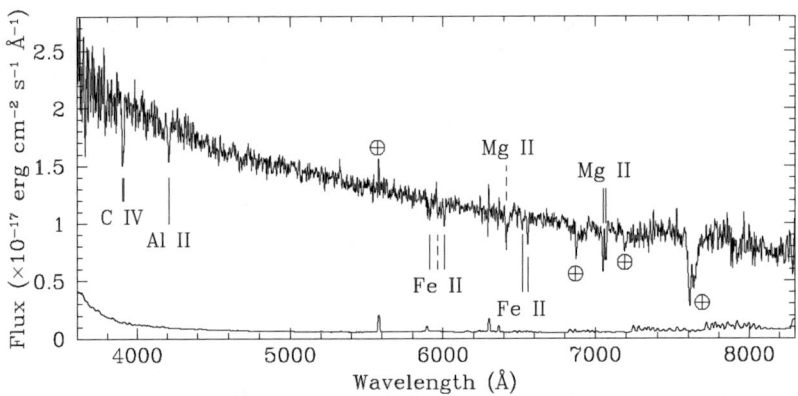

FIGURE 2. Optical spectrum (and corresponding Poisson error) of the afterglow of GRB 030328. Two systems at redshift $z = 1.5216$ and $z = 1.295$ are apparent: the former (solid hyphens) likely originates in the host galaxy, whereas the latter (dashed hyphens) is produced by a foreground absorber. The symbol ⊕ indicates atmospheric and telluric features.

yielded for the OT a linear polarization $P = (2.4 \pm 0.6)$ % and a polarization angle $\theta = 170° \pm 7°$, corrected for the polarization bias (Wardle & Kronberg 1974).

In order to check whether variations of P and θ occurred during the polarimetric run, we also separately considered each of the 5 single polarimetry cycles. Although with lower S/N, P and θ are consistent with being constant across the whole polarimetric observation run.

REFERENCES

12. Beuermann, K., Hessman, F.V., Reinsch, K., et al. 1999, A&A, 352, L26
12. Butler, N.R., Ricker, G. R., Ford, P.G., et al. 2005, ApJ, 629, 908
12. Di Serego Alighieri, S., 1997, in: Instrumentation for large telescopes, ed. J.M. Rodriguez Espinosa, A. Herrero, & F. Sánchez (Cambridge: Cambridge Univ. Press), p. 287
12. Gorosabel, J., Jelinek, M., de Ugarte Postigo, A., Guziy, S., Castro-Tirado, A.J. 2005, in: 4th Workshop on Gamma-Ray Bursts in the Afterglow Era, ed. L. Piro, L. Amati, S. Covino, & B. Gendre, Il Nuovo Cimento, in press [astro-ph/0504059]
12. Henden, A.A. 2003, GCN Circ. 2114
12. Maiorano, E., Masetti, N., Palazzi, E., et al., 2005, A&A, submitted
12. Mészáros, P., & Rees, M.J. 1999, MNRAS, 306, L39
12. Peterson, B.A., & Price, P.A. 2003, GCN Circ. 1974
12. Sari, R., Piran, T., & Narayan, R. 1998, ApJ, 497, L17
12. Sari, R., Piran, T., & Halpern, J.P. 1999, ApJ, 519, L17
12. Schlegel, D.J., Finkbeiner, D.P. & Davis, M. 1998, ApJ, 500, 525
12. Villasenor, J., Crew, G., Vanderspek, R., et al. 2003, GCN circ. 1978
12. Wardle, J.F.C., & Kronberg, P.P. 1974, ApJ, 194, 249

Colors of optical afterglows of GRBs in analysis of the dust in their environment

V. Šimon*, R. Hudec* and G. Pizzichini[†]

*Astronomical Institute, Academy of Sciences of the Czech Republic, 251 65 Ondřejov, Czech Republic
[†]INAF/IASF Bologna, via Gobetti 101, 40129 Bologna, Italy

Abstract. We find that the behaviour of the color indices of the optical afterglows (OAs) of GRBs can be divided into several intervals during the first several tens of days. We present an analysis of the color variations in the very early phases of the OAs and their relation to those of the ensemble of 25 OAs studied by [8]. Our analysis of the Swift/UVOT observations of the very early phase of the OA of GRB050922C reveals chromatic variations which possibly suggest a multi-component, rapidly varying spectral profile during $t - T_0 < 0.01$ days. The colors of this OA do not show any significant intrinsic extinction and put constraints on the changes of the extinction caused by the GRB and/or its OA. We argue that the ensemble of OAs studied here is not reddened. A very broad range of their redshifts ($0.17 < z < 3.5$) suggests that these GRBs presumably occur in the different times of the evolution of their host galaxies, and hence in the regions with different metallicity and dust content. Although several mechanisms can play a role, our finding shows that all of them must be able to provide a negligible extinction for the events analysed here. We briefly discuss also the implications for the orphan afterglows.

Keywords: radiation mechanisms: non-thermal; dust, extinction; galaxies, starburst; gamma-rays: bursts
PACS: 95.85.Kr; 95.85.Pw; 97.10.Zr; 98.54.Ep; 98.58.-w; 98.70.Rz

INTRODUCTION

Color indices of optical afterglows (OAs) of GRBs are a powerful and innovative approach to the study of such events. Among others, they help us to search for the common properties of the afterglows, resolve among the individual radiation mechanisms, search for the relations among colors, luminosities and the decay rates of the OAs (if the redshift z is known). Here we will concentrate on the analysis of the color variations in the very early phases of the OAs and on their impact on the investigation of the properties of the local interstellar medium of GRBs in their host galaxies.

COLLECTION AND ANALYSIS OF THE DATA

This analysis concentrates on the very early time intervals ($t - T_0 < 0.05$ days) of the OAs of the long GRBs and puts the findings into the context of the papers by [7, 8]. The magnitudes of the OA of GRB050502A were taken from the ground-based observations by [1]. Swift/UVOT data were used for the OA of GRB050922C [3, 4]. The color indices of the ensemble of 25 OAs of the long GRBs were taken from our earlier analyses [7, 8]. The magnitudes and color indices were corrected for

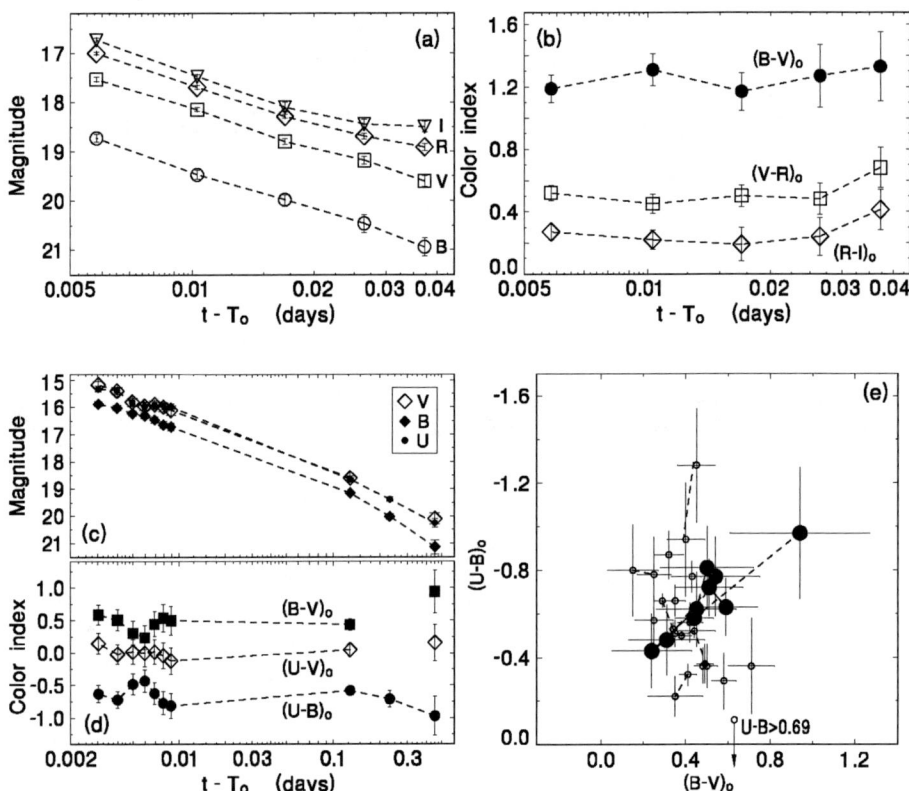

FIGURE 1. (a) The light curves of the OA of GRB050502A using the interpolated data of [1] to obtain the color indices (b). (c) The light curves of the OA of GRB050922C from Swift/UVOT data calibrated by [3, 4]. We interpolated them to obtain the color indices in (d); notice the anti-correlated variations in $(B-V)_0$ and $(U-B)_0$. (e) The color-color diagram for the OA of GRB050922C. The colors of the ensemble of the OAs (0.14 days $< t - T_0 <$ 10.2 days) [8] are displayed for comparison and show that, on average, the colors of the OA of GRB050922C are consistent with them even in the very early phase.

the Galactic reddening according to [6]. Because of space limitations, the reader is referred to http://gcn.gsfc.nasa.gov/gcn/gcn3_archive.html (GCN Circulars, ed. S. Barthelmy) and J. Greiner's Web page http://www.mpe.mpg.de/~jcg/grbgen.html for full bibliographic references on each OA. The details of the data analysis and the determination of the color indices for the OAs can be found in [7].

RESULTS

We find that the behaviour of the color indices of the OAs can be approximately divided in the following way.

Very early phases of the OAs ($t - T_0 < 0.05$ days): $(V-R)_0$ and $(R-I)_0$ of the OA of

GRB050502A are consistent with those of the ensemble of unreddened OAs already at $t - T_0 \approx 0.005$ days (Fig. 1ab). The effect of the Lyman break is visible only in $(B-V)_0$.

Swift/UVOT observations of GRB050922C show chromatic variations in the very early phase of the event, during $t - T_0 < 0.01$ days (Fig. 1cd). We find the anti-correlated variations in $(B-V)_0$ and $(U-B)_0$ at roughly constant $(U-V)_0$. The variations in the U and V magnitudes are caused by a transient increase of the decay rate of this OA. The color-color diagram in Fig. 1e shows the variations in $(B-V)_0$ and $(U-B)_0$ to be inconsistent with the changes of the reddening, because they occur roughly perpendicularly with respect to the reddening vector (see [7] for details). It can be seen that the colors of the OA of GRB050922C are consistent with the ensemble of the OAs from [8]. We interpret these changes in terms of a rapidly varying multi-component spectrum.

We thus argue that the above-mentioned three OAs do not show any sign of a variable (decreasing) extinction even in the very early phases of their evolution.

Medium phases of the OAs (0.14 days$< t - T_0 <$ 10.2 days): As we have shown [7, 8], the color indices of 25 OAs of the long GRBs form a distinct ensemble in the color-color diagrams in the observer frame in this time interval, which suggests that their spectral shape must be quite smooth, with no bumps or strong lines, between the observed B to I passbands (2000–5600 Å in the rest frame), and is consistent with a power-law spectral distribution, in accordance with the model by [5]. The shape of the spectrum of these OAs does not change significantly while their luminosity decreases by a large amount. We note that this ensemble appears to be limited by the increasing effect of the Lyman break at $z > 3.5$, with the increasing strength in the blue bands for the OAs at higher z, namely GRB050502A ($z = 3.793$) and GRB000131 ($z = 4.5$).

This time interval appears to be suitable for the investigation of the dust not destructed by the initial flash. We find no sign of any elongation of the pattern along the reddening vectors to be apparent for these OAs with $z < 3.5$.

Late epochs of the OAs ($t - T_0 >$ 10 days): We argue that the colors of the OA at these phases, caused by an emerging supernova (SN), can mimic those of a reddened afterglow or a dust condensation, and a careful analysis is necessary to resolve the correct mechanism. The shifts of the OA of GRB030329 in $(V-R)_0$ vs. $(R-I)_0$ and $(B-V)_0$ vs. $(V-R)_0$ diagrams, and especially the colors in $t - T_0 \approx 35$ days, can serve as an example [9]. In this case, the colors in $t - T_0 \approx 35$ days are caused by the underlying SN, but they also lie on the reddening vectors. We note that all the GRBs studied here and in our previous papers are long ones, so for example the contribution of the emerging SN may not apply to short GRBs.

Implication of the strong concentration of most colors of OAs: The intrinsic reddening of the GRBs studied here inside their host galaxies must be quite similar and relatively small. In case of a large reddening it would be quite unlikely to obtain such similar values of extinction in all cases. Several possibilities can be considered: (a) these GRBs lie on earth-watching side of a star-forming region; (b) the density and dust abundance of the local interstellar medium are substantially reduced by intense radiation (e.g. models by [11]); (c) the dust grain size distribution is biased toward large grains [10]; (d) low metallicity (e.g. [2]); (e) low dust-to-gas ratio [2]. As already stated, a very broad range of redshifts ($0.17 < z < 3.5$) suggests that these GRBs presumably occur in the different times of the evolution of their host galaxies, and hence in the regions with different metallicity and dust content. It is thus quite likely that several mechanisms can play a role but our finding shows that all of them must be able to provide a negligible extinction for the events analysed here. Although the color indices in the very early phase are available only for three OAs, they put strong constraints on any evolution of the extinction caused by the GRB and/or its OA.

Implications for the orphan afterglows: If the dust destruction is caused mainly by the high-energy flash, then the orphan afterglows can be still heavily absorbed due to the smaller opening angle of the gamma and X-rays in comparison with the opening angle of the optical emission. Such orphan OAs would be considerably fainter and redder, and significantly more difficult to detect in comparison with the OAs of the observed GRB.

ACKNOWLEDGMENTS

This study was supported by the grant A3003206 provided by the Grant Agency of the Academy of Sciences of the Czech Republic, the project ESA PRODEX INTEGRAL 90108, ESA PECS project 98023, and the CNR-AVČR collaborative project Investigation of GRBs (2004/2007).

REFERENCES

1. Guidorzi, C., Monfardini, A., Gomboc, A., et al., *ApJ*, **630**, L121 (2005).
2. Hjorth, J., Moller, P., Gorosabel, J., et al., *ApJ*, **597**, 699 (2003).
3. Li, W., Jha, S., Filippenko, A. V., et al., *GCN 4095* (2005).
4. Li, W., Jha, S., Filippenko, A. V., et al., *astro-ph/0505504* (2005).
5. Sari, R., Piran, T., and Narayan, R., *ApJ*, **497**, 17 (1998).
6. Schlegel, D. J., Finkbeiner, D. P., and Davis, M., *ApJ*, **500**, 525 (1998).
7. Šimon, V., Hudec, R., Pizzichini, G., and Masetti, N., *A&A*, **377**, 450 (2001).
8. Šimon, V., Hudec, R., Pizzichini, G., and Masetti, N., "Colors of Optical Afterglows of GRBs and Their Time Evolution," in **Gamma-Ray Bursts: 30 Years of Discovery: Gamma-Ray Burst Symposium**, edited by E. E. Fenimore and M. Galassi, AIP Conference Proceedings 727, American Institute of Physics, New York, 2004, pp. 487–490.
9. Šimon, V., Hudec, R., and Pizzichini, G., *A&A*, **427**, 901 (2004).
10. Stratta, G., Perna, R., Lazzati, D., et al., *A&A*, **441**, 83 (2005).
11. Waxman, E., and Draine, B. T., *ApJ*, **537**, 796 (2000).

LATE AFTERGLOWS & GRB/SN CONNECTION

Supernova and GRB connection: Observations and Questions

Massimo Della Valle

INAF-Arcetri Astrophysical Observatory, Largo E. Fermi, 5 Firenze, Italy
Kavli Institute for Theoretical Physics, University of California, Santa Barbara, CA 93106

Abstract. We review the observational status of the supernova/gamma-ray burst connection. Present data suggest that SNe associated with GRBs form a heterogeneous class of objects including both bright and faint hypernovae and perhaps also 'standard' Ib/c events. Evidence for association with other types of core-collapse SNe (e.g. IIn) is much weaker. After combining the local GRB rate with the local SN-Ibc rate and beaming estimates, we find the ratio GRB/SNe-Ibc in the range $\sim 0.5\% - 4\%$. In most SN/GRB associations so far discovered, the SN and GRB events appear to go off simultaneously. In some cases data do not exclude that the SN explosion may have preceded the GRB by a few days. Finally we discuss a number of novel questions started by recent cases of GRB-SN associations.

Keywords: Supernovae, Gamma Rays
PACS: 97.60

INTRODUCTION

Intensive optical, infrared and radio follow-up of GRBs, occurring in the last decade, has established that long-duration GRBs (Klebesadel 1990; Dezalay et al. 1992; Kouveliotou et al. 1993), or at least a significant fraction of them, are directly connected with supernova explosions. Most of the evidence arises from observations of supernova features in the spectra of a few GRB afterglows. Examples of the SN/GRB connection include SN 1998bw/GRB 980425 (Galama et al. 1998), SN 2003dh/GRB 030329 (Stanek et al. 2003, Hjorth et al. 2003), SN 2003lw/GRB 031203 (Malesani et al. 2004), SN 2002lt/GRB 021211 (Della Valle et al. 2003), XRF 020903 (Soderberg et al. 2005), SN 2005nc/GRB 050525A (Della Valle et al. 2006) and more recently SN 2006aj/GRB 060218 (Masetti et al. 2006; Modjaz et al. 2006, Campana et al. 2006; Sollerman et al. 2006; Pian et al. 2006; Mirabal et al. 2006; Cobb et al. 2006). In addition there are about a dozen afterglows which show, days to weeks after the gamma-ray events, rebrightenings and/or flattenings in their lightcurves (e.g. Zeh et al. 2004). These bumps are interpreted as SNe emerging out of their afterglows (Bloom et al. 1999, Castro-Tirado & Gorosabel 1999). The detection of star-formation features in the host galaxies of GRBs (Djorgovski et al. 1998, Fruchter et al. 1999) provided the earliest hint for the existence of a link between GRBs and the death of massive stars. Le Floc'h et al. (2003) and Christensen, Hjorth & Gorosabel (2004) have found that GRB hosts are galaxies with a fairly high (relative to the local Universe) star formation of the order of 10 $M_\odot$ yr$^{-1}/L^\star$ or more, while recent studies on the parent galaxies of GRBs (Conselice et al. 2005, Wainwright et al. 2005) show that a significant fraction of GRB hosts (∼ 50%) exhibit a merger/disturbed morphology.

GRB 980425/SN 1998BW

SN 1998bw was the first SN discovered spatially and temporally coincident with a GRB (GRB 980425; Galama et al. 1998). It was discovered in the nearby galaxy ESO 184-G82 at 40 Mpc. This implied that GRB 980425 was underenegetic by about 3–4 orders of magnitude with respect to the "standard" γ-energy budget of $\sim 10^{51}$ erg (Frail et al. 2001, Panaitescu & Kumar 2001). The associated SN was extremely energetic with expansion velocities 3-4 times higher than those exhibited by normal Ib/c SNe (Patat et al. 2001). The theoretical modeling of the light curve and spectra (Iwamoto et al. 1998; Woosley, Eastman, & Schmidt 1999) suggests that SN 1998bw originated in a $\sim 40 M_\odot$ star (on the main sequence), with a C+O core of about $\sim 10 M_\odot$. This picture is supported by the radio properties of SN 1998bw (Kulkarni et al. 1998, Weiler et al. 2002) that can be explained in terms of an interaction of a mildly relativistic ($\Gamma \sim 1.8$) shock with a dense circumstellar medium fed by the strong wind of the massive envelope-stripped progenitor. Recently, Maeda et al. (2006), after analyzing the Fe and [O I] line profiles in the nebular spectra of SN 1998bw, give support to the idea that SN 1998bw was the product of an asymmetric explosion (see also Höflich, Wheeler & Wang 1999) viewed relatively off-axis from the jet direction, $\theta < 30°$ (Maeda et al. 2006) and $\theta > 15°$ (Yamazaki et al. 2003). An asymmetric explosion decreases the kinetic energy input by a factor $\lesssim 3$ (in Maeda et al. models) with respect to spherically symmetric models.

However, the association between two peculiar astrophysical objects, such as SN 1998bw and GRB 980425 was not taken as proof of a general SN/GRB connection.

GRB 021211/SN 2002LT

GRB 021211 was detected by the HETE–2 satellite (Crew et al. 2003), allowing the localization of its optical afterglow (Fox et al. 2003) and the measurement of its redshift $z = 1.006$ (Vreeswijk et al. 2006). Figure 1 shows the result of the late-time photometric follow-up, carried out with the ESO VLT–UT4 (Della Valle et al. 2003), together with observations collected from literature. A rebrightening is apparent, starting ∼ 15 days after the burst and reaching the maximum ($R \sim 24.5$) during the first week of January. For comparison, the host galaxy has a magnitude $R = 25.22 \pm 0.10$, as measured in late-time images. The spectrum of the bump (Fig. 1, right panel) obtained during the rebrightening phase is characterized by a broad absorption, the minimum of which is measured at ∼ 3770 Å (in the rest frame of the GRB). The comparison with the spectra of other SNe supports the identification of the broad absorption with a blend of the Ca II H and K absorption lines. The blueshifts corresponding to the minimum of the absorption and to the edge of the blue wing imply velocities $v \sim 14\,400$ km/s and $v \sim 23\,000$ km/s, respectively. In Fig. 1 the light curve of SN 1994I (dereddened by $A_V = 2$ mag) is added to the afterglow and host contributions, after applying the appropriate K-correction (solid line). As it can be seen, this model reproduces well the shape of the observed light curve. It is interesting to note that SN 1994I (the spectrum of which provides the best match to the observations) is a "standard" type-Ic event (Filippenko et al. 1995) rather than a bright *hypernova* (HN) (a hypernova is a broad-line type-Ibc SN) as the ones proposed for association with other long-duration GRBs. However we note that even if the pre-maximum spectra of some HNe (e.g. 2002ap, Mazzali et al. 2002) show significantly broader lines than our case, this difference vanished after maximum, such that it may not be easy to distinguish at later stages between the two types of SNe.

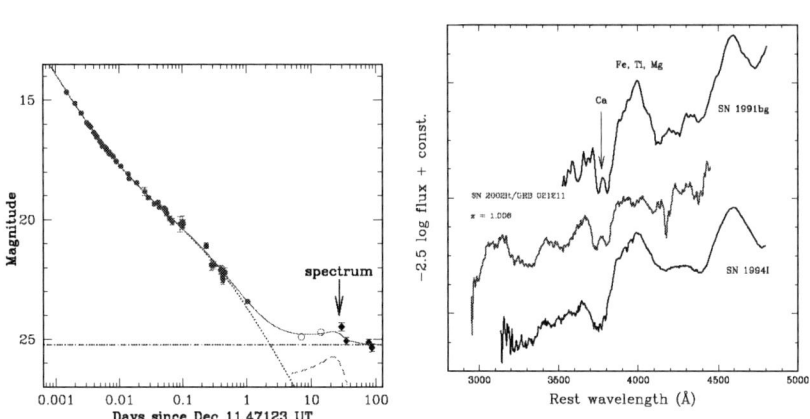

FIGURE 1. Left panel. Light curve of the afterglow of GRB 021211. Filled circles represent data from published works (Fox et al. 2003; Li et al. 2003; Pandey et al. 2003), open circles are converted from HST measurements (Fruchter et al. 2002), while filled diamonds indicate our data; the arrow shows the epoch of our spectroscopic measurement. The dotted and dot-dashed lines represent the afterglow and host contributions respectively. The dashed line shows the light curve of SN 1994I reported at $z = 1.006$ and dereddened with $A_V = 2$ (from Lee et al. 1995). The solid line shows the sum of the three contributions. **Right panel.** Spectrum of the afterglow+host galaxy of GRB 021211 (middle line), taken on 2003 Jan 8.27 UT (27 days after the burst). For comparison, the spectra of SN 1994I (type Ic, bottom) and SN 1991bg (peculiar type Ia, top) are displayed, both showing the Ca absorption. Plots from Della Valle et al. (2003, 2004).

THE "SMOKING GUN": GRB 030329/SN 2003DH

The breakthrough in the study of the GRB/SN association arrived with the bright GRB 030329. This burst, also discovered by the HETE–2 satellite, was found at a redshift $z = 0.1685$ (Greiner et al. 2003). SN features were detected in the spectra of the afterglow by several groups (Stanek et al. 2003, Hjorth et al. 2003; see also Kawabata et al. 2003; Matheson et al. 2003a) and the associated SN (SN 2003dh) looked strikingly similar to SN 1998bw (Fig. 2). The gamma-ray and afterglow properties of this GRB were not unusual among GRBs, and therefore, the link between GRBs and SNe was eventually established to be general.

The modeling of the early spectra of SN 2003dh (Mazzali et al. 2003) has shown that SN 2003dh had a high explosion kinetic energy, $\sim 4 \times 10^{52}$ erg (if spherical symmetry is assumed). However, the light curve derived from fitting the spectra suggests that SN

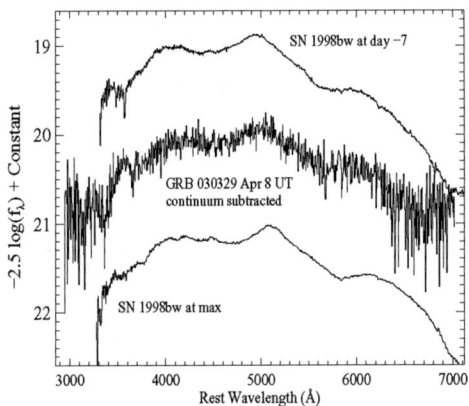

FIGURE 2. Spectrum of SN 2003dh taken on 2003 April 8, afterg sbutracting the spectrum of April 4 rescaled. The residual spectrum shows broad bumps at approximately 5000 and 4200 Å (rest frame), which is similar to the spectrum of the peculiar type-Ic SN 1998bw a week before maximum light (Patat et al. 2001). Plot from Stanek et al. 2003.

2003dh was probably fainter than SN 1998bw (but see Bloom et al. 2004), ejecting only ~ 0.35 $M_\odot$ of ^{56}Ni. The progenitor was a massive envelope-stripped star of $\sim 35 - 40 M_\odot$ on the main sequence. The spectral analysis of the nebular-phase emission lines carried out by Kosugi et al. (2004) suggests that the explosion of the progenitor of GRB 030329 was aspherical, and that its axis was well aligned with both the GRB relativistic jet and our line of sight.

GRB 031203/SN 2003LW

GRB 031203 was a 30s burst detected by the INTEGRAL burst alert system (Mereghetti et al. 2003) on 2003 Dec 3. At $z = 0.1055$ (Prochaska et al. 2004), it was the second closest burst after GRB 980425. The burst energy was extremely low, of the order of 10^{49} erg, well below $\sim 10^{51}$ erg of normal GRBs. In this case, a very faint NIR afterglow was discovered, orders of magnitude dimmer than usual GRB afterglows (Malesani et al. 2004). A few days after the GRB, a rebrightening was apparent in all optical bands (Thomsen et al. 2004; Cobb et al. 2004; Gal-Yam et al. 2004). For comparison, in Fig. 3 the VRI light curves of SN 1998bw are plotted (solid lines), placed at $z = 0.1055$, stretched by 1.1 and dereddened with $E_{B-V} = 1.1$. After assuming a light curve shape similar to SN 1998bw, which had a rise time of 16 days in the V band, data suggest an explosion time nearly simultaneous with the GRB. With the assumed reddening, SN 2003lw appears to be brighter than SN 1998bw by 0.5 mag in the V, R, and I bands. The absolute magnitudes of SN 2003lw are hence $M_V = -19.75 \pm 0.15$, $M_R = -19.9 \pm 0.08$, and $M_I = -19.80 \pm 0.12$. Fig. 3 also shows the spectra of the rebrightening on 2003 Dec 20 and Dec 30 (14 and 23 rest-frame days after the

GRB), after subtracting the spectrum taken on Mar 1 (81 rest-frame days after the GRB, Tagliaferri et al. 2004). The spectra of SN 2003lw are remarkably similar to those of SN 1998bw obtained at comparable epochs (shown as dotted lines in Fig. 3). Both SNe show very broad absorption features, indicating high expansion velocities. The analysis of early spectra of 2003lw (Mazzali et al. 2006) indicates that this HN produced about $\sim 0.5 M_\odot$ of Ni. The progenitor mass could be as large as 40-50 $M_\odot$ on the main sequence.

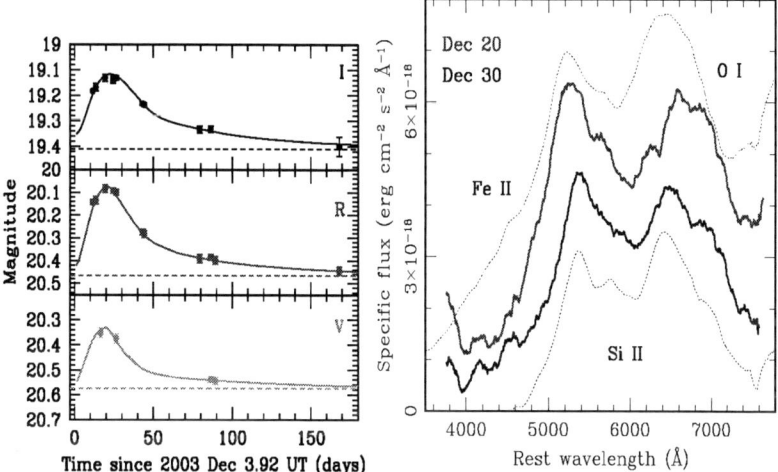

FIGURE 3. Left panel. Optical and NIR light curves of GRB 031203 (dots). The solid curves show the evolution of SN 1998bw (Galama et al. 1998; McKenzie & Schaefer 1999), rescaled at $z = 0.1055$, stretched by a factor 1.1, extinguished with $E(B-V) = 1.1$, and brightened by 0.5 mag. The dashed lines indicate the host galaxy contribution. **Right panel.** Spectra of SN 2003lw, taken on 2003 December 20 and 30 (solid lines), smoothed with a boxcar filter 250Å wide. Dotted lines show the spectra of SN 1998bw (from Patat et al. 2001), taken on 1998 May 9 and 19 (13.5 and 23.5 days after the GRB, or 2 days before and 7 days after the V-band maximum, respectively), extinguished with $E(B-V) = 1.1$ and a Galactic extinction law (Cardelli et al. 1989). The spectra of SN 1998bw were vertically displaced for presentation purposes. Plots from Malesani et al. 2004.

RATES OF SNE IB/C, HYPERNOVAE AND GRBS

GRB 980425, XRF 020903, GRB 031203, and GRB 060218 have γ-energy budgets between 2–4 orders of magnitude fainter than those exhibited by "standard" GRBs. The increasing number of discovery of these underenergetic events (with an associated SN component) can no longer considered a simple collections of peculiar, atypical cases. These bursts were so faint, that they would have been easily missed at cosmological distances, therefore it is very possible that they are the most frequent GRBs in the universe. These 4 events have been detected in 9 years of observations within a volume of ~ 4 Gpc^3 (XRF 020903 is the most far away under-energetic event so far observed,

TABLE 1. Hypernovae

SN	cz km/s	References
1997dq	958	Mazzali et al. 2004
1997ef	3539	Filippenko 1997b
1998bw	2550	Galama et al. 1998
1999as	36000	Hatano et al. 2001
2002ap	632	Mazzali et al. 2002, Foley et al. 2003
2002bl	4757	Filippenko et al. 2002
2003bg	1320	Filippenko & Chornack 2003
2003dh	46000	Stanek et al. 2003, Hjorth et al. 2003
2003jd	5635	Filippenko et al.2003; Matheson et al. 2003b
2003lw	30000	Malesani et al. 2004
2004bu	5549	Foley et al. 2004
2005kz	8117	Filippenko, Foley & Matheson 2005
2006aj	9000	Masetti et al. 2006

at $z = 0.25$). These figures may imply an "observed" rate of ~ 0.1 GRB Gpc^{-3} yr^{-1}. On the other hand the "observed" rate has to be corrected to account for the effective full-sky coverage of the satellites. This task is not so simple to be carried out because the correction should include different factors (effective field of view, downtimes...) which are not easy to be quantify. As a conservative estimate (based on the published technical reports) one can crudely assume a correction factor of $\gtrsim 10$. Then the "observable" local rate turns out to be: ~ 2 GRB Gpc^{-3} yr^{-1}, which is slightly larger than derived by Schmidt (2001) (~ 0.5 GRB Gpc^{-3} yr^{-1}) and consistent with 1.1 GRB Gpc^{-3} yr^{-1}) derived by Guetta et al. (2004).

A rate of $\sim 2.6 \times 10^4$ SNe-Ibc Gpc^{-3} yr^{-1} can be derived by combining the the local density of B-band luminosity of $\sim 1.2 \times 10^8 L_{B,\odot}$ Mpc^{-3} (e.g. Madau, Della Valle & Panagia 1998) with the local rate of 0.22 SNe-Ibc (in Irr and Sm Hubble types) per century and per 10^{10} $L_{B,\odot}$ (Cappellaro, Evans & Turatto 1999). This SN rate has to be compared with $\lesssim 2$ GRB Gpc^{-3} yr^{-1} after rescaling for the jet beaming factor[1]. There exist different estimates for this parameter: from ~ 500 (Frail et al. 2001) to ~ 75 (Guetta, Piran & Waxman 2005), corresponding to beaming angles $\sim 4°-10°$, respectively. Taking these figures at their face value, we find the ratio GRB/SNe-Ibc to be in the range: $\sim 4\% - 0.5\%$. Izzard et al. (2004) have modeled the stellar progenitors of type-Ibc SNe, selecting those capable to produce GRBs. They find ratios comparable with the numbers presented here. Radio and optical surveys give less severe constraints: Soderberg et al. (2006b) and Rau et al. (2006) find $f_b^{-1} \lesssim 10^4$ and $f_b^{-1} \lesssim 12500$, then implying $\theta \gtrsim 0.8°$ and GRB/SNe-Ibc $\lesssim 30\%$.

The computation of the ratio GRB/HN requires a further step. The measurement of the SN rate is based on the control-time methodology (Zwicky 1938) that implies the systematic monitoring of galaxies of known distances. Unfortunately all HNe so far discovered (see Tab. 1) have been not found during "time controlled" surveys. An

---

[1] The beaming factor is defined as $f_b^{-1} = 1 - \cos\theta$

TABLE 2. Supernova/gamma-ray burst time lag. A negative time lag indicates that the SN explosion precedes the GRB.

GRB	SN	+Δt(days)	−Δt(days)	references
GRB 980425	1998bw	0.7	2	Iwamoto et al. 1998
GRB 000911	bump	1.5	7	Lazzati et al. 2001
GRB 011121	2001ke	0	5	Bloom et al. 2002b
		−	a few	Garnavich et al. 2003
GRB 021211	2002lt	1.5	3	Della Valle et al. 2003
GRB 030329	2003dh	2	8	Kawabata et al. 2003
		−	2	Matheson et al. 2003a
GRB 031203	2003lw	0	2	Malesani et al. 2004
GRB 041006	bump	2.7	0.9	Stanek et al. 2005
GRB 050525A	2005nc	0	< 3.5	Della Valle et al. 2006
GRB 060218	2006aj	0	0	Campana et al. 2006

heuristic approach to derive the rate of HNe is to compute the frequency of occurrence of all SNe-Ib/c and HNe in a limited distance sample of objects and to assume that they have been efficiently (or inefficiently) monitored by the same extent. From an upgraded version of Asiago catalog we have extracted 91 SNe-Ib/c (8 of which are HNe) with $cz < 6000$ km/s. This velocity threshold is suitable to make the distance distribution of 'normal' Ib/c and HNe statistically indistinguishable (KS probability=0.42). After excluding SN 1998bw, because it was searched in the error-box of GRB 980425, one can infer that the fraction of HNe is about $7/91 \simeq 8\%$ of the total number of SNe Ib/c. Therefore the ratio GRB/HNe turns out to be $\sim 0.5 \div 0.06$ (cfr. ~ 1, Podsiadlowski et al. 2004).

Finally we like to stress two points: *a)* sub-energetic GRBs may be less collimated than classical events. Their low energy demand does not pose any problem for most progenitor models, thus the beaming factor does not need to be too large. Moreover, from the sparse data on their afterglows, it appears that their breaks are quite late, if existent. For example, interpreting the break in the X-ray light curve of GRB 031203 as due to a jet we obtain $\theta \sim 16° \div 30°$. Therefore it is likely that small GRB/SN ratios are favored; *b)* the estimate of the "local" rate of GRBs is seriously plagued by the small number of available events. For example, if we consider only the 3 nearest events (which have occurred within 0.4 Gpc^3) we cannot exclude values as high as ~ 15 GRB Gpc^{-3} yr^{-1}. In this case the ratio GRB/SNe-Ibc becomes $\sim 30\% - 4\%$ (for $f_b = 500$ and 75, respectively). The former value would imply that a significant fraction of "standard" SNe-Ibc contributes to originate GRBs. The latter value might be consistent with a ratio GRB/HNe ~ 1.

FACTS AND OPEN QUESTIONS

From the data presented in the previous sections, a number of both established facts and intriguing questions emerge:

1. Long duration GRBs are closely connected with the death of massive stars. This has been spectroscopically confirmed over a large range of redshifts:

GRB 980425/SN 1998bw at $z = 0.0085$ (Galama et al. 1998); GRB 060218/SN 2006aj at $z = 0.03$ (Campana et al. 2006); GRB 031203/SN 2003lw at $z = 0.1055$ (Malesani et al. 2003); GRB 030329/SN 2003dh at $z = 0.16$ (Stanek et al. 2003, Hjorth et al. 2003); XRF 020903/SN 1998bw-like at $z = 0.23$ (Soderberg et al. 2005); GRB 050525A/SN 2005cn at $z = 0.6$ (Della Valle et al. 2006); and GRB 021211/SN 2002lt at $z \sim 1$ (Della Valle et al. 2003). In spite of this tight connection with SN explosions, Fruchter et al. (2006) have demonstrated that GRBs and SNe do not occur in similar galactic environments, these authors argue that this unexpected behavior may be related to the low metallicity content observed in the GRB hosts.

2. It is not clear whether or not only HNe (i.e. broad-lines SNe-Ibc) are capable of producing GRBs or even "standard" Ib/c events. There is weak evidence that other type of core-collapse SNe, such as type IIn, can contribute to the SN population of GRBs (Germany et al. 2000; Turatto et al. 2000; Rigon et al. 2003). However, in a recent study, Valenti et al. (2005) (see also Bosnjak et al. 2006) were not able to corroborate, on statistical basis, the associations with core-collapse SNe different from SNe-Ibc. The best evidence for the case of an association between a type-IIn SN and a GRB has been provided by Garnavich et al. (2003), who found that the color evolution of the bump associated with GRB 011121 is consistent with the color evolution of an underlying SN (SN 2001ke) strongly interacting with a dense circumstellar gas due to the progenitor wind.

3. GRB-SN data (including the bumps: Della Valle et al. 2003, Fynbo et al. 2004, Levan et al. 2005, Masetti et al. 2003, Price et al. 2003, Soderberg et al. 2005, Gorosabel et al. 2005, Stanek et al. 2005, Soderberg et al. 2006a, Bersier et al. 2006) indicate that the magnitude at maximum of SNe associated with GRBs may span a range of about 5 magnitudes, which is similar to that exhibited by "standard" stripped-envelope stars (Richardson, Branch & Baron 2006). However all GRB-SNe which have so far been spectroscopically confirmed appear to belong to the bright tail of SNe-Ib/c population (all have $M_V \sim -18.5/-19$). If this is the effect of an observational bias (which favors the spectroscopic observations of bright SNe) operating on a small number of objects or it has a deeper physical meaning is not yet clear.

4. There are events, such as XRF 040701 (Soderberg et al. 2005), for which the SN has been unsuccessfully searched with HST, down to magnitude $M_V \sim -15.8 \div -13$ (according to different assumption on the host galaxy extinction). These observations may imply that some GRBs can be associated with very underluminous SNe-Ibc. On the other hand such very faint objects have never been observed (see Richardson, Branch & Baron 2006). However, it should be noticed that a few unusually faint core-collapse events (belonging to the type-II class, not the Ibc!) have been already observed at magnitudes $M_V \sim -13 \div -14.5$ (Turatto et al. 1998; Pastorello et al. 2004). It would be possible that SNe-Ibc of comparable low luminosity do exist but they have not been observed just because they are rarer objects than type II (SNe-Ibc are about 15-30% of type II in late spirals/Irr, e.g. Mannucci et al. 2005).

5. Several authors have reported the detection of Fe and other metal lines in GRB X-ray afterglows (e.g. Piro et al. 1999). If valid (see Sako et al. 2005 for a critical view) these observations would have broad implications for both GRB emission models and would strongly link GRBs with SN explosions. For example, Butler et al. (2003) have reported the detection in a Chandra spectrum of emission lines whose intensity and

blueshift would imply that a supernova occurred > 2 months prior to the γ event. This kind of observations can be accommodated in the framework of the *supranova* model (Vietri & Stella 1998), where a SN is predicted to explode months or years before the γ burst. In Tab. 2 we have reported the estimates of the lags between the SN explosions and the associated GRBs, as measured by the authors of the papers. After taking these data at their face value, one can conclude that most SN and GRB events occur simultaneously, and only in some case the SN may have preceded the GRB by a few days (at the most). However we note that Swift has not detected X-ray lines in any afterglows so far observed.

6. Only a very small fraction of all massive stars are capable of producing GRBs. SNe-Ibc appear to be the natural candidates because they have already lost the Hydrogen envelope when the collapse of the core occurs, then allowing the ultra-relativistic jets to escape from the progenitor star. Nevertheless this fact does not seem to be sufficient. According to the current SN and GRB rates and $\langle f_b \rangle$ estimates, only $\lesssim 4\%$ of type-Ibc SNe are able to produce GRBs. This implies that GRB progenitors must have some other special characteristic other than being just massive stars. Recent studies have extensively discussed the role that stellar rotation (Woosley & Heger 2006; Yoon & Langer 2005; Fryer & Heger 2005), binarity (Podsiadlowski et al. 2004; Mirabel 2004; Tutukov & Cherepashchuk 2003; Smartt et al. 2002), asymmetry (Maeda et al. 2005) and metallicity (e.g. Fruchter et al. 2006) play in the GRB phenomenon.

7. The "optical" properties (i.e. luminosity at peak and expansion velocities) of the 4 closest SNe associated with GRBs vary by at most $\sim \pm 50\%$, while the γ-budget covers about 4 order of magnitudes. These facts may be interpreted in at least 2 different ways: a) we may have observed intrinsically similar phenomena under different angles. GRB 030329/SN 2003dh may be viewed almost pole-on, GRB 980425/SN 1998bw relatively off-axis ($15° < \theta < 30°$), while GRB 031203/SN 2003lw may lie in between (Ramirez-Ruiz et al. 2005). A consequence of this scenario is that the γ-properties are strongly dependent upon the angle ($\sim \theta^4$), whereas the optical properties are affected much less by changing the viewing angle up to $\Delta\theta \sim 30°$; b) the recent event GRB 060218/SN 2006aj ($E_{\rm iso} \sim 6 \times 10^{49}$ erg), may suggest a different interpretation. This GRB may be an example of intrinsically fainter event (Campana et al. 2006; Amati et al. 2006). This might indicate that there exists an intrinsic dispersion in the properties of the relativistic ejecta for SNe having similar optical properties (e.g. peak of luminosity, velocity of the ejecta). This fact is not unconceivable after keeping in mind that the observed relativistic energies at play in the GRB phenomenon, at least in the local universe ($z < 0.1$), appear to be just tiny fluctuations ($10^{-2/-4}$) of the kinetic energy involved in the 'standard' SNe-Ibc ($\sim 10^{51}$ erg) or HN explosions ($\sim 10^{52}$ erg).

8. As for AGN, it has been proposed (Lamb et al. 2005, see also Kouveliotou et al. 2004 and Dado et al. 2004) a unification scheme where GRBs, XRRs, XRFs and SNe-Ibc are the same phenomenon, but viewed at different angles. Given the rates of GRBs and type Ibc SNe discussed in the section 6, the unification scenario would work for $\langle f_b^{-1} \rangle \sim 30000$, which would correspond to beaming angles of $\sim 0.5°$. On the other hand the measured f_b^{-1} factors are much smaller, likely in the range 75–500 (Guetta et al. 2005, Yonetoku et al. 2005; van Putten & Regimbau 2003, Frail et al. 2001) that corresponds to beaming angles of $\sim 10° - 4°$.

Acknowledgments

I wish to thanks Daniele Malesani, Maurice van Putten and Evan Scannapieco for the critical reading of the manuscript and all colleagues of the "Supernova-Gamma Ray Burst Connection" program at the KITP (UCSB) for useful discussions.
This research was partially supported by the National Science Foundation under Grant No. PHY99-0794.

REFERENCES

Amati, L., Frontera, F., Guidorzi, C., & Montanari, E. 2006, GCN 4846
Bersier, D., et al. 2006, ApJ, in press (astro-ph/0602163)
Bloom, J.S., Kulkarni, S.R., Djorgovski, S.G. et al. 1999, Nature, 401, 453
Bloom, J.S., Kulkarni, S.R., Price, P.A., et al. 2002, ApJ, 572, L45
Bloom, J.S., et al. 2004, AJ, 127, 252
Bosnjak, Z., Celotti, A., Ghirlanda, G., Della Valle, M., & Pian, E. 2006, A&A, 447, 121
Butler, N.R., Marshall, H.L., Ricker, G.R., Vanderspek, R.K., Ford, P.G., Crew, G. B., Lamb, D.Q., & Jernigan, J.G. 2003, ApJ, 597, 1010
Campana, S., et al. 2006, Nature, submitted (astro-ph/0603279)
Cappellaro, E., Evans, R., & Turatto, M. 1999, A&A, 351, 459
Cardelli, J.A., Clayton, G.C., & Mathis, J.S. 1989, ApJ, 345, 245
Castro-Tirado, A., & Gorosabel, J. 1999, A&AS, 138, 449
Christensen, L., Hjorth, J., & Gorosabel, J. 2004, A&A, 425, 913
Cobb, B.E., Baylin, C.D., van Dokkum, P.G., Buxton, M.M., & Bloom, J. S. 2004, ApJ, 608, L93
Cobb, B.E., Bailyn, C.D., van Dokkum, P.G., & Natarajan, P. 2006, ApJL, submitted (astro-ph/0603832)
Conselice et al. 2005, ApJ, 633, 29
Cox, A.N. 2000, in Allen's astrophysical quantities, 4th ed. Publisher: New York: AIP Press; Springer, 2000
Crew, G.B., Lamb, D.Q., Ricker, G.R., et al. 2003, ApJ, 599, 387
Dado, S., Dar, A., & De Rujula, A. 2004, A&A, 422, 381
Della Valle M., Malesani, D., Benetti, S., et al. 2003, A&A, 406, L33
Della Valle M., Malesani, D., Benetti, S., et al. 2004, in the Proceedings of the 2003 GRB Conference (Santa Fe, 2003 Sep 8-12), eds. E. Fenimore & M. Galassi, p. 403
Della Valle, M., Malesani, D., Bloom, J.S., et al. 2006, ApJ, in press (astro-ph/0604109)
Dezalay, J.P. et al. 1992, AIP Conf. Proc. 265, p. 304 (Huntsville, Oct. 16-18 1991, GRB Workshop, eds. W.S. Paciesas & G.J. Fishman)
Djorgovski, S.G., Kulkarni, S.R., Bloom, J.S., Goodrich, R., Frail, D.A., Piro, L., & Palazzi, E. 1998, ApJ, 508, L17
Filippenko, A.V. 1995, ApJ, 450, L11
Filippenko, A.V. 1997b, IAUC 6783
Filippenko, A.V., Leonard, D.C., & Moran, E.C. 2002, IAUC 7845
Filippenko, A.V., & Chornock, R. 2003, IAUC 8084
Filippenko, A.V., Foley, R.T., & Swift, B. 2003, IAUC 8234
Foley, R.J., Wong, D.S., Moore, M., & Filippenko, A.V. 2004, IAUC, 8353
Foley, R.J., Papenkova, M.S., Swift, B.J., et al. 2003, PASP, 115, 1220
Fox, D.W., Price, P.A., Soderberg, A.M., et al. 2003, ApJ, 586, L5

Frail, D.A., Kulkarni, S.R., Sari, R., et al. 2001, ApJ, 562, L55
Fryer, C.L., & Heger, A. 2005, ApJ, 623, 302
Fruchter, A.S., Thorsett, S.E., Metzger, M.R., et al. 1999, ApJ, 519, L13
Fruchter, A.S., Levan, A., Vreeswijk, P.M., Holland, S.T. & Kouveliotou, C. 2002, GCN 1781
Fruchter, A.S., et al. 2006, Nature, submitted (astro-ph/0603537)
Fynbo, J., Sollerman, J., Hjorth, J., et al. 2004, ApJ, 609, 962
Gal-Yam, A., Moon, D.S., Fox, D.B., et al. 2004, ApJ, 609, L59
Galama, T.J., Vreeswijk, P.M., van Paradijs, J., et al. 1998, Nature, 395, 670
Garnavich, P.M., Stanek, K.Z., Wyrzykowski, L., et al. 2003, ApJ, 582, 924
Germany L., Reiss, D.J., Sadler, E.M., Schmidt, B.P., & Stubbs, C.W. 2000, ApJ, 533, 320
Gorosabel, J., et al. 2005, A&A, 437, 411
Greiner, J., et al. 2003, GCN 2020
Guetta, D., Perna, R., Stella, L., & Vietri, M. 2004, ApJ, 615, L73
Guetta, D., Piran, T., & Waxman, E. 2005, ApJ, 619, 412
Hatano, K., Branch, D., Nomoto, K., Deng, J.S., Maeda, K., Nugent, P., & Aldering, G. 2001, 198th BAAS, 33, 838
Hjorth, J., Sollerman, J., Moller, P., et al. 2003, Nature, 423, 847
Höflch, P., Wheeler, J.C., & Wang, L. 1999, ApJ, 521, 179
Iwamoto, K., Mazzali, P.A., Nomoto, K., et al. 1998, Nature, 395, 672
Izzard, R.G., Ramirez-Ruiz, E., & Tout, C.A. 2004, MNRAS, 348, 1215
Li, W., Filippenko, A.V., Chornock, R., & Jha, S. 2003, ApJ, 586, L9
Kawabata, K.S., Deng, J., Wang, L., et al. 2003, ApJ, 593, L19
Klebesadel, R.W. 1990, Taos, July 29 - August 3 (Los Alamos Workshop on GRBs, Eds. Cheng H, Richard I. Epstein, Edward E. Fenimore, Cambridge University Press, 1992), 161
Kouveliotou, C., et al. 1993, ApJ, 413, L101
Kouveliotou, C., et al. 2004, ApJ, 608, 872
Kosugi, G., Mizumoto, Y., Kawai, N., et al. 2004, PASJ, 56, 61
Kulkarni, S.R., Frail, D.A., Wieringa, M.H., et al. 1998, Nature, 395, 663
Lamb, D.Q., Donaghy, T.Q., & Graziani, C. 2005, ApJ, 620, 355
Lazzati, D., Covino, S., Ghisellini, G., et al. 2001, A&A, 378, 996
Le Floc'h, E., Duc, P.-A., Mirabel, I.F., et al. 2003, A&A, 400, 499
Levan, A., Nugent, P., Fruchter, A., et al. 2005a, ApJ, 624, 880
Lee, M.G., Kim, E., Kim, S.C., Kim, S.L., Park, W., Pyo, T.S. 1995, JKAS, 28, 31L
Madau, P., Della Valle, M., & Panagia, N. 1998, MNRAS, 297, L17
Maeda, K., Nakamura, T., Nomoto, K., Mazzali, P., Patat, F., Hachisu, I. 2002, ApJ, 565, 405
Maeda, K., Mazzali, P., & Nomoto, K. 2006, ApJ, in press (astro-ph11389)
Malesani, D. Tagliaferri, G., Chincarini, G., et al.2004, ApJ, 609 L5
Mannucci, F., Della Valle, M., Panagia, N., Cappellaro, E., Cresci, G., Maiolino, R., Petrosian, A., & Turatto, M. 2005, A&A, 433, 807
Masetti, N., Palazzi, E., Pian, E., et al. 2003, A&A, 404, 465
Masetti, N., et al. 2006, GCN 4803
Matheson, T., Garnavich, P.M., Stanek, K.Z., et al. 2003a, ApJ, 599, 394
Matheson, T., Challis, P., & Kirshner, R. 2003b, IAUC, n. 8234
Mazzali, P., Deng, J., Maeda, K., et al. 2002, ApJ, 572, L61
Mazzali, P., Deng, J., Tominaga, N., et al. 2003, ApJ, 599, L95
Mazzali, P.A., Deng, J., Maeda, K., Nomoto, K., Filippenko, A.V., & Matheson, T. 2004, ApJ, 614, 858

Mazzali, P., et al. 2006, ApJ, in press (astro-ph 0603516)
McKenzie, E.H., & Schaefer, B.E. 1999, PASP, 111, 964
Mereghetti, S., & Götz, D. 2003, GCN Circ. 2460
Mirabal, N., Halpern, J.H., An, D., Thorstensen, J.R., Terndrup, D.M. 2006, ApJL, submitted (astro-ph/0603686)
Mirabel, I.F. 2004, RMxAC, 20, 14
Modjaz, M., et al. 2006, ApJL, submitted (astro-ph 0603377)
Panaitescu, A., & Kumar, P. 2001, ApJ, 560, L49
Pandey, S.B., Anupama, G.C., Sagar, R., Bhattacharya, D., Castro-Tirado, A.J., Sahu, D.K., Parihar, P., & Prabhu, T.P. 2003, A&A, 408, L21
Pastorello, A., et al. 2004, MNRAS, 347, 74
Patat, F., et al. 2001, ApJ, 555, 900
Pian, E., et al. 2006, Nature submitted (astro-ph/0603530)
Piro, L., Costa, E., Feroci, M., et al. 1999, ApJ, 514, L73
Podsiadlowski, P., Mazzali, P., Nomoto, K., Lazzati, D., & Cappellaro, E. 2004, ApJ, 607, L17
Price, P.A., Kulkarni, S.R., Schmidt, B.P., et al. 2003, ApJ, 584, 931
Prochaska, J.X., Bloom, J.S., Chen, H., Hurley, K.C., Melbourne, J., Dressler, A., Graham, J.R., Osip, D.J., & Vacca, W.D. 2004, ApJ, 611, 200
Rau, A., Greiner, J., & Schwarz, R. 2006, A&A, 449, 79
Ramirez-Ruiz, E., Granot, J., Kouveliotou, C., Woosley, S.E., Patel, S. K, & Mazzali, P. 2005, ApJ, 625, L91
Richardson, D, Branch, D., & Baron, E. 2006, AJ, in press)astro-ph/0601136)
Rigon, L., Turatto, M., Benetti, S., Pastorello, A., Cappellaro, E., Aretxaga, I., Vega, O., Chavushyan, V., Patat, F., Danziger, I.J., & Salvo, M. 2003, MNRAS, 340, 191
Sako, M., Harrison, F., & Rutledge, R. 2005, ApJ, 623, 973
Schmidt, M. 2001, ApJ, 552, 36
Smartt, S.J., et al. 2002, ApJ, 572, L147
Soderberg, A.M., et al. 2005, ApJ, 627, 877
Soderberg, A.M., et al. 2006a, ApJ, 636, 391
Soderberg, A.M., Nakar, E., Kulkarni, S.R., & Berger, E. 2006b, ApJ, 638, 930
Sollerman, J., et al. 2006, A&A, submitted (astro-ph/0603495)
Stanek, K.Z., Matheson, T., Garnavich, P. M., et al. 2003, ApJ, 591, L17
Stanek, K.Z., Garnavich, P.M., Nutzman, P.A., Hartman, J. D., & Garg, A. 2005, ApJ 626, L5
Tagliaferri, G., Covino, S. Fugazza, D., et al. 2004, IAU Circ. 8308
Thomsen, B., Hjorth, J., Watson, D., et al. 2004, A&A, 419, L21
Turatto M., Suzuki, T., Mazzali, P., Benetti, S., Cappellaro, E., Danziger, I.J., Nomoto, K., Nakamura, T., Young, T.R., & Patat, F. 2000, 534 L57
Turatto, M., et al. 1998, ApJ, 498, L129
Tutukov, A., & Cherepashchuk, A.M. 2003, Astron. Rep. 47, 386
Valenti, S., Cappellaro, E., Della Valle, M., Frontera, F., Guidorzi, C., & Montanari, E. 2005, Nuovo Cim. 28C, 633
van Putten, M.H.P.M., & Regimbau, T. 2003, ApJ, 593, L15
Vietri, M., & Stella, L. 1998, ApJ, 507, L45
Vreeswijk, P.M., Smette, A., Fruchter, A.S., et al. 2006, A&A, 447, 145
Wainwright, C., Berger, E., & Penprase, B.E. 2005, AAS, 207, 19.08
Weiler, K.W., Panagia, N., Montes, M.J., & Sramek, R.A. 2002, ARA&A, 40, 387

Woosley, S.E., & Heger, A. 2006, ApJ, 637, 914
Woosley, S.E., Eastman, R.G., & Schmidt, B.P. 1999, ApJ, 516, 788
Yamazaki, R., Yonetoku, D., & Nakamura, T. 2003, ApJ, 594, L79
Yonetoku, D., Yamazaki, R., Nakamura, T., Murakami, T. 2005, MNRAS, 362, 1114
Yoon, S.C., & Langer, N. 2005, A&A, 443, 643 Zeh, A., Klose, S., & Hartmann, D.H. 2004, ApJ, 609, 952
Zwicky, F. 1938, ApJ, 88, 529

A Broader Perspective on the GRB-SN Connection

Alicia M. Soderberg

Division of Physics, Mathematics and Astronomy,105-24, California Institute of Technology, Pasadena, CA 91125

Abstract. Over the last few years our understanding of local Type Ibc supernovae and their connection to long-duration gamma-ray bursts has been revolutionized. Recent discoveries have shown that the emerging picture for core-collapse explosions is one of diversity. Compiling data from our dedicated radio survey of SNe Ibc and our comprehensive HST survey of GRB-SNe together with ground-based follow-up campaigns, I review our current understanding of the GRB-SN connection. In particular, I compare local SNe Ibc with GRB-SNe based on the following criteria: (1) the distribution of optical peak magnitudes which serve as a proxy for the mass of ^{56}Ni produced in the explosion, (2) radio luminosity at early time (few days to weeks) which provides a measure of the energy coupled to on-axis relativistic ejecta, and (3) radio luminosity at late time (several years) which constrains the emission from GRB jets initially directed away from our line-of-sight. By focusing on these three points, I will describe the complex picture of stellar death that is emerging.

Keywords: Supernovae – Gamma-ray Bursts
PACS: 97.60.Bw

INTRODUCTION

Twenty years have passed since the class of Type Ibc supernovae (SNe Ibc) was initially recognized as a distinct population of core-collapse explosions [1, 2, 3]. Their lack of homogeneity and low event rate, ($\sim 10\%$ of locally discovered SNe), did not motivate focused observational programs.

In 1998, however, SNe Ibc enjoyed an explosion of new-found interest thanks to the discovery of Type Ic SN 1998bw ($d \approx 36$ Mpc) within the *BeppoSAX* localization error box of gamma-ray burst, GRB 980425 [4, 5]. While the γ-ray energy release, E_γ, of GRB 980425 was a factor of 10^4 below that of GRBs, SN 1998bw was (and still remains) the most luminous radio SN ever observed [6]. Two unusual features were noted from the radio data: significant energy ($E_{\rm radio} \sim 10^{49}$ erg) coupled to mildly relativistic (Lorentz factor, $\Gamma \sim 3$) ejecta and evidence for episodic energy injection [6, 7]. Moreover, the bright optical emission required production of $\sim 0.5 M_\odot$ ^{56}Ni, comparable to that inferred for Type Ia supernovae while the broad absorption lines (indicative of photospheric velocities above 30,000 km s^{-1}) implied a total kinetic energy of 3×10^{52} erg [8, 9].

These observations have been interpreted under the framework of the "collapsar model" (e.g. [10]) in which a central engine (accreting black hole) plays a significant role in exploding the star.

THE GRB-SN CONNECTION: AN OVERVIEW

In the seven years since the discovery of SN 1998bw/GRB 980425, about a dozen SNe Ibc have been reported in association with GRBs, all at $z \geq 0.1$ (see [11, 12] for recent compilations). Of these associations, three were unambiguously confirmed through spectroscopic identification of SN features (GRB 030329 [13]; GRB 031203 [14]; XRF 020903 [15]) which were observed to be unusually broad and similar to those seen in SN 1998bw. The majority of GRB-SN associations are inferred based on the emergence of a red "bump" in the afterglow light-curves approximately $20(1+z)$ days after the explosion and attributed to a thermal supernova component. These observations imply that at least some SNe Ibc are powered by a central engine.

At the same time, several broad-lined supernovae have been discovered locally and are currently estimated to represent $\sim 5\%$ of the Type Ibc population [16]. Given their spectral similarity to SN 1998bw and GRB-SNe, it has been argued that broad-lined SNe Ibc can be used as signposts for GRBs, even in the absence of observed gamma-ray emission (e.g. [17]).

The question has thus become, what is the connection (if any) between the engine-driven GRB-associated SNe and local SNe Ibc? Here I present optical and radio observations for these two samples in an effort to address this question and to offer a broader perspective on the GRB-SN connection.

AN OPTICAL PERSPECTIVE ON THE GRB-SN CONNECTION

In Figure 1 I compare the peak optical magnitudes and ^{56}Ni mass estimates for GRB-associated SNe and local SNe Ibc. The figure clearly shows that ^{56}Ni mass scales with peak optical luminosity. As a result, M_V can be used as a proxy for the synthesized ^{56}Ni mass in the cases where estimates are not available. Several striking conclusions can be drawn directly from this compilation:

- **SN 1998bw is not the most luminous event of either sample.** In fact several local SNe Ibc and GRB-associated SNe are actually brighter. This emphasizes the fact that not all GRB-SNe are like SN 1998bw.

- **The distributions of local and GRB-associated SNe show significant overlap.** We conclude that GRB-associated SNe are not necessarily more luminous nor do they produce more ^{56}Ni than local SNe Ibc. In fact, a K-S test on the two data samples shows a 53% probability that the two have been taken from the same parent population of events. This may indicate a similar ^{56}Ni production mechanism for both samples and thus imposes significant constraints on progenitor models.

- **SNe Ibc with broad optical absorption lines are not more luminous than other events.** In fact, they display a range of optical luminosities comparable to the spread observed for both the local and GRB-SN samples. This emphasizes that broad-optical absorption lines cannot be used as a proxy for a large ^{56}Ni mass.

These four points illustrate the fact that optical observations cannot be used to distinguish the class of GRB-SNe from the local SNe Ibc.

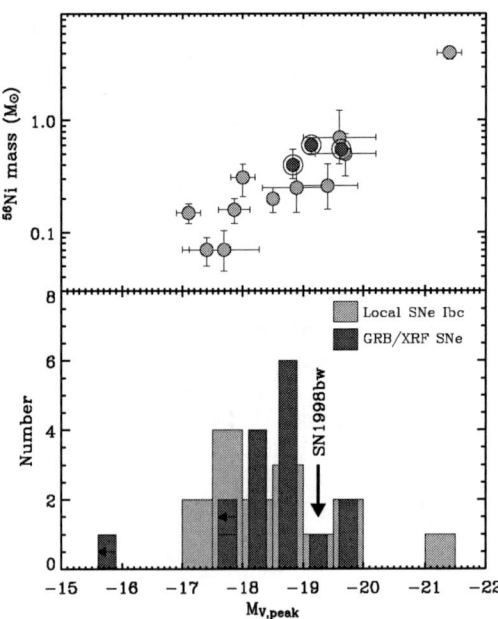

FIGURE 1. The compilation of peak optical magnitudes and ^{56}Ni mass estimates for GRB-SNe and local SNe Ibc from Soderberg *et al.* (2006a) has been extended to included SN 2003lw/GRB 031203 (Mazzali *et al.*, in prep) and local SNe 2003jd [17] and 2005bf [18]. The distributions for GRB-SNe and local SNe Ibc show significant overlap. A K-S test shows a 53% probability that the two samples are drawn from the same parent population of SNe.

A RADIO PERSPECTIVE OF THE GRB-SN CONNECTION

Radio observations offer a better way to distinguish between GRB-SNe and local SNe Ibc since they provide the best calorimetry of the explosion. Radio emission from SNe Ibc is produced by the dynamical interaction of the fastest ejecta with the circumstellar medium [19], in much the same way that GRB afterglows are produced. As the ejecta sweep up and shock the surrounding medium they produce synchrotron emission with a spectral peak near the radio band on a timescale of days to weeks. The emission is brightest for SNe with copious energy coupled to (mildly) relativistic ejecta, as in the case of SN 1998bw. Radio observations are therefore unique in that they provide a measure of the speed and energy of the fastest ejecta produced in the explosion.

Motivated thus, since 1999 we have been monitoring local SNe Ibc with the Very Large Array on a timescale of days to years after the explosion. Early observations are used to probe on-axis ejecta components while late-time data constrain components that were initially directed away from our line-of-sight. This six year effort has resulted in several key advances in our understanding of the GRB-SN connection.

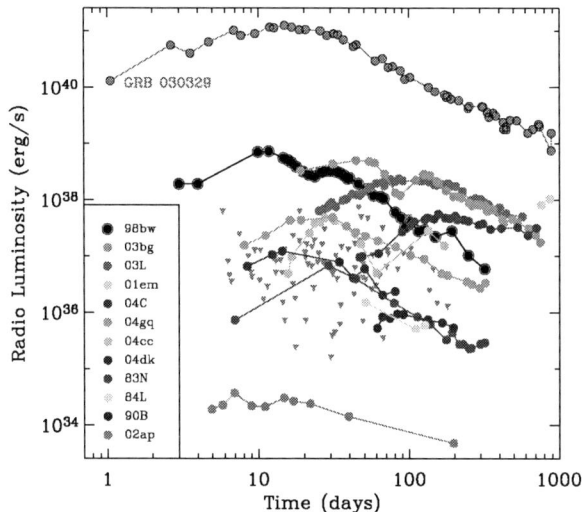

FIGURE 2. A compilation of SNe Ibc radio observations resulting mainly from our dedicated VLA survey (Soderberg et al. in prep). Upper limits are shown as inverted grey triangles. All detections have been studied as part of our VLA program with the exception of SNe 1990B, 1984L, 1983N, 2001em and 1998bw which were taken from the literature [20, 21, 22, 23, 6]. For comparison we show the radio lightcurve of GRB 030329 which had a radio luminosity typical of long duration GRBs [24]. These radio data show that there is a clear distinction between local SNe Ibc and GRB-SN explosions.

Early-Time Observations

First, through our extensive sample of 146 local SNe Ibc, we now know that only 10% have detectable radio emission on a timescale of a few days to years (Figures 2 and 3). These events typically peak in the radio band several weeks after the explosion with average luminosities a factor of 10^2 times fainter than SN 1998bw on a comparable timescale. We compare the optical and radio properties for this sample of SNe Ibc and find no strong correlations. In particular, broad-lined SNe Ibc are **not** more radio luminous than the rest of the sample and can be significantly fainter (e.g. SN 2002ap, [25]; SN 2003jd, [26]). We conclude that radio bright SN 1998bw-like events are rare: less than 2% of the local population ([27]; Soderberg et al., in prep).

Next, we find a clear distinction between local SN Ibc explosions and cosmological GRBs. As clearly shown in Figure 2, GRB-SN explosions are a factor of about 10^4 times more radio luminous than typical SNe Ibc. This is attributed to the fact that GRB-SN explosions couple the bulk of their energy to highly relativistic ejecta while SNe Ibc couple a relatively tiny fraction. These results strongly suggest that if central engines power the majority of local SNe Ibc, the engines must be weaker than those of GRB-SN explosions.

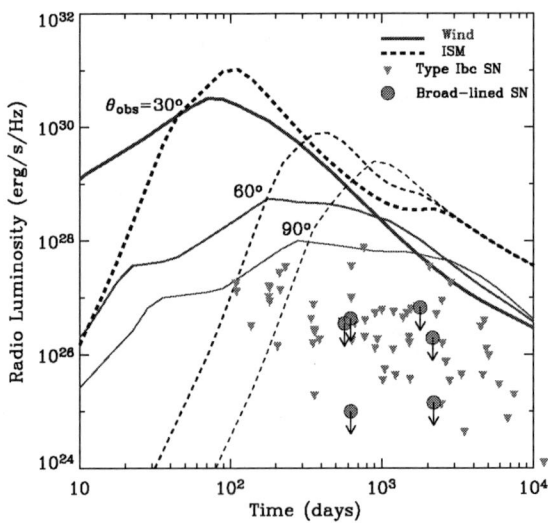

FIGURE 3. A compilation of late-time radio data for 68 local SNe Ibc as published in Soderberg *et al.* (2006b). Upper limits are shown as inverted grey triangles and we emphasize the limits for broad-lined SNe with green circles/arrows. For comparison we show the predicted radio light-curves for a typical GRB afterglow ($E = 10^{51}$ erg, $n = A_* = 1$, $\theta_{\rm jet} = 5°$) observed 30, 60 and 90 degrees away from the collimation axis expanding into a constant density medium (dashed blue lines) and a stellar wind environment (solid red lines). Nearly all of the upper limits are fainter than the predicted emission from a GRB, even viewed 90° off-axis. This holds in particular for the broad-lined SNe where we rule out the scenario that every broad-lined SN Ibc harbors a GRB.

Late-Time Observations

Finally, we use our late-time data obtained between 1 to 30 years after the explosion to search for evidence of relativistic GRB jets that were initially directed away from our line-of-sight. As the jets sweep up material and decelerate they spread sideways, eventually intersecting our viewing angle [28, 29]. At this point the GRB afterglow emission becomes visible and is most easily detected in the radio band. In Figure 3 we show that of the 68 events in our late-time sample, none show evidence for strong radio emission that can be attributed to an off-axis GRB jet.

We [26] compare these data to the radio luminosities of cosmological GRBs to limit the fraction of SNe Ibc hosting GRB jets to $\leq 10\%$ (90% confidence). This holds in particular for the broad-lined events: we rule out a scenario in which every broad-lined Ibc hosts a GRB (84% confidence). This result, taken together with the early-time radio data, reiterates that **broad optical absorption lines do not imply the presence of relativistic ejecta.**

CONCLUSIONS

We compare the optical and radio properties for local SNe Ibc and GRB-SNe. We show that the optical luminosities for GRB-SNe and local SNe Ibc are comparable and therefore cannot be used to distinguish between the two samples. From our comprehensive radio survey of local SNe, however, we are able to show a clear distinction between GRB-SNe and SNe Ibc: the radio luminosities of local events are typically 10^4 times fainter than those of typical GRB-SNe and 10^2 times fainter than SN 1998bw. We conclude that GRB-SN explosions couple the bulk of their energy to highly relativistic ejecta while local SNe couple a relatively tiny fraction. Finally, we use our late-time radio observations to constrain the fraction of SNe Ibc harboring off-axis GRBs to less than 10%. This holds in particular for the local broad-lined SNe Ibc which have been argued (based on their spectral similarity to GRB-SNe) to be associated with off-axis GRBs. In conclusion we find that while most GRB explosions have a supernova component, only a small fraction of SNe Ibc are capable of producing the copious relativistic ejecta characteristic of gamma-ray bursts.

REFERENCES

1. J. H. Elias, K. Matthews, G. Neugebauer, and S. E. Persson, *ApJ*, **296**, 379 (1985).
2. J. C. Wheeler, and R. Levreault, *ApJ*, **294**, L17 (1985).
3. A. Fillipenko, *AAR&A*, **35**, 309 (1997).
4. T. J. Galama et al., *Nature*, **395**, 670 (1988).
5. E. Pian et al., *ApJ*, **536**, 778 (2000).
6. S. R. Kulkarni et al., *Nature*, **395**, 663 (1998).
7. Z.-Y. Li and R. A. Chevalier, *ApJ*, **526**, 716 (1999).
8. K. Iwamoto et al., *Nature*, **395**, 672 (1998).
9. S. E. Woosley et al., *ApJ*, **516**, 788 (1999).
10. A. I. MacFadyen et al., *ApJ*, **550**, 410 (2001).
11. A. Zeh, S. Klose, and D. H. Hartmann, *ApJ*, **609**, 952 (2004).
12. A. M. Soderberg et al., *ApJ*, **636**, 391 (2006a).
13. T. Matheson et al., *ApJ*, **599**, 394 (2003).
14. D. Malesani et al., *ApJL*, **609**, L5 (2004).
15. A. M. Soderberg et al., *ApJ*, **627**, 877 (2005).
16. P. Podsiadlowski, P. A. Mazzali, K. Nomoto, D. Lazzati, and E. Cappellaro, *ApJL*, **607**, L17 (2004).
17. P. A. Mazzali et al., *Science*, **308**, 1284 (2005).
18. T. Nominaga et al., *ApJL*, **633**, L97 (2005).
19. R. A. Chevalier, *ApJ*, **499**, 810 (1998).
20. S. D. van Dyk, R. A. Sramek, K. W. Weiler, and N. Panagia, *ApJ*, **409**, 162 (1993).
21. N. Panagia, R. A. Sramek, and K. W. Weiler, *ApJL*, **300**, L55 (1986).
22. R. A. Sramek, N. Panagia, and K. W. Weiler, *ApJL*, **285**, L59 (1984).
23. C. J. Stockdale et al., *IAU Circulars*, **8472**, 2 (2005).
24. E. Berger et al., *Nature*, **426**, 154 (2003).
25. E. Berger, S. R. Kulkarni, and R. A. Chevalier, *ApJL*, **577**, L5 (2002).
26. A. M. Soderberg, E. Nakar, E. Berger, and S. R. Kulkarni, *ApJ*, in press (2006b), astro-ph/0507147.
27. E. Berger, S. R. Kulkarni, D. A. Frail, and A. M. Soderberg, *ApJ*, **599**, 408 (2003).
28. B. Paczynski, *ACTA Astronomica*, **51**, 1 (2001).
29. E. Waxman, *ApJ*, **602**, 886 (2004).

Late-Time X-ray Flares during GRB Afterglows: Extended Internal Engine Activity

A. D. Falcone*, D. N. Burrows*, P. Romano[†], S. Kobayashi*,**, D. Lazzati[‡], B. Zhang[§], S. Campana[¶], G. Chincarini[¶,||], G. Cusumano[††], N. Gehrels[‡‡], P. Giommi[§§], M. R. Goad[¶¶], O. Godet[¶¶], J. E. Hill[***,‡‡], J. A. Kennea*, P. Mészáros*,[†††], D. Morris*, J. A. Nousek*, P. T. O'Brien[¶¶], J. P. Osborne[¶¶], C. Pagani*, K. Page[¶¶], G. Tagliaferri[†] and the Swift XRT Team

*Department of Astronomy & Astrophysics, 525 Davey Lab., Penn. State University, University Park, PA 16802, USA
[†]INAF-Osservatorio Astronomico di Brera, Via Bianchi 46, 23807 Merate, Italy
**Astrophysics Research Institute, Liverpool John Moores University, Birkenhead CH41 1LD, UK
[‡]JILA, University of Colorado, Boulder, CO 80309, USA
[§]Department of Physics, University of Nevada, Las Vegas, NV
[¶]INAF – Osservatorio Astronomico di Brera, Merate, Italy
||Università degli studi di Milano-Bicocca, Dipartimento di Fisica, Milano, Italy
[††]INAF- Istituto di Fisica Spazialee Fisica Cosmica sezione di Palermo, Palermo, Italy
[‡‡]NASA Goddard Space Flight Center, Greenbelt, MD
[§§]ASI Science Data Center, via Galileo Galilei, 00044 Frascati, Italy
[¶¶]Department of Physics and Astronomy, University of Leicester, Leicester, UK
***USRA, 10211 Wincopin Circle, Suite 500, Columbia, MD, 21044-3432, USA
[†††]Department of Physics, Penn. State University, University Park, PA 16802, USA

Abstract. Observations of gamma ray bursts (GRBs) with *Swift* produced the initially surprising result that many bursts have large X-ray flares superimposed on the underlying afterglow. These flares were sometimes intense, rapid, and late relative to the nominal prompt phase. The most intense of these flares was observed by XRT with a flux $> 500\times$ the afterglow. This burst then surprised observers by flaring again after > 10000 s. The intense flare can be most easily understood within the context of the standard fireball model, if the internal engine that powers the prompt GRB emission is still active at late times. Recent observations indicate that X-ray flares are detected in $\sim 1/3$ of XRT detected afterglows. By studying the properties of the varieties of flares (such as rise/fall time, onset time, spectral variability, etc.) and relating them to overall burst properties, models of flare production and the GRB internal engine can be constrained.

Keywords: Gamma Ray Bursts, X-rays
PACS: 98.70.Rz, 95.85.Nv

INTRODUCTION

Since its launch on 2004 November 20, *Swift* [1] has provided detailed measurements of numerous gamma ray bursts (GRBs) and their afterglows with unprecedented reaction times. By detecting burst afterglows promptly, and with high sensitivity, the properties of the early afterglow and extended prompt emission can be studied in detail for the first time. This also facilitates studies of the transition between the prompt emission and the afterglow. The rapid response of the pointed X-ray Telescope (XRT) instrument [2] on *Swift* has led to the discovery that large X-ray flares are common in GRBs and occur at times well after the initial prompt emission.

While there are still many unknown factors related to the mechanisms that produce GRB emission, the most commonly accepted model is that of a relativistically expanding fireball with associated internal and external shocks [3]. In this model, internal shocks produce the prompt GRB emission. Observationally, this emission typically has a timescale of ~ 30 s for long bursts and ~ 0.3 s for short bursts [4]. The expanding fireball then shocks the ambient material to produce a broadband afterglow that decays quickly (typically as $\propto t^{-\alpha}$). When the Doppler boosting angle of this decelerating fireball exceeds the opening angle of the jet into which it is expanding, then a steepening of the light curve (jet break) is also predicted [5]. For a description of the theoretical models of GRB emission and associated observational properties, see Mészáros [6], Zhang & Mészáros [7], Piran [8], and Van Paradijs et al. [9]. For an alternative explanation that describes both the prompt emission and the afterglow emission with a forward shock, see Dermer & Mitman [10].

With the advent of recent *Swift*-XRT observations of many large flares at various times after the burst, it is clear that a new constraint on GRB models is available to us. We now know that the few previous observations of relatively small flux increases [11, 12] did not provide a complete picture of the X-ray flaring activity during and following GRBs. Recent observations by XRT indicate that flares are common, that they can have a fluence comparable to the initial prompt emission, and that they have various timescales, spectra, and relative flux increase factors [13, 14, 15, 16]. By studying the properties of these flares and by delving into the details of the GRB models, the nature of the X-ray flares, and possibly the GRB internal engine, may be elucidated.

OVERALL XRT OBSERVATIONS

As of 27 December 2005, *Swift*-BAT detected and imaged 95 GRBs, which extrapolates to a rate of $\sim 100/year$. *Swift* slewed to 80 of these bursts within 200 ks, and 74% of these observations resulted in detections of an X-ray afterglow. XRT slewed promptly to 59 of these bursts within 350 s, and 95% of these observations resulted in detections of an X-ray afterglow. From the sample of 56 bursts with prompt slews and detections, more than 24 of them have significant detections of X-ray flares at late times, relative to the nominal prompt emission time frame. In short, $> 25\%$ of all *Swift*-BAT detected bursts have significant X-ray flares, and $> 43\%$ of the bursts with a prompt XRT detection have significant X-ray flares.

A FEW REMARKABLE FLARING GRBS

The flaring GRBs discussed below are just a small subset of those observed so far. A more comprehensive sample will be published soon in two forthcoming papers.

XRF 050406

XRF 050406 was the first *Swift* burst with flaring that was clearly significant, independent of any supporting observations [15, 13]. It is worth mentioning that GRB050219a exhibited flaring, but confidence was not achieved until the higher significance detections of flaring from XRF 050406, and then GRB 050502B. XRF 050406 has a flare with a peak at about 210 s after the BAT trigger time. The flare rises above the underlying power law decay by a factor of ~ 6. When the underlying power law decay, which has a temporal decay index of 1.58 ± 0.17, is subtracted from the flare data, the rise and fall of the flare are nearly symmetric with temporal power law indices of ± 6.8. The $\delta t/t$ for this flare is ~ 0.2. The underlying decay curve before and after the flare are consistent with a single temporal power law decay. This flare did not provide enough photons to perform a detailed spectral analysis, but from plotting the band ratio, it could be seen that the flare had a harder spectrum at the onset which softened back to that of the underlying afterglow as the flare decayed [15]. XRF 050406 is also notable since it was an X-ray Flash, rather than a classic GRB. This common feature of XRFs and GRBs suggests a potential link between the two classes.

GRB 050502B

GRB 050502B is a prime example of a GRB with at least one large flare at late times after the cessation of the initial prompt emission detected by BAT [14, 13]. The light curve from XRT data is shown in Figure 1. A giant flare, with a flux increase by a factor of $\sim$500, was observed using XRT. The fluence during the giant flare, $(1.2 \pm 0.05) \times 10^{-6}$ erg cm^{-2} in the 0.2 – 10 keV band, was slightly above that during the initial prompt emission detected by BAT. The flare rises to a sharp peak at 743 ± 10 s, but this appears to be on top of a broader peak that extends from 640 ± 20 s to 790 ± 20 s. In the hard band (1 – 10 keV), there is significant time structure within the peak of the giant flare itself. During the flare, the spectrum can be fit best by an absorbed cutoff power law (or Band function) [17], rather than a simple absorbed power law, which fits the underlying afterglow nicely. For details, see Falcone et al. [14]. The spectral index hardens significantly during the flare (with a cutoff energy of $\sim$2.5 keV in the XRT band) before returning back to a softer and more typical afterglow spectrum after the flare has ended. Before and after the flare, the temporal decay of the underlying afterglow can be fit well with a single power law $\sim t^{-0.8 \pm 0.2}$. At much later times, between $(1.9 \pm 0.3) \times 10^4$ s and $(1.1 \pm 0.1) \times 10^5$ s, there are two broad bumps (or possibly one broad bump with some structure). These bumps are notable in themselves since they could be more flaring, or they could be due to a combination of flaring and energy injection into the forward shock.

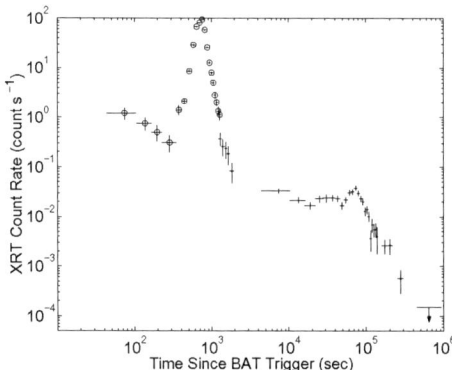

FIGURE 1. X-ray light curve of GRB 050502B. Open circles are window timing mode data, and dots are photon counting mode data. For details, see Falcone et al. [14].

GRB 050607

GRB 050607 is notable due to the fast rise of one of its multiple flares [18]. The second, and brightest flare, had a peak at ~310 s. To borrow a term from prompt GRB descriptions, this flare was FRED-like (fast-rise, exponential-decay), with a very steep rise. The temporal power law index was ~22 if one placed t_0 at the burst trigger time, and it was ~4.1 if one places t_0 at the time of the flare onset [18]. The $\delta t/t$ for this flare is ~ 0.2.

Flaring Short bursts

The exceptional short burst, GRB 050724, exhibited significant flaring detected by XRT (for details, see Barthelmy et al. [19]). There were several flare-like features. In particular, there is the broad bump detected with a peak at $\sim 5 \times 10^4$ s.

It is also possible that GRB 051227, which has a significant X-ray flare peaking at ~ 110 s, is a short burst [20]. However, there is some ambiguity in its characterization as short or long.

Flaring from High Redshift bursts

GRB 050904, at a redshift of 6.29, is the most distant GRB detected to date. This burst has a very interesting X-ray light curve (see Figure 2) with many flares superimposed on top of the underlying temporal decay, and on top of one another [21]. Even after a transformation of the light curve into the rest frame of the GRB, there is significant flaring at times as late as ~5000 s.

In addition to GRB 050904, there are other moderately high redshift bursts with multiple flares (e.g. GRB 050730 at $z \sim 4$). As more data arrive from flaring GRBs,

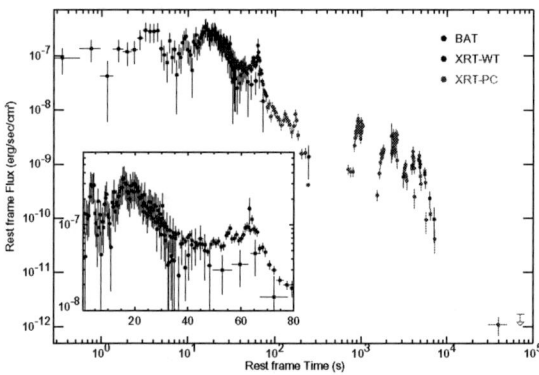

FIGURE 2. Background-subtracted X-ray light curve of GRB 050904, transformed into the rest frame for z=6.29. The BAT data points are from an extrapolation into the XRT energy range (black points, from 0-75 s) superimposed on the XRT light curve (20-10,000 s). For details, see Cusumano et al. [21].

it will be interesting to compare all of the redshift corrected rest frame light curves. While the high redshift certainly makes the emission more extended in the observer frame, these bursts are remarkably sporadic at late times, even in the rest frame.

DISCUSSION AND CONCLUSIONS

It is clear that we now have a recently realized characteristic of GRBs that can be used to probe their nature. The X-ray flares have myriad characteristics. Many have very fast rises and decays, whereas others are relatively gradual. Some occur at early times, along with the nominal prompt emission detected by BAT, whereas others occur at very late times ($\sim 10^5$ s). They occur during all of the underlying decay curve phases (see Nousek et al. [22], Zhang et al. [23] for discussion of decay phases), with the possible exception of post-jet-break times. Some of the flares are huge, whereas others are small bumps. Some GRBs exhibit many flares, whereas other GRBs have only one.

A large fraction of the flares have several characteristics that point towards continued internal engine activity. These characteristics include: 1) The temporal decay index before and after many (but not all) flares are identical, indicating that the afterglow had already begun before the flare, 2) the rise time and decay time of the flares are frequently very fast ($\delta t/t \ll 1$), thus the flare is difficult (although not impossible) to explain with mechanisms associated with the external shock (see Ioka et al. [24], Zhang et al. [23] for discussion), 3) there is even faster time structure near the peaks of some of the flares, 4) the spectra during some flares are represented better by a Band function or cutoff power law model, rather than a simple power law, similar to the nominal prompt emission, 5) the hardness before and after some flares is consistent with an afterglow that has already begun before the flare and continues with approximately the same spectral index after the flare, whereas the spectra during some flares are frequently harder than the underlying afterglow. A final piece of supporting evidence for the restarting of the central engine is that the decay parameters following flares (and BAT prompt emission) usually imply a t_0 that is consistent with the onset of the event, when the decay is interpreted as being dominated by the curvature effect; for details, see Liang et al. [25].

For at least some GRBs with flares, continued internal engine activity is likely, but some flares allow for the possibility of external shock processes, within the framework of the standard fireball model. It is beyond the scope of this paper to address particular extended internal engine models that can explain these flare observations. However, it is important to note that any such models must be capable of emission at very late times ($> 10^4$ s), sporadic and repeated emission to explain multiple flares, very fast rise/decay times, and total energy input comparable to that of the initial prompt emission.

Studies of the overall properties of a sample of many flaring GRBs are necessary to truly characterize their nature, and to determine if there are classes. Results from these studies are forthcoming.

ACKNOWLEDGMENTS

This work is supported at Penn State by NASA contract NAS5-00136; at the Univ. of Leicester by the Particle Physics and Astronomy Research Council under grant PPA/Z/S/2003/00507; and at OAB by funding from ASI under grant I/R/039/04.

REFERENCES

1. Gehrels, N., Chincarini, G., Giommi, P., et al., *Astrophys. Journ.*, **611**, 1005 (2004).
2. Burrows, D.N., Hill, J.E., Nousek, J.A., et al., *Space Science Reviews*, in press, (2005).; astro-ph/0508071
3. Mészáros, P. & Rees, M.J., *Astrophys. Journ.*, **476**, 232 (1997).
4. Meegan, C.A., et al., *Astrophys. Journ. Sup.*, **106**, 65 (1996).
5. Rhoads, J.E., *Astrophys. Journ.*, **525**, 737 (1999).
6. Mészáros, P., *Ann. Rev. Astron. & Astrophys.*, **40**, 137 (2002).
7. Zhang, B. & Mészáros, P., *Int. Journ. of Mod. Phys. A*, **19**, 2385 (2004).
8. Piran, T., *Rev. Mod. Phys.*, **76**, 1143 (2005).
9. van Paradijs, J., Kouveliotou, C., & Wijers, R.A.M.J., *Ann. Rev. Astron. & Astrophys.*, **38**, 379 (2000).
10. Dermer, C. D. & Mitman, K. E., *Astrophys. Journ. Lett.*, **513**, L5 (1999).
11. Piro, L., et al., *Astrophys. Journ.*, **623**, 314 (2005).
12. in't Zand, J.J.M., Heise, J., Kippen, R.M., et al., *Proceedings of 3rd Rome Workshop*, ASP **312**, 18 (2003); astro-ph/0305361
13. Burrows, D.N., Romano, P., Falcone, A., et al., *Science*, **309**, Issue 5742, 1833 (2005).
14. Falcone, A., et al., *Astrophys. Journ.*, accepted, (2006).; astro-ph/0512615
15. Romano, P., et al., *Astron. & Astrophys.*, accepted, (2006).; astro-ph/0601173
16. Burrows, D.N., et al., *Proceedings of the X-ray Universe 2005 Symposium, La Escorial*, in press, (2006).; astro-ph/0511039
17. Band, D., et al., *Astrophys. Journ.*, **413**, 281 (1993).
18. Pagani, C., et al., *Astrophys. Journ.*, submitted, (2006).
19. Barthelmy, S., et al., *Nature*, **438**, 994 (2005).
20. Barthelmy, S., et al., GCN Circular #4401
21. Cusumano, G., et al., *Nature*, accepted, (2006); astro-ph/0509737
22. Nousek, J. A., Kouveliotou, C., Grupe, D., et al., *Astrophys. Journ.*, accepted (2006).
23. Zhang, B., Fan, Y.Z., Dyks, J., et al., *Astrophys. Journ.*, accepted (2006).
24. Ioka, K., Kobayashi, S., & Zhang, B., *Astrophys. Journ.*, **631**, 429 (2005)
25. Liang, E. W., et al., in preparation, (2006).

Physical origin of X-ray flares following GRBs

Bing Zhang

Department of Physics, University of Nevada Las Vegas, Las Vegas, NV 89154

Abstract. One of the major achievements of *Swift* is the discovery of the erratic X-ray flares harboring nearly half of gamma-ray bursts (GRBs), both for long-duration and short-duration categories, and both for traditional hard GRBs and soft X-ray flashes (XRFs). Here I review the arguments in support of the suggestion that they are powered by reactivation of the GRB central engine, and that the emission site is typically "internal", i.e. at a distance within the forward shock front. The curvature effect that characterizes the decaying lightcurve slope during the fading phase of the flares provides an important clue. I will then discuss several suggestions to re-start the GRB central engine and comment on how future observations may help to unveil the physical origin of X-ray flares.

Keywords: gamma-ray bursts; gamma-ray; X-ray; radiation mechanism; polarization
PACS: 95.30.Gv, 95.30.Lz, 95.85.Nv, 95.85.Pw, 98.70.Rz

INTRODUCTION

The successful launch and operation of the *Swift* satellite offers the opportunity to unveil the final gap between the prompt emission and the late afterglow. With the co-operation of the Burst Alert Telescope (BAT) and the X-Ray Telescope (XRT), a canonical early X-ray afterglow lightcurve is emerging[1, 2, 3, 4], which includes five components beside the prompt emission itself (Fig.1), a steep decay component, a shallower-than-normal decay component, a normal decay component, a possible post-jet-break steep decay component, as well as one or more erratic X-ray flares. Not every segment exists in every burst, but all the lightcurves could be in principle understood within such a general framework. These lightcurves bring invaluable information to understand prompt emission - afterglow transition, GRB emission site, central engine activity, forward-reverse shock physics, and GRB immediate environment (for a full discussion, see Ref. [2]).

The most interesting component is erratic X-ray flares detected in nearly half *Swift* GRBs[5, 6, 7, 4, 8]. The general observational properties of these flares include (e.g. [9]): (1) flares typically have rapid rise and fall times, with $\delta t/t_{peak} \ll 1$; (2) many light curves have evidence for an underlying afterglow power law component with the same slope; (3) in many bursts, multiple early flares exist in a same burst; (4) in some cases, e.g. GRB 050502B [5, 6], the flux increases are very large (factors of tens to hundreds); (5) flares soften as they progress; (6) the durations of the flares are positively correlated with the epochs when the flares happen; (7) in some cases, very late flares occur at around days after the trigger (e.g. GRBs 050502B [6] and 050724 [10, 11]); (8) there is no apparent difference between long-duration and short-duration GRBs and between normal, hard GRBs and soft XRFs as far as the X-ray flare properties are concerned.

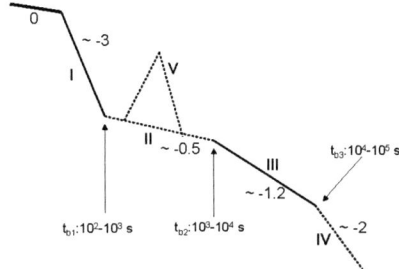

FIGURE 1. The canonical X-ray afterglow lightcurve. From Zhang et al. [2].

X-RAY FLARE MECHANISM: LATE CENTRAL ENGINE ACTIVITY AND INTERNAL ENERGY DISSIPATION

The data greatly constrained the possible models to interpret X-ray flares.

- The rapid rise and fall with $\delta t/t_{peak} \ll 1$ strongly disfavor the scenarios that invoke large angular scale external shock models[2, 12], including refreshed shocks, patchy-shell jets, multi-component jets, signatures induced by neutron decay, etc.
- The existence of multiple flares in a same GRB disfavors scenarios that can only account for one flare, e.g. the synchrotron self-Compton emission in the reverse shock[13], the deceleration of the blastwave[14], and the progenitor models invoking a companion[15].
- The comparable fluence (and hence, energy budget) of the giant X-ray flare of GRB 050502B[5, 6] disfavors models that interpret X-ray flares and prompt emission without additional energy input, including the clumpy medium model[2].
- The similar flaring behavior observed in both long and short bursts suggest that the flare mechanism is insensitive to the progenitor[16].

The leading model is that flares are due to the reactivation of the GRB central engine, and that they are produced from an "internal" dissipation radius within the external blastwave front[5, 2, 12, 17]. There are two advantages for such a model to interpret the flare data[2]. First, re-starting the GRB central engine effectively re-sets the starting-time. For example, should there be no prompt gamma-rays, the onset of a late flare would be effectively the "trigger" time. Each episode of the late central engine activity is equivalent to each other and should define its own time zero point (t_0). Such a shift is essential to define the most relevant temporal decay index in the $\log F_v - \log(t-t_0)$ lightcurves, especially when t is not much larger than t_0 (here all the numerical number of the time quantities are with respect to the trigger time). If t_0 is far from the trigger time, the conventional lightcurves ($\log F_v - \log t$, with time zero point at the trigger time) would show artificial very steep decays as observed (sometimes the decay index after the flare peak is steeper than -7). Second, this model is very economical energetically. This is because we invoke two different emission sites to interpret the X-ray flares and the background power-law decaying X-ray afterglow. The former are from an internal radius, while the latter is from the external forward shock. In order to produce the same X-ray flux level, the internal model usually only requires a small amount of energy budget with respect to that of the prompt emission, and the lightcurve including the prompt emission and the flares essentially reflects the time history of the central engine energy input. For the external shock model, on the other hand, one requires at least comparable energy budget with the intial blastwave in order to make any noticeable change of the afterglow flux level[18]. The flare amplitudes are usually a factor of several to several hundreds. This would require an unbelievably large energy input from the central engine without any observational signature. This is rather implausible. For external shock models that invoke very small density clumps[19], the energy budget problem could be eased, but so far no modeling could successfully reproduce the giant flares as observed in GRB 050502B.

Invoking an unknown central engine activity inevitably introduces difficulties to model the flares (as compared with the simply afterglow model). However, one has a clean signature that could be used to diagnose the above interpretation. This is the so-called "curvature effect" that dominates the decaying phase of the flare. Assuming that radiation stops abruptly at an internal radius from a conical shell, and that the cooling frequency is below the X-ray band before the cessation of the synchrotron emission, no fresh electrons would contribute to the emission in the X-ray band since the cessation of the emitter. What one detects is then the emission from higher latitudes with respect to the viewing direction that propagates to the observer with slight delay due to the extra distance the photons need to travel. There is a clear prediction for such a scenario: the temporal decay index is connected to the spectral index by $\alpha = \beta + 2$ with the convention $F_\nu \propto \nu^{-\beta} t^{-\alpha}$ [20, 21, 2, 17, 22, 23]. Such a prediction is rather robust regardless of the speed history of the shell[2, 17] and is largely insensitive to the jet structure as long as the viewing angle is not very far away from the bright core of the structured jets[23].

The clean $\alpha = \beta + 2$ characteristic is complicated by two additional effects [2]. One is the t_0 effect, i.e. one needs to identify the correct central engine starting time for the flare in investigation. If the internal-origin of the flares is correct, this t_0 should be at the beginning of the rising segment of the flare. The second effect is the superposition effect. There is an underlying afterglow level which tends to make the decay slope shallower, especially near the transition time between the steep decay component (segment I in Fig.1) and the shallow decay components from the afterglow (segments II or III in Fig.1). However, since both the spectral index (β) and the temporal decay indices in different segments (α_1, α_2 and α_3) could be directly inferred from the data, one has a clean strategy to test the correctness of the internal-curvature interpretation.

Liang et al. [8] have performed such a systematic test. The procedure is the following. (1) Identify a steep decay component with $\alpha > 2$ following a certain X-ray flare (or the steep decay component following the prompt emission). (2) Perform a fit to the afterglow lightcurve with two overlapping power laws. The late-time index is fitted by the data and the early-time index is assumed to be $\beta + 2$. (3) Search for the time zero point t_0 that

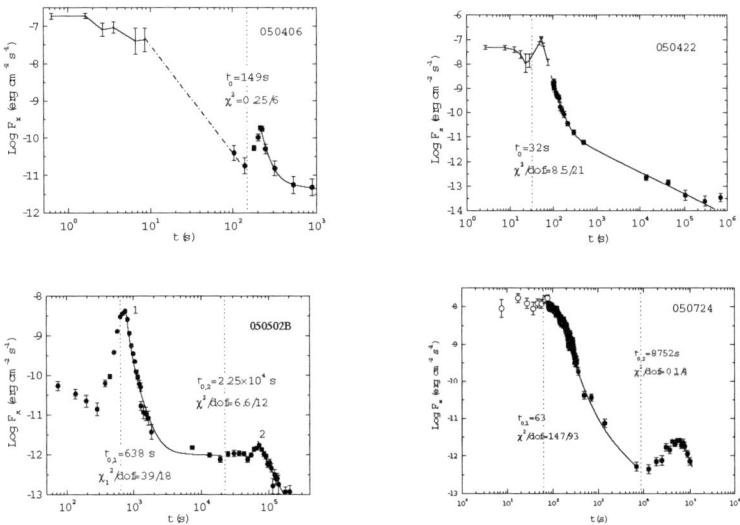

FIGURE 2. Several examples of the searched t_0 of X-ray flares. From Liang et al. [8].

gives the best fit to the data. The results are impressive. In many cases, the searched t_0 is exactly where it is expected, i.e. at the beginning of the corresponding flare (Fig.2). This gives strong support to our beginning hypothesis: that the flares are from internal dissipation of energy and that the decay is controlled by the curvature effect. In some cases the signature is not as clear as expected, but these cases also have overlapping flares or extended emission so that the rising part of the flare is usually burried beneath other emission components.

One interesting thing to note is that both GRB 050502B (a long-duration burst [6]) and GRB 050724 (a short-duration burst associated with non-star-forming galaxy [10]) have a late flare peaking around a day. The decay after the peak clearly follows the prediction of curvature effect, strongly suggesting that the central engine is turned on again (for an extended emission period) at such a late epoch. This poses great challenge to the possible central engine models for both long and short GRBs.

HOW TO RESTART THE CENTRAL ENGINE

Now that the X-ray flare data suggest that the GRB central engine must be re-started at a delayed time and that it works in a non-continuous manner, a straightforward question is then how a GRB central engine model would be able to give rise to what is observed.

This is still largely an open issue. Nonetheless a list of suggestions have been made since the discovery of the X-ray flares.

- King et al. [24] suggest that within the collapsar progenitor model for long GRBs, the rotating stellar core may undergo fragmentation during the collapse. The accretion of the fragmented blobs into the newly formed black hole offers a plausible mechanism of X-ray flares. Whether such a fragmentation can be realized calls for proofs from numerical simulations. The existence of similar X-ray flares in short GRB 050724 [10] suggests that the flare mechanism likely also work for progenitors without a heavy envelope. This makes this suggestion inconclusive.

- Based on energetics alone, Fan et al. [25] argue that the flares following the short GRB 050724 (and possibly also GRB 050709) must be of magnetic origin. This is because in various compact star merger models, the mass of the fuel to power the accretion is limited to be at most about 1 solar mass. The durations of the early flares typically last for hundreds seconds, suggesting that the accretion rate ($\leq 0.01 M_\odot/s$) is too low for the neutrino-annihilation mechanism to power the observed 10^{48} erg s^{-1} luminosity. As a result, the X-ray flares are expected to be somewhat linearly polarized, although the detailed degree of polarization is hard to predict. The argument may be also extended to flares in long GRBs
- Motivated by the similarity of the X-ray flare properties for both long and short bursts, Perna et al. [16] argue that the flaring mechanism must be related to the property of something in common for both types of bursts. They suggest the accretion disk is this common link, and the episodical flares may be caused by fragmentation of the accretion disk itself. Gravitational instability in the outer parts of the disk may cause the fragmentation. The model can naturally interpret some properties of the flares, including the duration-timescale correlation and the duration-peak luminosity anticorrelation.
- Based on results of previous MHD numerical simulations and theoretical analyses, Proga & Zhang [26] suggest that the accretion flow does not have to be chopped at large radii due to fragmentation. Rather, strong magnetic fields near the black hole can play the role to modulate the accretion flow. The magnetic barrier, which has been evident in previous numerical simulations [27, 28], can act as an agent to repeatedly stop and restart the accretion flow. This gives a plausible mechanism to power X-ray flares in both long and short GRBs. In this model, the launched jet is magnetized, which coincides the argument based on energetics [25].
- Dai et al. [29] argue that the postmerger product of double neutron star merger may well be a massive neutron star with milliseocnd rotation period. The differentially-rotating stellar body would wind up interior poloidal magnetic fields to form progressively stronger toroidal magnetic fields. These tangled fields then float up and break through the stellar surfaces, triggering magnetic-reconnection-driven explosive events. This is an attractive mechanism to produce X-ray flares in short GRBs.
- A two-step central engine model is discussed by Gao [30].

Entering the second-year operation, *Swift* will keep collecting flare data for a large sample of bursts. From data point of view, it is high time to systematically analyze the flare data and perform statistical analyses of the properties of flares. In the mean time, detailed theoretical modeling of the flare properties (e.g. [31]) would shed more light on the radiation mechanism and the dissipation site of X-ray flares. In long terms, coordinated observations of *Swift* and high energy detectors such as *GLAST* would reveal whether there are high-energy counterparts of X-ray flares, and hence, greatly constraint the emission physics of the flares [32]. An even far-reaching goal is to detect polarization of X-ray flares with future X-ray polarimeters, which would leads to more direct diagnoses of the GRB central engine [25].

ACKNOWLEDGMENTS

I thank stimulative collaborations with E. W. Liang, Y. Z. Fan, J. Dyks, D. N. Burrows, S. Kobayashi, P. Mészáros, R. Perna, P. J. Armitage, D. Proga, Z. G. Dai, X. Y. Wang, X. F. Wu, J. Nousek, N. Gehrels and many other members in the Swift team on various topics covered in this talk. This work is supported by NASA through grants NNG05GB67G, NNG05GH91G, and NNG05GH92G.

REFERENCES

1. J. A. Nousek, C. Kouveliotou, D. Grupe, et al. *Astrophys. J.*, in press (2006), astro-ph/0508332.
2. B. Zhang, Y. Z. Fan, J. Dyks, S. Kobayashi, P. Mészáros, D. N. Burrows, J. A. Nousek, and N. Gehrels, *Astrophys. J.*, in press (2006), astro-ph/0508321.
3. G. Chincarini, A. Moretti, P. Romano, et al. astro-ph/0506453
4. P. T. O'Brien, R. Willingale, J. Osborne, et al. *Astrophys. J.*, submitted (2006), astro-ph/0601125
5. D. N. Burrows, P. Romano, A. Falcone, et al. *Science*, **309**, 1833 (2005).
6. A. Falcone, D. N. Burrows, D. Lazzati, et al. *Astrophys. J.*, in press (2006), astro-ph/0512615.
7. P. Romano, A. Moretti, P. L. Banat, et al. *Astron. Astrophys.*, in press (2006), astro-ph/0601173.
8. E. W. Liang, B. Zhang, P. T. O'Brien, et al. *Astrophys. J.*, submitted (2006).
9. D. N. Burrows, P. Romano, O. Godet, et al. "Swift XRT Observations of X-ray flares in GRB afterglows", to appear in *X-Ray Universe 2005*, conference proceedings, (2005), astro-ph/0511039.
10. S. D. Barthelmy, G. Chincarini, D. N. Burrows, et al. *Nature*, **438**, 994 (2005).
11. S. Campana, G. Tagliaferri, D. Lazzati, et al. *Astron. Astrophys.*, submitted (2006).
12. K. Ioka, S. Kobayashi, B. Zhang, *Astrophys. J*, **631**, 429 (2005)
13. S. Kobayashi, B. Zhang, P. Mészáros, D. N. Burrows, *Astrophys. J.*, submitted (2005), astro-ph/0506157
14. L. Piro, M. De Pasquale, P. Soffitta, et al. *Astrophys. J.*, **623**, 314 (2005)
15. A. I. MacFadyen, E. Ramirez-Ruiz, W. Zhang preprint, (2005), astro-ph/0510192
16. R. Perna, P. J. Armitage, B. Zhang, *Astrophys. J.*, **636**, L29 (2006)
17. Y. Z. Fan, D. M. Wei, *Mon. Not. R. Astron. Soc.*, **364**, L42 (2005)
18. B. Zhang, P. Mészáros, *Astrophys. J.*, **566**, 712
19. C. D. Dermer, in preparation (2006)
20. P. Kumar, A. Panaitescu, *Astrophys. J.*, **541**, L51 (2000)
21. C. D. Dermer, *Astrophys. J.*, **614**, 284 (2004)
22. A. Panaitescu, P. Mészáros, N. Gehrels, D. Burrows, J. Nousek, *Mon. Not. R. Astron. Soc.*. in press (2006), astro-ph/0508340
23. J. Dyks, B. Zhang, Y. Z. Fan, *Asttrophys. J.*, submitted (2006), astro-ph/0511699
24. A. King, P. T. O'Brien, M. R. Goad, J. Osborne, E. Olsson, K. Page, *Astrophys. J.*, **630**, L113 (2005)
25. Y. Z. Fan, B. Zhang, D. Proga, *Astrophys. J.*, **635**, L129 (2006)
26. D. Proga, B. Zhang, *Astrophys. J.*, submitted (2006), astro-ph/0601272
27. D. Proga, Begelman, M. C. *Astrophys. J.*, **592**, 767 (2003)
28. I. V. Igumenshchev, R. Narayan, M. A. Abramowicz, *Astrophys. J.*, **592**, 1042 (2003)
29. Z. G. Dai, X. Y. Wang, X. F. Wu, B. Zhang, submitted (2006)
30. W. H. Gao, *Astrophys. J.*, submitted (2006), astro-ph/0512646
31. X. F. Wu, Z. G. Dai, X. Y. Wang, Y. F. Huang, L. L. Feng, T. Lu, *Astrophys. J.*, submitted (2006) astro-ph/0512555
32. X. Y. Wang, Z. Li, P. Mészáros, *Astrophys. J.*, submitted (2006), astro-ph/0601229

The Supernova Gamma-Ray Burst Connection

S. E. Woosley* and A. Heger*†

*Department of Astronomy and Astrophysics, UCSC, Santa Cruz CA 95064
†Theoretical Astrophysics Group, T-6, MS B227, Los Alamos National Laboratory, Los Alamos, NM 87545

Abstract. The chief distinction between ordinary supernovae and long-soft gamma-ray bursts (GRBs) is the degree of differential rotation in the inner several solar masses when a massive star dies, and GRBs are rare mainly because of the difficulty achieving the necessary high rotation rate. Models that do provide the necessary angular momentum are discussed, with emphasis on a new single star model whose rapid rotation leads to complete mixing on the main sequence and avoids red giant formation. This channel of progenitor evolution also gives a broader range of masses than previous models, and allows the copious production of bursts outside of binaries and at high redshifts. However, even the production of a bare helium core rotating nearly at break up is not, by itself, a sufficient condition to make a gamma-ray burst. Wolf-Rayet mass loss must be low, and will be low in regions of low metallicity. This suggests that bursts at high redshift (low metallicity) will, on the average, be more energetic, have more time structure, and last longer than bursts nearby. Every burst consists of three components: a polar jet (~ 0.1 radian), high energy, subrelativistic mass ejection (~ 1 radian), and low velocity equatorial mass that can fall back after the initial explosion. The relative proportions of these three components can give a diverse assortment of supernovae and high energy transients whose properties may vary with redshift.

Keywords: supernovae, gamma-ray bursts, stellar evolution
PACS: 98.70.Rz, 97.60.Bw, 26.50.+x

INTRODUCTION

As talks summarized elsewhere in this proceedings and papers in the literature have made clear [1], most GRBs of the long-soft variety (henceforth just GRBs) are a consequence of the deaths of massive stars. Evidence supporting this comes from: 1) the location of GRBs in regions of active star formation [2]; 2) the clear presence of supernovae of Type Ic-BL ("broad-lined Ic") in conjunction with three GRBs: 980425, 030329, and 031203; 3) the presence of supernova-like bumps in most other GRBs where they might be observed [3]; and 4) the similarity in energy between the beaming-corrected GRB energy and that of a supernova.

Accepting this as a starting point, an important question must be why some massive stars die as supernovae of the ordinary variety, while others die as GRBs. The fraction that do die as GRBs is apparently very small. Taking an event rate of core-collapse supernovae visible from the earth today as ~ 20 per 16 arc min squared per year [4], the integrated rate of supernovae on the sky is about 6 per second. BATSE saw about a burst a day. Correcting by a factor of 300 for beaming and another factor of three for Earth occultation and bursts that were missed for reasons other than beaming, the GRB sky rate is about 0.02 per second. That is, GRBs are a fraction of order 0.3% of all massive star deaths. This fraction could be larger if there are numerous sub-luminous events like

GRB 980425 or cosmic X-ray flashes (XRFs), and it might increase with redshift, but apparently GRBs are a rare channel of massive star death.

Another interesting question is whether GRBs and ordinary supernovae are the extrema of a continuum of events or separate classes of explosions. The existence of such diverse phenomena as XRFs, 980425, and GRBs with varying energy and supernova properties suggests a continuum. So, too, does the growing class of "hyper-energetic" or grossly asymmetric supernovae like SN 2005bf [5, 6, 7]. Add to this the fact that most broad-lined Type Ic supernovae do *not* harbor GRBs [8], and a compelling case might be made for some continuously variable parameter that dials between ordinary supernovae and energetic GRBs like 990123. Clearly, the difference is that GRBs concentrate a significant fraction of their 10^{51} - 10^{52} ergs into highly relativistic ($\Gamma > 200$) beamed ejecta while ordinary supernovae do not, but searches beyond that for a deeper physical cause take us into what is suspected, and away from what is certain.

GRB PROGENITORS

The two key quantities that determine how a massive star dies are its mass and the rotation rate of its inner few solar masses. Stars that are more massive when they die are more likely to make black holes. More massive stars also evolve more quickly and may more effectively preserve the large angular momentum they have at birth in their core [9]. Rotation is a key ingredient in any successful GRB model. If the rotational energy of a neutron star is to provide the $\sim 10^{52}$ erg inferred for some of the supernovae accompanying GRBs, it must have a rotational period ~ 1 ms. This implies a specific equatorial angular momentum, $j \approx 7 \times 10^{15}$ cm^2 s^{-1} $(P/1$ ms$)(R/10$ km$)^2$, about 20 or more times that of a typical pulsar. If a disk is to form around a black hole one needs more. At 3 $M_\odot$, the specific angular momentum of the last stable orbit is $j_{lso} = 2\sqrt{3} GM/c = 4.6 \times 10^{16}$ cm^2 s^{-1} $(M_{BH}/3M_\odot)$ for a Schwarzschild black hole, and $j_{lso} = 2/\sqrt{3} GM/c = 1.5 \times 10^{16}$ cm^2 s^{-1} $(M_{BH}/3M_\odot)$ for an extreme Kerr hole. Given that angular momentum increases monotonically with interior mass from 1.5 to 3 $M_\odot$, the necessary values for neutron star models and black hole models are qualitatively similar.

Studies of massive star evolution that include estimated magnetic torques [9] have shown that such large values of angular momentum are not easily achieved if the star spends much time as a red giant, or even if it loses its envelope early on, but has a high mass loss rate afterwords as a Wolf-Rayet (WR) star [10]. A frequently discussed scenario is a stellar merger, but the central region of a differentially rotating star that has lost its angular momentum is virtually impossible to spin up again. This limits viable models to those where the merger removes the envelope (with enough time still left for for its dispersal), produces a compact helium star that never becomes a blue or red giant itself, and still does not slow down the denser, inner regions of the star.

Several possibilities remain, however. One is that the magnetic torques used in the above study were simply too large, but then one must take care both to produce the large number of slowly rotating neutron stars that are seen, and not overproduce GRBs. Another possibility is a binary merger between two massive stars, both of which are burning helium in their centers [11]. The merger ejects both envelopes and the (now rapidly rotating) residual helium star evolves to produce the burst a few hundred thousand years later, after the envelope has left the vicinity. Even then, though, the new helium star must lose a very limited amount of mass before dying, so it helps if the merger occurs late during helium burning for one or both stars.

A third possibility is that GRBs come from the merger of a black hole or neutron star with either the helium core of a massive star [12, 13] or a white dwarf [14]. In the former case, the giant star's envelope must be completely dispersed before the final merger occurs. In the white dwarf case, one would not expect such a tight correlation with star forming regions, but a fraction of GRBs could still be made this way. In both of these compact merger models, however, there is so *much* angular momentum that the total duration of the event is quite long, longer than typical GRBs. It could be that the observed burst only reflects the epoch of maximum accretion in which case both enduring activity and long precursors might be present. Or perhaps this model only makes very long GRBs.

A final possibility, and one that has received a lot of attention lately, is that GRBs result from a rare channel of single star evolution in which red giant formation is avoided altogether. The possibilities here are exciting and warrant a more lengthy discussion.

A Single Star Model

It has been recently realized [15, 16] that stars which rotate *very* rapidly on the main sequence, with equatorial speeds near 400 km s^{-1} instead of 200 - 300 km s^{-1}, may experience nearly complete mixing on the main sequence [17, 18]. In fact, rotationally-induced mixing in such stars, chiefly by Eddington-Sweet circulation, keeps the star's composition nearly homogeneous until the end of helium burning. A red giant is never formed and the star goes straight from the main sequence to being a Wolf-Rayet star. Such stars will have more rapidly rotating iron cores when they die, though still not fast enough to be GRBs if the helium star loses a lot of mass. Because the progenitors are on the very high end of the observed distribution function for O- and B-star rotation [19], GRBs will be a rare channel of star death.

This model has several beneficial features for GRBs. First, the progenitor star is a compact WR star, mostly composed of oxygen with little helium at its surface - a WO star. This agrees well with the properties of those few GRB-supernovae that have been well studied spectroscopically. Second, it is possible to produce a wider range of GRB progenitor masses. Main sequence stars as light as 10 $M_\odot$ produce helium and heavy element cores that - *modulo* the mass loss - are about as large as that of a 25 $M_\odot$ star with slow rotation (both about 9 $M_\odot$). Such big cores evolve rapidly, retain their angular momentum, and develop big iron cores when they die. They are more likely to make black holes. Below about 10 $M_\odot$, larger, possibly unphysical rotation speeds are necessary on the main sequence to cause Eddington-Sweet circulation to mix the star efficiently. On the upper end, very massive GRB progenitors can be produced from main sequence stars of more moderate mass than previously thought. For slowly rotating, solar-metallicity stars, the largest helium core that can exist when the star dies is around 15 $M_\odot$, corresponding to a main sequence star of 35 $M_\odot$. Heavier stars lose

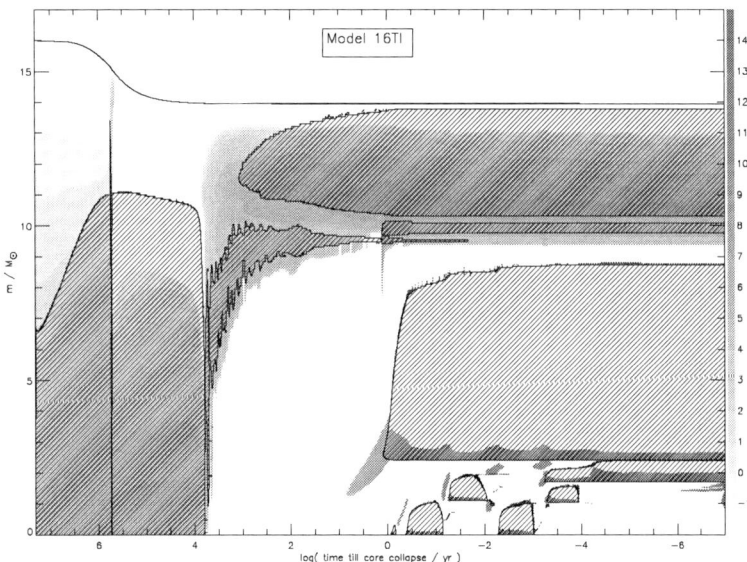

FIGURE 1. Convective history of a rapidly rotating 16 $M_\odot$ star that experiences nearly complete mixing on the main sequence and avoids red giant formation. The left axis is the interior mass, cross hatching shows regions that are convective, and gray shading indicates specific nuclear energy generation on a logarithmic scale (right gray-scale bar, in log ergs/g/s). The bottom axis is the time until death (on a logarithmic scale). This model had an initial composition of one per cent solar and an equatorial velocity half way through hydrogen burning of 380 km s^{-1}. The final mass was 14.0 $M_\odot$ and the iron core mass was 1.60 $M_\odot$. The abundances in the surface convection zone of the presupernova star were 9.5% He, 30% C, 57% O, and 2.7% Ne. The radius was 4.1×10^{10} cm.

their envelopes to winds and the resulting helium core shrinks by mass loss. But for these rapidly rotating, well-mixed stars, the upper bound on the helium core is, in principle, equal to the main sequence mass. Of course, mass loss, both on the main sequence and especially as a WR star, will still shrink the mass. As we shall see shortly however, turning the metallicity down can alleviate the mass loss of a WR star, so that GRB progenitors in low metallicity regions could have very big mass.

An upper limit to the helium core mass that comes from these rapidly rotating models is the first mass to encounter the pulsational pair instability [20], about 40 $M_\odot$ of helium and heavy elements. The evolved star experiences violent, repeated, nuclear-powered explosions when the star ignites oxygen burning. Each outburst ejects solar masses of surface material. This material surrounds the star when it finally dies (typically death happens months to years later) and prevents a GRB from getting out (though such explosions are interesting in their own right). For helium cores above about 65 $M_\odot$, the pair-instability becomes so violent that it leads to the complete disruption of the star and no GRB will be made, but above 140 $M_\odot$ a new regime is encountered where black holes are formed and GRBs of a more energetic, longer-lasting variety become possible [21]. In slowly rotating stars, without mass loss, the pulsational pair instability is first encountered for main sequence stars of $\sim$100 $M_\odot$ and black holes are made starting

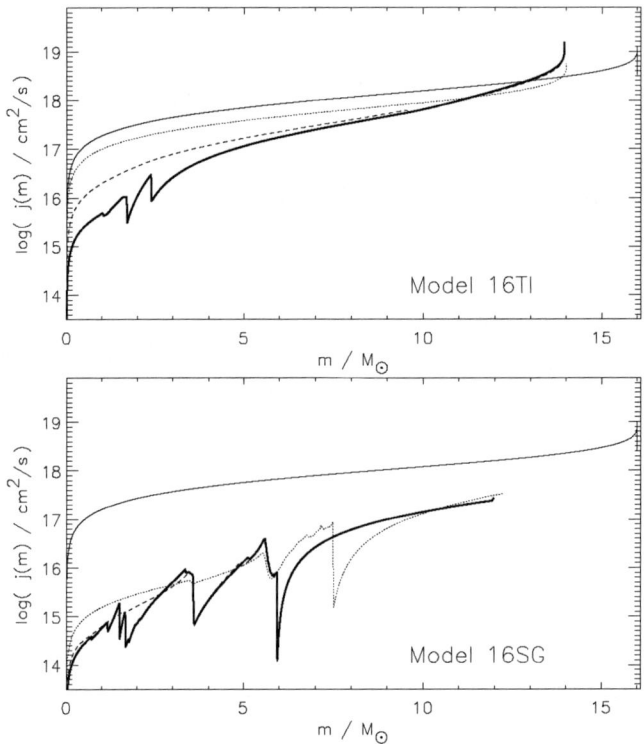

FIGURE 2. Specific angular momentum as a function of mass in two stars with initial mass 16 $M_\odot$ [15]. Both models included angular momentum transport due to rotationally induced mixing processes and magnetic torques. The angular momentum is evaluated at the zero age main sequence (top line, solid), central helium depletion (dotted line, next down), carbon depletion (dashed line), and presupernova (dark solid line). The top star had an equatorial rotation rate on the main sequence of 380 km s^{-1}, experienced complete mixing, and avoided ever making a red giant. The top star also had a Wolf-Rayet stellar mass loss rate corresponding to 1% solar metallicity. The bottom star had solar metallicity and a slower rotation rate on the main sequence, 215 km s^{-1}. The bottom star became a red supergiant, lost more mass, and produced a slowly rotating iron core that would give a 8 ms pulsar period after collapse and neutrino loss. The top star ended up with a higher mass and much greater rotation rate. This star would form a black hole accretion disk starting at 3.5 $M_\odot$. Avoiding red giant formation and reducing the mass loss greatly increases the probability of making a GRB.

at about 260 $M_\odot$, so these lower values for rapidly rotating stars are very significant changes. The numbers are currently uncertain, however, because few rapidly rotating models have been calculated and the effect of the rotation on the pair instability has not been included in a self-consistent way.

A final interesting consequence of rapid rotation and nearly homogeneous mixing, is that single stars, even at low metallicity, can end their lives as compact WR stars without the need of a binary companion to absorb the envelope. This has interesting implications for making GRBs at high redshift.

The Metallicity Dependence of GRBs

Whether the WR-star that serves as the progenitor of a GRB is made by merger or rotationally-induced mixing, it is essential that its mass loss, or more specifically, its angular momentum loss, be low. A massive WR star typically spends 0.5 to 1 million years burning helium. If, during that time, its mass loss rate is over 10^{-5} $M_\odot$ yr^{-1}, it will lose a significant fraction of its mass. Magnetic torques maintain nearly rigid rotation during helium burning, so the continued expansion of layers deep in the star to larger radius brakes the rotation of the inner core beyond what is needed later to make a GRB. Mass loss is the enemy of GRBs.

Fortunately, it has been recently determined [22] that reducing the metal content of a WR star even by a factor of a few lowers its mass loss very appreciably, approximately as $Z^{0.86}$. Once the metallicity has been reduced by a factor of 10, typical mass loss rates for 10 - 20 $M_\odot$ WR stars are $\sim 10^{-6}$ $M_\odot$ yr^{-1} and less. Losing less than a solar mass would not greatly alter the final angular momentum distribution of a massive WR star, so at metallicities $\sim$10% solar and less, GRBs should be plentiful. It is important to note that the metallicity employed in the scaling here is the initial concentration of iron in the star. It does not include the carbon and oxygen at the surface of a WC or WO star that was made in the star itself during helium burning, at least not until the iron abundance becomes so low that the mass loss is negligible anyway, roughly 0.01 solar.

This does not mean that GRBs cannot happen in regions with solar metallicity, only that it is harder. Measured WR mass loss rates show a considerable spread and some merger models might also still work. The magnetic torques estimated in the models are uncertain and, for stars that are highly deformed the angular dependence of the mass loss rate is an issue. Mass loss preferentially from the poles would reduce the amount of angular momentum carried away by each gram [23].

However, the key role of mass loss and its strong metallicity dependence *does* suggest that the fraction of massive stars dying while making GRBs may be larger in regions where less nucleosynthesis has occurred. As the metallicity declines it may also be possible to make more energetic, longer lasting bursts. Higher mass and angular momentum at death increase the reservoir of material that can accrete into a black hole. It may also lead to more rapidly rotating neutron stars in the millisecond magnetar model.

The low mass loss rate also has implications for the afterglow analysis, and a lower density may be more consistent with observations [24]. One must take care, however, since the wind that is sampled in the afterglows was ejected during the post-helium burning evolution of the star, during which the mass loss rate may have varied from what is observed on the helium-burning main sequence.

THE THREE COMPONENTS OF A GRB

A GRB with an energetic supernova accompanying it will have three components: 1) a highly relativistic, $\Gamma \gtrsim 200$, central jet with an opening angle $\sim$0.1 radian measured from the rotational axis; 2) a broader region of very energetic, but subrelativistic ejecta extending out to angles $\sim$1 radian; and 3) slower moving ejecta in the equator. In general, region 1 is responsible for the GRB, region 2 is necessary for the supernova and the

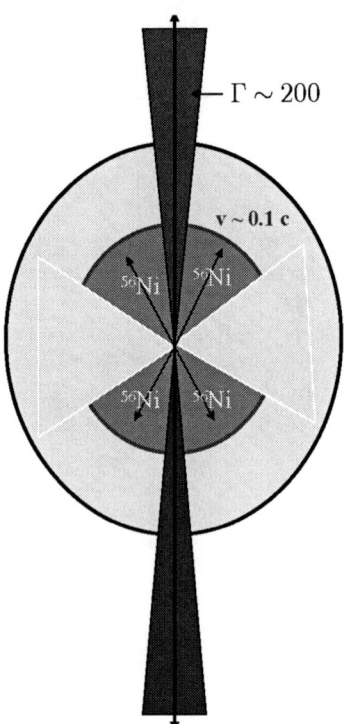

FIGURE 3. Schematic illustrating the three components of a typical GRB and its accompanying supernova. A core relativistic jet (~0.1 radian, $\Gamma \sim 200$, KE $\sim 10^{51}$ erg) is responsible for the GRB and its afterglows. A broader angle, energetic outflow (~ 1 radian, v/c ~0.1, KE ~10^{52} erg) is responsible for exploding the star and making the ^{56}Ni to power the light curve. A third component of low velocity material exists in the equatorial plane of those models in which central mass ejection is blocked by an accretion disk (e.g., collapsars). This component typically fails to achieve ejection on the first try and falls back to power a continuing explosion.

^{56}Ni to make it bright, and region 3 is naturally present in any model where outflow in the equator is blocked, e.g., by an accretion disk in the collapsar model. Region 2 probably contains most of the energy and is necessary because a narrow jet, by itself, is an ineffective way of blowing up a star or producing explosive nucleosynthesis. Region 3 may move out initially, but is largely responsible for the "fall-back" that may keep an accretion-powered source active long after the initial burst is over. One might also properly discuss a fourth region, the jet cocoon lying between regions 1 and 2, but here we count that as part of the jet.

In the collapsar model [25], the jet is produced in the near vicinity of the black hole by neutrinos or MHD processes, while the large angle ejecta come from the disk wind. In the millisecond magnetar model, theorists have yet to talk about the two components in any detail, but one might imagine a large angle component from filling a cavity with nearly isotropic (thermal?) radiation, while simultaneously producing a narrower, electromagnetically focused outflow. Without better knowledge of the physics of how the relativistic jet is launched and the efficiency of the disk-wind, it is not possible to say what fraction of the energy goes into each of these components, but in all likelihood that fraction varies with the mass and angular momentum distribution of the presupernova star. It is also worth keeping in mind that the central engine must not only maintain a powerful jet for at least the ~10 seconds it takes the jet to reach the surface of the star, but must hold the direction of that jet steady to better than 3 degrees [26]. If the jet wavers by more than that, it becomes contaminated with too many baryons on the way out to make a GRB.

If the three components can vary independently, and it is hard to see why they wouldn't, one expects a wide variety of phenomena resulting from essentially the same central engine. These could include a) ordinary GRBs [Regions 1 and 2 strongly active]; b) anisotropic broad-lined supernova without any GRB or bright afterglow [Region 1 trapped in the star, Region 2 dominant; e.g., ref. 8] c) GRBs with continuing activity after the burst [Regions 1 and 3 active]; d) XRFs and SN 1998bw [either the cocoon of an ordinary GRB seen somewhat off axis or an ordinary GRB with higher baryon loading in the jet]; and others. In this context, SN 1998bw and SN 2003dh need not be typical of all supernovae associated with GRBs or XRFs. There would be a continuum of events with the fainter GRBs and supernovae quite possibly the dominant case in a volume-limited sample.

It must be one of the observational goals of the future to gather a sufficient sample to determine if this is true. Are GRBs and core-collapse supernovae a continuum of events with a common central engine and a smoothly varying parameter - e.g., rotation - underlying them all. Or are GRBs and supernovae two discrete, different classes of high energy explosions. Personally, our underlying bias [27] is that rotation plays little role in most supernovae. These are the result of (slowly rotating) neutron star formation and neutrino transport. But "GRBs" are a diverse class of phenomena, overlapping common supernovae in observable properties. Perhaps rotation powers them all? This is a very old question, but still a critical one. Just how do massive stars die (and explode)?

GRBS AT HIGH REDSHIFT

If the properties of GRBs are sensitive to the metallicity as we have described, then one expects systematic differences in the appearance of GRBs "locally" and at high redshift (where the metallicity is presumably low). On the average, though perhaps not individually, GRBs in more metal-deficient regions will come from stars that are more massive and that have lost less angular momentum. Their disks will draw from a larger reservoir of matter and, if, as seems reasonable, the total burst of energy correlates with the total mass accreted, the GRB will last longer and have more energy. The supernova component may also be brighter if the disk wind lasts longer and carries more mass. Bigger helium stars are more tightly bound, however, and harder to explode [28], so the supernova could experience significantly more fall-back. Continuing accretion activity would be the norm.

It is also possible, in the collapsar model, to have too much angular momentum [25, 29, 30]. Angular momentum much in excess of 10^{17} cm^2 s^{-1} will lead to a pile up in the disk at such large radii that neutrino dissipation is negligible. Unable to dissipate its binding energy, the disk becomes unstable and perhaps dominated by an outflow, accompanied by very little accretion. Such behavior has been seen recently in unpublished calculations by Weiqun Zhang and Andrew MacFadyen. Since the outflow has too little energy to explode the whole star in one try, a limit cycle may operate (though this has yet to be followed on the computer). Since angular momentum increases monotonically outwards in the equator of a GRB progenitor, material accretes efficiently until a limiting angular momentum is reached. During that time it maintains a strong jet which can escape the star and make a GRB. Eventually though, the angular momentum becomes so large that the disk ceases to be an NDAF (neutrino-dominated accretion flow) and stagnates. Matter is still falling in, however, especially from high latitudes. Mixing and shear will eventually reduce the angular momentum to the point where accretion can begin again, launch a new jet, and repeat the cycle. The characteristic time scale would be given by mixing and fall back. A few hours is reasonable [31].

This physics is possibly reflected in the observed features of GRB 050904, the most distant GRB discovered [32] and localized [33] so far (z = 6.29). This was a long, multi-peaked, bright burst that lasted well over 205 seconds [34] and had an equivalent isotropic energy of 0.66 to 3.2×10^{54} erg [35, 36] and a beaming corrected energy of 4 to 12×10^{51} erg. Repeated flaring activity was seen from the burst for 1.5 hours in the rest frame [36]. While the properties of this burst are not dramatically different from some others seen closer by, it does lie at an extreme of energy, duration and variability. It will be very interesting to see if future bursts at this redshift and higher show these same characteristics.

ACKNOWLEDGMENTS

The authors appreciate helpful discussions with and access to the unpublished calculations of Andrew MacFadyen and Weiqun Zhang. This research was supported by NASA (NAG5 12036, MIT 292701, NNG05GG28G, SWIF03-0047-0037, and NAG5-13700) and the NSF (AST 0206111). AH is supported at LANL by DOE contract W-7405-ENG-36 to the Los Alamos National Laboratory.

REFERENCES

1. Woosley, S. E., & Bloom, J. 2006, ARAA, submitted
2. Fruchter, A. 2006, *Nature*, submitted
3. Zeh, A., Klose, S., & Hartmann, D. H. 2004, ApJ, 609, 952
4. Madau, P., della Valle, M., & Panagia, N. 1998, MNRAS, 297, L17
5. Folatelli, G. et al. 2006, ApJ in press, astroph-0509731
6. Tominaga, N., et al. 2005, ApJL, 633, L97
7. Anupama, G. C., et al. 2005, ApJL, 631, L125
8. Soderberg, A. M., Nakar, E., Kulkarni, S. R., & Berger, E. 2005, submitted to ApJ, astroph-0507147

9. Heger, A., Woosley, S. E., & Spruit, H. C. 2005, ApJ, 626, 350
10. Heger, A., & Woosley, S. E. 2003, AIPC, Vol. 662, p. 214
11. Fryer, C. L., & Heger, A. 2005, ApJ, 623, 302
12. Fryer, C. L., & Woosley, S. E. 1998, ApJL, 502, L9
13. Zhang, W., & Fryer, C. L. 2001, ApJ, 550, 357
14. Fryer, C. L., Woosley, S. E., Herant, M., & Davies, M. B. 1999, ApJ, 520, 650
15. Woosley, S. E., & Heger, A. 2006, ApJ in press, astroph-0508175
16. Yoon, S.-C., & Langer, N. 2006, A&A, in press, astroph-0508242
17. Fliegner, J., & Langer, N. 1995, IAUS, 163, 326
18. Heger, A., & Langer, N. 2000, ApJ, 544, 1016
19. Gies, D. R., & Huang, W. 2004, IAU Symposium, 215, 57
20. Heger, A., & Woosley, S. E. 2002, ApJ, 567, 532
21. Fryer, C. L., Woosley, S. E., & Heger, A. 2001, ApJ, 550, 372
22. Vink, J. S., & de Koter, A. 2005, A&A, 442, 587
23. Maeder, A., & Meynet, G. 2000, A&A, 261, 159
24. Chevalier, R. A., Li, Z.-Y., & Fransson, C. 2004, ApJ, 606, 369
25. MacFadyen, A. I., & Woosley, S. E. 1999, ApJ, 524, 262
26. Zhang, W., Woosley, S. E., & Heger, A. 2004, ApJ, 608, 365
27. Woosley, S. E., & Janka, T. 2005, Nature, Physics, 1, 147
28. Woosley, S. E., Heger, A., & Weaver, T. A. 2002, Reviews of Modern Physics, 74, 1015
29. Narayan, R., Piran, T., & Kumar, P. 2001, ApJ, 557, 949
30. Lee, W. H., Ramirez-Ruiz, E., & Page, D. 2005, ApJ, 632, 421
31. MacFadyen, A. I., Woosley, S. E., & Heger, A. 2001, ApJ, 550, 410
32. Cummings, J., et al. 2005, GCN Circ. 3910
33. Kawai, N., Yamada, T., Kosugi, G., Hattori, T., & Aoki, K. 2005, GCN, 3937
34. Sakamoto, T. et al 2005, GCN 3938
35. Tagliaferri, G., et al. 2005, A&A, 443, L1
36. Cusumano, G. et al. submitted to Nature, astroph-0509737

Evidence for intrinsic absorption in the Swift X–ray afterglows

S. Campana*, P. Romano*, S. Covino*, D. Lazzati[†], A. De Luca**, G. Chincarini[‡]*, A. Moretti*, G. Tagliaferri*, G. Cusumano[§], V. Mangano[§], V. La Parola[§], T. Mineo[§], P. Giommi[¶], M. Perri[¶], M. Capalbi[¶], L.A. Antonelli[∥], D.N. Burrows[††], J.E. Hill[††], J.L. Racusin[††], J.A. Kennea[††], D.C. Morris[††], C. Pagani[††], J.A. Nousek[††], J.P. Osborne[‡‡], M.R. Goad[‡‡], K.L. Page[‡‡], A.P. Beardmore[‡‡], O. Godet[‡‡], P.T. O'Brien[‡‡], A.A. Wells[‡‡], L. Angelini[§§] and N. Gehrels[§§]

*INAF - Osservatorio Astronomico di Brera, Via Bianchi 46, I-23807 Merate (LC), Italy
[†]JILA, Campus Box 440, University of Colorado, Boulder, CO 80309-0440, USA
**INAF - Istituto di Astrofisica spaziale e Fisica Cosmica, Via Bassini 15, I-20133, Milano, Italy
[‡]Universitá di Milano-Bicocca, Dip.lazz di Fisica, Piazza della Scienza 3, I-20126 Milano, Italy
[§]INAF - Istituto di Astrofisica spaziale e Fisica Cosmica, Via La Malfa 153, I-90146 Palermo, Italy
[¶]Agenzia Spaziale Italiana, Science Data Center, Via Galileo Galilei, I-00044, Frascati, Italy
[∥]INAF - Osservatorio Astronomico di Roma, Via di Frascati 33, I-00040 Monteporzio Catone (Roma), Italy
[††]Department of Astronomy and Astrophysics, 525 Davey Lab., Pennsylvania State University, University Park PA 16802, USA
[‡‡]Department of Physics and Astronomy, University of Leicester, Leicester LE1 7RH, UK
[§§]NASA/Goddard Space Flight Center, Greenbelt MD 20771, USA

Abstract. Gamma-ray burst (GRB) progenitors are observationally linked to the death of massive stars. X-ray studies of the GRB afterglows can deepen our knowledge of the ionization status and metal abundances of the matter in the GRB environment. Moreover, the presence of local matter can be inferred through its fingerprints in the X–ray spectrum, i.e. the presence of absorption higher than the Galactic value. A few studies based on BeppoSAX and XMM-Newton found evidence of higher than Galactic values for the column density in a number of GRB afterglows. Here we report on a systematic analysis of 17 GRBs observed by Swift up to April 15, 2005. We observed a large number of GRBs with an excess of column density. Our sample, together with previous determinations of the intrinsic column densities for GRBs with known redshift, provides evidence for a distribution of absorption consistent with that predicted for randomly occurring GRB within molecular clouds.

INTRODUCTION

Evidence has been accumulating in recent years that at least a subclass of Gamma–ray bursts (GRBs), the ones with a long ($\gtrsim 2$ s) burst event, are associated with deaths of massive stars (e.g. Woosley 1993; Paczyński 1998; MacFadyen & Woosley 1999). This evidence was initially based on the relatively small offset of the GRB location with respect to the center of the host galaxy (Bloom, Kulkarni & Djorgovski 2002). Moreover, decisive supernova features have been observed in the afterglow of a few nearby GRBs (Galama et al. 1998, Della Valle et al. 2003; Stanek et al. 2003; Malesani et al. 2004), directly linking long GRBs to massive stars. This also provides strong observational evidence for the connection of GRBs to star formation (Djorgovski et al. 1998; Fruchter et al. 1999; Prochaska et al. 2004). A study on the GRB host galaxies by Le Floc'h et al. (2003) found that these hosts have very blue colours, comparable to those of the faint blue star-forming sources at high redshift. The association of long GRBs with star forming regions supports the idea that a large fraction of the optically-dark GRBs (i.e. GRBs without an optical afterglow), as well, are due to high (dust) absorption (Lazzati, Covino & Ghisellini 2002; Rol et al. 2005; Filliatre et al. 2005).

Together with optical studies, which probe the dust content of the GRB environment, X–ray studies of the GRB afterglows can give insight on the ionization status and metal abundances of the matter in the GRB environment. This can be done by using either emission or absorption features (e.g., Böttcher et al. 1999; Ghisellini et al. 2002). Although emission features are more apparent, the cumulative effect of low-energy cutoff is easier to detect in the relatively low signal to noise spectra of X–ray afterglows. Moreover, if the absorbing material is located close to the GRB site ($\sim 0.1 - 10$ pc), it is expected that GRB photons may lead to a progressive photoionization of the gas, gradually reducing the effect of low energy absorption (Lazzati & Perna 2002).

X–ray absorption in excess of the Galactic value has been reported for a handful of GRB afterglows (Owens et al. 1998; Galama & Wijers 2001; Stratta et al. 2004; De Luca et al. 2005; Gendre, Corsi & Piro 2005). Evidence for a decrease of the intrinsic column density with time in the X–ray prompt emission of some GRBs has also been found (GRB980506, Connors & Hueter 1998; GRB980329, Frontera et al. 2000; GRB011211, Frontera et al. 2004). Lazzati & Perna (2002) interpreted them as evidence for GRBs occurring within overdense regions in molecular clouds similar to star formation globules in our Galaxy.

Stratta et al. (2004) presented a systematic analysis of a sample of 13 bright afterglows observed with BeppoSAX narrow field instruments. They found a significant detection of additional intervening material in only two cases (namely, GRB990123 and GRB010222), but, owing to the limited photon statistics, they could not exclude that intrinsic X–ray absorption is also present in the other bursts. Chandra observations of GRB afterglows have yielded a few detections and constraints of the presence of intrinsic X–ray absorption (Gendre et al. 2005). XMM-Newton observed 9 GRB afterglows (for a review see De Luca et al. 2005 and Gendre et al. 2005). These are mainly INTEGRAL GRBs, the large majority of which has been discovered close to the Galactic plane (i.e. are characterized by a relatively high Galactic column density). From XMM-Newton data one can gather evidence that at least several GRBs occur in high density regions within their host galaxies (e.g, De Luca et al. 2005).

Here we investigate the presence of intrinsic absorption in the complete set of all the 17 GRBs promptly observed by Swift up to April 15, 2005, for more details (especially on the data) see Campana et al. (2006) and Table 1.

TABLE 1. Results of the spectral analysis.

GRB	N_H Gal.* 10^{20} cm^{-2}	N_H obs.† 10^{20} cm^{-2}	$\chi^2_{\rm red}$ (dof)
041223	10.9 (9.9) [5.6]	$16.8^{+5.2}_{-4.2}$	0.8 (26)
050124	5.2 (2.6) [1.7]	$6.2^{+3.9}_{-2.5}$	1.3 (35)
050126	5.3 (3.2) [2.6]	$4.1^{+2.7}_{-2.6}$	0.9 (10)
050128	4.8 (4.9) [3.8]	$12.5^{+1.4}_{-1.3}$	1.3 (105)
050215b	2.1 (2.0) [0.9]	< 3.4	1.0 (4)
050219a	8.5 (10.1) [8.1]	$30.1^{+6.5}_{-5.9}$	1.0 (57)
050219b	3.8 (3.0) [1.7]	$23.8^{+4.0}_{-3.7}$	1.0 (98)
050223	7.1 (6.6) [4.4]	$9.5^{+23.8}_{-7.3}$	1.2 (3)
050306	3.1 (2.9) [3.5]	$46.1^{+35.6}_{-28.8}$	1.0 (3)
050315	4.3 (3.3) [2.5]	$14.9^{+3.9}_{-2.2}$	1.3 (42)
050318	2.8 (1.8) [0.9]	$4.2^{+1.9}_{-1.5}$	0.8 (39)
050319	1.1 (1.2) [0.5]	$3.0^{+0.9}_{-0.8}$	1.3 (29)
050326	4.5 (3.8) [1.8]	$18.9^{+7.1}_{-6.0}$	1.0 (25)
050401	4.8 (4.4) [3.3]	$21.1^{+2.2}_{-1.8}$	1.0 (297)
050406	2.8 (1.7) [1.1]	< 6.6	1.2 (10)
050408	1.7 (1.5) [1.3]	$30.7^{+5.5}_{-4.9}$	0.9 (45)
050412	2.2 (1.7) [1.0]	$26.4^{+14.9}_{-12.4}$	1.4 (11)

* Column density values are from Dickey & Lockman (1990). Values between parentheses are from Kalberla et al. (2005) and in square parentheses from Schlegel, Finkbeiner & Davis (1998), using the usual conversion $N_H = 5.9 \times 10^{21} E(B-V)$ cm^{-2}.

† The values of the column density have been computed at $z = 0$ since we do not have any knowledge of the GRB redshift for most of them (but see Table 2).

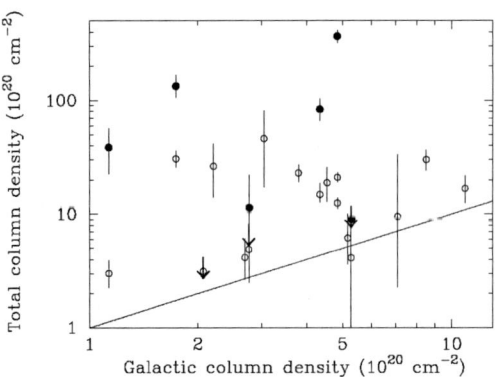

FIGURE 1. Galactic column density versus column density obtained from spectral fit of the X-ray afterglow. Open circles are values obtained without any redshift information. Filled circles indicate values for the six GRBs with known redshift at the redshift of the host galaxy. Upper limits are also indicated with filled and open circles, as above. The line represents the prints of equal values between the Galactic and the total column density (i.e. no intrinsic absorption).

TABLE 2. Results of the spectral analysis of GRB with known redshift. Refs.: 1) Berger, Cenko, Kulkarni 2005; 2) Kelson & Berger 2005; 3) Berger & Mulchaey 2005; 4) Fynbo et al. 2005; 5) Fynbo et al. 2005; 6) Berger, Gladders & Oemler 2005. Values in the second part of the table are from Stratta et al. (2004), De Luca et al. (2005) and Gendre et al. (2005).

GRB	Redshift (ref.)	N_H Gal. 10^{20} cm^{-2}	N_H obs.* 10^{20} cm^{-2}	$\chi^2_{\rm red}$ (dof)	N_H DLA (90%) 10^{20} cm^{-2}
050126	1.29 (1)	5.3	< 9.4	1.0 (10)	0.8 (6.4)
050315	1.95 (2)	4.3	$83.7^{+19.7}_{-17.4}$	1.4 (43)	1.4 (5.7)
050318	1.44 (3)	2.8	$11.3^{+10.7}_{-8.9}$	0.6 (39)	0.9 (12.9)
050319	3.24 (4)	1.1	$38.6^{+18.0}_{-16.0}$	1.4 (29)	3.6 (447)
050401	2.90 (5)	4.8	366^{+47}_{-46}	1.1 (294)	3.0 (272)
050408	1.24 (6)	1.7	134^{+34}_{-28}	1.0 (45)	0.8 (4.7)
980703	0.97	5.8	290^{+71}_{-27}		0.6 (1.7)
990123	1.60	2.1	30^{+70}_{-20}		1.1 (23.1)
990510	1.63	9.4	160^{+19}_{-13}		1.1 (25.1)
000210	0.85	2.5	50^{+10}_{-10}		0.0 (0.0)
000214	0.47	5.8	< 2.7		0.0 (0.0)
000926	2.07	2.7	40^{+35}_{-25}		1.6 (73.9)
010222	1.48	1.6	120^{+70}_{-60}		1.0 (15.1)
001025a	1.48	6.1	66^{+30}_{-30}		1.0 (15.1)
020322	1.80	4.6	130^{+20}_{-20}		1.3 (40.7)
020405	0.70	4.3	47^{+37}_{-37}		0.0 (0.0)
020813	1.25	7.5	< 36.5		0.8 (5.0)
021004	2.33	4.3	< 34		2.0 (118)
030226	1.98	1.6	68^{+41}_{-33}		1.5 (60.8)
030227	3.90	22	680^{+18}_{-38}		5.2 (649)
030328	1.52	4.3	< 44.3		1.0 (17.3)

* Column density values have been computed at the GRB redshift.

DISCUSSION

The evidence that a large fraction of GRBs is characterized by an absorbing column density larger than the Galactic one clearly points to a high density interstellar medium (ISM) in the proximity of the GRB (in fact with X–rays we directly probe the GRB line of sight, whereas in the optical the observations might be contaminated by the host galaxy contribution). Dense environments in the host galaxy, possibly associated with star forming regions, provide a further clear signature in favour of the association of long GRBs to the death of massive stars. Moreover, the effect of an intrinsic column density can be hidden either by a large Galactic absorber (as often occurs for INTEGRAL GRBs) or by a large redshift shifting the energy scale by $(1+z)$ and the column density effective value by $\sim (1+z)^{2.6}$.

We combine our sample of GRBs with known redshift with other intrinsic column densities available in the literature (Stratta et al. 2004; De Luca et al. 2005; Gendre et al. 2005), obtaining a sample of 21 GRBs (see Table 2). In principle the absorption excess found in most GRBs might not be local to the host galaxy but may come from a line-of-sight interlooper. This often occurs in optical studies of quasar with dumped Lyman absorber (DLA). Based on quasar studies (Wolfe, Gawiser & Prochaska 2005; Péroux et al. 2003) we simulated for each of our bursts a distribution of line-of-sights (10000 trials), evaluating a mean absorption (weighted as $(1+z)^{2.6}$) and 90% confidence value. These values are reported in column six of Table 2. The influence of DLA systems is marginal in our sample even if there are a few GRBs in which the observed absorption excess may come from intervening DLAs. Indeed, such systems have recently been found in few Swift GRBs (e.g. GRB050730 with $\log(N_H) = 22.3$, Starling et al. 2005, Chen et al. 2005, and GRB 050401, Watson et al. 2005, $\log(N_H) = 22.5$). However, they are due to the ISM in the GRB host. One of these GRBs is part of our sample, GRB 050401, and indeed we obtained a high value of the intrinsic column density.

In order to understand the origin of the absorption excess, we compared the distribution of measured intrinsic column densities with the distribution expected for bursts occurring in Galactic-like molecular clouds (Reichart & Price 2002) and with the one expected for bursts occurring following a host galaxy mass distribution using the Milky Way as a model (Vergani et al. 2004). For each of these two column density distributions we simulated 10000 GRBs and compared, by means of a Kolmogorov-Smirnov (KS) test, their intrinsic absorption distribution to the observed distribution. We found that the observed distribution is inconsistent with the galaxy column density distribution (Vergani et al. 2004), with a KS null hypothesis probability of 10^{-11}, but it is consistent with GRBs distributed randomly in molecular clouds (KS null hypothesis probability of 0.61). These results would support an origin of long GRBs within high density regions such as molecular clouds. We stress that our sample also contains upper limits and that we are sensitive to low values of column density, which however are found only in a small fraction of the total sample.

SUMMARY AND CONCLUSIONS

The main goal of the present paper is to investigate the presence of intrinsic absorption in the X–ray spectra of GRB afterglows. We analyzed a complete set of 17 afterglows observed by Swift XRT before April 15, 2005. In 10 of them we found clear signs of intrinsic absorption, i.e. with a column density higher than the Galactic value estimated from the maps by Dickey & Lockman (1990) at > 90% confidence level (and with low probability of contamination from intervening DLA systems). For the remaining 7 cases, the statistics are not good enough to draw firm conclusions. This clearly suggests that long GRBs are associated with high density regions of the ISM, supporting the idea that they are related to the deaths of massive stars.

For the 6 GRBs with known redshift, together with 15 already known, we can have an unbiased view of the intrinsic absorption in the host galaxy rest frame. We found a range of $(1-35) \times 10^{21}$ cm^{-2} for all GRBs. This range of values is consistent with the hypothesis that GRBs occur within giant molecular clouds, spanning a range of column density depending on their exact location (Reichart & Price 2002). In our rest frame this column density is then reduced by a factor $\sim (1+z)^{2.6}$, making it more difficult to determine the intrinsic column density, especially for distant GRBs or for GRBs occurring at large Galactic column densities.

Finally, we compared the distribution of GRB column densities with known redshift with theoretical predictions available in the literature finding good agreement with the expectation (Reichart & Price 2002) for bursts occurring in molecular clouds.

REFERENCES

. Berger, E., Cenko, S.B., Kulkarni, S.R. 2005, GCN 3088
. Berger, E., Gladders, M., Oemler, G. 2005, GCN 3201
. Berger, E., Mulchaey, J. 2005, GCN 3122
. Bloom, J.S., Kulkarni, S.R., Djorgovski S.G. 2002, AJ, 123, 1111
. Böttcher, M., Dermer, C.D., Crider, A.W., Liang, E.P. 1999, A&A, 343, 111
. Campana, S., Romano, P., Covino, S., et al. 2006, A&A in press (astro-ph/0511750)
. Chen, H.-W., Prochaska, J.X., Bloom, J.S., Thomson, I.B. 2005, ApJL in press (astro-ph/0508270)
. Connors, A., Hueter, G.J. 1998, ApJ, 501, 307
. Della Valle, M., Malesani, D., Benetti, S. et al. 2003, A&A, 406, L33
. De Luca, A., Melandri, A., Caraveo, P.A., et al. 2005, A&A, 440, 85
. Dickey, J.M., Lockman F.J. 1990, ARA&A, 28, 215
. Djorgovski, S.G., Kulkarni, S.R., Bloom, J.S., et al. 1998, ApJ, 508, L17
. Filliatre, P., D'Avanzo, P., Covino, S., et al. 2005, A&A, 438, 793
. Frontera, F., Amati, L., Costa, E., et al. 2000, ApJS, 127, 59
. Frontera, F., Amati, L.; in't Zand, J.J.M., et al. 2004, ApJ, 616, 1078
. Fruchter, A.S., Thorsett, S.E., Metzger, M.R., et al. 1999, ApJ, 519, L13
. Fynbo, J.P.U., Hjorth, J., Jensen, B.L., et al. 2005, GCN 3136
. Fynbo, J.P.U., Jensen, B.L., Hjorth, J., et al. 2005, GCN 3176
. Galama, T.J., Vreeswijk, P.M., van Paradijs, J., et al. 1998, Nat, 395, 670
. Galama, T.J., Wijers, R.A.M.J. 2001, ApJ, 549, L209
. Gehrels, N., Chincarini, G., Giommi, P., et al. 2004, ApJ, 611, 1005
. Gendre, B., Corsi, A., Piro, L. 2005, A&A submitted (astro-ph/0507710)
. Ghisellini, G., Lazzati, D., Rossi, E., Rees, M.J. 2002, A&A, 389, L33
. Kalberla, P.M.W., Burton, W.B., Hartmann, D., et al. 2005, A&A, 440, 775
. Kelson, D., Berger, E. 2005, GCN 3101
. Lazzati, D., Covino, S., Ghisellini, G. 2002, MNRAS, 330, 583
. Lazzati, D., Perna, R. 2002, MNRAS, 330, 383
. Le Floc'h, E., Duc, P.-A., Mirabel, I.F., et al. 2003, A&A, 400, 499
. MacFadyen, A.I., Woosley S.E. 1999, ApJ, 524, 262
. Malesani, D., Tagliaferri, G., Chincarini, G., et al. 2004, ApJ, 609, L5
. Owens, A., Denby, M. Wells, A., et al. 1997, ApJ, 476, 924
. Paczyński, B. 1998, ApJ, 494, L45
. Péroux, C., et al. 2003, MNRAS, 345, 480
. Prochaska, J.X., Bloom, J.S., Chen, H.-W., et al. 2004, ApJ, 611, 200
. Reichart, D.E., Price, P.A. 2002, ApJ, 565, 174
. Rol, E., Wijers, R.A.M.J., Kouveliotou, C., et al. 2005, ApJ, 624, 868
. Schlegel, D., Finkbeiner D.P., Davis, M. 1998, ApJ, 500, 525
. Stanek, K.Z., Matheson, T., Garnavich, P.M., et al. 2003, ApJ, 591, L17
. Starling, R.L.C., Vreeswijk, P.M., Ellison, S.L. et al. 2005, A&A Lett. in press (astro-ph/0508237)
. Stratta, G., Fiore, F., Antonelli, L.A., Piro, L., De Pasquale, M. 2004, ApJ, 608, 846
. Vergani S.D., Molinari, E., Zerbi, F.M., Chincarini, G. 2004, A&A, 415, 171
. Watson, D., Fynbo, J.P.U., Ledoux, C., et al. 2005, ApJ submitted (astro-ph/0510368)
. Wolfe, A.M., Gawiser, E., Prochaska, J.A. 2005, ARA&A, 43, 861
. Woosley, S.E. 1993, ApJ, 405, 273

Constraining the GRB Collimation with a Survey for Orphan Afterglows

Arne Rau[*,†], Jochen Greiner[*] and Robert Schwarz[**]

[*]*Max-Planck Institut für extraterrestrische Physik, Giessenbachstrasse, 85748 Garching, Germany*
[†]*INAF - Osservatorio di Trieste, via G.B. Tiepolo 11, 34131 Trieste, Italy*
[**]*Astrophysikalisches Institut Potsdam, An der Sternwarte 16, 14482 Potsdam, Germany*

Abstract. GRB afterglows without detected prompt-emission counterparts are powerful tools to estimate the collimation of GRB jets. Here we report on the analysis of a dedicated survey for the search of these orphan afterglows. The survey was obtained the Wide Field Imager at the 2.2 m MPI/ESO telescope at La Silla, Chile and includes monitoring of $\sim$12 deg^2 with a limiting magnitude of R=23 in up to 25 nights typically spaced by one to two nights. Four new optical transients were discovered, three of these associated with a flare star, a cataclysmic variable and a dwarf nova. The fourth source shows indications for an extra-galactic origin. We discuss the results in the context of the GRB collimation.

Keywords: gamma rays: bursts; surveys
PACS: 98.70.Rz; 95.80.+p

INTRODUCTION

Within the frame work of the internal external shock model of GRBs [1] ultra-relativistic collimated outflows are favored both theoretically as well as observationally. As a consequence of the collimation, only a fraction of all bursts in the Universe are detectable on Earth through their γ-ray emission. Assuming a universal jet structure, the total GRB rate will be higher than the observed rate by a factor of roughly $\theta_{jet}^{-2} \propto \Gamma^2$, where θ_{jet} is the jet opening angle and Γ the bulk Lorentz factor of the ejecta.

GRBs which are missed at high energies can in principle be discovered through their afterglow radiation with a larger solid angle. One distinguished between two classes of these so-called "orphan afterglows": (i) "on-axis" orphans are expected when the narrow γ-ray emission misses the observer by a small amount but a the cocoon of material with lower Γ surrounding the γ-ray emitting jet falls within the field of view [6]. On-axis orphan afterglows resemble afterglows of bursts with detected prompt emission. (ii) "off-axis" orphan afterglows are expected at larger viewing angles [2], [3], [4]. Their light curves will initially be faint with a brightening to a viewing angle dependent maximum and become similar to regular GRB afterglows after the jet break later on [5] ,[6].

Orphan afterglows are an important tool for the study of the initial opening angle of the jets and allow to constrain the collimation of the optical afterglow emission [2]. Especially on-axis orphans are suitable as they are substantially brighter than off-axis orphans and thus easier to detect in a dedicated survey [6]. While a small number of surveys dedicated to the search of untriggered optical GRB counterparts was performed over the past years, e.g. [7], [8], [9], no candidate event was reported so far.

OBSERVATIONS AND DATA REDUCTION

The survey was performed during three periods in 1999 with the Wide Field Imager (WFI, $34' \times 33'$ field of view) at the MPI/ESO 2.2 m telescope in La Silla, Chile [10]. We monitored 7 different sky fields covering $\sim 12\,\text{deg}^2$ in up to 25 nights each. Photometry was taken in the R-band with 420 s exposure, corresponding to a limiting magnitude of $R=23$ at 10σ under good conditions. In total 39 nights of observations were scheduled, but unfavorable weather condition caused interruptions of the planned observing schedule of each period and lead to a significant decrease of the detection sensitivity for orphan afterglows.

The primary strategy was to obtain repetitive deep monitoring of a number of selected sky fields. The observing scheme was selected in order to allow for the detection of a GRB orphan afterglow in at least two epochs together with earlier and later upper limits. To combine the availability of a large aperture telescope together with the observed brightness decay of GRB afterglows, we decided on consecutive observations of a given field in every 2nd night. During this time on-axis orphans are expected to be brighter than $R=21$ while off-axis orphans will be fainter [6].

A PERL based pipeline, including a number of well tested astronomical software packages, was developed for the automatic reduction and analysis of the survey data. The basic image reduction was performed using *IRAF/MSCRED* and astrometric solutions were derived with the *WIFIX/ASTROMETRIX*[1] package by comparing the positions of detected sources with those compiled in the USNO-A2.0 catalog [11] separately for each of the 8 CCDs. The photometry inside a Gaussian-shaped point-spread-function was obtained within *IRAF/DAOPHOT*.

Variability informations for all detected sources were derived via the differential photometry technique. Here, an ensemble of at least 20 local, non-saturated, non-variable reference stars was selected for each sub-field. By deriving the median brightness offset of these stars with respect to the brightness of the same stars obtained in a reference image of a given sub-field, the photometric offset between the two observations were estimated. This was successively done for all observations of a given sub-field and thus provided a common photometric zero-point for all pointings of the respective sub-field. The absolute photometric calibration was obtained using observations of the field SA113-1 which contains a number of Landoldt standard stars.

For all detected sources light curves were obtained and candidate transient objects were selected based on the deviation of their light curve from their mean light curve. For all candidates with one-time detections or $\Delta R > 0.75$ mag (~ 12000), light curve plots as well as thumbnail images of all pointings were produced. These were examined by eye in order to remove spurious transients arising from nearby bright stars, extended objects, stray light effects at the edge of the FoV, bad focus or detections which were consistent with faint stars at the limiting magnitude of the individual pointings. The strategy of the survey was aimed to catch a possible orphan afterglow in at least two consecutive observations. Therefore, we considered only sources with at minimum two detections for further investigations. Out of the ~ 12000 candidates four remained.

---

[1] http://www.na.astro.it/~radovich/wifix.htm

RESULTS

From the four new optical transients three were identified based on the shape of the light curve and amplitude as a candidate cataclysmic variable, a dwarf nova and a flare star. The results on the fourth object, a candidate extra-galactic transient superimposed on an underlying faint object, are described below.

The transient (RA(2000)=13:28:13.7, Dec(2000)=−21:42:37) was detected on June 20.07 UT 1999 at its maximum brightness of $R=19.9$. Unfortunately, the observing period finished after the observation of the outburst and the next R-band pointing was performed more than six weeks later (August 4). Thus, the fading could not be monitored. A faint ($R=21.3$) persistent point source was found $\sim 0.''8$ offset of the position of the transient in images taken before and after the outburst (Fig. 1). This suggests an extra-galactic origin of the transient assuming the persistent counterpart can be associated with a candidate host galaxy. Nevertheless, this vague assumption based on the available observational data requires a confirmation by an accurate distance measurement.

The ROSAT All-sky survey did not show a source at the position of transient. Similarly, no counterpart was found in the SIMBAD and NED databases. The object is visible near the limiting magnitude of the DSS in the B and R-band and not detected in the DSS I-band.

The lack of observations during the light curve decay leaves the identification of the origin of the source open. If the faint persistent source is indeed a galaxy, the transient could e.g., be associated with a supernova explosion occurring in this galaxy. The brightening of ~ 1.5 mag in ~ 2 days is very steep compared to observed supernovae though [12]. Other explanations, like a foreground flare star close to the line of sight towards the persistent background source, or the projection of two galactic objects or a orphan afterglow are also possible. In order to test the latter hypothesis we searched for triggered GRBs which occurred during the time between the preceding observation (June 18.05 UT) and the outburst. We found 5 cataloged[2] GRBs during this period (BATSE #7609 & 7610 and IPN #2066, 2067 & 2069; K.Hurley, private communication). None of those has a position consistent with the transient which allows to exclude an association with a triggered GRB. Altogether, the transient is the best candidate, although unconfirmed, found during our survey.

DISCUSSION

To constrain the collimation of GRBs requires to understand first the detection efficiency for on-axis optical afterglows. This was done using a set of Monte-Carlo simulations folded with the observing schedule of the survey. Here, afterglows with random sky coordinates, light curve parameters and explosion times were simulated distributed over the periods in which observations were taken. See [10] for more details. The light curves were described by a broken power law and parameterized by the pre-break slope,

---

[2] http://grbcat.gsfc.nasa.gov/grbcat/grbcat.html

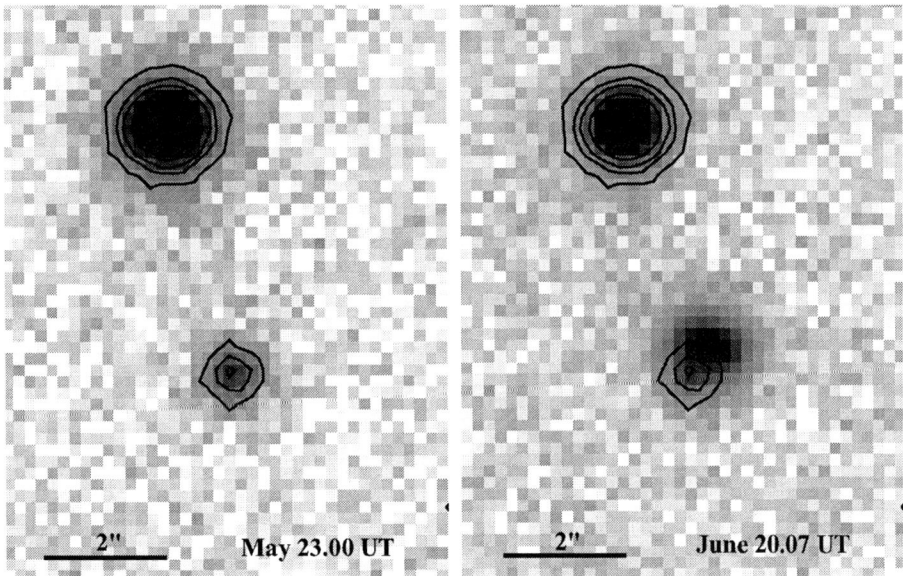

FIGURE 1. left: R-band image of the transient (center source) and a nearby bright star taken on May 23.00 UT when the source had a brightness of $R=21$. Flux contours are over-plotted for both objects. North is up and East to the left. right: Same field observed on June 20.07 UT together with the flux contours from May 23.00 UT. The flaring source shows an offset of $\sim 0.''8$ with respect to the quiescent counterpart.

α_1, break time, t_b, post-break slope, α_2 and initial R-band magnitude, R_{in}. Here, the initial magnitude corresponds to the observer frame R-band brightness with which an afterglow is created in the simulations. We used Gaussian distributions for $-1.8 < \alpha_1 < -0.4$ and $-2.8 < \alpha_2 < -1.4$ and an uniform distribution for t_b ranging from 0.4–4 days using observations collected in [13].

In Figure 2 we show the probability for the first detection of an afterglow in a simulation with three different distributions for R_{in} and one single event over a year and full sky. Naturally, the probability to detect an afterglow increases with the depth of the survey. More transients are expected to be detected above a certain magnitude for afterglow models with brighter R_{in}. Due to the lower number of bright afterglows and the increasing probability with diminishing initial brightness, the non-uniform distribution gives the lowest expectation rates. At the limiting magnitude of the survey of $R \sim 23$ the probability to detect a specific randomly selected afterglow in at least one observation is approximately 3×10^{-7}. The probability of identifying an event in two consecutive observations (N_{MC}; shown for the non-uniform distribution) is smaller by a factor of ~ 3 and can be interpreted as an upper limit on the true number of on-axis afterglows per year and full sky.

The number of detected orphan afterglows (in our case zero) together with the detection probability allow to constrain the collimation factor, f_c, which corresponds to the ratio of the true rates of on-axis optical afterglows, N_A, and long-duration GRBs which

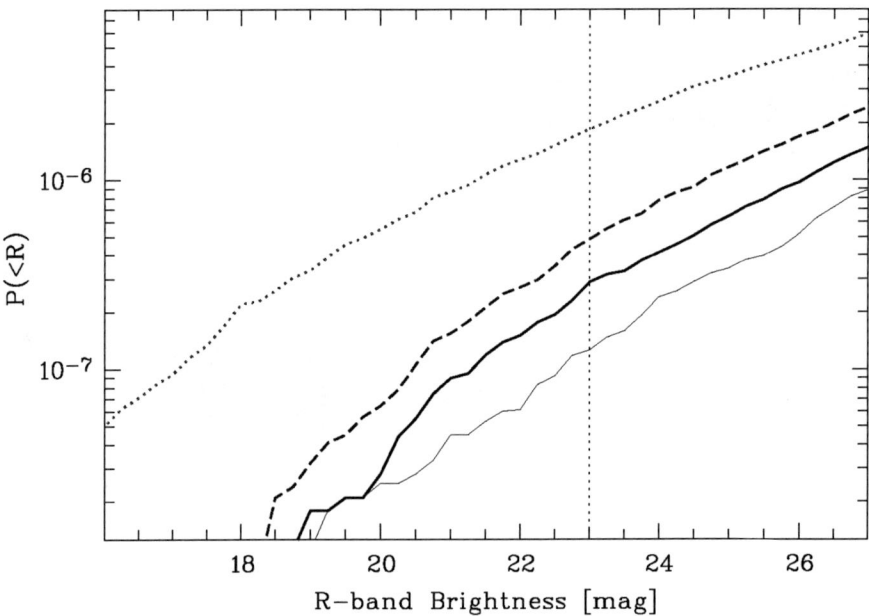

FIGURE 2. Probability distributions for the detection of a randomly selected afterglow for three different models of the initial brightness. The dotted and dashed functions represent uniform distributions of R_{in}=9–20 and R_{in}=13–23, respectively. The thick and thin solid lines correspond to the first-time and second-time observations for a distribution proportional to $0.2 \times R_{in}$ in the range of R_{in}=13–23. The vertical dotted line marks the limiting magnitude of the presented survey.

produce observable optical afterglows, N_γ, pointed at the observer. While an upper limit for N_A can be taken from N_{MC} derived above, N_γ depends multiplicative on a number of factors. This includes the observed rate of long-duration GRBs with a specific γ-ray instrument corrected for sky coverage, Earth-blockage and instrument down-time, e.g., $\sim 444 \, \text{yr}^{-1}$ using the full sky long-duration GRB rate of with BATSE [14]. A further correction factor of order 2 for events outside of the energy range of BATSE needs to be taken into account. In addition, approximately 10% of the afterglows are optically dim or dark ([15], [16], e.g., intrinsically faint, absorbed, high-z) and will therefore be to faint for a detection.

Simple assumptions for the described factors provide $N_\gamma \sim 800$. This ogether with $N_{MC}=1\times 10^7$ obtained from the Monte-Carlo simulations results in a collimation factor of $f_C<12500$. The rather conservative upper limit is significantly higher than the values for the beaming factor, the ratio of the overall rate of GRBs to the detected burst rate, derived by [17] and [18] of 75±25 and 500, respectively. The γ-ray beaming factor should be an upper limit for f_C as it corresponds to the case of isotropic afterglow radiation.

The large value for f_C suggests that the effective coverage of the performed observations was insufficient to provide a strong constraint for the collimation of the optically emitting GRB outflow. We performed a further Monte Carlo simulation in order to estimate the properties for an "ideal" survey assuming a schedule of one observation per field every two nights over 150 nights with a limiting magnitude of $R=23$. We find that $f_C<500$ (<75, <10) would be reached with such a configuration covering a field of $50\,\mathrm{deg}^2$ ($300\,\mathrm{deg}^2$, $2500\,\mathrm{deg}^2$). Existing (Palomar-Quest) and near-future instrumentation (e.g., VLT Survey Telescope, Visible & Infrared Survey Telescope for Astronomy, Pan-STARSS, Large Synoptic Telescope) will be able to breach into this region perform comprehensive searches for untriggered GRBs in the near future. However, rapid spectroscopic follow-up observations of candidate orphans will be crucial for the identification of orphan afterglows and other optical transients as well as for the detailed astrophysical studies.

ACKNOWLEDGMENTS

We thank the many observers who have assisted in the completion of this survey, including M. Braun, D. Clowe, J. Eislöffel, J. Fried, P. Heraudeau, R. Klessen, T. Kranz, I. Lehmann, R. Schmidt, C. Wolf, D. Woods and the helpful night assistants at La Silla. We thank Mario Radovich (INAF - Osservatorio Astronomico di Capodimonte) for making the WIFIX image reduction and astrometry package publicly available. The WFI is a joint project between the European Southern Observatory, the Max-Planck-Institut für Astronomie in Heidelberg (Germany) and the Osservatorio Astronomico di Capodimonte in Naples (Italy). AR acknowledges support and collaboration within the EU RTN Contract HPRN-CT-2002-00294.

REFERENCES

1. M. J. Rees, and P. Meszaros, *MNRAS*, **258**, 41 (1992)
2. J. Rhoads, *ApJ*, **487**, L1 (1997)
3. R. Perna, R., and A. Loeb, *ApJ*, **509**, L85 (1998)
4. N. Dalal, K. Griest, and J. Pruet, *ApJ*, **564**, 209 (2002)
5. J. Rhoads, *ApJ*, **525**, 943 (1999)
6. E. Nakar, and T. Piran, *New Astr.*, **8**, 141 (2003)
7. R. Kehoe, C. Akerlof, R. Balsano, et al. *ApJ*, **577**, L159 (2002)
8. W. T. Vestrand, K. Borozdin, D. J. Casperson, *AN*, **325**, 549 (2004)
9. E. Rykoff, F. Aharonian, C. W. Akerlof, et al. *ApJ*, **631**, 1032 (2005)
10. A. Rau, J. Greiner and R. Schwarz, *A&A in press*, 2006
11. D. B. A. Monet, B. Canzian, C. Dahn, et al. *VizieR Online Data Catalog*, 1252 (1998)
12. B. Leibundgut, G. A. Tammann, R. Cadonau, and D. Cerrito, *A&AS*, **89**, 537 (1991)
13. A. Zeh, S. Klose, D. A. Kann, *ApJ*, **637**, 889 (2006)
14. W. S. Paciesas, C. A. Meegan, G. N. Pendleton, et al. *ApJSS*, **122**, 465 (1999)
15. P. Jakobsson, J. Hjorth, J. P. U. Fynbo, et al. *ApJ*, **617**, L21 (2004)
16. E. Rol, R. A. M. J. Wijers, C. Kouveliotou, L. Kaper, and Y. Kaneko, *ApJ*, **624**, 868 (2005)
17. D. Guetta, T. Piran, E. Waxman, *ApJ*, **619**, 412 (2005)
18. D. A. Frail, S. R. Kulkarni, R. Sari, et al. *ApJ*, **562**, L55 (2001)

Long-term optical monitoring of GRB 030329

D. Bersier[*], K.Z. Stanek[†], P.M. Garnavich[**], T. Matheson[‡] and P. Mazzali[§]

[*]*Astrophysics Research Institute, Liverpool John Moores University, Birkenhead, CH41 1LD, UK*
[†]*Dept of Astronomy, Ohio State University, Columbus, OH 43210, USA*
[**]*Dept of Physics, University of Notre Dame, Notre Dame, IN 46556, USA*
[‡]*National Optical Astronomy Observatory, Tucson, AZ 85719, USA*
[§]*INAF-Osservatorio Astronomico, 34131 Trieste, Italy*

Abstract. We present photometric data for GRB 030329/SN 2003dh obtained over a year after the event. The decay rate of the SN is very similar to that of SN 1998bw. We discuss the "jitter" episode that occurred $\sim 50-65$ days after the explosion and argue that it was produced by the GRB afterglow and not by the supernova. A spectrum obtained eight months after the explosion, combined with the late-time light curve, allow us to determine the amount of nickel. We find that at least 0.35 $M_\odot$ of ^{56}Ni was synthesized during the explosion; the total mass of ^{56}Ni may be up to about 1 $M_\odot$.

Keywords: Gamma-ray bursts, supernovae, GRB 030329, SN 2003dh
PACS: 97.60.Bw, 98.70.Rz, 26.30.+k

1. DATA

The burst of 29 March 2003 is the best observed burst ever. It provided undisputable evidence for an association between GRBs and supernovae [1, 2, 3]. Its low redshift ($z = 0.1685$) means that we can follow the SN from the ground until very late. We report on such a late-time monitoring.

We obtained images with the 48" telescope at the Fred L. Whipple Observatory (Mt Hopkins, Arizona), up to 87 days after the burst. Data were also obtained at Kitt Peak National Observatory on the 4m telescope equipped with the MOSAIC camera, on the 6.5m Baade telescope with the wide-field instrument IMACS, on the 6.5m Clay telescope and the Magellan Instant Camera (MagIC), with the MiniMosaic camera at the WIYN telescope, on the Vatican Advanced Technology Technology Telescope, and on the MMT equipped with MegaCam. Most of the data were obtained in the *R*-band, occasionaly we also secured *B*-, *V*- or *I*-band data. The data were reduced in a standard manner with IRAF. We used Henden's list of secondary standards [4] to calibrate our photometry. The light curve is presented in Figure 1.

The light curve up to two months has been described in detail in Matheson et al. [3] and Lipkin et al. [5]. Focusing on the late-time behavior (after two months), we fit an exponential and a constant term (representing the host). We find that the decay rate is almost exactly identical to that of SN 1998bw in the *V* band (0.019 mag/day).

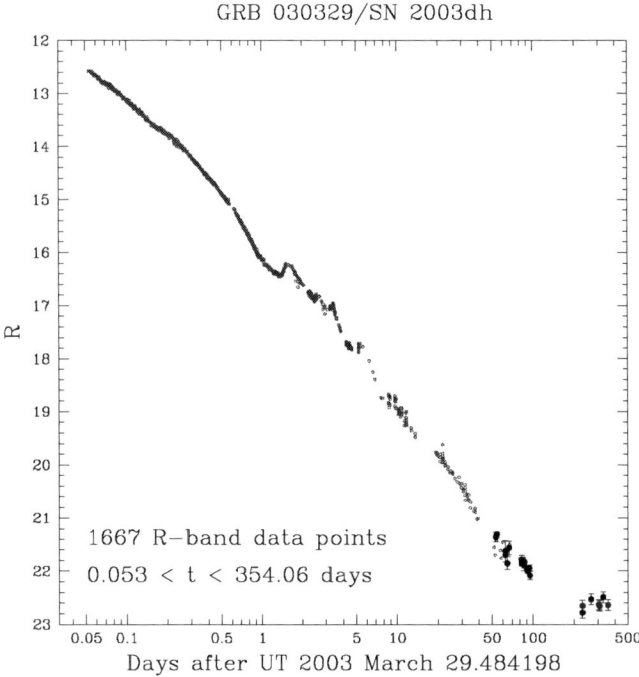

FIGURE 1. Complete *R* band light curve of GRB 030329/SN 2003dh, using data published in Matheson et al. and Lipkin et al. [3, 5], and new data presented here. Red (open) symbols are published data, black filled symbols with error bars are our new data.

2. JITTER EPISODE

Between 50 d and 70 d the photometric data show a large scatter, much larger than expected on the basis of measurements errors. Figure 2 shows the light curve during this particular time interval. The data from Ibrahimov et al. [6] follow the trend shown by our data: after 50 d, the OT became brighter during several days. It then faded snd reached a minimum around 64 d, when it started brightening again. In this last phase the data from Lipkin et al. [5] are in agreement with our data. Radio observations also show variability during this time interval. Frail et al. [7] state that "there are real variations at higher frequencies such as the 'bump' between 50 and 60 days seen at 15 GHz."

Local supernovae don't show this type of behavior. We thus attribute it the afterglow. To create this variability (of order 0.3 mag), the afterglow must have been significantly brighter than expected at that time.

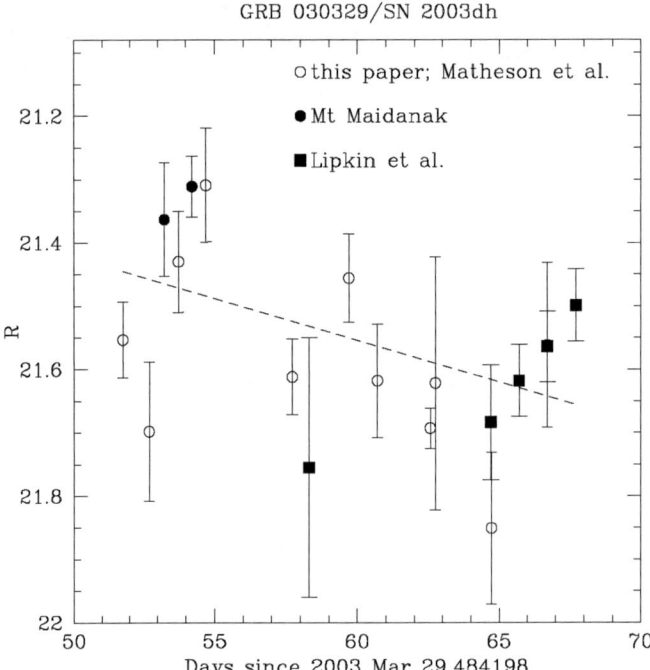

FIGURE 2. Light curve of GRB 030329 between 50 and 70 days. Open symbols are our data, dots are from Ibrahimov et al. [6], and filled squares are from Lipkin et al. [5].

3. LATE-TIME SPECTROSCOPY

We obtained a spectum in November 2003 with the Keck telescope and the Low Resolution Imaging Spectrometer. We subtracted the continuum from the host galaxy. The resulting spectrum is shown on Figure 3.

The most notable feature is the O I] 6300, 6363Å line, typical of type Ic supernovae. Other features that can be detected are broad emission lines near 4800 and 5300Å, which could be identified with the same broad Fe II]-dominated features observed in SN 1998bw.

The spectrum has been modeled using the code described in Mazzali et al. [8]. It was not possible to find a single model reproducing both the narrow O I] line *and* the broad Fe II] line. We thus computed two models: one where we try to fit the O I] line, and a "broad" spectrum where we fit the Fe II] line. These two models put lower, respectively upper, limits on the amount of ^{56}Ni synthesized during the explosion. These limits are 0.35 $M_\odot$ and 1.05 $M_\odot$. In any case, this supernova appears to have produced a fairly large amount of ^{56}Ni.

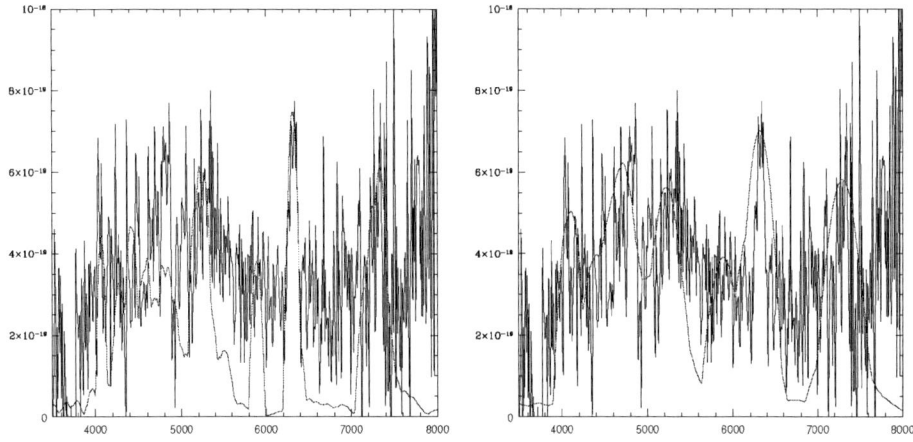

FIGURE 3. Spectrum of GRB 030329/SN 2003dh obtained in November 2003, compared to two models. *Left:* "Narrow-line" model. *Right* "Broad-line" model.

REFERENCES

1. Stanek, K. Z., et al., 2003, ApJ, 591, L17
2. Hjorth, J., et al. 2003, Nature, 423, 847
3. Matheson, T., et al. 2003, ApJ, 599, 394
4. Henden, A. A. 2003a, GCN Circ. 2082, 1
5. Lipkin, Y. M., Ofek, E. O., Gal-Yam, A., et al. 2004, ApJ, 606, 381
6. Ibrahimov, M. A., et al. 2003, GCN Circ. 2288, 1
7. Frail, D. et al., 2005, ApJ, 619, 994
8. Mazzali, P., et al. 2001, ApJ, 559, 1047

XRF 011030: the case of a late X-ray flare observed by BeppoSAX

A. Galli and L. Piro

Istituto Astrofisica Spaziale e Fisica Cosmica- Sezione di Roma, INAF, via del Fosso del Cavaliere, 100-00133 Rome, Italy

Abstract. The X-ray flash XRF 011030 observed by BeppoSAX has a light curve characterized by several outburst activities with the last detected flare occurring about 1300 s after the burst. We perform the temporal and spectral analysis of XRF 011030, study the late X-ray flare and compare it with the late afterglow emission observed by Chandra. We study the late X-ray flare in the contest of several variants of the external shock model and finally explain it as the beginning of the afterglow emission from a thick shell fireball emitted by a long duration central engine activity. We find that the late X-ray flare and the afterglow can be described both in a stellar wind and in a uniform interstellar medium (ISM).

Keywords: synchrotron,burst
PACS: 95.85.Pw

INTRODUCTION

The study of the temporal and and spectral evolution of Gamma-Ray Bursts (GRBs), and in particular the transition from the initial hard prompt emission to the following softer afterglow emission, is a powerful investigation tool. It gives information about the medium surrounding the central engine and the energy injection/activity duration and this help to constrain the nature of GRBs progenitor. Already during the BeppoSAX era became clear that the temporal evolution from the prompt to the afterglow emission shows different behaviours. Some bursts, e.g. GRB 970508 [1], GRB 971214 [2] and GRB 000210 [3] are characterized by a smooth connection between the tail of the prompt emission and the backward extrapolation of the late afterglow. Other bursts during the prompt-to-afterglow transition show the typical hard to soft spectral evolution but have a discontinuous temporal evolution. In fact, the main pulse is followed by a sudden re-emission occurring in the X-ray band, now usually called flare [2]. In particular BeppoSAX observed X-ray flares on different time scales. Short time scale flares (tens of seconds) were observed with BeppoSAX for example in GRB 970228 [4] and in GRB 980613 [2]. Theses event are characterized by a spectrum softer than that of the prompt and consistent with that of the afterglow and are well connected to the late afterglow with a power law. Long time scale flares (several minutes) were observed by BeppoSAX in GRB 011121 and in the X-Ray Rich XRR 011211 [5]. These two late X-ray flares show the same spectral behaviour of the short time scale flares, but they can be connected to the late afterglow only if the origin of the time t_0 is shifted to the instant of the flare, $F \propto (t - t_0)^{-\alpha}$.

More recently also the Swift satellite [6] observed long time scale X-ray flares in an increasing number of GRBs. The flares occurring in GRB 050126 and GRB 050219a [7] do not show a significant spectral evolution and their spectrum is consistent with that of the afterglow. In particular their light curves can be fitted with a $(t-t_0)^{-\alpha}$ power law. Also the late flare of GRB 050904 does not show spectral evolution, thus it seems to be similar to BeppoSAX flares. Instead in other bursts, e.g. GRB 050406, GRB 050421, GRB 050502B, GRB 050607 and GRB 050730, the flares show some evidence of hard-to-soft spectral evolution resembling the behaviour of the prompt emission [8].

These two different kinds of spectral behaviour can suggest the existence of two groups of flares with a different origin.

Several scenarios were proposed to explain the origin of X–ray flares, but we can distinguish mainly between internal shock models and external shock models. Burrows et al. [9] propose to explain late X–ray flares with late internal shocks, similar to that producing the GRB prompt emission, caused by a re-activation or a long duration activity of the central engine. Piro et al. [5] instead propose to explain the X–ray flares as the beginning of the afterglow emission due to the consistency of the spectrum of some flares with that of the late afterglow. In this model the flares are produced by a thick shell fireball coming out from a long duration central engine activity during the external shock. We underline that both the models require that the central engine remains active for a long time period.

TEMPORAL AND SPECTRAL PROPERTIES OF XRF 011030

The XRF 011030 light curve is characterized by a first bump (precursor), lasting about 300 s, followed by the main event, that has a duration of about 400 s and by a flare starting at about 1300 s. The comparison of the BeppoSAX data with the two Chandra observations also revealed the presence of a break in the light curve occurring between 10^4 and 10^6 s after the burst [10].

The event is an X-ray flash, consequently its spectrum is softer than that of a typical long GRB. In any case the spectral analysis points out that the flare spectrum is marginally softer than the main pulse and it is consistent with the late afterglow spectrum observed by Chandra [10].

DISCUSSION

In the previous section we have shown that the the flare spectrum is softer than the prompt and is consistent with the afterglow. We also know that late X–ray flares (e.g GRB 011121 and XRR 011211) are connected to the late afterglow if the origin of the time is shifted to the instant of the flare. The temporal and spectral properties of the flare give support to the idea that the it represents the beginning of the afterglow emission and is produced in the external shock with the circumburst medium. This lead us to study XRF 011030 in the framework of external shock.

The standard external shock model

In the standard external shock model the afterglow emission is produced by a thin shell fireball [11] expanding in a medium with a continuous or discontinuous density profile.

When a continuous density profile is assumed (ISM or wind) the flux rises and decays too slowly to describe the shape of the flare emission (see Fig. 5 of [10]).

A discontinuous density profile can affect the light curve if the observational frequency is below the electron cooling frequency. In fact, this condition is necessary so that the emitted flux is sensible to the density profile of the external medium. Also when this condition is satisfied, the observed flux increases too much and too fast to be consistent with the kinematical upper limits established by [12] on flares caused when the fireball interacts with density variations of the external medium.

Thick shell fireball from a long duration central engine activity

A shell is defined as thick if its thickness Δ satisfies the condition $\Delta > (E/nm_p c^2)^{1/3} \Gamma_0^{-8/3}$ [11]. The thickness Δ of the shell depends on the dynamical regime of the fireball, i.e on the duration of the activity of the central engine. In the thick shell scenario most of the energy is transferred to the external medium by the final layers of the shell when the engine turns off, that is about the instant of the late flare appearance. In this context the flare represents the beginning of the afterglow emission and comes out from the external shock. We develops our thick shell external shock model shifting the origin of the time t_0. We calculate the emitted flux also before the beginning of the deceleration regime carrying out the numerical integration of the basic equations of the so called standard fireball model [13]. This permit us to study the flare emission both during the phases of the rise and the decrease of the emission.

We apply this scenario to XRF 011030 shifting the origin of the time to the instant of the flare, $t_0 \simeq 1300$ s, and find solutions that can describe the late X-ray flare both when the fireball expands in a ISM or in a stellar wind [10]. These solutions also take into account for the late afterglow emission observed by Chandra. We first consider a fireball expanding in a wind. In this case we can explain the break as a spectral break. In fact, in a wind the electron cooling frequency increases with the time as $t^{1/2}$ and when it becomes greater than the observational frequency a break occurs in the light curve. When the fireball expand in a ISM the cooling frequency decreases with the time as $t^{-1/2}$. Also in this case a spectral break can occur, but after the break the light curve decays slower than in the wind case and it is difficult describe the two Chandra observations. Thus, in the ISM case we can explain the break as a jet break. When the fireball has jetted symmetry after the break the flux decays as t^{-p}. This temporal decay is faster than that expected in the case of a fireball with spherical symmetry and is in good agreement with the decay observed by Chandra.

CONCLUSIONS

XRF 011030 is one of the longest BeppoSAX event with the last detected flare occurring at about 1300 s and lasting about 200 s. The flare spectrum is consistent with that of the afterglow and this lead us to study the flare in the framework of the external shock. Standard external shock scenarios, based on a thin shell fireball expanding in a continuous (ISM or wind) or in a discontinuous density profile do not explain the sudden emission occurring during the flare. We find that the flare can be caused by a thick shell fireball coming out from a long duration central engine activity ($t_{eng} \sim t_{flare}$) when it catches up whit the external medium. In the case of a thick shell fireball we obtain good solutions both for a fireball expanding in a ISM and in a stellar wind.

ACKNOWLEDGMENTS

We thank E. Massaro, V. D'Alessio, B. Gendre and A. Corsi for their comments. This work is partially supported by the EU FP5 RTN Gamma Ray Burst: an enigma and a tool.

REFERENCES

1. Piro L. et al., 1998, A&A, 331, L41
2. Soffitta P., De Pasquale M., Piro L. & Costa E., 2002, *GRB in the Afterglow Era: 3rd Rome Workshop*, ASP Conference Series Edition Volume 312 pag.23
3. Piro L. et al., 2002, ApJ, 577, 680
4. Frontera F. et al., 1998, ApJ, 493, L67
5. Piro L. et al., 2005, ApJ, 623, 314P
6. Gehrels N. et al., 2004, ApJ, 611, 1005
7. Tagliaferri G. et al., 2005, Nature, 436, 1132
8. Burrows D. N. et al., 2005, astro-ph/0511039
9. Burrows D. N. et al., 2005a, Science, 309, 1833
10. Galli A. & Piro L., 2005, astro-ph/0510852
11. Sari R. & Piran T., 1999, ApJ, 520, 641
12. Ioka K., Kobayashi S. & Zhang B., 2005, ApJ, 631, 429
13. Panaitescu A. & Kumar P., 2000, ApJ, 543, 66

A catalog of X-ray afterglows observed by BeppoSAX, XMM-Newton and Chandra

B. Gendre*, A. Corsi*, L. Piro* and M. De Pasquale*,†

*IASF/INAF, via fosso del cavaliere 100, 00133 Roma, Italy
†MSSL, Holmbury St. Mary - Dorking - Surrey - RH5 6NT, U.K.

Abstract.
We present a catalog of X-ray afterglows observed by BeppoSAX, XMM-Newton and Chandra. We put constraints on the burst environment and geometry. We observe a temporal evolution of these constraints, by comparing the fast BeppoSAX and XMM-Newton follow-up and the late Chandra observations. We do not observe strong discrepancies when comparing the BeppoSAX sample to the XMM-Newton one and to some SWIFT bursts selected to be observed at late time.

Keywords: X-rays : general; Gamma-ray : bursts
PACS: 95.85.Nv : X-rays—98.70.Rz : gamma-rays sources, gamma-ray bursts

INTRODUCTION AND DATA ANALYSIS

The afterglow emission of Gamma-Ray Bursts (GRBs) is well described by the fireball model [1, 2, 3] and provides powerful diagnostics of the close environment of the burst. A large number of datasets available for a systematic study in X-rays is available. We have initiated a re-analysis of all X-ray afterglows observed so far, focusing on the burst environment properties [4, 5]. We retrieved all observations done by BepposAX, XMM-Newton and Chandra before the 1st of October 2004. Sixty six bursts observations were carried before this date. We obtained good constraints on the spectral and temporal indexes for 28 of these. For 13 bursts, the observed count rates did not allow to derive good constraints on the spectral and decay indexes. For 5 bursts, we do not detect any convincing afterglow. We finally discarded 3 burst observations. 52 afterglows were securely detected (out of 66 observations), and 10 observation gave a doubtfull afterglow detection. This turns to be a detection efficiency of 78 % (94 % if the doubtfull detection is indeed an afterglow). Except for the case of GRB 011030, all results reported here are consistent with the previous publications. When quoted, fluxes are unabsorbed and within the 2.0-10.0 keV X-ray band. The light curves and spectra will be made available through a local web page[1].

---

[1] http://grb.rm.iasf.cnr.it/catalog

TABLE 1. Mean parameters of each sub-sample from this paper. We include the decay index (δ), the spectral index (α), the flux at 11 hours after the burst and the observation start time.

	Chandra Imaging	Chandra Grating	XMM	BeppoSAX
δ	2.0 ± 0.3	1.4 ± 0.1	1.2 ± 0.2	1.3 ± 0.1
α	0.8 ± 0.1	1.0 ± 0.1	1.2 ± 0.2	1.2 ± 0.1
Log(Flux) @ 11 hours	-11.6 ± 0.3	-11.8 ± 0.3	-12.6 ± 0.2	-12.2 ± 0.1
Obs. time (ksec)	145.4	95.9	30.1	36.5

DISCUSSION

Comparison between afterglows

Our sample can be separated into the BeppoSAX, XMM-Newton, Chandra-grating and Chandra-imaging sub-samples. In the Chandra-grating sample, most of the observations were triggered because of a bright prompt emission detected in X-ray instruments. These bursts were therefore selected to be bright (see Table 1). The other samples are constituted by approved TOOs or observations made because it was possible to perform them. One may consider these samples as not too biased. In the remaining, we have assumed the BeppoSAX, Chandra-imaging and XMM-Newton samples not to present a selection bias.

We list the mean parameters of each subsample in Table 1. Discrepancies between the XMM-Newton and Chandra samples appear in both the flux and the decay index values. There is no strong difference between the XMM-Newton and BeppoSAX samples : there is no selection bias between these two samples.

Comparison with SWIFT bursts

[6] found that SWIFT X-ray afterglows are fainter compared to those observed by other instruments. They explain this discrepancy by a selection effect on the instrument that detected the burst (BeppoSAX, IPN, INTEGRAL, HETE-2 versus SWIFT). We tested this result using our catalog. We removed the Chandra observations and retained only XMM-Newton and BeppoSAX bursts because of the bias indicated above. The mean afterglow flux measured by SWIFT is $10^{-12.4}$ erg cm^{-2} s^{-1} at 12 hours after the burst, the BeppoSAX one is $10^{-12.2}$ erg cm^{-2} s^{-1}. We thus do not observe any strong discrepancy between the two samples.

Time evolution of the closure relationships

The values of the decay index (δ) and the spectral index (α) are linked through closure relationships. These relationships depend on the burst environment and the burst geometry [7, 8, 9]. The average closure relationship positions for each sub-samples are $\delta - 2.0\alpha = -0.93 \pm 0.36$, -1.14 ± 0.18, -0.36 ± 0.17, and 0.33 ± 0.32 for XMM-Newton, BeppoSAX, Chandra (grating) and Chandra (imaging) respectively for the Jet effects (theoretical expected values are 1.0 and 0.0, mean obervation times are 30.1, 36.5, 95.9, 145.4 respectively). Jet effects are excluded within BeppoSAX and XMM-Newton bursts, and the mean position value changes from one extreme position (XMM-Newton and BeppoSAX samples) to another one (Chandra sample). This evolution can be explained either by a passing through of the cooling frequency in the X-ray band or by a convergence toward the expected value in case of jet signature. The observed evolution of the decay index (from 1.2 ± 0.2 with XMM-Newton to 2.0 ± 0.3 with Chandra, $\Delta\delta = 0.8 \pm 0.5$) is consistent with a jet break (theoretical value : $\Delta\delta = 0.83 - 1.33$, depending on the model) but not with a cooling break ($\Delta\delta = 0.25$). We thus explain the evolution of the closure relationship values and the steep Chandra Imaging burst decays by a jet effect at late time. The observations that led to non detections are, with one exception, performed very late. At that time, one should observe a steep decay of the afterglow due to the jet effect, that could prevent the detection.

Burst geometry

The time of jet break can be linked to the jet opening angle [10]. We can rule out a jet signature within XMM-Newton bursts and cannot rule out this signature within Chandra bursts. Also, the Chandra bursts appear brighter than the XMM-Newton ones when extrapolated back to half a day after the burst. This may indicate that the crude extrapolation using a single power law is not correct, and that there is a break in the light curve *before* the observation. We can then set lower and an upper limits to be 0.09 rad $< \theta < 0.22$ rad, which is in agreement with previous works [e.g. 11, 12].

We have looked for a standard energy emission in the X-ray afterglow as in [12], using 14 bursts with known distance, spectral index, temporal decay and beaming angle. We have computed the luminosity at a common epoch of 11 hours after the burst using for each burst its spectral and temporal decay indices [note that 12, used a mean spectral index and mean values for the temporal decay index and redshift for unknown values]. We have fitted a Gaussian distribution to the isotropic luminosity distribution and the beaming corrected luminosity distribution. We obtained $\sigma = 0.5 \pm 0.2$ and $\sigma = 0.4 \pm 0.2$ respectively. The values obtained by [12] were $\sigma = 0.7 \pm 0.2$ and $\sigma = 0.3 \pm 0.1$ respectively. We observe the same (within error bars) distributions when one corrects for beaming. However, our isotropic luminosity distribution is narrower than the one observed by [12], while compatible within the error bars. In fact, its width is consistent with that of the beaming-corrected distribution. So, we do not observe a significant shrinking of the luminosity distribution when taking into account the beaming. We thus cannot confirm the [12] claim for a standard release of energy in the X-ray afterglow.

Surrounding medium

Most of the bursts can fit both a wind or an interstellar medium, due to the degenerations observed for 2 closure relationships. However, we can distinguish the medium type for GRB 040106. Using the X-ray afterglow data, we can constrain the environment of this burst to be a wind.

CONCLUSION

We have observed variations of the global properties of the X-ray afterglows within the samples that we explain by a jet effect. This is observed from both a variation of the closure relationships and the steepening of the mean decay index for bursts observed late. We have shown that jet effects are not present within one day after the burst. Taking advantage of the late observation time of Chandra, we find that the jet signature occurs between two and ten days after the burst. This allowed us to constrain the jet opening angle value. We have not noticed differences between the XMM-Newton and BeppoSAX samples that can be explained by an absence of a selection bias between these observatories. We have not observed a significant shrinking of the luminosity distribution when we corrected for beaming. This result needs to be confirmed by more burst observations, e.g. by SWIFT. In addition, we have found that the SWIFT X-ray afterglows are similar to those of XMM-Newton and BeppoSAX, while Chandra ones are biased toward higher luminosities. If this bias is not considered, the SWIFT bursts are fainter than the complete sample of all bursts listed here.

ACKNOWLEDGMENTS

We thank G. Garmire for his help during the analysis of the Chandra grating data. This work was supported by the EU FP5 RTN 'Gamma ray bursts: an enigma and a tool'. A.C. acknowledges support from an INFN grant. This work is based on observations obtained with XMM-Newton, an ESA science mission with instruments and contributions directly funded by ESA Member States and NASA.

REFERENCES

1. Rees, M.J., & Meszaros, P., 1992, MNRAS, 258, 41
2. Meszaros, P, & Rees, M.J., 1997, ApJ, 476, 232
3. Panaitescu, A., Meszaros, P., & Rees, M.J., 1998, ApJ, 503, 314
4. De Pasquale, M., Piro, L., Gendre, B., et al., 2005, A&A submitted, astro-ph/0507708
5. Gendre, B., Corsi, A., & Piro, L., 2006, A&A, in press
6. Berger, E., Kulkarni, S.R., Fox, D.B., et al., 2005, submitted to ApJ (astro-ph/0505107)
7. Rhoads, J.E., 1997, ApJ, 487, L1
8. Sari, R., Piran, T., & Narayan, N., 1998, ApJ, 497, L17
9. Chevalier, R.A., & Li, Z.Y., 1999, ApJ, 520, L29
10. Sari, R., Piran, T., & Halpern, J.P., 1999, ApJ, 519, L17
11. Frail, D.A., Kulkarni, S.R., Sari, R., et al., 2001, ApJ, 562, L55
12. Berger, E., Kulkarni, S. R., & Frail, D. A., 2003, ApJ, 590, 379

Swift Observations of GRB 050603

Dirk Grupe*, Peter J. Brown*, Alon Retter*, David N. Burrows*, John A. Nousek*, Peter Mészáros *,†, Bing Zhang** and Jay Cummings‡

*Department of Astronomy and Astrophysics, Pennsylvania State University, 525 Davey Lab, University Park, PA 16802, USA
†Department of Physics, Pennsylvania State University, University Park, PA 16802, USA
**Department of Physics, University of Nevada, Las Vegas, NV 89154, USA
‡NASA Goddard Space Flight Center, Greenbelt, MD 20771, USA

Abstract.
We report the results of *Swift* observations of the Gamma Ray Burst GRB 050603. With a V magnitude V=18.2 10 hours after the burst, the optical afterglow was the brightest so far detected by Swift and one of the brightest afterglows ever seen at late times. The Burst Alert Telescope (BAT) trigger light curves showed three fast-rise-exponential-decay spikes with $T_{90}=12\pm2$s and a fluence of 7.6×10^{-6} ergs cm^{-2} in the 15–150 keV band. The Swift spacecraft started the observations of the afterglow with the narrow-field instruments about 10 hours after the detection of the burst. The burst was bright enough to be detected by the Swift UV/Optical telescope (UVOT) for almost 3 days and by the X-ray Telescope (XRT) for a week after the burst. The X-ray light curve showed a rapidly fading afterglow with a decay index $\alpha=1.76^{+0.15}_{-0.07}$ This is one of the steepest late time slopes seen by Swift so far. The X-ray energy spectral index was $\beta_X=0.71\pm0.10$ with the absorption parameter in agreement with the Galactic value. The spectral analysis did not show an obvious change in the X-ray spectral slope over time. The optical UVOT light curve decayed with a slope of 1.97. The steepness and the similarity of the optical and X-ray decay rates suggest that the afterglow was observed after the jet break. Its redshift z=2.821 was measured by Berger & Becker (2005).

Keywords: GRB afterglow, individual (GRB 050603)
PACS: 98.70.Rz, 98.70.Qy

SWIFT OBSERVATIONS

GRB 050603 was detected by the BAT on 2005-June-03 at 06:29:05 UT (Retter et al. 2005). Due to engineering tests of the Swift satellite, the narrow-field instruments, UVOT and XRT, were unable to observe the GRB afterglow until about 9.5 and 11 hours after the burst, respectively. The UVOT began observing at 15:42:59 UT in the V-filter (Brown et al. 2005). The XRT started to take data at 17:19:27 UT (Racusin et al. 2005). GRB 050603 was observed over a period of about two weeks for a total of about 220 ks (Grupe et al. 2005).

BAT TRIGGER OBSERVATION

The BAT mask-weighted light curve (Figure 1) shows three fast-rise-exponential-decay (FRED) like spikes with peaks at 2.7 and 0.85 s before the trigger and 0.15 s afterwards. Each spike had a width of 0.6 s FWHM. The time-averaged spectrum between T_0 −3s and T_0 +18s in the 15–150 keV band is well-fitted by a single power law with an energy

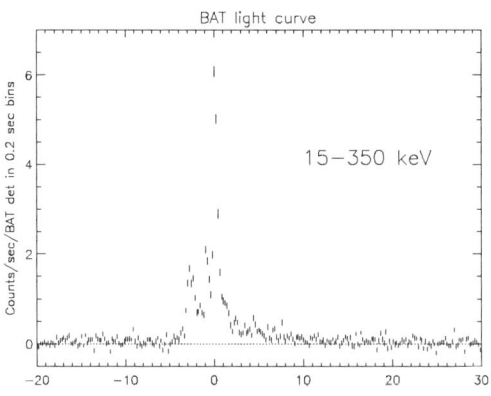

FIGURE 1. BAT Light curve of GRB 050603

spectral slope $\beta_\gamma = 0.17^{+0.07}_{-0.08}$. GRB 050603 had $T_{90}=12\pm2$s which classifies it as a long burst. The fluence in the 15–150 keV band was 7.6×10^{-6} ergs cm^{-2}.

ANALYSIS OF THE XRT DATA

The left panel of Figure 2 displays the 0.3-10 keV flux X-ray light curve of GRB 050603. The initial count rate at the beginning of the XRT observation was 0.06 counts s^{-1}. The decay slope $\alpha=1.76^{+0.15}_{-0.07}$ is unusually steep and compared to other afterglows at late time phases (Nousek et al. 2005) GRB 050603 is one of the most rapidly fading afterglows. The X-ray spectra of the first and second days after the burst are shown in the right panel of Figure 2. Both spectra are well fitted by absorbed power law with an energy spectral slope $\beta_X=0.71\pm0.10$. No additional absorber column above the Galactic value (1.2×10^{20} cm^{-2}; Dickey & Lockman 1990) is required. The spectral fits are consistent with no spectral variability between the two observations. The XRT position was measured by Moretti et al. (2005) to α_{2000}=02h39m56.73s, δ_{2000}=-25d10m54.36s.

UVOT V-FILTER ANALYSIS

Figure 3 displays the UVOT V-filter light curve (below) and the simultaneous XRT light curve. The UVOT decay slope is 1.8 ± 0.2, similar to the X-ray decay slope (see above). The magnitude at 9.5 hours after the burst was V=18.2, which is brighter than any other *Swift* optical afterglow and most other optical afterglows at this time, except

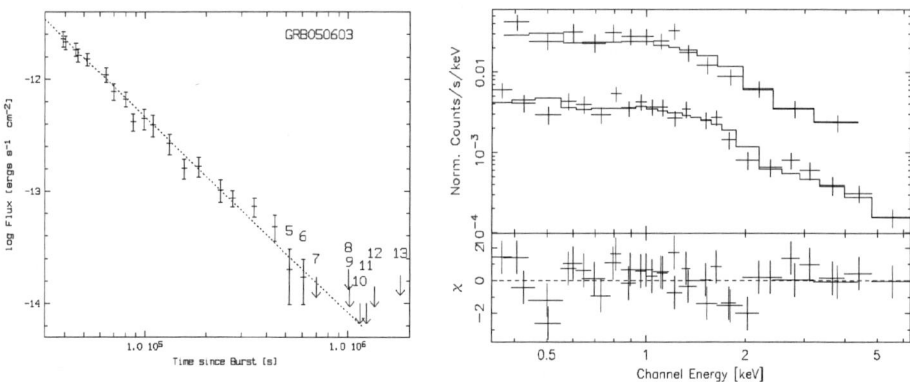

FIGURE 2. XRT Light curve and spectra of GRB 050603. The spectra were created from the data of segments 001 and 002

GRB 030329 (e.g. Berger et al. 2005).

DISCUSSION

With an optical magnitude V=18.2 10 hours after the burst, GRB 050603 was the brightest optical afterglow detected so far by *Swift*. Following the canonical afterglow light curve as shown in Nousek et al. (2005) and Zhang et al. (2006) the decay slope α=1.76 is steeper than the mean decay slope of the third phase of the afterglow in Nousek et al. (2005), but agrees with the fourth phase after the jet break in Zhang et al. (2006). The similar steep V-filter decay slope also argues for an early jet break. The X-ray spectral slope is rather flat compared to other afterglows (Nousek et al. 2005). The afterglow does not show significant evidence of spectral changes. The X-ray spectral analysis suggests that the absorption column is consistent with the Galactic value. A detailed description of this afterglow and the discussion can be found in Grupe et al. (2005). In this paper we also present a list of the serendipitous sources in the XRT field of view.

REFERENCES

1. Berger, E., et al. 2005, ApJ, 629, 328
2. Berger, E., & Becker, G. 2005, GCN 3520
3. Brown, P., et al. 2005, GCN 3516

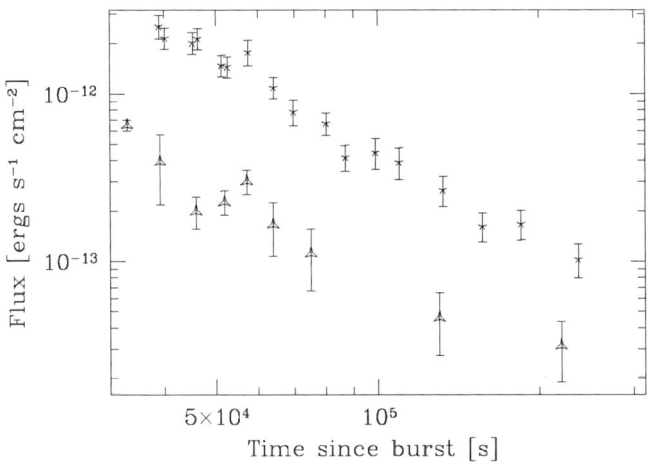

FIGURE 3. XRT Light curve of GRB 050603. The UVOT V fluxes were multiplied by a factor of 10^6 in order to plot them on the same scale as the XRT points. The XRT data are displayed by crosses and the UVOT V-band measurements by triangles.

4. Dickey, J.M., & Lockman, F.J. 1990, ARA&A, 28, 215
5. Grupe, D., et al. 2005, ApJ, submitted
6. Moretti, A., et al. 2005, ApJ, submitted, astro-ph/0511604
7. Nousek, J.A., et al. 2005, ApJ, accepted, astro-ph/0508332
8. Racusin, J.L., et al. 2005, GCN 3514
9. Retter, A., et al. 2005, GCN 3509
10. Zhang, B., et al. 2006, ApJ, in press, astro-ph/0508321

ACKNOWLEDGMENTS

This research was supported by NASA contract NAS5-00136 (D.G., D.B., & J.N.).

Swift Panchromatic Observations of the Bright Gamma-Ray Burst GRB 050525A

Stephen T. Holland*, Alex J. Blustin[†] and The Swift Science Team[**]

Code 660.1, NASA's GSFC, Greenbelt, MD 20771, USAd
[†]*Mullard Space Science Laboratory, Department of Space & Climate Physics, University College London, Holmbury St Mary, Dorking, Surrey RH5 6NT, UK*
[**]

Abstract.
The *Swift* observatory discovered and observed GRB 050525A. The X-ray and optical/ultraviolet afterglow decay curves exhibit an achromatic break approximately 0.15 days after the burst. This break, and the total gamma-ray energy of the burst, implies a jet opening angle of $2°\!.5$. Prior to the break the X-ray spectrum can be modelled by a power law with $\beta_X = -1.2$. A short X-ray flare occurred 300 s after the burst. The optical/ultraviolet data have a complex decay, with evidence of a reverse shock dominating in the first few minutes and a forward shock dominating at later times. The X-ray/ultraviolet/optical spectrum suggests that a cooling break moved through the optical between 800 and 25 000 s after the burst. The temporal decay and spectral indices agree with the standard fireball model, assuming a uniform interstellar medium.

Keywords: gamma rays: bursts
PACS: 98.70.Rz

INTRODUCTION

GRB 050525A was detected at 00:02:53 UT on 25 May 2005 by the *Swift*/BAT. X-ray, ultraviolet, and optical counterparts were found in the data sent to the the ground via the TDRSS link. The burst was optically bright with a redshift of $z = 0.606$ [1]. This burst occurred 15° above the Galactic Plane, and the Galactic extinction is $A_V = 0.32$ mag [2]

BAT DATA

The gamma-ray burst was a "typical" long–soft burst. The light curve shows two peaks in all four energy bands and $T_{90} = 8.8 \pm 0.5$ s. Both a Band spectrum [3] and a power-law spectrum with an exponential cut-off provide equally good fits to the burst spectrum. These best-fit models have a slope of $\beta = +0.013$ and $E_P = 78.8$ keV.

XRT DATA

The XRT's Photodiode and Photon Counting modes were used to monitor the X-ray decay from 133 s to 4.77 days after the burst. A rebrightening, or flare, occurs at ≈ 300 s, and a break is seen at 3.8 hours. The decay slopes are -1.20 before the break and -1.62 after. The spectrum shows no evidence for time evolution between 128 and 1048 s after the trigger. The X-ray data is well fit by an absorbed power law with $\beta_X = -0.98$ and a hydrogen-equivalent column density of 2.06×10^{21} cm^2.

UVOT DATA

Decay Curves

UVOT began observing 65 s after the BAT trigger and took images for the next 30 days. Data were taken with the UBV, ultraviolet, and White filters and calibrated using the HEASARC's Swift/UVOTA calibration database (CalDB)[1]. The optical counterpart was found in the initial 100 s V-band finding chart image at the ROTSE position [4].

The best fit to the decay before the break is the sum of two power laws with $\alpha_1 = -1.56$ and $\alpha_2 = -0.62$. These two components contribute equally to the flux at $t_0 + 7.2$ minutes. This is similar to what is seen in the early optical afterglows of GRB 021211 and GRB 9901123 [5]. The break occurs simultaneously in the optical/ultraviolet and in the X-ray data, which is consistent with a jet break. The post-break decay has $\alpha_j = -1.76$ at all frequencies. The data and the fits are shown in Figure 1.

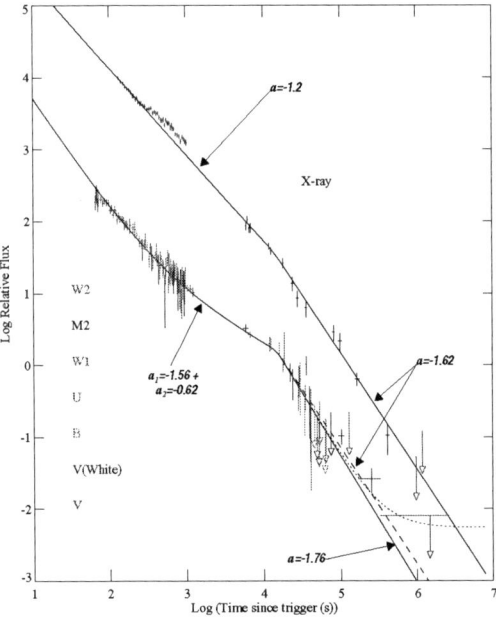

FIGURE 1. The upper curve shows the X-ray data (black crosses) with the best-fit broken power law. The lower curve shows the UVOT data with the model described above. The dashed line is the best post-break fit to the X-ray data superimposed onto the UVOT data.

---

[1] http://swift.gsfc.nasa.gov/docs/heasarc/caldb/swift/

Panchromatic Spectra

Figure 2 shows the combined X-ray/ultraviolet/optical spectrum of the afterglow of GRB 050525A. The observed spectrum at each epoch is well fit by a syncrotron spectrum that has been modified by absorption due to hydrogen gas in the Galaxy and in the host, dust in the host galaxy, and a Lyman cut-off. The best fit occurs using dust with an SMC extinction law [6] with $E_{B-V} = 0.43$ mag and $N_H = 1.8 \times 10^{21}$ cm^2 in the host. A cooling break at 2.42×10^{17} Hz (1 keV) is needed to fit the 250 s and 800 s spectra. No break is needed for the 25 000 s spectrum. This suggests that the cooling break evolved between 800 and 25 000 s.

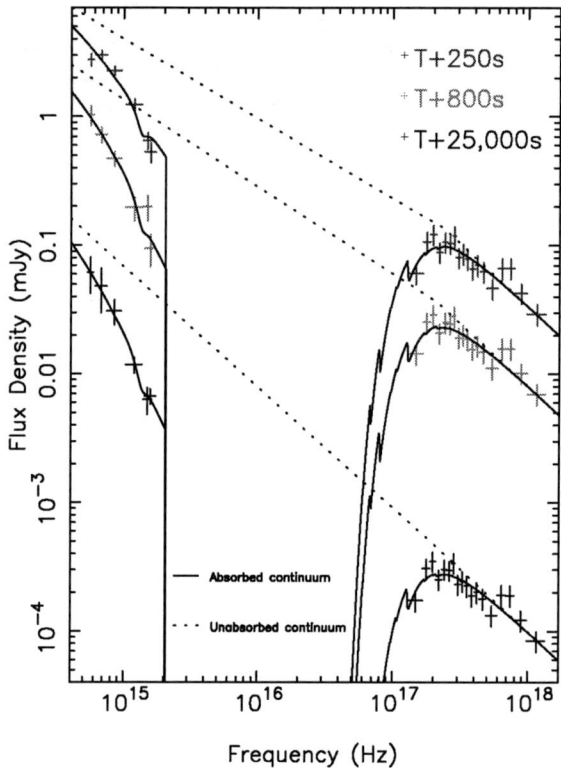

FIGURE 2. The observed panchromatic spectrum of GRB 050525A's afterglow at three epochs: 250, 800, and 25 000 s after the BAT trigger. The solid lines are the best fitting synchrotron spectra combined with SMC dust absorption in the host, hydrogen in the Galaxy and host, and a Lyman cut-off. The dashed lines are the corresponding pure synchrotron models.

INTERPERTATION

The early decay is a consistent with radiation from both a reverse shock and forward shock. The break at approximately 3.8 hr, and the post-break decay slope, is consistent with a jet break. *Swift* observations of the afterglow are consistent with the standard fireball model expanding into a uniform circumburst medium. The data are not consistent with a stellar wind environment. The observed spectral and decay slopes imply an electron index of $p = 2.2$, which is typical for GRBs.

The cooling break appears to have moved redward through the optical between ≈ 800 and 25 000 s after the BAT trigger. This is faster than the $v_c \propto t^{-1/2}$ evolution of the cooling frequency predicted by the standard afterglow model. This enhanced cooling may be due to the ratio of magnetic energy density to internal energy (ε_B) evolving with time.

The derived jet opening angle is $\approx 2°\!.5$, implying implies a low gamma-ray equivalent energy of $E_\gamma = 4.9 \times 10^{49}$ erg, assuming an ambient density of $n_0 = 1$ cm^{-2}. Despite being underluminous in gamma rays GRB 050525A had a "typical" optical afterglow.

ACKNOWLEDGMENTS

The Swift programme is supported by NASA, PPARC, and ASI (contract number I/R/039/04).

REFERENCES

1. R. J. Foley, H.-W. Chen, J. Bloom, and J. X. Prochaska, GCN Circular 3483 (2005).
2. D. J. Schlegel, D. P. Finkbeiner, and M. Davis, *ApJ*, **500**, 525–553 (1998).
3. D. Band, J. Matteson, L. Ford, B. Schaefer, D. Palmer, B. Teegarden, T. Cline, M., Briggs, W. Paciesas, G. Pendleton, G. Fishman, C. Kouveliotou, C. Meegan, R. Wilson, and P. Lestrade, *ApJ*, **413**, 281–292 (1993).
4. E. S. Rykoff, S. A. Yost, H. Swan, and R. Quimby, GCN Circular 3468 (2005).
5. S. T. Holland, D. Bersier, J. S. Bloom, P. M. Garnavich, N. Caldwell, P. Challis, R. Kirshner, K. Luhman, B. McLeod, K. Z. Stanek, *AJ*, **128**, 1955–1964 (2004).
6. Y. C. Pei, *ApJ*, **395**, 130–139 (1992).

Correlation of Color Behaviour of Blazars and Optical Afterglows of GRBs

René Hudec[1], Robert Filgas[2,1], Vojtech Simon[1], Milan Basta[1,2] and Martin Topinka[1,2]

[1] *Astronomical Institute of the Academy of Sciences of the Czech Republic, CZ-251 65 Ondrejov, Czech Republic rhudec@asu.cas.cz*
[2] *Astronomical Institute, Faculty of Mathematics and Physics, Charles University, Prague, Czech Republic*

Abstract. It has been demonstrated that the colors of optical afterglows (OAs) of GRBs are unique and almost identical for most of these events detected and analyzed in detail so far [1]. This enables conclusions toward physics of the source and physics of the source environment. In addition, OAs can be identified on CCD images taken at one time but in various filters. In this study, we compare the color behaviour of OAs of GRBs with those of other extragalactic objects. We will show that not only the physical processes involved as well as environment of the objects can be studied this way but also a classification of objects of various types is possible with color photometric data provided e.g. by deep CCD searches and experiments.

Keywords: gamma-ray astrophysics, gamma-ray bursts, active galactic nuclei - blazars
PACS: 95.55.Ka, 95.85.Pw

INTRODUCTION: THE METHOD OF COLOR INDICES

Color indices of optical afterglows (OAs) of gamma-ray bursts (GRBs) are a powerful and innovative approach to the study of such events. They help us to: (i) search for the common properties of the afterglows, (ii) understand the related physical processes, (iii) find out whether an optical event is related to a GRB even without observable gamma-ray detection by using the color indices of OAs, (iv) search for the orphan afterglows, (v) search for the relations among colors, luminosities and the decay rates of the OAs (if the redshift z is known), and (vi) constrain the properties of the local interstellar medium of GRBs.

Analogous analysis can be applied also to blazars and other extragalactic sources. In this study, we analyse the color indices of blazars and AGNs and compare them to the typical colors found for OAs of GRBs.

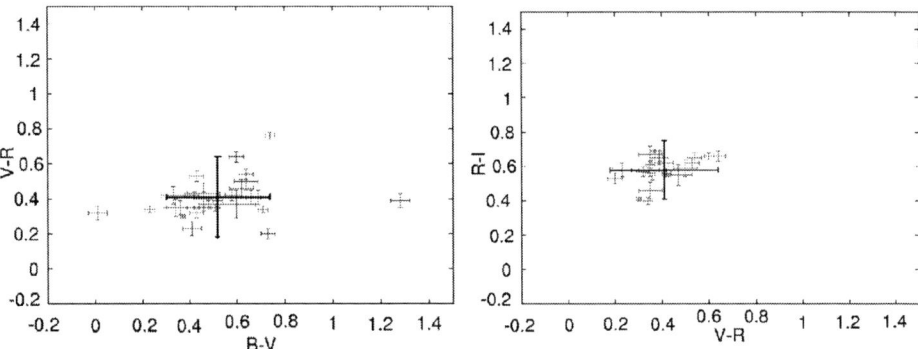

FIGURE 1. Color-color diagrams of blazars with corresponding error bars and centroids. Data from [2][3][4][5][6][7][8] with no additional correction applied.

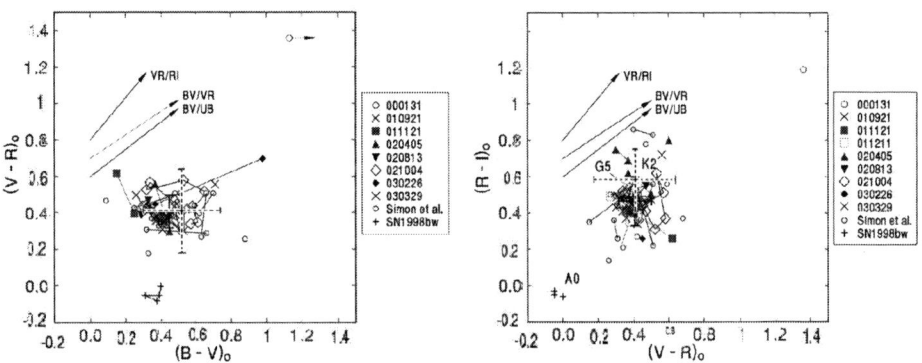

FIGURE 2. Color-color diagrams for the OAs of GRBs with centroids plus dashed centroids of blazars. Data from [1] and several blazar catalogues [2][3][4][5][6][7][8] are used.

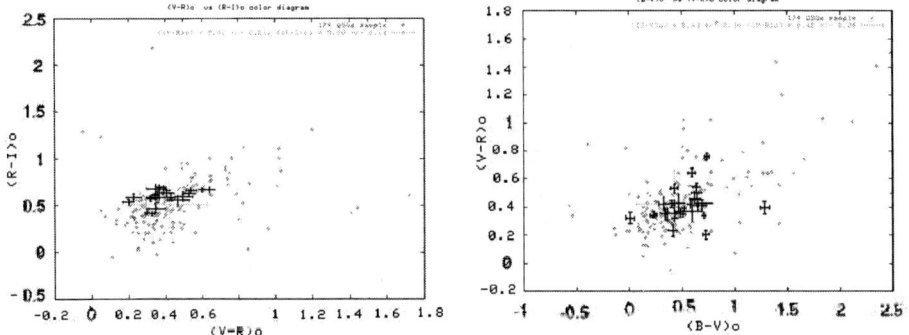

FIGURE 3. Color-color diagrams of AGN (from [9], small dots) plus crosses of blazar positions (with error bars).

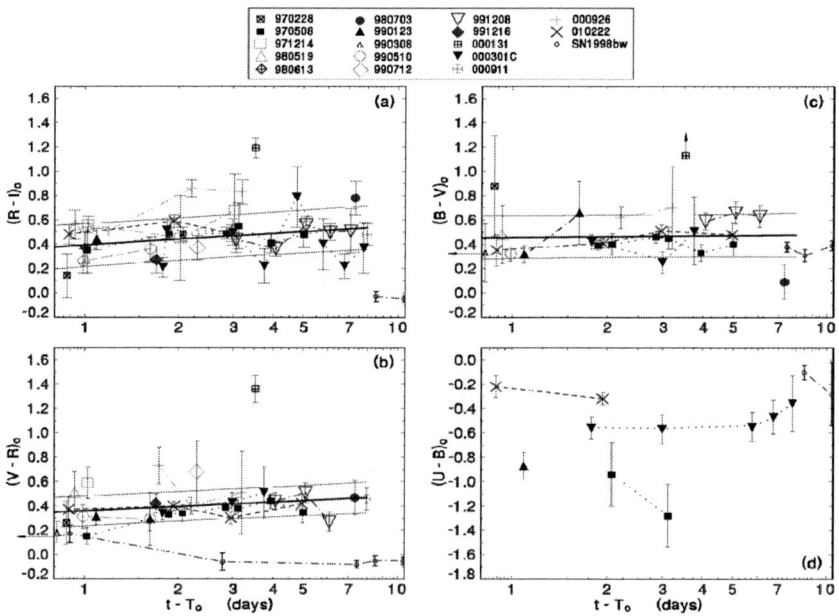

FIGURE 4. The curves of color indices of OAs during the first 10 days – the temporal evolution of color indices.

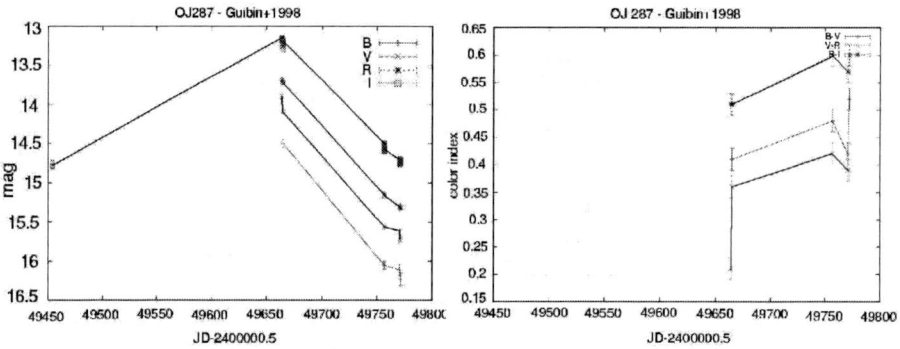

FIGURE 5. Left: Blazar OJ287 – light curve. Right: Blazar OJ287 – color index. Based on the observations [11].

RESULTS

Not only the OAs of GRBs, but also blazars exhibit specific colors. The main difference is that (i) the colors of blazars exhibit a somewhat larger scatter than those of the OAs of GRBs and (ii) the colors of blazars are similar to those of OAs of GRBs with only small systematic shifts less than 0.2 mag. An interesting feature is that both OAs of GRBs and blazars do not exhibit large color variations during the very large amplitude changes. This is in contrast to the (B-V) color evolution of type Ia SNe for wide range of initial decline rates and peak luminosities, with little or no reddening from dust in their host galaxies, where the color index (B-V) can vary by > 1 mag [10]. The comparison with the colors of a larger sample of AGN in general shows that the AGNs have the colors similar to the colors of blazars but the scatter is larger.

CONCLUSION

Both blazars and OAs of GRBs exhibit specific colors. The detailed analysis has shown that the typical color indices of OAs inside the time interval $0.14<t-T_0<10$ days are $(V-R)_0=0.41+/-0.09$, $(R-I)_0=0.46+/-0.13$, $(B-V)_0=0.45+/-0.15$ [1]. The typical colors of blazars and AGNs are shown in this contribution. The results of this analysis can be applied in automated searches for OAs of GRBs and blazars with photographic and CCDs surveys with filters. Also the physical conclusions can be drawn regarding the physics of the sources as well as the physics of their environment.

ACKNOWLEDGEMENTS

We acknowledge the support provided by the Grant Agency of the Academy of Sciences of the Czech Republic, grant A3003206.

REFERENCES

1. Simon, V. et al., *A&A*, **377**, 450 (2001).
2. Raiteri C.M., Ghisellini G., Villata M., De Francesco G., Lanteri L.. *Astron. Astrophys. Suppl. Ser.* **127**, 445 (1998).
3. Fiorucci M., et al., *Astron. Astrophys. Suppl. Ser.* 117, 475 (1996).
4. Villata M., Raiteri C.M., Ghisellini G., De Francesco G., Bosio S., Astron. Astrophys. Suppl. Ser. **121**, 119 (1997).
5. Bai J.M, Xie G.Z., Li K.H., Zhang X., Liu W.W., *Astron. Astrophys. Suppl. Ser.* **132**, 83 (1998).
6. Bai J.M, Xie G.Z., Li K.H., Zhang X., Liu W., Astron. Astrophys. Suppl. Ser. **136**, 455 (1999).
7. Xie G.Z., Li K.H., Zhang Y.H., Liu F.K., Fan J.H., Wang J.C., *Astron. Astrophys. Suppl. Ser.* **106**, 361 (1994).
8. Xie G.Z., Li K.H., Zhang X., Bai J.M., Liu W.W.., *Astrophys. J.* **522**, 846 (1999).
9. Veron-Cetty, M.-P.; Veron, P., *Astronomy and Astrophysics*, **412**, 399 (2003).
10. Phillips, M. M., et al., *AJ*, **118**, 1766 (1999).
11. Guibin, J. et al., *Astron. Astrophys. Suppl. Ser.*, **128**, 315 (1998).

Identification of Two Categories of optically bright Gamma-Ray bursts and A Model-Independent Luminosity Indicator

Enwei Liang[1,2] and Bing Zhang[1]

[1]*Physics Department, University of Nevada, Las Vegas, NV89154;*
lew@physics.unlv.edu, bzhang@physics.unlv.edu
[2]*Department of Physics, Guangxi University, Nanning 530004, China*

Abstract. We identify two categories of optically bright gamma-ray bursts (GRBs) with the intrinsic optical afterglow light curves for a complete sample of gamma-ray bursts (GRBs) detected from Feb. 1997 to Aug. 2005. The light curves follow two universal tracks at 2 hours in the rest frame since GRB triggers, and the optical luminosities at 1 day show a bimodal distribution. While the luminous group (~ 75% of the GRBs in our sample) has a wide range of redshift distribution, the bursts in the dim group all appear at a redshift lower than 1.1. We make a blind search for the relations of gamma-ray energy to observabales of prompt gamma-rays and optical afterglows by a multi-variable regression analysis method without imposing any theoretical models and assumptions for the bursts in our sample. We find a tight relation of the isotropic gamma-ray energy to the the peak energy and the optical light curve break time in the rest frame. Regarding this relation as a luminosity indicator, we explore the possible constraints on the cosmological parameters. An approach also proposed to calibrate the GRB luminosity indicators without introducing a low-redshift GRB sample based on the Bayesian theory.

Keywords: gamma-ray bursts; Observational cosmology
PACS: 98.70.Rz;98.80.Es

INTRODUCTION

Since the cosmological nature of the GRBs was identified and the association of the long GRBs with energetic core-collapsars was established (see review by [1]), they are expected to be new probes of cosmology and galaxy evolution[2]. With uncovered luminosity indicators, some authors have used GRBs as rulers to constrain the cosmography and cosmological dynamics[3, 4, 5, 6, 7, 8, 9, 10, 11].

In this work we systematically study the optical afterglows of GRBs. Over 8 years of optical afterglow hunting, more than 70 optically-bright GRBs have been detected, among which 44 bursts have well-sampled light curves and redshift measurements. We analyze these light curves in the cosmic rest frame and identify new subclasses of GRBs[12]. We further pursue the topic of GRB cosmology with the optical afterglow data. We find a tight relation of the isotropic gamma-ray energy (E_{iso}) to the peak energy of νf_ν spectrum (E'_p) and the break time of the optical afterglow light curves (t'_b), and analyzed possible constraints on the cosmological parameters[10]. Based on the Bayesian theory, we also propose an approach to calibrate the GRB luminosity indicators without introducing a low-redshift GRB sample based on the Bayesian theory [13].

IDENTIFICATION OF TWO CATEGORIES OF OPTICALLY BRIGHT GRBS

We make a complete search from the literature for the R-band afterglow light curves detected during the time period from Feb. 1997 to Aug. 2005. We obtain a GRB sample of 44 GRBs with well-sampled optical afterglow light curves and redshift measurement[1]. We make corrections of the Galactic extinction, the host galaxy extinction, and the $k-$correction and then calculate the intrinsic optical light curves $[L_R(t')$ vs. $t']$. Figure 1 shows the light curves and the 2-dimensional distribution of the luminosity at 1 day, $L_{R,1d}$, versus $E_{\gamma,\text{iso}}$, and the distributions of the two quantities, respectively. It is found that although the light curves at $t' < 0.1$ days vary significantly, they are clustered and follow two apparent universal tracks at $t' > 0.1$ days. While the $E_{\gamma,\text{iso}}$ distribution displays a power-law with sharp cutoff around $10^{51.5}$ ergs (due to the selection effect), $\log L_{R,1d}$ shows a well-defined bimodal distribution, which is well fitted by a two Gaussian model. These results indicate that within the optically bright GRBs there exist two well-separated sub-categories. The bursts in the luminous group ($\sim 75\%$ of the GRBs in our sample) are typically brighter than those in the dim group by a factor of ~ 30.

MODEL-INDEPENDENT MULTI-VARIABLE GAMMA-RAY BURST LUMINOSITY INDICATOR AND ITS POSSIBLE COSMOLOGICAL IMPLICATIONS

As a luminosity indicator it should be directly measured, i.e., it should be defined by observables only. We thus make a blind search for the relations of gamma-ray energy to observables by a multi-variable regression analysis method without imposing any theoretical models and assumptions. We find a tight relation of $E_{\gamma,\text{iso}}$ to E'_p and t'_b in the rest frame. It is shown in Figure 2. Regarding this relation as a luminosity indicator, we explore the possible constraints on the cosmological parameters. Since there is no low-redshift GRB sample to calibrate this relationship, we weigh the probability of using this relation in each cosmology to serve as a standard candle by χ^2 statistics, and then use this cosmology-weighed standard candle to evaluate cosmological parameters. The results are also presented in Figure 2. Through Monte Carlo simulations, we find that the GRB sample satisfying our relationship observationally tends to be a soft and bright one, and that the constraints on the cosmological parameters can be much improved either by the enlargement of the sample size or by the increase of the observational precision. Furthermore, the detections of a few high redshift GRBs satisfying the correlation could greatly tighten the constraints.

Unlike Type-Ia supernovae, calibration of GRB luminosity indicators using a low-redshift sample is difficult. This is because the GRB rate drops rapidly at low redshifts,

---

[1] A full version of the GRB sample with references to the observational data are available in the electronic version of [12]

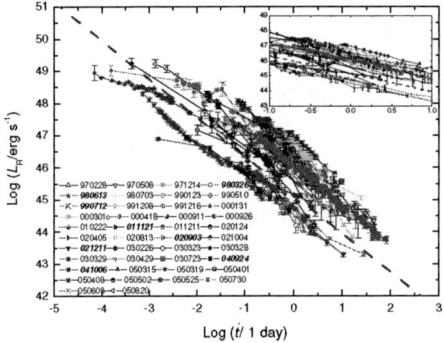
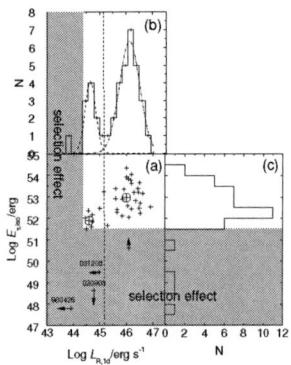

FIGURE 1. *Left*—The R-band light curves ($L_R(t')$ vs. t') in the cosmic proper rest frame. The dashed line is a division of the two groups of GRBs, $\log L_R = 45.15 - 1.2 \log t'$. The upper inset zooms in the light curves in the time regime from 0.1 days to 10 days. Those burst name in italic typeface in the figure legend belong to the dim group. *Right*—The 2-dimensional distribution of $L_{R,1d}$ and $E_{\gamma,\mathrm{iso}}$ (panel a), as well as the distributions of both quantities (panels b and c) for the bursts in our sample. The significant outliers, GRBs 030226, 970508, and 050408 have been excluded. The $E_{\gamma,\mathrm{iso}}$ has been corrected to the band pass 20 – 2000 keV in the rest frame according to the spectral parameters of prompt gamma-ray emission. The circled-crosses are the means of the two quantities for the two groups (excluding those bursts with limits). The grey area marks the parameter region in which the flux-threshold selection effect plays a dominant role. The dotted line in panel (b) is the best fit using a two Gaussian model, and the perpendicular dotted-line is the separation of the two groups.

and more importantly, some nearby GRBs appear to be different from their cosmological brethren. Based on the Bayesian theory, we propose an approach to calibrate the GRB luminosity indicators without introducing a low-redshift GRB sample. GRB luminosity indicators are generally written as $\hat{L} = c\prod x_i^{a_i}$, where c is the coefficient, x_i is the i-th observable, and a_i is its corresponding power-law index. The essential points of our approach include, (1) calibrate $\{a_i\}$ with a sample of GRBs in a narrow redshift range (Δz); and (2) marginalize the c value over a reasonable range of cosmological parameters. The selection of Δz for a particular GRB sample could be judged according to the size and the observational uncertainty of the sample. Our simulation suggests that with the current observational precisions of E_{iso}, E_p, and t_b, 25 GRBs within a redshift bin of $\Delta z \sim 0.30$ would give fine calibration to our luminosity indicator.

CONCLUSIONS

We classify the optically bright GRBs into two categories by using the intrinsic optical afterglow light curves at $t' > 2$ hours since GRB trigger. The luminous group ($\sim 75\%$ of the GRBs in our sample) is typically brighter than the dim one with a factor of ~ 30. While the luminous group ($\sim 75\%$ of the GRBs in our sample) has a wide range of redshift distribution, the bursts in the dim group all appear at a redshift lower

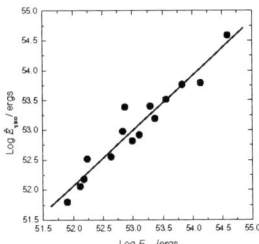 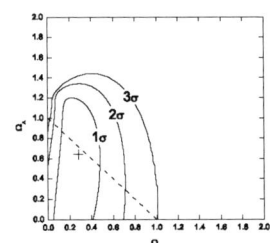 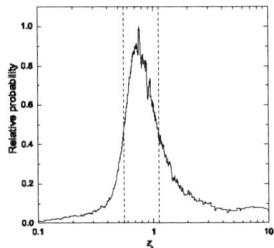

FIGURE 2. *Left*—Log $\hat{E}_{\gamma,\text{iso}}$ calculated by our relation as compared with log $E_{\gamma,\text{iso}}$ derived from the theoretical model with $\Omega_M = 0.28$ and $\Omega_\Lambda = 0.72$. The solid line is the regression line. *Middle*—The contours of likelihood interval distributions in the $(\Omega_M, \Omega_\Lambda)$-plane inferred from the current GRB sample using our method. The cross marks the most possible value of $(\Omega_M, \Omega_\Lambda)$, which is $(0.28, 0.64)$. *Right*—The smoothed likelihood function of the transition redshift from a decelerating Universe to an accelerating Universe inferred from the current GRB sample. The dashed lines mark the 1σ region, and the best value of $\hat{z}_t$ is $0.78^{+0.32}_{-0.23}$.

than 1.1. With the data at hand we find a tight relation of E_{iso} to E'_p and t'_b. Possible constraints on the cosmology parameters from the current GRB sample with this relation are analyzed. We also propose an approach to calibrate the GRB luminosity indicators without introducing a low-redshift GRB sample based on the Bayesian theory.

ACKNOWLEDGMENTS

This work is supported by NASA NNG04GD51G, a NASA Swift GI (Cycle 1) program, and the National Natural Science Foundation of China (No. 10463001).

REFERENCES

1. Zhang, B. & Mészáros, P. *International Journal of Modern Physics A*, **19**, 2385-2472 (2004)
2. Djorgovski, S. G., et al., in "Discoveries and Research Prospects from 6- to 10-Meter-Class Telescopes II", edited by Guhathakurta, Puragra. Proceedings of the SPIE 4834, pp. 238-247 (2003)
3. Bloom, J. S., Frail, D. A., & Kulkarni, S. R., *ApJ*, **594**, 674-683(2003)
4. Schaefer, B. E., *ApJ*, **583**, L67-L70(2003)
5. Dai, Z. G., Liang, E. W., & Xu, D., *ApJ*, **612**, L101-L104(2004)
6. Ghirlanda, G., et al., *ApJ*, **613**, L13-L16 (2004)
7. Friedman, A. S., & Bloom, J. S., *ApJ*, **627**, 1-25(2005)
8. Firmani, C.,Ghisellini, G., Ghirlanda,G.,& Avila-Reese, V., *MNRAS*, **360**, L1-L5 (2005)
9. Xu, D., Dai, Z. G., & Liang. E. W., *ApJ*, **633**, 603-610 (2005)
10. Liang, E. W. & Zhang, B., *ApJ*, **633**, 611-623 (2005)
11. Wang, F. Y. & Dai, Z. G., *MNRAS*, submitted (astro-ph/0512279) (2005)
12. Liang, E. W. & Zhang, B., *ApJ*, in press (astro-ph/0508510) (2006)
13. Liang, E. W. & Zhang, B., *MNARS*, submitted (astro-ph/0512177) (2005)

Optically Selected GRB Afterglows

F. Malacrino* and J-L. Atteia*

Laboratoire d'Astrophysique Observatoire Midi-Pyrénées, 14 avenue Edouard Belin, 31400 Toulouse, France

Abstract. Since November 2004, we attempt to detect GRB optical afterglows in near real-time on images taken at the Canada France Hawaii Telescope within the Very Wide Survey, component of the CFHT Legacy survey. To do so, a Real Time Analysis System automatically and quickly analyzes MegaCAM images and extracts from them a list of photometrically and astrometrically variable objects which is then validated by a member of the collaboration. Each month, we repeatedly observe 15 to 30 square degrees down to magnitude i' = 22.5. A few objects are classified as candidates and analyzed more deeply, and statistics are done showing the treatment's performance.

Keywords: gamma-ray bursts, observation and data reduction techniques, image processing
PACS: 98.70.Rz, 95.75.-z, 95.75.Mn

THE SURVEY

The Canada France Hawaii Telescope is a 3.6 m telescope located on the Mauna Kea in Hawaii. Built in the late 70's, it has been recently equipped with a new instrument, Megacam, a full square degree CCD imager.

The Very Wide Survey covers 1200 square degrees down to i' = 22.5, through 3 filters (r', g', i'). Initially conceived to discover and follow 2000 Kuiper Belt Objects, its observing strategy (Fig 1) is well suited to detect optical afterglows.

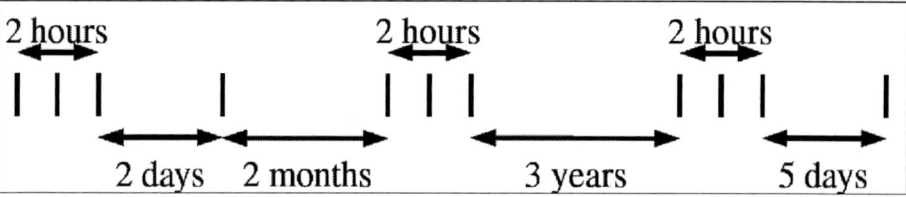

FIGURE 1. Observational strategy for one field (each vertical line is for one exposure, exposure time depends on the filter, but is typically of the order of one hundred seconds)

Each field is observed several times. This recurrence can be used to compare images between them in order to detect variable or transient objects.

---

[1] on behalf of the GRB RTAS collaboration

THE REAL TIME ANALYSIS

Catalog Process

This part of the process consists of reducing the useful information from 700 Mo to a few ten Mo and to prepare all the useful data for the comparison process. The pipeline automatically checks the presence of a new image as soon as it has been pre-processed by Elixir and starts the treatment for each of the 36 CCDs.

It consists of converting FITS images to JPEG format, extracting FITS headers and creating a catalog of objects, containing their main characteristics. These objects are then sorted according to their astrophysical properties (stars, galaxies, cosmic rays, etc..) and astrometically matched with the USNO catalog. Results are finally summarized on a web page to be checked and validated by a collaboration member. Images treated in this way can then be used in the comparison process.

Comparison Process

The goal of this process is first to list all the possible comparisons between images of the same field. To be possible, a comparison must involve images with the same filter and exposure time. In a second step, the process checks if the field has already been observed during this run, and in this case starts the comparison between the two best quality images of each night.

The comparison process classifies objects into two categories: matched and single objects. Matched objects are used for photometric inter-calibration. Asteroids are detected among single objects. At last, a search of matched objects having significant variation in luminosity is done.

All the information about variable objects and asteroids is gathered on a interactive web page. Then a member of the collaboration has to reject false detections or to validate objects as really variable. This validation process typically takes 2-3 hours for 1 run.

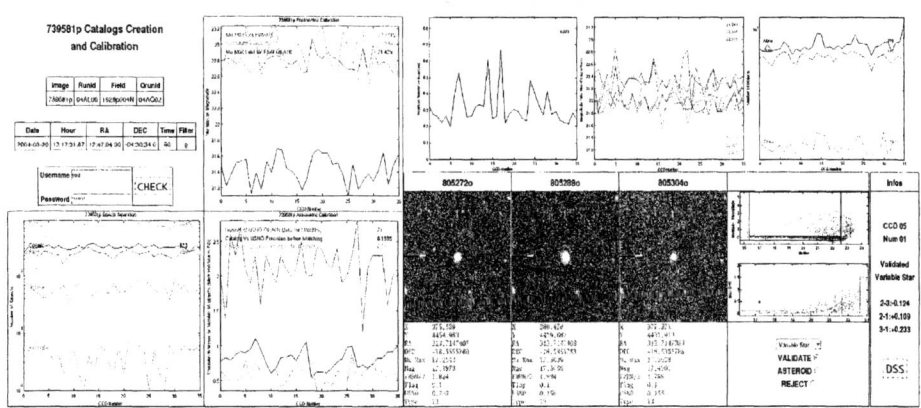

FIGURE 2. Example of a "catalog creation" web page (left) and a "triple comparison" web page (right)

THE POST ANALYSIS

At the end of each run, statistics are performed on catalogs, comparisons and validation. These statistics allow a more comprehensive view of the treatment and the possibility of comparing different runs. Ultimately the performances of the survey will be evaluated on a population of simulated GRB afterglows to constrain the rate of GRB afterglows.

Catalogs. Table 1 shows statistics for the catalog process of the MegaCAM run 05BQ07. The Very Wide survey usually works by sets of 16 or more fields, that's why images are classified by sets. We can notice that i' filter images contain twice more objects than r' filter ones, although the completness magnitudes are almost the same. It can also be noted that the last i' filter set is worse compared with the first two due to undefringe images.

TABLE 1. Statistics for 05BQ07 catalogs

Filter	Set	Date	Number of Images	Square Degrees	USNO Precision	Completness Magnitude	Objects per Square Degrees
r'	1	2005-10-25	18	16.018	0.60	22.53	15652
r'	2	2005-10-25	16	14.288	0.59	22.40	14415
r'	3	2005-10-25	17	15.165	0.60	22.34	14139
r'	4	2005-10-26	17	15.266	0.60	22.19	12613
i'	1	2005-10-26	16	14.438	0.62	22.32	28926
i'	2	2005-10-26	16	14.438	0.62	22.39	30221
i'	3	2005-10-26	17	15.341	0.62	22.10	28499

Comparisons. Using these catalogs, 32 triple comparisons have been processed, 16 for each filter. Triple comparisons involves images of the same night. r' filter triple comparisons are very impressive, with only 0.05 % of photometrically variable objects and many asteroids detected down to magnitude r' = 22.1. Unfortunately, the bad third image of the i' filter set leads to a huge number of photometrically variable objects, and this set of triple comparisons hasn't been validated by the team due to many false detections.

Triple Comparisons											
Exposure Time	Filter	Set	Date	Hour	Number of Comparisons	Square Degrees	Position Tolerance	Limit Magnitude	Objects per $°^2$	Dif Objects	Asteroids	Dif Objects for 10000 objects per $°^2$
110	r	1	2005-10-25	11:36:09.37	16	14.037	0.27	22.07	13186	93	736	5.02
180	i	2	2005-10-26	5:59:28.08	16	14.438	0.36	21.81	26010	814	446	21.68
Double Comparisons											
Exposure Time	Filter	Set	Date 1	Date 2	Number of Comparisons	Square Degrees	Position Tolerance	Limit Magnitude	Objects per $°^2$	Dif Objects	Alone Objects	Dif Objects for 10000 objects per $°^2$
110	r	1	2005-10-26	XXXX-XX-XX	1	0.902	0.35	21.99	10228	5	0	5.42
110	r	2	2005-10-26	2005-10-25	16	14.213	0.33	22.05	12289	91	11	5.21

FIGURE 3. Statistics for 05BQ07 comparisons

Double comparison statistics only concern r' filter images. The first set is insignificant, as it contains only one double comparison of 2 images of the same night. On the contrary, the second set looks like what we get in case of perfect double comparisons. Once again, the process selects approximatively 0.05% of luminosity variable objects down to magnitude r' = 22, and 11 objects which were classified as matched in the corresponding triple comparison, and are not present in the image taken one day later. The behaviour of these objects looks like the expected behaviour of GRB afterglows, showing that we can indeed detect GRB afterglows.

Validation. In the last part, we combine triple and double comparisons, and their corresponding validations, in order to compute statistics for asteroids and variable objects per field, as we consider that fields observed 2 days apart are not independent.

Exposure Time	Filter	Nature	Total Number	Rejected	Validated	Cosmic	CCD Defect	CCD Edge	Seeing	Contaminated	Faint	Other	Galaxy	Variable Star	TNO	Candidate
									Variable Objects							
110	r	Triple+Double	181	160	21	15	19	7	98	20	0	2	1	17	0	2

FIGURE 4. Statistics for 05BQ07 variable objects validation

Only 12% of objects classified as variable by the process are really variable, and most of them are variable stars. More than 50% of the detections are due to seeing variations, despite corrections applied by the process. 5 interesting candidates have been identified which were all discarded by comparison with previous images of the field.

CONCLUSION

We have shown that optical afterglow detection is possible using Megacam images from the Very Wide survey at the CFHT and a real time comparison pipeline. About 1 to 10 afterglows per year can be reasonably expected. The real time process has started one year ago, and 12 runs have been succesfully processed, but unfortunately without any strong afterglow candidate. Now, we have started the complete reprocessing of the whole Very Wide survey runs since the beginning of the legacy survey in April 2003. This study will allow us to compute a huge database of objects which will be very useful to plot luminosity function of the most interesting objects, and to strongly constrain the rate of orphan GRB afterglows and their beaming factor. Our work is continuing with new images taken every month, and we recently submitted a proposal for the next semester, which is specifically dedicated to the search of orphan GRB afterglows and will have a efficiency 2-3 times greater than the Very Wide Survey for this task.

Please visit http://www.cfht.hawaii.edu/~grb for more informations on this research.

ACKNOWLEDGMENTS

The authors gratefully acknowledge the support of the CFHT staff for the steady operation of the RTAS.

Afterglow calculation in the Blandford-Lyutikov electromagnetic model of GRBs

F. Genet, F. Daigne and R. Mochkovitch

Institut d'Astrophysique de Paris
UMR 7095 CNRS – Université Pierre et Marie Curie-Paris VI, 98 Bd Arago, 75014 Paris, France

Abstract. We compute the afterglow of GRBs produced by electromagnetic outflows as proposed by Lyutikov and Blandford (2003). Using a simple model for the injection of electromagnetic energy to the forward shock we obtain the afterglow evolution both during the period of activity of the central source and after. Our method applies even to a variable central source.

Keywords: Gamma-ray bursts, Shock waves
PACS: 98.70.Rz, 96.50.Fm

INTRODUCTION

Lyutikov and Blandford [1] have recently proposed an alternative to the standard fireball model where the central engine produces a purely electromagnetic outflow instead of a relativistic baryonic wind. Observationally this electromagnetic model (hereafter EMM) differs from the standard internal/external shock model by the absence of any reverse shock contribution, a different early afterglow evolution and a high polarization of the prompt emission [2]. Here we concentrate on the early afterglow (while the central source is still active) and compare the EMM to the standard model in X-rays and the visible for a uniform external medium or a wind environment.

DYNAMICS OF THE FORWARD SHOCK

We describe the evolution of the forward shock by writing the conservation of energy-momentum of the swept-up mass as it receives electromagnetic energy from the central source

$$E + Mc^2 = M\Gamma\Gamma_i c^2$$
$$E = M\beta\Gamma\Gamma_i c^2 \quad (1)$$

where M is the swept-up mass, β and Γ the velocity and Lorentz factor for the bulk motion of the shocked material and Γ_i the Lorentz factor for internal motions. The energy E received from the source reads

$$E = \int_0^t L_{\rm EM}(t) dt \quad (2)$$

where $L_{EM}(t)$ is the source electromagnetic power, t being the time in the source rest frame (which is also the observer time modulo the $(1+z)$ factor for time dilation). After elimination of Γ_i in (1) we get

$$2Mc^2\Gamma^2 = E + Mc^2 \sim E \tag{3}$$

Differentiation of (3) with respect to observer time yields

$$L_{EM}(t) = 4\mu c^2 \left(R^\varepsilon \Gamma \frac{d\Gamma}{dt} + \varepsilon c R^{\varepsilon-1} \Gamma^4 \right) \tag{4}$$

where we have written

$$M(R) = \mu R^\varepsilon \tag{5}$$

with $\mu = \frac{4\pi}{3}\rho$ and $\varepsilon = 3$ for a uniform medium of density ρ and $\mu = 4\pi A$ and $\varepsilon = 1$ for a wind environment with $\rho = A/R^2$. With the additional relation between observer time and shock radius

$$\frac{dR}{dt} = 2c\Gamma^2 \tag{6}$$

the problem can be solved for any law $L_{EM}(t)$. With a constant L_{EM} it can be easily shown that the solutions of (4) and (6) are

$$\Gamma = (Q/2)^{-1/2}(ct)^{-1/4} \quad \text{and} \quad R = 2Q(ct)^{1/2} \quad \text{with} \quad Q = \left(\frac{3L_{EM}}{32\pi\rho c^3}\right)^{1/4} \tag{7}$$

for a uniform medium while for a stellar wind

$$\Gamma = \left(\frac{L_{EM}}{16\pi A c^3}\right)^{1/4} \quad \text{and} \quad R = 2c\Gamma^2 t. \tag{8}$$

AFTERGLOW CALCULATION

Method

Using the dynamical evolution of the forward shock obtained above we have calculated afterglow lightcurves in X-rays and the visible for both the electromagnetic and the standard model. We obtain the two critical frequencies v_m and v_c following [3]. In the EMM, for a uniform medium and a constant source the ratio v_m/v_c is fixed

$$\frac{v_m}{v_c} = 0.065\, \varepsilon_{e,-1}^2\, \varepsilon_{B,-3}^2\, n L_{52} \tag{9}$$

where n is the density in cm^{-3}, L_{52} the electromagnetic power in units of 10^{52} erg.s^{-1}, $\varepsilon_e = 0.1 \times \varepsilon_{e,-1}$ and $\varepsilon_B = 0.001 \times \varepsilon_{B,-3}$. While the source is active, the cooling regime – fast or slow – therefore remains unchanged. We have adopted the following choice of parameters: $\varepsilon_{e,-1} = \varepsilon_{B,-3} = L_{52} = 1$ and $n = 1$ or $A_* = 0.1$ in the uniform medium or wind case respectively. With these parameters the afterglow is always in the slow

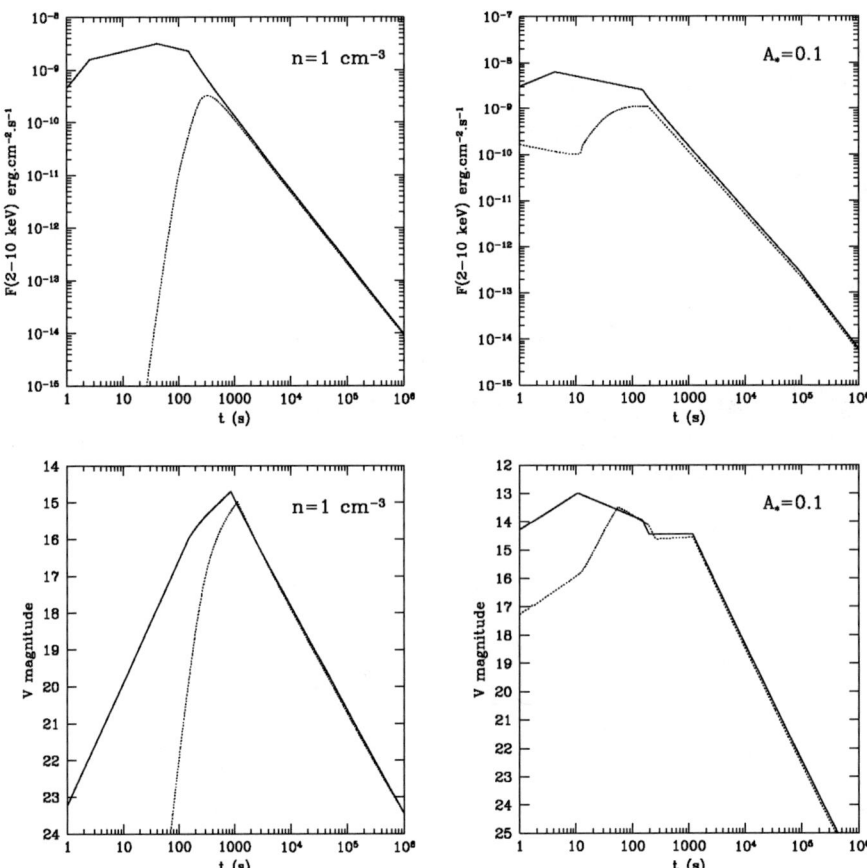

FIGURE 1. Afterglow comparison between the EMM (full line) and the standard model (dotted line). Top: X-ray band (2 - 10 keV); Bottom: V band; Left column: uniform medium of density $n = 1$ cm^{-3}; right column: stellar wind with $A_* = 0.1$.

cooling regime in the uniform medium. In the wind case, it starts in fast cooling while the source is active and moves to slow cooling shortly after.

Results

We have represented in Fig.1 afterglow light curves in X-rays (2 - 10 keV) and the visible for a uniform external medium and a stellar wind. For comparison we also show in Fig.1 standard afterglows computed in the internal/external shock model with the same total injected energy. A duration of 50 s for the phase of energy injection and a redshift $z = 2$ have been assumed. In all cases the afterglows are brighter at early times in the EMM while at late times (in the Blandford-McKee regime) the models coincide. The difference in early evolution is larger for a uniform external medium, especially in X-rays. In the EMM, the X-ray afterglow contribution is already bright at the beginning of the burst while it takes some 100 s to build up in the standard model.

CONCLUSION

The lightcurves in Fig.1 show that the EMM and the standard model notably differ at early times (essentially during the period of source activity). However relying on these differences alone to identify the physical origin of GRBs appears to be a difficult task which first requires a very early follow-up of the afterglow. In X-rays, SWIFT should be able to do that (at least in some cases) but the problem will come here from the mixing of the afterglow contribution with the brighter prompt emission component. In the visible, where the burst prompt emission is (probably) negligible, the EMM predicts a brighter afterglow (for a given set of parameters ε_e, ε_B, n or A_*). But in real afterglows these parameters are not known a priori and deciding between models will be tricky. Polarization properties of the burst prompt emission [2] when they become more easily accessible may provide a clearer evidence.

REFERENCES

1. Lyutikov, M. & Blandford, R.D., 2003, astro-ph/0312347.
2. Lyutikov, M., 2004, 35th COSPAR Meeting, astro-ph/0409489.
3. Sari, R., Piran, T. & Narayan, R., 1998, ApJ 497, L17.

Interaction of Electromagnetic Dominated Outflow with Inhomogeneous Ambient Medium

Koichi Noguchi[*,†] and Edison Liang[*]

[*]*Department of Physics & Astronomy, Rice University, Houston, TX 77005, USA*
[†]*T-15, Los Alamos National Laboratory, Los Alamos, NM 87545*

Abstract.
The effect of background plasma on particle acceleration via Poynting fluxes is studied in 3D PIC simulation of electron-positron and ion-electron plasmas. When strongly magnetized plasma at the center expands to background low-temperature electron-positron background, electromagnetic wave front accelerates particles in background ambient plasma and clump, and captures them in the Ponderomotive potential well. In electron-positron background case, the acceleration continues for the entire simulation, whereas strong charge separation between ions and electrons prevents particles from acceleration in the ion-electron case

Keywords: gamma rays:bursts—PIC—methods:numerical—relativity
PACS: PACS 52.65.-y, 52.65.Rr, 52.30.-q

INTRODUCTION

There are two competing paradigms for the origin of the prompt gamma-ray burst (GRB) emissions: hydrodynamic internal shocks [1, 2] versus Poynting fluxes[3]. Both pictures require the rapid and efficient acceleration of nonthermal electrons to high Lorentz factors in moderate magnetic fields to radiate gamma-rays. In the Poynting flux scenario, long-wavelength electromagnetic (EM) energy can be directly converted into gamma-rays using the electrons or electron-positron pairs as radiating agents. Such Poynting flux may originate as hoop-stress-supported magnetic jets driven by strongly magnetized accretion onto a nascent blackhole, or as transient millisecond magnetar winds, in a collapsar event[4] or in the merger of two compact objects[5]. The recent discovery of the diamagnetic relativistic pulse accelerator (DRPA)[6, 7], in which intense EM pulses imbedded in an overdense plasma (EM wavelength $\lambda \gg$ plasma skin depth c/ω_{pe}) capture and accelerate surface particles via sustained in-phase Lorentz forces when the EM pulses try to escape from the plasma, is particularly relevant to the Poynting flux scenario of GRBs. DRPA may be launched when the magnetic jet head or the magnetar stripe wind emerge from the stellar envelope surface due to the sudden deconfinement of the magnetic field and embedded e+e-plasma. Noguchi et. al.[8] recently showed that the mechanism is robust even with the radiation damping force.

In this article we report 3D PIC simulations of particle acceleration driven by electromagnetic expansion (Poynting flux) with background ambient medium and low-density clump with newly developed 3D PIC code, and we show the power spectrum and radiation strength from each particles.

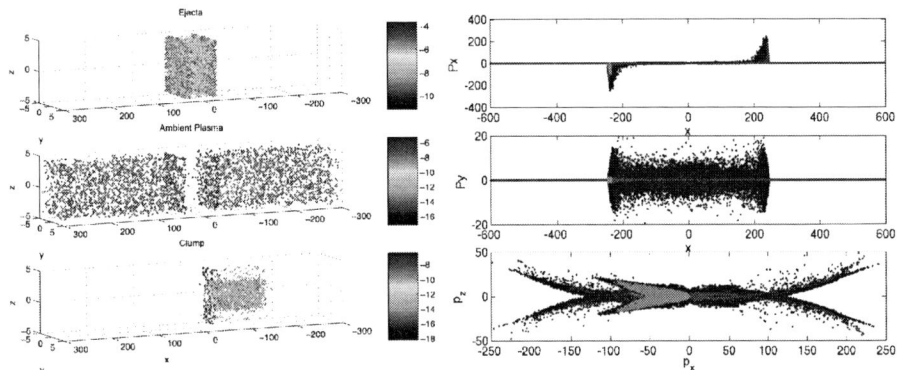

FIGURE 1. Left: The spatial distribution of particles in ejecta (top), ambient medium (middle) and clump (bottom) at $t\Omega_{ce} = 650$. Right: The phase plot of particles at $t\Omega_{ce} = 12000$ with $P_x - x$ (top), $P_y - x$ (middle) and $P_z - P_x$ (bottom).

INITIAL SETUP OF THE SIMULATION

We use the 3D explicit PIC simulation scheme based on the explicit leap-frog method for time advancing [9]. Spatial grids for the fields are uniform in both all directions, $\Delta x = \Delta y = \Delta z = c/\omega_{pe}$, where ω_{pe} is the electron plasma frequency. The simulation domain is $-600\Delta x \le x \le 600\Delta x$, $-5\Delta y \le y \le 5\Delta y$ and $-5\Delta z \le z \le 5\Delta z$ with periodic boundary conditions in each direction.

Following Noguchi et. al.[8], the initial electron-positron ejecta is uniformly distributed at the center of the simulation box, $-6\Delta x < x < 6\Delta x$, $-5\Delta y \le y \le 5\Delta y$ and $-5\Delta z \le z \le 5\Delta z$. The background uniform magnetic field in the y direction is applied only in the same region, so that the magnetic field freely expands toward the ambient plasma regions, $x > 6\Delta x$ and $x < -6\Delta x$ with accelerating plasma.

The clump and the ambient plasma consist of either electron-positron or electron-ion. The ratio between the number density of ejecta ρ_{ej} and the ambient plasma ρ_{am} is 0.01, whereas the clump ρ_{cl} is 0.1. The clump is a $100\Delta x \times 6\Delta y \times 6\Delta z$ column, whose center is located at $(-60\Delta x, 0, 0)$ so that the distance between the ejecta and the column is $4\Delta x$. The remained volume is filled with the ambient plasma.

The initial temperature of ejecta is assumed to be a spatially uniform relativistic Maxwellian, with 1MeV. The temperature of the clump and the ambient plasma is also uniform Maxwellian with the temperature 100eV.

We choose $m_i = 100m_e$ and $\omega_{pe}/\Omega_{ce} = 0.1$ in all simulations, where Ω_{ce} is the electron cyclotron frequency.

RESULTS

First, we study the electron-positron background case. Figure 1 shows the spatial distribution of particles at $t\Omega ce = 650$ and the phase plot at $t\Omega_{ce} = 12000$. The color of

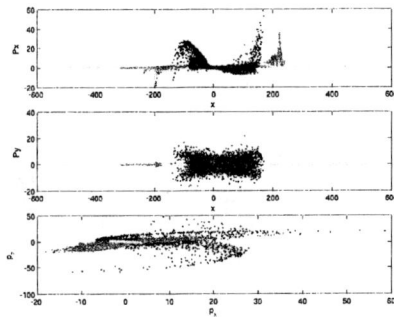

FIGURE 2. The phase plot of the ion-electron clump and ambient plasma case at $t\Omega_{ce} = 10000$ with $P_x - x$ (top), $P_y - x$ (middle) and $P_z - P_x$ (bottom).

each particles represents the magnitude of estimated radiation damping force using the relativistic dipole formula [10]

$$\langle P \rangle = \frac{2e^2}{3m^2c^3}(F_\parallel^2 + \gamma^2 F_\perp^2), \qquad (1)$$

where $F_\parallel$ and $F_\perp$ are the parallel and perpendicular components of the force with respect to the particle's velocity. As the ejecta expands, electric field is automatically generated in the z direction, and particles in the clump and ambient plasma are also accelerated relativistically in the direction of Poynting vector due to the ponderomotive force. Particles are also accelerated by the electric field, which expands the clump in the z direction.

At $t\Omega_{ce} = 12000$, the clump is swept by the EM field, and acceleration still continues. The highest γ in the ejecta is around 250, whereas $\gamma \simeq 100$ in the background and the clump. As we mentioned, there is no acceleration in the y direction, and the momentum distribution in the y direction does not change. Since all particles in this run are either electrons or positrons, there is no chage separation occurs and acceleration continues.

Figure 2 shows the phase plot of the ion-electron clump and ambient plasma case at $t\Omega_{ce} = 10000$. Different from the electron-positron case, the acceleration by the ponderomotive force is strongly reduced by the charge separation. Especially with the clump, the EM field is too weak to accelerate particles, and particles are bounced back to the center. However, particles in the wavefront ($x \simeq \pm 300$) are still get accelerated.

Next, we compare the power spectrum of electrons in both runs at $t\Omega_{ce} = 3000$. Figure 3 (a) shows the electron-positron case and (b) shows the ion-electron case with the clump ($x < 0$) and without the clump ($x > 0$). In the electron-positron case, the existence of clump slows down the acceleration of the ejecta, and shifts the peak of the ambient plasma from $1.3 m_e c^2$ to $0.9 m_e c^2$.

The ion-electron case also shows that the peak of the ambient plasma shifts to the lower energy, and the acceleration of the ejecta is severely reduced because of the charge separation between ions and electrons even though the ejecta consists of electrons and positrons.

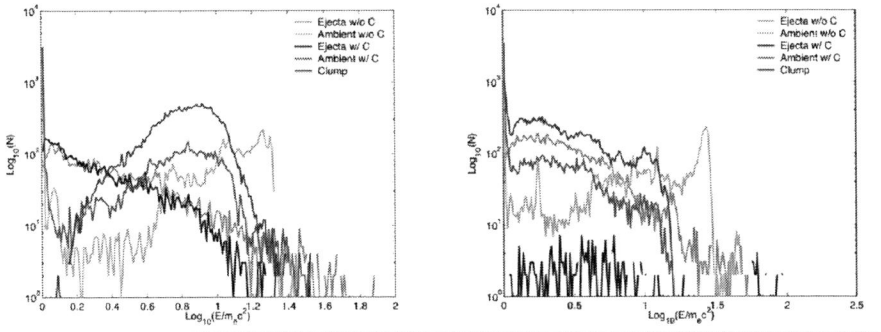

FIGURE 3. The power spectrum of electrons in electron-positron case (left) and ion-electron case (right).

SUMMARY

We studied the effect of electron-positron and ion-electron background ambient plasma on particle acceleration via Poynting fluxes. With electron-positron ambient plasma, the acceleration mechanism is still robust, and particles in the background and clump are also accelerated. With the ion-electron case, however, the acceleration is suppressed due to the charge separation in the background and clump plasma. If the density of interstellar medium near the ejecta is more than 10%, acceleration by the Poynting flux may not enough to create very high energy tail.

ACKNOWLEDGMENTS

This research is partially supported by NASA Grant No. NAG5-9223, NSF Grant No. AST0406882, and LLNL contract nos. B528326 and B541027. The authors wish to thank ILSA, LANL, B. Remington and S. Wilks for useful discussions.

REFERENCES

1. T. Piran, *Phys. Repts.* **33**, 529 (2000)
2. P. Mészáros, *Annual Review of Astronomy and Astrophysics*, **40**, 137 (2002)
3. M. Lyutikov, and E. G. Blackman, *MNRAS*, **321**, 177 (2001)
4. W. Zhang, S. E. Woosley, and A. I. MacFadyen, *ApJ*, **586**, 356 (2003)
5. M. Ruffert, and H-Th. Janka, *Gamma-Ray Burst and Aftergrow Astronomy*, AIP Conference Proceedings, **662**, 193 (2003)
6. E. Liang, K. Nishimura, H. Li, and S. P. Gary, *Phys. Rev. Lett.*, **90**, 085001 (2003)
7. E. Liang, and K. Nishimura, *Phys. Rev. Lett.*, **92**, 175005 (2004)
8. K. Noguchi, E. Liang, and K. Nishimura, *Nuovo Ciment C*, **028**, 381 (2005)
9. C. K. Birdsall,and A. B. Langdon, *Plasma Physics via Computer Simulation*, McGraw-Hill, 1985
10. G. B. Rybicki, and A. P. Lightman, *Raidiative Processes in Astrophysics*, Wiley-interscience, New York, 1979

Search for Correlations Between BATSE Gamma-Ray Bursts and Supernovae

J. Polcar*, M. Topinka*, D. Nečas†, F. Hroch**, R. Hudec*, V. Hudcová*, G. Pizzichini‡, N. Masetti‡ and E. Palazzi‡

*Astronomical Institute of the Academy of Sciences of Czech Republic
†Brno University of Technology, Faculty of Civil Engineering, Czech Republic
**Masaryk University Brno, Faculty of Science, Czech Republic
‡C.N.R. Instituto di Astrofisica Spaziale e Fisica Cosmica - I.A.S.F./CNR, Bologna, Italy

Abstract. We report on complex statistical research of space-time correlated supernovae and CGRO-BATSE gamma-ray bursts. We show that there exists a significantly higher abundance of core-collapse supernovae among the correlated supernovae, but the subset of all correlated objects does not seem to be physically different from the whole set.

Keywords: gamma ray bursts:supernovae:correlations:statistics
PACS: 95.85.Pw:97.60.Bw:97.10.Yp

INTRODUCTION

The origin and source of gamma-ray bursts (GRBs) still remain a puzzle. This tremendous energy could be released on a short time-scale during a collapse of a massive star in a supernova-like explosion. There are several pieces of observational evidence supporting the connection between GRBs and supernovae (SNe): some GRBs reveal an underlying SN in an optical afterglow lightcurve (e.g. GRB980326), sometimes supported by spectral and colour signatures (e.g. GRB030329 and SN2003dh), and there are a few objects with space-time coincidence (e.g. GRB980425 and SN1998bw).

We have used the current CGRO-BATSE GRB catalogue (up to date version to April 2004) [2] and combined data from Asiago Padova [4], Harvard [3] and Sternberg [1] catalogues of SNe.

PSEUDOTYPES

There is a wide range of the SN types existing in the input data. For easier manipulation we have divided all the SNe into several types, called *pseudotypes*, see Table 1. Note that the division is not disjunct. For the purposes of our study it is useful to simplify this division into three pseudotypes only: the `dwarf`, the `core` and the `unknown` SNe.

SPACE AND TIME CORRELATION

The situation is easy in the space domain, A GRB and a SN make a pair if the location of the SN falls into the GRB errorbox which yield 3σ error.

TABLE 1. Pseudotypes clasification.

pseudotype	catalogue clasification
unknown	'NULL', '?'
core	'ib', 'ib?', 'ic', 'ic?', 'ic', 'ib/c', 'ic-p', 'ii', 'ii?', 'iib', 'iib?', 'iic', 'iic?', 'ii-n', 'ii-n?', 'ii-l', 'ii-p', 'ii-p?', 'iipec'
dwarf	'ia' ,'ia?', 'ia-p', 'ia-p?', 'i', 'i?', 'i-p', 'i-l'

In the time domain it is not a priori theoretically clear what is the time delay between the SN and an eventual gamma emission. For our purposes we postulated the time of the optical emission to be the time of the maximum T_{max} of the optical lightcurve of the SN. Unfortunately only a fraction of the SNe provides an information about the date of T_{max}. We solved the problem statistically and we assumed the time delay between T_{max} and the time of the discovery T_{disc} to be the median $M_{max} = -4$ days. We get it from the distribution of all known time delays. Thus for the SNe which have the time of the maximum missing we computed it as

$$T_{max} = T_{disc} + M_{max}.$$

The time delay T_δ between the possible prompt gamma emission T_γ of the SN and its optical emission T_{max} may depend on the type of the SN. We respect it by the division into the pseudotypes. We represent the time delay by the shift in time T_δ. Generally it could be written as

$$T_\gamma = (T_{max} + \mu_\gamma) \pm 3\sigma_\gamma.$$

We tried to make the best possible estimate. The typical time delay T_δ can be estimated for the dwarf SN as a relatively narrow time interval $+20 \pm 7$ days [5]. For the core-collapse SNe we choose the typical values of the time delay to be 0 ± 30 days. For the SNe of unknow or uncertain type we used a statistical estimation of the time-shift and the width of the time error. Other uncertainty is the method in the T_{max} estimate. For each pseudotype we assume the Gaussian distribution of the time delays $T_\gamma - T_{max}$. We calculated the weighted average of the particular densities of the probability with respect to the level of uncertainty we had. The matching process itself reflects the effect of the selection of the SNe into the pseudotypes, it results into the different size of the time error used for each pseudotype. The larger time-errorbox of the particular pseudotype we have the greater number of matched pairs of this kind we expect. To compare this effect it is useful to do the matching separately for two different sets of settings: $\mathbb{A}$ for the one which is dependent on the pseudotype of the SN and $\mathbb{B}$ for which is not.

RESULTS OF THE MATCHING

We obtained 92 of possibly correlated pairs in the case $\mathbb{A}$ and 127 in the case $\mathbb{B}$. The sky maps illustrating the results are plotted in Figure 1.

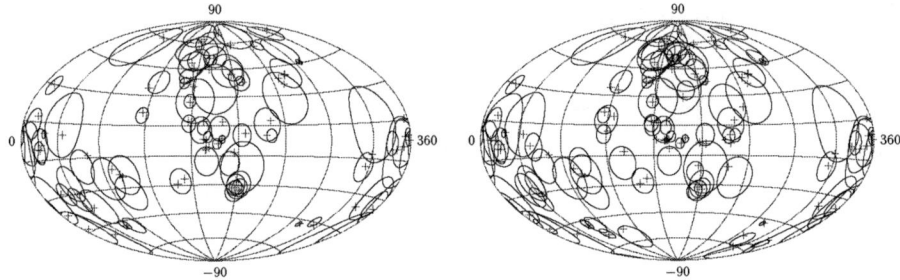

FIGURE 1. Results of the matching of possibly correlated pairs in the case A (left) and in the case B.

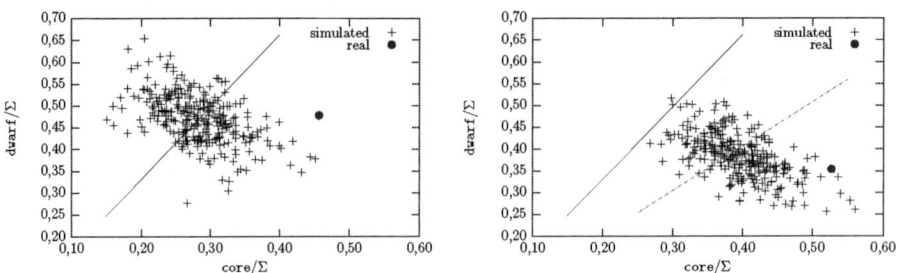

FIGURE 2. Fraction of pseudotype core for case A (left) and for case B.

RELATIVE ABUNDANCE OF PSEUDOTYPES

We studied the spectrum of SN pseudotypes in the set of possibly correlated SNe with respect to all SNe. We test if the relative abundance of the core SNe differs from the whole SN catalogue.

To exclude the various possible selection effect and to prove the higher number of the core SNe among the coinciding pairs we compare the results of the real match with artificial random SN and GRB distribution we can generate using the rotation method. We rotated the SN catalogue around arbitrary axes. We choose 295 different rotation both for A and B to get 295 different sets of artificial correlated pairs.

In the case of artificially build catalogues we see a significant difference between the A and B sets of parameters (Figure 2). It is due to the different size of the time window.

However the relative abundance of the core SNe in the real match still remains different from the distribution of the relative abundances in the random cases. The distribution of the abundances of the core SNe in the artificial data set is Gaussian up to 97.8% in the meaning of Kolmogorov–Smirnov test and the relative abundance of the core SNe in the real match is out of 3σ interval.

ISOTROPIC ENERGY EQUIVALENT

The peak of the distribution of GRB isotropic energy equivalent (IEE) radiated in gamma-rays for matched pairs is similar to the IEE of SN1998bw 8×10^{47} ergs. It could result that there are other unrevealed SN1998bw-like among the BATSE GRBs, but the IEE distribution for the real match is not significantly different from the one for the random match. It proves that the number of the pairs matched due to the physical link is below our level of resolution by such a test. It could also mean that coincidence between SN1998bw and GRB980425 was just accidental.

The IEE of matched GRBs is four orders of magnitude lower than the average GRB IEE, which means that assuming their connection with the matched SNe we most probably underestimate their distances and thus the connection is not real for most of the matched pairs.

CONCLUSION

We have carried out a complex analysis of space-time correlated SNe and CGRO-BATSE GBRs. We have compared count of matched possible correlated pairs SN-GRB with a random one and have estimated the detectable fraction of physically and random correlated pairs. There exists a significantly higher abundance of core-collapse supernovae among the correlated supernovae, but the subset of all correlated objects does not seem to be physically different from the whole set. Even the IEE of GRB980425 fits well into IEE distribution it is not relevant according to the tests we did.

ACKNOWLEDGMENTS

We acknowledge the support by the Grant Agency of the Academy of Sciences of the Czech Republic, grant A3003206 and 205/03/H144.

REFERENCES

1. http://www.sai.msu.su/sn/
2. http://cossc.gsfc.nasa.gov/batse/index.html
3. http://cfa-www.harvard.edu/iau/lists/Supernovae.html
4. http://merlino.pd.astro.it/~supern/snean.txt
5. Petschek, A. G., Supernovae, Springer Verlag New York 1996
6. Vietri, M. and Stella, L., 1998 ApJ, 507, L45
7. MacFayeden, A.I. and Woosley, S.E., 1999, ApJ, 524, 262

GRB Supernova Luminosities – Correcting for the Host Extinction

A. Zeh[*], C. Riddle[†], S. Klose[*], D. A. Kann[*] and D. H. Hartmann[†]

[*]*Thüringer Landessternwarte Tautenburg, Sternwarte 5, D-07778 Tautenburg, Germany*
[†]*Department of Physics & Astronomy, Clemson University, 118 Kinard Laboratory, Clemson, SC 29634-0978, USA*

Abstract. We present the results of a systematic analysis of the world sample of optical/near-infrared afterglow light curves observed in the pre-*Swift* era by the end of 2004. We analyzed this sample in an attempt to detect the predicted underlying supernova light component. To gain statistical insight on the phenomenological properties of these supernova, we have also corrected this sample now for the visual extinction in their host galaxies.

Keywords: Gamma-Ray Bursts, Supernovae
PACS: 98.70.Rz

INTRODUCTION

The physical association between long-duration Gamma-Ray Bursts and supernovae, observationally first revealed for GRB 980425/SN 1998bw [1] and GRB 980326 [2], received strong confirmation from spectroscopic observations of the afterglow of GRB 030329/SN 2003dh [e.g. 3, 4], GRB 021211 [5], GRB 031203/SN 2003lw [6], and XRF 020903 [7]. Here we report on further results of our study [8, 9] to search for and to analyze extra light in GRB afterglows in an attempt to detect the predicted underlying supernova light component and to gain statistical insight of its phenomenological properties.

Our input sample of afterglows through the end of 2004 with enough data points and established redshifts contains 37 bursts (GRB 970228 - GRB 041006). We fit the light curves as the sum of an afterglow, an underlying host galaxy, and a supernova component. The latter is modeled using published multicolor light curves of SN 1998bw [1] as a template by shifting them to the corresponding redshift of the burst (see [8]). We parameterized these light curves by introducing a parameter k that shifts the light curve up and down (in luminosity), and a parameter s that stretches the light curve in time (light curve shape). The fit equation for the magnitude of the optical transient is

$$m(t) = -2.5 \log\{10^{-0.4 m_c}[(t/t_b)^{\alpha_1 n} + (t/t_b)^{\alpha_2 n}]^{-1/n} + k \times 10^{-0.4 m_{\rm SN}(t/s)} + 10^{-0.4 m_{\rm host}}\}. \tag{1}$$

The parameters in Equation (1) are the prebreak decay slope α_1, the postbreak decay slope α_2, the break time t_b, the sharpness of the break n, the brightness of the host galaxy $m_{\rm host}$, the constant m_c which corresponds to the magnitude of the fitted light curve for the case $n = \infty$ at the break time t_b (the intersection point of the prebreak and the postbreak slope, without considering a smooth transition), as well as the supernova parameters k and s, which indicate the luminosity ratio and the stretch factor normalized to SN 1998bw. Before the fitting process the data were corrected for Galactic extinction using the maps provided in [10]. For a more detailed description of our procedure we refer to [8].

In the present study we take into account the host extinction derived by Kann et al. [11] from a fit of the spectral energy distribution of all afterglows with sufficient data in the pre-*Swift* era. In doing so, here we derive the GRB supernova luminosities corrected for extinction in the GRB host galaxies.

RESULTS AND DISCUSSION

By the end of 2004, there were 12 optical afterglows with known redshifts that showed evidence for extra light in the R band at later times (GRBs 970228, 990712, 991208, 000911, 010921, 011121, 020405, 021211, 020903, 030329, 031203, 041006). For eight afterglows with a known supernova bump (GRBs 991208, 000911, 010921, 011121, 020405, 021211, 030329, 041006) Kann et al. [11] were able to derive a host extinction value, A_V(host), which we used here in order to determine the extinction-corrected luminosities of these supernovae (Fig. 2). Note that this procedure could not be performed for all supernovae due to a lack of multicolor afterglow data.

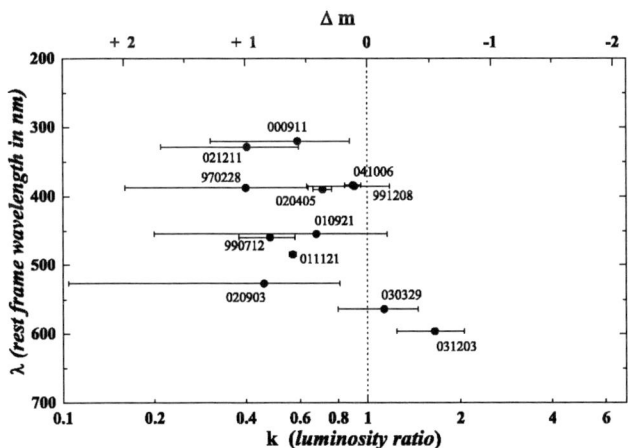

FIGURE 1. The luminosity ratio k of the individual GRB supernovae in the observer frame (R band) without correction for the host extinction. The ordinate shows the center of the R band in the corresponding supernova host frame. By definition, $k = 1$ for SN 1998bw. During the fit the stretch parameter s (Eq. 1) was assumed to be a free parameter.

While a correction for the host extinction is the only way to get insight of the real supernova luminosity distribution, the application of this procedure is limited by the requirement to have reliable multicolor afterglow data in order to extract a host extinction value. In addition, any such correction introduces additional errors to the

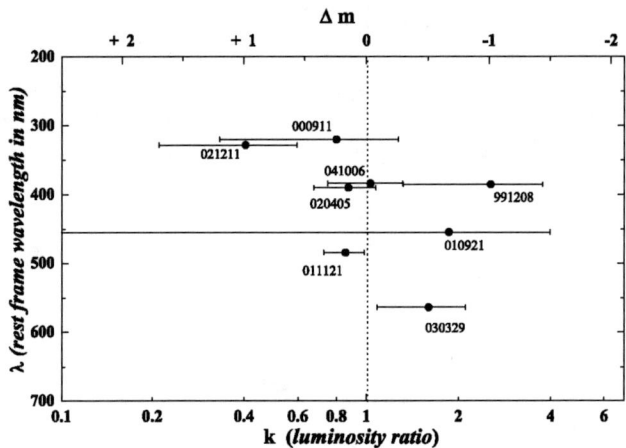

FIGURE 2. The same aus Figure 1 after correction for the host extinction. Note that for SN 1998bw we assumed A_V(host) = 0 mag.

derived individual supernova luminosities. This naturally limits any conclusions one can draw from the presently available small data set.

With some caution one tends to interprete Figure 2 as evidence for a concentration of the extinction-corrected supernova luminosities around $k=1$. In particular, in the present data base there is no strong evidence for events that are much more luminous than SN 1998bw, supporting numerical investigations of this class of supernovae [13]. However, more data are required in order to put this observational evidence on a reliable statistical basis.

REFERENCES

1. Galama, T. J., et al. 1998, Nature, 395, 670
2. Bloom, J. S., et al. 1999, Nature, 401, 453
3. Hjorth, J., et al. 2003, Nature, 423, 847
4. Matheson, T., et al. 2003, ApJ 599, 394
5. Della Valle. M., et al. 2003, A&A 406, L33
6. Malesani, D., et al. 2004, ApJ 609, L5
7. Soderberg, A. M., et al. 2005, ApJ, 627, 877
8. Zeh, A., Klose, S., Hartmann, D. H. 2004, ApJ, 609, 952
9. Zeh, A., Klose, S., Hartmann, D. H. 2005, 22nd Texas Symp. (astro-ph/0503311)
10. Schlegel, D., Finkbeiner, D., & Davis, M. 1998, ApJ, 500, 525
11. Kann, D. A., Klose, S. Zeh, A. 2006 ApJ, accepted (astro-ph/0512575)
12. Stanek, K. Z., et al. 2005, ApJ, 626, L5
13. Woosley, S. 2005, priv. comm.

The frontier of darkness: the cases of GRB 040223, GRB 040422, GRB 040624

P. D'Avanzo[*,†], P. Filliatre[**,‡], P. Goldoni[**,‡], L. A. Antonelli[§], S. Campana[*], G. Chincarini[¶], S. Covino[*], A. Cucchiara[∥], M. Della Valle[††], A. De Luca[‡‡], S. Foley[§§], D. Fugazza[*], N. Gehrels[¶¶], D. Götz[‡‡], L. Hanlon[§§], G. L. Israel[§], D. Malesani[***], B. McBreen[§§], S. McBreen[†††], S. McGlynn[§§], S. Mereghetti[‡‡], L. Moran[‡‡‡], J. A. Nousek[∥], R. Perna[§§§], L. Stella[¶¶¶¶] and G. Tagliaferri[*]

[*]*INAF, Osservatorio Astronomico di Brera, via E. Bianchi 46, I-23807 Merate (LC), Italy*
[†]*Università Insubria, Dipartimento di Fisica e Matematica, via Valleggio 11, I-22100 Como, Italy*
[**]*Laboratoire Astroparticule et Cosmologie, UMR 7164, 11 place Marcelin Berthelot, F-75231 Paris Cedex 05, France*
[‡]*Service d'Astrophysique, CEA/DSM/DAPNIA/SAp, CE-Saclay, Orme des Merisiers, Bât. 709, F-91191 Gif-sur-Yvette Cedex, France*
[§]*INAF, Osservatorio Astronomico di Roma, via Frascati 33, Monteporzio Catone, I-00040 Roma, Italy*
[¶]*Università degli studi di Milano-Bicocca, Dipartimento di Fisica, Piazza delle Scienze 3, I-20126 Milano, Italy*
[∥]*Department of Astronomy and Astrophysics, Pennsylvania State University, 525 Davey Laboratory, University Park, PA 16802*
[††]*INAF, Osservatorio Astrofisico di Arcetri, largo E. Fermi 5, I-50125 Firenze, Italy*
[‡‡]*INAF - Istituto di Astrofisica Spaziale e Fisica Cosmica di Milano, via E. Bassini 15, I-20133 Milano, Italy*
[§§]*Department of Experimental Physics, University College Dublin, Dublin 4, Ireland*
[¶¶]*NASA Goddard Space Flight Center, Code 661, Greenbelt, MD 20771*
[***]*International school for advanced studies (SISSA/ISAS), via Beirut 2-4, I-34014 Trieste, Italy*
[†††]*Astrophysics Missions Division, Research Scientific Support Department of ESA, ESTEC, Noordwijk, The Netherlands*
[‡‡‡]*Department of Physics & Astronomy, University of Southampton, Southampton, SO17 1BJ, United Kingdom*
[§§§]*Department of Astrophysical and Planetary Sciences, University of Colorado at Boulder, 440 UCB, Boulder, CO, 80309, USA*
[¶¶¶¶]*INAF, Osservatorio Astronomico di Roma, via Frascati 33, Monteporzio Catone, 00040 Rome, Italy*

Abstract. Understanding the reasons for the faintness of the optical/near-infrared afterglows of the so-called dark bursts is essential to assess whether they form a subclass of GRBs, and hence for the use of GRBs in cosmology. With VLT and other ground-based telescopes, we searched for the afterglows of the *INTEGRAL* bursts GRB 040223, GRB 040422 and GRB 040624 in the first hours after the triggers. A detection of a faint afterglow and of the host galaxy in the K band was achieved for GRB 040422, while only upper limits were obtained for GRB 040223 and GRB 040624, although in the former case the X-ray afterglow was observed. A comparison with the magnitudes of a sample of afterglows clearly shows the faintness of these bursts, which are good examples of a population that an increasing usage of large diameter telescopes is beginning to unveil.

Keywords: gamma-ray sources; gamma-ray burst

CP836, *Gamma-Ray Bursts in the Swift Era*,
edited by S. S. Holt, N. Gehrels, and J. A. Nousek
© 2006 American Institute of Physics 0-7354-0326-0/06/$23.00

PACS: 98.70.Rz

INTRODUCTION

The study of GRB afterglows is a promising tool for cosmology, since their absorption spectra convey information on the distance and the chemical composition of a new set of galaxies (e. g. [4]), with the possibility of exploration up to the reionization epoch [9]. However, a debated fraction of GRBs – from less than 10% [10] to 60% [11] – did not show any detectable afterglow in the optical band. Popular and non-mutually exclusive explanations are: these bursts have intrinsically faint afterglows in the optical band (e. g. [5]; [11]); their decay is very fast [1]; the optical afterglow is extinguished by dust in the vicinity of the GRB or in the star-forming region in which the GRB occurs (e. g. [8]; [15]); their redshift is above 6, so that the Lyman-α absorption by neutral hydrogen in the host galaxy and along the line of sight damps the optical radiation of the afterglow [8]. To these physical explanations, one must add the possibility that the search techniques are neither accurate nor quick enough [10]. The possibility that some afterglows are intrinsically faint has of course a big impact on the modelling of the GRBs themselves, as well as on their application in cosmology. On the other hand, if one or several of the other explanations are correct, a substantial reduction of the fraction of dark bursts can be achieved by quick observations in the infrared. In any case, a fast and multiwavelength follow-up campaign of observations is mandatory for the study of GRB afterglows. To this aim, the ESA's International Gamma-Ray Astrophysics Laboratory *INTEGRAL* [17], launched in October 2002, has a burst alert system called IBAS (*INTEGRAL* Burst Alert System, [12]). IBAS carries out rapid localizations for GRBs incident on the IBIS detector with a precision of a few arcminutes [13]. The public distribution of these coordinates enables multi-wavelength searches for afterglows at lower energies. *INTEGRAL* data on the prompt emission in combination with the early multi-wavelength studies, such as those presented in this work for GRB 040223, GRB 040422 and GRB 040624, can probe these high energy phenomena.

THE AFTERGLOW OF GRB 040422

GRB 040422 was detected by the *INTEGRAL* satellite at an angle of only 3 degrees from the Galactic plane. We observed the afterglow of GRB 040422 with the ISAAC and FORS 2 instruments at the VLT less than 2 hours after the burst. Such a prompt reaction, together with careful inspection of the crowded field of this GRB, led to the discovery of its near-infrared afterglow, for which we obtained the astrometry and photometry. We measured for the afterglow a magnitude $K = 18.0 \pm 0.1$ (1.9 hours after the burst), a value which is below the limit of the 2MASS catalogue. No detection could be obtained in the R and I bands, partly due to the large extinction in the Milky Way. We imaged the position of the afterglow again two months later in the K band, and detected a likely bright ($K \sim 20$) host galaxy (Fig. 1, for more details see [2]). We compare the magnitude of the afterglow with those of a compilation of promptly

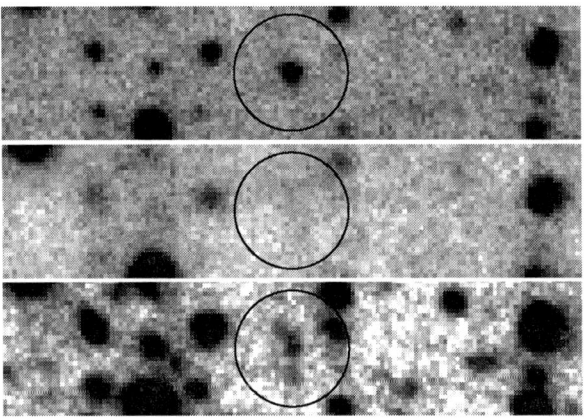

FIGURE 1. The region of the afterglow of GRB 040422. From top to bottom: 2004 Apr 22.37, 2004 May 05.33, 2004 Jun 26.11 UT.

observed counterparts of previous GRBs, and show that the afterglow of GRB 040422 lies at the very faint end of the distribution, after accounting for Milky Way extinction (Fig. 2). This observation suggests that the proportion of dark GRBs can be lowered significantly by a more systematic use of 8-m class telescopes in the infrared in the very early hours after the burst.

THE DARK GRB 040223 AND GRB 040624

GRB 040223 was detected by *INTEGRAL* close to the Galactic plane while GRB 040624 was at high Galactic latitude. The two GRBs have long durations, slow pulses and are weak. The γ-ray spectra of both bursts are best fit with steep power-laws, implying they are X-ray rich. GRB 040223 is among the weakest and longest of *INTEGRAL* GRBs. The X-ray afterglow of this burst was detected 10 hours after the prompt event by *XMM-Newton*. The measured spectral properties are consistent with a column density much higher than that expected from the Galaxy, indicating strong intrinsic absorption. We carried out near-infrared observations 17 hours after the burst with the ESO-NTT, which yielded upper limits. Given the intrinsic absorption, we find that these limits are compatible with a simple extrapolation of the X-ray afterglow properties. For GRB 040624, we carried out optical observations 13 hours after the burst with FORS 1 and 2 at the VLT, and DOLoRes at the TNG, again obtaining upper limits. As for GRB 040422, we compare these limits with the magnitudes of a compilation of promptly observed counterparts of previous GRBs and find again that they lie at the very faint end of the distribution (Fig. 2, for a more detailed analysis see [3]). Together with GRB 040422, these two bursts are good examples of a population of bursts with dark or faint afterglows that are being unveiled through the increasing usage of large diameter telescopes engaged in comprehensive observational programmes.

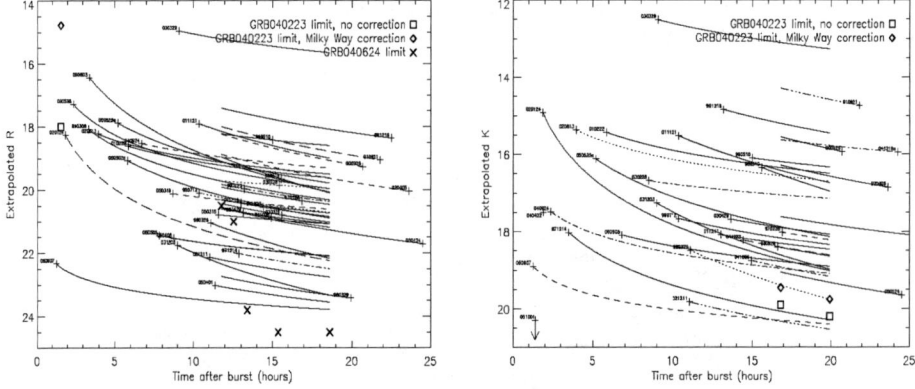

FIGURE 2. Light curves of a set of 39 afterglow data, observations of GRB 040422 and limits of GRB 040223 and GRB 040624. For clarity, different line styles are used to indicate the temporal power-law decay extrapolations, which only cover our observation epochs. (*Left*) Magnitudes extrapolated to the R band (when necessary) after correction for Galactic absorption. The cross is placed at the time of observation. The diamond and the square indicate the magnitude limits for GRB 040223 with and without correction for Galactic extinction, respectively, at 2-σ ([6]). The crosses indicate the magnitude limits for GRB 040624. The first two come from the GCNs ([14, 7]). The last three points are the 3-σ limits reported in [3]. (*Right*) Magnitudes extrapolated to the K band (when necessary) after correction for Galactic absorption. The cross is placed at the time of observation. The diamonds and the squares indicate our 3-σ magnitude limits with and without correction from Galactic extinction, respectively. The K magnitude for GRB 040422 ([2]) and the limit for GRB 051001 ([16]) are also reported.

REFERENCES

1. Berger, E., Kulkarni, S. R., Bloom, J. S., et al. 2002, *ApJ*, 581, 981
2. Filliatre, P., D'Avanzo, P., Covino, S., et al. 2005, *A&A*, 438, 793
3. Filliatre, P., Covino, S., D'Avanzo, P., et al. 2005, *A&A accepted, astro-ph/0511722*
4. Fiore, F., d'Elia, V., Lazzati, D., et al. 2005, *ApJ*, 624, 853
5. Fynbo, J. U., Jensen, B. L., Gorosabel, J., et al. 2001, *A&A*, 369, 373
6. Gomboc A., Marchant J.M., Smith R.J., Mottram C.J., Fraser S.N. 2004, *GCN 2534*
7. Gorosabel J., Casanova V., Verdes-Montenegro L., et al. 2004, *GCN 2615*
8. Lamb, D. Q. 2000, *Phys. Rep.*, 505, 333
9. Lamb, D. Q., & Reichart, D. E. 2000, *ApJ*, 536, 1
10. Lamb, D. Q., Ricker, J. R., Atteia, J.-L., et al. 2004, *New A Rev.*, 48, 423
11. Lazzati, D., Covino, S., & Ghisellini, G. 2002, *MNRAS*, 330, 583
12. Mereghetti, S., Götz, D., Borkowski, J., Walter, R., & Pedersen, H. 2003, *A&A*, 411, L291
13. Mereghetti, S., Götz, D., Borkowski, J., et al. 2004a, in Proceedings of the 5th *INTEGRAL* Workshop: The *INTEGRAL* Universe (Munich), ESA Special Publication SP-552, ed. V. Schönfelder, G. Lichti & C. Winkler, *astro-ph/0404019*
14. Piccioni A., Bartolini C., Guarnieri A., et al. 2004, *GCN 2623*
15. Reichart, D. E. & Price, P. A. 2002, *ApJ*, 565, 174
16. Rol, N., Levan, A., E., Tanvir, et al. 2005b, *GCN 4053*
17. Winkler, C., Courvoisier, T. J., Di Cocco, G., et al. 2003, *A&A*, 411, L1

PROGENITORS, COSMOLOGY AND X-RAY FLASHES

Constraints on the Diverse Progenitors of GRBs from the Large-Scale Environments

J. S. Bloom[*] and J. X. Prochaska[†]

[*]*Astronomy Department, University of California, Berkeley, 601 Campbell Hall, Berkeley, CA 94720*
[†]*Department of Astronomy and Astrophysics, UCO/Lick Observatory; University of California, 1156 High Street, Santa Cruz, CA 95064*

Abstract.
The pursuit of the progenitors of short duration-hard spectrum gamma-ray bursts (SHBs) draws strongly upon similar quests for the origin of supernovae (SNe) and long duration-soft spectrum GRBs (LSBs). Indeed the notion that, in the absence of smoking guns, the progenitors of cosmic explosions betray their identities both on the global and local scale, motivates the study of SHB redshifts, host galaxies, and locations with respect to hosts. To this end, we suggest both a historical and emergent physical analogy of GRBs with SNe: long-soft GRBs are to core-collapsed supernovae as short-hard GRBs are to Type Ia supernovae ("LSB:CC::SHB:Ia"). Still, the SHB progenitor pursuit is just beginning and we caution that while there are some substantive differences between observations of LSBs and SHBs on large-scales, particularly in host demographics, neither the offset nor the redshift distributions of SHBs are yet (statistically) distinct with those of LSBs.

Keywords: Gamma-ray Bursts — Short-Hard Subclass; Galaxies
PACS: 98.70.Rz

INTRODUCTION

The subclassification of high-energy bursting phenomena relies, for good reason, on the properties of the prompt emission. Yet whether the diversity of burst duration and spectral hardness reflects a true diversity in the physics of the explosions and the nature of the progenitors, was largely a leap of faith for the vast majority of GRBs. Soft-gamma ray repeaters aside, there was until 2005 little evidence to suggest a real difference between long-soft (LSB) and short-hard (SHB) GRBs [1]. Following the first rapid locations of SHBs by Swift [2] and *HETE-2* [3], the first X-ray afterglows were found [2], allowing for an unprecedented localization of five SHBs. K. Hurley gave an excellent and thorough overview of high-energy properties of short-hard bursts in this conference.

What has emerged with the limited set of SHBs are both commonalities and striking differences between the two populations. Broadband modeling [4, 5, 6, 7] has shown a broad consistency of afterglows with the external shock model. As predicted [8], because the luminosity of the afterglow scales with the input energy of the explosion, the afterglows of SHBs appear to be fainter analogues of LSBs. Moreover, there is now evidence that short bursts are weakly collimated with half angle jetting of more than 10% (see [7]). Yet unlike with the relatively coöperative LSBs, it may be some time before a definitive observation reveals the origin of short bursts: the best smoking gun for the merger hypthothesis is a concident detection of a burst of gravitational waves.

TABLE 1. Basic Properties of Short-Hard GRB Hosts

	redshift	host classification	in cluster?	SFR ($M_\odot$ yr^{-1})	Min Age Since Starburst (Gyr)	Metallicity ($Z/Z_\odot$)
050509b	0.225	E	Yes	<0.1	~1	~1
050709	0.160	Irr/late-type dwarf	No	>0.3	~ongoing	0.25
050724	0.258	early (E+S0)	No	<0.05	8	0.2
050813	0.722 (1.7?)*	E?	Likely	<0.2		~1
051221	0.5459	late-type dwarf?	?	~0.2		~1

Source: [4, 2, 14, 15, 5, 16, 7, 17] and Covino (this conference)

* Berger, this conference

Thankfully, the progenitors of cosmic explosions are observationally manifested on length scales ranging from circumburst to galaxy cluster. Indeed, a census of the *types of host galaxies* of SNe beginning in the 1950s yielded insights into the various progenitors: Reaves [9] noticed that Type I (now subclassed as Ia) SNe occurred in galaxies of all sorts whereas Type II arise from essentially only in late-types. This observation is fully borne out in modern studies [10]. Likewise, systematic surveys of the *locations of SNe around galaxies* further informed the progenitor question. Van Dyk [11] concluded that the location of SNs "suggest that type Ia supernovae do not arise from massive short-lived stellar populations. Type Ib/Ic and type II supernovae, however, are very likely to be associated with H II regions and therefore with massive stellar progenitors." Recently, measurements of the *redshift distribution* of Ia SNe have shown evidence for delay since starburst [12, 13]. All such observations of SNe on large scales support the established belief from detailed modeling of the transient emission that massive young stars are responsible for Type II SNe and degenerate old stars are responsible for Ia SNe.

As reviewed below, the quest for an understanding of the progenitors of LSBs benefited from a similar analysis of observations on large scales. Now, thanks to the breakthrough localization of afterglows and motivated by past work with SNe and LSBs, the search for the origin of SHBs is following parallel (productive) tracks.

PROGENITORS REVEALED ON LARGE SCALES

Host Galaxies

The population of LSB hosts appear to be best classified as "faint blue galaxies." (see reviews by N. Tanvir and E. Le Floc'h, this conference). To summarize, they are underluminous (median $L \sim 0.1 L_*$) with disturbed/irregular morphologies. Indeed, of the more than 50 LSBs that have been well-localized to date, no LSB has been definitively associated with an early-type host. The hosts do appear to be especially ripe environments for making collapsars [18, 19, 20, 21, 22, 23, 24, 25]. In particular, the star formation per unit mass appears high — with half of well-studied hosts showing more than 10 $M_\odot$ yr^{-1} L_*^{-1} — indicating recent star burst activity ([26, 20, 22, 27] and

Prochaska, this meeting). There is also anecdotal evidence that the ratio of [Ne III]/[O II] is high, indicative of hot H II regions, and suggesting a propensity for making massive stars. While the hosts do appear to be low in metallicity, it is not clear whether the hosts are, in general, lower in metallicity than galaxies at similar redshifts.

In contrast to the LSB hosts, the galaxies putatively associated with the SHB population exhibit a wide range of morphology, star-formation rate, and metallicity. The SHB host population currently includes both solar metallicity elliptical galaxies and metal-poor late-type dwarfs [15, 7]. The discovery of the early-type galaxies hosts argue for a greater than 1 Gyr delay between the formation of the progenitor star(s) and the burst. It may also be notable that even the putative hosts with current star-formation exhibit evidence for old stars (Covino, this conference; [7]), although this is a common characteristic of dwarf galaxies in the Local Group (e.g., [28]).

If SHBs share a common progenitor, it may be possible to constrain the lifetime of SHB progenitors based on the morphological mix of host galaxies ([29]; Ramirez-Ruiz, this conference). For example, the morphological types for the first four bursts in Table 2 reflect a higher incidence of early-type galaxies than Type Ia SN and suggest a progenitor lifetime significantly exceeding 1 Gyr. These conclusions, however, are not robust and are subject to the small size of the current sample (for example, the inclusion of GRB 051221 found after the conference increased the late-type fraction by a factor of two!). On the other hand, a large progenitor lifetime would help explain the apparent high incidence of cluster membership. This could be naturally explained by the fact that galaxies in overdense regions form earlier in hierarchical cosmologies [29].

Redshift Distributions

If LSBs arise from the death of massive stars, then a natural ramification is that the long-soft bursting rate should trace the universal star formation rate (SFR). The brightness of the gamma-ray data alone has shown broad consistency (e.g., [30]) but it was not until relatively recently, with the advent of spectroscopic redshifts of LSBs, that details of the bursting rate could be inferred. The difficulty in making the comparison is two-fold. First, the "true" universal star formation rate is not known a priori and is subject to observational selections such as the suppression of UV starlight due to dust. The canonical model rate used for GRB litmus tests, parameterized in Porciani and Madau [31], has the SFR rising rapidly from $z=0$ to $z=1$, then flat out to large redshifts. Second, there are inherent biases in redshift discovery of GRBs which are often difficult to quantify, such as translating non-detections of spectroscopic redshifts into limits and using complex trigger efficiencies near the detection level to infer rates at the faint end of the flux distribution. Redshifts found mostly from emission lines in the host – the majority of pre-Swift redshifts – are especially subject to complex biases due to the finite bandpass of optical spectrographs, emission from night sky lines, and limited number of SF emission lines; nevertheless, pre-Swift LSBs were shown consistent with the SFR (e.g. [32]). Redshifts discovered by means of absorption-line spectroscopy, how most Swift burst redshifts are found thanks to rapid localizations, are significantly less subject to observational biases. P. Jakobsson and B. Gendre presented results on the true

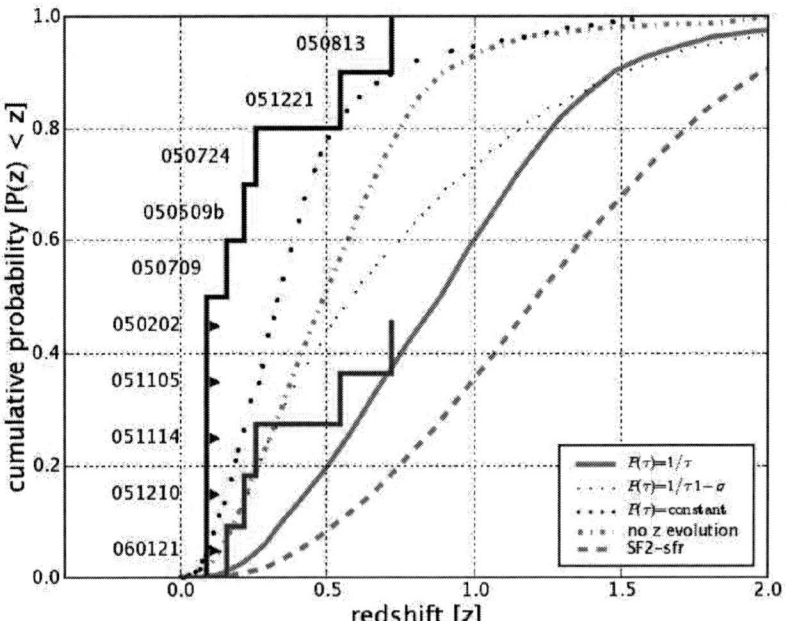

FIGURE 1. A "complete" redshift distribution of those SHBs localized rapidly by *HETE-2* and Swift, not associated with SGRs, compared with progenitor models (from [34]) with different temporal delays from starburst. As of 21 Jan 2006, there are 6 SHBs without redshift (shown in the leftmost cumulative histogram with right-pointing arrows) and 5 with redshift in this sample. The truncated histogram at right, shows how the first 5 SHBs with redshift would lie if all those SHBs without redshift arose from z greater than 0.7. Aside from tracing the universal SFR (SF2, dashed line), the current sample does not exclude any of the Guetta & Piran models. If a redshift for SHB 050813 of $z = 1.7$ is adopted (Berger, this conference), then the SF2 model is statistically allowed by the data.

redshift rate in the Swift era, finding a reasonable consistency with the universal SFR (see also [33]).

As the individual *a posteriori* identification of the 5 SHBs with afterglows with specific redshifts range from $P \approx 10^{-4}$ (e.g., 050904) to $P \approx 10^{-3}$ (e.g., 050509b), there is now little doubt that the population short-hard GRBs are of extragalactic origin. Still, as of writing, there has been no absorption line redshift measured for a *bone fide* short-hard GRB: the inference of SHB redshifts relies at present on the statistical connection to a putative host galaxy and spectroscopy of that host.

A cursory comparison of the redshift distribution of SHBs with LHBs reveals what appears to be a significantly different population. Augmenting those 5 SHBs (with known afterglows) in Table 1 with 790613 ($z = 0.09$, inferred by statistical arguments of burst-only position [29]), the population indeed appears to be heavily skewed to lower redshifts. Whereas the median redshift of LSBs is $z = 2.8$ [33], of these 6 reasonably secure redshifts of SHBs, the median is $z = 0.24$. Guetta & Piran [34] and Nakar et al. [35] have suggested that the difference could be only partly due to the relative faintness of the prompt burst emission brightness resulting in only the relatively nearby SHBs causing a trigger. Despite the brightness bias, the Guetta & Piran claim is that the intrinsic rate of SHBs are more skewed to lower z than the universal SFR. Nakar et al. further claimed (before the $z = 1.7$ hypothesis by E. Berger was advanced at the conference and 051221 was discovered) that the true SHB redshift distribution requires a progenitor population which goes boom more than a few Gyr after starburst. If true, this stretches even the longest-timescale merger scenarios.

Given the instrumental and environmental biases against detection of higher redshift SHBs, the non-detections of SHB redshifts in a complete sample must be properly accounted for statistically when drawing conclusions about the true rate. As an illustration, if we define the sample as "all *HETE-2* and Swift short-hard bursts which were radpily localized but not associated with an SGR" the list includes those in Table 1 plus GRBs 060121, 051210, 051114, 051105, and 050202; that is, 5 with redshifts and 5 without[1]. Applying survival analysis on the sample[2], we find that the data are consistent with having been drawn (and censored) at random by all of the temporal delay models proposed by Guetta & Piran. Even the universal star-formation rate (SF2) is allowed (e.g., under the Peto & Peto Generalized Wilcoxon Test) with a few percent chance. The Kaplan Meier mean is $z = 0.38 \pm 0.10$. If SHB 050813 arose from $z = 1.7$ (as suggested by Berger at this conference) the Kaplan Meier mean redshift is $z = 0.58 \pm 0.26$ and the data are consistent statistically with the SF2 model. The conclusions, which implicitly ignore all the short bursts too faint and distant to trigger on prompt emission, are clearly still sensitive to individual bursts; this reflects the danger of drawing strong conclusions from the current small sample.

Locations In and Around Galaxies

From the first sub-arcsec localization of LSBs, showing an offset of GRB 970228 from the centroid of the optical light (e.g. [37]), it was clear that the phenomena was unlikely related to activity of the central black hole [38]. Unfortunately, due to the relatively large redshifts of LSBs, resolved imaging of the immediate environment of bursts (say at the 100 pc level) is simply not possible even with diffraction limited imaging of HST and 10-m groundbased telescopes. Inferences about what the locations of GRBs imply for the progenitors must be made in a statistical manner. A comparison of the first 20 GRB localizations with apparent centers of their hosts revealed two surprising results in the context of the then paradigm of LSBs as merger products. First,

---

[1] At the conference, we showed similar results using small IPN error boxes as the defining sample of SHBs. Bursts 051227, 050911, 051211A were not included because their (current) ambiguous classifcation as SHB.

[2] Since the error circles of those without redshift were searched for nearby galaxies to depths exceeding the DSS limit, I have assumed a lower limit of $z = 0.09$ (the lowest redshift of an IPN burst found in Gal-Yam et al.; this is between the 1 and 2 σ lower limits of Gal-Yam et al. of other fields).

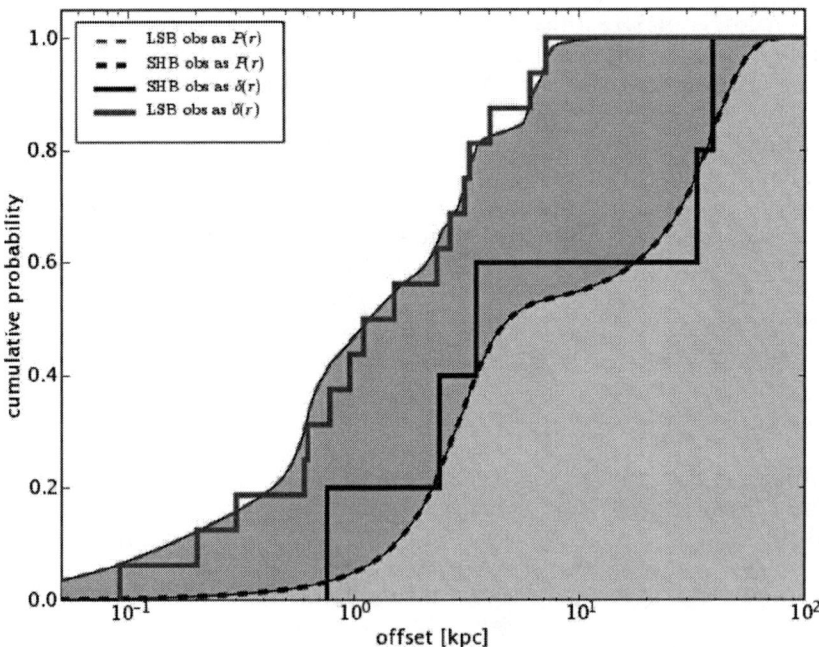

FIGURE 2. Comparison of the cumulative offset distribution of long-soft (left) and short-hard GRBs (right). The SHB sample is taken from Table 2 and the LSB sample comprises the first 16 bursts with known redshifts and offsets. The histograms are made assuming that the offsets are known precisely (ie., a δ-function at the measured offset), whereas the smooth curves account for the uncertainty in the offset measurements following the formalism of [36]. While SHBs *appear* qualitatively to be more diffusely located with respect to their putative hosts, the Kolomogorov-Smirnoff probability that the observed SHB population was drawn at random from the observed LSB population is 38%. That is, the data do not yet support the qualitatively assessment that the locations of SHBs are substaintially different than LSBs.

the locations are relatively concentrated towards the centers of galaxies, all less than 10 kpc from the apparent centers. This tight concentration is statistically inconsistent with nominal merger locations. [The standard NS-NS formation scenarios usually lead to median merger times between 0.1 – 1 Gyr from starburst with systemic kick velocities of order 100 km s^{-1}. The expected offsets from mergers is of course dependent on the host properties[3].] Second, the locations of LSBs appeared to trace the location of the UV light of hosts, suggesting an intimate connection of LSBs with star formation. A. Fruchter presented results at the conference, making use of HST imaging of new GRBs, and showing a continued connection of LSBs with the light of their hosts. In particular, the locations of LSBs appear to prefer some of the highest surface brightness

---

[3] Since the hosts are relatively low mass, suggesting that double NSs could escape the host potenitial before coalescence, a tight concentration is especially disfavored. This inconsistency is entirely dependent upon the dominant channel for double NS production. Other production channels do posit small offsets of GRBs from star formation locations (e.g., [39]).

TABLE 2. Offsets of Short Gamma-Ray Bursts from Their Putative Host Galaxies

	Angular Offset (arcsec)	Projected (kpc)
050509B	10.9 ± 3.6	39 ± 13
050709	1.29 ± 0.48	3.5 ± 1.3
050724	0.61 ± 0.23	2.4 ± 0.9
050813*	4.66 ± 2.5	33.2 ± 17.6
051221	0.12 ± 0.04	0.760 ± 0.030

Source: [15, 7]

* Here, we have assumed that source "B" in Prochaska et al. is the host. This is by no means the only possible offset for this burst. The deprojection is made assuming $H_0 = 71$ km s^{-1} Mpc^{-1}, $\Omega_m = 0.3$, $\Omega_\Lambda = 0.7$.

components of their hosts.

In light of the LSB location results, one of the startling observations of SHB 050509b was not only the statistical connection to a low-redshift giant elliptical galaxy, but, if the connection were true, the relatively large offset from that galaxy. Whereas no LSB had been found more than 10 kpc in projection from a putative host, the first SHB was well-localized at 39 ± 13 kpc in projection. This, in and of itself, is suggestive of a different population of progenitors for SHBs. In contrast, the location of SHB 050709, on the outskirts of a faint star-burst galaxy bears a striking resemblance to the burst-host configuration of LSB 970228. As with LSBs, analysis of *individual* locations would clearly lead to diverse conclusions about the nature of the progenitors.

Thankfully, while there are only 5 SHBs with ∼arcsec localizations (see Table 2), we are now in the position to begin to test various progenitor scenarios[4]. What can be stated with certainty is SHBs do not all arise from the central activity of galaxies. Following Figure 2 SHBs *appear* to be more diffusely positioned around galaxies than LSBs. But, owing to the small numbers and sometimes large uncertainties in the offset measurements, we find that the locations of SHBs are still statistically significant ($P_{K-S} = 0.38$) with having been drawn from the same population of as LSBs.

DISCUSSIONS AND CONCLUSIONS

The search for the origin(s) of SHBs presents a substantial challenge relative to discovery of long-burst progenitors: there is no obvious smoking gun for short burst progenitors that have not already been looked for. Whereas an obvious manifestation of the collapsar

---

[4] It is a popular misconception that NS–NS merger locations are unlikely to be close (less than ∼5 kpc) to their hosts based on the cumulative offset distribution presented in Figure 22 of Fryer, Woosley, Hartmann [40]. It appears that the scale of the x (distance) axis in that figure is incorrect by an order of magnitude with respect the differential distributions shown in Figure 21. To be sure, the median offset for an M_* galaxy is ∼10 kpc, which is consistent with other simulations from other groups.

model was the connection to star forming hosts and star forming regions within hosts, viable progenitor models predict offsets and locations within hosts that are far from robust. So what may we conclude about how the current host observations of SHBs relative to LSBs reflect the progenitors? A secure statement is that "SHB hosts contain a generally older population of stars than LSB" and a reasonable, but somewhat more directed statement is that "SHBs come from an old stellar population whereas LSBs do not." What we think is still a matter of debate is whether "the frequency of early-type to late-type hosts for SHB is consistent with NS–NS" (E. Ramirez-Ruiz discussed his views on this in the conference). Similar statements can be made of the offset distribution of SHBs: yes, there appears to be some differences between LSBs and SHBs but nothing stastistically significant at this time. What appears relatively secure, based on the small offsets of three of 5 SHBs from low-mass galaxies is that:

The progenitors of SHBs cannot have both large systematic kicks (> 100 km s^{-1}) at formation AND inhere large delay times from starburst (> 1 Gyr).

Making even stronger statements about the nature of the progenitors is not only hampered by small number statistics but in the lack of robust predictions from the models. But the above statement can be reworked in the form of a generic set of predictions:

- In Long delay (>1 Gyr) progenitors scenarios with kicks the offsets should anti-correlate with host mass and correlate with average stellar age
- In Short delay (< 1 Gyr) progenitors the offsets should correlate with host mass and anti-correlate with average stellar age.

If the progenitor lifetime of the SHBs is long and kicks are small, then the bursts should correspond spatially to the oldest stellar populations in a given galaxies. For early-type galaxies, the distribution would presumably follow the light of the galaxy. In contrast, the distribution in star-forming galaxies might be more concentrated in the spheroid (e.g., bulge of the Milky Way).

Those of us that have worked on creating *ab initio* predictions for NS–NS progenitors all basically agree on the offsets and delays provided we use a production channel dominated by binaries where the helium stars do not fill their respective Roche lobe before exploding as a SN. But other productions might dominate, leading to predictions of relatively short merger times and small offsets [39, 41]. Both T. Piran and J. Grindlay discussed other production channels in the conference. For example, if SHBs arise from NS binaries made in the centers of globular clusters [42], then the offsets should scale with the halo size of the host (rather than exponential disk size) and the systemic kicks could be small. This results in diverse predictions of offsets and host demography for the same progenitors.

Fostered by the presence of a curious 100 sec timescale for prompt emission and lacking the detection of a Li-Paczyński minisupernova [43], the theory community is rapidly developing other progenitor models for SHBs such as WD–WD mergers [44] and accretion induced collapse of a NS [45, 46]. In this respect the connection of GRBs with SNe may not only be a historical one, there may be a deeper analogy: long-soft GRBs are to core-collapased supernovae as short-hard GRBs are to Type Ia supernovae

("LSB:CC::SHB:Ia"). Not only does this ring true for the hosts and locations (like Type Ia SNe, short bursts can occur outside star forming regions in galaxies of any type, whereas, like core-collapsed SNe, long bursts only occur near star forming regions of galaxies actively forming stars) but could also reflect a symmetry of the progenitors: LSBs result from a special (extreme?) version of core-collapased massive young stars whereas SHBs might arise in the NS to BH transition (and Ia SNs in WD to NS transition).

Whether the LSB:CC::SHB:Ia analogy will hold at the physical level remains to be seen. But as progenitor predictions for SHBs are honed, new SHB offsets plus inferences of the host mass will continue to constrain. Still, a direct comparison with model predictions will also be hampered by the possibly of an "ambient density bias," where afterglows, the brightnesses of which scale as the circumburst density to the half power, are more likely to be discovered if the burst occurs in regions of larger density. Are we missing a population of bursts at large offsets that occur in the intergalactic medium? If the internal shock paradigm for short burst holds then ∼arcsecond location of the prompt emission in future missions should be immune to this bias. Last, we note that while SHB afterglow follow-up tends to be more difficult than LHBs, one positive result of the low z nature of detected SHBs is that we may have the possibility of resolving the sub-structure of the birthsites within galaxies (e.g., spiral arms, H II regions) using HST and groundbased adaptive optics.

ACKNOWLEDGMENTS

We are thankful to Daniel Perley, Enrico Ramirez-Ruiz, Jonathan Granot, and Hsiao-Wen Chen for their respective insights and work on the nature of short bursts. We thank Avishay Gal-Yam and Alexander Kann for constructive comments on a version of this text. JSB and JXP acknowledge support from NASA/Swift grant NNG05GF55G. JSB thanks the organizing committee for partial support of students Perley and Katey Alatalo.

REFERENCES

1. C. Kouveliotou, C. A. Meegan, G. J. Fishman, N. P. Bhat, M. S. Briggs, T. M. Koshut, W. S. Paciesas, and G. N. Pendleton, *ApJ (Letters)* **413**, L101–L104 (1993).
2. N.Gehrels, et al., *Nature* **437**, 851–854 (2005).
3. J. S. Villasenor, et al., *Nature* **437**, 855–858 (2005).
4. J. S. Bloom, J. X. Prochaska, et al., Closing in on a Short-Hard Burst Progenitor: Constraints from Early-Time Optical Imaging and Spectroscopy of a Possible Host Galaxy of GRB 050509b (2006), ApJ; astro-ph/0505480.
5. D. B. Fox, et al., *Nature* **437**, 845–850 (2005).
6. A. Panaitescu, The Energetics and Environment of the Short-GRB Afterglows 050709 and 050724 (2005), astro-ph/0511588.
7. A. M. Soderberg, et al., The Afterglow and Host Galaxy of the Energetic Short-Hard Gamma-Ray Burst 051221 (2006), astro-ph/0601455.
8. A. Panaitescu, P. Kumar, and R. Narayan, *ApJ (Letters)* **561**, L171–L174 (2001).
9. G. Reaves, *PASP* **65**, 242 (1953).
10. E. Cappellaro, M. Turatto, D. Y. Tsvetkov, O. S. Bartunov, C. Pollas, R. Evans, and M. Hamuy, *A&A* **322**, 431–441 (1997).

11. S. D. van Dyk, *AJ* **103**, 1788–1803 (1992).
12. A. Gal-Yam, and D. Maoz, *mnras* **347**, 942–950 (2004).
13. L.-G. Strolger, et al., *ApJ* **613**, 200–223 (2004).
14. E. Berger, et al., A Merger Origin for Short Gamma-Ray Bursts Inferred from the Afterglow and Host Galaxy of GRB 050724 (2005), astro-ph/0508115.
15. J. X. Prochaska, J. S. Bloom, H.-W. Chen, R. J. Foley, D. A. Perley, E. Ramirez-Ruiz, J. Granot, W. H. Lee, D. Pooley, K. Alatalo, and K. Hurley, The Galaxy Hosts and Large-Scale Environments of Short-Hard γ-ray Bursts (2005), submitted to ApJ, astro-ph/0510022.
16. S. D. Barthelmy, et al., *Nature* **438**, 994–996 (2005).
17. J. Gorosabel, et al., The short-duration GRB 050724 host galaxy in the context of the long-duration GRB hosts (2005), astro-ph/0510141.
18. V. V. Sokolov, T. A. Fatkhullin, A. J. Castro-Tirado, A. S. Fruchter, V. N. Komarova, E. R. Kasimova, S. N. Dodonov, V. L. Afanasiev, and A. V. Moiseev, *A&A* **372**, 438–455 (2001).
19. S. G. Djorgovski, et al., "The GRB Host Galaxies and Redshifts," in *Gamma-ray Bursts in the Afterglow Era*, 2001, p. 218.
20. J. P. U. Fynbo, et al., *A&A* **406**, L63–L66 (2003).
21. E. Le Floc'h, et al., *A&A* **400**, 499–510 (2003).
22. L. Christensen, J. Hjorth, and J. Gorosabel, *A&A* **425**, 913–926 (2004).
23. C. J. Conselice, P. M. Vreeswijk, A. S. Fruchter, A. Levan, C. Kouveliotou, J. P. U. Fynbo, J. Gorosabel, N. R. Tanvir, and S. E. Thorsett, *ApJ* **633**, 29–40 (2005).
24. C. Wainwright, E. Berger, and B. E. Penprase, A Morphological Study of Gamma-Ray Burst Host Galaxies (2005), astro-ph/0508061.
25. A. S. Fruchter, A. J. Levan, L. Strolger, P. M. Vreeswijk, et al., Positions of GRBs (2005), submitted to ApJ.
26. P. M. Vreeswijk, et al., *ApJ* **546**, 672–680 (2001).
27. P. Jakobsson, et al., *MNRAS* **362**, 245–251 (2005).
28. T. A. Smecker-Hane, A. A. Cole, J. S. Gallagher, and P. B. Stetson, *ApJ* **566**, 239–244 (2002).
29. A. Gal-Yam, E. Nakar, E. Ofek, D. B. Fox, S. B. Cenko, S. R. Kulkarni, A. M. Soderberg, F. Harrison, P. A. Price, B. E. Penprase, D. Frail, E. Berger, M. Gladders, and J. Mulchaey, The Progenitors of Short-Hard Gamma-Ray Bursts from an Extended Sample of Events (2005), astro-ph/0509891.
30. T. J. Loredo, and I. M. Wasserman, *ApJ* **502**, 75 (1998).
31. C. Porciani, and P. Madau, *ApJ* **548**, 522–531 (2001).
32. J. S. Bloom, *AJ* **125**, 2865–2875 (2003).
33. P. Jakobsson, et al., A mean redshift of 2.8 for Swift gamma-ray bursts (2005), astro-ph/0509888.
34. D. Guetta, and T. Piran, *A&A* **435**, 421–426 (2005).
35. E. Nakar, A. Gal-Yam, and D. B. Fox, The Local Rate and the Progenitor Lifetime of Short-Hard Gamma-Ray Bursts: Synthesis and Predictions for LIGO (2005), astro-ph/0511254.
36. J. S. Bloom, S. R. Kulkarni, and S. G. Djorgovski, *AJ* **123**, 1111–1148 (2002).
37. K. C. Sahu, et al., *Nature* **387**, 476 (1997).
38. B. Carter, *ApJ (Letters)* **391**, L67–L70 (1992).
39. R. Perna, and K. Belczynski, *ApJ* **570**, 252–263 (2002).
40. C. L. Fryer, S. E. Woosley, and D. H. Hartmann, *ApJ* **526**, 152–177 (1999).
41. K. Belczynski, R. Perna, T. Bulik, V. Kalogera, N. Ivanova, and D. Q. Lamb, A Study of Compact Object Mergers as Short Gamma-ray Burst Progenitors (2006), astro-ph/0601458.
42. J. Grindlay, S. Portegies Zwart, and S. McMillan, Short gamma-ray bursts from binary neutron star mergers in globular clusters (2005), astro-ph/0512654.
43. L.-X. Li, and B. Paczyński, *ApJ (Letters)* **507**, L59–L62 (1998).
44. A. J. Levan, et al., Short GRBs in old populations: Magnetars from WD-WD mergers (2006), astro-ph/0601332.
45. C. D. Dermer, and A. Atoyan, Collapse of Neutron Stars to Black Holes in Binary Systems: A Model for Short Gamma Ray Bursts (2006), astro-ph/0601142.
46. A. I. MacFadyen, E. Ramirez-Ruiz, and W. Zhang, X-ray flares following short gamma-ray bursts from shock heating of binary stellar companions (2005), astro-ph/0510192.

The Electromagnetic Model of Gamma ray Bursts

Maxim Lyutikov

University of British Columbia, 6224 Agricultural Road, Vancouver, BC, V6T 1Z1, Canada
Department of Physics and Astronomy, University of Rochester, Bausch and Lomb Hall, P.O. Box 270171, 600 Wilson Boulevard, Rochester, NY 14627-0171, USA

Abstract. The electromagnetic model assumes that rotational energy of a relativistic, stellar-mass central source (black-hole–accretion disk system or fast rotating neutron star) is converted into magnetic energy through unipolar dynamo mechanism, propagated to large distances in a form of relativistic, subsonic, Poynting flux-dominated wind and is dissipated directly into emitting particles through current-driven instabilities. Collimating effects of magnetic hoop stresses lead to strongly non-spherical expansion and formation of jets. Long and short GRBs may develop in a qualitatively similar way, except that in case of long burst ejecta expansion has a relatively short, non-relativistic, strongly dissipative stage inside the star.

PACS: 98.70.Rz

PRINCIPAL ISSUES: ELECTROMAGNETIC AND FIREBALL MODELS

We describe electromagnetic model of GRBs which assumes that the energy that will power a GRB comes from rotational kinetic energy of a central source. Whether magnetic fields play an important dynamical role at any stage in the outflow remains, in our view, one of the principal issues in GRBs physics. Currently, the overwhelming point of view, advocated by the fireball model (FBM) is that magnetic fields do not play any major dynamical role (except, perhaps, at a very early stage; after which fields are dissipated quickly). FBM advocates that in the emission region magnetic field are re-created locally (*e.g.* through development of Weibel instability [1]), with energy density typically much smaller than plasma energy density. Fields are small scale, with correlation length l_c much smaller than the "horizon" length $l_c \ll R/\Gamma$ (R is the radius of the outflow in the laboratory frame and Γ is its bulk Lorentz factor). Alternative approach, advocated by MHD and electromagnetic models [*e.g.* 2, 3, 4] is that there are dynamically important large scale fields with "super-horizon" correlation length $l_c \geq R/\Gamma$, which are created at the source and which may play a major role in driving the whole outflow in the first place.

To quantify the dynamical importance of large scale magnetic fields, it is useful to introduce magnetization parameter σ as a ratio of Poynting $F_{\rm Poynting}$ to (cold) particle $F_{\rm p}$ fluxes (or as a ratio of rest frame energy densities)

$$\sigma \equiv \frac{F_{\rm Poynting}}{F_{\rm p}} = \frac{B^2}{4\pi\Gamma\rho c^2} = \frac{b'^2}{8\pi\rho' c^2} \tag{1}$$

where B and ρ are magnetic field and plasma density in the lab frame, b' and ρ' are magnetic field and plasma density in the plasma frame (where electric field is zero). For $\sigma \ll 1$ magnetic fields are dynamically unimportant (this is assumed within a framework of a conventional FBM), while for $\sigma \geq 1$ magnetic fields start to play important dynamical role. For $\sigma \gg 1$, there is an important qualitative change in the dynamical behavior of the flow at σ_{crit} $\Gamma^2/2$. For $\sigma < \sigma_{crit}$ the flow is super-Alfvenic, while for $\sigma > \sigma_{crit}$ the flow is sub-Alfvenic (and sub-fastmagnetosonic). The different is somewhat analogous to the difference between sub-sonic and supersonic flows in hydrodynamics.

Thus, depending on the parameter σ three qualitatively different regimes for expansion of the ejecta may be identified, which I will call (i) fireball model (FBM below) $\sigma \ll 1$, (ii) MHD models $1 \leq \sigma < \sigma_{crit}$ and (iii) electromagnetic model (EMM below) $\sigma > \sigma_{crit}$. These three possibilities leads to *a qualitatively different dynamic behavior of flows*.

SOURCE FORMATION AND ENERGY RELEASE IN EMM

EMM assumes that the GRB "prime mover" is relativistic, fast rotating, near stellar-mass source. For numerical estimates I will assume that a central source generates luminosity L $L_{50} \times 10^{50}$ erg/s for a time t_s, where $t_s \sim 100$ s for long bursts and $t_s \sim 1$ s for short ones (there are indications that both long and short bursts are powered by the same luminosity, but for different time [5]). The mass of the central source is $\sim 0.01 M_\odot$ for short bursts and $\sim M_\odot$ for long bursts (this is the total mass passing through the disk, not in the black hole). The source is assumed to rotate with a spin frequency $\sim$ kHz. In addition, it is assumed that a source possesses a large magnetic field of $B_s \sim 10^{14} L_{50}^{1/2}$ G. If initially the core is fast rotating [*e.g.* 6], the total rotational energy in the disk, $\sim M_{disk} R_{disk}^2 \Omega^2$, in case of core collapse is much larger, $\sim 7.9 \times 10^{52} M_{disk}/M_\odot R_{disk}/ \times 10^6 \mathrm{cm}^2 \Omega/6.28 \times 10^3 \mathrm{rad/s}^2$ erg, than in the case of mergers, $\sim 7.9 \times 10^{50}$ erg with $M_{disk} \sim 0.01 M_\odot$ in the above estimate. The source is expected to be active for $t_s \sim E/L \sim 100$ s for longs and ~ 1 s for shorts.

Rotational energy of the central object is extracted by magnetic fields through unipolar induction mechanism, similar to prevailing model of AGN jets [*e.g.* 7]. Magnetic fields both launches the jet (*e.g.* through Blandford-Znajek mechanism) and collimates it by hoop stresses. Latest full relativistic MHD numerical simulations of accretion disk–black hole systems do show formation of the strongly-magnetized axial funnel [*e.g.* 8]. Thus, *large scale, energetically dominant magnetic fields may be expected in the launching region of GRB jets*.

Qualitatively, in the immediate vicinity of the source, the plasma is separated into two phases: an internal matter-dominated phase in which large currents are flowing and external magnetically dominated phase. Strong magnetic fields and magnetic flux are generated in the dense medium (a disk or a differentially rotating neutron star-like object), while relativistic outflow is generated in the magnetically dominated phase. In this case matter loading may be expected to be small (*e.g.* analogous to pulsar wind). As the source remains active for $\sim$ thousand to million dynamical times the flow will be able to settle down quickly to a quasi-steady state evolving slowly as the hole or neutron star slows down. The separation into matter- and magnetic field-dominated phases is somewhat similar to the Sun, where dynamo operates in tachocline, deep below the surface, while magnetic energy, and most importantly magnetic flux, are dissipated outside the star.

The form of expanding bubble depends on lateral distribution of source luminosity, which within a framework of EMM mode is given by lateral distribution of current. At relativistic stage expansion is nearly ballistic while at a non-relativistic stage inside a star a flow may be collimated both through the action of magnetic hoop stresses *and* interaction with the surrounding gas. One particular stationary outflow configuration which captures the essential features of the outflow, is that the outgoing current is confined to the poles and the equatorial plane and closes along the surface of the bubble. The magnetic field in the bubble is inversely proportional to the cylindrical radius, $B_\phi \propto 1/r\sin\theta$. Accompanying this magnetic field is a poloidal electrical field so that there is a near radial Poynting flux carrying energy away from the source. Thus, as the magnetic field strength is strongest close to the symmetry axis, the bubble will expand fastest along the polar direction. The internal structure of an outflow then corresponds to a "structured jet" with $L_\theta \sim \theta^{-2}$, so that the central source releases an equal amount of energy per decade of θ.

RELATIVISTIC EXPANSION

In case of long GRBs inflating a bubble inside a star, eventually the bubble will break free out of the star forming two axial jets along which Poynting flux will flow until the central source slows down on the time scale $t_s \sim 100$ s. Outside the star, the bubble will expand ultra-relativistically and bi-conically. For short GRBs, presumably associated with merger of neutron stars in a low density environment, there is no preceding non-relativistic stage so expansion of the bubble is relativistic from the beginning. In case of relativistic motion the effective load can consist of the performance of work on the expanding blast wave. This is where most of the power that is generated by the central magnetic rotator ends up.

After the bubble has expanded beyond a radius $r_{sh} \sim ct_s \sim 3 \times 10^{12} t_{s,2}$cm ($\sim 10^{10}$ cm for short bursts) the electromagnetic energy will be concentrated within an expanding, electromagnetic shell with thickness $\sim r_{sh}$ and with most of the return current completing along its trailing surface. The global dynamics of this shell and its subsequent expansion are set in place by the electromagnetic conditions at the light cylinder and within the collimation region. An important property of ultra-relativistic outflows is that they are hard to collimate, so that any collimation should be achieved close to the source, within a star, where the flow is only mildly relativistic.

Relativistic expansion of the magnetized shell may be separated into two stages, which we will call "early" and "late", depending on whether or not most of the fast waves emitted by the central source have caught up with the CD and their energy has been given to the circumburst medium. The transition between two stages occurs at the moment which is similar to the deceleration radius in fireball model, except that in the case of EMM the shell is decelerating all the time, but with different laws before and after the transition. Keeping with the tradition we will still call the transition radius as deceleration radius.

"Early" Stage.

At $r > r_{ph}$ the outflow becomes a relativistically expanding shell of thickness $\sim ct_s \sim 3 \times 10^{12}$ cm for long GRBs and $\sim 3 \times 10^{10}$ cm for short GRBs. The shell contains toroidal magnetic field; the current now detaches from the source and completes along the shell's inner surface. At this stage the CD is constantly re-energized by the fast-magnetosonic waves propagating from the central source. The average motion of the CD Rt is determined by the average luminosity at the retarded time t':

$$L_\Omega t' \sim \rho c^3 \Gamma^4 Rt^2 \beta^3 \qquad (2)$$

which for constant luminosity gives $\Gamma \sim L_\Omega/\rho c^{3\,1/4} r^{-1/2}$ (in a constant density medium) or $\Gamma \sim L_\Omega/4\kappa\rho_0 r_0^2 c^{3\,1/4}$ const (in a ρr $\rho_0 r_0/r^2$ wind). If the central source releases most of the current along the axis and the equatorial plane, then $\Gamma \propto 1/\sqrt{\sin\theta}$ at this stage. If source's luminosity varies, this will be reflected in the "jitter" of the CD. Development of instabilities at the CD, like impulsive Kruskal-Schwarzschild instability [4] may lead to dissipation and particle acceleration. The internal structure of the magnetic shell is a messy mixture of the outgoing waves from the source and the ingoing waves reflected from the CD, similar to a pre-Sedov phase in hydrodynamical explosions. Unlike the case of a hydrodynamic blast wave with energy supply, no internal discontinuities form inside magnetic shell.

Early stage lasts for $ct_s < r < r_{\text{dec}}$, where

$$r_{\text{dec}} \; L_\Omega t_s^2/\rho c^{1/4} \sim 3 \times 10^{16} L_{50}^{1/4} t_{s2}^{1/2} n^{-1/4} \text{cm}, \qquad (3)$$

for long bursts (in the observer frame this phase lasts $\sim t_s \sim 100$ s) and $r_{\text{dec}} \sim 3 \times 10^{15}$ cm for shorts (similarly, in the observer frame this phase lasts $\sim t_s \sim 1$ s). Radius r_{dec} (3) is somewhat similar to the deceleration radius in case of FBM; at this moment most energy of the shell in given to the circumburst medium, $L_\theta t_s \sim E\theta \sim \rho c^2 r_{\text{dec}}^3 \Gamma r_{\text{dec}}, \theta^2$ (note that here Γ Γr_{dec}, not Γ_0 as in case of FBM, since there is no formal definition of Γ_0 in case of EMM).

"Late Stage" ($r_{\text{dec}} < r < r_{\text{NR}} \equiv L_\Omega t_s/\rho c^{2\,1/3}$).

At distances $r > r_{\text{dec}}$ most of the waves reflected from the CD have propagated throughout the shell, so that all the regions of the shell come into causal contact. Most of the energy of the explosion will reside in the blast wave which will eventually settle down to follow a self-similar expansion. As the expanding shell performs work on the surrounding medium its total energy decreases; the amount of energy that remains in the ejecta shell during the late stage is small, $\sim E_\Omega/\Gamma^2$. Most of the energy is still concentrated in a thin shell with $\Delta R \sim R/\Gamma^2$ near the surface of the shell which is moving according to $\Gamma \sim \sqrt{E_\Omega/\rho c^2}\, r^{-3/2}$ (in a constant density medium), or $\Gamma \sim r^{-1/2}$ (in a $\rho \sim r^{-2}$ wind). If the central source releases most of the current along the axis and the equatorial plane, then $\Gamma \propto 1/\sin\theta$ at this stage.

"Late stage" of magnetic shell expansion corresponds to the conventional afterglow phase when synchrotron and inverse Compton radiation is emitted throughout the electromagnetic spectrum. The initially aspheric expansion will give the appearance of a jet with the "achromatic break" occurring when the fastest Lorentz factor of the spine $\Gamma\theta\ 0$ becomes comparable with the reciprocal of the observer's inclination angle with respect to the symmetry axis, $\Gamma\theta\ 0 \sim 1/\theta_{ob}$. When $r > r_{NR} \sim Lt/\rho c^{21/3} \sim 2 \times 10^{18} L_{50}^{1/3} t_{s2}^{1/3} n^{-1/3}$ cm, the blast wave become non-relativistic and will become more spherically symmetric, while evolving towards a Sedov solution.

PRODUCTION OF GRBS AND AFTERGLOWS

By the time the shell radius expands to r_{dec} most of the electromagnetic Poynting flux from the source will have caught up with the CD and been reflected by it, transferring its momentum to the blast wave.

We propose that the *γ-ray-emitting electrons are accelerated near* $r_{dec} \sim 10^{15} - 10^{16}$ *cm (for long bursts) and* $\sim 10^{15}$ *cm (for short bursts) due to development of electromagnetic current-driven instabilities* (conventional model of particle acceleration - acceleration at internal shocks - cannot work in this model since in the limit $\sigma \gg 1$ fast shocks are either weak or do not form at all). The development of current instabilities usually results in enhanced or anomalous plasma resistivity which leads to an efficient dissipation of the magnetic field. The magnetic energy is converted into heat, plasma bulk motion and, most importantly, into high energy particles, which, in turn, are responsible for the production of the prompt γ-ray emission. The conversion of magnetic energy into particles may be very efficient. For example, recent RHESSI observations of the Sun indicate that, in reconnection regions, most magnetic energy goes into non-thermal electrons.

Except at early stage (as discussed in §), afterglows are generated in a similar way both in the FBM and EMM. As the magnetic shell expands, its energy is gradually transferred to the preceding forward shock wave. Relativistic particles are accelerated in the blast wave producing the observed afterglow in a manner which is similar to what is proposed for fluid models, except that contact discontinuity itself may be an important source of magnetic flux through impulsive Kruskal-Schwarzschild instability [4], so that afterglows may result from a mixture of relativistic particles, derived from the shock with magnetic field derived from the shell.

At late times, well beyond r_{dec} (which in observer frame is nearly coincident with the prompt phase), the temporal behavior of proper afterglow (as opposed to tails of prompt emission, see §) is determined by the total energy release E_Ω and not by the form of that energy. In case of EMM, the preferred lateral distribution of the magnetic field, energy fluxes and luminosity correspond to the line current, so that $L \sim 1/\sin^2\theta$.

TESTS OF GRB MODELS

Testing the Fireball Model: Reverse Shock Emission

Perhaps the simplest test of GRB models could have come from observations of emission from the reverse shock propagating in the ejecta, which typically falls into the optical range. FBM predicts strong reverse shock emission, so that absence of nearly contemporaneous optical emission in most GRBs would be a strong argument against FBM. In MHD models with $\sigma > 1$ reverse shock is weak, while in EMM reverse shock is absent altogether.

In the Swift era, *not a single GRB has shown the predicted behavior*. This is despite the fast on-board optical telescope and a large number of ground based robotic telescopes (RAPTOR, ROTSEE, TAROT and others). *Reverse shock emission is virtually an unavoidable prediction of the fireball model, so that absence of predicted reverse shocks emission in the Swift era argues against the fireball model.* Naturally, there is a number of ways that through which optical flashes can be suppressed (*e.g.* cooling of optically emitting electrons on photons of prompt emission, "thin shell case", when the reverse shock emission is spread over longer times, producing weaker signal. *A possible explanation of an absence of clear reverse shock signal is that ejecta plasma is strongly magnetized.* In the case when energy density of magnetic field dominates the total energy density ($\sigma \geq 1$) reverse shock becomes very inefficient in dissipating flow energy [9].

Electromagnetic Model: Bright Early Afterglows

Fireball and electromagnetic models make very different prediction for the properties of early afterglows [10]. According to EMM, at the early afterglow stage the Lorentz factor and peak frequency are larger (and falling with time) than in the FBM (constant Lorentz factor and peak frequency). *Early afterglow in the EMM are more energetic than in FBM, and can blend with the prompt phase.*

Emission Radius of Prompt Photons and Early Swift Afterglows

One of the surprising early results from Swift satellite was detection of X-ray spikes and breaks in light curves at intermediate times, much longer than burst duration but well before the conventional jet break [*e.g.* 11, 12, 13]. A typical behavior includes fast-slow-fast decay with transitions near $100 - 1000$ seconds and $\sim 10^4$ seconds. This presents a real challenge to GRB models, since if the emission is seen "head on", within angle $\theta \leq 1/\Gamma$, the radii at which these features should be produced correspond to radii much larger than deceleration radius. At these times most of the energy is in the forward shock which should produce smooth light curve.

The initial fast decaying part of afterglows was argued to be a "sideways" prompt emission, coming from angles $\theta > 1/\Gamma$ [14]. If this interpretation is correct, one can determine emission radii of the prompt emission and compare them with model predictions. (We remind that FBM predicts radii of emission $r_{em} \sim 2\Gamma_0^2 c \delta t \sim 10^{12} - 10^{13}$ cm, while EMM predicts $r_{em} \leq r_{dec} \sim 10^{16}$ cm, Eq. 3). If emission is generated at r_{em} and is coming to observer from large angles, $\theta > 1/\Gamma$, its delay with respect to the start of the prompt pulse is $\Delta t \sim r_{em}/c\theta^2/2$. For typical observer angle $\theta \sim 0.1$ and first break of a light curve at $\Delta t \sim 1000$ seconds, the implied emission radius is $r_{em} \sim 6 \times 10^{15}$ cm. *This is at least two orders of magnitude larger than is assumed in the fireball model, but is close to the assumption of the electromagnetic model.* [To be consistent with FBM and variability on short times scales, the Lorentz factor of the flow should be $\Gamma_0 \sim 3000$, but this would imply that emission is strongly de-boosted, $\Gamma_0 \theta \sim 300 \gg 1$.] *Interpretation of light curves breaks at $\sim 10^3$ s as been due to prompt emission seen at large angles, $\theta > 1/\Gamma$, is inconsistent with the fireball model.*

Fast Variability from Large Radii

If prompt emission is produced at distances $\sim 10^{15} - 10^{16}$ cm, how can fast variability, on times scales as short as milliseconds, be achieved? One possibility, is that emission is beamed in the outflow frame, for example due to relativistic motion of "fundamental emitters" [4]. Possible origin of relativistic motion of "fundamental emitters" may be the fact that in case of relativistic reconnection occurring in plasma with $\sigma \gg 1$, the outflowing matter reaches relativistic speeds with $\gamma_{out} \sim \sigma$ [15]. Internal synchrotron emission by such jets, or Compton scattering of ambient photons will then be strongly beamed in the frame of the outflow.

Consider an outflow moving with a bulk Lorentz factor Γ with randomly distributed emitters moving with respect to the shell rest frame with a typical Lorentz factor γ_T. Highly boosted emitters, moving towards an observer, have Lorentz factor $\gamma \sim 2\gamma_T \Gamma$, so that modest values of $\gamma_T \sim 5 - 10 \ll \Gamma \sim 100 - 300$ suffice to produce short time scale variability from large distances. As the burst progresses, larger angles and more of internal jets producing prompt emission become visible. Most of them will be seen from large angles $> 1/\gamma_T$ in the bulk frame, producing smooth curves. Occasionally, a jet at large viewing angle, $\theta > 1/\Gamma$, but directed towards an observer will be seen, producing an X-ray flare. One expects a break in the light curve at $\Delta t \sim r_{em}/c\theta^2/2$, where θ is a viewing angle (in a structured jet model, this is the angle between the jet axis and direction to the observer). Afterglow should start to blend with prompt emission at later times. In Fig. 1 we plot an example of a prompt light curve in this model (Lyutikov, in prog.). The model readily explains many unusual properties of early afterglows: (i) X-ray flares and light curve breaks at late times, much longer than conventional prompt GRB duration (extended source activity is not needed!), (ii) fast variability, (iii) gradual softening of the spectrum, (iv) hardemig of a spectrum during X-ray flares.

Observational Implications of the Electromagnetic Model

In this section we give a short discussion of how the main GRB phenomena are (or may be) explained within a framework of EMM. (i) Jet break in afterglow. GRB outflows have large opening angles, but do not have a jet in a proper sense. Outflows are non-isotropic so an achromatic break is inferred when the viewing angle is $\theta_{ob} \sim 1/\Gamma$. (ii) Structured jet. The model predicts and gives a theoretical foundation for the "structured jet" profile of the external shock. (iii) XRF flashes. Another testable prediction of the model is that much more numerous X-ray flashes (XRFs) should be observed, which may be coming "from the sides" of the expanding shell, where the flow is less energetic and the Lorentz boosting is weaker. In addition, the total *bolometric* energy inferred for XRFs (from observations of afterglows before radiative losses become important) should be comparable to the total bolometric energy of γ-ray bursts. Generally, the distributions of parameters of XRFs should continuously match those of GRBs. (iv) Weak thermal precursor. If a fraction $1/\sigma \sim 0.01 - 0.1$ of the magnetic energy is dissipated near the source, this should produce a thermal precursor with luminosity $\sim 0.01 - 0.1$ of the main GRB burst. (v) Hard-soft evolution . The trend of GRB spectra to evolve from hard to soft during a pulse is explained as a synchrotron radiation in an expanding flow with magnetic field decreasing with radius $B \propto \sqrt{L}/r$ (later in a pulse emission is produced further out where magnetic field is weaker, so that the peak energy will be lower; this is similar to "radius-to-frequency mapping" in radio pulsars and AGNs). (vi) Amati $E_{peak} - L$ correlation. A correlation between peak energy and total luminosity, $E_{peak} \sim \sqrt{L}$ follows from the assumption of a fixed typical emission radii and fixed minimum particle energy since $B \sim \sqrt{L_\Omega}$. (vii) Variability. Variability of the prompt emission reflects the statistical properties of dissipation (and not the source activity as in the FBM). Magnetic fields are non-linear dissipative dynamical system which often show bursty behavior with power law PDF. [For example, solar flares show variability on a wide range of temporal scales, down to minutes, which are unrelated to the time scale of 22 years of magnetic field generation in the tachocline.] (viii) Prompt and afterglow polarization. Claims of high polarization [16] if confirmed, may provide a decisive test of GRB models [see though 17]. The best way to produce polarization in the range $10\% \leq \Pi \leq 60\%$ is through synchrotron emission in large scale magnetic fields [18]. (Larger polarization can only be produced with inverse Compton mechanism, smaller polarization can be produced by small scale magnetic fields.) Large scale field structure in the ejecta emission may also be related to polarization of afterglows if fields from the magnetic shell are mixed in with the shocked circumstellar material. In this case, *the position angle should not change through the afterglow* while if polarization is observed both in prompt and afterglow emission the position angle should be the same. Also, polarization should be most independent of the "jet break" moment.

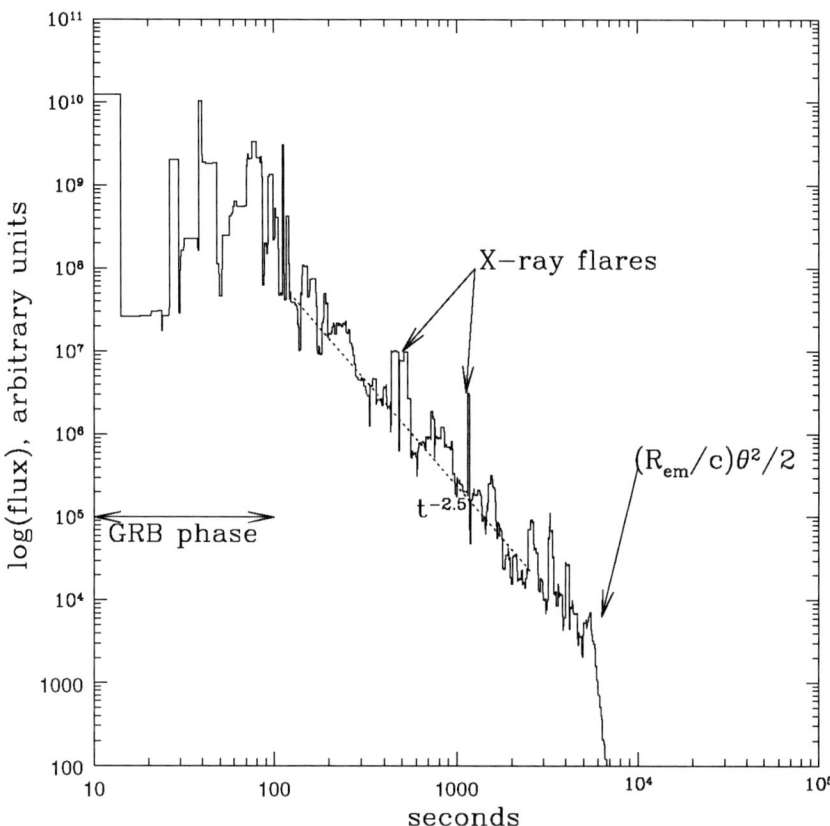

FIGURE 1. Prompt emission produced by emitters moving randomly in the bulk frame. Emission is generated within a shell of thickness $t_s c$ 3×10^{12} cm moving with Γ 100 at distance r_{em} $\Gamma^2 t_s c$ by randomly distributed jets with random orientation moving with random Lorentz factors $1 < \gamma_T < \gamma_{T,max}$ 5. Each emitter is active for random time $0 < t'_{em} < 0.5 t_s c \Gamma$ $t_{pulse,max}$ in its rest frame. Homogeneous jet centered on an observer with opening angle θ 0.1, dimensionless parameters $N\pi/\Gamma \gamma_{T,max} \theta^2$ 1.2 (probability of seeing one sub-jet "head-on" from angles $< 1/\Gamma$) and $Nct_{pulse,max}/2^2/r_{em}^2 \theta^2 t_s c \Gamma$ 0.19 (efficiency of energy conversion), where N is total number of emitters. Intensity of emission is $\propto \delta^{3\alpha}$, where δ is Doppler factor and α 0.5 is spectral index. As the burst progresses, the average Doppler factor $\delta \approx t_s \Gamma / t$ and the average flux decays as $t^{-2\alpha}$ $t^{-2.5}$ (Lyutikov, in prog.)

REFERENCES

Medvedev, M. V. and Loeb, A., 1999, , 526, 697

Usov, V. V., 1992, Nature, 357, 472

Blandford, R. D. 2002 Lighthouses of the Universe ed. R. Sunyaev Berlin:Springer-Verlag

Lyutikov M., Blandford R., 2003, astro-ph/0312347

Fox, D. B, *et al.*, Nature in press, astro-ph/0510110

Woosley, S. and Heger, A., 2005, submitted to ApJ, astro-ph/0508175

Ferrari, A., 2004, , 293, 15

De Villiers, J.-P. and Hawley, J. F. and Krolik, J. H. and Hirose, S., 2005, , 620, 878

Kennel, C. F., & Coroniti, F. V. 1984a, ApJ, 283, 694

Lyutikov, M., 35th COSPAR Scientific Assembly, p237, astro-ph/0409489

Tagliaferri, G. *et al.*, 2005, , 436, 985

Nousek, J. A. *et al.*, 2005, astro-ph/0508332

Chincarini, G. *et al.*, 2005, astro-ph/0511107

Kumar, P., & Panaitescu, A. 2000, ApJ Lett., 541, 51

Lyutikov, M., Uzdensky D., 2003, ApJ, 589, 893

Coburn, W., & Boggs, S.E. 2003, , 423, 415

Rutledge, R., Fox, D., 2003, astro-ph/0310385

Lyutikov, M., Pariev, V., Blandford, R., 2003, ApJ, 597, 998

The Progenitors of Short Gamma-Ray Bursts

Enrico Ramirez-Ruiz

Institute for Advanced Study, Einstein Drive, Princeton, NJ 08540;enrico@ias.edu

Abstract. Although they were discovered more than 30 years ago, short gamma-ray bursts are still a mystery. All that we can be confident about is that they involve compact objects and highly relativistic dynamics. Most theoretical discussions attribute their energy production either to an accreting stellar mass black hole or to a precursor stage whose inevitable end point is a stellar mass black hole. Current ideas and prospects are briefly reviewed. There are, fortunately, several new observations that could help clarify the issues.

A BURST OF PROGRESS

Until recently, short GRBs were known predominantly as bursts of γ-rays, largely devoid of any observable traces at any other wavelengths. However, a striking development in the last several months has been the measurement and localization of fading X-ray signals from several short GRBs, making possible the optical and radio detection of afterglows, which in turn enabled the identification of host galaxies at cosmological distances[3, 5, 7, 25]. The presence in old stellar populations e.g., of an elliptical galaxy for GRB 050724, rules out a source uniquely associated with recent star formation[2]. In addition, no bright supernova is observed to accompany short GRBs[3, 5, 10], in distinction from most nearby long-duration GRBs[9]. There is now a stronger motivation to develop models in fuller detail. This paper outlines some of the key issues.

BESTIARY

As is well known, one of the first proposals that was made, soon after the discovery of quasars in 1963, was that they were powered by accretion onto massive black holes[38, 33]. The fundamental reason why this proposal was made was that quasars were known to be prodigiously powerful, with luminosities equivalent to hundreds of galaxies, and that up to $\sim 0.1c^2 \equiv 10^{20}$ erg g^{-1} of energy per unit mass could be released by lowering matter close to a black hole. This efficiency could be over a hundred times that traditionally associated with nuclear power. Since this time, we have also learned about black holes with masses $\sim 5-10$ $M_\odot$ in Galactic binary systems and ultra-luminous X-ray sources which, with decreased confidence, we also associate with black holes, primarily on energetic grounds. In these objects, accretion (and the accompanying radiation) is usually thought to be limited by the Eddington rate, a self–regulatory balance imposed by Newtonian gravity and radiation pressure. The standard argument gives a maximum luminosity $L_{\rm Edd} = 1.3 \times 10^{38}(M/M_\odot)$ erg s^{-1}. Although this may not be strictly the case in reality – as in the current argument concerning the

TABLE 1. Estimated rates of short GRBs and plausible progenitors in yr^{-1} Gpc^{-3}

Progenitor	Rate*($z=0$)
NS–NS	80
BH–NS	10–300
BH–WD	10
BH–He	1000
NS AIC	50
SN Ib/c	60000
SN Ia	150000
SGRBs	$0.2(4\pi/\Omega)$

* derived from [6, 23, 34]

nature of ULXs – it does exhibit the qualitative nature of the effect of radiation pressure on accreting plasma, in the limit of large optical depth.

The photon luminosity, for the few-second duration of a typical short burst, is of course colossal: it exceeds by many thousands the most extreme output from any active galactic nucleus (thought to involve super massive black holes), and is 12 orders of magnitude above the Eddington limit for a stellar-mass object. The total energy, however, is not out of line with some other phenomena encountered in astrophysics - indeed it is reminiscent of the energy released in the core of a supernova. The Eddington photon limit is circumvented if the main cooling agent is emission of neutrinos rather than electromagnetic waves. This regime requires correspondingly large accretion rates, of the order of one solar mass per second, and is termed hypercritical accretion. Such high accretion rates are never reached for black holes in XRBs or AGN, where characteristic rates are below the Eddington rate. They can, however, be achieved in the process of forming neutron stars and solar-mass black holes in the core collapse of massive stars. In such a situation, the densities and temperatures are so large ($\rho \simeq 10^{12}$ g cm^{-3}, $T \simeq 10^{11}$ K) that photons are completely trapped, and neutrinos are emitted copiously.

The current view is that short GRBs arise in a very small fraction $\sim 10^{-6}$ of stars which undergo a catastrophic energy release event toward the end of their evolution. One conventional possibility is the coalescence of binary neutron stars [13, 22, 4, 21, 27, 30]. Double neutron star (NS) binaries, such as the famous PSR1913+16, will eventually coalesce due to angular momentum and energy losses to gravitational radiation. When a neutron star binary coalesces, the rapidly-spinning merged system could be too massive to form a single neutron star; on the other hand, the total angular momentum is probably too large to be swallowed immediately by a black hole [28]. The expected outcome would then be a spinning hole, orbited by a torus of NS debris.

Other types of progenitor have been suggested - e.g. a NS-BH merger, where the neutron star is tidally disrupted before being swallowed by the hole [21, 12, 11]; the merger of a WD with a black hole [6]; or accretion induced collapse (AIC) of a NS [36, 19], where the collapsing neutron star has too much angular momentum to collapse quietly into a black hole. Table 1 provides a summary of the various rate estimates for some of these short GRB progenitors, while Figure 1 illustrates their different production

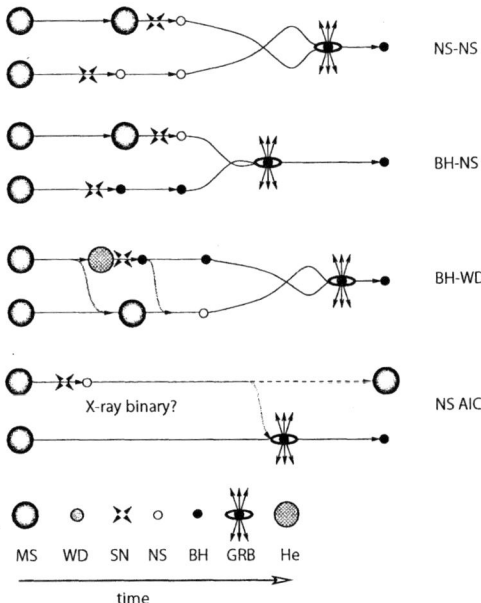

FIGURE 1. Schematic scenarios for plausible short GRB progenitors. The dominant production channel for each scenario is depicted, where MS denotes the primary main sequence star. The (rough) relative in-spiral times due to gravitational radiation for compact mergers are shown.

channels. Aside from the rate of SNe events, the rate of short GRBs and plausible progenitors in Table 1 are uncertain by at least a factor of a few. All of the progenitor scenarios listed roughly scale with the rate of star formation; therefore, the rates at redshift of $z = 1$ are a factor of ~ 10 higher than locally.

METABOLIC PATHWAYS

It has become increasingly apparent in the last few years that most plausible short GRB progenitors suggested so far (e.g. NS-NS or NS-BH, WD-BH mergers, and NS AIC) are expected to lead to a black hole plus debris torus system – although a possible exemption includes the formation from accretion[35] or from a WD-WD merger[17] of a fast rotating neutron star with an ultrahigh magnetic field. In this case there is no external agent feeding the accretion disk, and thus the event is over roughly on an accretion timescale (which would be on the order of one second). The binding energy of the orbiting debris, and the spin energy of the BH are the two main reservoirs in the former case. The first provides up to 42% of the rest mass energy of the torus, while the latter can grant up to 29% (for a maximal spin rate) of the mass of the black hole itself.

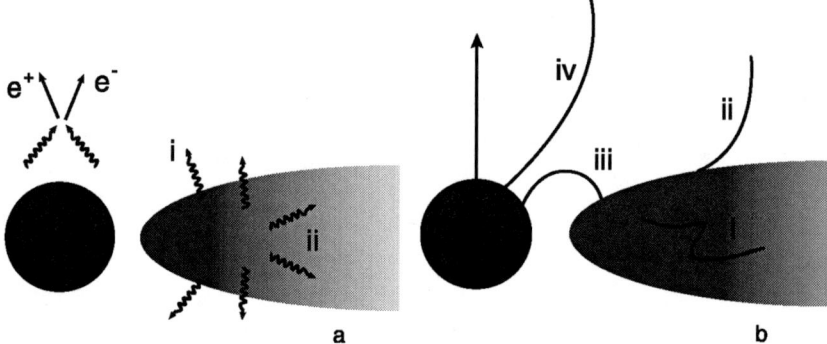

FIGURE 2. Metabolic pathways for the extraction of energy. *Panel a:* Energy released as neutrinos is reconverted via collisions outside the dense core into electron-positron pairs or photons (i). The neutrinos that are emitted from the inner regions of the debris deposit part of their energy in the outer parts of the disc (ii), driving a strong baryonic outflow. This wind may be responsible for collimating the jet. *Panel b:* Strong magnetic fields anchored in the dense matter can convert the binding and/or spin energy into a Poynting outflow. A dynamo process of some kind is widely believed to be able to operate in accretion discs, and simple physical considerations suggest that fields generated in this way would have a canonical length-scale of the order of the disc thickness (i). Open field lines can connect the disc outflow and may drive a hydromagnetic wind (ii). The above mechanism can tap the binding energy of the debris torus. A rapidly rotating hole could contain a larger energy reservoir. This energy could be extracted in principle through MHD coupling to the rotation of the hole (iii;iv).

How can the energy be transformed into outflowing relativistic plasma? There seem to be two options. The energy released as thermal neutrinos is expected to be reconverted, via collisions outside the dense core, into electron-positron pairs and photons. Alternatively, strong magnetic fields anchored in the dense matter could convert the gravitational binding energy of the system into a Poynting-dominated outflow. A brief summary of the various metabolic pathways is presented in Figure 2.

SUPERNOVA FAMILY TIES

Current observational limits[10, 3] indicate that any supernova-like event accompanying short GRBs would have to be over 50 times fainter than normal Type Ia SNe or Type Ic hypernovae, 5 times fainter than the faintest known Ia or Ic SNe, and fainter than the faintest known Type II SNe. These limits strongly constrain progenitor models for short GRBs.

It should be noted that the merger of compact objects does not necessarily imply the absence of optical or other long-wavelength phenomena after the GRB. Neutron-rich material may be dynamically ejected from a NS-NS or a NS-BH merger. Its subsequent decompression may synthesize radioactive elements through the r- process, whose radioactive decay could power an optical transient[18]. Figure 3 shows the predicted emission of the simple, analytic model of [18] for different parameters along with upper limits for the GRB 050509b (assuming the redshift $z = 0.2248$ of its tentative host galaxy).

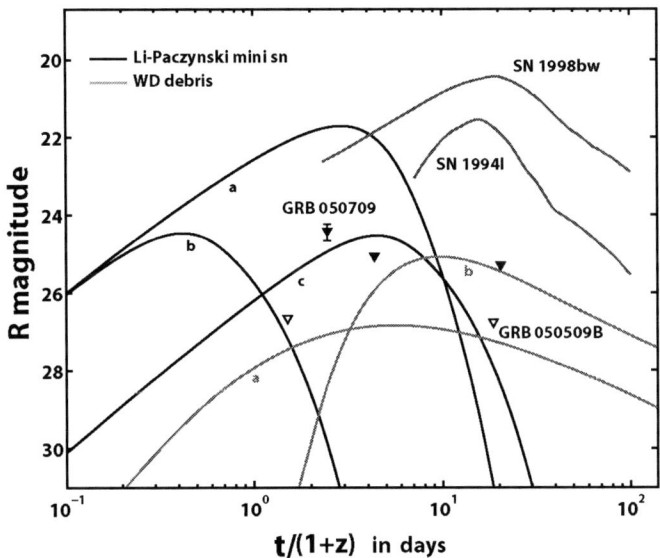

FIGURE 3. Magnitudes and upper limits of short GRB afterglows (GRB 050709-filled triangles- and GRB 050509b -empty triangles), together with the predictions (black lines a, b and c) of the 'mini-SN' model of [18] scaled to the redshift of GRB 050509b (z=0.225). The three curves are for different values of the model parameters: the total ejected mass, its velocity, and the fraction of the rest energy that is converted to heat through radioactive decay. The light curve is expected to peak after approximately a day, when the ejecta becomes optically thin. A quasi thermal spectrum is expected, as in a SN, which peaks around the optical or near UV, with a reasonably large theoretical uncertainty. The gray lines show the expected emission from the debris disk in (a) a BH-WD and (b) WD-WD merger with 0.3 and 0.9 solar masses of ejecta, respectively. For comparison, two Type Ic supernovae are also shown: SN 1998bw was associated with the long GRB 980425.

The optical flux from a 'mini-SN' associated with GRB 050509b should have been a factor of $\sim 10^3 (f/10^{-3})(M/0.01 M_\odot)^{1/2}(3v/c)^{1/2}$ higher than the upper limit derived by [10] at $t = 1.85$ days, where M and v are the mass and velocity of the ejected material, and f is the fraction of its rest energy that goes into radioactive decay. For a kinetic energy of $10^{51} E_{51}$ erg, where $E_{51} = (M/0.01 M_\odot)(3v/c)^2 \sim 1$, varying M and v within a reasonable range ($0.003 < M/M_\odot < 1$ and $0.03 < v/c < 0.5$) would not change the optical luminosity by more than one order of magnitude. A larger uncertainty is the value of f, which reflects the amount of radioactive material synthesized in the accompanying NS-NS or NS-BH wind. From the above simple arguments we derive an approximate upper limit of $f < 10^{-5}$.

We note here that the most efficient conversion of nuclear energy to the observable luminosity is provided by the elements with a decay timescale comparable to t_τ. In reality, there is likely to be a large number of nuclides with a very broad range of decay timescales. Our limits thus place interesting constraints on the abundances and the lifetimes of the radioactive nuclides that form in the rapid decompression of nuclear-density matter – they should be either very short or very long when compared to t_τ so that radioactivity is inefficient in generating a large luminosity.

Scenarios involving a massive WD have been considered as triggers for SHBs, and can in principle be also tested for. As illustrated in Figure 3, if a short GRB arises following the disruption of WD, a large and massive remnant debris disk could power a substantial energy release through nuclear reactions.

CONGENITAL TIME SCALES AND ENERGETICS

A question that has remained largely unanswered so far is what determines the typical duration of a short GRB, which can extend to tenths of seconds. This is of course long in comparison with the dynamical or orbital time scale for the models described above. While bursts can easily be derived from a very short, impulsive energy input, this is generally unable to account for a large fraction of short bursts which show complicated light curves. This hints at the desirability for a central engine lasting much longer than a typical dynamical time scale[26].

As we argue above, the various progenitors differ only slightly in the mass of the hole and that of the debris torus they produce, however, they may differ more markedly in the time the gas mass remains orbiting. In a coalescing compact binary (e.g. NS-NS or BH-NS mergers), the draining time is largely determined by the viscous timescale of the accreting gas[15, 32], while in a WD-BH merger, the large potential energy available when the WD material accretes into the hole provides a longer lasting supply of gas. If the trigger of a short burst involves a black hole, then an acceptable model requires that the surrounding torus should not completely drain into the hole, or be otherwise dispersed, on too short a time scale.

Once a disk is formed, the energy output depends on its initial mass, M_t and temperature. We have recently calculated [16] a realistic set of time-dependent models for their dynamical evolution, covering the typical duration time scales of short GRBs (a few tenths of a second). From the resulting neutrino luminosities we have computed the total energy deposition that could drive a relativistic outflow through $\nu\bar{\nu}$ annihilation, assuming a 1% efficiency at $L_\nu = 10^{53}$ erg s^{-1} [24] and its duration. The results for various disk masses and effective α-disk viscosities are shown in Figure 4 (joined square symbols). The total output $E_{\nu\bar{\nu}} \simeq 10^{49}[M_t/0.03M_\odot]^2$ erg, is very roughly independent of the inferred duration, which increases with decreasing disk viscosity since the overall evolution is slower. The strong dependence on disk mass is simply a reflection of the sensitivity of the neutrino emission rates on temperature ($e^\pm$ capture on free nucleons dominates the cooling rate, with emissivity $\dot{q} \propto \rho T^6$).

Assuming full equipartition of the magnetic field energy density with the internal energy in the fluid in the inner disk gives the estimates shown in Figure 4 (round symbols). The energy flux is sensitive primarily to the equatorial density in the accretion flow. It is thus initially roughly constant, and then drops on an accretion (i.e. viscous) timescale. This explains the energy–duration correlation in Figure 4. Since the observed flux sets the threshold for burst detection, neutrino powered events will enhance the relative importance of shorter events (since $E_{\nu\bar{\nu}} \sim$ const, then $L_{\nu\bar{\nu}} \propto t^{-1}$), while magnetically dominated short GRBs more truthfully reflect the underlying intrinsic distribution (since $L_{BZ} \sim$ const, then $E_{\nu\bar{\nu}} \propto t$). Relaxing the assumption of full equipartition will

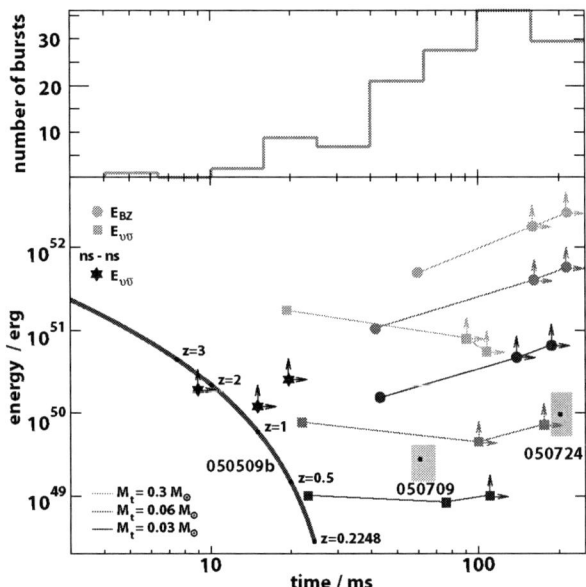

FIGURE 4. *Top:* Histogram of observed short burst durations taken from BATSE. *Bottom:* Energy–duration relation[25] for GRB 050509b (black line), GRB 050709 and GRB 050724. The square (round) joined symbols show the total isotropic energy release (assuming collimation of the outflow into $\Omega_b = 4\pi/10$) and duration (t_{50}) for $\nu\bar{\nu}$ annihilation (Blandford–Znajek)–powered bursts, as computed from our 2D disk evolution models. The range in initial disk mass covers one order of magnitude and the effective disk viscosities are $\alpha = 10^{-1}, 10^{-2}, 10^{-3}$ (left to right). Many of the estimates are lower limits because at the end of our calculations not enough mass had been drained from the disk for the luminosity to drop appreciably. The stars correspond to $\nu\bar{\nu}$–driven outflows in NS-NS mergers, as calculated by [29].

lower the total energy budget accordingly. The dependence on disk mass is different, with $E_{BZ} \simeq 5 \times 10^{50} [M_t/0.03 M_\odot][\alpha/10^{-1}]^{-0.55}$ erg. Our estimates assume that whatever seed field was present has been amplified to the correspondingly high values extremely rapidly. Whether this will actually occur is unclear, particularly for the shortest events, as the field can grow only in a time scale associated with proto–neutron star–like convection or differential rotation in the case of the MRI.

In brief, we have shown here that the energetics and durations of short GRBs can be accounted for by small, dense disks around stellar mass black holes, based on dynamical modeling of such systems. However, long duration (~ 100 s) X-ray flares have been recently observed to follow several short GRBs[37, 1], e.g., GRB 050724, after a delay of ~ 30 s. There is also independent support that X-ray emission on these timescales is detected when light curves of many bursts are stacked[14]. This is hard to reconcile with current models, which suggest that the mergers of compact objects will lead to brief energy input episodes, typically of the order of the duration of a short burst[15]. The flare emission could, however, be produced, for example, by the interaction of an extended -

possibly magnetically dominated - ejecta emanating from a short GRB with the envelope of a giant companion star[19]. This more isotropic GRB ejecta component need not necessarily dominate the total burst energetics, but it can be efficiently reprocessed by the envelope of the companion star at distances $\sim 10^{12}$ cm, into an X-ray flare with a luminosity and timescale comparable to the observed values. In this model, the burst is triggered by the collapse of a neutron star after accreting matter from the companion.

LIFETIMES, HOSTS AND COSMOLOGICAL SETTING

The distribution of time delays for short GRB is not yet well understood. It can be constrained using, for example, the burst rate as a function of redshift [20, 8]. This method makes use of the luminosity function of short GRBs and its redshift evolution, which are uncertain. In [39], we propose to use the local rates of short GRBs in different types of galaxies as an alternative and complementary method to constrain the time delay distribution.

In the local universe, about 55–70% of the stellar mass is in early-type galaxies and the corresponding stars mainly formed about 10 Gyr ago. Three of the four host galaxies of short GRBs found so far are associated with old and massive galaxies with little current or recent star formation, which makes it unlikely that short bursts are associated with massive stars. The lifetime of the progenitor systems can be estimated by using the star formation history of elliptical galaxies from a galaxy formation model. This allow us to separate the early- and late-type galaxy contributions to the overall cosmic star formation history. Presently available data suggests, but not yet prove, a long time delay between the formation of the progenitor system and the short GRB outburst — for progenitors that can outburst within a Hubble time, about half of them have lifetime longer than $\sim$10 Gyr. The progenitors of short GRBs thus appear to be longer lived than those of Type Ia supernovae. Models of the probability distribution of time delays, here parametrized as $P(\tau) \propto \tau^n$, with $n > 3/2$ are favored.

Two of the four short GRB host galaxies are found in cluster environments[3, 25]. To study the association of short GRBs with clusters, it would be useful to separate the stellar mass function into that for field galaxies and that for cluster galaxies in addition to early- and late-type galaxies. More promising for the immediate future, the preponderance of cluster environments can be investigated observationally. Important information may be gained by studying the local stellar mass inventory. Approximately 50% of the stellar mass contents in early-type galaxies are in galaxies with $M_{\text{star}} > 10^{11} M_\odot$ that typically reside in clusters. Since it is likely that in clusters galaxies shut off their star formation process early on, a long progenitor lifetime further increases the tendency for short GRBs to happen in cluster galaxies. It is fair to conclude that the observed preponderance of cluster environments for short GRBs is consistent with an old stellar population that preferentially resides in early-type galaxies.

Detailed observations of the astrophysics of individual GRB host galaxies may be essential before stringent constraints on the lifetime of short GRB progenitors can be placed. If confirmed with further host observations, this tendency of short GRB progenitors to be relatively old can help differentiate between various ways of forming a short GRB.

PROSPECTS

Much progress has been made in the past year in understanding the nature of cosmological short GRBs. Still, various alternative ways of triggering the explosions responsible for short GRBs remain: NS-NS, NS-BH, BH-WD or WD-WD binary mergers, recycled magnetars, spun-down supra-massive NS and accretion induced collapse of a NS. Can we decide between the various alternatives? The progenitors of short GRBs are essentially masked by afterglow emission, largely featureless synchrotron light, which reveals little more than the basic energetics and micro physical parameters of relativistic shocks. In the absence of a supernova-like feature (Figure 3), the interaction of GRB ejecta with a stellar binary companion may be the only observable signature in the foreseeable future shedding light on the identity of the progenitors. A definitive understanding will, however, come with the observations of concurrent gravitational radiation or neutrino signals arising from the dense, opaque central engine.

We must also remain aware of other possibilities. There still remain a number of mysteries, especially concerning the identity of their progenitors, the nature of the triggering mechanism, the transport of the energy and the time scales involved. It could be that the accretion-induced collapse of a WD, or (for some equations of state) the merger of two neutron stars, could give rise to a rapidly-spinning magnetar, temporarily stabilized by rapid rotation. The afterglow could then, at least in part, be due to a magnetar continuing power output. It could also be that mergers of unequal mass NS, or NS with other compact companions, lead to the delayed formation of a black hole. Such events might also lead to repeating episodes of accretion[31] and orbit separation, or to the eventual explosion of a NS which has dropped below the critical mass, all of which would provide a longer time scale, episodic energy output. And there could be more subclasses of GRBs than just short ones and long ones. And if the class of models that we have advocated here turns out to be irrelevant, the explanation of short GRBs will surely turn out to be even more remarkable and fascinating.

ACKNOWLEDGMENTS

I am especially grateful to M. Rees, W. Lee, S. Rosswog, A. MacFadyen, J. Granot, Z. Zheng, J. Hjorth, J. Bloom, Dave Pooley and J. Prochaska for extended collaboration, and to Stan Woosley and N. Gehrels for discussions.

REFERENCES

1. S. D. Barthelmy, et al., Nature **438**, 994
2. E. Berger, et al., Nature **438**, 988 (2005)
3. J. S. Bloom, et al., ApJ **638**, 354 (2006)
4. D. Eichler, M. Livio, T. Piran, and D. N. Schramm, Nature **340**, 126 (1989)
5. D. B. Fox, et al, Nature **437**, 845 (2005)
6. C. L. Fryer, S. E. Woosley, and D. H. Hartmann, ApJ **526**, 152 (1999)
7. N. Gehrels, et al., 2005, Nature **437**, 851 (2005)
8. D. Guetta, and T. Piran, A&A **435**, 421 (2005)
9. J. Hjorth, et al., 2003, Nature **423**, 847 (2003)

10. J. Hjorth, et al., 2005, ApJ **630**, L117
11. H.-T. Janka, T. Eberl, M. Ruffert, and C. Fryer, ApJ **527**, L39 (1999)
12. W. Kluźniak, and W. H. Lee, ApJ **494**, L53 (1998)
13. J. M. Lattimer, D. N. Schramm, ApJ **210**, 549 (1976)
14. D. Lazzati, E. Ramirez-Ruiz, and G. Ghisellini, A&A **379**, L39 (2001)
15. W. H. Lee, E. Ramirez-Ruiz, and D. Page, ApJ **608**, L5 (2004)
16. W. H. Lee, E. Ramirez-Ruiz,and D. Page, ApJ **632**, 421 (2005)
17. A. J. Levan, et al., (astro-ph/0601332)
18. L. X. Li, and B. Paczyński, ApJ, **507**, L59 (1998)
19. A. I. MacFadyen, E. Ramirez-Ruiz, and W. Zhang, (astro-ph/0510192)
20. E. Nakar, A. Gal-Yam, and D. B. Fox, (astro-ph/0511254)
21. R. Narayan, B. Paczynski,and T. Piran, ApJ **395**, L83 (1992)
22. B. Paczyński, ApJ **308**, L43 (1986)
23. E. S. Phinney, ApJ **380**, L17 (1991)
24. R. Popham, S. E. Woosley, and C. L. Fryer, ApJ **518**, 356 (1999)
25. Prochaska, J. X. et al., (astro-ph/0510022)
26. M. J. Rees, A&AS **138**, 491 (1999)
27. S. Rosswog et al., A&A **341**, 499 (1999)
28. S. Rosswog, and E. Ramirez-Ruiz, MNRAS **336**, L7 (2002)
29. S. Rosswog, and M. Liebendorfer, MNRAS **342**, 673 (2003)
30. S. Rosswog, E. Ramirez-Ruiz, and M. B. Davies, MNRAS **345**, 1077 (2003)
31. S. Rosswog, ApJ **634**, 1202 (2005)
32. S. Setiawan, M. Ruffert, and H.-T. Janka, MNRAS **352**, 753 (2004)
33. E. E. Salpeter, ApJ **140**, 796 (1964)
34. M. Schmidt, ApJ **523**, L117 (2001)
35. H. C. Spruit, A&A **341**, L1 (1999)
36. M. Vietri, and L. Stella, ApJ **527**, L43 (1999)
37. J. S. Villasenor, et al., Nature **437**, 855 (2005)
38. Zel'dovich Ya. B., Novikov I. D., Sov. Phys. Dok. **158**, 811 (1964)
39. Z. Zheng, and E. Ramirez-Ruiz, (astro-ph/0601622)

GRB Cosmology and the First Stars

Volker Bromm* and Abraham Loeb†

*Astronomy Department, University of Texas, Austin, TX 78712
†Astronomy Department, Harvard University, Cambridge, MA 02138

Abstract.
Gamma-Ray Bursts (GRBs) are unique probes of the cosmic star formation history and the state of the intergalactic medium up to redshifts of the first stars. In particular, the ongoing *Swift* mission might be the first observatory to detect individual Population III stars, provided that the massive, metal-free stars were able to trigger GRBs. *Swift* will empirically constrain the redshift at which Population III star formation was terminated, thus providing crucial input to models of cosmic reionization and metal enrichment.

Keywords: cosmology: theory – early universe – gamma rays: bursts
PACS: 95.30.Lz; 97.10.Bt; 97.20.Wt; 98.70.Rz

INTRODUCTION

Gamma-Ray Bursts (GRBs) are believed to originate in compact remnants (neutron stars or black holes) of massive stars. Their high luminosities make them detectable out to the edge of the visible universe [1, 2]. GRBs offer the opportunity to detect the most distant (and hence earliest) population of massive stars, the so-called Population III (Pop III henceforth), one star at a time. In the hierarchical assembly process of halos which are dominated by cold dark matter (CDM), the first galaxies should have had lower masses (and lower stellar luminosities) than their low-redshift counterparts. Consequently, the characteristic luminosity of galaxies or quasars is expected to decline with increasing redshift. GRB afterglows, which already produce a peak flux comparable to that of quasars or starburst galaxies at $z \sim 1-2$, are therefore expected to outshine any competing source at the highest redshifts, when the first dwarf galaxies have formed in the universe.

The first-year polarization data from the *Wilkinson Microwave Anisotropy Probe* (*WMAP*) indicates an optical depth to electron scattering of $\sim 17 \pm 4\%$ after cosmological recombination [3, 4]. This implies that the first stars must have formed at a redshift $z \sim 20$, and reionized a substantial fraction of the intergalactic hydrogen around that time [5, 6, 7, 8, 9]. Early reionization can be achieved with plausible star formation parameters in the standard ΛCDM cosmology; in fact, the required optical depth can be achieved in a variety of very different ionization histories since *WMAP* places only an integral constraint on these histories [10]. One would like to probe the full history of reionization in order to disentangle the properties and formation history of the stars that are responsible for it. GRB afterglows offer the opportunity to detect stars as well as to probe the ionization state [11] and metal enrichment level [12] of the intervening intergalactic medium (IGM).

GRBs, the electromagnetically-brightest explosions in the universe, should be detectable out to redshifts $z > 10$ [1, 2]. High-redshift GRBs can be identified through infrared photometry, based on the Lyα break induced by absorption of their spectrum at wavelengths below $1.216\,\mu\mathrm{m}\,[(1+z)/10]$. Follow-up spectroscopy of high-redshift candidates can then be performed on a 10-meter-class telescope. Recently, the ongoing *Swift* mission [13] has detected a GRB originating at $z \simeq 6.3$ (e.g., [14]), thus demonstrating the viability of GRBs as probes of the early universe.

There are four main advantages of GRBs relative to traditional cosmic sources such as quasars:

(i) The GRB afterglow flux at a given observed time lag after the γ-ray trigger is not expected to fade significantly with increasing redshift, since higher redshifts translate to earlier times in the source frame, during which the afterglow is intrinsically brighter [1]. For standard afterglow lightcurves and spectra, the increase in the luminosity distance with redshift is compensated by this *cosmological time-stretching* effect.

(ii) As already mentioned, in the standard ΛCDM cosmology, galaxies form hierarchically, starting from small masses and increasing their average mass with cosmic time. Hence, the characteristic mass of quasar black holes and the total stellar mass of a galaxy were smaller at higher redshifts, making these sources intrinsically fainter [15]. However, GRBs are believed to originate from a stellar mass progenitor and so the intrinsic luminosity of their engine should not depend on the mass of their host galaxy. GRB afterglows are therefore expected to outshine their host galaxies by a factor that gets larger with increasing redshift.

(iii) Since the progenitors of GRBs are believed to be stellar, they likely originate in the most common star-forming galaxies at a given redshift rather than in the most massive host galaxies, as is the case for bright quasars [11]. Low-mass host galaxies induce only a weak ionization effect on the surrounding IGM and do not greatly perturb the Hubble flow around them. Hence, the Lyα damping wing should be closer to the idealized unperturbed IGM case and its detailed spectral shape should be easier to interpret. Note also that unlike the case of a quasar, a GRB afterglow can itself ionize at most $\sim 4 \times 10^4 E_{51} M_\odot$ of hydrogen if its UV energy is E_{51} in units of 10^{51} ergs (based on the available number of ionizing photons), and so it should have a negligible cosmic effect on the surrounding IGM.

(iv) GRB afterglows have smooth (broken power-law) continuum spectra unlike quasars which show strong spectral features (such as broad emission lines or the so-called "blue bump") that complicate the extraction of IGM absorption features. In particular, the continuum extrapolation into the Lyα damping wing (the Gunn-Peterson absorption trough) during the epoch of reionization is much more straightforward for the smooth UV spectra of GRB afterglows than for quasars with an underlying broad Lyα emission line [11].

Although the nature of the central engine that powers the relativistic jets of GRBs is still unknown, recent evidence indicates that long-duration GRBs trace the formation of massive stars (e.g., [16, 17, 18, 19, 20, 21]) and in particular that long-duration GRBs are associated with Type Ib/c supernovae [22]. Since the first stars in the universe are predicted to be predominantly massive [23, 24, 25], their death might give rise to large numbers of GRBs at high redshifts. In contrast to quasars of comparable brightness, GRB afterglows are short-lived and release ~ 10 orders of magnitude less energy into the surrounding IGM. Beyond the scale of their host galaxy, they have a negligible effect on their cosmological environment[1]. Consequently, they are ideal probes of the IGM during the reionization epoch. Their rest-frame UV spectra can be used to probe the ionization state of the IGM through the spectral shape of the Gunn-Peterson (Lyα) absorption trough, or its metal enrichment history through the intersection of enriched bubbles of supernova (SN) ejecta from early galaxies [12]. Afterglows that are unusually bright (> 10mJy) at radio frequencies should also show a detectable forest of 21 cm absorption lines due to enhanced HI column densities in sheets, filaments, and collapsed minihalos within the IGM [27].

Another advantage of GRB afterglows is that once they fade away, one may search for their host galaxies. Hence, GRBs may serve as signposts of the earliest dwarf galaxies that are otherwise too faint or rare on their own for a dedicated search to find them. Detection of metal absorption lines from the host galaxy in the afterglow spectrum, offers an unusual opportunity to study the physical conditions (temperature, metallicity, ionization state, and kinematics) in the interstellar medium of these high-redshift galaxies. A small fraction (~ 10) of the GRB afterglows are expected to originate at redshifts $z > 5$ [28, 29]. This subset of afterglows can be selected photometrically using a small telescope, based on the Lyα break at a wavelength of $1.216\,\mu\mathrm{m}\,[(1+z)/10]$, caused by intergalactic HI absorption. The challenge in the upcoming years will be to follow-up on these candidates spectroscopically, using a large (10-meter class) telescope. GRB afterglows are likely to revolutionize observational cosmology and replace traditional sources like quasars, as probes of the IGM at $z > 5$. The near future promises to be exciting for GRB astronomy as well as for studies of the high-redshift universe.

IMPORTANT OPEN QUESTIONS

Using GRBs to probe the high redshift universe has great promise. In the following, we discuss some key open questions.

Cosmic Star Formation at High Redshifts

It is of great importance to constrain the Pop III star formation mode, and in particular to determine down to which redshift it continues to be prominent. The extent of the Pop III star formation will affect models of the initial stages of reionization (e.g., [8, 6, 30, 9, 31]) and metal enrichment (e.g., [32, 12, 33, 34, 35]), and will determine whether planned surveys will be able to effectively probe Pop III stars (e.g., [36]). The constraints on Pop III star formation will also determine whether the first stars could have contributed a significant fraction to the cosmic near-IR background (e.g., [37, 38, 39, 40, 41]).

---

[1] Note, however, that feedback from a single GRB or supernova on the gas confined within early dwarf galaxies could be dramatic, since the binding energy of most galaxies at $z > 10$ is lower than 10^{51} ergs [26].

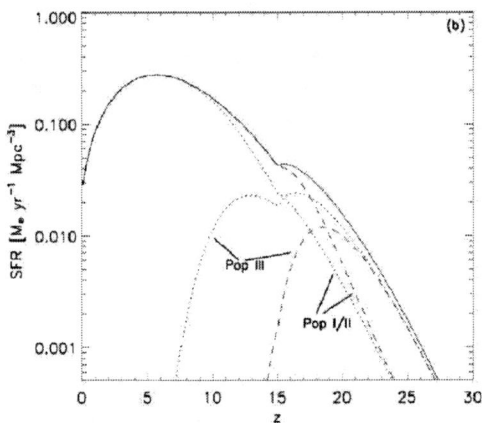

FIGURE 1. Cosmic comoving star formation rate (SFR) in units of $M_\odot$ yr^{-1} Mpc^{-3}, as a function of redshift (from [29]). We assume that cooling in primordial gas is due to atomic hydrogen only, a star formation efficiency of $\eta_* = 10\%$, and reionization beginning at $z_{\text{reion}} \approx 17$. *Solid line:* Total comoving SFR. *Dotted lines:* Contribution to the total SFR from Pop I/II and Pop III for the case of weak chemical feedback. *Dashed lines:* Contribution to the total SFR from Pop I/II and Pop III for the case of strong chemical feedback. Pop III star formation is restricted to high redshifts, but extends over a significant range, $\Delta z \sim 10-15$.

To constrain high-redshift star formation, one has to carry out a two-step approach: *(1) What is the signature of GRBs that originate in metal-free, Pop III progenitors?* Simply knowing that a given GRB came from a high redshift is not sufficient to reach a definite conclusion as to the nature of the progenitor. Pregalactic metal enrichment was likely quite inhomogeneous, and we expect normal Pop I and II stars to exist in galaxies that were already metal-enriched at these high redshifts [29]. Pop III and Pop I/II star formation is thus predicted to have occurred concurrently at $z > 5$. How is the predicted high mass-scale for Pop III stars reflected in the observational signature of the resulting GRBs? Our preliminary results indicate that circumburst densities are systematically higher in Pop III environments. GRB afterglows will then be much brighter than for conventional GRBs. In addition, due to the systematically increased progenitor masses, the Pop III distribution may be biased toward long-duration events.

(2) The modelling of Pop III cosmic star formation histories has a number of free parameters, such as the star formation efficiency and the strength of the chemical feedback. The latter refers to the timescale for, and spatial extent of, the distribution of the first heavy elements that were produced inside of Pop III stars, and subsequently dispersed into the IGM by supernova blast waves. Comparing with theoretical GRB redshift distributions one can use the GRB redshift distribution observed by *Swift* to calibrate the free model parameters. In particular, one can use this strategy to measure the redshift where Pop III star formation terminates.

In Figure 1 and 2, we illustrate this approach (based on [29]). Figure 2 leads to the robust expectation that $\sim 10\%$ of all *Swift* bursts should originate at $z > 5$. This prediction is based on the contribution from Population I/II stars which are known to

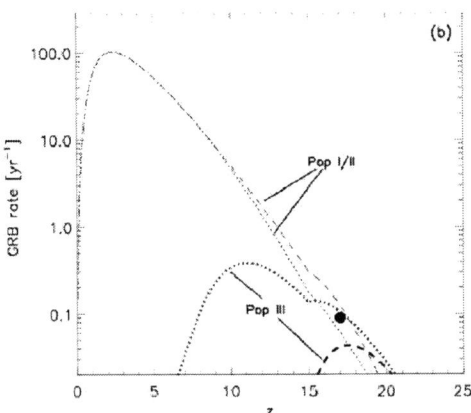

FIGURE 2. Predicted GRB rate to be observed by *Swift* (from [29]). Shown is the observed number of bursts per year, $dN^{\rm obs}_{\rm GRB}/d\ln(1+z)$, as a function of redshift. All rates are calculated with a constant GRB efficiency, $\eta_{\rm GRB} \simeq 2 \times 10^{-9}$ bursts $M_\odot^{-1}$, using the cosmic SFRs from Fig. 1. *Dotted lines:* Contribution to the observed GRB rate from Pop I/II and Pop III for the case of weak chemical feedback. *Dashed lines:* Contribution to the GRB rate from Pop I/II and Pop III for the case of strong chemical feedback. *Filled circle:* GRB rate from Pop III stars if these were responsible for reionizing the universe at $z \sim 17$.

exist even at these high redshifts. Additional GRBs could be triggered by Pop III stars, with a highly uncertain efficiency. Assuming that long-duration GRBs are produced by the collapsar mechanism, a Pop III star with a close binary companion provides a plausible GRB progenitor. We have estimated the Pop III GRB efficiency, reflecting the probability of forming sufficiently close and massive binary systems, to lie between zero (if tight Pop III binaries do not exist) and ~ 10 times the empirically inferred value for Population I/II (due to the increased fraction of black hole forming progenitors among the massive Population III stars).

A key ingredient in determining the underlying star formation history from the observed GRB redshift distribution is the GRB luminosity function, which is only poorly constrained at present. The improved statistics provided by *Swift* will enable the construction of an empirical luminosity function. With an improved luminosity function, we will be able to re-calibrate the theoretical prediction in Figure 2 more reliably.

Physical Properties of GRB Hosts

In order to predict the observational signature of high-redshift GRBs, it is important to know the properties of the GRB host systems. Within variants of the popular CDM model for structure formation, where small objects form first and subsequently merge to build up more massive ones, the first stars are predicted to form at $z \sim 20$–30 in minihalos of total mass (dark matter plus gas) $\sim 10^6 M_\odot$ [42, 26, 9]. These objects are the sites for the formation of the first stars, and thus are the potential hosts of the highest-redshift GRBs. *What is the environment in which the earliest GRBs and their*

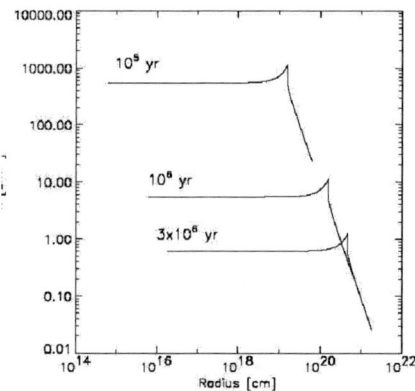

FIGURE 3. Effect of photoheating from a Population III star on the density profile in a high-redshift minihalo. The curves, labeled by the time after the onset of the central point source, are calculated according to a self-similar model for the expansion of an HII region. Numerical simulations closely conform to this analytical behavior. Notice that the central density is significantly reduced by the end of the life of a massive star, and that a central core has developed where the density is constant.

afterglows did occur? This problem breaks down into two related questions: (i) what type of stars (in terms of mass, metallicity, and clustering properties) will form in each minihalo?, and (ii) how will the ionizing radiation from each star modify the density structure of the surrounding gas? These two questions are fundamentally intertwined. The ionizing photon production strongly depends on the stellar mass, which in turn is determined by how the accretion flow onto the growing protostar proceeds under the influence of this radiation field. In other words, the assembly of the Population III stars and the development of an HII region around them proceed simultaneously, and affect each other. As a first step (see Figure 3), we describe the photo-evaporation as a self-similar champagne flow [43], with parameters appropriate for the Population III case.

Notice that the central density is significantly reduced by the end of the life of a massive star, and that a central core has developed where the density is nearly constant. Such a flat density profile is markedly different from that created by stellar winds ($\rho \propto r^{-2}$). Winds, and consequently mass-loss, may not be important for massive Population III stars [44, 45], and such a flat density profile may be characteristic of GRBs that originate from metal-free Population III progenitors.

We have carried out idealized simulations of the protostellar accretion problem, allowing us to estimate the final mass of a Population III star [47]. Using the smoothed particle hydrodynamics (SPH) method, we have included the chemistry and cooling physics relevant for the evolution of metal-free gas (see [24] for details). Improving on earlier work [48, 24] by initializing our simulations according to the ΛCDM model, we have focused on an isolated overdense region that corresponds to a 3σ−peak [47]: a halo containing a total mass of $10^6 M_\odot$, and collapsing at a redshift $z_{\rm vir} \simeq 20$.

We have found that one high-density clump has formed at the center of the minihalo, possessing a gas mass of a few hundred solar masses. Soon after its formation, the

FIGURE 4. Supernova explosion in the high-redshift universe (from [46]). The snapshot is taken $\sim 10^6$ yr after the explosion with total energy $E_{\rm SN} \simeq 10^{53}$ ergs. We show the projected gas density within a box of linear size 1 kpc. The SN bubble has expanded to a radius of ~ 200 pc, having evacuated most of the gas in the minihalo. *Inset:* Distribution of metals. The stellar ejecta (*gray dots*) trace the metals and are embedded in pristine metal-poor gas (*black dots*).

clump becomes gravitationally unstable and undergoes runaway collapse. Once the gas has exceeded a threshold density of 10^7 cm^{-3}, a sink particle is inserted into the simulation. This choice for the density threshold ensures that the local Jeans mass is resolved throughout the simulation. The clump (i.e., sink particle) has an initial mass of $M_{\rm Cl} \simeq 200 M_\odot$, and grows subsequently by ongoing accretion of surrounding gas. High-density clumps with such masses result from the chemistry and cooling rate of molecular hydrogen, H$_2$, which imprint characteristic values of temperature, $T \sim 200$ K, and density, $n \sim 10^4$ cm^{-3}, into the metal-free gas [24]. Evaluating the Jeans mass for these characteristic values results in $M_J \sim$ a few $\times 10^2 M_\odot$, which is close to the initial clump masses found in the simulations.

The high-density clumps are clearly not stars yet. To probe the subsequent fate of a clump, we have re-simulated the evolution of the central clump with sufficient resolution to follow the collapse to higher densities (see [49, 50] for a description of the refinement technique). Our refined simulation enables us to study the three-dimensional accretion flow around the protostar (see also [51, 52, 53, 54]). We now allow the gas to reach densities of 10^{12} cm^{-3} before being incorporated into a central sink particle. At these high densities, three-body reactions [55] have converted the gas into a fully molecular form. We follow the growth of the molecular core over the first $\sim 10^4$ yr after its formation, making the idealized assumption that the protostellar radiation does not affect the accretion flow. The accretion rate is initially very high, $\dot{M}_{\rm acc} \sim 0.1 M_\odot$ yr^{-1}, and subsequently declines roughly as a power law of time. The mass of the molecular core, taken as a crude estimate for the protostellar mass, grows approximately as: $M_* \sim \int \dot{M}_{\rm acc} dt \simeq 0.8 M_\odot (t/1 \text{ yr})^{0.45}$. A robust upper limit for the final mass of the star is then: $M_*(t = 3 \times 10^6 \text{yr}) \sim 500 M_\odot$. In deriving this upper bound, we have conservatively assumed that accretion cannot go on for longer than the total lifetime of a massive star.

Our numerical results can be understood within the general theoretical framework of how stars form [56]. Star formation typically proceeds from the 'inside-out', through the accretion of gas onto a central hydrostatic core. Whereas the initial mass of the hydrostatic core is very similar for primordial and present-day star formation [57], the accretion process – ultimately responsible for setting the final stellar mass – is expected to be rather different. On dimensional grounds, the accretion rate is simply related to the sound speed cubed over Newton's constant (or equivalently given by the ratio of the Jeans mass and the free-fall time): $\dot{M}_{\rm acc} \sim c_s^3/G \propto T^{3/2}$. A simple comparison of the temperatures in present-day star forming regions ($T \sim 10$ K) with those in primordial ones ($T \sim 200 - 300$ K) already indicates a difference in the accretion rate of more than two orders of magnitude.

Can a Population III star ever reach this asymptotic mass limit? The answer to this question is not yet known with any certainty, and it depends on whether the accretion from a dust-free envelope is eventually terminated by feedback from the star (e.g., [51, 52, 53, 58, 54]). The standard mechanism by which accretion may be terminated in metal-rich gas, namely radiation pressure on dust grains [59], is obviously not effective for gas with a primordial composition. Recently, it has been speculated that accretion could instead be turned off through the formation of an HII region [58], or through the radiation pressure exerted by trapped Lyα photons [54]. The termination of the accretion process defines the current unsolved frontier in studies of Population III star formation.

The first galaxies may be surrounded by a shell of highly enriched material that was carried out in a SN-driven wind (see Fig. 4). A GRB in that galaxy may show strong absorption lines at a velocity separation associated with the wind velocity. Simulating these winds and calculating the absorption profile in the featureless spectrum of a GRB afterglow, will allow us to use the observed spectra of high-z GRBs to directly probe the degree of metal enrichment in the vicinity of the first star forming regions (see [12] for a semi-analytic treatment).

As the early afterglow radiation propagates through the interstellar environment of the GRB, it will likely modify the gas properties close to the source; these changes could in turn be noticed as time-dependent spectral features in the spectrum of the afterglow and used to derive the properties of the gas cloud (density, metal abundance, and size). The UV afterglow radiation can induce detectable changes to the interstellar absorption features of the host galaxy [60]; dust destruction could have occurred due to the GRB X-rays [61, 62], and molecules could have been destroyed near the GRB source [63]. Quantitatively, all of the effects mentioned above strongly depend on the exact properties of the gas in the host system.

Population III Progenitors

Most studies to date have assumed a constant efficiency of forming GRBs per unit mass of stars. This simplifying assumption may either lead to an overestimation or underestimation of the frequency of GRBs. Metal-free stars are thought to be massive [23, 24] and their extended envelopes may suppress the emergence of relativistic jets out of their surface (even if such jets are produced through the collapse of their core to a spinning black hole). On the other hand, low-metallicity stars are expected to have weak winds with little angular momentum loss during their evolution, and so they may preferentially yield rotating central configurations that make GRB jets after core collapse.

What kind of metal-free, Pop III progenitor stars may lead to GRBs? Long-duration GRBs appear to be associated with Type Ib/c supernovae [22], namely progenitor massive stars that have lost their hydrogen envelope. This requirement is explained theoretically in the collapsar model, in which the relativistic jets produced by core collapse to a black hole are unable to emerge relativistically out of the stellar surface if the hydrogen envelope is retained [64]. The question then arises as to whether the lack of metal line-opacity that is essential for radiation-driven winds in metal-rich stars, would make a Pop III star retain its hydrogen envelope, thus quenching any relativistic jets and GRBs.

Aside from mass transfer in a binary system, individual Pop III stars could lose their hydrogen envelope due to either: (i) violent pulsations, particularly in the mass range $100-140M_\odot$, or (ii) a wind driven by helium lines. The outer stellar layers are in a state where gravity only marginally exceeds radiation pressure due to electron-scattering (Thomson) opacity. Adding the small, but still non-negligible contribution from the bound-free opacity provided by singly-ionized helium, may be able to unbind the atmospheric gas. Therefore, mass-loss might occur even in the absence of dust, or any heavy elements.

REFERENCES

1. Ciardi, B., & Loeb, A. 2000, ApJ, 540, 687
2. Lamb, D. Q., & Reichart, D. E. 2000, ApJ, 536, 1
3. Kogut, A., et al. 2003, ApJS, 148, 161
4. Spergel, D. N., et al. 2003, ApJS, 148, 175
5. Cen, R. 2003, ApJ, 591, L5
6. Ciardi, B., Ferrara, A., & White, S.D.M. 2003, MNRAS, 344, L7
7. Somerville, R. S., & Livio, M. 2003, ApJ, 593, 611
8. Wyithe, J. S. B., & Loeb, A. 2003, ApJ, 588, L69
9. Yoshida, N., Bromm, V., & Hernquist, L. 2004, ApJ, 605, 579
10. Haiman, Z., & Holder, G. P. 2003, ApJ, 595, 1
11. Barkana, R., & Loeb, A. 2004, ApJ, 601, 64
12. Furlanetto, S. R., & Loeb, A. 2003, ApJ, 588, 18
13. Gehrels, N., et al. 2004, ApJ, 611, 1005
14. Haislip, J., et al. 2005, Nature, submitted (astro-ph/0509660)
15. Wyithe, J. S. B., & Loeb, A. 2002, ApJ, 581, 886
16. Totani, T. 1997, ApJ, 486, L71
17. Wijers, R.A.M.J., Bloom, J.S., Bagla, J.S., & Totani, T. 1997, ApJ, 486, L71 1998, MNRAS, 294, L13
18. Blain, A. W., & Natarajan, P. 2000, MNRAS, 312, L35
19. Kulkarni, S. R., et al. 2000, Proc. SPIE, 4005, 9
20. Bloom, J. S., Kulkarni, S. R., & Djorgovski, S. G. 2002, AJ, 123, 1111
21. Natarajan, P., Albanna, B., Hjorth, J., Ramirez-Ruiz, E., Tanvir, N., & Wijers, R.A.M.J. 2005, MNRAS, 364, L8
22. Stanek, K. Z., et al. 2003, ApJ, 591, L17
23. Abel, T., Bryan, G., & Norman, M. L. 2002, Science, 295, 93
24. Bromm, V., Coppi, P. S., & Larson, R. B. 2002, ApJ, 564, 23
25. Bromm, V., & Larson, R. B. 2004, ARAA, 42, 79

26. Barkana, R., & Loeb, A. 2001, Phys. Rep., 349, 125
27. Furlanetto, S. R., & Loeb, A. 2002, ApJ, 579, 1
28. Bromm, V., & Loeb, A. 2002, ApJ, 575, 111
29. Bromm, V., & Loeb, A. 2006, ApJ, in press (astro-ph/0509303)
30. Sokasian, A., Yoshida, N., Abel, T., Hernquist, L., & Springel, V. 2004, MNRAS, 350, 47
31. Alvarez, M. A., Bromm, V., & Shapiro, P. R. 2006, ApJ, in press (astro-ph/0507684)
32. Mackey, J., Bromm, V., & Hernquist, L. 2003, ApJ, 586, 1
33. Furlanetto, S. R., & Loeb, A. 2005, ApJ, 634, 1
34. Schaye, J., Aguirre, A., Kim, T.-S., Theuns, T., Rauch, M., & Sargent, W. L. W. 2003, ApJ, 596, 768
35. Simcoe, R. A., Sargent, W. L. W., & Rauch, M. 2004, ApJ, 606, 92
36. Scannapieco, E., Madau, P., Woosley, S., Heger, A., & Ferrara, A. 2005, ApJ, 633, 1031
37. Santos, M. R., Bromm, V., & Kamionkowski, M. 2002, MNRAS, 336, 1082
38. Salvaterra, R., & Ferrara, A. 2003, MNRAS, 339, 973
39. Kashlinsky, A., Arendt, R. G., Mather, J., & Moseley, S. H. 2005, Nature, 438, 45
40. Madau, P., & Silk, J. 2005, MNRAS, 359, L37
41. Dwek, E., Arendt, R. G., & Krennrich, F. 2005, ApJ, 635, 784
42. Tegmark, M., Silk, J., Rees, M. J., Blanchard, A., Abel, T., & Palla, F. 1997, ApJ, 474, 1
43. Shu, F. H., Lizano, S., Galli, D., Cantó, J., & Laughlin, G. 2002, ApJ, 580, 969
44. Baraffe, I., Heger, A., & Woosley, S. E. 2001, ApJ, 550, 890
45. Kudritzki, R. P. 2002, ApJ, 577, 389
46. Bromm, V., Yoshida, N., & Hernquist, L. 2003, ApJ, 596, L135
47. Bromm, V., & Loeb, A. 2004, NewA, 9, 353
48. Bromm, V., Coppi, P. S., & Larson, R. B. 1999, ApJ, 527, L5
49. Kitsionas, S., & Whitworth, A. P. 2002, MNRAS, 330, 129
50. Bromm, V., & Loeb, A. 2003, ApJ, 596, 34
51. Omukai, K., & Palla, F. 2001, ApJ, 561, L55
52. Omukai, K., & Palla, F. 2003, ApJ, 589, 677
53. Ripamonti, E., Haardt, F., Ferrara, A., & Colpi, M. 2002, MNRAS, 334, 401
54. Tan, J. C., & McKee, C. F. 2004, ApJ, 603, 383
55. Palla, F., Salpeter, E. E.., & Stahler, S. W. 1983, ApJ, 271, 632
56. Larson, R. B. 2003, Rep. Prog. Phys., 66, 1651
57. Omukai, K., & Nishi, R. 1998, ApJ, 508, 141
58. Omukai, K., & Inutsuka, S. 2002, MNRAS, 332, 59
59. Wolfire, M. G., & Cassinelli, J. P. 1987, ApJ, 319, 850
60. Perna, R., & Loeb, A. 1998, ApJ, 501, 467
61. Waxman, E., & Draine, B. T. 2000, ApJ, 537, 796
62. Fruchter, A., Krolik, J. H., & Rhoads, J. E. 2001, ApJ, 563, 597
63. Draine, B. T., & Hao, L. 2002, ApJ, 569, 780
64. MacFadyen, A. I., Woosley, S. E., & Heger, A. 2001, ApJ, 550, 410

Long Gamma-Ray Bursts as standard candles

Davide Lazzati*, Giancarlo Ghirlanda†, Gabriele Ghisellini†, Lara Nava†, Claudio Firmani†, Brian Morsony* and Mitchell C. Begelman*

*JILA, University of Colorado, 440 UCB, Boulder CO 80309-0440, USA
†Osservatorio Astronomico di Brera, via Bianchi 46, 22807 Merate, LC, Italy

Abstract. As soon as it was realized that long GRBs lie at cosmological distances, attempts have been made to use them as cosmological probes. Besides their use as lighthouses, a task that presents mainly the technological challenge of a rapid deep high resolution follow-up, researchers attempted to find the Holy Grail: a way to create a standard candle from GRB observables. We discuss here the attempts and the discovery of the Ghirlanda correlation, to date the best method to standardize the GRB candle. Together with discussing the promises of this method, we will underline the open issues, the required calibrations and how to understand them and keep them under control. Even though GRB cosmology is a field in its infancy, ongoing work and studies will clarify soon if and how GRBs will be able to keep up to the promises.

Keywords: gamma-ray sources; gamma-ray bursts; cosmology; dark energy
PACS: 98.70.Rz, 98.80.-k, 95.36.+x

INTRODUCTION

At first glance, gamma-ray bursts (hereafter GRBs), are all but standard candles. We can quantify this statement by computing the isotropic equivalent energy, i.e. their energy output in photons assuming they radiate in every direction with the same properties:

$$E_{\mathrm{iso}} = \frac{4\pi D_L^2}{1+z} \mathscr{F}_{\mathrm{bol}} \qquad (1)$$

where $\mathscr{F}_{\mathrm{bol}}$ is the burst bolometric fluence, z its redshift and D_L its luminosity distance. This quantity is easy to measure, provided that the burst has been detected and its redshift measured. A basic knowledge of its spectrum is necessary in order to perform a bolometric correction, unless a statistical approach is adopted [1]. Compilation of E_{iso} for burst samples show that this quantity spans, at least, three orders of magnitude [8, 3]. The isotropic equivalent energy is therefore all but a good standard candle. Type Ia supernovæ (SNe), as a test bench, have a RMS scatter of 0.075 decades.

Despite this apparent failure, the interest in GRBs as cosmological probes did not wane. Another related research branch was the effort to find a redshift estimate for bursts without optical afterglows. The real requirement of a cosmological test is to measure the source distance without measuring its redshift. Such a measure leads to an estimate of a luminosity distance or angular distance that allows for comparison with cosmological models.

In this paper I will review the attempts made to use GRBs as standard candles (§ 2). I will then discuss the Ghirlanda relation (§ 3), that seem to finally provide us with a

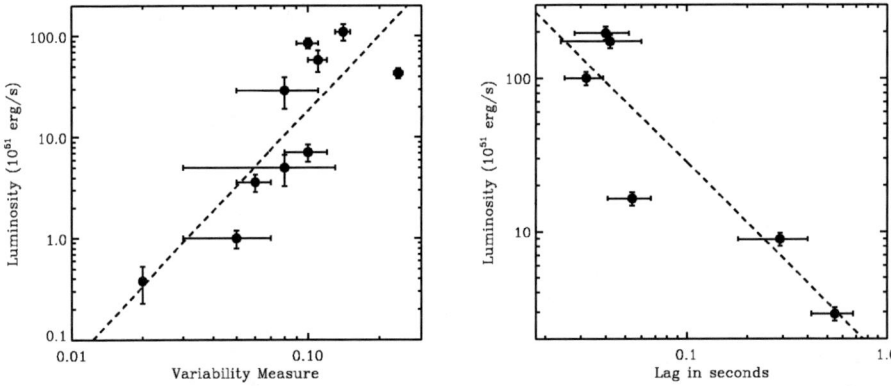

FIGURE 1. The variability-luminosity (left panel) and lag-luminosity (right panel) correlations for a sample of *Beppo*SAX GRBs. Data taken from [5, 6].

candle that is standard enough to be used to test cosmological models (§ 4). I will finally discuss future prospects and developments of this idea (§ 5).

CORRELATIONS

The first attempt to find a correlation between GRB observables that could eventually lead to a distance measurement independent of redshift was the so-called variability-luminosity correlation [4, 5] (see Fig. 1). It was found, for a small sample of *Beppo*SAX bursts with measured redshifts, that the intrinsic peak luminosity is correlated with the amount of variability present in the light curve. Therefore, if the light curve of a GRB can be observed with enough accuracy, the amount of variability can be converted into an intrinsic luminosity that, compared to the flux, yields a distance measurement.

Almost simultaneously, a correlation was discovered also between the luminosity and the time lag between light curve features at different frequencies [6] (see Fig. 1). The time lag is measured by cross-correlating the light curve of a GRB in two different energy bands (usually BATSE channels 1 and 3). It is found that the longer the lag between the two channels, the smaller the peak luminosity of the event. An attempt to use these correlations for cosmological purposes yielded poor results, due to the large scatter of both relations. Only formal limits to the cosmological parameters could be obtained [7].

A different correlation was discovered after the realization that GRBs are most likely beamed in a cone rather than isotropic explosions. Frail et al. [8] and Panaitescu and Kumar [9] noted independently that the more luminous is the GRB, the earlier is its break time (and therefore the smaller its beaming angle). If the isotropic equivalent energy E_{iso} is corrected for the beaming angle, a remarkable clustering of the "true" energies is obtained. Given the need to know the break time of the afterglow light curve, it is more difficult to obtain all the necessary information for a given GRB. Analogously

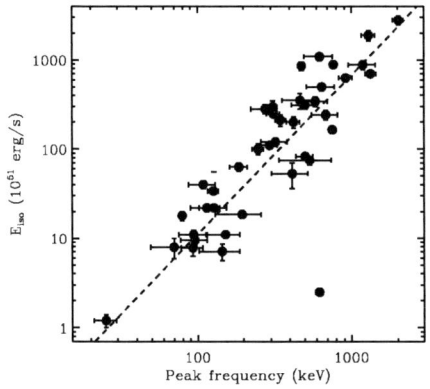

FIGURE 2. The Amati correlation. Data from [12].

to the other correlations described above, this correlation is affected by a large scatter that does not allow its use for cosmological tests [10].

Finally, a new insight into the parameter space was opened by the systematic analysis of spectra of the prompt emission of *BeppoSAX* GRBs [11]. Amati et al. showed that a remarkable correlation is present between the typical photon frequency (measured as the peak of the spectrum in $\nu F(\nu)$ units) and the burst isotropic equivalent energy (see Fig. 2). Again, the scatter of the correlation is too big to allow for any cosmological use. A debate is open on whether there is a selection effect increasing the narrowness of the correlation [13].

THE GHIRLANDA CORRELATION

A dramatic improvement over the previous attempts was achieved by merging the standard energy results [8, 9] with the spectral results [11]. Ghirlanda and collaborators [3] noticed that if the Amati plot (Fig. 2) is modified by using the beaming corrected energy rather than the isotropic equivalent one, a very narrow relation between points is obtained. The scatter of points around the correlation is impressively small. A Gaussian fit yielded a $1-\sigma$ dispersion of 0.15 decades, approximatively twice as large as the scatter of Type Ia SNe. To counterbalance the small scatter of the correlation, placing a GRB on the Ghirlanda plot requires a complicated set of observations. The measure of the typical frequency of photons in the prompt GRB emission requires broadband coverage, from X-rays to soft γ-rays. If such a broad coverage was provided in the past by BATSE and *BeppoSAX*, most of the Swift bursts do not have an adequately large band, unless observed simultaneously by HETE-2. On the other hand, the measure of the beaming corrected energy requires a knowledge of the opening angle of the GRB. This can be obtained only with continuous sampling of the afterglow light curve, possibly in multiple bands, to unambiguously identify an achromatic break that can be associated to the geometry of the fireball. Finally, a good measure of the opening angle of the GRB

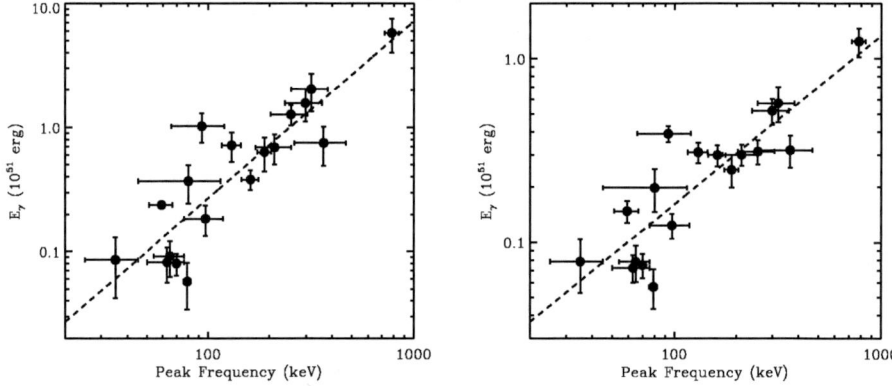

FIGURE 3. The Ghirlanda correlation. The left panel shows the results assuming a uniform environment for all GRBs, while the right panel shows the result under the assumption of a wind environment. Data from [14].

requires the knowledge of the density profile of the medium surrounding the GRB. The association of GRBs with the death of massive stars and simultaneous explosion with Type Ic SNe [15, 16, 17] suggests a stratified medium surrounding the GRB. On the other hand, external shock fits generally suggest a flat distribution of density. It can be shown [14] (see the right panel of Fig. 3) that changing the assumption on the environment profile does not destroy the correlation, but changes its slope. A tighter correlation is found for a wind environment.

THE GHIRLANDA CORRELATION AS A COSMOLOGY TOOL

Thanks to the narrowness of the Ghirlanda correlation, a GRB Hubble diagram can be produced yielding promising results [18]. A deep knowledge of the systematic errors is however necessary before a meaningful constraint on cosmological parameters can be derived [19]. Besides the uncertainty on the environment stratification and the difficulty in the measurement of the peak frequency with a relatively narrow band instrument, a major issue is to understand the relation between the break time in the light curve and the opening angle of the jet. So far toy models for which the jet is uniform inside a well-defined opening angle and absent outside have been used. However, it is expected that this model is not extremely accurate. If the jet is hydrodynamically confined inside the progenitor star [20, 21], the jet reaching th surface of the star is expected to be complex (see Fig. 4). The jet-star interaction creates a high-pressure cocoon that surrounds and confines the jet. The jet structure evolves then toward a stable solution with a freely expanding jet core surrounded by a boundary layer that flows parallel to the jet-cocoon discontinuity and is in pressure equilibrium with the cocoon. The jet that emerges on the surface of the star is therefore complex as can be seen in Fig. 5, where density, pressure and entropy of the jet are shown few seconds after the jet breakout (Morsony et al. in

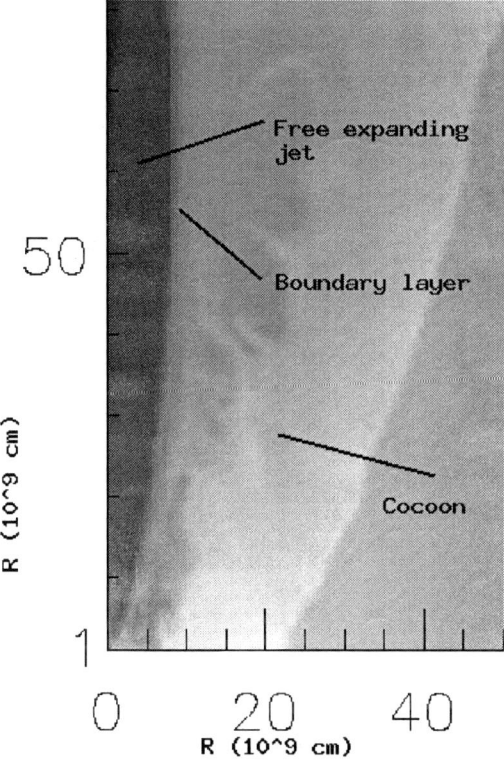

FIGURE 4. The structure of a relativistic jet expanding inside a massive star core.

preparation).

Besides the structure of the jet, another aspect to consider is the viewing angle of the observer with respect to the jet axis. Rossi et al. [22] showed that, in the simplest assumption of a conical jet that does not expand sideways, the viewing angle has a strong influence on the time of the break (Fig. 6). This uncertainty should propagate and affect the dispersion of the points around the Ghirlanda correlation. The observed dispersion is too small and therefore a smaller dependence of the break time from the viewing angle has to be relevant in the real case. One possibility to avoid such a large dependence of the break time on the viewing angle is to assume a so-called structured jet [22]. In this case, all the jets are assumed to be alike, but with an energy distribution in solid angle,

$$\frac{dE}{d\Omega} \propto \theta^{-2} \qquad (2)$$

It can be shown that with such an energy configuration the afterglow light curve is similar to that of a uniform jet, with an early-time shallow decay, a late-time fast decay, and a break time that connects the two regimes. The break time, however, is not related to the opening angle of the jet but to the viewing angle. As a consequence, to any break time is

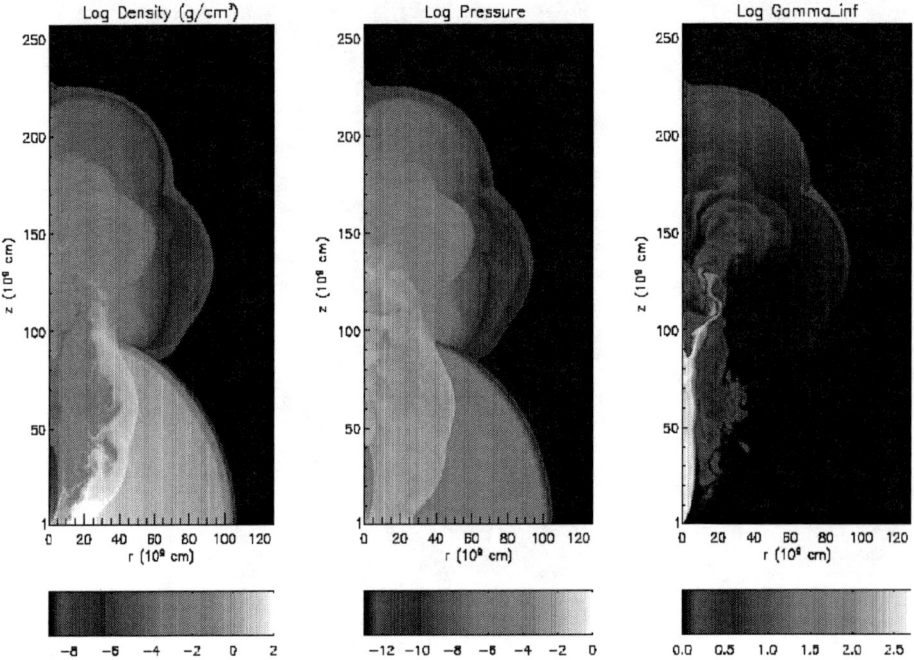

FIGURE 5. The structure of a relativistic jet as it exits from the surface of the progenitor massive star. The left panel shows density, the central panel shows pressure and the right panel shows Lorentz factor at infinity, or dimensionless entropy.

associated a single viewing time and a single value of E_{iso}. This configuration mimicks a wide low-energy jet for oservers far from the axis of the beam and mimicks a narrow powerful jet for observers close to the jet axis. This solution explains successfully the small scatter of the Ghirlanda relation. Attempts to unveil the true structure of the jet have so far been inconclusive. Theoretically, the jet structure may be due to the turbulent interaction of the jet with the star or to the spreading of the jet as the star confinement progressively wanes and disappear [21].

Despite these theoretical and observational uncertainties one can blindly apply the Ghirlanda correlation to derive constraints on the cosmological parameters [18, 23]. Figure 7 shows the result that can be derived with the present sample of GRBs. The allowed region is still wide, mainly due to the paucity of the sample. A fairer comparison can be obtained by simulating a sample of 150 bursts. Figure 8 shows the result of such an effort. GRBs indeed provide strong constraints on the cosmological parameters and on the Dark Energy evolution.

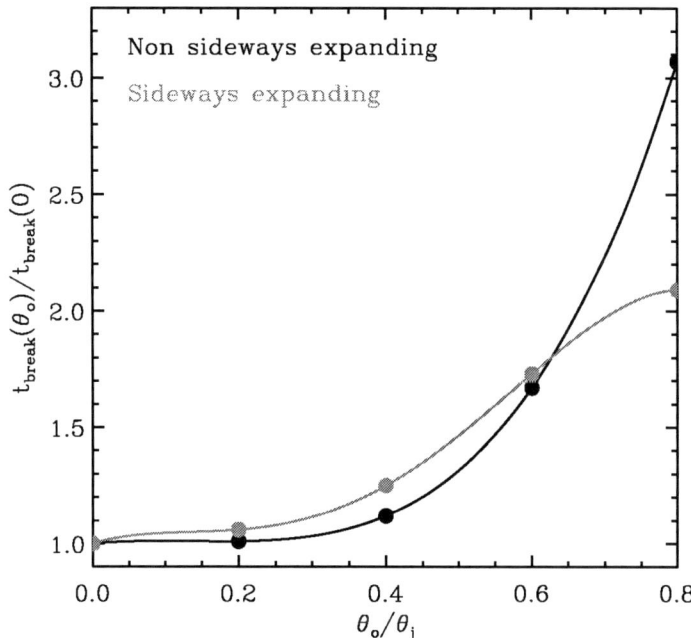

FIGURE 6. The dependence of the break time on the viewing angle of the jet. The black line and points show the dependence for a non sideways expanding jet while the gray line and points show the behavior for a sideways expanding jet.

DISCUSSION

The attempt to standardize the GRB candle has a long history. Since it was realized that GRBs explode at large redshifts, it became of great interest to understand whether GRBs could be used as distance indicators. It was soon clear that at first glance GRBs are useless as distance indicators, since their apparent luminosity spans at least three orders of magnitude. Soon after, however, it was realized that correlations between observables exist and that these correlations can lead to a more standard GRB candle. Attempts with the variability-luminosity and lag-luminosity correlations yielded inconclusive results. The large scatter of the Amati relation hampered, in a similar way, all attempts to use it for cosmological purposes.

The recent discovery of the Ghirlanda correlation [3] opened new possibilities. The scatter of points around that correlation is only approximatively twice as large as that of Type Ia SNe around the stretching-luminosity correlation. With a suitably large sample (comparable in number to that of Ia SNe) GRBs could provide interesting new constraints on cosmological parameters and the evolution (if any) of the Dark Energy. However, more insight in the origin of the correlation and its systematic errors has to

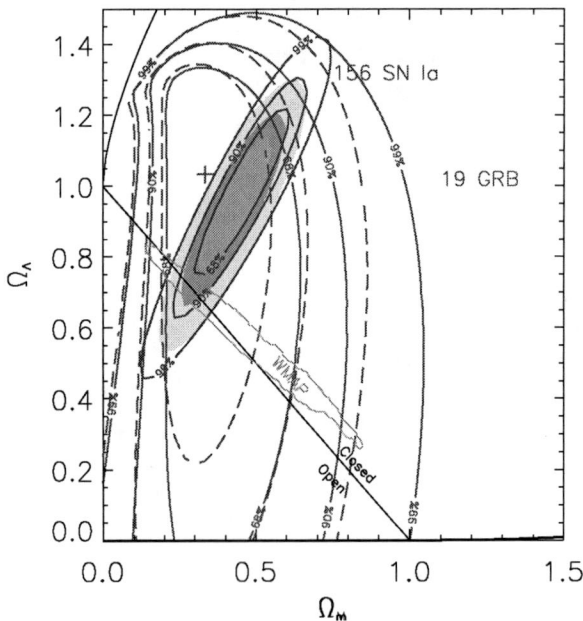

FIGURE 7. Constraints in the $\Omega_M - \Omega_\Lambda$ plain that can be obtained with the application of the Ghirlanda relation to the present sample of 19 GRBS (data from [23]). The shaded contours show the constraints obtained by a simultaneous fit of Ia SN data with GRBs.

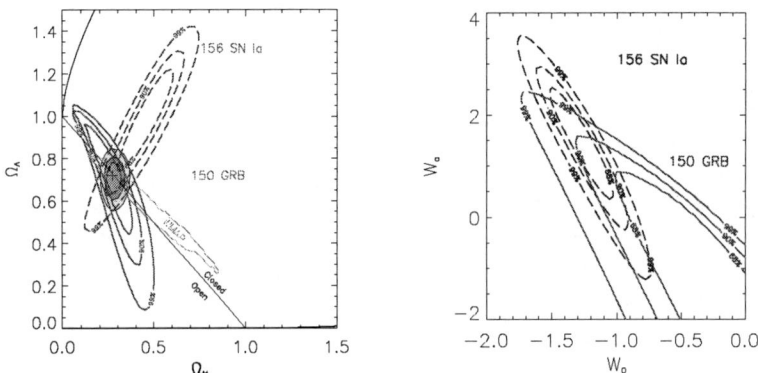

FIGURE 8. Constraints on the cosmological parameters and on the evolution of dark energy from a simulated set of 150 GRBs compared to analogous constraints obtained from the present Ia SN sample.

be obtained before strong claims can be made. In particular, we still have to understand clearly the structure of the external medium, the jet structure and the relation of the jet break time to the opening angle (or structure) of the jet.

ACKNOWLEDGMENTS

This work was supported by NSF grant AST-0307502 and NASA Astrophysical Theory Grant NAG5-12035.

REFERENCES

1. J. S. Bloom, D. A. Frail, and R. Sari, *The Astronomical Journal* **121**, 2879–2888 (2001).
2. D. A. Frail, et al, *The Astrophysical Journal Letters* **562**, L55–L58 (2001).
3. G. Ghirlanda, G. Ghisellini, and D. Lazzati, *The Astrophysical Journal* **616**, 331–338 (2004).
4. E. E. Fenimore and E. Ramirez-Ruiz, *astro-ph/0004176* (2000).
5. D. E. Reichart, et al., *The Astrophysical Journal* **552**, 57–71 (2001).
6. J. P. Norris, G. F. Marani, and T. J. Bonnell, *The Astrophysical Journal* **534**, 248–257 (2000).
7. B. E. Shaefer, *The Astrophysical Journal Letters* **583**, L67–L70 (2003).
8. D. A. Frail, et al., *The Astrophysical Journal Letters* **562**, L55–L58 (2001).
9. A. Panaitescu and P. Kumar, *The Astrophysical Journal Letters* **560**, L49–L53 (2001).
10. J. S. Bloom, D. A. Frail, S. R. Kulkarni *The Astrophysical Journal* **594**, 674–683 (2003).
11. L. Amati, et al. *Astronomy & Astrophysics* **390**, 81–89 (2002).
12. L. Amati, *Submitted to MNRAS*, astro-ph/0601553 (2006).
13. E. Nakar and T. Piran, *Monthly Notices of the Royal Astronomical Society* **360**, L73–L76 (2005).
14. L. Nava, G. Ghisellini, G. Ghirlanda, F. Tavecchio, and C. Firmani, *Astronomy & Astrophysics in press* astro-ph/0511499 (2006).
15. J. Hjorth, et al., *Nature* **423**, 847–850 (2003).
16. K. Stanek, et al., *The Astrophysical Journal Letters* **591**, L17–L20 (2003).
17. D. Malesani, et al., *The Astrophysical Journal Letters* **609**, L5–L8 (2004).
18. G. Ghirlanda, G. Ghisellini, D. Lazzati, and C. Firmani, *The Astrophysical Journal Letters* **613**, L13–L16 (2004).
19. A. S. Friedman and J. S. Bloom, *The Astrophysical Journal* **627**, 1–25 (2005).
20. A. I. MacFadyen and S. E. Woosley, *The Astrophysical Journal* **524**, 262–289 (1999).
21. D. Lazzati and M. C. Begelman, *The Astrophysical Journal* **629**, 903–907 (2005)
22. E. Rossi, D. Lazzati, J. D. Salmonson, and G. Ghisellini, *Monthly Notices of the Royal Astronomical Society* **354**, 86–100 (2004).
23. G. Ghirlanda, G. Ghisellini, C. Firmani, L. Nava, F. Tavecchio, and D. Lazzati, *Astronomy & Astrophysics submitted* astro-ph/0511559 (2006).

Variability in GRB Host Galaxy Line Fluxes

Aybüke Küpcü Yoldaş*, Jochen Greiner* and Arne Rau*,†

*Max-Planck-Institut für extraterrestrische Physik, Giessenbachstrasse 1, D-85748 Garching, Germany
†INAF - Osservatorio Astronomico di Trieste, via G.B. Tiepolo 11, 34131 Trieste, Italy

Abstract. We present the results of the spectral analysis of the host galaxies of gamma-ray bursts (GRB) 031203 and 990712 observed up to ∼1.5 and ∼6 years after their original outburst, respectively. We compare the emission line fluxes with that of the previous observations of the same GRBs, in search for photoionization signatures. The energy output in the GRB prompt emission and afterglow phase is expected to photoionize the surrounding medium similar to the case observed in supernovae. Investigating the variability of the emission line fluxes helps to constrain the recombination time scales and gas density of the circumburst environment. We further discuss the implications of variable line fluxes on the host galaxy parameters infered using emission lines, i.e. star-formation rate (SFR), metallicity etc.

Keywords: gamma-ray sources; gamma-ray bursts, galaxies
PACS: 98.70.Rz

IONIZATION OF THE CIRCUMBURST MATERIAL

In the fireball model of GRBs [1], the energy released from the collapse of the massive star is converted into the kinetic energy of thin baryonic shells which expand collimated at ultra-relativistic speed as a jet like structure. After producing the prompt γ-ray emission by internal shocks between different shells, the residual impacts on the surrounding gas and drives an ultra-relativistic shock into the ambient medium. The shock accelerates relativistic electrons leading to the observed radio to X-ray afterglow radiation through synchrotron emission. The emission of the X-ray afterglow, integrated over the first 7–10 days, typically contains the same energy as the primary γ-ray burst itself [2].

If the GRB resides in a region of star formation, photoionization of the circumburster material by the prompt γ/X-ray emission and also by the X-ray and UV afterglow emission will be inevitable. If the circumburst density is high, it will lead to time dependent (on hour timescale) absorption [3] and emission line [4, 5] features, such as claimed to be seen in X-rays. On longer timescales, the GRB photoionization may lead to indicative recombination line features, which allow the identification of remnants of GRBs in nearby galaxies [6, 7].

Photoionization of the ambient medium is well studied in the case of supernovae, being the most similar cases to that of GRBs. For SN 1987A IUE observations showed that the prominent UV lines started to increase simultaneously after 60–80 days and stayed on a constant level until 400 days after the initial exciting supernova outburst [8]. After 400 days most lines decreased fastly and had reached the noise level by day 1500 [9]. A similar behavior was observed for [NII] and [OIII]. Detailed modelling of the ionization zones and the line emission of the circumstellar gas [10] showed that the matter which had to be optically thick for the ionizing radiation had gas densities ranging between 6×10^3 cm^{-3} and 3×10^4 cm^{-3}. A total gas mass up to a distance of 10^{16} cm of $\sim4.5\times10^{-2}$ solar masses had to be ionized in order to account for the observed line fluxes which e.g. reached 2×10^{-12} erg/cm^2/s (L$\sim3\times10^{35}$ erg/s) for [OIII].

In the case of a GRB, ionization by prompt emission, the afterglow (photon field) and by the blast wave (shock-ionization) will appear additionally to the SN component and most likely even dominate. However, the influence of the blast wave on the ionization is uncertain. For the following, we assume that the photoionization is the dominating ionization mechanism. Most part of the energy (e.g. kinetic, magnetic) stored in the GRB-jet ($\sim10^{50}$ erg) will be released in the afterglow. This radiation will ionize the ambient medium while the jet decelerates and expands sidewards, e.g. the afterglow of GRB 030329 had an approximate size of $\sim10^{18}$ cm after 250 days [11]. The volume which such afterglow might ionize is thus $\sim10^4$ times larger than in the case of SN 1987A. This leads to a $\sim10^4$ higher amount of ionized gas mass.

The strength and timescale of the recombination emission depends strongly on the ambient density. While the modelling of the broad-band SED of afterglows has led to densities in the range 1-10cm^{-3}, there are also observational indications for much higher densities: (i) observed variable X-ray line features require densities of $\sim10^5$-10^6cm^{-3} [12, 13], (ii) some GRB afterglow data require a dense ($\sim10^4$cm^{-3}) shell around some nearby low-density medium (1-10cm^{-3}) [14]. The SN-GRB connection is now clearly proved for three GRBs [15, 16, 17, 18], indicating the link between long-duration GRBs and deaths of massive stars. Observations of massive, Wolf-Rayet like stars, have shown that they lose matter via strong stellar winds. A Wolf-Rayet (WR) stellar wind, interacting with a circumstellar medium of a typical density n=1 cm^{-3} leads to the formation of a shell of density $\sim10^3$ cm^{-3}, extending from 10^{16} to 10^{18} cm [19, 14]. With a density of n=10^3 cm^{-3} and plasma temperatures of $\sim10^4$ K, the shell volume will be ionized in a few seconds with a recombination time scale e.g. for [OIII] of ~1.9 years.

All the afterglow photons are available for ionizing a region with radius of up to 10^{18}-10^{19}cm (depending on density). First order estimates show that only a small fraction ($\sim10^{-6}$) of the ionizing fluence is converted into line emission. In case of higher gas densities and/or lower plasma temperatures the recombination time scale will be shorter. Then, only a small ionized volume will be visible and thus the total line intensitiy will be weaker and the temporal evolution of the observable line flux be different.

GRB 031203

GRB 031203 triggered IBIS on INTEGRAL and was initially localized with a 2.5 arcmin accuracy [20]. X-ray and radio afterglows of the GRB were detected and the host galaxy was identified optically. The redshift was then determined spectroscopically as z = 0.1055 [21].

The spectra of the host galaxy of GRB 031203 were obtained on April 09, 2005 with FORS2 on VLT, using 300V/GG375 grism/filter. The slit width was 1.0 arcsec and the total exposure time was 3×20 minutes. The spectra were reduced using standard

TABLE 1. The emission line flux ratios for GRB 031203

Line	$F_\lambda/F_{H\beta}$*	
	06 Dec 2003[†]	09 Apr 2005
NeIII	0.60±0.05	0.668±0.034
Hδ	0.25±0.02	0.289±0.014
Hγ	0.50±0.04	0.479±0.021
OIII (4364)	0.111±0.008	0.086±0.006
OIII (4959)	2.1±0.1	2.172±0.023
OIII (5007)	6.4±0.3	6.462±0.089
Hα	2.8±0.1	3.019±0.054
NII	0.151±0.012	0.167±0.007
ArIII	0.068±0.005	0.078±0.006

* The flux ratios were corrected for an extinction of $E(B-V) = 1.17$. The H_β flux, without the extinction correction, is 5.18±0.08 and 5.82±0.04 ×10^{-16} erg/s/cm^2 for 2003 and 2005 observations, respectively (see the text for further explanation).
[†] Flux ratio errors are assumed to be ∼5 – 8%.

IRAF packages and flux calibrated using the observations of the standard star LTT 6248 obtained in the same night. We then corrected the spectra for foreground extinction assuming $E(B-V) = 1.17$ as used by Prochaska et al. [21]. Line fluxes were determined by fitting a Gaussian to the line using the SPLOT task of IRAF. The continuum level was determined locally.

The only previous spectroscopic observations of the GRB 031203 are those used for determining the redshift and were obtained 3 days after the burst using the 6.5m Magellan/Baade telescope with a total exposure time of 2400s [21]. However, the slit width was 0.75 arcsec for these observations, which resulted in an underestimate of the fluxes as the galaxy overfilled the slit [21]. Therefore, instead of directly comparing the emission line fluxes, we compared the line flux ratios, i.e. $F_\lambda/F_{H\beta}$. For the line flux ratios given by Prochaska et al. [21], we assumed the error percentage is ∼ 5 – 8% based on the error bars of individual line fluxes given in their paper. As shown in Table 1, the line flux ratios at day 3 and day 492, are consistent within the errors.

GRB 990712

GRB 990712 was discovered by GRBM and WFC onboard BeppoSAX on July 12.69655 UT, 1999. The duration of the burst was 30s and it was first localized by WFC at R.A. = 22:31:50, Dec. = -73:24:24 with an error radius of 2 arcmin [22]. Follow-up observations led to the discovery of the GRB afterglow. The redshift of the burst is z=0.4331 [26]. The host galaxy of GRB 990712 is one of the brightest GRB host galaxies with V = 22.3 and R = 21.8 [25].

The spectra of GRB 990712 were obtained on July 05 and 06, 2005, approximately 6 years after the GRB using VLT/FORS2 under good seeing conditions (∼0.6 arcsec). The

TABLE 2. GRB 990712: Log of observations

Date	Instrument	Grism	Exposure time
14 July 1999	FORS1	150I	1200 s
06 June 2002	FORS1	600R	4320 s
17 July 2004	FORS2	600RI	3600 s
05-06 July 2005	FORS2	300V	7200 s

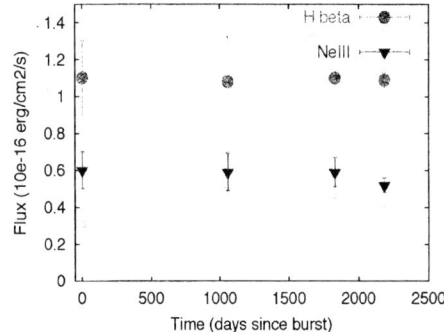 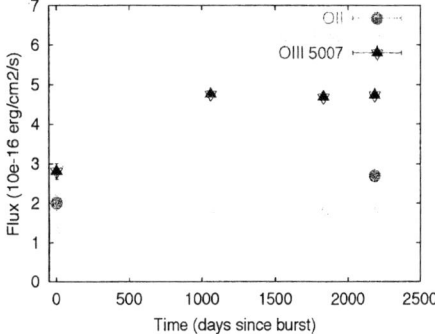

FIGURE 1. *Left:* NeIII, Hγ, Hβ, and *Right:* OII and OIII 4959, 5007 line flux light curves for GRB 990712. Error bars are smaller than the symbol size, if not visible.

300V/GG435 grism/filter was used with a slit width of 1.0 arcsec. The total exposure time was 2 hours (4×30 minutes). The spectra were reduced using standard IRAF packages and calibrated using standard star G158-100 observed on July 05, 2005 with the same grism/filter. We corrected the spectra for the foreground extinction given as E(B − V) = 0.03 [25]. The line fluxes were determined as described above in the case of GRB 031203.

The line fluxes are compared with those derived using VLT archival data of observations obtained on July 14, 1999 (PI: Galama), June 06, 2002 (PI: Mirabel) and on July 17, 2004 (PI: Le Floc'h) (see Tab.2). July 1999 data was reduced using standard IRAF tasks and flux calibrated using observations of the standard star EG 274 obtained on September 15, 1999. The spectrum was corrected for foreground extinction of E(B − V) = 0.03 and line fluxes were determined as in the case of GRB 031203. The flux values we obtain are different than the published values obtained by flux calibrating the same data using the afterglow brightness extrapolated in time [23] [1]. The previously unpublished 2002 and 2004 data were similarly reduced using IRAF tasks and flux calibrated using observations of the standard stars LTT 3854 taken on June 05, 2002 and LTT 377 taken on July 15, 2004, respectively. The spectra were foreground extinction corrected and line fluxes were determined as in the case of GRB 031203.

The NeIII, Hγ, and Hβ lines are observed to have constant fluxes (see Fig.1 left panel).

---

[1] We are in contact with Paul Vresswijk to verify the results.

On the other hand, OIII fluxes are observed to increase after the burst till July 2002, and then stayed constant thereafter as shown in Figure 1 on the right. The OIII (4959) line has increased by 25±13% whereas the OIII (5007) line has increased by 41±5%. The OII line has increased by 26±9% in ∼6 years (see Fig.1 right panel). However the lack of observations between July 1999 and July 2005, prevents us to further constrain the timescale of the increase.

SUMMARY AND DISCUSSIONS

For GRB 031203, the comparison of the line flux ratios at day 3 and day 492 revealed that they are the same within the uncertainities. Line flux ratios are not absolute measures of changes in individual line fluxes. Nevertheless they are reliable under the assumptions that the spatial profile is the same for all lines and the temporal change of fluxes due to photoionization would be different given the different cooling/recombination timescales of individual lines. The second assumption would implicitly include a change in the line flux ratios in time, since i.e. Hβ would change differently than OIII, it would not be possible to observe the same line flux ratio in time. Therefore we conclude that the line fluxes did not change within our errors. Note however that we cannot exclude the possibility that some of the line fluxes rise up and fall back to the same value between these two observations obtained at day 3 and day 492.

In the case of GRB 990712, we observed a change in Oxygen line fluxes within the first ∼3 years after the burst, whereas the Balmer lines stayed constant over ∼6 years after the burst. This is the first time that a change with a timescale of years is observed in emission line fluxes of a GRB host galaxy. This implies that indeed the GRB changes its environment drastically so that it is recognizable against the total emission of the rest of the galaxy.

Independent of the detailed interpretation of these variable line fluxes, this result has two consequences of misleading calculations of the galaxy properties, namely the calculations of the SFR and of the metallicity. Currently the SFR calculations based on emission lines are relying on observations obtained at a single epoch. Similarly, metallicity calculations rely on the line flux ratios derived based on single epoch observations. However, if the variability seen in the case of GRB 990712 is confirmed, current strategies would lead to overestimated SFRs and wrong values of metallicities.

Furthermore, this relative behaviour of constant Balmer line fluxes and increasing Oxygen line fluxes was also obtained by Perna et al. [7] as a result of their simulations of the emission spectrum due to the cooling and recombination of a dense environment ($n = 10^2$ cm^{-3} to a radius of 10 pc) photoionized by the GRB. These values are consistent with the simulated Wolf-Rayet environments [19, 14]. Nonetheless we cannot rule out a wind environment or a more complex environmental structure without further detailed modelling.

ACKNOWLEDGMENTS

We thank to Rosalba Perna and Markus Böttcher for insightful comments and discussions. AKY acknowledges support from the International Max-Planck Research School (IMPRS) on Astrophysics. AR acknowledges support and collaboration within the EU RTN Contract HPRN-CT-2002-00294.

REFERENCES

1. M. J. Rees and P. Meszaros, *MNRAS* **258**, 41 (1992).
2. M. J. Rees and P. Meszaros, *ApJ* **496**, L1 (1998).
3. R. Perna and A. Loeb, *ApJ* **501**, 467 (1998).
4. M. Böttcher et al., *A&A* **343**, 111 (1999).
5. G. Ghisellini et al., *ApJ* **517**, 168 (1999).
6. D. L. Band and D. H. Hartmann, *ApJ* **386**, 299 (1992).
7. R. Perna, J. Raymond and A. Loeb, *ApJ* **533**, 658 (2000).
8. C. Fransson et al., *ApJ* **341**, 429 (1989).
9. G. Sonneborn et al., *ApJ* **477**, 848 (1997).
10. P. Lundqvist and C. Fransson, *ApJ* **464**, 924 (1996).
11. G. B. Taylor et al., *ApJ* **609**, L1 (2004).
12. D. Watson et al., *A&A* **393**, L1 (2002).
13. J. N. Reeves et al., *A&A* **403**, 463 (2003).
14. R. Chevalier et al., *ApJ* **606**, 369 (2004).
15. T. J. Galama et al., *Nature* **395**, 670 (1998).
16. J. Hjorth et al., *Nature* **423**, 847 (2003).
17. K. Z. Stanek et al., *ApJ* **591**, L17 (2003).
18. D. Malesani et al., *ApJ* **609**, 5 (2004).
19. E. Ramirez-Ruiz et al., *MNRAS* **327**, 829 (2001).
20. D. Gotz et al., *GCN Circ.* 2459 (2003).
21. J. X. Prochaska et al. *ApJ* **611**, 200 (2004).
22. J. Heise et al., *IAU Circ.* 7221 (1999).
23. P. Vreeswijk et al., *ApJ* **546**, 672 (2001).
24. K. Sahu et al., *ApJ* **540**, 74 (2000).
25. D. J. Schlegel, D. P. Finkbeiner and M. Davis, *ApJ* **500**, 525 (1998).

Missing GRB host galaxies in deep mid-infrared observations: implications on the use of GRBs as star formation tracers

Emeric Le Floc'h*, Vassilis Charmandaris†, Bill Forrest**, Félix Mirabel‡, Lee Armus§ and Daniel Devost¶

*University of Arizona, Tucson, AZ 85721, USA
†University of Crete, GR-71003, Heraklion, Greece
**University of Rochester, Rochester, NY 14627, USA
‡European Southern Observatory, Santiago 19, Chile
§Spitzer Science Center, Pasadena, CA 91125, USA
¶Cornell University, Ithaca, NY 14853, USA

Abstract.
We report on the first mid-infrared observations of 16 GRB host galaxies performed with the *Spitzer Space Telescope*, and investigate the presence of evolved stellar populations and dust-enshrouded star-forming activity associated with GRBs. Only a very small fraction of our sample is detected by *Spitzer*, which is not consistent with recent works suggesting the presence of a GRB host population dominated by massive and strongly-starbursting galaxies (SFR $\gtrsim 100\,M_\odot\,yr^{-1}$). Should the GRB hosts be representative of star-forming galaxies at high redshift, models of galaxy evolution indicate that $\gtrsim 50\%$ of GRB hosts would be easily detected at the depth of our mid-infrared observations. Unless our sample suffers from a strong observational bias which remains to be understood, we infer in this context that the GRBs identified with the current techniques can not be directly used as unbiased probes of the global and integrated star formation history of the Universe.

Keywords: GRB host galaxies - Infrared - Luminous Infrared Galaxies - Observational Cosmology
PACS: 98.54.Ep – 98.70.Lt – 98.70.Rz – 98.80.Es

INTRODUCTION

In the past few years, the connection between long Gamma-Ray Bursts (hereafter GRBs) and the activity of massive star formation in distant galaxies has been established in a robust way [e.g., 1, 2, 3, 4]. Furthermore, GRBs are very little affected by dust extinction and they are likely detectable up to very high redshift. They could thus be used as probes of the whole star formation history of the Universe independently of all the usual biases affecting the current deep surveys. This statement is however based on the assumption that the production rate of GRBs as a function of redshift is strictly proportional to the amount of massive stars which are formed, with no redshift evolution of the parameters that may influence the trigger of these catastrophic events. In this context, the GRB host galaxies in a given redshift bin should be representative of the sources responsible for the bulk of the star-forming activity at this redshift. Comparing the properties of the GRB hosts with those of field sources is thus one way to test how well GRBs can signpost the sites of massive star formation in the distant Universe.

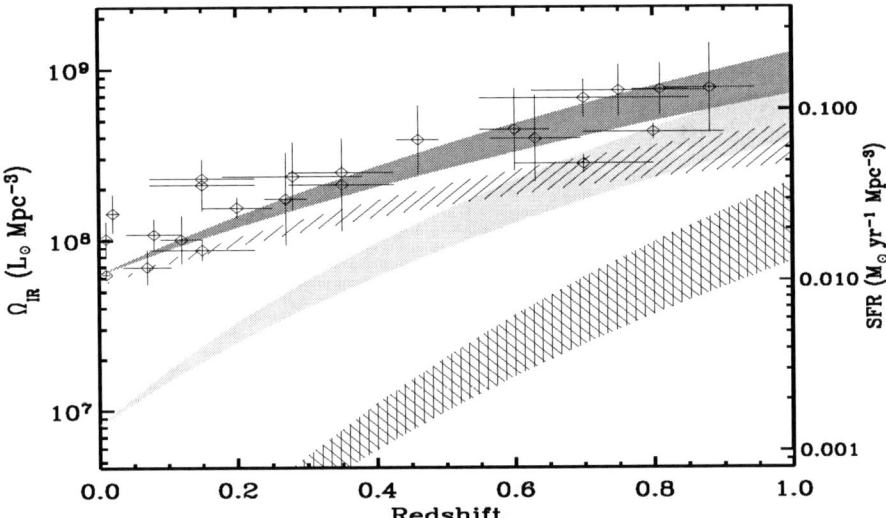

FIGURE 1. Evolution of the comoving IR energy density up to $z=1$ (dark shaded region) and the respective contributions from low luminosity galaxies (i.e., $L_{IR} < 10^{11} L_\odot$, lined area), "infrared luminous" sources (i.e., $L_{IR} \geq 10^{11} L_\odot$, light shaded region) and ULIRGs (i.e., $L_{IR} \geq 10^{12} L_\odot$, cross-hatched region). At $z \sim 1$ IR luminous galaxies represent $70\pm15\%$ of the comoving IR energy density and dominate the star formation activity. From Le Floc'h et al. [7].

From the multi-wavelength deep surveys that were carried out in the last decade we know that the activity of star formation has been progressively shifting from very massive and very luminous starbursts at redshifts $z \sim 2$–3 from low-mass sub-luminous star-forming dwarves in the local Universe [e.g., 5, 6]. This is often called the "downsizing" of the cosmic evolution and it is illustrated in Fig. 1 which reports on the star formation history (SFH) of the Universe from $z=0$ to $z=1$. In this figure the integrated SFH has been decomposed into the contribution of galaxies classified as a function of their infrared (IR) luminosity (hence as a function of their star formation rate). We see that beyond $z \sim 0.7$, the so-called Luminous Infrared Galaxies and Ultra-Luminous InfraRed Galaxies (respectively LIRGs: $10^{11} L_\odot \leq L_{IR} = L[8-1000\mu m] \leq 10^{12} L_\odot$, and ULIRGs: $L_{IR} \geq 10^{12} L_\odot$) dominate the SFH [7]. These IR-luminous starbursts are also intermediate or high-mass objects. If the long GRBs are tracing the star formation history, a significant fraction of these cosmic explosions should thus be observed in luminous and massive sources.

The most efficient facility currently accessible to the astronomical community for tracking the properties of massive, dusty and luminous starbursts at high redshift is the *Spitzer Space Telescope*. *Spitzer* was launched in August 2003 as the final mission of the NASA Great Observatories Program. This satellite is an infrared cold telescope with an 80 cm-diameter mirror, equipped with two imagers called IRAC and MIPS and one spectrometer called IRS. IRAC observes in broad-band filters centered at 3.6, 4.5, 5.8

and 8.0 μm, MIPS images the sky at 24, 70 and 160 μm and the IRS instrument obtains mid-IR spectra of sources between 5 and 35 μm. *Spitzer* is now providing spectacular results which allow us better insights into the physical processes of dust emission in Galactic star-forming regions and nearby galaxies. It is also capable to detect very high redshift galaxies that are completely invisible in the deepest optical data ever taken by the *Hubble Space Telescope* [e.g., 8]. It routinely obtains mid-IR spectra of the SCUBA submillimeter sources at $z \sim 2$–3 [e.g., 9] and it has recently resolved up to $\sim 90\%$ of the infrared background thanks to the unprecedented sensitivity of the MIPS instrument [e.g., 10, 11].

We thus used *Spitzer* to study the properties of several GRB host galaxies. Here we report on our observations and we analyze our data in the goal of testing whether GRBs are detected in massive and luminous IR starbursts at high redshift.

OBSERVATIONS

We targeted a sample of 16 GRB host galaxies using *Spitzer* as part of the IRS GTO program (PI: J.Houck). Each object was imaged with IRAC at 4.5 and 8 μm down to 3.5 μJy and 20 μJy (3σ) respectively, as well as with MIPS at 24 μm down to 85 μJy (3σ). These sensitivity limits are slightly shallower than those typically reached in the GTO surveys undertaken by *Spitzer* (e.g., $5\sigma \sim 80$ μJy, [7]) but still reasonably deep. Assuming a typical conversion between the star formation rate (SFR) and the IR continuum emission (e.g., [12]), we infer that our 24 μm data are sensitive to SFR $\gtrsim 15$ M$_\odot$ yr^{-1} at $z \sim 1$ and SFR $\gtrsim 150$ M$_\odot$ yr^{-1} at $z \sim 2$.

Our sample is composed of the host galaxies of all GRBs that were localized with a sub-arsecond accuracy between 1997 and July 1999, with the exception of the host of GRB 970228 that was replaced with the host of GRB 010222. A persistent submillimeter emission was in fact detected at the location of this burst [13], which makes it an obviously interesting target for IR observations. There was no other pre-selection using an *a priori* knowledge of e.g., redshifts, optical or near-infrared magnitudes, detections at other long wavelengths...

RESULTS

Most of the sources from our sample (i.e., $\sim 80\%$) could not be detected with *Spitzer* (with neither IRAC nor MIPS), including the hosts of GRB 970508, GRB 980703 and GRB 010222 that have been proposed as potential ultra-luminous IR galaxies by [14], [15] and [13] respectively. One noticeable exception is the host of GRB 980613 presented as a merger-induced starburst by [16]. As seen in Fig. 2 this host galaxy is composed of several interacting knots. Two of them are clearly detected at 4.5 μm with IRAC, and there is also evidence for a detection at 8.0 and 24 μm. These two components have very red $R - K$ colors, likely pointing to a dust-obscured starburst. Nonetheless, we note that the afterglow of the GRB was identified in another region located 2" away to the North, and which is not detected by *Spitzer*. This is therefore an interesting

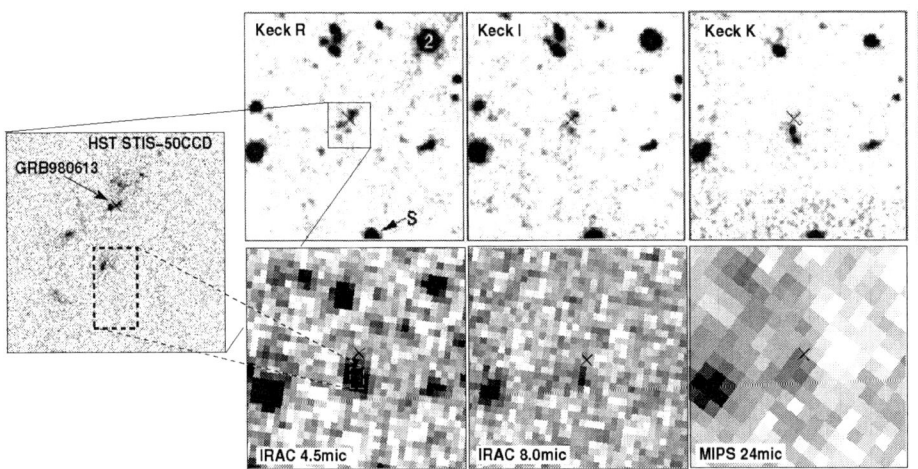

FIGURE 2. An optical and infrared view of the host of GRB 980613, an interacting system at $z = 1.1$. In each panel the position of the GRB is indicated with a cross. This burst did not occur within the most IR-luminous (thus the most starbursting) component of the merger. This may reveal the influence of other physical parameters favoring the formation of such events (metallicity, efficiency of the star-forming activity, IMF, ...). From Djorgovski et al. [16], Holland et al. [17] and Le Floc'h et al. (submitted).

example of a GRB that did not occur in the most active star-forming environment of its host (see also [18]).

In summary, our sample of GRB host galaxies observed with *Spitzer* is not consistent with a population of massive and dusty luminous starbursts. This actually confirms previous optical and near-IR studies of the GRB hosts, performed either on a case-by-case basis (e.g. [1, 19]) or following a more statistical approach (e.g., [2, 20, 21, 22], and that led to the conclusion that these objects are mostly blue, sub-luminous, low-mass and young galaxies with a modest amount of star formation and a low level of dust extinction.

IMPLICATIONS

At 24 μm, the differential source number counts derived from the *Spitzer* deep surveys present a turn-over between 200 and 300 μJy when they are normalized to the Euclidean slope (e.g., [10]). This reveals that most of the IR background is produced by galaxies that are much brighter than the 3σ sensitivity limit of our GRB host observations with MIPS. Using phenomenological models of IR galaxy evolution (e.g., [23]) we inferred that more than 50% of our sample should have easily been detected by MIPS if the GRBs effectively trace the whole activity of star formation at high redshift. Our large fraction of non-detections at 24 μm thus implies either that our sample is strongly biased or that the long GRBs preferentially occur in faint objects.

There is clearly one selection effect that may potentially affect our current sample. In fact these objects were identified mostly using the sub-arcsec location of optically-bright

GRB afterglow transients. This type of selection may thus induce a bias against dusty sources. In favor of this interpretation, a fraction of the so-called "dark bursts" (i.e., GRBs with no detectable afterglows despite rapid and deep optical follow-ups) appears to be truly enshrouded behind dusty material. An example of these dusty dark GRBs is the GRB 970828, which was localized thanks to its radio afterglow [24]. Its host galaxy is actually one of the very few sources that were detected in our 24 μm data, which clearly supports the scenario of a burst accompanied by an afterglow extinguished by dust. On the other hand, another large fraction of dark GRBs is also due to intrinsically faint bursts and GRBs with steep time decays. Furthermore it has been suggested that dust grains can be destroyed by hard X-ray emission along the line of sight of GRBs, which may substantially reduce the effect of extinction in the selection of star-forming environments with GRBs. Finally, we did not detect any IR emission toward the hosts of two other dark bursts (i.e., GRB 990506 and GRB 981226). This shows that these dark GRBs are not systematically associated with dusty star-forming galaxies, an interpretation already proposed by Barnard et al. [25] based on SCUBA observations of four dark GRB host galaxies. We infer that if this bias does exist it can not be important enough to explain all our non-detections by *Spitzer*.

Consequently, our results may reflect the influence of parameters more physically related to the environments where GRBs are produced, and which may explain why these GRBs preferentially take place in young, sub-luminous and blue objects rather than luminous and massive starbursts. For instance, the potential effect of a low metallicity in the GRB progenitor enveloppe is now intensively discussed by theorists (e.g, Woosley, MacFadyen et al., these proceedings) as it may clearly favor the trigger of such events. In fact, low metallicites have already been measured from the integrated spectra of several GRB hosts [26, 27], and this may also explain the statistically low luminosity of this GRB-selected population. Rotation effects and the implication of GRB progenitors within binary systems are also currently explored.

SUMMARY

As previously stated, our *Spitzer* observations reveal that long GRBs are statistically not observed in the massive and luminous infrared galaxies that dominate the activity of star formation in the early Universe. This actually confirms previous claims arguing for a population rather dominated by blue, young and low-mass objects, and it shows that the hosts of long GRBs are not representative of the sources that produced the bulk of stellar mass throughout the lifetime of the Universe. This tells us that the relation between massive star formation and long GRBs is likely much more complex than previously assumed, and it strongly suggests that the history of the GRB production rate can not be directly converted into the integrated star formation history. This disagreement might reflect the influence of specific parameters in the trigger of GRBs (e.g., metallicity, rotation effects, binarity, ...), that we need to understand if we want to control the use of GRBs as star formation tracers.

ACKNOWLEDGMENTS

This conference in Washington DC was a really fruitful and pleasant meeting. The first author would like to thank all the people who contributed in making this event such a success.

REFERENCES

1. Fruchter, A. S., Thorsett, S. E., Metzger, M. R., et al. 1999, ApJ 519, L13
2. Sokolov, V. V., Fatkhullin, T. A., Castro-Tirado, A. J., et al. 2001, A&A 372, 438
3. Bloom, J. S., Kulkarni, S. R., & Djorgovski, S. G. 2002, AJ 123, 1111
4. Stanek, K. Z., Matheson, T., Garnavich, P. M., et al. 2003, ApJ 591, L17
5. Chary, R. & Elbaz, D. 2001, ApJ 556, 562
6. Chapman, S. C., Blain, A. W., Ivison, R. J., & Smail, I. R. 2003, Nature 422, 695
7. Le Floc'h, E., Papovich, C., Dole, H., et al. 2005, ApJ 632, 169
8. Mobasher, B., Dickinson, M., Fergusson, H., et al. 2005, ApJ in press (astro-ph/0509768)
9. Lutz, D., Valiante, E., Sturm, E., et al. 2005, ApJ 625, L83
10. Papovich, C., Dole, H., Egami, E., et al. 2004, ApJS 154, 70
11. Dole, H., Lagache, G., Puget, J.-L., et al. 2006 (submitted to A&A)
12. Kennicutt, R. C. 1998, ARA&A 36, 189
13. Frail, D. A., Bertoldi, F., Moriarty-Schieven, G. H., et al. 2002, ApJ 565, 829
14. Hanlon, L., Laureijs, R. J., Metcalfe, L., et al. 2000, A&A 359, 941
15. Berger, E., Kulkarni, S. R., & Frail, D. A. 2001, ApJ 560, 652
16. Djorgovski, S. G., Bloom, J. S., & Kulkarni, S. R. 2003, ApJ 591, L13
17. Holland, S., Fynbo, J., Thomsen, B., et al. 2000, GRB Circular Network, 698
18. Hjorth, J., Thomsen, B., Nielsen, S. R., et al. 2002, ApJ 576, 113
19. Bloom, J. S., Odewahn, S. C., Djorgovski, S. G., et al. 1999, ApJ 518, L1
20. Le Floc'h, E., Duc, P.-A., Mirabel, I. F., et al. 2003, A&A 400, 499
21. Courty, S., Björnsson, G., & Gudmundsson, E. H. 2004, MNRAS 354, 581
22. Christensen, L., Hjorth, J., & Gorosabel, J. 2004, A&A 425, 913
23. Lagache, G., Dole, H., Puget, J.-L., et al. 2004, ApJS 154, 112
24. Djorgovski, S. G., Frail, D. A., Kulkarni, S. R., et al. 2001, ApJ 562, 654
25. Barnard, V. E., Blain, A. W., Tanvir, N. R., et al. 2003, MNRAS 338, 1
26. Prochaska, J. X., Bloom, J. S., Chen, H.-W., et al. 2004, ApJ 611, 200
27. Soderberg, A. M., Kulkarni, S. R., Berger, E., et al. 2004, ApJ 606, 994

GRB Afterglows as a New ISM and IGM Probe

Hsiao-Wen Chen*, Jason X. Prochaska† and Josh S. Bloom**

*Department of Astronomy & Astrophysics, University of Chicago, Chicago, IL 60637, U.S.A.
†UCO/Lick Observatory, University of California at Santa Cruz, Santa Cruz, CA 95064, U.S.A.
**Department of Astronomy, University of California at Berkeley, Berkeley, CA 94720, U.S.A.

Abstract.
We summarize results from a study of the metallicity, relative abundances, gas density, and kinematics of dense media in the host environment of two *Swift* bursts, GRB 050730 at $z = 3.968$ and GRB 051111 at $z = 1.549$. Both GRB hosts exhibit strong absorption features from excited Si^+ and Fe^+ ions, indicating an extreme ISM environment that is similar to what is found around massive stars like luminous blue variables (LBV) and Wolf-Rayet stars. The extreme ISM properties have never been observed in intervening quasar absorption line systems beyond the local universe.

Keywords: gamma rays: bursts—quasars: absorption lines
PACS: 98.70.Rz

1. UNVEILING EXTREME STAR-FORMING REGIONS

Mounting evidence has demonstrated that long-duration gamma-ray bursts (GRBs) arise in active star-forming regions [e.g. 1, 2], supporting a physical connection between the bursts and the catastrophic death of massive stars [3, 4]. Studies of the circumburst environment together with the progenitor stars therefore bear directly on our understanding of star formation and metal production in the early universe.

The extreme brightness of GRB optical afterglows, albeit brief, offers a unique means to unveil the physical conditions of the ambient medium around the burst progenitors. For example, low-resolution afterglow spectra have yielded a few diagnostic measurements such as the H I column density, the gas metallicity, and the dust-to-gas ratio of the GRB hosts [e.g. 8, 9]. High-resolution spectroscopy of the optical afterglows has further uncovered detailed kinematic signatures, population ratios of excited ions, and chemical compositions of the interstellar medium (ISM) and the circumstellar medium (CSM) of the progenitor star of the GRB [e.g. 5, 6, 7].

The majority of GRB host galaxies are known to have large neutral gas content with neutral hydrogen column density much beyond $\log N(H I) = 20.3$ [e.g. 9], the standard threshold for selecting damped Lyα absorbers (DLAs). DLAs selected along random lines of sight toward background quasars dominate the mass density of neutral gas in the universe and represent a uniform sample of distant galaxies with known $N(H I)$. GRB host DLAs ($z_{DLA} = z_{GRB}$), which are presumably selected by vigorous star formation and therefore probe deep into the center regions of distant galaxies, presents a nice complement to classical DLA system ($z_{DLA} < z_{GRB}$), which arises preferentially at large galactocentric radii due to a larger cross-section of the outskirts than the inner regions. In this article, we summarize our analysis of the circumstellar medium observed in two GRB host galaxies, and compare with what is known for classical DLAs.

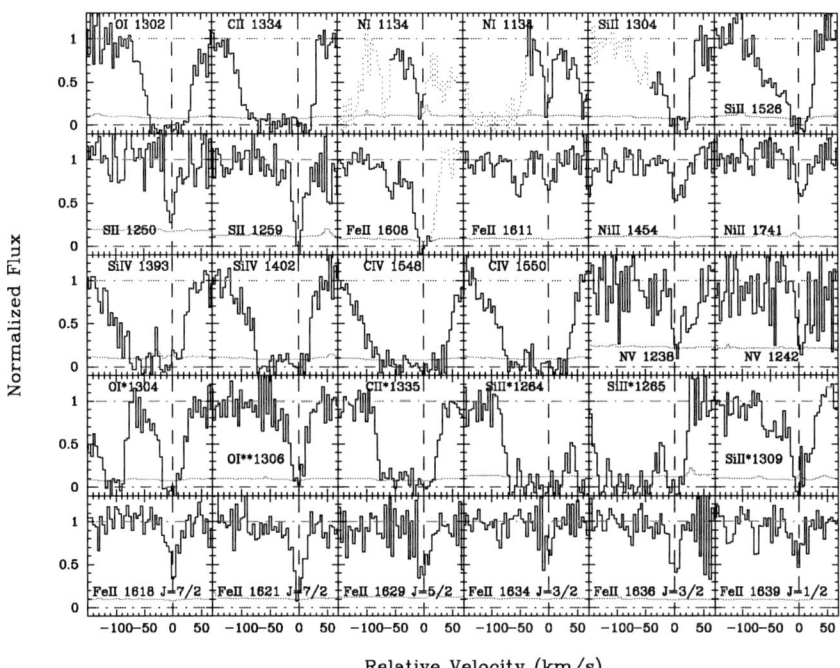

FIGURE 1. Absorption profiles of various ionic transitions found at the host redshift of GRB 050730. Resonance transitions are presented in the top three rows, while absorption features from excited O^0, C^+, Si^+, and Fe^+ are presented in the bottom two rows. Dotted curves indicate contaminating features that are not related to the marked transitions. The zero relative velocity corresponds to redshift $z = 3.96855$.

2. ABSORPTION-LINE PROPERTIES OF GRB HOSTS

High-resolution echelle spectra of GRB 050730 were obtained using the MIKE echelle spectrograph on the Magellan Clay telescope. The spectra cover a wavelength range from $\lambda = 3300$ Å to 9000 Å with a spectral resolution of FWHM ≈ 10 km s^{-1} at wavelength $\lambda = 4500$ Å and ≈ 12 km s^{-1} at $\lambda = 8000$ Å. We have identified a suite of metal-absorption lines at the redshift of the GRB $z = 3.96855$, including neutral species such as O I, low-ionization transitions such as C II, Si II, S II, Ni II, and Fe II, and high-ionization transitions such as C IV, Si IV, and N V. We have also identified strong fine-structure lines such as O I*, O I**, Si II*, C II*, and Fe II*.

Absorption profiles of these transitions are presented in Figure 1. The presence of strong fine-structure transitions indicate an extreme ISM environment with high gas density that is rarely observed in classical DLA systems. Furthermore, the profiles of well resolved lines (e.g. S II 1250) show that $> 90\%$ of the neutral gas is confined to a velocity width 20 km s^{-1}, which is considerably smaller than the median value of intervening DLA systems and implies a quiescent environment. The asymmetry of these line profiles is also suggestive of an organized velocity field, e.g. rotation or outflow.

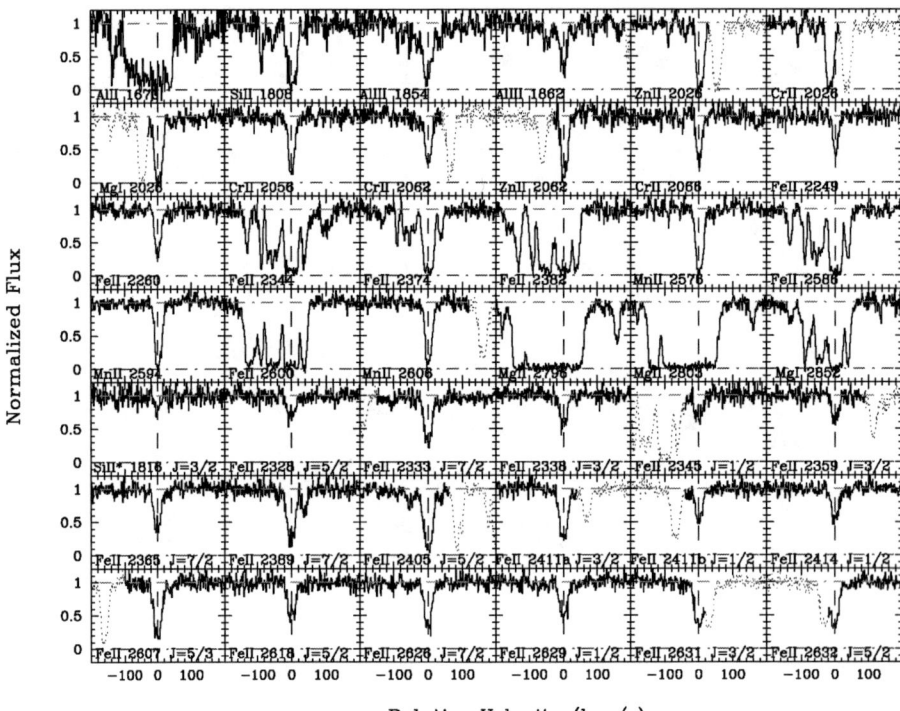

FIGURE 2. Absorption profiles of various ionic transitions found at the host redshift of GRB 051111. Resonance transitions are presented in the top four rows, while absorption features from excited Si^+, and Fe^+ are presented in the bottom three rows. The zero relative velocity corresponds to redshift $z = 1.54948$.

Additional properties measured for the host of GRB 050730 are summarized as follows. (1) We measure $\log N(H\,I) = 22.15 \pm 0.05$ and $[S/H] = -2.0 \pm 0.1$. Because S is non-refractory, its gas-phase abundance gives a direct measurement of the gas metallicity. (2) We find $[S/Fe] = +0.3$, consistent with the gas phase $[\alpha/Fe]$ measurements of low-metallicity DLA systems. Even if we adopt an intrinsic solar abundance pattern, the dust-to-gas ratio in the host ISM is very low [c.f. 8]. (3) We measure $[N/S] = -1.0 \pm 0.2$, again consistent with low-metallicity DLA systems. (4) We detect no molecular lines in the Lyα forest with a 3-σ upper limit to the molecular fraction $f_{H_2} \equiv 2N(H_2)/[2N(H_2) + N(H\,I)] < 10^{-8}$. The lack of H_2 suggests a warm gas phase, consistent with the implication from the detections of excited Si^+ and Fe^+.

High-resolution echelle spectra of GRB 051111 were obtained using the High Resolution Echelle Spectrometer (HIRES) on the Keck I telescope. The spectra cover a wavelength range from $\lambda = 4165 - 8720$Å with a spectral FWHM resolution of $\approx 5\,\text{km s}^{-1}$. We have identified a suite of metal absorption lines at the redshift of the GRB $z = 1.54948$, including resonance transitions such as Si I, Mg I, Zn II, Cr II, Mg II, Fe II, Mn II, and Al III, as well as fine-structure transitions due to excited Si^+ and Fe^+.

Absorption profiles of these transitions are presented in Figure 2. We find that while saturated lines such as Fe II, Mg I and Mg II exhibit multiple components over a velocity range from -140 km s^{-1} to 40 km s^{-1}, well resolved lines such as Cr II and weak Fe II show that > 90% of the neutral gas is confined to a velocity width 30 km s^{-1}. In addition, the Si II 1808 and Zn II 2026,2062 transitions are saturated, indicating log N(Zn) > 13.5 and log N(Si) > 15.9. These measurements represent by far the largest Zn and Si column densities observed in QSO absorption line systems. The presence of Fe$^+$ in all four excited ^6D_J states with $J = 7/2, 5/2, 3/2$, and $1/2$ suggests a large gas density in the absorbing medium. The weak Mg I 2026 transition appears to be saturated, indicating a warm gas of temperature $T > 1000$ K.

The hydrogen Lyα transition of GRB 051111 is not observed in the ground-based data. Adopt the observed lower limit to the Zn abundance log N(Zn) > 13.5, we estimate log N(H I) > 20.8 for gas of solar metallicity. Lower metallicity would lead to higher N(H I) for the neutral gas content in the host ISM. In addition, the observed Zn to Fe ratio is [Zn/Fe] > 1.2. This large ratio suggests significant differential depletion in the host of GRB 051111, contrary to what is observed for the host of GRB 050730.

3. CIRCUMBURST ENVIRONMENT OF GRB PROGENITORS

An emerging feature of GRB progenitor environments is the presence of strong fine-structure transitions from excited states of C$^+$, Si$^+$, O^0, and Fe$^+$. In contrast to resonance transitions of the dominant ions in neutral gas, these absorption lines can reveal the temperature and density of the gas, as well as the ambient radiation field. In particular, detections of Fe II fine structure transitions have only been reported in rare places such as broad absorption-line (BAL) quasars, η Carinae and the circumstellar disk of β Pictoris. Identifications of strong Fe II fine-structure transitions therefore suggest extreme gas density and temperature in the GRB progenitor environment.

We examine the excitation mechanism and gas density of the absorbing medium through comparisons of the observed relative abundances between different excited states of the Fe$^+$ ion. The population ratios between different states are determined based on the balance between excitation and de-excitation rates. When the density is sufficiently large that the collisional de-excitation rates exceed the spontaneous decay rate, the excited states are populated according to a Boltzmann distribution,

$$\frac{n_i}{n_j} = \frac{g_i}{g_j} \exp[-(E_{ij}/k)/T_{Ex}], \quad (1)$$

where g_i is the degeneracy of state i, $E_{ij} \equiv E_i - E_j$ is the difference in energy between the two states, and T_{Ex} is the excitation temperature.

Figure 3 shows the observed column density N_i scaled by the corresponding degeneracy g_i for each of the Fe$^+$ excited states as a function of the energy E_{ij} above the ground state $J = 9/2$. The error bars reflect 1-σ uncertainty in N_i. The solid (blue) curve indicates the best-fit Boltzmann function with a best-fit excitation temperature $T_{Ex} = 6100$ K for GRB 050730 and $T_{Ex} = 2600$ K for GRB 051111. The inset shows the minimum χ^2 values as a function of T_{Ex}. The reduced χ^2 is nearly unity, $\chi^2_\nu = 1.02$, supporting our assumption that the Boltzmann function is a representative model.

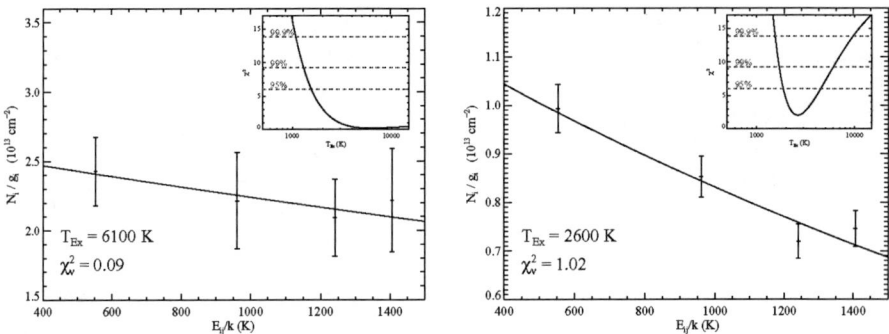

FIGURE 3. Relative abundance ratios between different excited Fe II states, N_i/g_i where g_i represents the degeneracy of the corresponding level, as a function of their energy E_{ij} above the ground state. We present measurements for the host of GRB 050730 in the left panel and GRB 051111 in the right panel. The solid curve in each panel represents the best-fit function $A\exp[-(E_{ij}/k)/T_{Ex}]$ for a minimum χ^2 value. We find a best-fit excitation temperature $T_{Ex} = 6100$ K for GRB 050730 and $T_{Ex} = 2600$ K for GRB 051111. The insets show the reduced χ^2 values for a range of T_{Ex} values.

Adopting the best-fit temperature for the absorbing medium, we can further constrain the electron density based on the observed column density ratios. We calculate the expected population ratio between excited fine-structure states i and ground state $J = 9/2$ for gas of $T = 6100$ K and $T = 2600$ K, using the PopRatio software package [10]. We find that $n_e > 1000$ cm^{-3} for both systems. To estimate the total gas density n_H based on the estimated n_e requires knowledge of the ionization state of the absorbing gas traced by the excited Fe$^+$. The large N(H I) estimated for both systems implies an optical depth to ionizing radiation ($h\nu > 1$ Ryd) $\tau > 1000$. While an O or B star (or the GRB itself) could significantly ionize a skin-layer on the outside of the cloud, the majority of the gas would remain neutral. Therefore, the hydrogen gas must be predominantly neutral and we can safely assume an ionization fraction $x < 10^{-2}$. We therefore infer $n_H > 10^5$ cm^{-3} for the absorbing medium, which further leads to an upper limit to the size of the excited medium $\ell = N(\text{H I})/n_H < 10^{17}$ cm, i.e., less than 1/3 pc.

This small dimension implies that it must arise near the GRB event (at $r \sim \ell$) or have a nearly unity covering fraction at large radii (e.g. a thin shell). Additional constraints can be derived based on absorption features due to neutral species and the afterglow light curves. A detailed discussion is presented in Prochaska, Chen, & Bloom (2006).

4. CLASSICAL DLAS VERSUS GRB HOSTS

A striking feature is that both GRB hosts exhibit Fe$^+$ and Si$^+$ fine-structure transitions which have never been observed in intervening quasar absorption line systems [e.g. 11]. Figure 4 presents a sharp contrast in the line strengths of different excited ions observed in a classical DLA with $\log N(\text{H I}) = 21.3$ at $z = 2.6264$ and in the host of GRB 050730. Our study shows that aside from having a much higher neutral gas density and temper-

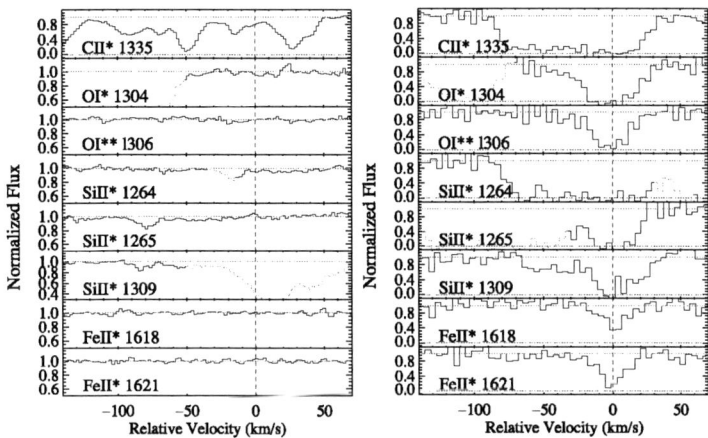

FIGURE 4. Comparisons of line strengths due to different excited ions observed in a classical DLA with log N(H I) = 21.3 at $z = 2.6264$ (left panel) and in the host of GRB 050730 (right panel).

ature as derived from the observed population ratios between different excited ions, the GRB hosts have very similar characteristics to known DLA systems at $z > 2$, such as low dust content, low metallicity, and α-element enhanced chemical composition.

As bright lighthouses for high-resolution optical spectroscopy, GRBs are now fulfilling their promise as detailed complementary probes of the distant universe. It is only a matter of time when high- resolution spectroscopy at **infrared** wavelengths will afford an picture of star formation and chemical enrichment in the epoch of reionization.

ACKNOWLEDGMENTS

The authors are grateful to Ian Thompson for obtaining the Magellan/MIKE data of GRB 050730. We thank Grant Hill, Derek Fox and Barbara Schaefer for their roles in obtaining the Keck/HIRES data of GRB 050111. H.-W.C., J.X.P., and J.S.B. are partially supported by NASA/Swift grant NNG05GF55G.

REFERENCES

1. Bloom, J. S. et al. 2002, ApJ, 572, L45
2. Stanek, K. Z. et al. 2003, ApJ, 591, L17
3. Woosley, S. E. 1993, ApJ, 405, 273
4. Paczyński, B. 1998, ApJ, 494, 45
5. Fiore, F. et al. 2005, ApJ, 624, 853
6. Chen, H.-W., Prochaska, J. X., Bloom, J. S., & Thompson, I. B. 2005, ApJ, 634, L25
7. Prochaska, J. X., Chen, H.-W., & Bloom, J. S. 2006, submitted to ApJ (astro-ph/0601057)
8. Savaglio, S., Fall, S. M., & Fiore, F. 2003, ApJ, 585, 638
9. Vreeswijk, P. M. 2004, A&A, 419, 927
10. Silva, A. I. & Viega, S. M. 2002, MNRAS, 329, 135
11. Howk, J. C., Wolfe, A. M., & Prochaska, J. X. 2005, ApJ, 622, L81

GRB Host Studies (GHostS)

S. Savaglio*, K. Glazebrook* and D. Le Borgne[†]

*Johns Hopkins University, Baltimore, USA
[†]CEA/Saclay, France

Abstract. The GRB Host Studies (GHostS) is a public archive collecting observed quantities of GRB host galaxies. At this time (January 2006) it contains information on 32 GRB hosts, i.e. about half of the total number of GRBs with known redshift. Here we present some preliminary statistical analysis of the sample, e.g. the total stellar mass, metallicity and star formation rate for the hosts. We found that these are generally low-mass objects, with 79% having $M_* < 10^{10}$ $M_\odot$. The total stellar mass and the metallicity for a subsample of 7 hosts at $0.4 < z < 1$ are consistent with the mass-metallicity relation recently found for normal star-forming galaxies in the same redshift interval. At least 56% of the total sample are bursty galaxies: their growth time-scale (the time required to form the observed stellar mass assuming that the observed SFR is constant over the entire life of the galaxy) is shorter than 400 Myr.

Keywords: gamma-ray bursts; high redshift galaxies
PACS: 98.54.Ep; 98.58.-w; 98.62.Ai; 98.70.Rz

INTRODUCTION

Although the typical nature of GRB hosts is still heavily under discussion, it is clear that they differ from normal high-z galaxies, as they are generally low-luminosity and young objects [1, 2]. We still cannot tell whether they form a galaxy population by themselves, or they are just much easier to detect than normal low-luminosity galaxies because they are associated with transient, but very luminous events.

To help investigate this issue, we have initiated a database dedicated to GRB host galaxies, called GRB Host Studies[1] (GHostS). Thanks to the advent of the Swift mission, the amount of results related to GRB hosts is in rapid ascent. The goal of GHostS is to gather, classify and synthesize GRB host information, and derive meangful parameters for a new statistically significant sample. At the present, it is the largest public archive of its kind. For each host, the optical-NIR photometry is provided, together with emission line fluxes, originating in the star-forming regions. So far, GHostS uses results coming from more than 70 different publications.

In this first work, we focus on the determination of the total stellar mass of the host galaxies, a parameter that has been hardly investigated in the past, for a number of good reasons [3]. We relate stellar masses to SFRs and metallicities.

---

[1] GHostS can be accessed at the URL http://www.pha.jhu.edu/~savaglio/ghosts

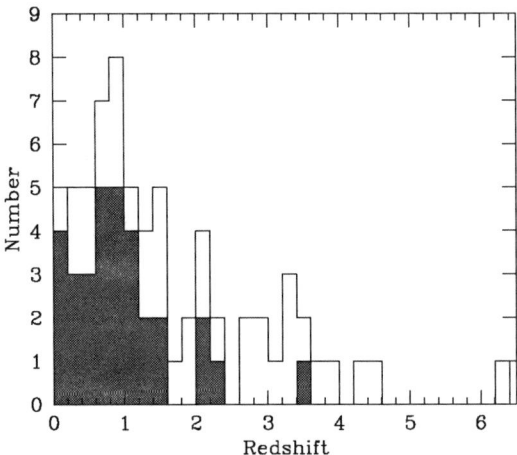

FIGURE 1. Redshift histogram of all GRBs with known redshift (67 in total - *open histogram*) and of the 32 GRBs with detected host, currently included in the GHostS archive (*filled histogram*). The median redshift of the two samples is $z \simeq 1.12$ and $z \simeq 0.84$, respectively.

THE GRB HOST SAMPLE

The median redshift of our 32 GRB hosts is $z \simeq 0.84$ (Figure 1), i.e. lower than the present median redshift for all GRBs with measured redshift ($z \simeq 1.12$). A couple of them are associated with a short-duration GRB, the remaining are long-duration GRBs. The objects are selected according to the requirement that optical-NIR photometry (necessary to estimate the total stellar mass) is available. Optical-NIR photometry is hard to measure for $z > 2$ galaxies in general. Our sample contains 4 objects with $z > 2$.

Among the 32 hosts, fluxes of [OII], [OIII] and Hβ emission lines are available for 19, 10 and 9 of them, respectively. These are used to derive metallicities and SFRs.

Total stellar masses

The stellar mass of the hosts is derived using a procedure which will be described in detail in a future paper (Le Borgne, Glazebrook & Savaglio, in preparation). Briefly, our technique uses SED fitting to the multi-band optical-NIR photometry. The observed NIR light, which in high-z galaxies samples rest-frame light above the 4000Å break, is closely related to the galaxy's total stellar mass and provides good stellar mass estimates up to $z \sim 2 - 3$ [4]. It is also insensitive to dust because most stars in a galaxy are not in birth clouds, and because the redder bands are less affected by extinction. The stellar mass derived this way is a much more meaningful physical variable than the luminosity,

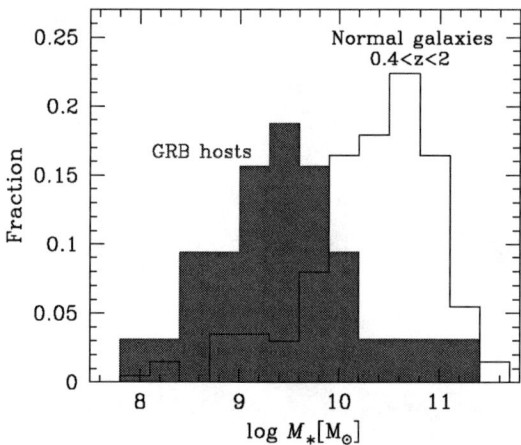

FIGURE 2. Fraction of GRB hosts per stellar-mass bins (*filled histogram*) and the comparison with 201 normal $0.4 < z < 2$ galaxies (*open histogram*) from the $K < 20.6$ Gemini Deep Deep Survey (see also [4]). In the GRB host sample, 79% have stellar masses below $M_* = 10^{10.0}$ M$_\odot$. The GDDS is complete for stellar masses above $M_* = 10^{10.8}$ M$_\odot$ and $M_* = 10^{10.1}$ M$_\odot$, for all galaxies and star-forming galaxies, respectively.

because it represents the integral of the past star-formation and merger history, and, in contrast for instance to UV light, can only increase with time.

In Figure 2 we show the stellar mass histogram for the 32 GRB hosts. The median/average mass and 1σ dispersion are $M_* = 10^{9.5}$ M$_\odot$ and 0.9 dex, respectively. This is compared to the same histogram obtained for normal $0.4 < z < 2$ galaxies from the Gemini Deep Deep Survey (GDDS; [5]). The GDDS is a deep optical-NIR ($K < 20.6$) survey and is complete at $0.4 < z < 2$ for all galaxies and for star-forming galaxies down to stellar masses $M_* = 10^{10.8}$ M$_\odot$ and $M_* = 10^{10.1}$ M$_\odot$, respectively. The comparison shows that GRB observations are much more efficient in detecting low-mass galaxies at high redshift than traditional high-z surveys.

Metallicities and SFRs

Metallicities are derived with the R_{23} calibrator, which uses [OII], [OIII] and Hβ line fluxes [6]. We adopted the formulation recently proposed by Kobulnicky & Kewley (2004) [7]. This set of emission lines allows the metallicity measurement for the largest possible number of GRB hosts. The $12 + \log(\text{O/H})$ value is derived for 9 hosts, 7 of which are in the redshift interval $0.4 < z < 1$ (Figure 3). Figure 3 also shows the same

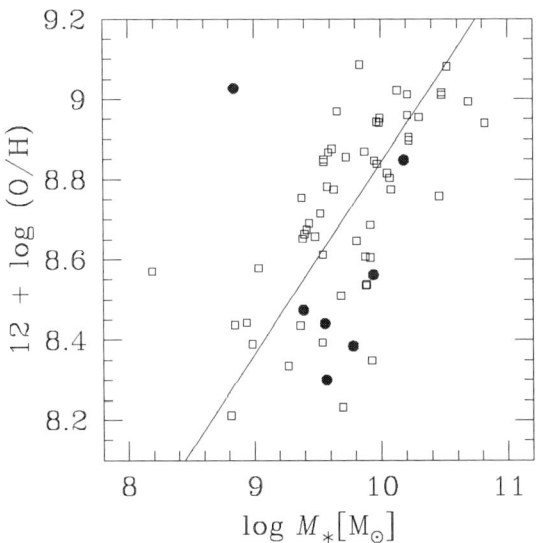

FIGURE 3. Total stellar mass and metallicity for $0.4 < z < 1$ GRB hosts (*filled circles*). Metallicities are derived assuming a modest (10%) Balmer stellar absorption correction and $A_V = 1$ dust extinction (Milky Way extinction law). The outlier at $\log M_* = 8.8$ is the GRB 991208 host galaxy at $z = 0.706$ [9]. *Open squares*: results for $0.4 < z < 1$ galaxies from GDDS and CFRS [10]. The straight line is the bisector fit for this sample.

parameters derived for GDDS and Canada-France Redshift Survey[2] (CFRS) galaxies, in the same redshift interval [10].

We estimate SFRs using the [OII] emission, the most common line measured in GRB hosts. The Hα emission flux provides a more robust SFR estimate, however this is measured in 5 GRB hosts only. Moustakas et al. (2006) [11] have shown that the dust-corrected [OII]-to-Hα flux ratio in local galaxies is on average one (0.12 dex dispersion) over a large range of B luminosities (from 10^7 to 10^{11} L$_\odot$). We adopt this relation to derive SFRs, after assuming a Milky Way extinction law with $A_V = 1$. The median SFR and the range spanned by the 19 GRB hosts are SFR = 12 M$_\odot$ yr^{-1} and $1 - 100$ M$_\odot$ yr^{-1}, respectively (we also apply a correction of a factor of 2 for slit-aperture loss).

Another indicative parameter is the SFR per mass unit, or the inverse of this, also called growth time-scale. This is defined as $\rho_* \equiv M_*/\text{SFR}$ [12], and is the time that a galaxy needs to build the observed stellar mass, if the SFR is assumed to be constant and is given by the observed value. The result as a function of redshift is shown in Figure 4, together with the comparison with normal star-forming $0.4 < z < 1$ galaxies from GDDS and CFRS [10].

---

[2] Emission line fluxes for the CFRS galaxies are taken from Lilly et al. (2003) [8]

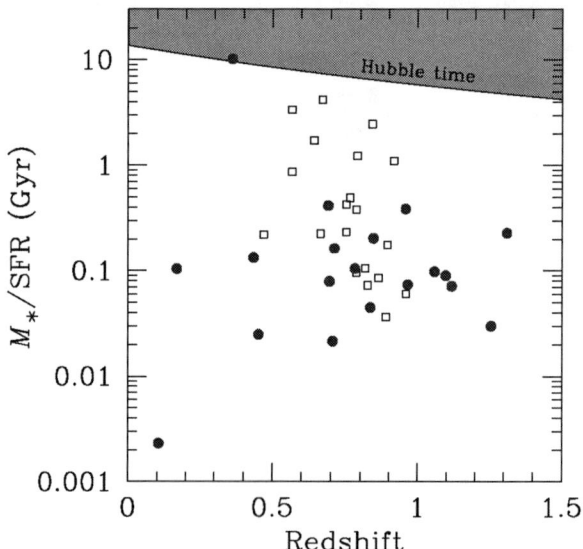

FIGURE 4. Growth time-scale (total stellar mass divided by observed SFR) as a function of redshift for 19 GRB hosts (*filled dots*). These are compared to the $0.4 < z < 1$ star-forming galaxies of GDDS and CFRS of Figure 3 (*open squares*). The curve marks the age of the universe (Hubble time) as a function of redshift, and indicates the transition from the quiescent star-formation mode to the bursty star-formation mode.

RESULTS

We presented the first results of the GHostS project. We focused on the stellar mass of 32 GRB hosts, SFRs for a subsample of 19, and metallicities for a subsample of 9, and compared these to normal high-z galaxies. We found that the total stellar mass is $< 10^{10.0}$ $M_\odot$ for 79% of the GRB hosts (Figure 2). At this mass, most of the high-z deep spectroscopic surveys are highly incomplete. The median stellar mass of the sample is $M_* = 10^{9.5}$ $M_\odot$, i.e. comparable to the stellar mass content of the Large Magellanic Cloud. Recently, Chary et al. (2002) [3] have derived stellar masses for 7 hosts and found a similar median value.

The median observed and dust-corrected (for $A_V = 1$) SFR in 19 hosts are 3 and 12 $M_\odot$ yr^{-1}, respectively. Given the generally low stellar masses for these GRB hosts, we conclude that a large fraction are bursty galaxies, with growth time-scales that are shorter than 400 Myr, and on average 100 Myr (Figure 4). If we consider the whole GRB population with measured redshift (67 in total), the host is a bursty galaxy in at least 1/4 of the cases.

The median metallicity in 9 GRB hosts is 0.6 solar, with values ranging from half to twice solar. These values are not far from expectations, given the stellar masses and redshifts of the galaxies. Moreover, they behave as predicted by the mass-metallicity relation observed at high redshifts in normal star-forming galaxies (Figure 3).

In summary, we quantified some of the known statements regarding GRB hosts, according to which a large fraction of them are low-mass starbursts. Although low-mass starbursts at high redshifts are hard to identify if no GRB event occurs, there is no evidence that GRB hosts represent a different population of galaxies that existed in the young universe.

ACKNOWLEDGMENTS

The authors thank the conference organizers for offering such an interesting event and enjoyable time, and the Swift team for their large contribution to our understanding of GRB phenomenology.

REFERENCES

1. Le Floc'h, E., et al. 2003, A&A, 400, 499
2. Christensen, L., Hjorth, J., & Gorosabel, J. 2004, A&A, 425, 913
3. Chary, R., Becklin, E. E., Armus, L. 2002, ApJ, 566, 229
4. Glazebrook, K., et al. 2004, Nature, 430, 181
5. Abraham, R. G. et al. 2004, AJ, 127, 2455
6. Pagel, B. E. J., Edmunds, M. G., Blackwell, D. E., Chun, M. S., & Smith, G. 1979, MNRAS, 189, 95
7. Kobulnicky, H. A., & Kewley L. J. 2004, ApJ, 617, 240
8. Lilly, S. J., Carollo, C. M., & Stockton, A. N. 2003, ApJ, 597, 730
9. Castro-Tirado, A. J., et al. 2001, A&A, 370, 398
10. Savaglio, S., et al. 2005, ApJ, 635, 260
11. Moustakas, J., Kennicutt, R. C., Tremonti, C. A. 2006, ApJ, in press (astro-ph/0511730)
12. Juneau, S., et al. 2005, ApJL, 619, 135

The redshift distribution of long gamma-ray bursts

Frédéric Daigne*, Elena Rossi[†,**] and Robert Mochkovitch*

*Institut d'Astrophysique de Paris
UMR 7095 CNRS – Université Pierre et Marie Curie-Paris VI, 98 bd Arago, 75014 Paris, France.
[†]JILA, University of Colorado, Boulder, CO 80309-0440, USA
**Chandra Fellow

Abstract. We study the redshift distribution of Gamma-Ray Bursts (GRBs) with Montecarlo simulations, assuming a comoving rate proportional to the star formation rate. We then use the energy band and flux threshold of several instruments (BATSE VI, HETE2 and SWIFT) to compute the observed distributions. Model parameters are constrained by fitting simultaneously (i) the log(N)-log(P) diagram of BATSE bursts; (ii) the Epeak distribution of bright BATSE bursts; (iii) the proportion of X-ray rich GRBs and X-ray flashes observed by HETE-2. This allows us to predict the redshift distribution of GRBs detected by SWIFT. We compare it with actual data, in the aim to discriminate between different possible SFRs at high redshift.

Keywords: Origin, formation, evolution, age, and star formation; gamma-ray sources; gamma-ray bursts; Cosmology
PACS: 98.35.Ac; 98.70.Rz; 98.80.-k

INTRODUCTION

Our study is motivated by two questions, related to the role of GRBs as cosmological tools : (i) do long GRBs trace the star formation rate ? (ii) what is the expected rate of long GRBs at high redshift ? On the one hand, a better understanding of the first issue would allow us to use GRBs to investigate the star formation history of the universe. Observations give strong indications in favor of an association of long GRBs with massive stars. However, the actual physical conditions (e.g. mass, metallicity, rotation, binarity) for a star to trigger a burst are not currently known. On the other hand, knowing the GRB rate at high redshift is also very important, as the afterglow emission of such bursts may be used to probe the ionization and metal enrichment histories of the intervening intergalactic medium. We use Montecarlo simulations to investigate these questions. Such an approach allows a realistic parametrization of the intrinsic GRB properties, a realistic treatment of detection criteria for several instruments and a study of the impact of the uncertainties in the GRB physics on the predicted GRB rate.

MODEL

In our Montecarlo simulations, each generated GRB is given a redshift, a luminosity and a spectrum, according to specific probability distributions. Then the peak flux is computed in different energy bands and compared with the detection sensitivity of

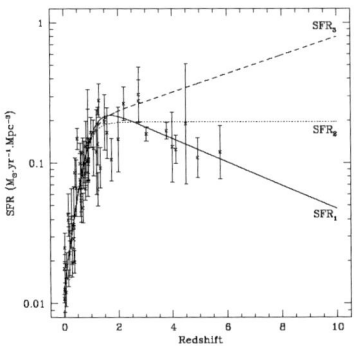

FIGURE 1. The three star formation rates considered in this study. Observed data are taken from [2].

BATSE, HETE-2 and SWIFT. This allows us to derive the expected distributions of observed redshifts, peak fluxes and peak energies. Finally, a comparison with current data puts constraints on the model parameters.

Comoving rate. Following [1], the GRB rate is assumed to be proportional to the type II Supernova rate, and therefore to the Star Formation Rate (SFR) if the initial mass function is constant in time:

$$\mathscr{R}_{GRB} = k \times \mathscr{R}_{SN}, \qquad (1)$$

where k is a free parameter of the model. We adopt $\mathscr{R}_{SN} = 0.0122\ M_\odot^{-1} \times$ SFR as in [1]. Concerning the SFR, we took a simple fit of observed data up to $z \sim 6$ [2]. This case is called SFR_1. We also considered two alternative rates: SFR_2 which is constant above $z \sim 2$ and SFR_3 which increases above $z \sim 2$. These three SFRs are plotted in figure 1.

Luminosity function. The luminosity function of long GRBs is assumed to follow a power-law:

$$p(L) \propto L^{-\delta} \text{ for } L_{min} \leq L \leq L_{max}. \qquad (2)$$

There are therefore three more model parameters: the slope δ and the minimum and maximum luminosities L_{min} and L_{max}.

Intrinsic spectrum. The intrinsic spectrum is assumed to be a broken power-law with a break at energy E_p [3]. The probability distributions of the low-energy slope α and the high-energy slope β follow the observed distribution of α and β in a set of long bright GRBs [4]. We have checked that the simulated distributions of the observed low- and high-energy slopes in long bright GRBs were very close to the intrinsic distributions. The peak energy E_p is computed according to two possible rules: (i) either we assume an intrinsic Amati-like relation

$$E_p \propto L^{0.43}, \qquad (3)$$

TABLE 1. Best models : parameters.

Rate	$\log L_{min}$	$\log L_{max}$	δ	$\log k$	$\log E_{p,0}$
	Intrinsic peak energy : Amati-like relation $E_p \propto L^{0.43}$				
SFR1	49.3 ± 0.8	53.6 ± 0.4	1.67 ± 0.08	-5.1 ± 0.5	
SFR2	49.4 ± 0.8	53.6 ± 0.5	1.65 ± 0.09	-5.2 ± 0.5	
SFR3	49.6 ± 1.1	53.5 ± 0.4	1.51 ± 0.17	-5.7 ± 0.4	
	Intrinsic peak energy : log-normal distribution with mean $E_{p,0}$				
SFR1	49.7 ± 1.1	53.2 ± 0.5	1.58 ± 0.18	-5.4 ± 0.5	2.73 ± 0.07
SFR2	49.8 ± 1.3	53.4 ± 0.7	1.57 ± 0.24	-5.5 ± 0.4	2.74 ± 0.09
SFR3	50.3 ± 1.3	53.4 ± 0.7	1.47 ± 0.38	-6.1 ± 0.3	2.76 ± 0.10

with $E_p = 380$ keV at $L = 1.6 \times 10^{52}$ erg s^{-1} and with a normal dispersion $\sigma = 0.2$ dex [5]; (ii) or the luminosity and the peak energy are independent intrinsic quantities. Then we assume that the peak energy follows a log-normal distribution centered on $E_{p,0}$ and with a dispersion $\sigma = 0.3$ dex. In this second case, $E_{p,0}$ is a new free parameter.

Detection criteria. We consider the detection by three instruments: BATSE, HETE-2 and SWIFT. For BATSE, we apply two different criteria : (i) we follow [6, 7, 8] and compare with their results for the observed $\log N - \log P$ diagram; (ii) we apply a threshold $P_{50-300 \text{ keV}} > 5$ ph cm^{-2} s^{-1} (thereafter called "bright BATSE bursts"), to compare with the results of the BATSE spectroscopic catalog [4]. This threshold typically corresponds to 5 % to 10 % of the brightests GRBs, in agreement with the BATSE GRB distribution. For HETE-2, we adopt $P_{2-10 \text{ keV}} > 1$ ph cm^{-2} s^{-1} (resp. $P_{30-400 \text{ keV}} > 1$ ph cm^{-2} s^{-1}) for WXM (resp. FREGATE) (J.L. Atteia, private communication). Finally, for SWIFT, we use $P_{15-150 \text{ keV}} > 0.2$ ph cm^{-2} s^{-1} (see the contribution of D.L. Band in these proceedings). We also define a sub-sample of "bright SWIFT bursts" by $P_{15-150 \text{ keV}} > 1$ ph cm^{-2} s^{-1}.

Observational constraints. Depending on the assumption for the intrinsic E_p distribution, our model has four (Amati-like relation) or five (log-normal E_p distribution) free parameters. To determine these parameters, we use three observational constraints : (1) the BATSE observed $\log N - \log P$ diagram [6, 7, 8]; (2) the observed peak energy distribution of long bright GRBs [4]; (3) the observed fraction of X-Ray Rich GRBs (XRR) and X-Ray Flashes (XRFs) in the sample of GRBs detected by HETE-2. We adopt the same definitions as the HETE-2 team for these sub-classes [9].

RESULTS

The use of the three observational constraints listed above allows to eliminate some degeneracies that appear when using the $\log N - \log P$ diagram only. Precisely, the

TABLE 2. Best models : mean redshift.

Rate	All	SWIFT	Bright SWIFT
	Intrinsic peak energy : Amati-like relation $E_p \propto L^{0.43}$		
SFR1	3.1	1.6	1.4
SFR2	8.0	1.7	1.4
SFR3	10.5	3.4	2.0
	Intrinsic peak energy : normal distribution with mean $E_{p,0}$		
SFR1	3.1	1.9	1.6
SFR2	8.0	2.2	1.7
SFR3	10.5	3.9	2.3

$\log N - \log P$ diagram alone can fix the slope of the luminosity function but is not enough to determine L_{min} and L_{max}. The observed E_p distribution then helps to fix L_{max} and allows also to fix L_{min} if combined with the constrain on the XRF-XRR fraction. In practice we make a joint-fit of the three constraints. In table 1, we list the best parameters that we obtain for each case, including 1 σ error bars. For each assumption on the intrinsic peak energy, and for each SFR, table 2 gives the mean redshift for all GRBs, all GRBs detected by SWIFT and bright GRBs detected by SWIFT. Figure 2 shows the corresponding redshift distribution, as well as the observed distribution for SWIFT bursts [10].

Peak energy distribution. As can be seen in Figure 2, the redshift distribution depends strongly on the assumed SFR but is almost insensitive to the rule adopted for the intrinsic peak energy distribution. The Amati-like relation case differs however from the log-normal distribution case regarding other properties of the long GRB distribution: (i) we find that selection effects do not create an observed Amati-like relation in the second case. This means that if the observed relation is real, it is probably intrinsic; (ii) When the intrinsic peak energy distribution is log-normal, the observed peak energy distribution for all simulated GRBs (not only bright long BATSE bursts) is also close to log-normal, with a mean value around 100 keV. On the other hand, if an intrinsic Amati-like relation is assumed, we get a peak energy distribution for the whole GRB population which favors lower peak energies (mean value of a few keV). This is due to the adopted luminosity function, which favors low-luminosity GRBs. Therefore, it would be only due to detection threshold effects that BATSE bright bursts happen to be clustered around 100 keV. This means that if the Amati relation is confirmed, there is probably a large population of low luminosity-low peak energy bursts which is still undetected.

Redshift distribution. Our results show that the present redshift distribution of SWIFT bursts can already put severe constraints on the comoving GRB rate (Fig. 2). For SFR_1, the redshift distribution of all GRBs is very close to the observed SWIFT distribution. However, due to the detection threshold, the simulated SWIFT distribution is very far from it. Analogously, the SFR_2 gives a mean redshift (see table 2) lower than

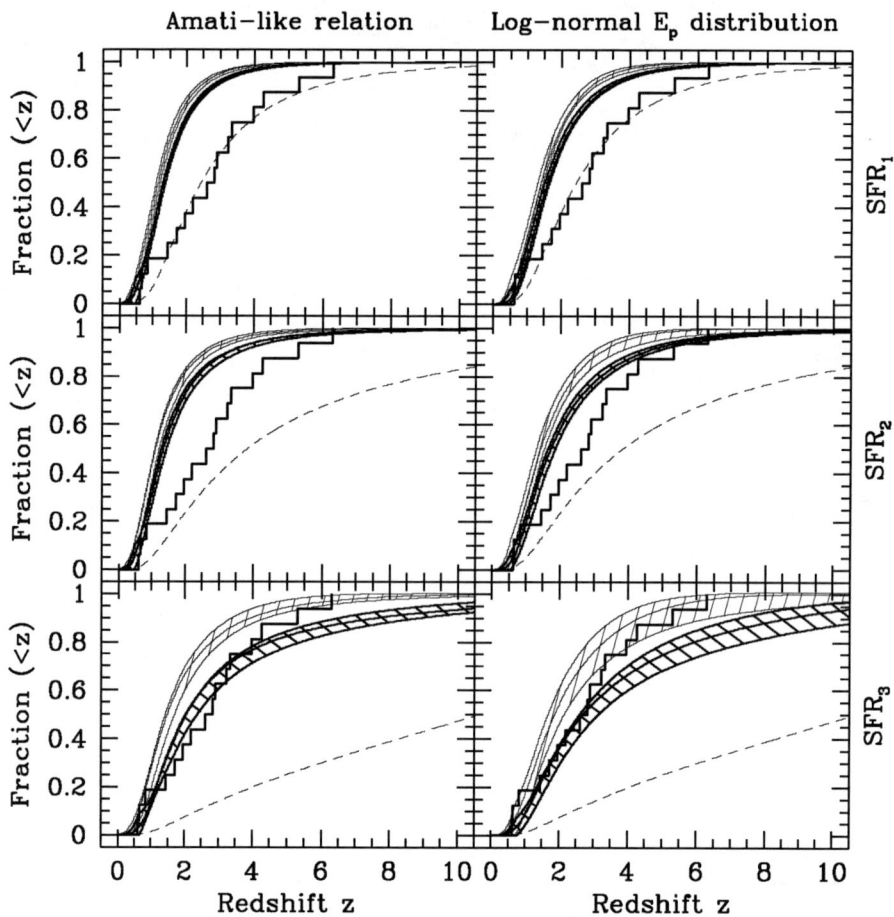

FIGURE 2. Best models : redshift distribution. The thick line shows the redshift distribution of GRBs detected by SWIFT [10]. The dotted line shows the redshift distribution of all simulated GRBs, with no detection threshold. The two strips show the observed redshift distribution (with 1 σ dispersion) of simulated GRBs that are detected by SWIFT (all bursts : thick line; bright bursts only : thin line).

the observed SWIFT bursts. Instead, the SFR_3 gives a much better agreement with data. The observed SWIFT distribution lies just between the simulated redshift distribution of all SWIFT bursts and that of bright SWIFT bursts. It is exactly what one could expect, if selection effects leading to the measurement of the redshift are related to the brightness of the source. Unfortunately, the ignorance on the selection effects involved in the redshift measurement prevents us from performing a more quantitative comparison with data.

DISCUSSION

Using Montecarlo simulations of the long GRB population under the assumptions that (i) the GRB rate follows the SFR; (ii) the GRB luminosity functions is a power-law; (iii) the peak-energy is either determined by an Amati-like intrinsic relation or by a log-normal distribution, we get the following results:

1. **Luminosity function**: the slope is well constrained: $\delta \sim 1.5 - 1.7$. The maximum luminosity is also determined within a factor of two, $L_{max} \sim 2 - 4 \times 10^{53}$ erg s^{-1}, while the minimum luminosity is not so well fixed, $L_{min} \sim 0.2 - 2 \times 10^{50}$ erg s^{-1}.
2. **Intrinsic peak energy distribution**: If the observed Amati relation is confirmed, we find that the relation needs to be intrinsic and cannot be obtained by selection effects only. A consequence of the relation is that the observed peak energy distribution is not representative of the distribution for the whole GRB population, which has a lower mean value of a few keV. Therefore, there would be many undetected XRRs and XRFs. On the other hand, a log-normal peak energy distribution can be also in very good agreement with all the observational constraints, but does not reproduce the observed Amati relation. In this case, the mean peak energy of the whole GRB population is close to 100 keV and the observed distribution is much more representative.
3. **GRB rate**: in agreement with [1], we find that one GRB pointing towards us is produced for about 10^5 to 10^6 supernovae in the Universe. When correcting for the beaming, the real GRB rate in the Universe could be multiplied by a factor ~ 500 [11]. The present redshift distribution of SWIFT bursts strongly favors SFR$_3$, which increases above $z \sim 2$. The expected rates of bright SWIFT bursts ($P > 1$ ph cm^{-2} s^{-1}) in this case is 2.1 yr^{-1} above $z = 5$, 1.2 yr^{-1} above $z = 6$ and 0.75 yr^{-1} above $z = 7$. However, such a SFR is unrealistic, as so many stars at high redshift would over-produce metals with respect to observations. This surprising result is probably an indication that *GRBs do not really trace the SFR* (see E. Le Floch et al., this conference). One simple way to reconcile our results with a more plausible SFR such as SFR$_1$ is to assume that the efficiency of GRB production by stars evolves with time, and is about $6 - 7$ times larger at $z = 7$ than at $z = 2$. This evolution could be related to various factors, such as the metallicity.

REFERENCES

1. C. Porciani, and P. Madau, *ApJ* **548**, 522–531 (2001).
2. A. M. Hopkins, *ApJ* **615**, 209–221 (2004).
3. D. Band, et al., *ApJ* **413**, 281–292 (1993).
4. R. D. Preece, et al., *ApJS* **126**, 19–36 (2000).
5. G. Ghirlanda, G. Ghisellini, C. Firmani, A. Celotti, and Z. Bosnjak, *MNRAS* **360**, L45–L49 (2005).
6. J. M. Kommers, et al., *ApJ* **533**, 696–709 (2000).
7. B. E. Stern, et al., *ApJL* **540**, L21–L24 (2000).
8. B. E. Stern, J.-L. Atteia, and K. Hurley, *ApJ* **578**, 304–309 (2002).
9. T. Sakamoto, et al., *ApJ* **629**, 311–327 (2005).
10. P. Jakobsson, et al., *to appear in A&A* (2005), astro-ph/0509888.
11. D. A. Frail, et al., *ApJL* **562**, L55–L58 (2001).

GRB 050814 at $z = 5.3$ and the Redshift Distribution of *Swift* GRBs

P. Jakobsson[*], A. Levan[†], J. P. U. Fynbo[*], R. Priddey[†], J. Hjorth[*], N. Tanvir[†], D. Watson[*], B. L. Jensen[*], J. Sollerman[*], P. Natarajan[**], J. Gorosabel[‡], J. M. Castro Cerón[*] and K. Pedersen[*]

[*]*Dark Cosmology Centre, Niels Bohr Institute, University of Copenhagen, Juliane Maries Vej 30, 2100 Copenhagen, Denmark*
[†]*Centre for Astrophysics Research, University of Hertfordshire, College Lane, Hatfield, Herts, AL10 9AB, UK*
[**]*Department of Astronomy, Yale University, PO Box 208101, New Haven CT 06520-8101, USA*
[‡]*Instituto de Astrofísica de Andalucía (CSIC), Apartado de Correos, 3004, E-18080 Granada, Spain*

Abstract. We report optical, near-infrared and X-ray observations of the afterglow of GRB 050814, which was seen to exhibit very red optical colours. By modelling its spectral energy distribution we find that $z = 5.3 \pm 0.3$. We next present a carefully selected sample of 19 *Swift* GRBs, intended to estimate in an unbiased way the GRB redshift distribution, including the mean redshift (z_{mean}) as well as constraints on the fraction of high-redshift bursts. We find that $z_{\mathrm{mean}} = 2.7$ and that at least 5% of the GRBs originate at $z > 5$. The redshift distribution of the sample is qualitatively consistent with models where the GRB rate is proportional to the star formation rate in the Universe. The high mean redshift of this GRB sample and the wide redshift range clearly demonstrates the suitability of GRBs as efficient probes of galaxies and the intergalactic medium over a significant fraction of the history of the Universe.

Keywords: dust, extinction – early Universe – galaxies: high redshift – gamma rays: bursts
PACS: 95.85.Kr – 95.85.Nv – 98.70.Rz

INTRODUCTION

The immense luminosities of the gamma-ray bursts (GRBs), coupled with their origin in the core collapse of massive stars [1, 2] and their γ-ray penetration through dust, open up a variety of intriguing cosmological applications. Much effort has been directed into the use of GRBs for studying star formation (e.g. [3]), as backlights for exploring high-redshift galaxies and the intergalactic medium (e.g. [4, 5]), and even as probes of cosmological parameters (e.g. [6, 7]). Although the GRB population observed until the end of 2004 had enabled much progress in the field, it was widely expected that the launch of *Swift*, and the subsequent order of magnitude increase in the number of GRBs open to detailed study, would allow further insight into the high-redshift Universe [8]. Indeed, the ability of *Swift* to locate and follow-up a fainter burst population than was previously possible [9] has allowed the study of more distant bursts. The mean redshift of pre-*Swift* bursts was $z_{\mathrm{mean}} = 1.4$, while we show here that bursts discovered by *Swift* now have $z_{\mathrm{mean}} = 2.7$, including the first burst to have been discovered with $z > 6$, GRB 050904 at $z = 6.295$ (e.g. [10]).

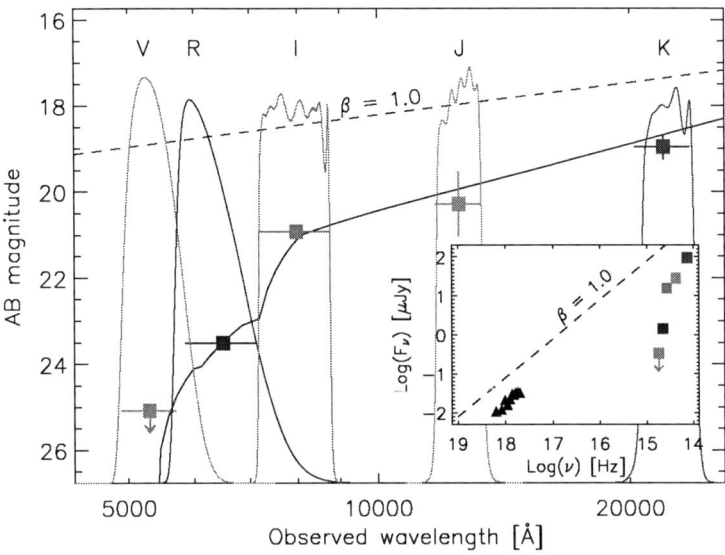

FIGURE 1. The spectral energy distribution of the GRB 050814 afterglow at $\Delta t = 14$ hr. The strong break blueward of the I-band is too strong to be readily explained by reddening alone and is best fit by the presence of the Lyα break at $z = 5.3$. The solid curve is a fit to the data at that redshift. The dashed line shows the spectral slope expected from a synchrotron emission in the fireball model with $\beta = 1$. The filter response functions are also shown. The horizontal error bars represent the FWHM of each filter. The V-band upper limit is 2σ. The inset shows the $VRIJK$ observations (filled squares) along with the X-ray spectrum (filled triangles) at $\Delta t = 14$ hr. The dashed line is the same $\beta = 1$ slope as in the main panel.

SPECTRAL ENERGY DISTRIBUTION OF THE AFTERGLOW

Our multiband observations of the GRB 050814 afterglow are presented in [11]. They allowed the construction of its spectral energy distribution (SED), displayed in Fig. 1, where we have corrected the observed data points for foreground (Galactic) extinction. The SED has a strong break blueward of the I-band, exhibiting colours of $I - K = 3.44 \pm 0.29$ mag and $R - I = 2.87 \pm 0.10$ mag, corresponding to spectral slopes of $\beta_{IK} = 1.78 \pm 0.12$ and $\beta_{RI} = 11.70 \pm 0.04$, respectively ($F \propto \nu^{-\beta}$). The latter value is unreasonable for GRB afterglows, implying an electron energy power-law index more than ten times higher than normally observed. Even in the case of high local extinction (A_V), such steep slopes cannot be obtained (see also [12]).

The most likely explanation for the steep break observed is due to the presence of the Lyα break at a redshift of $5 < z < 6$. To provide a more robust estimate of the GRB 050814 redshift we fit the available photometry at different redshifts, allowing for a range in β and A_V modelled using the parametrization of [13]. The models of [14] provide the average hydrogen opacity as a function of redshift. The minimum χ^2 is obtained for $z = 5.3 \pm 0.3$. However, we are only able to obtain weak constraints on β and A_V. Fixing $\beta = 1.0$, a typical value for GRB afterglows, results in a best fit of a restframe $A_V = 0.9$ mag and an unchanged redshift. This A_V is marginally higher than

TABLE 1. An update to the list of 28 long-duration GRBs from table 2 in [11]; these additional nine bursts were detected after 30 September 2005. Here θ_{Sun} is the Sun-to-field distance, θ_{Moon} the Moon-to-field distance and I_{Moon} the Moon illumination at the time the burst occurred. For a burst detected in the optical but without a reported redshift, an upper redshift limit is estimated based on the filter it is detected in.

GRB	z	A_V^{Gal} [mag]	θ_{Sun} [deg]	θ_{Moon} [deg]	I_{Moon} [%]	Ref.
051001		0.05	142	156	4	
051006		0.22	83	121	12	
051016A		0.29	76	116	98	
051016B	0.94	0.11	73	117	99	[15]
051117B		0.18	130	48	97	
060108	< 8.5	0.05	146	96	69	[16]
060111A	< 5.0	0.09	61	111	90	[17]
060115	3.53	0.44	121	72	99	[18]
060124	2.30	0.44	121	132	30	[19]

has been inferred from the SEDs of pre-*Swift* bursts with bright optical afterglows (OAs), but is a necessary consequence of the red $I - K$ colour. The extrapolated $\beta = 1.0$ line, normalized for $A_V = 0.9$ mag, slightly overestimates the predicted X-ray flux (inset of Fig. 1), indicating that β is a bit steeper; $\beta = 1.1$ would make the X-ray data fall on the line. Since the best fit X-ray spectral index is consistent with the assumed optical/NIR one, a cooling break between the optical and X-rays can be ruled out.

THE REDSHIFT DISTRIBUTION OF *SWIFT* BURSTS

In order to study the redshift distribution of GRBs, it is important to carefully select a sample containing bursts which have "observing conditions" favorable for redshift determination. In [11] we introduced four criteria with the aim of constructing such a sample. Here we recap those criteria and add a fifth one: (1) Small error circles, hence the bursts have to be localised with the XRT. (2) The Galactic extinction in the direction to the burst has to be sufficiently small or $A_V^{Gal} < 0.5$ mag. (3) The XRT error circle should be distributed quickly (within 12 hours) for a relatively rapid follow-up. Although the automatic slewing of *Swift* was enabled in the middle of January 2005, part of the following month was dedicated to calibration which could not be interrupted. Therefore, we have only included bursts occurring after 1 March 2005. (4) Rejection of bursts with a declination unsuitable (above $+70°$ or below $-70°$) for follow-up observations. (5) The Sun-to-field distance has to be large enough, with $\theta_{Sun} \gtrsim 55°$.

The first 28 bursts in the sample were listed in table 2 in [11]. Nine additional bursts are presented in Table 1. For each burst we have also listed the Moon-to-field distance (θ_{Moon}) and the Moon illumination at the time of the burst. This is done to examine if these parameters affect the redshift determination significantly, e.g. a full Moon close to a burst location. This is of course difficult to quantify as the OA brightness also plays

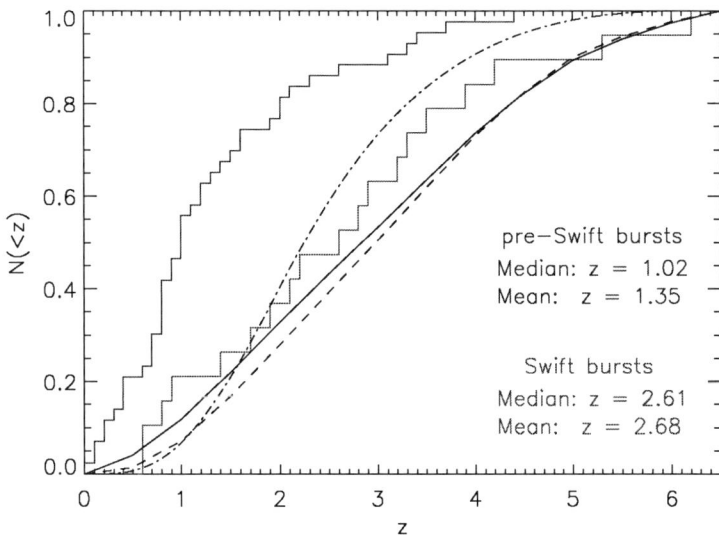

FIGURE 2. The cumulative fraction of GRBs as a function of redshift for 43 pre-*Swift* bursts (upper stepwise curve) and 19 *Swift* bursts (lower stepwise curve). Overplotted are three simple models for the expectation of the redshift distribution of GRBs: model II from [20] in which the GRB rate is proportional to the star formation rate (solid curve), model IV from [20] in which the GRB rate increases with decreasing metallicity (dashed curve) and a model from [23] in which the GRB rate is proportional to the star formation rate (dash-dotted curve). All three models fold in the *Swift*/BAT flux sensitivity.

a role. For example, GRB 050820A (table 2 in [11]) has a measured redshift even if it occurred during a full Moon and $\theta_{Moon} = 34°$. Therefore, we decided not to limit the sample further. This relatively "clean" sample of 37 bursts has a redshift recovery rate of roughly 50% (19/37).

Figure 2 shows the redshift distribution of the 19 bursts with a reported redshift in our *Swift* sample. Both the mean and the median is $z \approx 2.65$, more than twice as large as the corresponding numbers for pre-*Swift* bursts. A natural explanation for this increase is the lower trigger threshold of *Swift* compared to previous missions, giving rise to fainter (*Swift* events are on average 1.7 mag fainter in the *R*-band at a similar epoch: [9]) and higher redshift bursts. This is complemented by the accurate positions provided by *Swift* and the rapid response of a variety of telescopes aimed at redshift determinations.

This *Swift* sample is the most uniform to date and it is of interest to compare its redshift distribution to models predicting the fraction of GRBs expected to occur at a given redshift. [20] have modelled the expected redshift distribution for GRBs, utilising several models including those which follow the globally averaged star formation rate (model II), and those which scale according to the average metallicity of the Universe at a given redshift (model IV, see e.g. [21, 22]). [23] have also carried out a similar exercise, where the GRB rate is assumed to be proportional to the star formation rate. These models are plotted in Fig. 2.

It is remarkable how similar the observed *Swift* redshift distribution is to the model predictions; we can now reason that GRBs indeed trace star formation (see also [24]). However, with the available sample and the limited flux sensitivity of the *Swift*/BAT for $z > 5$ bursts, it is currently not possible to determine if GRBs are unbiased tracers of star formation. For example, models II and IV from [20] are nearly indistinguishable when comparing to the relatively small sample of 19 bursts. Note that although model II from [20] and the [23] model both presuppose that the GRB rate is proportional to the star formation rate, they use different assumptions regarding the poorly determined GRB luminosity function and the intrinsic spectral shape, explaining their difference in Fig. 2.

By including all the bursts in Table 1 and table 2 in [11], we are able to constrain the number of bursts above a specific redshift. For example, 5%–40% of the bursts are located at $z > 5$. The [20] and [23] predictions are 10% and 2%, respectively, suggesting that the GRB luminosity function parameters and/or the GRB spectral index assumed in [20] might be more appropriate. [25] also predict that 10% of the *Swift* GRBs should originate at $z > 5$.

DISCUSSION & CONCLUSIONS

The mean redshift of our relatively unbiased *Swift* sample ($z_{mean} = 2.7$) is larger than the median redshift of sub-mm galaxies ($z_{median} = 2.2$: [26]) and is similar to that of Type 2 AGNs ($z_{mean} \sim 3$: [27]). With two $z > 5$ GRBs discovered within a space of a month, and primarily due to the spectroscopic redshift of $z = 6.295$ for GRB 050904 [10], we are finally accessing the GRB high-redshift regime. Are we starting to probe the era of Pop III stars? If the transition between the dark ages and the era of reionization occurred around $z \approx 6$–7 (see e.g. [28] for a review), the answer might be positive. However, [29] have calculated that at most one massive metal-free star forms per pre-galactic halo, and since the GRB progenitor may need to be a member of a close binary system in the collapsar scenario (e.g. [30, 31, 32]), it seems unlikely that Pop III stars could end their lives as GRBs. [33] have proposed a non-binary possibility in the collapsar scenario, introducing unusually rapidly rotating massive stars. It is therefore possible that Pop III stars are GRB progenitors, although the number of unknowns is currently too large to arrive at a concrete conclusion.

The sample contains GRB 050814, whose OA was particularly faint in the *R*-band; the observed optical to X-ray spectral slope is flatter ($\beta_{OX} = 0.36$) than expected for the fireball model. Hence, GRB 050814 is classified as a dark burst as defined by [34]. We have argued that this is most likely due to the high-redshift nature ($z = 5.3$) of this burst; the $R - I$ colour is extremely red which is impossible to explain by strong extinction given the observed $I - K$ colour. Indeed, a similar conclusion was proposed for GRB 980329 [35].

It is clear that GRBs have now opened up a window to the very high-redshift Universe. The emerging GRB redshift histogram (Fig. 2) strongly indicates that GRBs can be used to trace the star formation in the Universe over a wide redshift range ($0 \lesssim z \lesssim 7$). Future instrumentation, such as the X-shooter [36], will hopefully shed light on the end of the dark ages and the possible GRB/Pop III connection.

ACKNOWLEDGMENTS

PJ, JF, BLJ and KP acknowledge support from the Instrument Centre for Danish Astrophysics (IDA). RP acknowledges PPARC for support, while AL and NRT thank PPARC for support through postdoctoral and senior research fellowships, respectively. JS acknowledges Danmarks Nationalbank for lodging. The research of JG is supported by the Spanish Ministry of Science and Education through programmes ESP2002-04124-C03-01 and AYA2004-01515. JMCC gratefully acknowledges partial support from IDA and the NBI's International Ph.D. School of Excellence. The Dark Cosmology Center is funded by the Danish National Research Foundation. The authors acknowledge benefits from collaboration within the EU FP5 Research Training Network "Gamma-Ray Bursts: An Enigma and a Tool".

REFERENCES

1. J. Hjorth, J. Sollerman, P. Møller, et al., *Nature*, **423**, 847–850 (2003).
2. K. Z. Stanek, T. Matheson, P. M. Garnavich, et al., *ApJ*, **591**, L17–L20 (2003).
3. L. Christensen, J. Hjorth, and J. Gorosabel, *A&A*, **425**, 913–926 (2004).
4. P. M. Vreeswijk, S. L. Ellison, C. Ledoux, et al., *A&A*, **419**, 927–940 (2004).
5. P. Jakobsson, J. Hjorth, J. P. U. Fynbo, et al., *A&A*, **427**, 785–794 (2004).
6. G. Ghirlanda, G. Ghisellini, D. Lazzati, and C. Firmani, *ApJ*, **613**, L13–L16 (2004).
7. E. Mörtsell, and J. Sollerman, *JCAP*, **6**, 9–18 (2005).
8. N. Gehrels, G. Chincarini, P. Giommi, et al., *ApJ*, **611**, 1005–1020 (2004).
9. E. Berger, S. R. Kulkarni, D. B. Fox, et al., *ApJ*, **634**, 501–508 (2005).
10. N. Kawai, G. Kosugi, K. Aoki, et al., *Nature*, in press (astro-ph/0512052) (2006).
11. P. Jakobsson, A. Levan, J. P. U. Fynbo, et al., *A&A*, in press (astro-ph/0509888) (2006).
12. D. E. Reichart, *ApJ*, **553**, 235–253 (2001).
13. D. Calzetti, L. Armus, R. C. Bohlin, et al., *ApJ*, **533**, 682–695 (2000).
14. P. Madau, *ApJ*, **441**, 18–27 (1995).
15. A. M. Soderberg, E. Berger, and E. Ofek, *GCN Circulars* 4186 (2005).
16. A. Monfardini, C. G. Mundell, C. Guidorzi, et al., *GCN Circulars* 4502 (2006).
17. A. J. Blustin, S. Zane, and S. R. Rosen, *GCN Circulars* 4482 (2006).
18. S. Piranomonte, V. D'Elia, F. Fiore, et al., *GCN Circulars* 4520 (2006).
19. N. Mirabal, and J. P. Halpern, *GCN Circulars* 4591 (2006).
20. P. Natarajan, B. Albanna, J. Hjorth, et al., *MNRAS*, **364**, L8–L12 (2005).
21. J. P. U. Fynbo, P. Jakobsson, P. Møller, et al., *A&A*, **406**, L63–L66 (2003).
22. C. L. Fryer, and A. Heger, *ApJ*, **623**, 302–313 (2005).
23. J. Gorosabel, N. Lund, S. Brandt, et al., *A&A*, **427**, 87–93 (2004).
24. P. Jakobsson, G. Björnsson, J. P. U. Fynbo, et al., *MNRAS*, **362**, 245–251 (2005).
25. V. Bromm, and A. Loeb, *ApJ*, in press (astro-ph/0509303) (2006).
26. S. C. Chapman, A. W. Blain, I. Smail, and R. J. Ivison, *ApJ*, **622**, 772–796 (2005).
27. P. Padovani, M. G. Allen, P. Rosati, and N. A. Walton, *A&A*, **424**, 545–559 (2004).
28. J. Miralda-Escudé, *Science*, **300**, 1904–1909 (2003).
29. T. Abel, G. L. Bryan, and M. L. Norman, *Science*, **295**, 93–98 (2002).
30. C. L. Fryer, S. E. Woosley, and D. H. Hartmann, *ApJ*, **526**, 152–177 (1999).
31. A. I. MacFadyen, and S. E. Woosley, *ApJ*, **524**, 262–289 (1999).
32. W. Zhang, S. E. Woosley, and A. Heger, *ApJ*, **608**, 365–377 (2004).
33. S. E. Woosley, and A. Heger, *ApJ*, submitted (astro-ph/0508175) (2006).
34. P. Jakobsson, J. Hjorth, J. P. U. Fynbo, et al., *ApJ*, **617**, L21–L24 (2004).
35. A. S. Fruchter, *ApJ*, **512**, L1–L4 (1999).
36. S. D'Odorico, M. I. Andersen, P. Conconi, et al., *SPIE*, **5492**, 220–229 (2004).

The true redshift distribution of Pre-SWIFT gamma-ray bursts

B. Gendre* and M. Boër[†]

*IASF/INAF, via fosso del cavaliere 100, 00133 Roma, Italy
[†]Observatoire de Haute Provence, 04870 Saint Michel l'Observatoire, France

Abstract. SWIFT bursts appear to be more distant than previous bursts. We present the Boer & Gendre relation that link redshift and afterglow luminosities. Taking advantage of the XMM-Newton, Chandra and BeppoSAX catalogs, and using this relation, we have investigated the redshift distribution of GRBs. We find that XMM burst sources with unknown redshift appear to be more distant than those with a known redshift. We propose that this effect may be due to a selection effect of pre-SWIFT optical observations.

Keywords: X-rays : general; Gamma-ray : bursts
PACS: 95.85.Nv : X-rays–98.70.Rz : gamma-rays sources, gamma-ray bursts

INTRODUCTION

The observations of long Gamma-Ray Burst (GRB) afterglows allowed the emergence of the fireball model [1, 2, 3]. In this model an isotropic blast wave propagates into a surrounding uniform interstellar medium (ISM). Two refinements were made: first, the isotropic assumption was relaxed. This model was called the "jet model" [4]. Second, observations showed that long GRBs may be linked with the explosion of a massive star (hypernova, [5]). In such a case, the surrounding medium is not uniform [6] because of the wind from the progenitor of the GRB. This model is referred as the "wind model" [7, 3, 6].

As GRBs are distant events, one may use them for cosmological studies. They may trace the star formation history and constrain the cosmological parameters. With distant bursts, such as GRB 050904 (z=6.29, [8]), observed by SWIFT and TAROT [9, 10], one can also observe the death of the first stars and the re-ionization period. Recently, it appears that SWIFT bursts were observed to be more distant than previous bursts [11]. This is rather obvious when looking at Fig. 1, which displays the cumulative distribution of redshifts derived from pre-SWIFT observations versus those measured from SWIFT detections.

We reported earlier that GRB X-ray afterglows with known redshifts have a bimodal luminosity evolution : the faintest GRB afterglows appear to decay more slowly than the brighter ones [12]. Bright and faint X-ray afterglows are separated by one order of magnitude in flux one day after the burst. We can use this relation as a redshift estimator. From the distances we compute, we are able to explain the observed difference in the cumulative distributions of redshift presented in Fig. 1, as a selection bias.

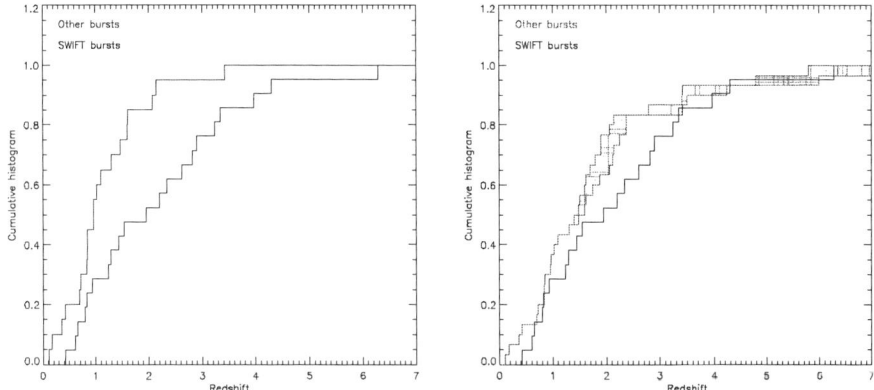

FIGURE 1. The redshift distributions of pre-SWIFT (color line) and SWIFT (black line) bursts. Left : using only bursts with a redshift measured by optical studies. Right : using bursts with a redshift measured either by an optical study or by the use of the Boër & Gendre relation. The shaded area represents the uncertainties of the redshift distribution.

TABLE 1. Bursts used to derive the Boër & Gendre relation. We indicate for each burst its redshift and the satellite that observed the X-ray afterglow.

Burst name	Redshift	X-ray satellite	Burst name	Redshift	X-ray satellite
GRB 970228	0.695	BeppoSAX	GRB 011121	0.36	BeppoSAX
GRB 970508	0.835	BeppoSAX	GRB 011211	2.14	XMM-newton
GRB 971214	3.42	BeppoSAX	GRB 020405	0.69	Chandra
GRB 980425	0.0085	BeppoSAX	GRB 020813	1.25	Chandra
GRB 980613	1.096	BeppoSAX	GRB 021004	2.3	Chandra
GRB 980703	0.966	BeppoSAX	GRB 030226	1.98	Chandra
GRB 990123	1.60	BeppoSAX	GRB 030328	1.52	Chandra
GRB 990510	1.619	BeppoSAX	GRB 030329	0.168	XMM-Neton
GRB 991216	1.02	Chandra	GRB 031203	0.105	XMM-Newton
GRB 000210	0.846	BeppoSAX	GRB 050401	2.90	SWIFT
GRB 000214	0.42	BeppoSAX	GRB 050525A	0.606	SWIFT
GRB 000926	2.066	BeppoSAX	GRB 050904	6.29	SWIFT
GRB 010222	1.477	BeppoSAX	GRB 050908	3.344	SWIFT

THE BOËR & GENDRE RELATION

Derivation of the relation

Our sample is listed in Table 1. We used only GRBs with known redshifts that exhibit an X-ray afterglow observed either by BeppoSAX, XMM-Newton, Chandra or SWIFT. The detail of data analysis is presented in [13].

We have corrected the fluxes for distance, time dilation, and energy losses due to the

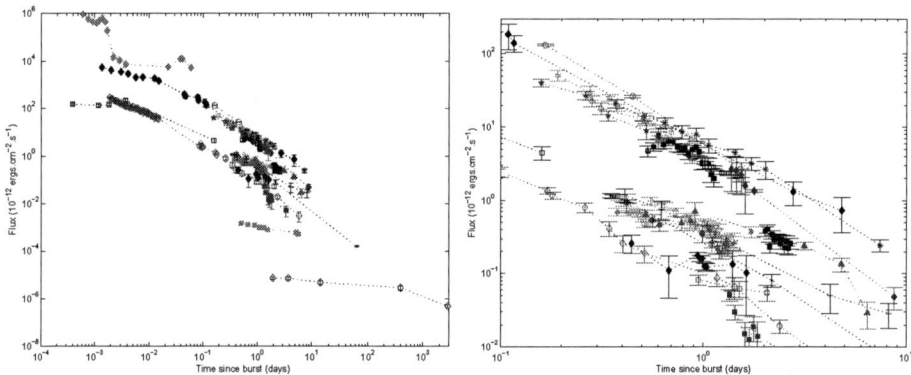

FIGURE 2. The Boër & Gendre relation made with 26 bursts. Left panel: full scale to show all bursts. Right panel: we display the same relation restricted to the time interval from 0.1 to 10 days after the GRB.

cosmological energy shift. To compute these corrections, we used a flat universe model, with an Ω_m value of 0.3. We normalized the flux to a common distance of $z=1$ rather than using the luminosity. We corrected the cosmological energy shift as in [14]. In order to reduce uncertainties, we did not correct for the time dilation effect by interpolating the flux as in [14]; instead, we computed the time of the measurement in the burst restframe. Finally, we restricted the light curves to the 2.0–10.0 keV X-ray band, where the absorption is negligible. This allowed us to neglect any other corrections for absorption by the ISM. We do not take into account any beaming due to a possible jet.

The two groups reported in [12] are still present (see Fig. 2). All but two bursts lie in one of the two groups. The only exceptions are GRB 980425 and GRB 031203. In the following we call *group I* the set of GRB afterglows with the brightest luminosity, and *group II* the dimmer ones. The probability that a power law luminosity distribution (letting the index be a free parameter) represent the observed distribution is, at the maximum, 1.2×10^{-6} (index value : -1), thus the observed clustering in two groups is very significant.

We computed the mean decay index of the groups. We find $\delta = 1.6 \pm 0.2$ for *group I*. If we take into account all bursts of *group II*, we find $\delta = 1.5 \pm 0.9$. However, if we take into account only the bursts with a good decay constraint (hence ignoring GRB 011121 and GRB 030226), we get $\delta = 1.1 \pm 0.2$. Using a Kolmogorov-Smirnov test to check if this repartition is due to a single population of GRBs, we obtain a probability of 0.13 : this distribution of decay indexes may be due to only one population.

Validity of the relation as a distance estimator

Before using this relation as a distance estimator, one may check its validity.

1. It has been reported by [15] a weak clustering in two groups. This effect is probably due to the time dilatation correction computed by these authors: As noted above,

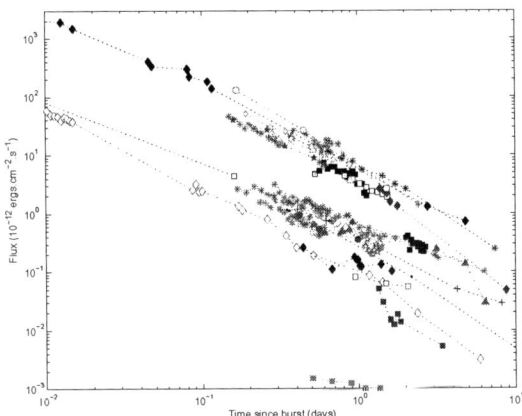

FIGURE 3. The rescaled bursts.

and also stated by [16], this correction needs to be computed without extrapolations (as done in [13]), otherwise the uncertainties on the decay index and on the mean flux level might broaden the true distribution.

2. Several discrepancies between SWIFT bursts and other bursts have been reported [17]. This might be the result of a selection effect. While this claim is still valid for the optical afterglows, [18] has shown that the X-ray afterglows of Beppo-SAX, XMM-Newton and SWIFT are similar. Thus, no selection effect can be objected if one uses X-ray afterglows, as we do.

3. Both nearby and distant bursts may be used to check the validity domain of the relation. At large distances ($z>5$) only one burst (GRB 050904) is present in our sample. It presents strong flaring activity [19, 20] which may make conclusions difficult to draw at first glance; however this burst agrees with the relation (see Fig. 2). At small distances ($z<0.1$), two bursts (GRB 980425 and GRB 031203) are present in the sample. As can be seen in Fig. 2, they do not follow the relation. These events are the only outliers we found, and this might be due to a distance effect : as a conservative hypothesis we prefer to restrict the validity of the Boër & gendre relation to redshifts larger than 0.5.

We emphasize that the bursts were sorted in each groups according only to their X-ray properties (the decay index), and that we did not use any redshift estimate. When the classification was ambiguous (e.g. for a decay index of 1.4)we systematically privileged the group that gave the lowest redshift value, i.e. disfavoring high redshifts (where we want to check the relation).

THE REDSHIFT DISTRIBUTION OF PRE-SWIFT BURSTS

We used the Boër & Gendre relation to derive the distance of the bursts listed in Table 2. As one may note, only one burst does not reach the lower threshold, and thus we do

TABLE 2. Redshift estimates derived from the Boër & Gendre relation

GRB name	Spectral index	Observing satellite	Redshift estimate
GRB 001025A	1.8	XMM-Newton	5.8
GRB 020322	1.1	XMM-Newton	1.8
GRB 040106	0.49	XMM-Newton	3.4
GRB 040223	1.6	XMM-Newton	1.9
GRB 040827	1.3	XMM-Newton	1.9
GRB 980329	1.44	BeppoSAX	1.5
GRB 980519	2.43	BeppoSAX	4.8
GRB 990704	1.68	BeppoSAX	1.4
GRB 990806	1.31	BeppoSAX	1.7
GRB 001109	1.29	BeppoSAX	2.8
GRB 020410	1.3	BeppoSAX	0.5

not consider it in the following discussion. The mean redshift for SWIFT bursts is 2.7, while the mean pre-SWIFT mean redshift was 1.2. We took into account the bursts from table 2 to recomputed the redshift distributions; the result is displayed on the right panel of Fig. 1. As it can be seen, with this method the pre-SWIFT and SWIFT distributions do agree both at low and high redshifts. The observed difference at intermediate redshift can be explained by a lack of SWIFT bursts at these distances, and is compatible with the expectation derived from Poisson counting statistics. We note also the interesting result of GRB 001025A. This burst was observed by XMM-Newton but its optical afterglow was never detected, and thus it was classified as dark burst [21]. With a calculated redshift of 5.8 ± 0.8, this can be easily explained by the Lyman alpha cutoff.

As stated above, there is a bias observed with the SWIFT optical afterglows : they appear fainter than the pre-SWIFT ones [17]. This has strong consequences on the distance estimation of the bursts. This estimate is based on spectroscopic observations of the optical afterglow which is detected mostly after an X-ray observation gave the precise position of the transient. Before SWIFT X-ray observations were made hours after the burst. Hence the optical follow-up occurred at least hours, or even days, after the event, when the optical transient had significantly faded. Thus the OT detection implied a bright source: they are not common as discovered by SWIFT. Because of the cosmological effects, a burst will have its flux decreasing with the distance if the afterglow spectral index is larger than 1 (as observed in most of the afterglows by [7, 16]) : bright bursts will be on average nearby, biasing the pre-SWIFT distribution against high redshifts. Since our analysis is unbiased in that sense, we do not see any difference in distance in the two samples.

CONCLUSIONS

We have presented the Boër & Gendre relation which links the X-ray afterglow luminosity and the GRB source distance. We use it to derive the distance of burst sources with unknown redshift. We observe significantly more distant bursts by using our method, than selecting them only on their optical properties. We show that the redshift distribution of GRB sources computed from the pre-SWIFT and SWIFT sample agree when using the Boër and Gendre relation.

We infer that the difference previously reported is due to a selection bias due to the optical measurements of the redshift of pre-SWIFT bursts.

ACKNOWLEDGMENTS

This work was supported by the EU FP5 RTN 'Gamma ray bursts: an enigma and a tool'.

REFERENCES

1. Rees, M.J., & Meszaros, P., 1992, MNRAS, 258, 41
2. Meszaros, P., & Rees, M.J., 1997, ApJ, 476, 232
3. Panaitescu, A., Meszaros, P., & Rees, M.J., 1998, ApJ, 503, 314
4. Rhoads, J.E., 1997, ApJ, 487, L1
5. Meszaros, P., 2001, Science, 291, 79
6. Chevalier, R.A., & Li, Z.Y., 1999, ApJ, 520, L29
7. Dai, Z.G., & Lu, T., 1998, MNRAS, 298, 87
8. Kawai, N., Kosugi, G., Aoki, K., et al., 2005, Nature, submitted
9. Cusumano G., et al., 2005, these proceedings
10. Boër, M., Atteia, J.L., Damerdji, Y., et al., 2005, these proceedingsd
11. Jakobsson, P., Levan, A., Fynbo, J. P. U., et al., 2005, ApJ, submitted
12. Boër, M. & Gendre, B. 2000, A&A, 361, L21
13. Gendre, B., Boër, M., 2005, A&A, 430, 465
14. Lamb & Reichart, 2000, ApJ, 536, L1
15. Nardini, M., Ghisellini, G., Ghirlanda, G., et al., 2005, A&A, submitted
16. Liang, E. & Zhang, B., 2005, Nature, submitted
17. Berger, E., Kulkarni, S.R., Fox, D.B., et al., 2005, ApJ, 634, 501
18. Gendre, B., Corsi, A., Piro, L., 2005, A&A, in press
19. Watson, D., Reeves, J. N., Hjorth, J., et al., 2005, ApJ, submitted
20. Boër, M., Atteia, J.L., Damerdji, Y., et al., 2005, ApJ, submitted
21. Pedersen, K., Hurley, K., Hjorth, J., et al., 2005, ApJ, in press

GRB 050904: the oldest cosmic explosion ever observed in the Universe

G. Cusumano*, V. Mangano*, G. Chincarini[†,**], A. Panaitescu[‡], D.N. Burrows[§], V. La Parola*, T. Sakamoto[¶,‖], S. Campana[†], T. Mineo*, G. Tagliaferri[†], L. Angelini[¶], S.D. Barthelemy[¶], A.P. Beardmore[††], P.T. Boyd[¶], L.R. Cominsky[‡‡], C. Gronwall[§], E.E. Fenimore[§§], N. Gehrels[¶], P. Giommi[¶¶], M. Goad[††], K. Hurley[***], J.A. Kennea[§], K.O. Mason[†††], F. Marshall[¶], P. Mèszàros[§,‡‡‡], J.A. Nousek[§], J.P. Osborne[††], D.M. Palmer[‡], P.W.A. Roming[§], A. Wells[††], N.E. White[¶] and B. Zhang[§§§]

*INAF-Istituto di Astrofisica Spaziale e Fisica Cosmica di Palermo, Via Ugo La Malfa 153, 90146 Palermo, Italy
[†]INAF – Osservatorio Astronomico di Brera, Via Bianchi 46, 23807 Merate Italy
[**]Università degli studi di Milano-Bicocca, Dip. di Fisica, Piazza delle Scienze 3, I-20126 Milan, Italy
[‡]Los Alamos National Laboratory, P.O. Box 1663, Los Alamos, NM 87545, USA
[§]Department of Astronomy & Astrophysics, Pennsylvania State University, PA 16802, USA
[¶]NASA/Goddard Space Flight Center, Greenbelt, MD 20771, USA
[‖]National Research Council, 2001 Constitution Avenue, NW, TJ2114, Washington, DC 20418, USA
[††]Department of Physics and Astronomy, University of Leicester, University Road, Leicester LE1 7RH, UK
[‡‡]Department of Physics and Astronomy, Sonoma State University, Rohnert Park, CA 94928-3609, USA
[§§]4Los Alamos National Laboratory, P.O. Box 1663, Los Alamos, NM 87545, USA
[¶¶]ASI Science Data Center, via Galileo Galilei, 00044 Frascati, Italy
[***]UC Berkeley Space Sciences Laboratory, Berkeley, CA 94720-7450
[†††]MSSL, University College London, Holmbury St. Mary, Dorking, RH5 6NT Surrey, UK
[‡‡‡]Department of Physics, Pennsylvania State University, PA 16802, USA
[§§§]Department of Physics, University of Nevada, Box 454002, Las Vegas, NV 89154-4002, USA

Abstract. Swift discovered the high redshift (z=6.29) GRB 050904 with the Burst Alert Telescope and began observing with its narrow field instruments only 161 s after the burst onset. GRB 050904 was a long, multi-peaked, bright GRB with a presence of flaring activity lasting up to 1-2 hours after the burst onset. The spectral energy distribution shows a clear softening trend along the burst evolution with a photon index decreasing from -1.2 up to -1.9. The observed variability is more dramatic than the typical Swift afterglow, the amplitude and rise/fall times of the flares are consistent with the behavior of nearby ($z \leq 1$) long GRBs and suggest the interpretation of the BAT and XRT data as a single continuous observation of long lasting prompt emission.

Keywords: Gamma Ray Bursts, X-rays
PACS: 98.70.Rz, 95.85.Nv

INTRODUCTION

The Swift [1] X-ray Telescope (XRT, [2]) is providing a growing number of unprecedented observations of the early stages of GRB afterglows in the 0.2-10 keV X-ray band.

The XRT rapid ($\leq$ 2 min) response to the Swift Burst Alert Telescope (BAT, [3]) triggers has already led to the discovery of rapid early declines followed by the smoother "standard" afterglow components and dramatic flaring in the early X-ray light curves [4, 5, 6, 7], as well as simultaneous peaks in the final part of the soft gamma-ray (15-350 keV) light curve observed by BAT and in the initial part of the X-ray light curve for several bursts. Thanks to its fast response, and precise (about 5″) source localization Swift is able to alert immediately the ground-based telescopes to locate the optical counterpart and get redshift measurements before the object becomes too faint.

Here we report on the gamma-ray and X-ray observation of GRB 050904. The GRB triggered the BAT on 2005 September 4 at 01:51:44 UT [8]. The burst was located onboard at RA_{J2000}=00h54m41s, Dec_{J2000}=+14° 08' 17" with an uncertainty of 3' radius (90% confidence level) and was quickly pointed towards by Swift. Early photometry indicated a high redshift (z>5, [9]). A photometric redshift $z = 6.1^{+0.37}_{-0.12}$ was measured by the MISTICI collaboration [10] and confirmed by a Subaru spectroscopic measurement of 6.29± 0.01 [11]. A break at T_b = 2.6±1.0 days was also found in the J-band light curve by the MISTICI collaboration [10]. Such a high redshift translates to a distance of 12.8 billion light-years from Earth. This gave GRB 050904 the distinction of being the most distant cosmic explosion ever observed.

GRB 050904 was observed by the XRT from 161 seconds up to 10 days after the burst onset, overlapping the BAT observations for about 300 seconds, before the end of the high energy prompt emission. During the first 598 seconds, data were accumulated in WT mode, while all the other data were accumulated in PC mode (see Hill et al. 2005 for an exhaustive discussion on XRT operative modes). Hereafter, errors are reported with a 90% single parameter confidence level. The afterglow position derived with *xrtcentroid* (v0.2.7) and including the boresight correction [12] is RA_{J2000} = 00h54m50s.8, Dec_{J2000} = +14°05'08"2, with an uncertainty of 3"2.

For the measured redshift $z = 6.29$, the 15-350 keV BAT band corresponds to a 109-2551 keV band while the 0.2-10 keV XRT band corresponds to a 1.4-73 keV band. The observed timescales are instead stretched by a factor $(1+z)$ with respect to the rest frame ones. In the following the GRB phenomenology is presented and discussed from the point of view of the source rest frame and referred to the GRB050904 onset T =2005 Sep 4, 01:51:44.3 UT.

DATA ANALYSIS

Timing Analysis

Fig. 1 (top panel) shows the evolution of the GRB flux and luminosity in the source rest frame. The observed BAT count rates were extrapolated into the XRT 0.2-10 keV band using a conversion factor evaluated from the BAT best fit spectral model (Table 1). The observed XRT count rates were converted into flux using the best fit spectral parameters listed in Table 1.

The BAT light curve displays three main peaks: two ~2-seconds long peaks at T+3.8 and at T+7.7 seconds, respectively, and a main long-lasting peak at ~T+13.7 seconds, where T is the time of the burst onset. Emission in the BAT energy range continues up to

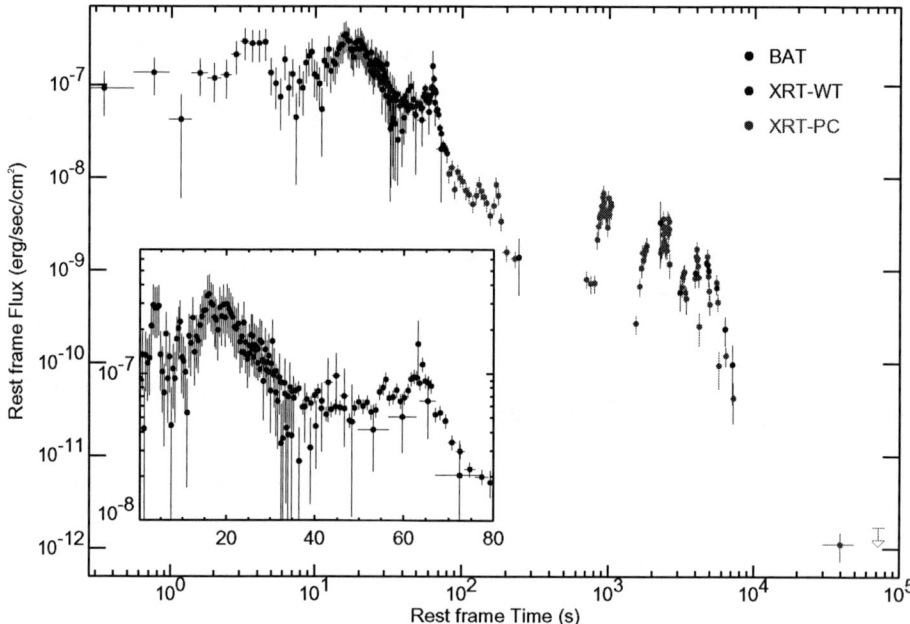

FIGURE 1. (Top panel) Light curve of GRB 050904 as observed by the BAT and XRT. Fluxes were then converted to rest frame by multiplying by $(1+z)^2$ with $z = 6.29$, and corresponds to flux emitted in the 1.4-73 keV energy band. (Bottom panel) Spectral evolution of GRB 050904 shows a photon index Γ changing during the observation.

almost T+69 seconds with a weak peak at ∼T+65 seconds, coincident with the first peak of the XRT light curve. The BAT and XRT light curves overlap between T+23 and T+69 seconds. The early XRT light curve shows a steep decay with a slope $\alpha = -2.07 \pm 0.03$ with three flares superimposed at T+65 seconds, T+126 seconds and T+171 seconds. These flares can be modeled by a linear rise lasting 26.6, 5.3 and 4.7 seconds, plus an exponential decay with decay time of 4.5, 10.98 and 5.2 seconds, respectively. Although interrupted by low Earth orbit observing constrains, the light curve from GRB 050904 reveals highly irregular rate variations likely due to the presence of flares up to T+1.5 hours. At later times the flaring activity is not detected and only a residual emission, 10^5 times lower than the initial intensity, is visible.

Spectral Analysis

The spectral analysis of GRB 050904 was performed by selecting two sets of time intervals for the BAT and XRT observations, corresponding to characteristic phases of the light curve evolution. The BAT spectra were accumulated in the 14-150 keV observed band in six time intervals up to 42 seconds from the burst onset. The XRT spectra were accumulated in twelve time intervals. Instrumental energy channels below 0.2 keV and 0.7 keV for PC and WT spectra, respectively, were ignored and the background was evaluated in regions free of contamination from other sources in the field of view. The BAT spectra were modeled with a power law with photon index Γ ($F(E) \propto E^{\Gamma+1}$) while the XRT spectra were modeled with a power law plus two absorption components: one for the intrinsic absorption in the host galaxy and one for the Galactic absorption. The latter was fixed to the line-of-sight value of 4.93×10^{20} cm^{-2} [13]. As a preliminary step, the intrinsic absorption column was evaluated from the PC spectra by leaving the redshifted N_H as a free parameter. We obtained a mean value of $(2.30 \pm 0.50) \times 10^{22}$ cm^{-2}. The spectrum of each interval was then fitted with an absorbed power law with both the Galactic column density fixed to 4.93×10^{20} cm^{-2} and the intrinsic absorption column fixed to 2.30×10^{22} cm^{-2}. More complex models, such as a Band function [14], cannot be constrained by the data. Fig. 1 (bottom panel) shows the evolution with time of the photon index Γ. The BAT spectra have $\Gamma \sim -1.2$, consistent with typical values of the α_{Band} parameter of the Band model [15]. This strongly suggests that the BAT observes the low energy part of the Band function and that the peak energy of the GRB spectrum is above $150 \times (1+z)$ keV in the source rest frame. If we exclude the spectrum of the first XRT flare at T+65 seconds, the XRT photon indices show a clear decreasing trend from about -1.2 to about -1.9 in the first T+200 seconds. No further spectral evolution is present in later XRT data, in agreement with the hardness ratio curve. The BAT and XRT photon indices are in good agreement in the overlapping region. Table 1 shows the best fit parameters for each of the selected time intervals.

We also evaluated the contribution to the total fluence in the 1.4-73 keV band of the three flares (T+65, T+126 and T+171 seconds) superimposed on the early XRT light curve. The fluence over the continuum is $(1.2 \pm 0.08) \times 10^{-6}$, $(4.7 \pm 0.5) \times 10^{-8}$ and $(5.8 \pm 0.6) \times 10^{-8}$ erg cm^{-2}, respectively. The fluence of the XRT continuum over the first orbit (i.e. from 23.2 to 244.4 s) is $(4.9 \pm 0.3) \times 10^{-6}$ erg cm^{-2}. The extrapolated 1.4-73 keV BAT fluence in the time interval from the burst onset to the start of the XRT observation is $(4.1 \pm 0.3) \times 10^{-6}$ ergs cm^{-2}. The three XRT flares are 5%, 1% and 1% of the total 1.4-73 keV emission observed up to T+244 s, respectively. The 1.4-73 keV fluence in the remaining part of the XRT observation is 1.8×10^{-6} ergs cm^{-2}. This value is only a lower limit because of the observational gaps.

DISCUSSION

GRB 050904 was a long, multi-peaked, bright GRB with strong X-ray flaring activity lasting up to 1-2 hours in the source rest frame. While the variability is more dramatic than the typical Swift afterglow, the amplitude and rise/fall times of these flares are consistent with the behavior of nearby ($z \lesssim 1$) long GRBs and suggest the interpretation of the BAT and XRT data as a single continuous observation of long lasting prompt emission. The flares in the XRT light curve could be interpreted as late internal shocks related to central engine activity. In this scenario they would have the same origin as the prompt gamma-ray emission [16, 17, 18]. This would require that the central engine remains active up to at least 5000 seconds, consistently with the collapsar model [19], which allows central engine activity for up to a few hours.

In the time interval from T+23 to T+ 244 seconds, the observed intensity underlying the XRT flares decays as t^α with $\alpha \sim -2$. An initial steep decay of the X-ray emission has been observed in many other GRBs detected by Swift [4, 16]. The measured decay slope together with the XRT energy index $\beta = \Gamma + 1 \sim -0.2$ are in good agreement with the interpretation of the observed emission as due to high-latitude emission [20]. This effect arises because of the Doppler delay of radiation emitted at large angles with respect to the observer's line of sight. Radiation observed as the tail of a peak is expected to be the off-axis emission of the shocked surface arriving at the observer at later times, and would decay as t^α with $\alpha = \beta - 2$. After T+50 seconds, due to the decrease of β to about -1, the predicted slope would be steeper than the measured -2. This deviation could be reconciled with the high-latitude emission assuming that the delayed radiation from the outer parts of the emitting curved shell is softer than the radiation along our line of sight and that a second soft component contributes to compensate for the expected steeper decay [17]. This additional component might be an emergent afterglow. The observational gaps after T+250 seconds do not allow us to confirm the presence of an underlying continuum at later times. The decrease of the photon index after T+50 seconds could be interpreted as an indication of a shift of E_p towards lower energies, but poor statistics and the narrowness of the XRT energy range do not allow us to verify this hypothesis.

Our lack of knowledge concerning the peak energy of the BAT and XRT spectra does not allow a precise estimate of the total energy released by GRB 050904. However, we can calculate lower and upper limits to the isotropic-equivalent radiated energy E_{iso} up to 244 seconds from the burst onset, i.e. including contributions from the first three XRT flares. To evaluate the lower limit to E_{iso} we integrated the best fit power law spectral energy distributions in the $(1-200)\times(1+z)$ keV band and in the $(1-10)\times(1+z)$ keV band for BAT and XRT, respectively, instead of the standard energy range $1\text{-}10^4$ keV (rest frame). The upper limit was obtained in the full $1\text{-}10^4$ keV band. We obtained 6.6×10^{53} ergs $< E_{iso} < 3.2 \times 10^{54}$ ergs. Additional contributions from the later flare portions are only a few percent. The large X-ray and gamma-ray isotropic equivalent energy of this burst is in agreement with the Amati relation [21] with an E_p of about 1500 keV in the rest frame.

The break observed in the optical and infrared afterglows by the MISTICI collaboration [10] at $T_b = 2.6\pm1.0$ days (observer frame) implies the half opening of the jet to be in the 3-4 degrees range. This corresponds to a collimation-corrected energy E_γ between 1.2 and 4×10^{51} ergs. This is well within the E_γ distribution of GRBs with known redshift [22, 23]. Consistency with the Ghirlanda relation [24] constraints the rest frame peak energy of the average spectrum to be within 560 and 1300 keV.

ACKNOWLEDGMENTS

The authors acknowledge support from ASI, NASA and PPARC.

TABLE 1. BAT and XRT spectral analysis results.

	Interval	Time Start	Time End	Γ	χ_ν^2
BAT	1	-1.43	2.69	-1.2 ±0.4	1.2 (57)
	2	2.69	4.89	-1.05±0.16	0.86 (57)
	3	4.89	10.1	-1.36±0.21	0.97 (57)
	4	10.1	20.4	-1.17±0.08	0.95 (57)
	5	20.4	30.6	-1.22±0.10	0.93 (57)
	6	30.6	41.6	-1.5 ±0.3	0.88 (57)
XRT	1	23.2	28.7	-1.13±0.07	0.70 (59)
	2	28.7	36.9	-1.31±0.06	1.13 (87)
	3	36.9	50.6	-1.34±0.06	0.81 (81)
	4	50.6	58.8	-1.78±0.07	1.23 (56)
	5	58.8	67.1	-1.50±0.07	1.18 (66)
	6	67.1	79.8	-1.88±0.12	0.78 (30)
	7	79.8	159.4	-1.80±0.10	1.00 (25)
	8	159.4	244.4	-1.96±0.19	0.82 (9)
	9	628	848	-1.81±0.22	1.51 (9)
	10	848	1040	-1.82±0.08	1.07 (37)
	11	1452	1863	-1.91±0.13	0.68 (14)
	12	2275	2618	-1.72±0.08	1.40 (35)
	13	3045	8173	-1.96±0.09	1.12 (36)

REFERENCES

1. N., Gehrels, G., Chincarini, P., Giommi, et al., *Astrophys. J.* **611**, 1005-1020 (2004)
2. D.N., Burrows, J.E., Hill, A., Nousek, et al., *Space Sci. Rev.*, in press, astro-ph/0508071 (2005)
3. S., Barthelmy, L.M., Barbier, J.R., Cummings, et al., *Space Sci. Rev.*, in press, astro-ph/0507410 (2005)
4. G., Tagliaferri, M., Goad, G., Chincarini, et al., *Nature* **436**, 985-988 (2005)
5. S.D., Barthelmy, J.K., Cannizzo, N., Gehrels, et al., *Astrophys. J.* submitted (2005)
6. Cusumano, G., Mangano, V., Angelini, L., et al., *Astrophys. J.*, in press, astro-ph/0509689 (2005)
7. D.N., Burrows, P., Romano, A., Falcone, et al., *Science* **309**, Issue 5742, 1833-1835 (2005)
8. J., Cummings, L., Angelini, S., Barthelmy, et al., *GCN Circular* **3910** (2005)
9. D.E., Reichart, *GCN Circular* **3915** (2005)
10. G., Tagliaferri, L.A., Antonelli, G., Chincarini, et al., *A&A* submitted, astro-ph (2005)
11. N., Kawai, T., Yamada, G., Kosugi, et al., *GCN Circular* **3937** (2005)
12. A., Moretti, M., Perri, M., Capalbi al., *Astron. Atrophys*, submitted (2005)
13. J.M., Dickey & F.J., Lockman, F.J., *ARA&A*, **28**, 215 (1900)
14. D., Band, J., Matteson, L., Ford, et al., *Astrophys. J.* **413**, 281-292 (1993)
15. R.D., Preece, M.S., Briggs, R.S, Mallozzi, et al., *Astrophys. J.S.* **126**, 19-36 (2000)
16. G., Chincarini, A., Moretti, P., Romano, et al., *Astrophys. J.* submitted, astro-ph/0506453, (2005)
17. B., Zhang, Y.Z., Fan, J., Dyks, et al., *Astrophys. J.* submitted, astro-ph/0508321 (2005)
18. J.A., Nousek, C., Kouvalioutu, D., Groupe, et al., *Astrophys. J.* submitted, astro-ph/0508332 (2005)
19. A.I., MacFadyen, S.E., Woosley, A., Heger, *Astrophys. J.* **550**, 410-425 (2001)
20. P., Kumar, A. & A. Panaitescu, *Astrophys. J.* **541**, L9-L12 (2000)
21. L., Amati, F., Frontera, M., Tavani, et al., *Astron. Atrophys* **390**, 81-89 (2002)
22. D.A., Frail, S.R., Kulkarni, R., Sari, et a l., *Astrophys. J.* **562**, L55-L58 (2001)
23. J.S., Bloom, D.A., Frail, S.R., Kulkarni, e t al., *Astrophys. J.* **594**, 674-683 (2003)
24. G., Ghirlanda, G., Ghisellini, D., Lazzati, *Astrophys. J.* **616**, 331-338 (2004)

X-ray flare in XRF 050406: evidence for prolonged engine activity

P. Romano*, A. Moretti*, P.L. Banat*, D.N. Burrows[†], S. Campana*, G. Chincarini*, S. Covino*, D. Malesani**, G. Tagliaferri*, A.D. Falcone[†], M. Capalbi[‡], G. Cusumano[§], P. Giommi[‡], V. La Parola[§], V. Mangano[§], M. Perri[‡] and C. Pagani*

*INAF–Osservatorio Astronomico di Brera, Via E. Bianchi 46, I-23807 Merate (LC), Italy
[†]Department of Astronomy & Astrophysics, Pennsylvania State University, PA 16802, USA
**International School for Advanced Studies (SISSA-ISAS), Via Beirut 2-4, I-34014 Trieste, Italy
[‡]ASI Science Data Center, via G. Galilei, I-00044 Frascati (Roma), Italy
[§]INAF–IASF, Via U. La Malfa 153, I-90146 Palermo, Italy

Abstract.
We present observations of XRF 050406, an X-ray flash with a relatively low fluence ($\sim 10^{-7}$ erg cm^{-2} in the 15–350 keV band), a soft spectrum (photon index $\Gamma_\gamma = 2.65$), no significant flux above ~ 50 keV and a peak energy $E_p < 15$ keV. XRF 050406 is the first burst detected by Swift clearly showing a flare in its X-ray light curve. The flare peaks 210 s after the BAT trigger, presents a flux variation $\delta F/F \sim 6$ in a timescale $\delta t/t_{\rm peak} \ll 1$ and a measured fluence of 1–15% of the prompt one. We argue that the producing mechanism is late internal shocks, which implies that the central engine is still active at 210 s, though with a reduced power with respect to the prompt emission. The X-ray light curve flattens to a more shallow slope with a decay index of ~ 0.5 after ~ 4400 s, also supporting continued central engine activity.

Keywords: Gamma rays: bursts; X-rays: bursts; X-rays: individuals (XRF 050406)
PACS: 98.70.Rz,98.70.Qy

DATA ANALYSIS

We present the XRT observations of the first Swift burst where a flare is clearly detected in its X-ray light curve [1, 2], during which the source count rate increased by a factor of ≥ 6. This feature had never been observed before in Swift data, and had rarely been observed before in any X-ray afterglow [3].

On 2005 Apr 6 at 15:58:48.40 UT, the Swift-BAT [4] triggered on GRB 050406 [5]. The Swift-XRT [6] observations of XRF 050406 started on 2005 Apr 6 at 16:00:12 UT, only 84 s after the trigger, and ended on 2005 Apr 22, thus summing up a total net exposure (in PC mode) of ~ 163 ks spread over a ~ 16 d baseline. Full details (especially about the data) can be found in Romano et al. [2].

Figure 1 shows the background-subtracted 0.2–10 keV light curve with the BAT trigger as origin of time. It clearly shows a complex behavior, with a power law decay underlying a strong flare which peaks at ≈ 210 s after the BAT trigger. Using the BAT trigger as reference time and excluding the data taken during the flare ($180\,{\rm s} < t < 300\,{\rm s}$), a fit with a broken power law yields slopes $\alpha_1 = 1.58^{+0.18}_{-0.16}$ and $\alpha_2 = 0.50 \pm 0.14$, and a break at ~ 4200 s. As a reference, the 0.2–10 keV unabsorbed flux at t_b is $(4 \pm 1) \times 10^{-13}$

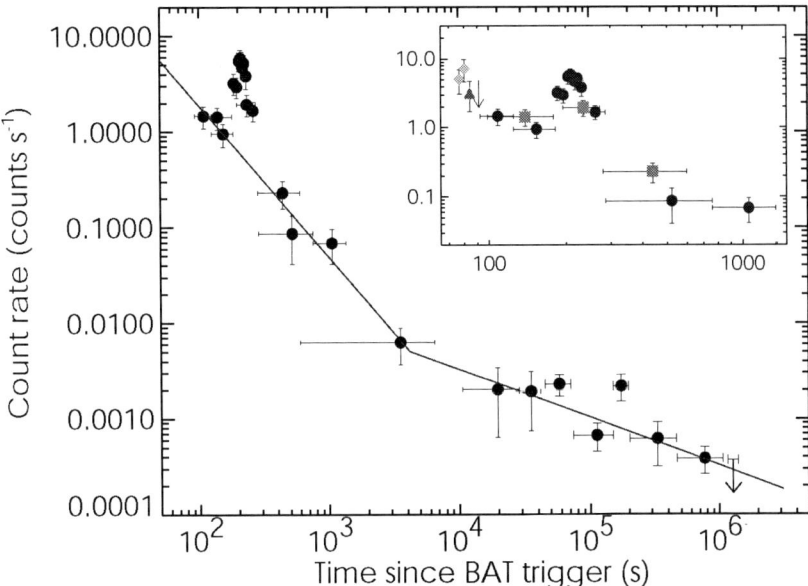

FIGURE 1. Background-subtracted X-ray light curve of the XRF 050406 afterglow in the 0.2–10 keV energy band, with time referred to the BAT trigger, 2005 Apr 06 at 15:58:48.4 UT. For $t < 4 \times 10^4$ s we binned the source counts with a minimum of 30 counts per time bin, and dynamically subtracted the normalized background counts in each bin. Afterwards, we used XIMAGE with the option SOSTA, which calculates vignetting- and PSF-corrected count rates within a specified box, and the background in a user-specified region. The last point after 10^6 s is a 3-σ upper limit. The inset shows the details of the first $\sim$ 1000 s, including data in all XRT modes. The diamonds represent LrPD mode data taken during the latter portion of the slewing phase; the triangle is the initial IM point (84 s after the trigger), the downward-pointing arrow is a LrPD limit (pointing, 91 s after the trigger), the circles are WT mode data (starting from 92 s after the trigger), and the squares are PC mode data (starting from 99 s after the trigger). The data have been corrected for pile-up (where appropriate) and PSF losses. The solid line represents the best-fit broken power-law model to the light curve (excluding the flare).

erg cm^{-2} s^{-1} and the luminosity in the 0.7–34.4 keV band is $(1.9 \pm 0.9) \times 10^{46}$ erg s^{-1} for $z = 2.44 \pm 0.36$ [7]. Both the rising and the falling part of the flare had very steep slopes that, when fit with a simple power law, yield $\alpha_{1,\text{flare}} = -6.8^{+2.4}_{-2.1}$ and $\alpha_{2,\text{flare}} = 6.8^{+3.6}_{-2.0}$ and the peak is at 213 ± 7 s from the BAT trigger (after the underlying power-law afterglow is subtracted). The flare can also be parameterised as a Gaussian with a peak at $211.1^{+5.4}_{-4.4}$ s and a width of $17.9^{+12.3}_{-4.6}$ s. The ratio of the characteristic time-scale and the peak time is $\delta t/t_{\text{peak}} \ll 1$, which puts severe constraints on the emission mechanisms that can produce the flare. Integration of the Gaussian best-fitting function yields an estimate of the fluence of the flare, $(1.4 \pm 1.0) \times 10^{-8}$ erg cm^{-2}, corresponding to an energy of $(2.0 \pm 1.4) \times 10^{50}$ erg.

The photon index is $\Gamma_X = 2.1 \pm 0.3$ and does not vary throughout the observation. Since the afterglow of XRF 050406 was very faint, it was not possible to perform time-

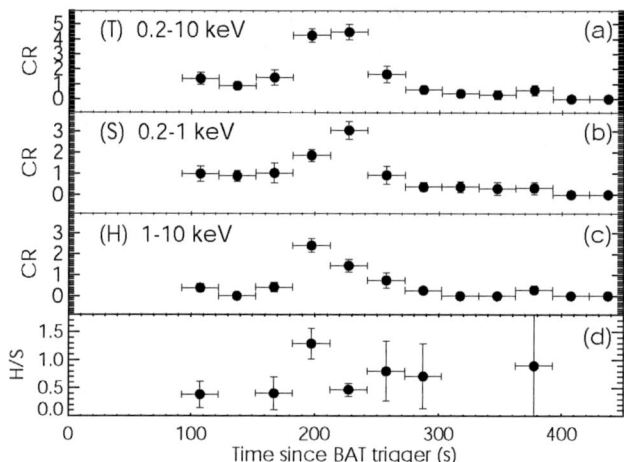

FIGURE 2. Windowed-timing background-subtracted light curves: total band (**T**, 0.2–10 keV, panel **(a)**), soft band (**S**, 0.2–1 keV, **(b)**), and hard band (**H**, 1–10 keV, **(c)**). The last panel **(d)** is the ratio of hard to soft count rates.

resolved spectroscopy to distinguish the spectral properties of the afterglow proper from the ones of the flare. Therefore, to test for spectral evolution we extracted events from the WT data in two more energy bands, 0.2–1 keV (soft, S) and 1–10 keV (hard, H), as well as the total band, 0.2–10 keV (Figure 2). During the rising portion of the flare the hard band flux increases by a factor of ≥ 6 while the soft band flux only increases slightly, so that the spectrum of the flare starts off harder than the underlying afterglow, and then evolves into a softer state as its flux decreases; this can be seen in the following time bin, when the soft band flux peaks with a flare to pre-flare flux ratio of ~ 3.5. Indication of spectral evolution during the flare comes as a $\sim 3\text{-}\sigma$ excess over a constant fit to the hardness ratio H/S. It should be noted that this behavior is reminiscent of that observed in the prompt emission, with the harder band peak preceding the softer band peak. At $t \sim 1.7 \times 10^5$ s a second faint bump is detected as a 2-σ excess over the underlying afterglow.

DISCUSSION

XRF 050406 is classified as an X-ray flash, with a 15–350 keV fluence $\sim 1 \times 10^{-7}$ erg cm^{-2}, a soft spectrum ($\Gamma_\gamma = 2.65$), no significant flux above ~ 50 keV and a peak energy $E_p < 15$ keV. The isotropic-equivalent gamma-ray energy of this event is $E_{\text{iso}} = (1.4^{+1.6}_{-0.6}) \times 10^{51}$ erg, and this effectively puts XRF 050406 in the low-energy tail of GRB energies [8]. Its main characteristics are however not qualitatively different from those of normal GRBs. The observed X-ray photon index ($\Gamma_X = 2.1$) is common among X-ray afterglows [9]. The light curve shows a break from a relatively steep decay ($\alpha_1 = 1.58$) to a flatter one ($\alpha_2 = 0.50$). Its overall shape is similar to the one typically observed by the XRT [9, 10], even though the initial slope is less steep than average. As observations accumulate, it becomes clear that these two classes of phenomena share many properties, and both have afterglows with similar characteristics. This is a clue that both events may have a common origin.

XRF 050406 is the first Swift-detected burst that showed a prominent flare in its X-ray light curve. A promising mechanism to produce the flare is late internal shocks [11], which implies that the central engine is still active at $t = 210$ s, even though the prompt emission ended after $t \sim 6$ s. The late-time activity in this case must have a reduced power with respect to the prompt emission, as the relative fluences indicate. Such a mechanism would naturally explain the steep rise and decay slopes. The indications of spectral evolution throughout the flare further support this interpretation. The flare appears to be harder than the underlying afterglow, which suggests a distinct origin for this emission. Furthermore, there are indications of spectral evolution, which shows the typical hard-to-soft pattern. Such a behavior is commonly observed in the prompt emission spikes of GRBs, which are produced in internal shocks. Further evidence of late engine activity comes from both the flat part of the light curve and possibly by the presence of the late-time bump observed at $t \sim 1.7 \times 10^5$ s.

We now know that flaring is quite a common behaviour, since $\sim 50\%$ of the bursts detected by XRT which were promptly observed showed flares. All the characteristics of the XRF 050406 flares have now been observed in most flaring GRBs, as well [12]. For example, highly significant spectral evolution throughout the flare has been reported in GRB 050502B, which was the brightest observed so far [13] and GRB 050724 [14]. In several cases the flares present large amplitudes and occur on short timescales. Furthermore, several flares are often observed in the same event, at times ranging from ~ 100 to 10^4–10^5 s after the burst. Finally, in most cases the afterglow is clearly present *before* the onset of the flare, and has consistent decay slope and flux levels with after the flare. The present case shows that flares are present both in XRFs and in GRBs indicating that flares are linked to some common properties of both kinds of bursts, and probably tied to their central engine.

REFERENCES

1. D. N. Burrows, P. Romano, A. Falcone, and et al., *Science* **309**, 1833–1835 (2005).
2. P. Romano, A. Moretti, P. L. Banat, and et al., *Astronomy and Astrophysics* **in press** (2006).
3. L. Piro, M. De Pasquale, P. Soffitta, and et al., *The Astrophysical Journal* **623**, 314–324 (2005).
4. S. D. Barthelmy, and et al., *Space Science Review* **120**, 143 (2005).
5. A. Parsons, S. Barthelmy, J. Cummings, and et al., *GRB Coordinates Network* **3180**, 1 (2005).
6. D. N. Burrows, J. E. Hill, Nousek, and et al., *Space Science Review* **120**, 165 (2005).
7. P. Schady, K. A. Mason, J. P. Osborne, and et al., *The Astrophysical Journal* **in press** (2006).
8. J. S. Bloom, D. A. Frail, and S. R. Kulkarni, *The Astrophysical Journal* **594**, 674–683 (2003).
9. G. Chincarini, A. Moretti, P. Romano, and et al. (2005), arXiv:astro-ph/0506453.
10. J. A. Nousek, C. Kouveliotou, D. Grupe, and et al., *The Astrophysical Journal* **in press** (2006).
11. B. Zhang, Y. K. Fan, J. Dyks, and et al., *The Astrophysical Journal* **in press** (2006).
12. D. N. Burrows, P. Romano, O. Godet, and et al. (2005), arXiv:astro-ph/0511039.
13. A. D. Falcone, D. N. Burrows, D. Lazzati, and et al., *The Astrophysical Journal* **in press** (2006).
14. S. Campana, G. Tagliaferri, D. Lazzati, and et al., *Astronomy and Astrophysiscs* **submitted** (2006).

The very long X-ray afterglow of XRF 050416A

V. Mangano*, G. Cusumano*, V. La Parola*, T. Mineo*, S. Campana[†], M. Capalbi**, G. Chincarini[†,‡], P. Giommi**, A. Moretti[†], M. Perri**, P. Romano[†], G. Tagliaferri[†], D.N. Burrows[§], O. Godet[¶], J. A. Kennea[§], K. Page[¶] and J.L. Racusin[§]

*INAF-Istituto di Astrofisica Spaziale e Fisica Cosmica di Palermo, Via Ugo La Malfa 153, 90146 Palermo, Italy
[†]INAF – Osservatorio Astronomico di Brera, Via Bianchi 46, 23807 Merate Italy
**ASI Science Data Center, via Galileo Galilei, 00044 Frascati, Italy
[‡]UniversitÃă degli studi di Milano-Bicocca, Dip. di Fisica, Piazza delle Scienze 3, I-20126 Milan, Italy
[§]Department of Astronomy & Astrophysics, Pennsylvania State University, PA 16802, USA
[¶]Department of Physics and Astronomy, University of Leicester, University Road, Leicester LE1 7RH, UK

Abstract.
XRF 050416A was discovered by the Swift Burst Alert Telescope and re-pointed with the Swift narrow field instruments only 64.5 s after the burst onset. The 15 – 150 keV BAT average spectrum has a photon index of $\Gamma \sim 3.0$ which classifies the bursts as an X-ray flash. The afterglow X-ray emission was monitored by the Swift X-Ray Telescope up to 74 days after the burst. The X-ray light curve shows a decay with three different phases: an initial steep decay with a decay slope of ~ 2.4 (phase A), then, starting at ~ 172 s from the burst onset, a second phase with a flat decay slope of ~ 0.44 (phase B), and finally, after ~ 1450 s from the burst onset, a third long-lasting phase with a decay slope of ~ 0.88 (phase C). We find evidence of spectral evolution from a softer emission in the phase A of the afterglow decay, with $\Gamma \sim 3.0$, to a harder emission with $\Gamma \sim 2.0$ in the phases B and C. A redshift of 0.6535 was measured for the source. The spectra show intrinsic absorption in the host galaxy of $\sim 6.8 \times 10^{21}$ cm^{-2}.

The consistency of the phase A photon index with the BAT photon index suggests that the initial fast decaying phase of the XRT afterglow might be the low energy tail of the prompt emission. The lack of jet break signatures in the X-ray afterglow light curve suggests very low collimation of the expanding fireball.

Keywords: Gamma Ray Bursts,X-rays
PACS: 98.70.Rz,95.85.Nv

INTRODUCTION

The Swift Burst Alert Telescope (BAT) detected and located a gamma-ray burst on 2005 April 16, 11:04:44.5 UT [1, 2]. The light curve showed a single peak followed by a small bump with $T_{90}=2.4\pm0.2$ s, with most of the energy emitted in the 15 – 50 keV band. The time-averaged energy distribution was modeled with a power law ($N(E) \propto E^{-\Gamma}$) with photon index $\Gamma = 3.1 \pm 0.2$ (90% confidence level) and gave a fluence of $(3.2 \pm 0.3) \times 10^{-7}$ erg cm^{-2} in the 15 – 50 keV band and $(3.6 \pm 0.4) \times 10^{-7}$ erg cm^{-2} in the 15 – 350 keV band [3]. The soft spectrum and the fact that the fluence in the X-ray energy band (15 – 30 keV) is larger than the fluence in the gamma-ray band

(30 – 400 keV) classifies this event as an X-ray flash (XRF; 4). The exhaustive ground analysis of the BAT data is presented in [3]. The satellite executed an immediate slew and began collecting data at 11:05:49 UT (64.5 s after the trigger) with the Ultraviolet and Optical Telescope (UVOT) and at 11:06:00.6 UT (i.e. 76.1 s after the trigger) with the X-Ray Telescope (XRT).

In the first 100 s of observation UVOT revealed a new source in the V filter at $RA_{J2000}= 12^h33^m54^s.56$, $Dec_{J2000}=+21° 03' 27''3$, with magnitude V=19.38 *mag* [5]. On ground analysis of XRT data revealed that at the same location a fading X-ray source was present [6].

Ground based follow-up optical, NIR and radio observations were performed with several instruments. In particular, a spectroscopic redshift of 0.6535 was measured with the Keck telescope [7].

Here we present first results on the long lasting and very well sampled X-ray afterglow of XRF 050416A.

XRT LIGHT CURVE AND SPECTRA

XRT was on target 76.1 s after the BAT trigger. It was operating in AUTO state and went through the standard sequence of observing modes. After the slew, operated in Low Rate PhotoDiode (LR) mode, XRT took a 2.5 s frame in Image (IM), 8 initial frames in Windowed Timing (WT) mode, and then correctly switched to Photon Counting (PC) mode for the rest of the orbit. XRF 050416A was then observed for 29 consecutive orbits for a total exposure time of 57454 s. XRF 050416A was further observed several times up to 74 days later in PC mode. This extraordinary observational campaign has allowed us to extract one of the longest and best sampled Swift X-ray light curves. Details on XRT data reduction and extraction of the light curve are given in [8]. Here we just want to stress that during on-ground analysis the source appeared to have been detected already during the slew and in the image frame (though the the on-board centroiding algorithm failed to converge). It was then possible to add to the light curve both an IM point and a LR point, where the latter had enough statistics for spectral analysis too.

The XRT light curve can be modeled by a doubly broken power law, with an initial slope $\alpha_A = 2.4 \pm 0.5$, a first break at the time $T_{break,1} = 172 \pm 36$, a second flat slope $\alpha_B = 0.44 \pm 0.13$, a second break at the time $T_{break,2} = 1.450 \pm 0.013$, and a final uninterrupted decay with slope $\alpha_C = 0.88 \pm 0.02$. Hereafter, with phase A, B and C we will refer to the time period before the first break at 172 s, the period between the two breaks and the period after the second break at 1450 s, respectively.

Average spectra extracted for the time intervals corresponding to the three phases were fitted with an absorbed power law model. The best fit results show an evidence for spectral variation among phases: the emission in phase A (with a photon index of $3.0^{+0.3}_{-0.4}$) is significantly softer than in the phases B and C, both consistent with a photon index of $2.04^{+0.11}_{-0.05}$ obtained by the joint fit of phase B and C spectra. The fit gave a value of $6.8^{+1.0}_{-1.2} \times 10^{21}$ cm^{-2} for the column density in excess with respect to the Galactic absorption (equal to 0.21×10^{21} cm^{-2}).

The complete light curve of the X-ray afterglow of XRF 060123A in flux units (0.2-10

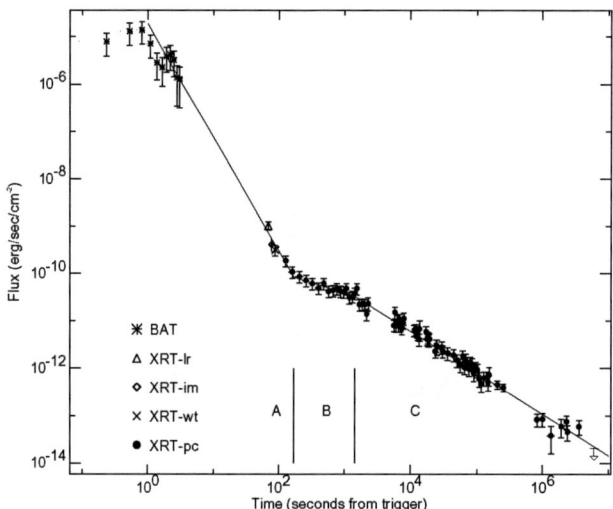

FIGURE 1. XRT light curve of XRF 050416A in flux units together with the BAT light curve extrapolated to the same 0.2-10 keV energy range used for the the XRT light curve. Phases A, B and C discussed in the text are marked. Note that the late extrapolation of the phase C decay is consistent with the flux upper limit measured 65 – 74 days after the prompt emission.

keV energy range) is shown in figure 1, together with the extrapolation of the BAT light curve to the XRT energy band. The XRT count rate light curve was converted into flux units by applying a conversion factor derived from the spectral analysis. The BAT light curve was extrapolated into the XRT energy band by converting the BAT count rate with the factor derived from the BAT spectral parameters obtained by a power law fit.

DISCUSSION

We have presented results of the analysis of the X-ray afterglow of XRF 050416A. XRT monitored the XRF 050416A X-ray emission from ∼64.5 s after the BAT trigger up to 74 days and observed its afterglow light curve evolving through three distinct phases corresponding to distinct decay slopes. The interpretation of these phases can be summarized as follow.

Phase A: The early steep and soft X-ray afterglows observed by Swift are generally interpreted as the tail of the gamma-ray burst emission due to high latitude emission (e.g. 9). For XRF 050416A we found that the best fit photon index determined for the phase A spectrum is consistent within the errors with the value obtained in the prompt burst emission fitting the BAT spectrum with a single power-law model. This suggests that the prompt burst emission and the phase A X-ray emission represent the time evolution of the same phenomenon observed in different energy ranges. In this scenario the first break in the X-ray light curve should be due to the emergence of the afterglow after fading of the GRB. To be consistent with the high latitude effect phase A decay slope should be $\alpha = 2 + \beta$ where β is the energy index measured during the decay. The decay slope of phase A ($\alpha_A = 2.4 \pm 0.5$) is definitely lower than the $\sim 4.0 \pm 0.4$ slope predicted by the high latitude effect for the observed $\beta \sim 2 \pm 0.4$ but can be reconciled with the model if we assume the shell emission does not stop instantaneously.

Phases B: The standard interpretation of the flat decay slope during phase B and the second temporal break in the afterglow is based on refreshed shocks (10). In the initial stages of the fireball evolution the forward shock, whose emission produces the X-ray afterglow, may be continuously refreshed with the injection of additional energy. Within this scenario, a flat decay of the afterglow is expected as the refreshed forward shock decelerates less rapidly than in the standard case. A transition to the standard afterglow evolution (i.e. a break) with no remarkable spectral changes is also expected when the additional energy supply ends. This is consistent with our findings.

Phase C: The phase C decay slope and spectral index are roughly consistent with $\alpha_C = (3p-2)/4$ and $\beta_C = p/2$ for $p \sim 2$. This is what is expected for fireball expansion in a uniform ISM when $v_c < v_X$ (here v_X represents the typical X–ray frequency and v_c is the synchrotron cooling frequency) and before the jet break. No other closure relation is satisfied by the phase C spectral and temporal indices. Since phase C remarkably continues uninterrupted to the end of the XRT observation 74 days after the burst, this interpretation would imply the absence of jet breaks in the X-ray afterglow.

Spherical expansion then becomes a distinct possibility for this afterglow. We note that up to now the detection of jet breaks in XRF afterglows is null, suggesting that XRFs in general may be less collimated than GRBs, in agreement with this result.

ACKNOWLEDGMENTS

The authors acknowledge support from ASI, NASA and PPARC.

REFERENCES

1. Sakamoto, T. et al., *Gamma-ray burst Coordinates Network* **3264** (2005)
2. Sakamoto, T. et al., *Gamma-ray burst Coordinates Network* **3273** (2005)
3. Sakamoto, T. et al., *Astrophys. J. Letters* in press, astro-ph/0512149 (2005)
4. Lamb, D.Q., Donaghy T.Q., Graziani C., *Astrophys. J.* **620**, 355 (2005)
5. Schady, P. et al., *Gamma-ray burst Coordinates Network* **3276** (2005)
6. Kennea, J.A. et al, *Gamma-ray burst Coordinates Network* **68** (2005)
7. Fox, D.B., *Gamma-ray burst Coordinates Network* **3408** (2005)
8. Mangano, V. et al., in preparation
9. Zhang, B. et al., *Astrophys. J.*, in press, astro-ph/0508321
10. Sari, R., & Mészáros, P., *Astrophys. J. Letters* **535**, 33 (2000)

Confirmation of the $E^{src}_{peak} - E_{iso}$ (Amati) relation from the X-ray flash XRF 050416A observed by Swift/BAT

T. Sakamoto[*,†], L. Barbier[*], S. Barthelmy[*], J. Cummings[*,†], E. Fenimore[**],
N. Gehrels[*], D. Hullinger[*,‡], H. Krimm[*,§], C. Markwardt[*,§], D. Palmer[**],
A. Parsons[*], G. Sato[¶] and J. Tueller[*]

[*]*NASA Goddard Space Flight Center*
[†]*National Research Council*
[**]*Los Alamos National Laboratory*
[‡]*University of Maryland*
[§]*Universities Space Research Association*
[¶]*Institute of Space and Astronautical Science*

Abstract. We report Swift Burst Alert Telescope (BAT) observations of the X-ray Flash (XRF) XRF 050416A. The fluence ratio between the 15-25 and 25-50 keV bands of this event is 1.1, thus making it the softest gamma-ray burst (GRB) observed by BAT so far. The spectrum is well fitted by the Band function with E^{obs}_{peak} of $15.6^{+2.3}_{-2.7}$ keV. Assuming the redshift of the host galaxy (z=0.6535), the isotropic-equivalent energy E_{iso} and the E_{peak} energy at the GRB rest frame (E^{src}_{peak}) of XRF 050416A are not only consistent with the correlation found by Amati et al. and extended to XRFs by Sakamoto et al., but also fill-in the gap of this relation around 30–80 keV range of E^{src}_{peak}. This result tightens the validity of the E^{src}_{peak}-E_{iso} relation form XRFs to GRBs.

Keywords: Prompt gamma-ray emission, X-ray flash
PACS: 43.35.Ei, 78.60.Mq

INTRODUCTION

The observations of X-ray flashes (XRF) are providing important information for understanding the nature of Gamma-Ray Bursts (GRBs). About 36% of the bright bursts observed by *Ginga* have E^{obs}_{peak} energy, which is the photon energy at which the νF_ν spectrum peaks, around a few keV and also show large X-ray to γ-ray fluence ratios [1]. The Wide Field Cameras (WFC) on-board the *Beppo*SAX satellite observed 17 XRFs in five years [2]. Kippen et al. [3] searched for GRBs and XRFs which were observed in both WFC and BATSE. The WFC and BATSE joint spectral analysis of XRFs shows that their E^{obs}_{peak} energies are significantly lower than those of the BATSE E^{obs}_{peak} distribution [4]. The systematic study of the spectral properties of XRFs observed by *HETE*-2 also supports this result [5].

The afterglow detection and the redshift measurement from the host galaxy of XRF 020903, which is one of the softest XRF observed by *HETE*-2, shows the dramatic progress in understanding the nature of XRFs. The prompt emission of XRF 020903 has $E^{obs}_{peak} < 5.0$ keV which is two orders of magnitude smaller than that of typical

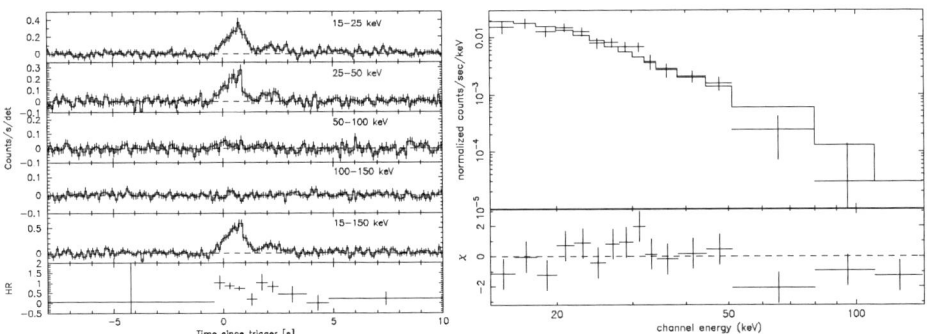

FIGURE 1. Left: The energy resolved light curves of BAT. The hardness ratio between 25-50 and 15-25 keV band is shown in a bottom panel. Right: The time-averaged spectrum with a simple power-law model.

TABLE 1. The time-averaged spectral parameters of XRF 050416A

Model	α	β	E_{peak} [keV]	K_{30} [ph cm^{-2} s^{-1} keV^{-1}]	χ^2/d.o.f.
PL	-3.1 ± 0.2			$(4.3 \pm 0.3) \times 10^{-2}$	50.74 / 57
Band	-1 (fixed)	< -3.4	$15.6^{+2.3}_{-2.7}$	$3.5^{+1.7}_{-0.8} \times 10^{-1}$	42.99 / 56

GRBs. The optical transient and the host galaxy of XRF 020903 were detected. Further spectroscopic observation of the host galaxy suggests that the redshift is 0.25 ± 0.01 [6]. Sakamoto et al. [7] calculated the isotropic-equivalent energy E_{iso} and the peak energy at the source frame E_{peak}^{src} using the redshift of the host galaxy, and found that XRF 020903 follows an extension of the empirical relationship between E_{iso} and E_{peak}^{src} found by Amati et al. [8] for GRBs (a.k.a. Amati relation). This result provides the observational evidence that XRFs and GRBs form a continuum and are a single phenomenon.

The X-ray flash, XRF 050416A, was detected and localized by the *Swift* Burst Alert Telescope (BAT) at 11:04:44.5 UTC on 2005 April 16 [9, 10]. *Swift* autonomously slewed to the BAT on-board position, and both *Swift* X-Ray Telescope (XRT) and UV-Optical Telescope (UVOT) detected the afterglow [11, 12]. Cenko et al. [13] reported that the host galaxy is faint and blue with large amount of the star formation and its redshift is z = 0.6535 ± 0.0002.

BAT DATA ANALYSIS

The left panel of figure 1 shows the energy resolved BAT light curves of XRF 050416A. It is clear that the signal of the burst is only visible below 50 keV. The burst signal is composed of two peaks. The first peak has a triangular shape with the rise time longer than the decay time. The t_{90} and t_{50} in the 15-150 keV band are 2.4 and 0.8 seconds, respectively. This t_{90} belongs to the shortest part of the "long GRB" classification based on the BATSE duration distribution [14]. The fluence ratio between the 15-25 keV band and the 25-50 keV band of 1.1 makes this burst one of the softest GRBs observed by

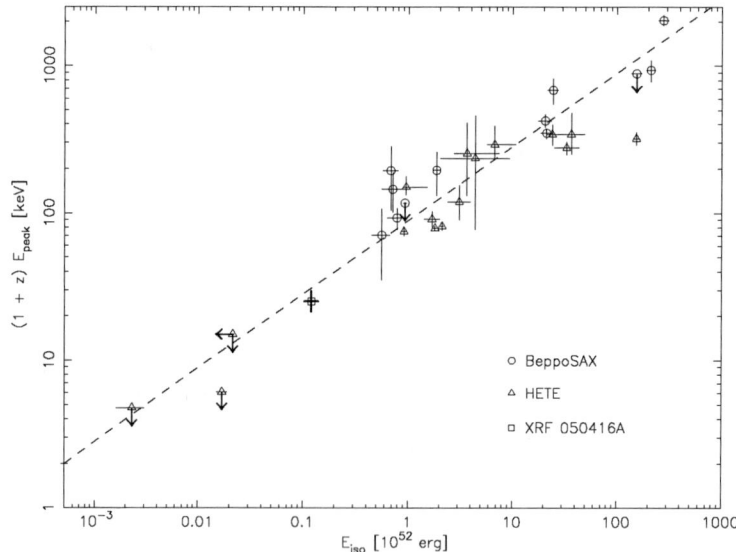

FIGURE 2. The isotropic-equivalent energy, E_{iso}, versus the peak energy at the GRB rest frame, $E_{\text{peak}}^{\text{src}}$, for XRF 050416A (square) and the known redshift GRBs from *Beppo*SAX (circle) and *HETE*-2 (triangle)

BAT so far. The bottom panel of figure 1 left shows the count ratio between the 25-50 keV and 15-25 keV bands. The spectral softening is clearly visible during the first and the second peak.

The right panel of figure 1 shows the time-averaged spectrum, accumulated over the time interval from -0.5 seconds to 3 seconds since the BAT trigger time, was fitted with a simple power-law model. The photon index β which is much steeper than -2 strongly indicates that the BAT observed the higher energy part of the Band function [15]. Motivated by this result, we tried to fit the spectrum with the Band function assuming the low energy photon index α to be fixed at -1, which is the typical value for both GRBs [4] and XRFs [3, 5]. The fitting shows a significant improvement from a simple power-law model to the Band function of $\Delta\chi^2$ of 7.75 for 1 degree of freedom. The observed E_{peak} energy, $E_{\text{peak}}^{\text{obs}}$, is well constrained at $15.6^{+2.3}_{-2.7}$ keV, and it confirms the soft nature of this burst.

One of the most important discoveries related to XRF 050416A is the confirmation of the $E_{\text{peak}}^{\text{src}} - E_{\text{iso}}$ relation [8]. We calculate the E_{peak} energy at the GRB rest frame, $E_{\text{peak}}^{\text{src}}$, and the isotropic-equivalent energy ($1 - 10^4$ keV at the rest frame), E_{iso}, using the redshift of the host galaxy (z=0.6535). Assuming $\alpha = -1$, $E_{\text{peak}}^{\text{src}}$ and E_{iso} of XRF 050416A are $25.1^{+4.4}_{-3.7}$ keV and $(1.2 \pm 0.2) \times 10^{51}$ erg, respectively. Figure 2 shows the data point of XRF 050416A with the known redshift GRBs of *Beppo*SAX and $HETE-2$ sample [16, 17, 7]. XRF 050416A not only follows the $E_{\text{peak}}^{\text{src}} \propto E_{\text{iso}}^{0.5}$ relation, but also fills in the gap of the relation around $E_{\text{peak}}^{\text{src}}$ of $30 - 80$ keV. This result tightens the validity of this relation at five orders of magnitude in $E_{\rm iso}$ and at three orders of magnitude in $E_{\rm peak}^{\rm src}$. XRF 050416A bridges the gap between XRFs which have $E_{\rm peak}^{\rm src}$ of less than 10 keV and GRBs in the $E_{\rm peak}^{\rm src} - E_{\rm iso}$ relation.

DISCUSSION

According to the XRT afterglow observation of XRF 050416A, the decay slope of the afterglow emission is ~ -0.9 from 0.015 days to ~ 34.7 days after the GRB trigger without any signature of a jet break [18].

Using $E_{\rm peak}^{\rm src}$ and $E_{\rm iso}$ of XRF 050416A measured by BAT, we can estimate the jet break time using the relation between $E_{\rm peak}^{\rm src}$ and the jet collimation-corrected energy E_γ found by Ghirlanda et al. [19] (Ghirlanda relation). When we use the empirical relation between $E_{\rm iso}$, $E_{\rm peak}^{\rm src}$, and the jet break time at the rest frame, $t_{\rm jet}^{\rm src}$, derived by Liang & Zhang [20] which is purely based on observational properties, the jet break time in the observer's frame is estimated to be ~ 1.5 days after the GRB on-set time. Thus, the estimated jet break time using the empirical $E_{\rm peak}^{\rm src}$-$E_{\rm iso}$-$t_{\rm jet}^{\rm src}$ relation is inconsistent with the null detection of a jet break until more than 34.3 days after the trigger by XRT. This non-jet break feature in the XRT afterglow light curve might be a further challenging for GRB jet emission, models and XRF/GRB unification scenarios.

REFERENCES

1. Strohmayer, T.E., Fenimore, E.E., Murakami, T., Yoshida, A. 1998, ApJ, 500, 873
2. Heise, J., in't Zand, J., Kippen, R.M., & Woods, P.M. 2000, in Proc. Second Rome Workshop: Gamma-Ray Bursts in the Afterglow Era, Ed. E. Costa, F. Frontera, & J. Hjorth (Berlin: Springer), 16
3. Kippen, R.M., Woods, P.M., Heise, J., in't Zand, J., Briggs, M.S., & Preece, R.D. 2002, in Gamma-Ray Bursts and Afterglow Astronomy, ed. G.R. Ricker & R. Vanderspek (Melville: AIP), 244
4. Preece, R.D., et al. 2000, ApJS, 126, 19
5. Sakamoto, T., et al. 2005, ApJ, 629, 311
6. Soderberg, A.M., et al. 2004, ApJ, 606, 994
7. Sakamoto, T., et al. 2004, ApJ, 602, 875
8. Amati, L., et al. 2002, A&A, 390, 81
9. Sakamoto, T., et al. 2005, GCN Circ. 3264
10. Sakamoto, T., et al. 2005, GCN Circ. 3273
11. Kennea, J., et al. 2005, GCN Circ. 3268
12. Schady, P., et al. 2005, GCN Circ. 3276
13. Cenko, S.B., et al. 2005, GCN Circ. 3542
14. Paciesas, W.S., et al. 1999, ApJS, 122, 465
15. Band, D.L., et al. 1993, ApJ, 413, 281
16. Amati, L., 2003, ChJAA, Vol. 3, Supplement, pp. 455-460
17. Lamb, D.Q. et al., NewAR, 48, 423
18. Nousek, J.A., et al. 2005 submitted to ApJ (astro-ph/0508332)
19. Ghirlanda, G., Ghisellini, G., Lazzati, D., 2004, ApJ, 616, 331
20. Liang, E., Zhang, B., 2005, ApJ, 633, 611

The X-ray spectrum and lightcurve of the redshift 6.29 γ-Ray Burst GRB 050904

D. Watson*, J. N. Reeves†, J. Hjorth*, J. P. U. Fynbo*, P. Jakobsson*, K. Pedersen*, J. Sollerman*, J. M. Castro Cerón*, S. McBreen** and S. Foley‡

*Dark Cosmology Centre, Niels Bohr Institute, University of Copenhagen, Juliane Maries Vej 30, DK-2100 Copenhagen Ø, Denmark.
†Laboratory for High Energy Astrophysics, Code 662, NASA Goddard Space Flight Center, Greenbelt, MD 20771, USA.
**Astrophysics Missions Division, Research Scientific Support Department of ESA, ESTEC, Noordwijk, The Netherlands
‡School of Physics, University College, Dublin 4, Ireland

Abstract. A γ-ray burst (GRB) has been found with a redshift comparable to the most distant quasars and galaxies: GRB 050904 at $z = 6.29 \pm 0.01$, making it the most distant X-ray source known. The X-ray lightcurve is not a power-law like many afterglows, but is dominated by large amplitude variability from a few minutes to at least half a day. The spectra soften during this time from a power-law with photon index $\Gamma = 1.2$ to 1.9. The spectra are well-described by an absorbed power-law with possible evidence of very large intrinsic absorption. There is no evidence for discrete features. This is in spite of the spectrum's very high signal-to-noise ratio, since GRB 050904 was extraordinarily bright in X-rays. In the first days after the burst, it was by far the brightest known X-ray source at $z > 4$. In the first minutes after the burst, the X-ray flux was $> 10^{-9}$ erg cm^{-2} s^{-1} in the 0.2–10 keV band, corresponding to an apparent luminosity between 10^5 and 10^6 times greater than the brightest X-ray quasars at similar distances. More photons were acquired in the first minutes with *Swift*-XRT than XMM-*Newton* and *Chandra* have obtained in ~ 300 ks of pointed observations of $z > 5$ AGN. The huge X-ray fluence detected from GRB 050904 is a clear demonstration of concept for efficient X-ray studies of the high-z IGM with new large area, high-resolution X-ray detectors, and shows that GRBs in their early phases are the only backlighting bright enough for X-ray absorption studies of the intervening matter at high redshift.

Keywords: <gamma-ray sources; gamma-ray bursts>
PACS: <98.70.Rz>

OBSERVATIONS AND DATA REDUCTION

GRB 050904 triggered *Swift*-BAT at 01:51:44 UT; early analysis of the BAT data indicate $T_{90} = 225 \pm 10$ s, a power-law photon index $\Gamma = 1.34 \pm 0.06$ and a fluence of $5.4 \pm 0.2 \times 10^{-6}$ erg cm^{-2} in the 15–150 keV band [1, 2]. The *Swift*-XRT began taking data less than three minutes after the trigger and localised a bright, flaring X-ray source [3]. The XRT began observations in windowed timing (WT) mode at ~ 170 s after the BAT trigger, photon counting (PC) mode observations began at ~ 580 s. The XRT data were reduced in a standard way using the most recent calibration files.

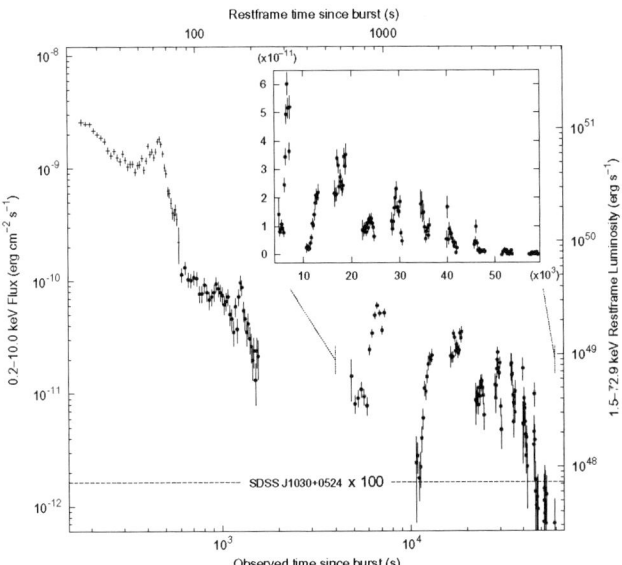

FIGURE 1. *Swift*-XRT 0.2–10.0 keV lightcurve of GRB 050904 (∼ 1.5–72.9 keV in the restframe). The equivalent isotropic luminosity at $z = 6.29$ is plotted on the right axis. WT and PC mode data are indicated by crosses and dots respectively. The flux of one of the most luminous X-ray sources known, the AGN SDSS J1030+0524, is plotted for comparison. We have had to multiply its flux by 100 to get it on the plot. SDSS J1030+0524 was the most distant known X-ray source before GRB 050904 and is, conveniently, at nearly the same redshift ($z = 6.28$). *Inset:* Linear blow-up of the data from ∼ 10 − 70 ks to illustrate the variability of the source at late times. The very hard spectral index at early times ($\Gamma \sim 1.2$) and the long BAT T_{90} for this burst, indicate that most of the WT mode data is dominated by prompt emission. However, the continued large amplitude variability more than one and a half hours after the GRB trigger (in the restframe), and the still relatively hard spectrum ($\Gamma \sim 1.9$), suggests that the XRT lightcurve is dominated by emission from the central engine during the first twelve hours of observations.

RESULTS

The XRT lightcurve (Fig. 1.) fades over the first day by more than three orders of magnitude. The lightcurve is not a simple power-law (or a three-segment power-law) decay (possibly with small amplitude variations) as observed in many other GRB afterglows [eg 4]. The early afterglow in WT mode contains a flare that doubles the flux prior to the rise, peaking at 446 ± 3 s (fitting a single peak to the full spectral band). This flaring activity is similar to that observed in other GRBs at early times [e.g. GRBs 050502B and 050406, 5], however, the lightcurve does not settle down to a power-law decay in the first day of observations, but continues to be dominated by large amplitude (up to a factor of ten) variability.

The early lightcurve in the soft and hard bands and hardness radios are shown in Figure 2.

The spectra (Fig. 3.) can be fit by a hard power-law with absorption at the Galactic level [4.9×10^{20} cm^{-2}, 6]. The spectrum softens appreciably during the observation,

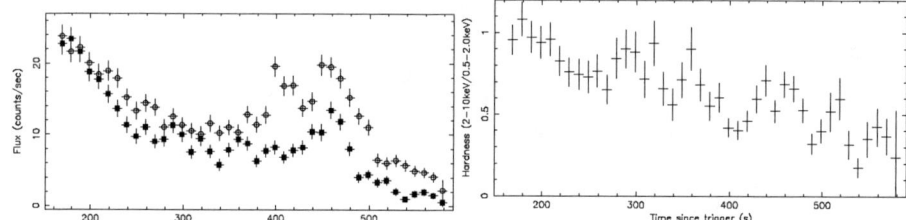

FIGURE 2. *Left:* Soft (0.5–2.0 keV, open circles) and hard (2–10 keV, filled squares) early lightcurve of GRB 050904. *Right:* Hardness ratio of the early lightcurve. The hard to soft evolution, observed in most GRB prompt emission, is fairly monotonic outside the flares, where small deviations are discernible.

reaching $\Gamma \sim 1.9$ in the 10–50 ks period after the BAT trigger (Table 1. and Fig. 3.).

There is no significant evidence for discrete features in the spectrum in emission or absorption, specifically, Fe^{26} at 6.97 keV and Ni^{28} at 8.10 keV have respective equivalent widths (EW) < 43 and < 44 eV in the WT mode spectra and < 27 and < 137 eV in the PC mode spectra in the restframe.

There is some evidence of excess absorption above the Galactic level: the best fit gives $N_H = 8.3 \pm 0.8 \times 10^{20}$ cm^{-2}. This excess ($N_H = 3.4 \times 10^{20}$ cm^{-2}) is statistically required (significant at a level $> 5\sigma$ using the f-test). With no discrete features, the redshift of the absorption is essentially unconstrained. Because of the redshift, to observe even a modest column at $z = 0$ requires a very high column at $z = 6.29$; in this case the best-fit excess column density at $z = 6.29$ is $2.8 \pm 0.8 \times 10^{22}$ cm^{-2}. However, the combination of the uncertainties in the Galactic column density due to the large beamsize, the current calibration uncertainty of the XRT response at low energies, and the possibility of convex curvature in the spectrum, renders one cautious about the reality of the detection of excess absorption in this case.

High-z Warm IGM Studies with GRBs

Access to the edge of the reionisation epoch using GRBs has begun with the observation of GRB 050904 at $z = 6.29$, pushing the age of the universe at which the most distant GRB has been detected down by 35%. The rapid response of *Swift* to GRB 050904 yielded high signal-to-noise ratio X-ray spectra in spite of the relatively modest aperture

TABLE 1. Spectral evolution of GRB 050904

Mode	Time since trigger (s)	Γ	N_H at $z = 6.29$ (10^{22} cm^{-2})
WT	174–374	1.23 ± 0.05	3.3 ± 1.5
WT	374–594	1.62 ± 0.06	3.6 ± 1.4
PC	594–1569	1.68 ± 0.08	< 1.6
PC	9080–63480	1.88 ± 0.04	2.9 ± 0.8

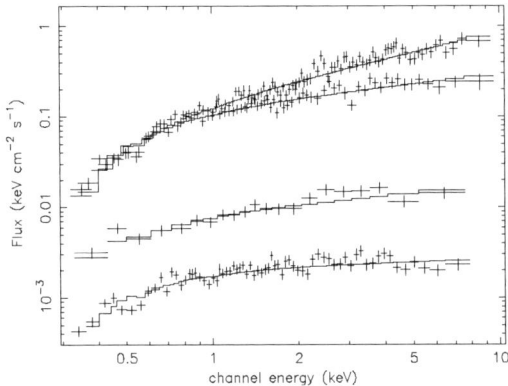

FIGURE 3. *Swift*-XRT $E^2F(E)$ (equivalent to νF_ν) spectra of GRB 050904 with the detector response removed. The spectra are fit with absorbed power-laws and show a clear hard to soft evolution, but the photon power-law indices are consistently $\Gamma < 2$, suggestive of a decreasing peak energy that is above the bandpass in the restframe). The best fit parameters are listed in Table 1.

of the XRT.

The $> 10^{-9}\,\mathrm{erg\,cm^{-2}\,s^{-1}}$ X-ray continuum detected in the first few hundred seconds after GRB 050904, demonstrates unequivocally the awesome power of GRBs to probe the universe in X-rays to the highest redshifts. A rapid response, similar to *Swift*'s, to a GRB like GRB 050904 with a large area detector with good spectral resolution and fast readout times (e.g. *Constellation-X* or *XEUS*) would yield between one and several million photons in an exposure of only a few minutes. This short observation would allow a study of the high-redshift universe that would otherwise require months of *effective* exposure time observing a similar redshift AGN with one of these large detectors. A sample of such observations could allow us not only to fix the fraction of baryonic dark matter, but to determine the metallicity and density evolution of the IGM, and put strong constraints on structure formation at high redshifts. Finally, this burst has been the subject of many publications [eg 7, 8]. Cusumano et al. (2005) have also recently examined this burst and found similar results. A small difference in the PC mode flux level is probably related to the spectrum used to convert the counts lightcurve to a flux lightcurve.

REFERENCES

1. Cummings, J., et al. 2005, GCN Circ., 3910
2. Palmer, D., et al. 2005, GCN Circ., 3918
3. Mineo, T., et al. 2005, GCN Circ., 3920
4. Nousek, J. A., et al. 2005, astro-ph/0508332
5. Burrows, D. N., et al. 2005, Sci, 309, 1833
6. Dickey, J. M., & Lockman, F. J. 1990, ARA&A, 28, 215
7. Tagliaferri, G. et al. 2005 A&A, 443, L1
8. Watson, D., et al. 2005, ApJL, in press (astro-ph/0509640)
9. Cusumano, D., et al. 2005 (astro-ph/0509737)

NON-PHOTONIC AND HIGH ENERGY EMISSION FROM GRBs

Non-photonic emission from γ-ray bursts

E. Waxman

Physics Faculty, Weizmann Inst. of Science, Rehovot 76100, Israel

Abstract.
γ-ray bursts (GRBs) are likely sources of ultra-high energy, $> 10^{19}$ eV, protons and high energy, > 1 TeV, neutrinos. Large volume detectors of ultra high energy cosmic rays (UHECRs) and high energy neutrinos, which are already operating and are being expanded, may allow to test in the coming few years the predictions of the GRB model for high energy proton and neutrino production. Detection of the predicted signals will allow to identify the sources of UHECRs and will provide a unique probe, which may allow to resolve open questions related to the underlying physics of GRB models. Moreover, detection of GRB neutrinos will allow to test for neutrino properties (e.g., flavor oscillations for which τ's would be a unique signature, and coupling to gravity) with an accuracy many orders of magnitude better than is currently possible.

Keywords: gamma ray bursts, cosmic rays, neutrinos
PACS: 98.70.Rz, 98.70.Sa, 95.85.Ry, 14.60.Pq

1. INTRODUCTION AND SUMMARY

The cosmic-ray spectrum extends to energies $\sim 10^{20}$ eV, and is likely dominated beyond $\sim 10^{19}$ eV by extra-Galactic sources of protons [1] (see, however, [2]). The origin of the highest energy, $> 10^{19}$ eV, cosmic rays (UHECRs) is a mystery. As explained in §2.1, the stringent constraints, which are imposed on the properties of possible UHECR sources by the high energies observed, rule out almost all source candidates, and suggest that γ-ray bursts (GRBs) and active galactic nuclei (AGN) are the most plausible sources. In § 2.2 we show that the energy loss of UHECR protons due to interaction with microwave background photons, which limits the proton propagation distance, disfavors AGNs as UHECR sources. It is also shown that both the energy production rate and the spectrum of UHECRs are consistent with those expected in the GRB model for UHECR production. Unique predictions of the GRB model, which may be tested with large UHECR detectors, are discussed in § 2.3 (for a pedagogical review of UHECR production in GRBs see [3]).

High energy neutrino production is discussed in § 3. It is first shown in § 3.1 that cosmic ray observations set an upper limit of $E_\nu^2 \Phi_\nu \leq E_\nu^2 \Phi_\nu^{\rm WB} = 5 \times 10^{-8} {\rm GeV/cm^2 s\, sr}$ on the diffuse extra-Galactic high energy neutrino intensity produced by sources which, like GRBs and AGN jets, are optically thin for high-energy nucleons to $p\gamma$ and $pp(n)$ interactions. The implications of the upper bound to the detector size required for detecting extra-Galactic high energy neutrinos are briefly discussed. In §3.2 we show that production of 100 TeV neutrinos in the region where GRB γ-rays are produced is a generic prediction of the GRB fireball model (for detailed reviews of the fireball model see [4]; for a pedagogical review of high energy neutrino production in GRBs see [3]). 100 TeV neutrino production is a direct consequence of the *assumptions* that energy is carried from the underlying engine, most likely a (few) solar mass black hole, as kinetic energy of protons and that γ-rays are produced by synchrotron emission of shock accelerated electrons. The detection of the predicted neutrino signal will therefore provide strong support for the validity of underlying model assumptions, which is difficult to obtain using photon observations (due to the high optical depth in the vicinity of the GRB "engine"). The predicted neutrino intensity, $\approx 0.2\Phi_\nu^{WB}$, implies a detection of ~ 20 muon induced neutrino events per yr in a km-scale neutrino detector. Since these events should be correlated in time and direction with GRB γ-rays, the search for GRB neutrinos is essentially background free. Neutrinos may be produced also in other stages of fireball evolution, at energies different than 100 TeV. The production of these neutrinos is dependent on additional model assumptions. As an example, we discuss in §3.3 the production of TeV neutrinos expected in the "collapsar" scenario, where GRB progenitors are associated with the collapse of massive stars.

The discussion of GRB neutrino emission demonstrates that in addition to identifying the sources of UHECRs, high-energy neutrino telescopes can also provide a unique probe of the physics of these sources. Moreover, detection of high energy neutrinos from GRBs may also provide information on fundamental neutrino properties [5]. High energy neutrinos are expected to be produced in astrophysical sources by the decay of charged pions, which lead to the production of neutrinos with flavor ratio $\Phi_{\nu_e} : \Phi_{\nu_\mu} : \Phi_{\nu_\tau} = 1 : 2 : 0$ (here Φ_{ν_l} stands for the combined flux of ν_l and $\bar{\nu}_l$). Neutrino oscillations then lead to an observed flux ratio on Earth of $\Phi_{\nu_e} : \Phi_{\nu_\mu} : \Phi_{\nu_\tau} = 1 : 1 : 1$ [6]. Upgoing τ's, rather than μ's, would be a distinctive signature of such oscillations, and it has been pointed out [6, 7] that searching for deviations from the standard flavor ratio $1 : 1 : 1$ may enable one to probe new physics. Detection of neutrinos from GRBs could, moreover, be used to test the simultaneity of neutrino and photon arrival to an accuracy of ~ 1 s, checking the assumption of special relativity that photons and neutrinos have the same limiting speed. These observations would also test the weak equivalence principle, according to which photons and neutrinos should suffer the same time delay as they pass through a gravitational potential. With 1 s accuracy, a burst at 1 Gpc would reveal a fractional difference in limiting speed of 10^{-17}, and a fractional difference in gravitational time delay of order 10^{-6} (considering the Galactic potential alone). Previous applications of these ideas to supernova 1987A (see [8] for review), yielded much weaker upper limits: of order 10^{-8} and 10^{-2} respectively.

In order to illustrate the possibilities that may be opened by high energy neutrino observations, we show in § 3.4 that electromagnetic energy losses of π's and μ's modify the flavor ratio (measured at Earth) of neutrinos produced by π decay, $\Phi_{\nu_e} : \Phi_{\nu_\mu} : \Phi_{\nu_\tau}$, from $1 : 1 : 1$ at low energy to $1 : 1.8 : 1.8$ at high energy. For GRBs the transition is expected at ~ 100 TeV, and may be detected by km-scale ν telescopes. While the detection of τ neutrinos would test flavor oscillations, a measurement of the flavor transition energy and energy-width will provide a unique probe of the physical conditions in the γ-ray production region.

2. ULTRA HIGH ENERGY COSMIC RAYS

2.1. The acceleration challenge

Most models of particle acceleration in astrophysical sources involve the acceleration of charged particles by the electro-motive force produced through the motion of a magnetized plasma. General phenomenological considerations imply that, independent of the details of the acceleration mechanism, the minimum energy output from a source capable of accelerating in this manner a proton to energy E_p is [9]

$$L > \frac{\Gamma^2}{\beta}\left(\frac{E_p}{e}\right)^2 c = 10^{45.5}\frac{\Gamma^2}{\beta}\left(\frac{E_p}{10^{20}\text{eV}}\right)^2 \text{erg/s}. \quad (1)$$

Here, $u = \beta c$ is the characteristic plasma velocity, and $\Gamma = (1-\beta^2)^{-1/2}$. Only two types of sources are known to satisfy this requirement. The brightest steady sources are active galactic nuclei (AGN). For them Γ is typically between 3 and 10, implying $L > 10^{47}$erg/s, which may be satisfied by the brightest AGN [10]. The brightest transient sources are GRBs. For these sources $\Gamma \simeq 10^{2.5}$ implying $L > 10^{50.5}$erg/s, which is generally satisfied since the typical observed MeV-photon luminosity of these sources is $L_\gamma \sim 10^{52}$erg/s.

The constraint of eq. (1) is essentially obtained by requiring the acceleration time to be shorter than the proton confinement time. A second constraint is imposed by requiring the proton acceleration time to be smaller than its energy loss time. For a relativistic wind, where acceleration takes place in internal shocks arising from variability of the underlying source driving the wind on time scale Δt, the proton acceleration time is smaller than its synchrotron energy loss time provided [11]

$$\Gamma > 130 \left(\frac{E_p}{10^{20}\text{eV}}\right)^{3/4} \left(\frac{\Delta t}{10\text{ms}}\right)^{-1/4}. \quad (2)$$

This constraint is remarkably similar to that inferred from γ-ray observations on the relativistic dissipative winds which are assumed, within the context of the fireball model (see [4] for reviews), to produce GRBs: $\Gamma > 300$ is implied by the γ-ray spectrum and by the short variability time, $\Delta t \sim 10$ms, through the requirement to avoid high pair-production optical depth. This, combined with the constraint of eq. (1), was one of the two main arguments suggested in [11] in support of an association between UHECR and GRB sources.

2.2. The GZK effect

The energy loss of UHECR protons due to interaction with microwave background photons (the "GZK effect") limits the proton propagation distance to < 100 Mpc for $E_p > 10^{20}$ eV [12]. This disfavors AGNs as UHECR sources, since no AGN powerful enough to satisfy the constraint of eq. (1) is known to lie within 100 Mpc from Earth.

Since most GRBs are known to reside at redshift $z > 1$, it may appear that they too are disfavored by the short, 100 Mpc, propagation distance of the protons. This is, however, not the case. While GRB γ-rays arrive as a burst, slight deflections of protons by an inter-galactic magnetic field would cause large delays in their arrival time (compared to the photon arrival time). The arrival time delay is accompanied by a comparable arrival time spread [11],

$$\tau(E_p, D) \approx 10^7 \left(\frac{E_p}{10^{20}\text{eV}}\right)^{-2} \left(\frac{D}{100\text{Mpc}}\right)^2 \frac{\lambda B^2}{10^{-8}\text{Mpc G}}\text{yr}, \quad (3)$$

where D is the source distance, B and λ are the field strength and correlation length, and a limit $B\lambda^{1/2} \leq 10^{-8}$G Mpc$^{1/2}$ is set by Faraday rotation measurements [13]. The number of GRBs contributing to the flux above energy E_p at any given time is [11, 14] $N_{\text{GRB}}(>E_p) \approx (4\pi/5) R_{\text{GRB}} D_c^3(E_p) \tau[E_p, D_c(E_p)]$, where R_{GRB} is the local ($z=0$) GRB rate density and $D_c(E_p)$ is the proton propagation distance. Using $R_{\text{GRB}}(z=0) \approx 0.5 \times 10^{-9}$Mpc^{-3} yr^{-1} [15], we find that a large number of GRBs,

$$N_{\text{GRB}}(>E_p) \approx 10^4 \left(\frac{E_p}{10^{20}\text{eV}}\right)^{-2} \left[\frac{D_c(E_p)}{100\text{Mpc}}\right]^5 \frac{\lambda B^2}{10^{-8}\text{Mpc G}}, \quad (4)$$

may contribute to the flux at any given time.

Figure 1, adapted from [16], presents a comparison of available UHECR data with the predictions of a model, where extra-galactic protons in the energy range $E_p \leq 10^{21}$ eV are produced by cosmologically-distributed sources at a rate and spectrum given by

$$E_p^2 \frac{d\dot{N}_p}{dE_p} = 0.65 \times 10^{44} \text{erg Mpc}^{-3} \text{yr}^{-1} \phi(z). \quad (5)$$

Here, $\phi(z)$ accounts for redshift evolution and $\phi(z=0) = 1$. The spectrum above 10^{19} eV is only weakly dependent on $\phi(z)$ since proton energy loss limits their propagation distance. For the heavy nuclei component dominating at lower, $< 10^{19}$ eV, energy the Fly's Eye experimental fit [17], $dN/dE \propto E^{-3.50}$, was used. The power-law spectrum of accelerated particles, $dN/dE \propto E^{-2}$, has been observed for both non-relativistic and relativistic shocks, and is believed to be due to Fermi acceleration in collisionless shocks [18].

Model predictions are in good agreement with the data of all experiments in the energy range 10^{19} eV to 10^{20} eV (As explained in detail in [16], the various experiments are consistent with each other when systematic errors in the absolute energy scale of the events are taken into account). The suppression of the flux above $\sim 10^{19.7}$ eV is the manifestation of the GZK effect. Above 10^{20} eV the Fly's Eye, HiRes and Yakutsk experiments are in agreement with each other and with the model, while the AGASA experiment reports a flux higher by a factor ~ 3. The origin of this discrepancy is unclear. The Auger UHECR experiment [20], which is currently under construction, is expected to significantly reduce the systematic uncertainty in cosmic-ray energy determination and to dramatically increase the number of detected UHECRs (thus reducing statistical

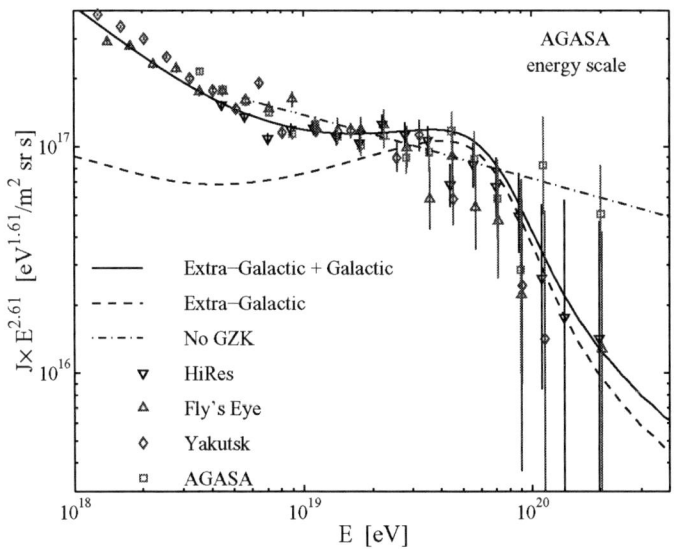

FIGURE 1. The solid curve shows the energy spectrum derived from the two-component model discussed in § 2.2 (with $\phi(z) \propto (1+z)^3$ up to $z = 2$, following the evolution of star formation rate). The dashed curve shows the extra-Galactic component contribution. The "No GZK" curve is an extrapolation of the $E^{-2.75}$ energy spectrum derived for the energy range of 6×10^{18} eV to 4×10^{19} eV [1]. Data taken from [17, 19] (AGASA's energy scale was chosen).

errors). It will provide an accurate determination of the UHECR spectrum to energies $> 10^{20}$ eV.

The local UHECR energy production rate, eq. (5), is remarkably similar to the local γ-ray energy production rate by GRBs, $\approx 10^{44}$ erg Mpc^{-3} yr^{-1} [21]. The similarity of the energy production rates was the second main argument suggested in [11] in support of an association between UHECR and GRB sources.

2.3. GRB model predictions

The rapid decrease of D_c with E_p implies, for $B\lambda^{1/2} \leq 10^{-8}$G Mpc$^{1/2}$ and $R_{\text{GRB}} \leq 1/$ Gpc^{3}yr, that for some energy E_c in the range 10^{20}eV $\leq E_c < 4 \times 10^{20}$eV the number of GRBs contributing to the UHECR flux is (on average) 1, $N_{\text{GRB}}(>E_c) = 1$ (see eq. 4). Fig. 2 presents the flux obtained for $E_c = 1.4 \times 10^{20}$eV in one realization of a Monte-Carlo simulation described in [14]. For each realization the positions and times at which cosmological GRBs occurred were randomly drawn. Most of the realizations gave an overall spectrum similar to that obtained in the realization of Fig. 2 when the brightest source of this realization (dominating at 10^{20}eV) is not included. At $E_p < E_c$, the number of sources contributing to the flux is very large, and the UHECR flux received at any given time is near the average (the average flux is that obtained when the UHECR

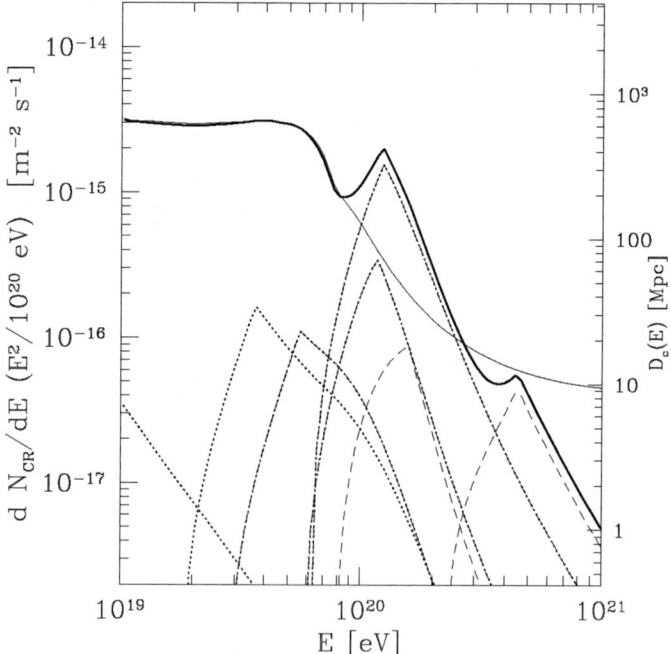

FIGURE 2. Results of a Monte-Carlo realization of the bursting sources model, with $E_c = 1.4 \times 10^{20}$ eV: Thick solid line- overall spectrum in the realization; Thin solid line- average spectrum, this curve also gives $D_c(E_p)$; Dotted lines- spectra of brightest sources at different energies.

emissivity is spatially uniform and time independent). At $E_p > E_c$, the flux will generally be much lower than the average, because there will be no burst within a distance $D_c(E_p)$ having taken place sufficiently recently. There is, however, a significant probability to observe one source with a flux higher than the average. A source similar to the brightest one in Fig. 2 appears $\sim 5\%$ of the time.

At any fixed time a given burst is observed in UHECRs only over a narrow range of energy, because if a burst is currently observed at some energy E_p, then UHECRs of much lower energy from this burst have not yet arrived, while higher energy UHECRs reached us mostly in the past. Thus, bursting UHECR sources should have narrowly peaked energy spectra, and the brightest sources should be different at different energies. For steady state sources, on the other hand, the brightest source at high energies should also be the brightest one at low energies. The large exposure expected for the Auger experiment [20] may allow to distinguish between the two cases, and, if the signature of bursting sources is detected, to determine the bursting source properties (e.g. rate, energy production).

3. HIGH ENERGY NEUTRINOS

3.1. An upper bound

The energy production rate, eq. 5, sets an upper bound to the neutrino intensity produced by sources which, like GRBs and AGN jets, are optically thin for high-energy nucleons to $p\gamma$ and $pp(n)$ interactions. For sources of this type, the energy generation rate of neutrinos can not exceed the energy generation rate implied by assuming that all the energy injected as high-energy protons is converted to pions (via $p\gamma$ and $pp(n)$ interactions). The resulting upper bound (for muon and anti-muon neutrinos, neglecting mixing) is [22]

$$E_\nu^2 \Phi_\nu < 2 \times 10^{-8} \xi_z \left[\frac{(E_p^2 d\dot{N}_p/dE_p)_{z=0}}{10^{44} \text{erg/Mpc}^3 \text{yr}} \right] \text{GeV cm}^{-2} \text{s}^{-1} \text{sr}^{-1}. \quad (6)$$

ξ_z is (a dimensionless parameter) of order unity, which depends on the redshift evolution of $E_p^2 d\dot{N}_p/dE_p$ (see eq. 5). In order to obtain a conservative upper bound, we adopt $(E_p^2 d\dot{N}_p/dE_p)_{z=0} = 10^{44} \text{erg/Mpc}^3 \text{yr}$ and a rapid redshift evolution, $\Phi(z) = (1+z)^3$ up to $z = 2$, following the evolution of star formation rate. This evolution yields $\xi_z \approx 3$. The WB upper bound is compared in fig. 3 with current experimental limits, and with the expected sensitivity of planned neutrino telescopes. The figure indicates that km-scale (i.e. giga-ton) neutrino telescopes are needed to detect the expected extra-Galactic flux in the energy range of ~ 1 TeV to ~ 1 PeV, and that much larger effective volume is required to detect the flux at higher energy.

3.2. "Generic" 100 TeV fireball neutrinos

Protons accelerated in the fireball internal shocks, where GRB γ-rays are expected to be produced, lose energy through photo-production of mesons in interactions with fireball photons. The decay of charged pions produced in this interaction results in the production of high energy neutrinos. The key relation is between the observed photon energy, E_γ, and the accelerated proton's energy, E_p, at the threshold of the Δ-resonance. In the observer frame,

$$E_\gamma E_p = 0.2 \, \text{GeV}^2 \, \Gamma^2. \quad (7)$$

For $\Gamma \approx 10^{2.5}$ and $E_\gamma = 1$ MeV, we see that characteristic proton energies $\sim 10^{16}$ eV are required to produce pions. Since neutrinos produced by pion decay typically carry 5% of the proton energy, production of $\sim 10^{14}$ eV neutrinos is expected [5].

The fraction of energy lost by protons to pions, f_π, is $f_\pi \approx 0.2$ [5, 3]. Assuming that GRBs generate the observed UHECRs, the expected GRB muon and anti-muon neutrino flux may be estimated using eq. 6 [5, 22],

$$E_\nu^2 \Phi_\nu \approx 0.8 \times 10^{-8} \frac{f_\pi}{0.2} \text{GeV cm}^{-2} \text{s}^{-1} \text{sr}^{-1}. \quad (8)$$

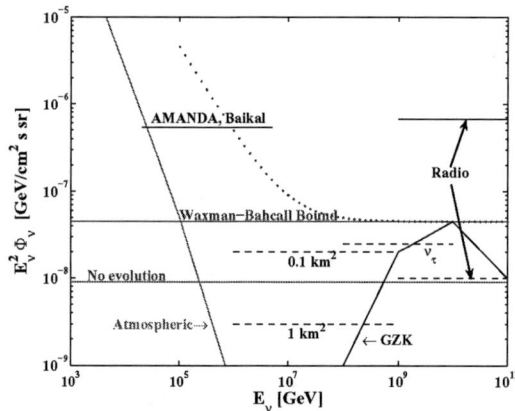

FIGURE 3. The upper bound on the extra-Galactic muon and tau neutrino flux (lower-curve: no source evolution, upper curve: source evolution following star formation rate), assuming oscillations lead to $\Phi_{\nu_e} : \Phi_{\nu_\mu} : \Phi_{\nu_\tau} = 1 : 1 : 1$, compared with experimental upper bounds (solid lines) of optical Cerenkov experiments (BAIKAL [23], AMANDA [24]), and of coherent Cerenkov radio experiments (RICE [25], GLUE [26]; see also [27]). The curve labelled "GZK" shows the intensity due to interaction with micro-wave background photons. Dashed curves show the expected sensitivity of 0.1 Gton (AMANDA, ANTARES, NESTOR) and 1 Gton (IceCube, NEMO) optical Cerenkov detectors [28], of the coherent radio Cerenkov (balloon) experiment ANITA [27] and of the Auger air-shower detector (sensitivity to ν_τ) [29]. Space air-shower detectors (OWL-AIRWATCH) may also achieve the sensitivity required to detect fluxes lower than the WB bound at energies $> 10^{18}$ eV [29].

This neutrino spectrum extends to $\sim 10^{16}$ eV, and is suppressed at higher energy due to energy loss of pions and muons [5, 30, 22]. Eq. 8 implies a detection rate of ~ 20 neutrino-induced muon events per year (over 4π sr) in a cubic-km detector [5, 31]. Since GRB neutrino events are correlated both in time and in direction with gamma-rays, their detection is practically background free.

3.3. TeV neutrinos

The 100 TeV neutrinos discussed in the previous section are produced at the same region where GRB γ-rays are produced. Their production is a generic prediction of the fireball model. It is a direct consequence of the assumptions that energy is carried from the underlying engine as kinetic energy of protons and that γ-rays are produced by synchrotron emission of shock accelerated particles. Neutrinos may be produced also in other stages of fireball evolution, at energies different than 100 TeV. The production of these neutrinos is dependent on additional model assumptions. We discuss below some examples related to the GRB progenitor. For a more detailed discussion see [32, 3] and references therein.

The most widely discussed progenitor scenarios for long-duration GRBs involve core collapse of massive stars. In these "collapsar" models, a relativistic jet breaks through the stellar envelope to produce a GRB. For extended or slowly rotating stars, the jet may be

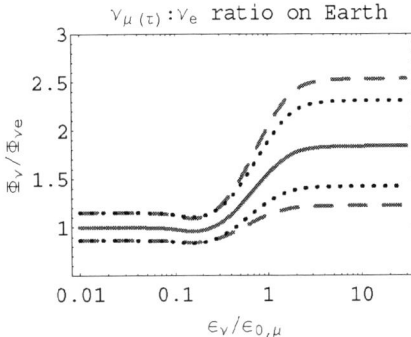 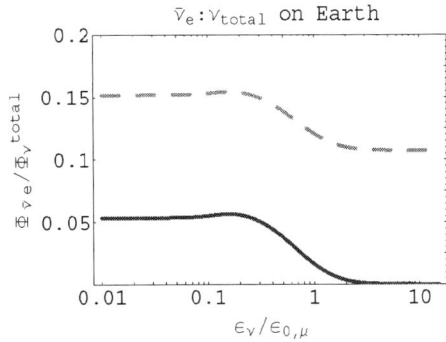

FIGURE 4. Flavor and anti-particle content of the flux of astrophysical neutrinos produced by pion decay, for pion energy spectrum (at production) $dn/dE_\pi \propto E_\pi^{-2}$ and electromagnetic pion energy loss rate $dE_\pi/dt \propto E_\pi^2$, as expected for internal shocks in GRB fireballs. Left: The ratio between $\Phi_{\nu_\mu(\nu_\tau)}$ and Φ_{ν_e} (solid line), with 90% CL lines of ν_μ (dashed) and ν_τ (dotted) fluxes (here, Φ_{ν_l} stands for the combined flux of ν_l and $\bar\nu_l$). Right: The ratio of $\bar\nu_e$ to total ν flux on Earth, solid (dashed) line for neutrinos produced by $p\gamma$ (pp) interactions. $E_{0,\mu}$ is the muon energy for which the muon life time is comparable to its electromagnetic energy loss time. $E_{0,\mu} \approx 10^3$ TeV for internal shocks in GRB fireballs.

unable to break through the envelope. Both penetrating (GRB producing) and "choked" jets can produce a burst of ~ 5 TeV neutrinos by interaction of accelerated protons with jet photons, while the jet propagates in the envelope [33, 34]. The estimated event rates may exceed $\sim 10^2$ events per yr in a km-scale detector, depending on the ratio of non-visible to visible fireballs. A clear detection of non-visible GRBs with neutrinos may be difficult due to the low energy resolution for muon-neutrino events, unless the associated supernova photons are detected. In the two-step "supranova" model, interaction of the GRB blast wave with the supernova shell can lead to detectable neutrino emission, either through nuclear collisions with the dense supernova shell or through interaction with the intense supernova and backscattered radiation field [35].

3.4. Energy dependent neutrino flavor ratio

Although a $1:1:1$ flavor ratio appears to be a robust prediction of models where neutrinos are produced by pion decay, energy dependence of the flavor ratio is a generic feature of models of high energy astrophysical neutrino sources [36]. Pions are typically produced in environments where they may suffer significant energy losses prior to decay, due to interaction with radiation and magnetic fields [5, 30]. Since the pion life time is shorter than the muon's, at sufficiently high energy the probability for pion decay prior to significant energy loss is higher than the corresponding probability for muon decay. This leads to suppression at high energy of the relative contribution of muon decay to the neutrino flux. The flavor ratio is modified to $0:1:0$ at the source, implying $1:1.8:1.8$ ratio on Earth [36].

Figure 4 presents the expected energy dependence of flavor and anti-particle content for internal shocks in GRB fireballs. Since the transition is expected at ~ 100 TeV, it may be detected by km-scale ν telescopes.

REFERENCES

1. M. Nagano, A.A. Watson, Rev. Mod. Phys. **72**, 689 (2000); T. Abu-Zayyad et al., Astrophys. J. **557**, 686 (2001).
2. A. A. Watson, astro-ph/0408110.
3. E. Waxman, Lect. Notes Phys. **576**, 122 (2001) (astro-ph/0103186).
4. Piran, T. 2000, Phys. Rep. **333**, 529; Mészáros, P. 2002, ARA&A **40**, 137; Waxman, E. 2003, Lect. Notes Phys. **598**, 393 (astro-ph/0303517).
5. E. Waxman & J. N. Bahcall Phys. Rev. Lett. 78, 2292 (1997).
6. J. G. Learned and S. Pakvasa, Astropart. Phys. 3, 267 (1995).
7. H. Athar, M. Jezabek & O. Yasuda, Phys. Rev. D62, 103007 (2000).
8. J. N. Bahcall, *Neutrino Astrophysics*, Cambridge University Press (NY 1989), pp. 438–460.
9. E. Waxman, Proc. Nobel Symp. 129: Neutrino Physics (World Scientific, Enköping, 2004), [arXiv:astro-ph/0502159].
10. Lovelace, R. V. E., Nature 262, 649 (1976).
11. Waxman, E. 1995, Phys. Rev. Lett. **75**, 386.
12. K. Greisen, Phys. Rev. Lett. 16, 748 (1966); G.T. Zatsepin, V.A. Kuzmin, JETP 4, 78 (1966).
13. Kronberg, P. P. 1994, Rep. Prog. Phys. **57**, 325.
14. Miralda-Escudé, J. & Waxman, E. 1996, Ap. J. **462**, L59.
15. Guetta, D., Piran, T. & Waxman, E. 2003, to appear in ApJ (astro-ph/0311488).
16. J. N. Bahcall, E. Waxman, Phys. Lett. B 556, 1 (2003).
17. D.J. Bird et al., Astrophys. J. 424, 491 (1994).
18. R. Blandford, D. Eichler, Phys. Rep. 154, 1 (1987); J. Bednarz & M. Ostrowski, Phys. Rev. Lett. 80, 3911 (1998); U. Keshet & E. Waxman, Phys. Rev. Lett. 94, id. 111102 (2005).
19. N. N. Efimov et al., in the Proceedings of the International Symposium on Astrophysical Aspects of the Most Energetic Cosmic-Rays, edited by M. Nagano and F. Takahara (World Scientific, Singapore 1991), p. 20; N. Hayashida et al., Astrophys. J. 522, 225 (1999), and astro-ph/0008102; T. Abu-Zayyad et al., Phys. Rev. Lett. 92, 151101 (2004).
20. http://www.auger.org/admin/
21. E. Waxman, Ap. J. 606, 988 (2004).
22. E. Waxman & J. N. Bahcall, Phys. Rev. D59, 023002 (1999); J. N. Bahcall & E. Waxman, Phys. Rev. D64, 023002 (2001).
23. V. Aynutdinov et al., Astropar. Phys. in press (astro-ph/0508675).
24. M. Ackermann et al., Astropart. Phys. 22, 127 (2004).
25. I. Kravchenko et al., Astropar. Phys. 19, 15 (2003).
26. P. W. Gorham et al., Phys. Rev. Lett. 93, 041101 (2004).
27. A. Silvestri et al., astro-ph/0411007.
28. F. Halzen, Proc. Nobel Symp. 129: Neutrino Physics (World Scientific, Enköping, 2004), [arXiv:astro-ph/0501593].
29. D. Saltzberg, Proc. Nobel Symp. 129: Neutrino Physics (World Scientific, Enköping, 2004), [arXiv:astro-ph/0501364], and references therein.
30. J. P. Rachen & P. Mészáros, Phys. Rev. D58, 123005 (1998).
31. J. Alvarez-Muñiz, F. Halzen, F., D. W. Hooper, Phys. Rev. D62, id. 093015 (2000); D. Guetta, D. Hooper, J. Alvarez-Muñiz, F. Halzen, E. Reuveni, Astropar. Phys. 20, 429 (2004).
32. P. Mészáros, ARA&A 40, 137 (2002).
33. P. Mészáros & E. Waxman, Phys. Rev. Lett. 87, 171102 (2001); S. Razzaque, P. Mészáros & E. Waxman, Phys. Rev. D69, 023001 (2004).
34. S. Ando & J. F. Beacom, Phys. Rev. Lett. 95, id. 061103 (2005).
35. D. Guetta, & J. Granot, Phys. Rev. Lett. 90, 201103 (2003); S. Razzaque, P. Mészéros & E. Waxman, Phys. Rev. Lett. 90, 241103 (2003); C. D. Dermer & A. Atoyan, Phys. Rev. Lett. 91, 071102 (2003).
36. T. Kashti & E. Waxman, Phys. Rev. Lett.. 95, .id. 181101 (2005).

A New Search Paradigm for Correlated Neutrino Emission from Discrete GRBs using Antarctic Cherenkov Telescopes in the Swift Era

Michael Stamatikos for the IceCube Collaboration[*] and David L. Band[†]

[*]*Department of Physics, University of Wisconsin, Madison, WI 53706, USA*[1]
[†]*NASA/GSSC, Greenbelt, MD 20771, USA and JCA/UMBC, Baltimore, MD 21250, USA*

Abstract. We describe the theoretical modeling and analysis techniques associated with a preliminary search for correlated neutrino emission from GRB980703a, which triggered the Burst and Transient Source Experiment (BATSE GRB trigger 6891), using archived data from the Antarctic Muon and Neutrino Detector Array (AMANDA-B10). Under the assumption of associated hadronic acceleration, the expected observed neutrino energy flux is directly derived, based upon confronting the fireball phenomenology with the discrete set of observed electromagnetic parameters of GRB980703a, gleaned from ground-based and satellite observations, for four models, corrected for oscillations. Models 1 and 2, based upon spectral analysis featuring a prompt photon energy fit to the Band function, utilize an observed spectroscopic redshift, for isotropic and anisotropic emission geometry, respectively. Model 3 is based upon averaged burst parameters, assuming isotropic emission. Model 4 based upon a Band fit, features an estimated redshift from the lag-luminosity relation, with isotropic emission. Consistent with our AMANDA-II analysis of GRB030329, which resulted in a flux upper limit of $\sim 0.150\,\mathrm{GeV/cm^2/s}$ for model 1, we find differences in excess of an order of magnitude in the response of AMANDA-B10, among the various models for GRB980703a. Implications for future searches in the era of Swift and IceCube are discussed.

Keywords: Gamma-ray: bursts, radiation mechanisms: nonthermal, neutrinos
PACS: 13.20.Cz, 14.60.Pq, 95.30.Cq, 95.55.Vj, 98.70.Rz, 98.70.Sa, 98.80.Es

1. GRB030329 & THE CASE FOR A NEW PARADIGM

Canonical fireball phenomenology, in the context of hadronic acceleration, predicts correlated MeV to EeV neutrinos from gamma-ray bursts (GRBs). Ideal for detection are $\sim$ TeV-PeV muon neutrinos, which arise as the leptonic decay products of photomeson interactions ($p + \gamma \to \Delta^+ \to \pi^+ + [n] \to \mu^+ + \nu_\mu \to e^+ + \nu_e + \bar{\nu}_\mu + \nu_\mu$) within the internal shocks of the relativistic fireball. Since the prompt γ-rays act as the ambient photon target field, these neutrinos are expected to be in spatial and temporal coincidence, with inverted energy spectra (see equation 2), which trace the photon energy spectra (see equation 1), due to the intrinsic threshold requirement that $4(1+z)^2 \varepsilon_p \varepsilon_\gamma \geq \left(m_{\Delta^+}^2 - m_p^2\right)\Gamma_{\mathrm{Bulk}}^2$. Constraints imposed by coincidence are tantamount to nearly[2] background-free searches in neutrino observatories such as AMANDA and IceCube (see table 1). The former, which has been calibrated with atmospheric neutrinos, has demonstrated the viability

---
[1] For recent results and a complete author list, please refer to [1].
[2] Non-trivial background is required to evaluate the statistical significance of on-time events (see table 1).

of high energy neutrino astronomy using the ice at the geographic South Pole as a Cherenkov medium since 1997. The latter, AMANDA's km-scale successor, is currently under construction with an anticipated completion date of 2010.

A positive detection of such high energy neutrinos would confirm hadronic acceleration in the relativistic GRB-wind, providing critical insight to the associated microphysics of the fireball, while possibly revealing an astrophysical acceleration mechanism for the highest energy cosmic rays. AMANDA analyses have used a diffuse formulation [2], predicated on an ensemble of average GRBs, to produce the most stringent upper limits upon multi-flavored correlated neutrino emission [1]. However, over 30 years of ground-based and satellite electromagnetic observations have documented the following facts: (i) the electromagnetic parameters of GRBs are characterized by distributions, often spanning multiple orders of magnitude, which differ both among and within bursts of different classes (i.e. short, long, x-ray rich, etc.), whose deviation from averaged values is often not accommodated by the inherent uncertainty of measurement; (ii) GRB satellite detectors exhibit a large dynamic range of thresholds and sensitivities which impart statistical sample bias via selection effects. These facts render the notion of an *average* GRB incompatible with the observational record. Furthermore, it has been argued that electromagnetic variations may lead to variations in the number of expected neutrino events associated with GRBs [3]. This reasoning has led to a new modeling paradigm for correlated neutrino emission searches, which is based upon the notion of a *discrete* GRB, i.e. one that is described by a unique set of electromagnetic parameters.

The quantitative effects on the expected neutrino number, energy and subsequent constraints upon astrophysical models (due to null detection), based upon multiple models of emission geometry and electromagnetic characterization, were initially illustrated in an analysis of GRB030329 with AMANDA-II [4]. For GRB030329, a peak effective area for muon neutrinos of $\sim 80\,\mathrm{m}^2$ and $\sim 700\,\mathrm{m}^2$ at ~ 2 PeV for AMANDA-II and IceCube, respectively, and an effective area for muons of $\sim 100,000\,\mathrm{m}^2$ and $\sim 1\,\mathrm{km}^2$ at ~ 200 TeV for AMANDA-II and IceCube, respectively, were achieved. Principle results, including neutrino flux upper limits for each model tested, are summarized in table 1. Further details regarding this analysis may be found elsewhere [4]. Here, supplementary to our conference presentation, we extend the paradigm by taking a first look at GRB980703a, one of the BATSE bursts currently under investigation using AMANDA archived data from 1997-2000 [5, 6].

TABLE 1. Summary of Results for GRB030329 [4].

Model	n_b, n'_b*	N_s, n_s, n'_s†	n_{obs}, n'_{obs}**	Flux Upper Limit‡ $\frac{\mathrm{GeV}}{\mathrm{cm}^2 \cdot \mathrm{s}}$
1	17.44, 0.06	0.1308, 0.0202, 0.0156	15, 0	0.150
2	17.44, 0.06	0.0691, 0.0116, 0.0092	15, 0	0.039
3	17.44, 0.06	0.0038, 0.0008, 0.0006	15, 0	0.035

* Number of background events expected during a 40 second on-time search window before (n_b) and after (n'_b) quality selection (optimized for discovery [7]), which resulted in a maximum search bin radius ($\Psi \equiv$ the space angle between the reconstructed muon trajectory and the GRB's position) of 11.3°, eliminating $\sim 99.6\%$ of the background while retaining $\sim 77.2\%$ of the signal for all models.

† Number of neutrino signal events expected on-time for IceCube (N_s) and AMANDA-II (n_s, n'_s).

** The number of observed events in AMANDA-II before (n_{obs}) and after (n'_{obs}) quality selection.

‡ Based upon null detection in AMANDA-II. The effects of neutrino flavor oscillations have been included. For more details on GRB030329's electromagnetic and neutrino parameterization, see [4].

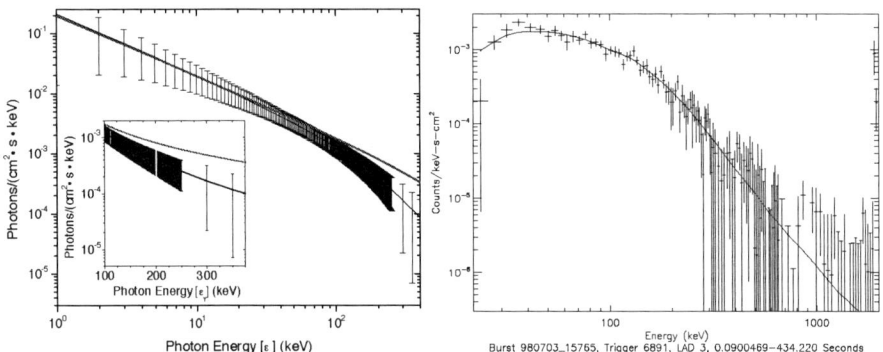

FIGURE 1. *Left Plate* - Prompt photon energy spectral fit to the Band function (see equation 1) for GRB980703a using the observed discrete electromagnetic parameters (black - used in models 1, 2, and 4), and average values (red - used in model 3). Inset illustrates the regime of the photon break energy (ε_γ^b) for the discrete parameter fit. Note that the disagreement between the curves is not resolved by the uncertainty in the discrete fit, illustrated as 1σ (propagated) error bars (assuming trivial covariance). See tables 2 and 3 for fit parameters. *Right Plate* - The prompt photon energy spectral Band fit, convolved with BATSE detector response (solid line), is compared to the measured BATSE photon count rate (data points with error bars) for the entire (fluence) emission of GRB980703a.

2. GRB980703A: EXTENDING THE DISCRETE GRB PARADIGM

Since the expected neutrino emission relies on the microphysics associated with the prompt γ-ray emission phase, a complete electromagnetic characterization of the GRB is compulsory. Spectral analysis, resulting in a prompt γ-ray photon energy spectrum, illustrated in figure 1, was performed via convolving an assumed spectral model with the BATSE detector response and comparing the fit with observed data. Our spectral assumption is characterized by an empirical model known as the *Band function* [8]:

$$N_{\varepsilon_\gamma}(\varepsilon_\gamma) = \begin{cases} A_\gamma \left(\dfrac{\varepsilon_\gamma}{100\,\text{keV}}\right)^\alpha exp\left(-\dfrac{\varepsilon_\gamma}{\varepsilon_\gamma^o}\right) & \varepsilon_\gamma \leq \varepsilon_\gamma^b \equiv (\alpha-\beta)\varepsilon_\gamma^o \\ A_\gamma' \left[\dfrac{(\alpha-\beta)\varepsilon_\gamma^o}{100\,\text{keV}}\right]^{\alpha-\beta} exp(\beta-\alpha) \left(\dfrac{\varepsilon_\gamma}{100\,\text{keV}}\right)^\beta & \varepsilon_\gamma \geq \varepsilon_\gamma^b \equiv (\alpha-\beta)\varepsilon_\gamma^o \end{cases} \quad (1)$$

Multi-wavelength afterglow observations have resulted in a more comprehensive electromagnetic characterization of GRB980703a, such as an observed spectroscopic redshift, z_{obs}, via (doppler) analysis of the optical transient. Since the blast wave evolution is sensitive to intrinsic burst properties such as the explosion energy, emission geometry and circumstellar density, n, the afterglow has been used to deduce anisotropic emission from a jet break time, t_{jet}, the relative energy fractions imparted to electrons, ε_e, and the magnetic field, ε_B. The best fit solution is consistent with the canonical fireball model, with a jet emission geometry that is inferred by assuming a relativistic outflow, beamed into an ambient medium of constant density, within a collimated jet of half angle, θ_{jet} [9]. A beaming fraction correction, f_B, is then used to correct isotropic values for luminosity, L_γ^{iso}, and energy, E_γ^{iso}, into beam-corrected (jet) values of L_γ^{jet} and E_γ^{jet}, respectively. Electromagnetic observables are summarized in table 2, while the electromagnetic parameterization of models 1 (discrete-isotropic), 2 (discrete-jet), 3 (average-isotropic) and 4 (lag-isotropic) are given in table 3. Model 4, based upon the lag-luminosity relation [10], may be used in the absence of an observed spectroscopic redshift.

TABLE 2. Observed Electromagnetic (EM) Properties of GRB980703a (BATSE Trigger #: 6891).

Parameter(s)	Value	Reference
RA, DEC, σ_R (°) [J2000]	359.07, 12.01, 1.67	[11]
T (UTC^s), T_{90} (SI^s)* [50-350 keV]	15765.22, 411.65 ± 9.27	[11]
F_γ^{Total} (ergs/cm^2) [50-350 keV]	$(6.22 \pm 0.43) \times 10^{-5}$	[11]
Φ_γ^{Peak} (ergs/cm^2/s) [20-1000 keV]†	$(1.56 \pm 0.07) \times 10^{-6}$	[6]
$\alpha, \beta, \varepsilon_\gamma^o$ (keV) [50-350 keV]**	$-1.02 \pm 0.20, -2.33 \pm 0.57, 223 \pm 96$	[6]
$\varepsilon_\gamma^b, \varepsilon_\gamma^p$ (keV) [50-350 keV]	$293 \pm 185, 219 \pm 104$	[6]
$\bar{z}_{obs}{}^\ddagger$	0.9660 ± 0.0006	[12]
d_L §	6219 ± 2843 Mpc $\sim (1.92 \pm 0.88) \times 10^{28}$ cm	[6]
$\varepsilon_e, \varepsilon_B$	$0.27 \pm 0.03, 0.0018^{+0.0004}_{-0.0003}$	[13]
t_{jet} (days)	3.43 ± 0.50	[9]
n (cm^{-3})	28 ± 10	[9]
θ_{jet}¶	$\sim (10.10 \pm 1.36)° \approx (0.176 \pm 0.024)$ rad	[6]
$f_B \equiv 1 - \cos\theta_{jet}$	0.016 ± 0.004	[6]

* The start of the T_{90} interval, with respect to the trigger time (T), was -6.14 seconds [11].

¶ $\theta_{jet} \approx 0.101$ rad $\left(\frac{t_{jet}}{1\,day}\right)^{\frac{1}{8}} \left(\frac{\xi}{0.2}\right)^{\frac{1}{8}} \left(\frac{n}{10\,cm^{-3}}\right)^{\frac{1}{8}} \left[\frac{(1+z)}{2}\right]^{-\frac{3}{8}} \left(\frac{E_\gamma^{iso}}{10^{53}\,ergs}\right)^{-\frac{1}{8}}$, with $\xi \sim 0.20^{+0.80}_{-0.20}$ (see [14]).

† We take the energy band pass of 20-1000 keV as bolometric.

** $A_\gamma = (1.97 \pm 0.43) \times 10^{-3}$ photons/cm^2/keV/s, $\chi_\nu^2 \sim 0.80$, and signal/noise ~ 4.11 (see equation 1).

‡ Based upon an average from emission and absorption redshift lines.

§ Λ_{CDM} cosmology: H$_o$ = 72 ± 5 km/Mpc/s, Ω_m = 0.29 ± 0.07, Ω_Λ = 0.73 ± 0.07 [15], is utilized throughout.

TABLE 3. Electromagnetic Parameterization for Neutrino Models of GRB980703a.

Parameter(s)	Model 1*	Model 2†	Model 3**	Model 4‡
z	See table 2	See table 2	~ 1	$0.6378^{+0.2222}_{-0.1398}$
d_L (Mpc)	See table 2	See table 2	6491 ± 2887	3730^{+2688}_{-2522}
L_γ (10^{51} ergs/s) [20-1000 keV]	7.21 ± 6.60	0.11 ± 0.11	10.08 ± 8.97	$2.38^{+2.29}_{-1.20}$
E_γ (10^{52} ergs) [20-2000 keV]	13.77 ± 12.68	0.21 ± 0.21	1.51 ± 1.35	$6.57^{+5.26}_{-2.67}$
ε_e	See table 2	See table 2	$0.33^{+0.67}_{-0.33}$	$0.33^{+0.67}_{-0.33}$
ε_B	See table 2	See table 2	$0.33^{+0.67}_{-0.33}$	$0.33^{+0.67}_{-0.33}$

* Based upon discrete electromagnetic parameters, assuming isotropic emission.

† Based upon discrete electromagnetic parameters, assuming beamed (jet) emission.

** Using averaged GRB values: $\alpha \sim -1, \beta \sim -2, \varepsilon_\gamma^o \sim \varepsilon_\gamma^b \sim \varepsilon_\gamma^p \sim 1$ MeV, $F_\gamma^{Total} \sim 6 \times 10^{-6}$ ergs/cm^2, $z \sim 1$, and $\Phi_\gamma^{Peak} \sim 2 \times 10^{-6}$ ergs/cm^2/s [2]. Under the assumption of isotropic emission.

‡ Isotropic emission and estimated z via the lag-luminosity method, with $\tau_o = 0.39^{+0.14}_{-0.20}$ seconds [10].

The neutrino spectral parameterization is given in equation 2 [4]. Values from tables 2 and 3 have been substituted into equation 2 to produce the values of the neutrino parameters and the number of neutrino events expected in AMANDA-B10 (via simulation) for models 1-4, given in table 4. The responses of AMANDA-B10 to models 1-4 for GRB980703a, prior to event quality selection, and AMANDA-II/IceCube to models 1-3 for GRB030329, are illustrated in the left and right plates of figure 2, respectively.

$$\varepsilon_{\nu_\mu}^2 \Phi_{\nu_\mu} \approx A_{\nu_\mu} \times \begin{cases} \left(\frac{\varepsilon_{\nu_\mu}}{\varepsilon_\nu^b}\right)^{-\beta-1} & \varepsilon_{\nu_\mu} < \varepsilon_\nu^b \\ \left(\frac{\varepsilon_{\nu_\mu}}{\varepsilon_\nu^b}\right)^{-\alpha-1} & \varepsilon_\nu^b < \varepsilon_{\nu_\mu} < \varepsilon_\pi^b \\ \left(\frac{\varepsilon_{\nu_\mu}}{\varepsilon_\nu^b}\right)^{-\alpha-1}\left(\frac{\varepsilon_{\nu_\mu}}{\varepsilon_\pi^b}\right)^{-2} & \varepsilon_{\nu_\mu} > \varepsilon_\pi^b \end{cases} \quad (2)$$

TABLE 4. Prompt Muon Neutrino Flux Parameterization of GRB980703a.

Parameter*	Model 1	Model 2	Model 3[†]	Model 4
$\Gamma_{\text{bulk}} \gtrsim 276 \left[L_{\gamma,52} t_{\nu,-2}^{-1} \varepsilon_{\gamma,MeV}^{max}(1+z)\right]^{\frac{1}{6}}$ **	293 ± 45	146 ± 23	310 ± 46 [300]	243^{+33}_{-17}
$f_\pi \simeq 0.2 \times \frac{L_{\gamma,52}}{\Gamma_{2.5}^4 t_{\nu,-2} \varepsilon_{\gamma,MeV}^b (1+z)}$	0.34 ± 0.43	0.09 ± 0.11	0.11 ± 0.12 [0.20]	$0.34^{+0.40}_{-0.28}$
$A_{\nu_\mu} \approx \frac{F_\gamma f_\pi}{8\varepsilon_\epsilon \ln(10) T_{90}}$ (10^{-6} GeV/cm^2/s)	6.49 ± 8.27	1.62 ± 2.13	$0.16^{+0.37}_{-0.24}$	$5.27^{+12.36}_{-6.81}$
$\varepsilon_\nu^b \approx \frac{7 \times 10^5}{(1+z)^2} \frac{\Gamma_{2.5}^2}{\varepsilon_{\gamma,MeV}^b}$ (10^5 GeV)	5.29 ± 3.71	1.32 ± 0.93	1.68 ± 0.50 [1]	$5.25^{+3.88}_{-3.51}$
$\varepsilon_\pi^b \approx \frac{10^8 \sqrt{\varepsilon_e} \Gamma_{2.5}^4 t_{\nu,-2}}{(1+z)\sqrt{\varepsilon_B}\sqrt{L_{\gamma,52}}}$ (10^8 GeV)	$5.37^{+4.15}_{-4.13}$	$2.68^{+2.16}_{-2.15}$	$0.46^{+0.75}_{-0.47}$ [0.1]	$0.40^{+0.64}_{-0.32}$
$n_{\nu_\mu}^{\text{AMANDA-B10}}$ (10^{-3})[‡]	2.18	0.96	0.11	1.77

* Where $L_\gamma \equiv L_{\gamma,52} \cdot 10^{52}$ ergs/s, $\Gamma \equiv \Gamma_{2.5} \cdot 10^{2.5}$, $t_\nu \equiv t_{\nu,-2} \cdot 10$ ms, $\varepsilon_\gamma^b \equiv \varepsilon_{\gamma,MeV}^b \cdot 1$ MeV, and $\varepsilon_\gamma^{max} \equiv \varepsilon_{\gamma,MeV}^{max} \cdot 100$ MeV.
[†] Bracketed values indicate nominal values used in average GRB parameterization [2].
** Super-Eddington luminosity within a compact source requires a lower bound to ensure transparent optical depth.
[‡] Events expected in AMANDA-B10, prior to quality selection (most optimistic case), for an on-time search window of 420.92 seconds. Event quality selection for AMANDA-B10 may systematically attenuate these values by $\sim 70\%$.

3. DISCUSSION & FUTURE OUTLOOK IN THE SWIFT ERA

Consistent with our results for GRB030329, we find that the most critical parameters, which translate into an observable variation in detector response, are the electromagnetic fluence, F_γ, and spectral characterization in the vicinity of the neutrino spectral break, ε_ν^b. The former is related to the number of neutrinos expected in the detector, while the latter affects the mean neutrino energy of the events. These effects on neutrino energy flux, which exceed one and two orders of magnitude in the comparison of models 1 and 3, in GRB980703a (see table 4) and GRB030329 (see figure 2), respectively, may be directly traced back to the variance in their electromagnetic characterization, via equation 2. The consequences of this reality are both clear and apropos, since the hallmark of the Swift era is the acquisition of a more complete electromagnetic characterization of fewer bursts (relative to the age of BATSE). With an order of magnitude increase in effective area for GRBs [4], realized by km-scale detectors such as IceCube, it is most likely that either the first evidence for correlated neutrino emission or the first real constraints on associated hadronic acceleration will come from the analysis of an exceptional (local)

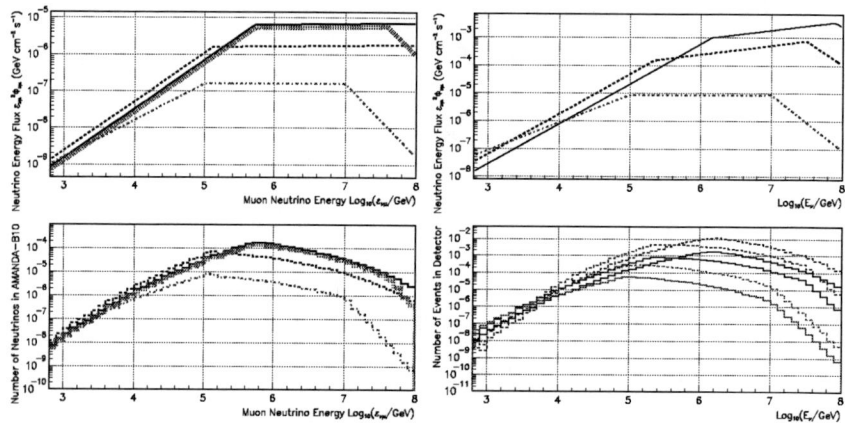

FIGURE 2. *Left Plate* - Prompt neutrino energy flux (*Upper panel*) and AMANDA-B10 detector response (*Lower panel*), for GRB980703a models 1, 2, 3, and 4, indicated by solid black, dashed blue, dot-dashed red, and hatched purple curves, respectively. *Right Plate* - *Upper panel* - Prompt neutrino energy flux for GRB030329 models 1 (solid black), 2 (dashed blue) and 3 (dot-dashed red). *Lower panel* - Detector response for models 1 (black), 2 (blue) and 4 (red) in AMANDA-II (solid) and IceCube (dashed).

discrete GRB, rather than an aggregate of hundreds of average emission. It is interesting to note that model 4, based upon an estimated redshift from the lag-luminosity relation, was consistent with model 1, which was based upon the observed redshift. Future work includes the analysis of a subset of BATSE GRBs from 1997-2000 [6], using Band function fits and lag-luminosity relation estimated redshifts. Ultimately, the synergy of γ-ray and neutrino astronomy may be realized via multi-wavelength and multi-messenger correlative GRB searches in an era of superior instruments such as IceCube and Swift.

REFERENCES

1. A. Achterberg, et al., *astro-ph/0509330* (See also K. Kuehn et al., these proceedings.).
2. E. Waxman, and J. Bahcall, *Phys. Rev. D* **59**, 023002 (1999).
3. D. Guetta, et al., *Astrop. Phys.* **20**, 429–455 (2004).
4. M. Stamatikos, J. Kurtzweil, and M. Clarke, *preprint (astro-ph/0510336)* (2005).
5. M. Stamatikos, et al., "Gamma-Ray Bursts: 30 Years of Discovery," AIPC 727, 2004, pp. 146–149.
6. M. Stamatikos, D. Band, D. Hooper, and F. Halzen, *In preparation, to be submitted to ApJ*. (2006).
7. G. Hill, J. Hodges, B. Hughey, A. Karle, and M. Stamatikos, "PHYSTAT," Oxford, 2005.
8. D. Band, et al., *ApJ* **413**, 281–292 (1993).
9. D. Frail, et al., *ApJ* **590**, 992 (2003).
10. D. Band, et al., *ApJ* **613**, 484–491 (2004).
11. (2003), URL http://www.batse.msfc.nasa.gov/batse/grb/catalog/current/.
12. S. Djorgovski, et al., *ApJ* **508**, L17–L20 (1998).
13. S. Yost, et al., *ApJ* **597**, 459–473 (2003).
14. A. Friedman, and J. Bloom, *ApJ* **627**, 1–25 (2005).
15. D. Spergel, et al., *ApJ Supp. Ser.* **148**, 174–194 (2003).

Searching for Cataclysmic Cosmic Events with a Coincident Gamma-ray Burst and Gravitational Wave Signature

Szabolcs Márka and Luca Matone

Columbia University, Pupin Laboratories, New York, NY 10027

Abstract. Coincident observation of gamma-ray bursts and gravitational waves will help us to dramatically improve our understanding of energetic processes in the universe while opening a new window on compact, and often difficult to study, astronomical objects. One of the major goals of interferometric gravitational wave detectors is to develop and exploit gravitational wave detection in conjunction with astrophysical observations. The collaboration among gravitational wave detectors and gamma-ray burst observatories is ongoing and flourishing. The present status of the collaborative research and the future plans are summarized and illustrated through practical experience with the Laser Interferometer Gravitational Wave Observatory (LIGO) detectors.

Keywords: gravitational wave detectors, Data analysis, Gravitational radiation, gamma-ray burst, GRB, HETE, SWIFT, LIGO
PACS: 04.80.Nn, 07.05.Kf, 95.85.Sz

BACKGROUND

Interferometric gravitational wave (GW) detectors LIGO, GEO, TAMA, and VIRGO are already searching for GW signals with unprecedented sensitivity [1, 2, 3, 4, 5, 6, 7]. For LIGO, the noise levels (see [1] Figure 1) are already reaching design specifications, giving a possibility for detection of the most intensive GW signals. It is very important to analyze the detection capabilities of these detectors and to estimate the event rates of potentially detectable GW signals. There is a considerable list of possible detection candidates [8]: the inspiral of neutron star (NS) or black hole (BH) binaries, the tidal disruption of NS by BH in NS/BH binaries, BH/BH merger and ringdown, low-mass X-ray binaries, pulsars, centrifugally hung-up proto neutron stars in white dwarf accretion-induced collapse, supernova core collapse, gamma-ray bursts (GRBs) [9], and the stochastic background. Direct detection of gravitational waves of cosmic origin will provide complementary data to the information obtainable via traditional observational methods.

Gamma Ray Bursts (GRBs) are short but very energetic pulses of gamma rays from astrophysical sources, with duration ranging between 10 ms and 100 s. GRBs are historically divided into two classes [10, 11] based on their duration: "short" ($<$ 2 s) and "long" ($>$ 2 s). Both classes are isotropically distributed and their detection rate can be as large as one event per day. The present consensus is that long GRBs [11] are the result of the

---

[1] http://www.ligo.caltech.edu/docs/G/G060010-01/G060010-01.eps

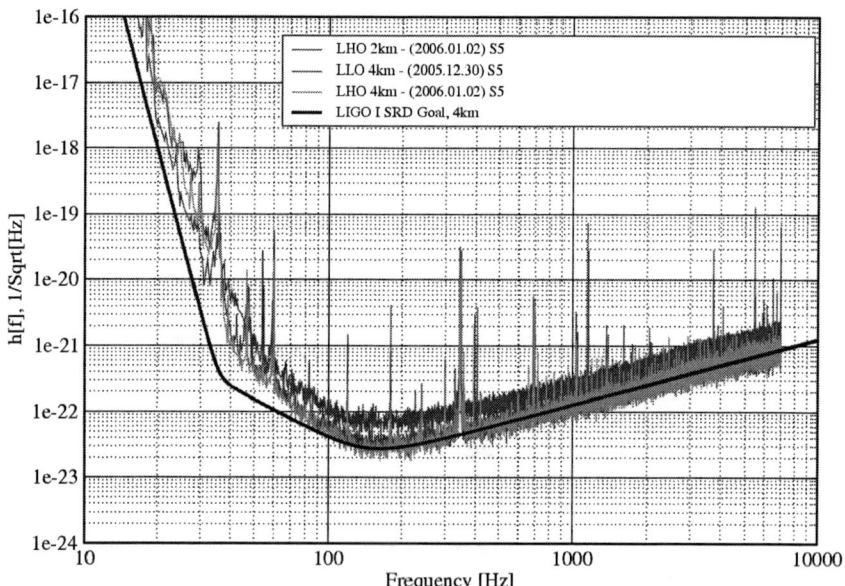

FIGURE 1. LIGO sensitivity curves during the S5 Scientific Run [strain Hz$^{-1/2}$]. The LIGO design sensitivity goal (SRD) is also indicated.

core collapse of massive stars resulting in black hole formation. The violent formation of black holes has long been proposed as a potential source of gravitational waves. Therefore, we have a reason to expect strong association between GRBs and gravitational waves. There is a large number of publications available describing possible associations between GRBs and gravity waves [12, 13, 14, 15, 12, 16, 17, 18, 19, 20, 21, 22].

A number of long GRBs have been associated with X-ray, radio and/or optical afterglows, and the cosmological origin of the host galaxies of their afterglows has been unambiguously established by their observed redshifts, which are usually of order of unity [11]. The smallest observed redshift of an optical afterglow associated with a detected GRB (GRB980425 [23, 24, 25]) is z=0.0085 ($\simeq$35 Mpc), which also pointed to a possible hypernova association of long GRBs.

There is mounting observational evidence [22, 26] supporting the theoretical argument that the short-hard GRBs are the result of a compact binary coalescence, which are also expected to be strong GW emitters [8, 14].

There are significant differences between the electromagnetic, neutrino and GW signatures of a given GRB event. In contrast to electromagnetic radiation, GWs do not interact strongly with matter, making the universe transparent for them. The electromagnetic radiation, the neutrinos and the GWs are usually emitted at different times by different sub-processes during a cataclysmic event, thus bringing complementary information about various aspects of the same cosmic source.

Due to strong beaming [27, 28] and limited detector sensitivity, present GRB observatories only detect a small fraction ($\sim$O(1/500) [27]) of the GRB events of our universe.

A detected GRB event can supply priceless information for GW searches, such as event time, event type and sky position. Therefore GW searches are performed through two fundamentally different ways:

1. The untriggered method searches for GW signals with arbitrary arrival times and sky position [2]. Untriggered searches primarily target undetected GRBs at a price of somewhat lower sensitivity.
2. Triggered search methods [3] have only access to a tiny subset of the GRB events of the universe (namely the well detected ones). However, the additional information available through the detection of the GRBs warrants the execution of specialized GW searches with potentially higher sensitivity and/or lower false detection rate. Triggered search for GW signals is usually restricted to an astrophyscially motivated range of arrival times around the trigger time. The non-GW observation is also used as a guide in selecting specific astrophysical models and optimizing the GW search algorithm. An additional future advantage is that a network of optical, neutrino and gravity wave detectors might be able to alert each other when an energetic cosmic event is detected.

The gravitational wave signal of a GRB is anticipated to have either a chirp signature due to the binary inspiral process (short GRBs, [14, 17, 18, 19, 20, 21, 22]) or a transient ($\sim O(10\text{-}100ms)$) burst like signal [3] (long GRBs [11, 12]). Accordingly, GW searches for GRB associated signals are also performed through different search methods depending on the GRB classification. Finn, Mohanty and Romano proposed that multiple GRB triggers can be combined to improve the collective signal-to-noise ratio [29]. Variants of this method are widely used today.

Exciting science can be derived from galactic sources, which lie in the overlapping range of neutrino [30], electromagnetic and gravity wave detectors. This is the range where correlated measurements may be the most profitable. Unfortunately the predicted event rate ($\sim$1 in 50 years)[31] for our galaxy is so low that this approach can be classified as opportunistic at best.

INTERFEROMETRIC GRAVITATIONAL WAVE DETECTORS

The new generation of GW detectors rely on interferometric monitoring of the relative separation of mirrors, which play the role of test masses. They respond to space-time distortions induced by the passage of GWs as they traverse the detectors. GWs with normal incidence to the plane of the detector, and polarized along the arms of the detector, cause the mirrors to move differentially. Waves incident from other directions and/or polarizations can also induce differential motion, albeit at a smaller level.

Presently, there is an operational international network of first generation interferometric GW detectors: LIGO, VIRGO, TAMA, and GEO. Advanced terrestrial detectors (such as Advanced LIGO [2]) and space detectors (such as LISA [3]) have advanced designs

---

[2] http://www.ligo.caltech.edu/advLIGO/scripts/ref_des.shtml
[3] http://www.srl.caltech.edu/$\sim$shane/sensitivity/

and a precise schedule.

Numerous future projects are in their planning phase: Next Generation LISA (NGLISA), Decihertz Interferometric Gravitational-wave Observatory (DECIGO [32], the Advanced Laser Interferometer Antenna (ALIA) [33] and the Big Bang Observer (BBO) [33]. Their sensitivities, detection frequency bands and capabilities will be quite different.

The present LIGO detectors are orthogonal arm enhanced Michelson laser interferometers. The LIGO Hanford Observatory (LHO) operates two identically oriented interferometric detectors, which share a common vacuum envelope: one having 4 km long arms (H1), and one having 2 km long arms (H2). The LIGO Livingston Observatory operates a single 4 km long detector (L1). The two sites are separated by $\simeq 3002$ km, representing a maximum arrival time difference of $\simeq \pm 10$ ms. A complete description of the LIGO interferometers as they were configured during LIGO's first Science Run (S1) can be found in [34].

SEARCHES

Overview of Past Results

Several studies [35, 36, 37, 38, 39, 40, 41] were published on the coincident analysis of GRBs and GWs by collaborations using traditional bar detectors [42, 43, 44]. None of the articles declared coincident detection or statistical association [29], however, they consistently improved their upper limits on the strength of the gravitational waves as technologies and techniques improved.

Baggio et al. [45] put upper limit on the strength of the initial GW burst of the December 27/2004 giant flare of SGR1806.

The GRB030329 trigger occurred during the Second Science Run (S2) of the LIGO detectors. No candidates with gravitational wave signal strength larger than a predetermined threshold were observed [3] and upper limits were set on the energy emitted in GWs for a number of possible scenarios.

Outlook

As their sensitivity improved, interferometric GW detectors became competitive partners of bar detectors. LIGO pursues an active program on coincidence studies between external triggers and GWs. LIGO presently uses data from SNEWS and major networks specializing in distribution of GRB observational data via the Internet (GCN [4], IPN). Pipelines were developed for both GW signal searches in conjunction with individual GRB triggers and searches aimed at finding statistical association between GRBs and GWs.

---

[4] Description of GCN is available at their homepage at: http://gcn.gsfc.nasa.gov/gcn/

Due to the apparent association between the short hard GRBs and compact binary inspirals, the traditional matched filtering analysis approach [46, 47] is a desirable choice. Accordingly, the sensitive search methods developed for untriggered binary inspiral searches are used to search for individual and also for statistical association between short GRBs and GWs.

Core collapse [12], black hole formation [16, 48] and black hole ringdown [49, 50] may each produce gravitational wave emissions. Unfortunately, there are no known gravitational waveforms that could be reliably used as templates for a matched filter search for long GRBs. Therefore present coincident searches between long GRBs and GWs are designed to be as waveform independent as practical. Predictions of gravitational wave amplitudes [51, 52, 53] are also uncertain by several orders of magnitude, making it difficult to predict the probability to observe the gravitational wave signature of distant GRBs. LIGO has ongoing triggered search program to search for the direct GW counterpart of GRBs [54, 55, 3] and statistical association between GRBs and GWs. A wide range of untriggered GW burst monitoring tools [56, 57] are used to also identify GW burst signals.

Besides the short and long GRBs, Soft Gamma-ray Repeaters (SGRs) may also prove to be an interesting source [45, 58] population in the future. Recently, close analysis of the associated X-ray lightcurve [59, 60, 61] revealed the possibility of massive starquakes in two sources. The consecutive Quasi-Periodic Oscillations in the lightcurve's tail were interpreted as a signature of excited toroidal modes of the star's crust. These in turn can emit GWs [62, 63], albeit these might prove too weak to be detected with today's GW detectors. There is an ongoing analysis effort at LIGO to take advantage of this discovery.

CONCLUSION

The international network of GW detectors is searching for GWs with unprecedented sensitivity. Coincident observation of astronomical events by GW detectors and traditional observatories shall improve our understanding of energetic cosmic processes, opening a new window on compact and difficult-to-study astronomical objects.

Currently, within the community of interferometric gravitational wave detectors, there are ongoing efforts to make use of the information provided by the GRB observations. Cooperation presents its own technical and scientific advantages and challenges. GRB triggers are fairly straightforward to use, however, the significant cosmological distances of the sources pose a great challenge for the earth-bound GW detectors.

Operating close to their target sensitivity, today the LIGO detectors are more sensitive than during the previous S2 data run by factors of 10 - 100, depending on frequency. This implies an improvement of a factor of ~ 1000 in sensitivity to the energy emitted in gravitational waves relative to S2 [3]. Detection of GRBs with measured redshifts significantly smaller than GRB030329's, as was 1998bw, is certainly possible.

Coincident detection of GRBs and GWs still remain an opportunistic enterprize, but not for long, as improved detectors are taking us closer to the ultimate goal of discovery.

ACKNOWLEDGMENTS

The authors are grateful for the support of the United States National Science Foundation under cooperative agreement PHY-04-57528 and Columbia University in the City of New York. The authors gratefully acknowledge the support of the United States National Science Foundation for the construction and operation of the LIGO Laboratory. We are grateful to Scott Barthelmy and the GCN network and Kevin Hurley and the IPN network for providing us with near real time GRB triggers and to the SWIFT, Konus, Integral and HETE experiments, who detect and generate the events distributed by GCN and IPN. We thank Kate Scholberg and the Supernova Early Warning System (SNEWS).

REFERENCES

1. S. A. Hughes, et al., *ArXiv Astrophysics e-prints* (2001), arXiv:astro-ph/0110349.
2. B. Abbott, et al., *Physical Review D* **72**, 062001–+ (2005).
3. B. Abbott, et al., *Physical Review D* **72**, 042002–+ (2005).
4. B. Abbott, et al., *Physical Review Letters* **94**, 181103–+ (2005).
5. H. Grote, et al., *Classical and Quantum Gravity* **22**, 193–+ (2005).
6. M. Ando, and the TAMA Collaboration, *Classical and Quantum Gravity* **22**, 881–+ (2005).
7. F. Acernese, et al., *Classical and Quantum Gravity* **22**, 869–+ (2005).
8. C. Cutler, and K. S. Thorne, *ArXiv General Relativity and Quantum Cosmology e-prints* (2002), arXiv:gr-qc/0204090.
9. T. Piran, *ArXiv Astrophysics e-prints* (2001), arXiv:astro-ph/0102315.
10. C. Kouveliotou, et al., *Astrophysical Journal, Letters* **413**, L101–L104 (1993).
11. B. Zhang, and P. Mészáros, *Int. J. Mod. Phys. A* (2003), in press, astro-ph/0311321.
12. M. B. Davies, et al., *Astrophysical Journal* **579**, L63–L66 (2002).
13. M. H. P. M. van Putten, *Astrophysical Journal, Letters* **575**, L71–L74 (2002).
14. E. Nakar, A. Gal-Yam, and D. B. Fox, *ArXiv Astrophysics e-prints* (2005), arXiv:astro-ph/0511254.
15. C. L. Fryer, S. E. Woosley, and A. Heger, *Astrophysical Journal* **550**, 372–382 (2001).
16. M. H. van Putten, et al., *Physical Review D* **69**, 044007–+ (2004).
17. W. H. Lee, E. Ramirez-Ruiz, and J. Granot, *Astrophysical Journal, Letters* **630**, L165–L168 (2005).
18. J. S. Villasenor, et al., *Nature* **437**, 855–858 (2005).
19. N. Gehrels, et al., *Nature* **437**, 851–854 (2005).
20. J. Hjorth, et al., *Nature* **437**, 859–861 (2005).
21. E. Nakar, and A. Gal-Yam, *American Astronomical Society Meeting Abstracts* **207**, –+ (2005).
22. D. B. Fox, et al., *Nature* **437**, 845–850 (2005).
23. S. R. Kulkarni, et al., *Nature* **395**, 663–669 (1998).
24. K. Iwamoto, et al., *Nature* **395**, 672–674 (1998).
25. T. J. Galama, et al., *Nature* **395**, 670–672 (1998).
26. K. Belczynski, et al., *ArXiv Astrophysics e-prints* (2006), arXiv:astro-ph/0601458.
27. D. A. Frail, et al., *Astrophysical Journal, Letters* **562**, L55–L58 (2001).
28. M. H. P. M. van Putten, and T. Regimbau, *Astrophysical Journal, Letters* **593**, L15–L18 (2003).
29. L. S. Finn, S. D. Mohanty, and J. Romano, *Physical Review D* **60**, 121101–+ (1999).
30. The Snews Group, *Nuclear Physics B Proceedings Supplements* **143**, 543–543 (2005).
31. R. Diehl, et al., *Nature* **439**, 45–47 (2006).
32. N. Seto, S. Kawamura, and T. Nakamura, *Physical Review Letters* **87**, 221103–+ (2001).
33. J. Crowder, and N. J. Cornish, *Physical Review D* **72**, 083005–+ (2005).
34. B. Abbott, et al., *Nuclear Instruments and Methods in Physics Research A* **517**, 154–179 (2004).
35. P. Tricarico, A. Ortolan, and P. Fortini, *Classical and Quantum Gravity* **20**, 3523–3531 (2003).
36. P. Tricarico, et al., *Physical Review D* **63**, 082002–+ (2001).
37. G. Modestino, and A. Moleti, *Physical Review D* **65**, 022005–+ (2002).

38. P. Astone, et al., *Astronomy and Astrophysics, Supplement* **138**, 603–604 (1999).
39. P. Astone, et al., *Physical Review D* **66**, 102002–+ (2002).
40. L. Amati, et al., *Astronomy and Astrophysics, Supplement* **138**, 605–606 (1999).
41. P. Astone, et al., *ArXiv Astrophysics e-prints* (2004), `astro-ph/0408544`.
42. G. A. Prodi et al., "Initial operation of the gravitational wave detector AURIGA," in *Second Edoardo Amaldi Conference on Gravitational Wave Experiments*, 1998, pp. 148–158, (CERN - Switzerland), edited by E. Coccia, G. Veneziano, G. Pizzella, World Scientific, Singapore.
43. P. Astone, et al., *Physical Review D* **47**, 362–375 (1993).
44. P. Astone, et al., *Astroparticle Physics* **7**, 231–243 (1997).
45. L. Baggio, et al., *Physical Review Letters* **95**, 081103–+ (2005).
46. C. W. Helstrom, *Statistical Theory of Signal Detection*, Pergamon Press, London, 1968, 2nd ed.
47. B. Abbott, et al., *Physical Review D* **69**, 122001–+ (2004).
48. K. Belczynski, T. Bulik, and B. Rudak, *Astrophysical Journal, Letters* **608**, L45–L48 (2004).
49. É. É. Flanagan, and S. A. Hughes, *Physical Review D* **57**, 4535–4565 (1998).
50. J. D. E. Creighton, *Physical Review D* **60**, 022001–+ (1999).
51. E. o. Müller, *The Astrophysical Journal* **603**, 221–230 (2004).
52. C. L. Fryer, and M. S. Warren, *The Astrophysical Journal* **601**, 391–404 (2004).
53. C. D. Ott, A. Burrows, E. Livne, and R. Walder, *The Astrophysical Journal* **600**, 834–864 (2004).
54. S. D. Mohanty, et al., *Classical and Quantum Gravity* **21**, 1831–+ (2004).
55. S. D. Mohanty, et al., *Classical and Quantum Gravity* **21**, 765–+ (2004).
56. L. Cadonati, and S. Márka, *Classical and Quantum Gravity* **22**, 1159–+ (2005).
57. F. Beauville, The Joint LIGO/Virgo Working Group, et al., *Classical and Quantum Gravity* **22**, 1293–+ (2005).
58. J. E. Horvath, *Modern Physics Letters A* **20**, 2799–2804 (2005).
59. G. L. Israel, et al., *Astrophysical Journal, Letters* **628**, L53–L56 (2005).
60. T. E. Strohmayer, and A. L. Watts, *Astrophysical Journal, Letters* **632**, L111–L114 (2005).
61. A. L. Watts, and T. E. Strohmayer, *ArXiv Astrophysics e-prints* (2005), `arXiv:astro-ph/0512630`.
62. L. Stella, et al., *Astrophysical Journal, Letters* **634**, L165–L168 (2005).
63. B. L. Schumaker, and K. S. Thorne, *Monthly Notices of the RAS* **203**, 457–489 (1983).

Spectroscopy of the Brightest Bursts up to Energies of 200MeV.

M.M. González[*], M. Carrillo-Barragán[*], B.L. Dingus[†], Y. Kaneko[**] and R.D. Preece[‡]

[*]*Instituto de Astronomía-UNAM*
[†]*Los Alamos National Laboratory*
[**]*Universities Space Research Association*
[‡]*Physics Department, University of Alabama in Huntsville, National Space Science and Technology Center*

Abstract. The EGRET calorimeter, TASC, was triggered by BATSE and thus observed several bursts in the energy range of 1-200 MeV. The analysis of combined LAD-TASC data has been developed and discussed previously. Spectroscopy of GRB941017 based on this analysis, uncovered a MeV-component that would be hidden in the brightness of the synchrotron component in a time integrated spectrum of the whole burst, while separation into shorter time intervals allowed a unique identification of a new component different from the synchrotron component. This paper performs the same analysis used for GRB941017 to study the spectral temporal evolution for 21/68 of the brightest BATSE bursts in the broadest energy range yet reported. We consider the Band function plus a power law to describe the spectra at keV and MeV energies respectively. The higher energy spectral component is a major fraction of the energy in GRB941017, but also is significant in one time interval of the spectrum of GRB980923. Other relevant results from this analysis are given.

Keywords: BATSE, TASC, spectroscopy, GRB, gamma-ray burst
PACS: 95.75.Fg, 95.55.Ka, 95.85.Nv, 95.85.Pw

Description of bursts spectra constrains the GRB models. Empirical laws in a large set of BATSE bursts have been found to support the most "natural" and accepted emission process, the synchrotron emission in either internal or external shocks. Some facts, such as the violation of the "line of death" [1], the narrow E_{peak} [2, 3] distribution and the temporal evolution of β [2, 3] seem to contradict the model. These facts could be either physical or intrinsic to the bursts or a consequence of the low BATSE detector sensitivity at the lowest (10keV) and highest (1MeV) energies. An analysis of the burst spectra in a broader energy range will help to solve these problems, especially for bursts with E_{peak} above 700keV.

We combined and analyzed LAD-TASC data for 57 bursts selected from the BATSE catalog because they were bright above 300keV. We followed the analysis described by [4, 5, 6]. For each burst, we used the Band function [7] plus a power law to fit jointly the photon flux observed by both detectors. Results for a similar analysis using only Band function refer to [5, 6]. The time intervals and the maximum energy included in the analysis were determined by EGRET-TASC data and varies from burst to burst.

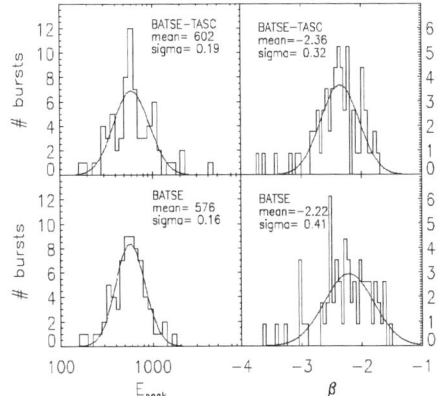 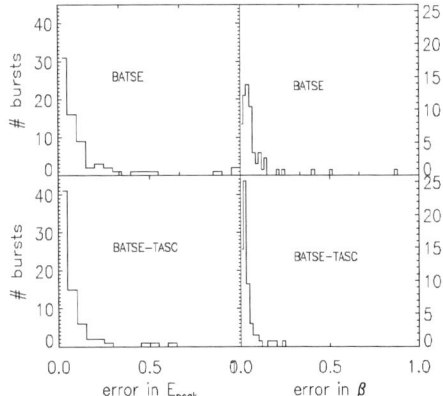

FIGURE 1. Distributions of E_{peak}, β (left four windows) and their uncertainties (right four windows) are shown for BATSE (bottom) and BATSE-TASC (top) data. Solid lines represent gaussian distributions that fit the E_{peak} and β distributions, mean and sigma are given in the figure.

RESULTS AND CONCLUSIONS

As shown in figure 1, BATSE values for E_{peak} close to 1MeV were corrected to values even higher than 2MeV resulting in a symmetric E_{peak} distribution around 602keV. β distribution became softer and narrower than the reported by BATSE. Significant improvement in the determination of E_{peak} and β has been achieved as shown by the error distributions of E_{peak} and β in figure 1. Two thirds of the bursts showed hard to soft evolution in single episodes. No generalized temporal evolution was found though the whole burst.

No cutoff in the spectra up to 10MeV and in some cases up to 100MeV was observed. Except for GRB941017 [8, 9] and GRB980923 (presented here), the spectra up to 10MeV is consistent with a smooth continuation of the spectra at keV energies and described only with a Band function.

For GRB980923, the TASC spectrum covering from 13s to 46s has an excess over the background of 8.6σ and 6.9σ in the energy ranges of 1-10MeV and 10MeV-200MeV. A power law to described the spectrum above 10MeV was added. Results are given in the table 1 and figure 2.

Limits to the higher energy component described by a power law function were calculated. Results are given in the table 2 below for the most significant GRBs. GRB910503 shown to be the most significant. Observations with the EGRET spark chamber [6] are consistent with the existence of such component.

REFERENCES

1. R.D. Preece, M.S. Briggs, R.S. Mallozzi, G.N. Pendleton, W.S. Paciesas, D.L. Band, *ApJ*,506,L23 (1998).

TABLE 1. Spectral fitting parameters of the differential photon flux. The first six rows contain the best spectral fit parameters for the two time intervals shown in Figure 2. The seventh row shows the probability that the improvement in χ^2 from the addition of the high-energy power law in the fit is due to chance, as determined by the χ^2 test. The last three rows contain the energy fluxes for three different energy ranges. In last column, a higher-energy component similar to the one observed in GRB941017 with spectral index -1 is assumed.

Time from BATSE trigger (s)	13 to 46	13 to 46	13 to 46
Band GRB function parameters			
$A(\frac{\times 10^{-2} \text{ph}}{\text{s cm}^{-2} \text{keV}})$	16.2 ± 0.1	15.9 ± 0.2	16.0 ± 0.1
E_{peak}(keV)	388 ± 4	397 ± 4	397 ± 4
α	-0.66 ± 0.01	-0.67 ± 0.01	-0.67 ± 0.01
β	-2.93 ± 0.06	-3.51 ± 0.25	-3.31 ± 0.12
High-energy power law parameters			
$A_{PL}(\frac{\times 10^{-7} \text{ph}}{\text{s cm}^{-2} \text{keV}})$	—	5.6 ± 0.7	4.9 ± 0.7
γ	—	-1.32 ± 0.25	$-1\,fixed$
Probability	—	1.1×10^{-6}	3.4×10^{-7}
Energy Flux ($\times 10^{-6} \frac{\text{erg}}{\text{s cm}^{-2}}$)			
0.03-2 MeV	11.1 ± 0.03	10.9 ± 0.7	10.9 ± 0.7
2-10 MeV	1.0 ± 0.003	0.8 ± 0.05	0.8 ± 0.05
10-200 MeV	0.26 ± 0.0008	2.7 ± 0.2	2.9 ± 0.2

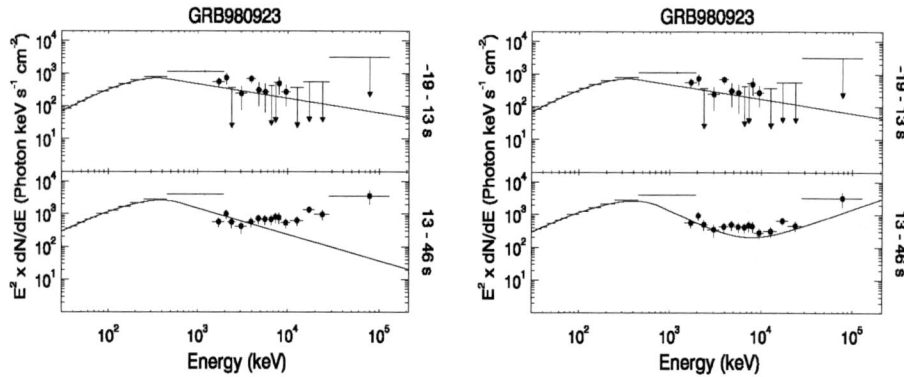

FIGURE 2. Energy fluxes for GRB980923, during the time intervals shown in Table 1. Crosses and circles correspond to BATSE-LAD and EGRET-TASC data respectively. For the purpose of the plot, but not for the spectral fit, the TASC data are binned in energy to give at least 2σ significance over the background. The upper limits correspond to 2σ deviation from the background. The solid lines represents the best fit to the joint LAD-TASC data using a Band function. The best fit parameters are shown in Table 1. The normalization factor between the data sets was fixed to 0.32 through the whole burst. A higher-energy component similar to the one observed in GRB941017 with spectral index -1 is assumed in the right window.

TABLE 2. The 66 percent confidence level upper and lower limits for the amplitude of a higher-energy component are given in the third column. The corresponding burst and time interval are given in first two columns. The probabilities of improving the fit by chance when adding the power law function are given in last column.

Burst	Time period (s)	$A_{PL}(\frac{\times 10^{-7} \text{ph}}{\text{s cm}^{-2} \text{keV}})$	Probability
GRB910503	3 - 7	4.6 - 7.3	3.5×10^{-5}
GRB920622	3 - 7	2.8 - 5.8	4.5×10^{-3}
GRB940703	32 - 64	1.9 - 4.1	2.7×10^{-2}
GRB950425	42 - 75	1.6 - 2.7	1.1×10^{-4}
GRB970223	7 - 23	0.5 - 2.2	1.5×10^{-1}
GRB970411	39 - 72	1.0 - 4.8	1.1×10^{-1}
GRB981021	1 - 3	1.1 - 9.9	2.6×10^{-4}

2. R.D. Preece, G.N. Pendleton, M.S. Briggs, R.S.Mallozzi, W.S. Paciesas, D.L. Band, J.L. Matteson and C.A. Meegan, *ApJ*,**496**, 849 (1998).
3. R.D. Preece, M.S. Briggs, R.S. Mallozzi, G.N. Pendleton, W.S. Paciesas and D.L. Band, *ApJ*,**126**, 19 (2000).
4. M.M. González, Y. Kaneko, R. Preece and B.L. Dingus, GAMMA-RAY BURST AND AFTERGLOW ASTRONOMY 2001 edited by G.R. Ricker and R.K. Vanderspek, AIP Conference Proceedings 662, American Institute of Physics, New York, 2003, pp.267.
5. M.M. González, B.L. Dingus, Y. Kaneko, R.D. Preece and M.S.Briggs, Gamma-Ray Bursts, edited by E.E. Fenimore and M. Galassi, AIP Conference Proceedings 727, American Institute of Physics, New York, 2004, pp.236.
6. M.M. González, Ph.D. thesis, University of Wisconsin-Madison, 2005.
7. D. Band, J. Matteson, L. Ford, B. Schaefer, D. Palmer, B. Teegarden, T. Cline, M. Briggs, W. Paciesas, G. Pendleton, G. Fishman, C. Kouveliotou, C. Meegan, R. Wilson and P. Lestrade, *ApJ*, **413**, 281 (1993).
8. González M.M., Dingus B.L., Kaneko Y., Preece R.D. Dermer C.D. and Briggs M.S., Gamma-Ray Bursts, edited by E.E. Fenimore and M. Galassi, AIP Conference Proceedings 727, American Institute of Physics, New York, 2004, pp.203.
9. M.M. González, B.L. Dingus, Y. Kaneko, R.D. Preece, C.D. Dermer and M.S.Briggs, *Nature*, **424**, 749 (2003).

The Search for Neutrinos from Gamma Ray Bursts with AMANDA

Kyler Kuehn, for the IceCube Collaboration and the IPN Collaboration

Department of Physics and Astronomy, University of California, Irvine CA 92697-4575

Abstract.
We report on the combined analysis of over 400 GRB time periods that occurred during seven years of AMANDA observations. AMANDA has seen no neutrinos correlated with these bursts, thus we report a neutrino flux limit that is the most stringent observational limit to date. In light of the new observational opportunities afforded by Swift, we also discuss the future potential for GRB neutrino detection with AMANDA's successor, IceCube. Finally, we briefly discuss the expansion of AMANDA's transient point-source search to other phenomena, such as jet-driven supernovae and gamma-ray dark bursts.

Keywords: Neutrino astronomy, Gamma-Ray Bursts, AMANDA, IceCube
PACS: 95.85.Ry, 98.70.Rz, 98.70.Sa

INTRODUCTION

The Antarctic Muon and Neutrino Detector Array (AMANDA) consists of 677 photomultiplier tubes housed in optical modules placed beneath the surface of the ice at the South Pole. This array of detectors is capable of observing the Čerenkov radiation from neutrino-induced muon events above energies of approximately 50 GeV [1]. AMANDA has been searching the heavens since 1997 for high-energy neutrinos from a variety of astrophysical sources, including Gamma-Ray Bursts (GRBs). This GRB neutrino search relies on spatial and temporal correlations with numerous ground- and space-based observatories, such as BATSE, HETE, and the other instruments of the Interplanetary Network [2].

A variety of different models exist for neutrino emission from GRBs. Neutrinos may be emitted in coincidence with or as a precursor (up to 100 seconds prior) to photon emission. Depending on the characteristics of the central engine and the circumburst environment, the predicted neutrino flux may vary significantly [3, 4, 5]. The search for neutrino emission will help to test models of hadronic acceleration in the fireball or other GRB scenarios, and the search for precursor neutrinos may constrain models of GRB progenitors [6, 7].

ANALYSIS PROCEDURE

AMANDA's observation procedure initially retains blindness during each GRB, allowing for optimization of data selection criteria while minimizing the possibility of false positive detections. For each GRB, a two-hour period is examined around the burst, excluding the 10-minute window immediately surrounding the burst. The two-hour "off-time" period is examined to ensure detector stability and to determine the background rate for each burst. Selection criteria are then determined by comparing Monte Carlo simulations of muon neutrino signals with the observed background events.

While spatial and temporal information for each burst serve as the primary criteria for data selection, secondary criteria include the number of hit optical modules (i.e. those participating together in a single event), the angular resolution of the reconstructed event track, the uniformity of the hit optical modules along the reconstructed track, and the likelihood of each reconstructed track. These criteria are selected to minimize the Model Rejection Factor (MRF) [8], which is defined as the 90% event upper limit divided by the expected number of signal events derived from simulations.

RESULTS

New results from the combination of seven years of AMANDA observations are shown in Table 1. From the observation of zero events for all of these bursts, we are able to derive neutrino flux limits for representative theoretical models[1]. The observed limits are approaching the predictions of several important models, including the supranova model and the Waxman-Bahcall broken power-law model (Figure 1A).

In addition to the specific models detailed here, the AMANDA results can be applied to any desired theoretical spectrum by means of the Green's Function Fluence method (as detailed in [11]). Based on the subset of 139 bursts searched by AMANDA from 2000 to 2003, we are able to set a (spectrum-independent) fluence limit that is a significant improvement over previously-reported results (Figure 1B). To calculate the flux limit for any desired spectrum, one needs only to fold that spectrum into the fluence limit shown here.

CONCLUSIONS AND OUTLOOK

We report here on the analysis of over 400 GRBs occurring over seven years of AMANDA observations. The detection of zero events during the relevant time periods results in a flux limit that is close to several theoretical predictions for neutrino emission from GRBs. While these results represent the bulk of the GRB time periods contained within the AMANDA data, there are approximately 100 additional GRBs from 2001-2003 that have not yet been examined. These bursts are taken from the archives of the InterPlanetary Network; most have localization errors much larger than the angular resolution of AMANDA, so a different procedure is currently being developed to analyze the time periods surrounding these bursts.

As AMANDA's successor IceCube [12] continues its observations, there are two primary benefits to the search for neutrinos from GRBs in the Swift era. First, localizations are much more accurate than those derived from BATSE or other IPN satellite

---

[1] We similarly obtained a null result from a search of twenty-four non-triggered BATSE GRBs [9, 10]; however, these bursts are not incorporated into the models and thus are excluded from our flux limits.

TABLE 1. Preliminary Results of GRB Analysis 1997-2003

	N_{Bursts}	$N_{BG,Exp}$	N_{OBS}	Event Upper Limit	MRF * (Sensitivity)	MRF (Observed)
1997-1999[†]	268	0.46	0	1.98	20	14
2000	88	1.02	0	1.61	20	10
2001	15	0.06	0	2.38	66	64
Precursor	15	0.05	0	2.39		
2002	17	0.08	0	2.36	54	54
Precursor	17	0.06	0	2.38		
2003	19	0.10	0	2.34	54	52
Precursor	18	0.06	0	2.38		
01-03	51	0.24	0	2.19	20	16
Precursor	50	0.16	0	2.28		
97-03	407	1.71	0	1.27	7	3

* for Waxman-Bahcall, modified for oscillations
[†] using AMANDA B-10

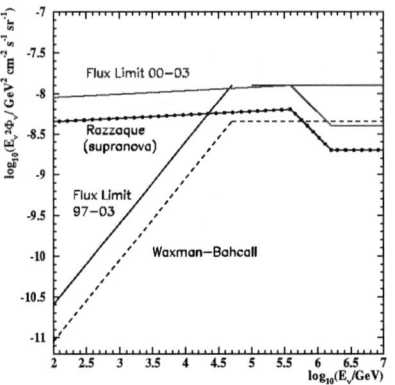 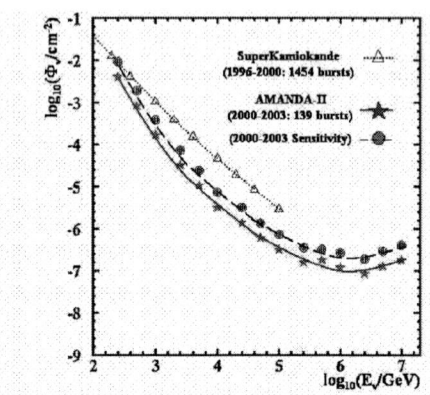

FIGURE 1. (A) Flux limits for AMANDA-II observations (solid lines) and theoretical predictions (dashed lines) for representative neutrino spectra. (B) Green's Function Fluence Limit for AMANDA-II observations of BATSE and IPN triggered bursts. The sensitivity is determined based upon the expected background and signal events prior to observations.

observations. Though there are approximately 400 bursts with precise localizations in seven years of AMANDA data, precise localizations for the additional (poorly-localized) bursts would have improved AMANDA's flux limits considerably. As Swift continues to observe bursts and provide precise localizations, the bursts will be incorporated with much greater efficiency into AMANDA/IceCube observations.

Second, distance determinations for GRB hosts provide an important new tool for neutrino flux predictions. Here we have discussed results based on models with averaged burst properties. However, burst-to-burst variation in total flux, redshift, peak energy, and other variables can significantly impact the expected neutrino flux (the expected flux for GRB030329, for example, is nearly a factor of 10 higher than the average burst [13]). While the averaged burst properties served as a useful tool when redshift estimates were missing or inadequate, IceCube's observations will lead to much more precise results when the predicted flux for each burst can be determined individually.

Finally, the transient point source search described here will also be broadened in scope as IceCube continues the legacy of AMANDA. While all of the bursts included in these results were observed by various photon observatories, gamma-ray dark or failed GRBs are hypothesized to occur at as much as $100\times$ the rate of standard bursts [7]. Even if the photons from these bursts are not observed, the neutrino signature may still be detectable by this method if spatial and temporal localization information can be derived from other observations (such as afterglows or correlated supernovae). Additionally, some supernovae are hypothesized to emit relativistic jets of material in a manner similar to GRBs. These GRB-like supernovae are also a promising candidate for neutrino emission detectable by IceCube or other km-scale detectors [14].

ACKNOWLEDGMENTS

We acknowledge the support of the following agencies: National Science Foundation Office of Polar Programs, National Science Foundation Physics Division, University of Wisconsin Alumni Research Foundation, Department of Energy, and National Energy Research Scientific Computing Center (supported by the Office of Energy Research and Department of Energy), the NSF-supported TeraGrid systens at the San Diego Supercomputing Center (SDSC), and the National Center for Supercomputing Applications (NCSA); Swedish Research Council, Swedish Polar Research Secretariat, and Knut and Alice Wallenberg Foundation, Sweden; German Ministry for Education and Research, Deutsche Forschungsgemeinschaft (DFG), Germany; Fund for Scientific Research (FNRS-FWO), Flanders Institute to encourage scientific and technological research in industry (IWT), and Belgian Federal Office for Scientific, Technical, and Cultural Affairs (OSTC); The Netherlands Organisation for Scientific Research (NWO).

REFERENCES

1. Ahrens, J., et al., ApJ **583** (2003) 1040
2. Hurley, K., Astron Telegram #19 (1998)
3. Waxman, E., and J. Bahcall, PRL **78** (1997) 2292
4. Alvarez-Muñiz, J., et al., PRD **62** (2000) 3015
5. Dermer, C., and A. Atoyan, PRL **91** (2003) 1102
6. Razzaque, S., et al., PRD **68** (2003) 3001
7. Mészáros, P., and E. Waxman, PRL **87** (2001) 1102
8. Hill, G.C., and K. Rawlins, Astropart Phys **19** (2003) 393
9. Stern, B.E., ct al., ApJ **563** (2001) 80
10. Schmidt, M., ApJ **523** (1999) L117
11. Fukuda, S., et al., ApJ **578** (2002) 317
12. Karle, A., et al., Nuc Phys B Proc Supp **118** (2003) 388
13. Stamatikos, M., these proceedings
14. Ando, S. et al., PRL **95** (2005) 171101

A Search for Short Duration Very High Energy Emission from Gamma-Ray Bursts

David Noyes for the Milagro Collaboration

University of Maryland, College Park

Abstract. Milagro is a water-Cherenkov detector capable of observing air showers produced by gamma rays with energies of approximately 100 GeV and higher. The wide field of view (~2 sr) and high duty cycle (>90) of Milagro make it ideal for searching for transient emission from GRBs.

Measurements have been made of GRB spectra up to a few GeV with no sign of a cutoff, however much is still unknown about the nature and existence of this VHE component. While many models predict VHE emission from GRBs, gamma/gamma absorption from infrared background photons or the optically thick region of the burst source complicate observations.

In three years of searching the Milagro data set for GRBs with durations ranging from $250\mu s$ to 40s, no significant evidence was found for VHE emission from GRBs. Models for GRB redshift and isotropic energy distributions are used to significantly constrain the VHE spectrum of GRBs.

INTRODUCTION

The search for Very High Energy (VHE) emission from GRBs is motivated by observations and theoretical models. While processes including inverse-Compton emission and proton-synchrotron emission could give rise to a VHE component to the GRB spectrum, they have mainly been observed in the keV to few MeV energy range. A number of observations suggest emission at higher energies. EGRET observations indicate that the GRB spectrum extends at least to 1 GeV with no sign of a cutoff [1]. GRB941017, was found to have a distinct high energy component which evolved independently of the low energy component [2]. In addition to satellite observations, Milagrito, the proto-type for Milagro, observed evidence for VHE emission from GRB970417a [3]. This high energy emission can be difficult to explain with standard models of the GRB emission, indicating the importance of more observations at these energies.

One obstacle to detecting VHE emission from GRBs is pair production with low energy photons in the GRB environment and in the Extra-Galactic Background Light (EBL). This requires large bulk Lorentz factors for GRBs and effectively limits observations of photons with energies above a few hundred GeV to $z < 0.5$.

Milagro [4] is a water-Cherenkov detector sensitive to air showers with primary energies above ~ 100 GeV. The detector consists of a 6-million gallon artificial pond instrumented with 723 photomultiplier tubes (PMTs) arranged in two layers. The pond is surrounded by an array of 175 individual Cherenkov detectors covering $10,000 m^2$. Milagro's large field of view (FOV ~ 2 sr) and high duty cycle (>90%), make it ideal for searching for transient emission produced by GRBs. While Milagro does perform coincident and follow-up observations of satellite-localized GRBs (see [5]). these make up a small fraction of the total number of bursts that occur in Milagro's FOV. For a rate

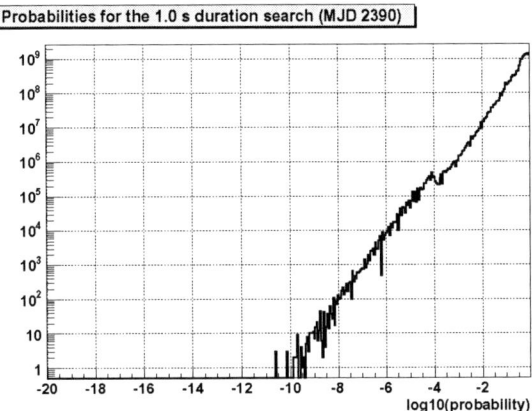

FIGURE 1. Probability distribution for the 1 s duration search for an entire day of data. After accounting for trials, none of the entries in this histogram are statistically significant.

of 700 GRBs/year in the universe, $700 \times (2sr/4\pi sr) \times 0.90 = 100$ GRB/year, occur in Milagro's FOV compared to 20 satellite-localized GRBs in the past year. This is why it is important to search for emission from GRBs even when there is no localization from another instrument.

THE SHORT DURATION BURST SEARCH

Lacking a localization from another instrument, a blind search of the Milagro data is performed. In contrast to a search for a known source, the start time, duration, and coordinates of a potential GRB are not known *a priori*, making it necessary to search over the entire sky, at all times, over multiple durations. This search is done on 27 durations ranging from $250\mu s$ to 40 s. Longer durations searches are described in [6][7].

The bulk of the events detected by Milagro are cosmic-ray induced air showers. This background is measured using a long period of time (relative to the search duration) and then compared to the number of events in a signal map. The signal map contains the number of events observed in a time window specified by the start time and duration of the search. Then the probability that the number of observed signal events was due to a random fluctuation of the background is calculated for each point on the sky. Fig. 1 shows the probability distribution for the 1 s duration search for an entire day of data. A GRB candidate would appear on this figure as a low probability value, away from the main distribution. What constitutes a low probability depends on the number of trials incurred by the search method. After properly accounting for the number of trials, none of the entries in this histogram are statistically significant (at the 5σ level).

The Milagro data was searched for VHE emission from GRBs from MJD 52353 to MJD 53372 (March 19, 2002 to January 1, 2005), 1020 days in total. After accounting for detector dead time and data quality cuts, the number of live days searched was 836.6 (2.3 years). In this period, no evidence for VHE emission from a GRB was observed.

CONSTRAINING THE VHE SPECTRUM OF GRBS

Since no evidence for VHE emission from GRBs was observed, this emission may be constrained. This requires a number of assumptions to be made in order to create simulations of GRB populations. In these simulations, it is assumed that the VHE spectrum is an E^{-2} power law from 100 GeV to 10 TeV, and that GRB durations follow the BATSE T90 distribution. Models are used to account for the attenuation due to the EBL, and burst redshift (z) and isotropic energy (E_{iso}) distributions.

To simulate a burst population, a z, duration, E_{iso}, and zenith angle are randomly drawn from their respective distributions. Given these parameters the number of photons expected in Milagro from each GRB is calculated. The number of photons expected (N_γ) is compared to the measured Milagro background (N_{bkg}) at the same zenith angle and duration. If the Poisson probability of observing $N_{bkg} + N_\gamma$ events when expecting N_{bkg} is greater than 5σ post-trials it is considered that the burst would be detectable by Milagro.

A large number of bursts are simulated in order to determine the fraction detectable by Milagro. Since the models for the E_{iso} distribution apply to the keV/MeV portion of the spectrum, it is assumed that the energy radiated in the form of VHE photons is a constant times that radiated in the form of keV/MeV photons. This energy ratio ($E_{iso}(VHE)/E_{iso}(keV)$) is varied in order to determine the fraction of the simulated burst population that would be detectable as a function of the energy ratio. Since no evidence for VHE emission was observed by Milagro in the 2.3 years that were searched, the 90% confidence level (CL) upper limit on the number of GRBs/year in this data set is 2.3/2.3 =1 GRB/year. The value of the energy ratio that gives this many events expected per year is the 90% CL upper limit on the ratio. Given a model of the redshift and E_{iso} distribution of GRBs, this limit depends on how many bursts in the model are bright enough and close enough for Milagro to detect.

A simple model is considered here that uses a model of the binary merger redshift distribution for short duration bursts ($< 2s$), and a theoretical Star Formation Rate (SFR) distribution for long duration bursts ($> 2s$). A double Gaussian (parameterized by two means and two widths) is used for the E_{iso} distribution. Given the redshift distribution above, the parameters of the E_{iso} distribution are determined by comparison to the measured BATSE fluence distribution (in the 20 keV to 300 keV band). The parameters of the E_{iso} distribution which most closely reproduce the measured BATSE fluence distribution are selected as those to be used in the model.

Using this model, a large number of GRBs are simulated in the manner described above. The results of this simulation are shown in Fig. 2. In this figure, the top three plots show the distributions used for this model, the bottom left plot shows the redshift distribution of the bursts that would have been detectable by Milagro assuming an energy ratio of 1, and the bottom right plot shows the number of bursts per year detectable by Milagro as a function of the energy ratio (assuming a total rate of GRBs in the universe of 700/year). If the total energy radiated in the form of 100 GeV to 10 TeV photons was equal to the total energy radiated in the form of 20 keV to 300 keV photons, Milagro would expect to have observed about 4 GRBs per year. The 90% CL upper limit on the ratio is 0.04 for this model. In other words, the Milagro result implies that the VHE emission is less than 4% of the keV emission at the 90% CL. This assumes that the energy ratio does not vary from burst to burst, and that the bulk Lorentz factor of GRBs

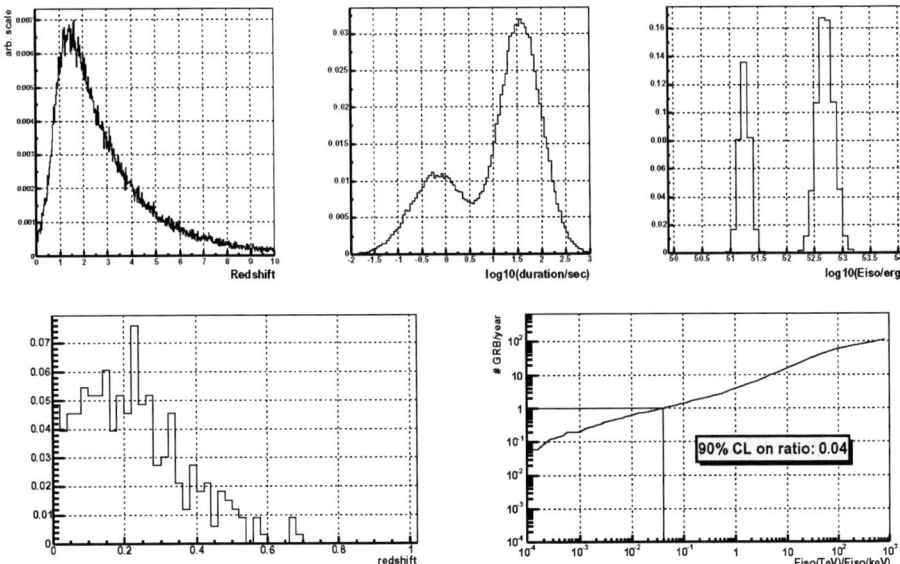

FIGURE 2. Results of the simulation simple model considered here. The top three plots show the model distributions used (z, duration, E_{iso}). The bottom left shows the redshift distribution of GRBs detectable by Milagro for an energy ratio of 1. The bottom right plot shows the number of GRBs/year detectable by Milagro as a function of the energy ratio. The 90% CL upper limit on the energy ratio is 0.04.

is sufficiently large that the VHE emission escapes the source, and only holds for the particular model considered here.

ACKNOWLEDGMENTS

We thank Scott Delay and Michael Schneider for their dedicated efforts on the Milagro experiment. We also gratefully acknowledge the financial support of the National Science Foundation (under grants PHY-0075326,-0096256,-0097315,-206656,-0245143,-0245234,-0302000, and ATM-0002744, the Department of Energy (Office of High Energy Physics), Los Alamos National Laboratory, the University of California, and the Institute for Geophysics and Planetary Physics at Los Alamos National Laboratory.

REFERENCES

1. Dingus, B. L. in High-Energy Gamma Ray Astronomy (eds Aharonian, F. A. & Volk, H. J.) 383 391
2. Gonzales, M. M., et al., Nature 424, 847 (2003).
3. Atkins, R., et al., ApJ., 533, L119 (2000).
4. Atkins, R. et al, Astrophys. J., 595, 803 (2003).
5. Saz Parkinson, Pablo Proceedings of Gamma Ray Bursts in the Swift Era (2006).
6. Atkins, R., et al., ApJ., 604, L25 (2004).
7. Hays, E. Ph.D Thesis, University of Maryland, College Park (2004).

Milagro Search for Very High Energy Emission from Gamma-Ray Bursts in the *Swift Era*

P. M. Saz Parkinson for the Milagro Collaboration[1]

Santa Cruz Institute for Particle Physics, University of California, 1156 High Street, Santa Cruz, CA 95064

Abstract.
The recently launched *Swift* satellite is providing an unprecedented number of rapid and accurate Gamma-Ray Burst (GRB) localizations, facilitating a flurry of follow-up observations by a large number of telescopes at many different wavelengths. The Very High Energy (VHE, >100 GeV) regime has so far been relatively unexplored. Milagro is a wide field of view (2 sr) and high duty cycle (> 90%) ground-based gamma-ray telescope which employs a water Cherenkov detector to monitor the northern sky almost continuously in the 100 GeV to 100 TeV energy range. We have searched the Milagro data for emission from the most recent GRBs identified within our field of view. These include three *Swift* bursts which also display late-time X-ray flares. We have searched for emission coincident with these flares. No significant detection was made. A 99% confidence upper limit is provided for each of the GRBs, as well as the flares.

INTRODUCTION

Some of the most important contributions to our understanding of gamma-ray bursts have come from observations of afterglows over a wide spectral range [1]. Very little, however, is known about the broadband spectra of GRBs in the prompt phase, due to its short duration. Many GRB production models predict a fluence at TeV comparable to that at MeV scales [2, 3, 4]. Almost all GRBs are detected in the energy range between 20 keV and 1 MeV, though several have been observed above 100 MeV by EGRET, indicating that the spectrum of GRBs extends at least out to 1 GeV [5]. A second component was also found in one burst which extended up to at least 200 MeV and had a much slower temporal decay than the main burst [6]. It is unclear how high in energy this component extends to and whether it is similar to the inverse Compton peak seen in many TeV sources. At very high energies there has been no conclusive emission detected for any single GRB, though a search for counterparts to 54 BATSE bursts with Milagrito, a prototype of Milagro, found evidence for emission from one burst, with a significance slightly greater than 3σ [7]. At these high energies, gamma rays suffer from attenuation due to the extra-galactic background light (EBL), which is redshift-

---

[1] A. Abdo, B. T. Allen, R. Atkins, D. Berley, E. Blaufuss, S. Casanova, D. G. Coyne, B. L. Dingus, R. W. Ellsworth, L. Fleysher, R. Fleysher, M. M. Gonzalez, J. A. Goodman, E. Hays, C. M. Hoffman, L. A. Kelley, C. P. Lansdell, J. T. Linnemann, J. E. McEnery, A. I. Mincer, M. F. Morales, P. Nemethy, D. Noyes, J. M. Ryan, F. W. Samuelson, P. M. Saz Parkinson, A. Shoup, G. Sinnis, A. J. Smith, G. W. Sullivan, V. Vasileiou, G. P. Walker, D. A. Williams, X. W. Xu and G. B. Yodh

dependent [8, 9, 10], making GRBs above z>0.5 very difficult to observe. Here, we describe our search for VHE emission from 20 GRBs which have occurred within the field of view of Milagro[2] between December 2004 and December 2005. Recently, *Swift* has also observed bright X-ray flares from GRB afterglows, sometimes of comparable energy to the burst itself [11, 12]. Although the most significant of these flares (from GRB 050502B) was unfortunately outside the field of view of Milagro, six other flares from three different GRBs were observable [13]. We present the upper limits derived from our observations.

DATA ANALYSIS AND RESULTS

A search for an excess of events above those due to the background was made for each of the 20 bursts in our sample, listed in Table 1, as well as for the flares listed in Table 2. The number of events falling within a 1.6 degree bin was summed for the relevant duration (column 2 of Table 1 or column 4 of Table 2). An estimate of the number of background events was made by characterizing the angular distribution of the background using two hours of data surrounding the burst, using a technique known as "direct integration" [14]. For those bursts whose redshift is known, we compute the effect of the absorption, according to the model of [10] and print the upper limits in bold. No significant emission was detected. Our upper limits are given in column seven of Table 1 (for the GRBs) and Table 2 (for the flares).

Some of the more interesting bursts in Table 1 include GRB 050509b, the second *short/hard* burst detected by *Swift*, with a reported duration of 30 ms and a relatively low fluence of 2.3×10^{-8} erg cm^{-2} in the 15–350 keV range [15]. Although Milagro detected no emission from this burst [16], the very favorable zenith angle (10°) and possible low redshift of 0.226 provide the opportunity to set interesting upper limits for TeV emission from this burst. Another burst that deserves mention is GRB 051103, a short burst detected by the IPN, possibly originating from the nearby (< 4 Mpc) galaxy M81 [17]. Although this burst was not at a favorable zenith angle for Milagro (θ >45°), we chose to analyse the data, due to its potential interest. If this burst were associated with M81, as has been suggested, the almost complete absence of absorption would make the Milagro upper limit on the fluence about 20% of the measured fluence in the X-ray band [18]. Finally, GRB 051109 occurred at a zenith angle of less than 10° for Milagro, making it the burst with the best location of our sample. Unfortunately, this burst had a measured redshift of 2.346, making the Milagro limit, once absorption is taken into account, several orders of magnitude larger than the measured fluence in the X-ray band [19].

In Table 2 we list 6 flares detected by *Swift* from 3 GRBs which were in Milagro's field of view [12]. Milagro detected no significant emission from any of these flares. The brightest of these flares, from GRB 050607, had a fluence in the X-ray range of roughly 1.5×10^{-7} erg cm^{-2} [20], placing the Milagro limit about an order of magnitude higher.

---

[2] We have included GRB 051103 in our sample, despite being at a relatively large zenith angle, given its potential interest and possible proximity.

TABLE 1. List of GRB in the field of view of Milagro in the Swift Era (December 2004 – December 2005), with preliminary upper limits.

GRB	T90/Dur.*	$\theta^\dagger$	z^{**}	Instrument[‡]	$\sigma^\S$	UL(fluence)[¶]
041219a	520	26.9	...	INTEGRAL	+1.7	5.8e-6
050124	4	23.0	...	Swift	-0.8	3.0e-7
050319	15	45.1	3.24	Swift	+0.6	...
050402	8	40.4	...	Swift	+0.6	2.1e-6
050412	26	37.2	...	Swift	-0.6	1.7e-6
050502	20	42.7	3.793	INTEGRAL	+0.6	...
050504	80	27.6	...	INTEGRAL	-0.8	1.3e-6
050505	60	28.9	4.3	Swift	+1.2	...
050509b	0.128	10.0	0.226?	Swift	-0.9	**1.1e-6**
050522	15	22.9	...	INTEGRAL	-0.6	5.1e-7
050607	26.5	29.3	...	Swift	-0.9	8.9e-7
050712	35	38.8	...	Swift	-0.1	2.5e-6
050713b	30	44.2	...	Swift	-0.3	4.0e-6
050715	52	36.9	...	Swift	-1.5	1.7e-6
050716	69	30.3	...	Swift	-0.5	1.6e-6
050820	20	21.9	2.612	Swift	+0.2	...
051103	0.17	49.9[‖]	0.001?	IPN	-0.2	**4.2e-6**
051109	36	9.7	2.346	Swift	-1.1	**4.3e-3**
051111	20	43.7	1.55	Swift	+0.7	**3.8e-2**
051211b	80	33.3	...	INTEGRAL	+0.4	2.6e-6

* Duration of burst.
[†] Zenith angle (degrees).
[**] redshift (when known).
[‡] Instrument reporting the first detection.
[§] Significance of the signal.
[¶] 99% upper limit on the fluence (0.2–20 TeV), in ergs cm^{-2}. The numbers in bold take into account absorption by the EBL (using the Primack 05 model). Those with three dots are at redshifts so high that all the emission is expected to be absorbed.
[‖] This burst was analyzed despite its large zenith angle due to its potential interest as a nearby short burst.

TABLE 2. Bright X-ray Flares in Swift GRB Afterglows

GRB	Burst Time*	Flare Time[†]	Dur.**	$\theta^\ddagger$	$\sigma^\S$	UL(fluence)[¶]
050607	33,083	110	120	29.1	-1.3	1.5e-6
050607	33,083	260	340	28.9	-1.3	2.4e-6
050712	50,427.5	170	265	38.5	+1.8	9.7e-6
050712	50,427.5	435	255	38.1	+2.3	1.0e-5
050716	45,364	135	65	30.7	-1.1	1.3e-6
050716	45,364	330	120	31.4	-0.4	2.3e-6

* BAT trigger time (UTC second of day).
[†] Time of onset of flare (seconds after BAT trigger time).
[**] Approximate duration of flare (s).
[‡] Zenith angle (degrees).
[§] Significance of the signal.
[¶] 99% upper limit on the fluence (0.2–20 TeV), in ergs cm^{-2}. No absorption is taken into account as the redshift of these bursts is unknown.

CONCLUSION

A search for VHE emission from GRBs was performed with the Milagro observatory in the range of 100 GeV to 100 TeV. A total of 20 satellite-triggered GRBs were well localized and fell within Milagro's field of view in the year since the launch of Swift. In addition, six bright X-ray flares from GRB afterglows were searched for VHE emission. No significant emission was detected from either the bursts or the flares. 99% confidence upper limits on the fluence are presented.

ACKNOWLEDGMENTS

Many people helped bring Milagro to fruition. In particular, we acknowledge the efforts of Scott DeLay, Neil Thompson and Michael Schneider. This work has been supported by the National Science Foundation (under grants PHY-0075326, -0096256, -0097315, -0206656, -0245143, -0245234, -0302000, and ATM-0002744) the US Department of Energy (Office of High-Energy Physics and Office of Nuclear Physics), Los Alamos National Laboratory, the University of California, and the Institute of Geophysics and Planetary Physics.

REFERENCES

1. van Paradijs, J., Kouveliotou, C. & Wijers, R. A. M. J. 2000, *Annual Review of Astronomy and Astrophysics* 38, 379
2. Dermer, C. D., Chiang, J., & Mitman, K. E. 2000, *ApJ* 537, 785
3. Pilla, R. P. & Loeb, A. 1998, *ApJL* 494, L167
4. Zhang, B. & Mészáros, P. 2001, *ApJ* 559, 110
5. Dingus, B. L. 2001, in Aharonian, F. A. and Volk, H. J. (eds), *High Energy Gamma Ray Astronomy*, AIP Conference Proceedings, 558, 383
6. Gonzalez, M. M. et al. 2003, *Nature* 424, 749
7. Atkins, R. et al. 2000, *ApJL* 533, L119
8. Stecker, F. & de Jager, O. C. 1998, *Astronomy and Astrophysics* 334, L85
9. Primack, J. R., Bullock, J. S., Somerville, R. S. & Macminn, D. 1999, *Astroparticle Physics*, 11, 93
10. Primack, J. R. et al. 2005, in Aharonian, F. A., Volk, H. J., and Horns, D. (eds) *Gamma 2004 Heidelberg*, AIP Conference Proceedings, 745, 23-33
11. Burrows, D. N. et al. 2005, *Science*, 309, 1833
12. Falcone, A. et al. 2005, *ApJ*, submitted
13. Falcone, A. *personal communication (2005)*
14. Atkins, R. et al. 2003, *ApJ* 595, 803
15. Barthelmy, S. et al. 2005, GCN Circular No. 3385
16. Saz Parkinson, P. M. 2005, GCN Circular No. 3411
17. Golenetskii, S. et al. 2005, GCN Circular No. 4197
18. Saz Parkinson, P. M. 2005, GCN Circular No. 4249
19. Saz Parkinson, P. M. 2005, GCN Circular No. 4265
20. King, A. et al. 2005, *ApJL*, L113-L115

FUTURE MISSIONS AND INSTRUMENTATION

Post-*Swift* Gamma-ray Burst Science and Capabilities Needed to *EXIST*

Jonathan E. Grindlay and the *EXIST* Team

Harvard-Smithsonian Center for Astrophysics, 60 Garden St., Cambridge, MA 02138

Abstract. The exhilerating results from *Swift* in its first year of operations have opened a new era of exploration of the high energy universe. The surge to higher redshifts of the Gamma-ray bursts now imaged with increased sensitivity establishes them as viable cosmic probes of the early universe. Wide-field coded aperture imaging with solid-state pixel detectors (Cd-Zn-Te) has been also established as the optimum approach for GRB discovery and location as well as to conduct sensitive full-sky hard X-ray sky surveys. I outline the current and future major science questions likely to dominate the post-*Swift* era for GRBs and several related disciplines and the mission requirements to tackle these. The *EXIST* mission, under study for NASA's Black Hole Finder Probe (BHFP) in the Beyond Einstein Program, could achieve these objectives as the Next Generation GRB Mission with 'ultimate' sensitivity and wide-field survey capability. Analysis tools for processing *Swift*/BAT slew data are under development at CfA and will both test *EXIST* scanning imaging and provide new data on GRBs and transients.

Keywords: Gamma-ray bursts, black holes, black hole surveys
PACS: 95.55Ka, 97.20Wt, 97.60Lf, 98.70Rz

INTRODUCTION

In less than a decade, Gamma-ray bursts (GRBs) have moved from being one of the outstanding cosmic mysteries for which the remarkable discovery of isotropy [1] with *BATSE* laid waste earlier views of their Galactic origin, to cosmological objects, as enabled by *BeppoSAX* prompt imaging [2] which in turn allowed followup optical identifications [3] and redshifts. With *Swift* [4], GRB studies have now entered a qualitatively new phase with wide-field γ-ray imaging for prompt GRB locations accurate ($\sim 2'$) and rapid (~ 10sec) enough to allow $\sim 10''$ source locations from X-ray afterglows typically only ~ 50-100sec later. As described in numerous conributions at this meeting, the questions now being asked of the *Swift* GRBs have raised the bar still further on the physics and astrophysics of GRBs, with discussion not just of star formation rates vs. redshift (SFR(z)), but metallicity vs. z (Z(z)) and – with the *Swift* first localization of a short GRB[5] (SGRB[1]) – even the rate of formation (and eventual merger) of double neutron star (DNS) systems vs. z.

Thus at each milestone in GRB research, major technological advances have, not surprisingly, spurred the next step. *BATSE* provided a large area, well-controlled (for

---

[1] Both SGRB and SHB (short hard bursts) have been proposed as the de facto required acronym. I suggest SGRBs to allow for the inevitable discovery of a short soft burst – it is likely we can predict time durations better than spectral signatures for NS-NS or NS-BH merger events.

threshold, sensitivity variations, and particle background-induced events) experiment yielding a large GRB sample, coarsely located, to provide a definitive isotropy result. *BeppoSAX* gave the first $\lesssim 10'$ GRB positions on timescales $\lesssim 0.3$d which allowed X-ray and then optical afterglows to be discovered which enabled identifications. And *Swift* has combined both prompt $\lesssim 3'$ positions and rapid re-pointing for the earliest-afterglow detections to achieve its breakthroughs for SGRB localization and identification (for which the ~ 10-$100\times$ fainter energetics makes rapid detection essential) as well as a significantly increased mean redshift[6] (to z ~ 2.7) enabled by its greater sensitivity to long GRBs (LGRBs).

Although it is early in the *Swift* mission to make predictions for what is needed next, I nevertheless attempt to do so here so that we (collectively) may have some hope for continuing this exciting quest to the next phase(s) of discovery. I begin by outlining some of the major questions (most dealt with in far more detail by other contributions to these Proceedings) and goals for GRB studies and what they could enable in opening up the high z and high energy universe. I then outline some of the results drawn from the ongoing concept study for the Energetic X-ray Imaging Survey Telescope, *EXIST*, which could be the ultimate-sensitivity and resolution Next Generation GRB Observatory as the Black Hole Finder Probe in NASA's Beyond Einstein Program.

OPEN QUESTIONS FOR SCIENCE OF AND ENABLED BY GRBS

In the post-*Swift* era, I suspect many of the "big" questions surrounding GRBs will still be at best only partly understood. This is likely if we take the understanding of core collapse SNe as a guide: after nearly 40y of trying, a mechanism has only just been reported[7] (but not yet confirmed) to launch a "successful" SN-II explosion and blast wave. In the following sub-sections, I describe briefly several key problems, from the (relatively) local to the most cosmologically extreme. This list is very incomplete, and represents some GRB challenges/opportunities that may have broadest impact on other disciplines. In each, I mention necessary (though perhaps not sufficient) requirements for next-generation GRB studies to make significant breakthroughs.

Short GRBs: Unravelling the mix

While the *Swift* [5] and *HETE-2* [8] detections and first localizations of SGRBs have very likely established that they are not core collapse events (like LGRBs) and that their hosts are in elliptical or cluster galaxies within limited (December, 2006) statistics of 3-4 of 5 plausible associations (see[9] for a current summary), a number of key issues remain.

First, is the longstanding possible mixing of distant SGR giant flare events and SGRBs. The plausible optical hosts for the *Swift* and *HETE-2* SGRBs are too distant for SGRs (even the most luminous SGR1806-20 event of December 27, 2004, is basically undetectable by *Swift*/BAT beyond ~ 70Mpc[10]). However SGRs could well be mixed into the *BATSE* sample of SGRBs, as suggested[11] for GRB970110. We note, however, this 13.8sec periodicity could also be due to a low frequency QPO from CygX-1 (just outside the *BATSE* position) since this would be consistent with QPOs seen at the $v_b \sim 0.1$Hz frequency characteristic[12] of CygX-1 in its low/hard state which described the source in early January 1997. Detection of the SGR component to SGRBs (which must be non-zero) will require enough sensitivity to detect the "smoking gun" magnetar pulsations, post-flash, which would not have been detectable with *BATSE* given the flux ratio for the SGR1806 event but which would be out to $\sim$10-40Mpc with *Swift* [10]. Extragalactic magnetar surveys would be enhanced by greater sensitivity at $\lesssim 10$ keV.

Second, and perhaps most interesting, are the questions now raised about SGRBs as, predominantly, DNS merger events in old stellar systems. We have suggested[13] that $\sim$10-30% of the SGRB population could arise from DNS systems formed by dynamical exchanges in globular cluster cores of NSs into compact binaries composed of NS-main sequence or, most often, millisecond pulsars with HeWD companions as now seen in great numbers in globular clusters. Two key observations from *Swift* and *HETE-2* , as well as the long-known DNS in a Galactic globular (M15-C) which *will* produce a SGRB, drove us to this model: 1) that the hosts are ellipticals or in galaxy clusters (with a high preponderance of ellipticals), and 2) that there is evidence[9] the DNS mergers (if indeed this is the underlying SGRB source) occurred in the (near-) haloes of their parent galaxies. What seems to not be appreciated by many in the GRB community is that offset SGRBs are *not* expected from dynamical kicks imparted to DNS systems by their NS formation events. Why? Because the 7 DNS systems known in the Galactic disk all appear to have velocities $\lesssim 50$ km s-1, which is consistent with recent understanding of binary evolution of DNS systems from massive binaries in which the second NS forms from a He star[14]. Although SGRBs must surely be produced by DNS mergers of these disk-bound systems, the globular cluster origin of some SGRBs will be strongly favored if *Swift* (and future missions) continue to find a signficant fraction offset from their host galaxies. Given the early evidence that SGRB afterglows are faint (no absorption line redshifts are yet detected from the SGRB afterglow events; rather all have come from the (nearby) putative host galaxy[9]), the requirement is for rapid and very accurate location of the SGRBs themselves. With higher sensitivity, energy bandwidth and resolution, the globular cluster component of the SGRB population can be isolated and used to trace the formation epoch of globulars as well as their dynamical evolution.

Third, and equally interesting, is the admixture of NS-NS vs. NS-BH merger events. Can the delayed "flares" in some events (e.g. the longer/softer but luminous afterpulse[8] in GRB050709 in a blue dwarf galaxy) be delayed inspiral from BH disruption of a NS? Or is it simply an afterglow from a local ISM, which for other SGRBs in or near ellipticals appear to be faint. Are the NS-BH binaries, which must exist, given kicks (despite their larger total mass than DNS systems) and thus sometimes expelled from galactic disks? Isolating the GRB signatures of NS-BH events would demonstrate that such systems exist, with their unequaled opportunity for testing GR with a precise clock (a millisecond pulsar) in close orbit around a stellar mass BH. A MSP-BH system remains a holy grail, and may be found first by MSP surveys in the radio, but the SGRB population will provide important constraints. Obviously a much larger sample of events and positions relative to their host galaxies is needed, together with $\sim$3-30 keV prompt spectra to look for softer afterpulses.

Highest redshift GRBs from Pop III stellar mass BHs?

The *Swift* discovery[15] of the current record redshift GRB, at z = 6.29, is a harbinger of things to come. As discussed many times, the rapid rise of the highest z value over the past few years makes it plausible that GRBs will soon be the highest z probe of stellar objects (including galaxies themselves and their central AGN) directly observed in the universe. The long-standing goal remains to detect Pop III stars, which are likely to be massive and may produce LGRBs without quenching their jets[16]. As pointed out by Bromm and Loeb, a "successful" Pop III GRB is even more likely if the progenitor is a binary that undergoes mass loss prior to collapse. Although the predicted rate of such GRBs is very uncertain, and dependent on timescales for reionization, the Pop III GRB rate predicted for *Swift* might be ~ 0.8 y^{-1}. This may be measurable with *Swift*, but clearly a larger area and field of view (FoV) future mission could increase this dramatically. The prospect of directly detecting the very first stars remains perhaps the most exciting long-term GRB objective.

Regardless of the uncertainties of Pop III GRBs, the "guaranteed" detection of Pop II GRBs (i.e. GRBs from collapsars which themselves were Pop II stars) is a prime objective. The prospect of GRBs as the "backlight" for surveying the IGM (and structures) at $z \gtrsim 7$ (as practically now reached) is potentially the most dramatic use of GRBs. To make this the powerful tool that it could be requires, once again, the sample size be increased dramatically from the estimate[16] that $\sim 10\%$ of *Swift* GRBs (or ~ 10 y^{-1}) will originate from $z \gtrsim 5$. Given that every such GRB is crucial, and that only $\sim 40\%$ of all *Swift* GRBs thus far have redshift measures, it is important to to confront the z *determination efficiency problem, zdep*, to exploit the potential of GRBs for cosmology in planning for future GRB science and missions. A partial answer is contained in the reasons to measure a very large sample of GRBs, as contained in our third "Open question".

Measuring cosmic physics with GRBs

Although much has been deciphered about the basic physics of GRBs, we really only have broad outlines of understanding of these remarkable events. Given their inherent interest, as well as great utility as probes of cosmic structure, there is great incentive to understand them by vastly larger numbers of GRBs observed with broader bandwidth, higher resolution measurements. A few examples can make the point:

Photometric redshifts: Much would be gained if we could remove remaining systematics from the currently-best correlative relations for GRB luminosity vs. measureable quantities – e.g. the Ghirlanda[17] relation for E_{peak} vs. E_{iso}. By calibrating this relation, or similar variants, with a large sample (several thousand) vs. the small numbers of (sometimes differing, or at least uncertain) objects presently, we might break the bonds of *zdep*. After all, our optical colleagues did this long ago with narrow-band photometry to identify "Lyman-break" galaxies. The imperative for distance measure, and "prompt" redshift estimates cannot be overstated: the rare Pop III GRB must be observed with moderate resolution $\gtrsim 2\text{-}10$micron spectra (i.e., ideally rest-frame Lyman-α to near-IR, to counter absorption in the IGM) to measure – ideally – clustering of absorption systems revealing cosmic structure along the line of sight. All of this would best be done within the first ~20min of the GRB, or during its luminous prompt phase, when it could be done with perhaps a dedicated space-borne ~1m telescope. If the prompt emission is missed, absorption line redshift measures from the much fainter afterglow would require signficant exposure time with JWST at the likely $\gtrsim 0.5$d delay needed to re-point JWST. Given this (very expensive) overhead, the requirement for near-certain photometric z values for every GRB becomes obvious: there must be $\gtrsim 90\%$ confidence that a JWST re-point is justified! Far better to hope that a co-orbiting (with *EXIST* ; see below) modest mid-IR telescope of small aperture could do prompt GRB redshift observations on *every* GRB.

Role of magnetic fields: The fundamental questions surrounding the GRB emission processes, and whether magnetic fields are dominant or GRB jets are Poynting flux dominated, could be revealed by polarimetry measures. The controversial claim [18] for polarization in GRB021606 does not mean that polarization is ruled out; simply that it has not yet been measured with sufficient sensitivity. Prompt vs. afterglow emission models (synchrotron vs. inverse Compton) would be possible to test and new constraints on beaming derived from polarization vs. time in the burst decay. Polarization can be measured from finely-pixellated detectors recording the spatial anisotropy of Compton scattered (~100-500 keV) photons, for a broad range of E_{peak}, in high time resolution spectra.

Nature of particle acceleration: High energy and high time resolution spectra of the fast spikes resolved in some GRBs would reveal the nature of particle acceleration in the internal shocks thought to comprise the prompt emission. This in turn would further constrain models[19] for production of ultra-high energy cosmic rays in GRBs and is possible to measure with sufficiently large area detectors to achieve good S/N within individual spikes.

Tests of Lorentz violations with GRBs: The same high time resolution spectra allow tests of fundamental physics – namely models which violate Lorentz invariance and would predict dispersive or an energy dependent speed of light. Such considerations, as recently discussed by Martinez and Piran[20], would again require very sensitive high time and spectral resolution measures over a broad energy band of high-z GRBs and could be achieved with the same requirements noted above.

EXIST : A NEXT GENERATION GRB CONCEPT

To answer or constrain the open questions outlined above requires a next generation GRB mission with very large collecting area and sensitivity over a large field of view and energy band. These requirements are well matched to those of the *EXIST* (Energetic X-ray Imaging Survey Telescope) all-sky deep hard X-ray imaging survey for black holes, which was recommended as one of the 3 high energy missions in the 2000 Decadal Survey Report. Over the past two years, a large collaboration has undertaken a concept study for *EXIST* as the implementation of the Black Hole Finder Probe (BHFP), one of three *Einstein Probe* missions proposed for NASA's Beyond Einstein Program. This ongoing study follows earlier studies in which the *EXIST* concept was first formulated

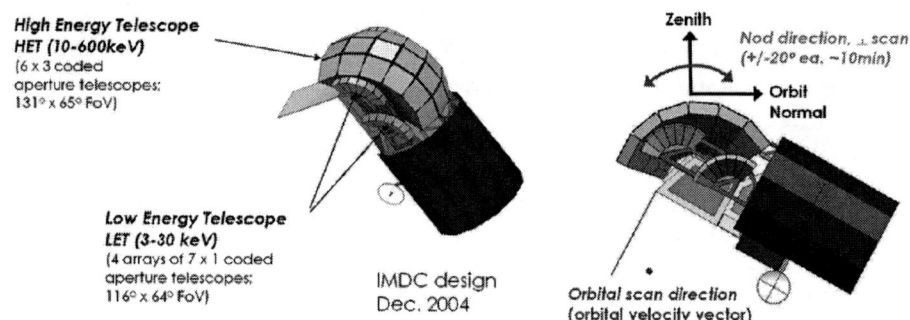

FIGURE 1. *Left:* Baseline concept for *EXIST* as developed at GSFC in December 2004. The large (blue) fixed solar panel array and yellow radiator panel are folded down after deployment on orbit. Scale: solar panel cylinder diameter is 5m, and the individual coded masks for the HET are 1.2m × 1.2m. *Right:* Orbital scanning configuration.

(see [21] and references therein). A detailed description of *EXIST* science generally and the full mission concept is given elsewhere[22]

The current baseline mission concept for *EXIST* is shown in Figure 1 and is centered around a very large area ($\sim$5.6m^2) array of imaging Cd-Zn-Te (CZT) detectors in 18 coded aperture sub-telescopes imaging a fully-coded 131° × 65° FoV. This High Energy Telescope (HET) images the full sky each orbit with 5$'$ angular resolution and $\lesssim$1$'$ source locations over the 10-600 keV band. Each HET sub-telescope includes a 56cm × 56cm array of imaging CZT (1.2mm pixels) collimated by surrounding (and rear) active shields (CsI) to a 22° × 22° fully-coded FoV defined by a radial-hole coded aperture mask. A co-aligned array of 1.1m^2 (total) of imaging Si detectors distributed in 28 coded-aperture sub-telescopes (20cm × 20cm each; 0.2mm pixels) constitutes the Low Energy Telescope (LET). With nearly the same FoV, the LET surveys the sky at 3-30 keV with 5× finer angular resolution and source location precision (1$'$ and $\sim$10$''$, respectively) to allow (nearly) unambiguous identifications for sources at the survey sensitivity limits. Sources a factor of (only) 3 brighter than the survey limits would be located to aspect-constrained limits of 3$''$ for unique identifications. The 5σ sensitivity limits for *EXIST* are 2mCrab and 0.05mCrab (2×10^{-11} and 5×10^{-13} erg cm^{-2} s^{-1}) for exposure times of 1 orbit (20min on source each 95min) and a 1 year survey, respectively, in any factor of 2 energy band up to 200 keV. The HET sensitivity falls off by a factor of 10 at the high energy limit (600 keV) due to both the 5mm thick CZT detectors and coded masks becoming transparent.

EXIST achieves full sky coverage each orbit by its zenith-pointed fan beam nodding ±20-25° each $\sim$10min (see Figure 1) so that a given source is scanned in both the orbit-velocity and orbit-normal directions for the $\sim$20min that it traverses the FoV of the telescope array. This scanning is the key to much larger dynamic range and limiting sensitivity that can extend significantly the limiting flux sensitivity below the systematic limits previously achieved in coded aperture imaging[23], as shown by simulations conducted[24] and on going as well as scanning experiments and analysis of *Swift* /BAT

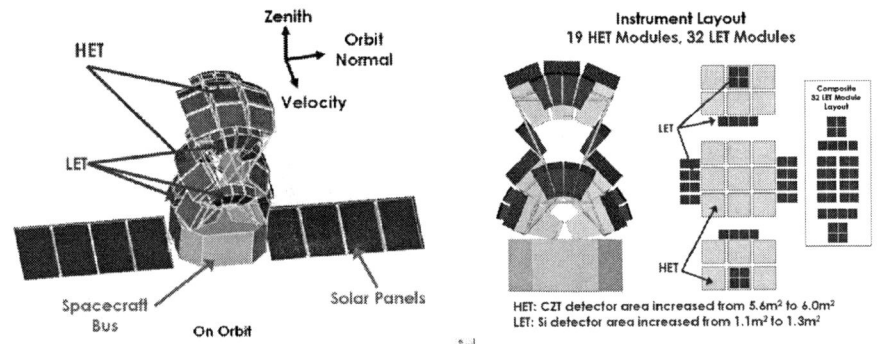

FIGURE 2. *Left:* New layout concept for *EXIST* as developed at General Dynamics in December 2005. The HET and LET telescope arrays are now symmetric on the S/C. *Right:* Increased fully-coded fields of view of HET (154° × 65°) and LET (160° × 64°) and total detector areas achieve more uniform sky coverage at the cost of a modest increase in total instrument mass.

slew data (see below). An even more uniform sky coverage (not as overexposed at the orbit poles) is achieved by a more symmetric telescope layout, which also allows much smaller articulated solar panels, as shown in Figure 2. This new design also better matches the LET total area, and thus sensitivity, to that of the HET. The ratio is nearly 1:4, or what is required for comparable S/N per unit time for a source with a Crab-like power law spectrum (with photon index 2) in the presence of a (much) flatter cosmic diffuse background, which dominates the aperture flux below 100 keV.

GRB sensitivities for *EXIST*

With this factor of ∼12× increase in total imaging CZT detector area over that of *Swift* /BAT, extended energy band (to both significantly lower (3keV) and higher (600 keV) energies), and larger total FoV (nearly ∼6sr, or half-sky, for $\gtrsim$50% coding response, or about 3× *Swift*), *EXIST* would achieve ∼5× greater sensitivity for GRBs and ∼10-20× greater sensitivity for its full-sky survey. Detailed simulations are being carried out over the coming year to include full background on orbit as well as effects of bright sources in the FoV (near the Galactic plane). Initial estimates by D. Band for the GRB sensitivity for *EXIST* , *Swift* and *BATSE* are shown in Figure 3 for GRBs with differing E_{peak} and redshift (all referenced to the flux values shown for a GRB at z = 1). The *EXIST* sensitivity curve is plotted for a *single* sub-telescope (i.e. 1 of the 19 in the new HET configuration). A given GRB is effectively imaged over 4 sub-telescopes, so the full sensitivity reaches a factor of 2 lower flux than shown.

The active shields (1cm CsI on 4 sides of each sub-telescope around the CZT array; 2cm planar shield beneath the CZT array) are read out for GRB spectroscopy. Simulations[25] show that their sensitivity for GRB spectra nicely extends that of the CZT array to much higher energies, as shown in Figure 3. E_{peak} values, and thus constraints on GRB luminosities, could be measured up to 5-10 MeV.

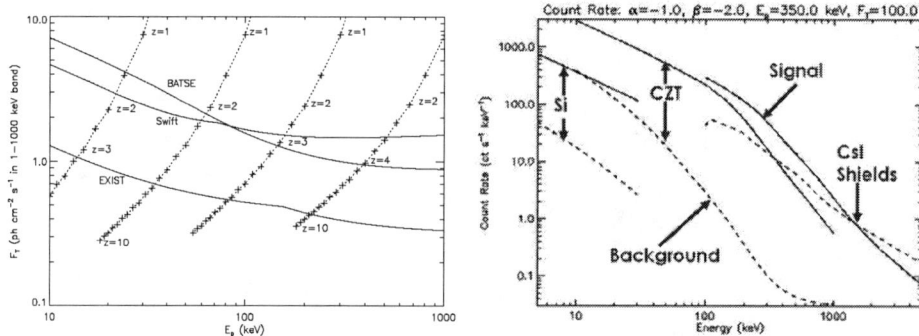

FIGURE 3. *Left:* Approximate sensitivity for one *EXIST* sub-telescope vs. *Swift* and *BATSE* for GRBs of E_{peak} and redshift shown. Full *EXIST* GRB sensitivity would be factor of 2 better (lower flux). *Right:* Matched GRB sensitivities in the LET, HET and CsI shields for the HET for GRB with spectral parameters shown.

What does this mean for the total number of GRBs that should be detected per year by *EXIST*, or the redshift distribution of GRBs expected? Using the original analysis of Bromm and Loeb[26] gives the relative GRB numbers vs. z shown in Figure 4 for *BATSE*, *Swift* and *EXIST*. The updated estimates[16] would move the *Swift* curve closer to the *BATSE* curve; and additional corrections are indicated for the different assumptions of reionization epoch as well as Pop I/II contributions. A more complete accounting of the *EXIST* sensitivities is also warranted. Nevertheless it appears *EXIST* would detect nearly half of all cosmic GRBs at z $\sim$10 and a quarter at z $\sim$15. Thus it appears feasible to design a next generation GRB observatory to achieve close to ultimate sensitivity.

The actual rate of GRBs anticipated for *EXIST* would be $\sim$600 y^{-1}. Again, this estimate is being refined and may increase for more accurate treatment of full vs. partial coding in the burst imaging. GRB triggers would be derived from rate increases as well as on-board imaging. The latter would be, as for *Swift*, essential for prompt triggers on high z bursts and the sought Pop III events. This objective, as well as the general black hole sky survey objectives, require that full sky-orbit images are derived continuously each orbit. The duty cycle for recording the long-duration high z GRBs would be limited by the time over which any source is fully or partially (50%) coded: $\sim$17min or 23min, respectively, each 95min orbit, or duty cycles of 18% and 24%. Thus the scanning geometry of *EXIST* is well suited to maximizing the detection probability of very long GRBs over the full sky.

BAT SLEW SURVEY: SCIENCE AND TESTING FOR *EXIST*

By continuous scanning its large FoV over the full sky each orbit, *EXIST* is effectively averaging over pixel to pixel systematics (e.g. gain variations, dead pixels, and changing backgrounds on any given pixel from other source variations or particle background

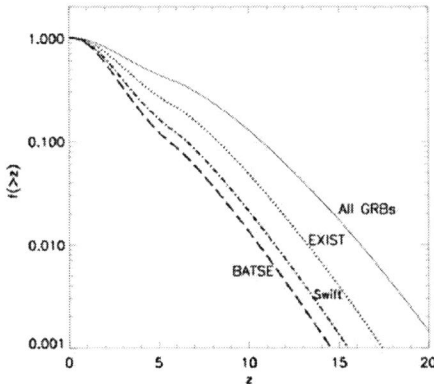

FIGURE 4. Estimated relative fractions of GRBs detected vs. redshift for *BATSE*, *Swift*, and *EXIST* as derived from the original formalism of Bromm and Loeb[26].

fluctuations). Our original suggestion[23] for scanning coded apertures with radial mask holes to both minimize systematics and flatten the wide-field imaging response has been borne out by simulations[24] and laboratory tests[27]. However an actual demonstration or test of the effects of scanning can be conducted with the *Swift* /BAT coded aperture telescope by analysis of its data obtained during slews (e.g. from a fixed pointing to a GRB; or between pointed targets). We have initiated a program to develop an analysis system for BATslew data to both test coded aperture scanning and to extend the *Swift* all sky survey for persistent sources, transients, and additional GRBs. A full report on the BATslew analysis system and initial results is in preparation[28], and an early look at some first results is given here.

Figure 5 shows a particularly interesting BAT slew image (right) and the 76 sec pointed observation (left) that preceeded this before a final slew onto SGR050904, the $z = 6.29$ GRB[15]. The portion of the slew shown in the image (right) is truncated to be the same exposure time (76sec) as in the immediately preceeding pointing, which had (fortuitously) the bright source CygX-1 moderately near the center of the FoV of the BAT (with 46% coding fraction). The slew image has been analyzed by co-adding images each 0.2sec (the *Swift* aspect readout resolution) with proper tangent plane projections (into $10'$ sky pixels) to be centered on CygX-1. The coding fraction is larger for CygX-1 in this combined slew image, with mean value 76% vs. 47% for the pointed image. However each of the images co-added has been corrected for its (differing) coding fraction so that the total may be compared directly for source strength and S/N with the pointed image. To facilitate a BATslew survey, *Swift* is now downloading $\sim$50 slews per day. This rich survey dataset to be explored will complement and extend the RXTE/ASM survey and recently inaugurated XMM slew survey[29].

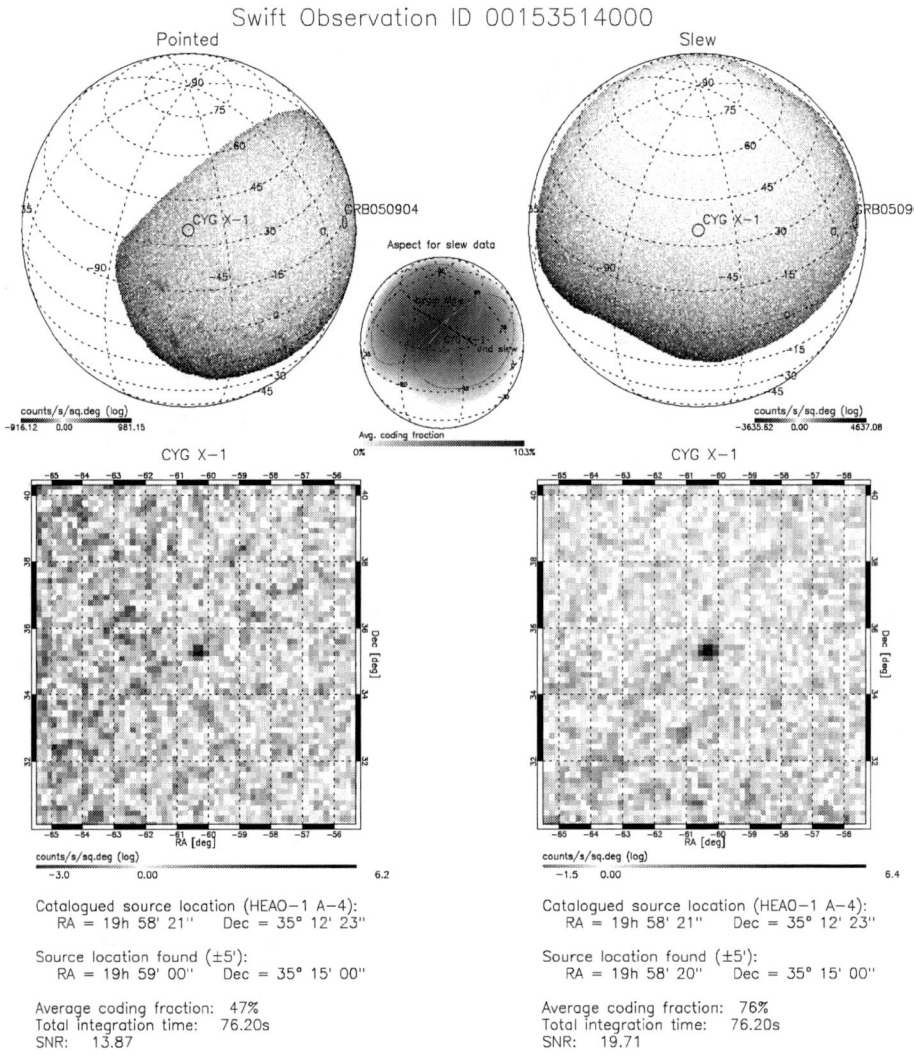

FIGURE 5. *Left:* Pointed image with CygX-1 near center of FoV showing full-sky view (top) and a 10° × 10° image around CygX-1 (bottom). *Right:* Same but for equal-exposure (76sec) of slew data, for which 380 × 0.2sec separate images have been co-added and weighted by their individual pixel variance values. For the slew image (right), pixels are included in the summed image only if their net exposure was $\gtrsim 10\%$ of the maximum (i.e. $\gtrsim 7.6$sec) in order to limit the effects of large fluctuations from low exposure pixels at the edge of the FoV. The resulting S/N is improved by ~42% due to the reduced noise in the slew image vs. the pointed image (compare both the full sky as well as zoomed images; the zoomed image shows negative fluctuations only ~half as large as the pointed observation) while the measured source flux (approximately constant between the two images) is preserved. The scan slew track is shown in the small plot centered at top: this slew was onto a target observed for only 76sec just before the subsequent slew to GRB050904, the *Swift* GRB at record z= 6.29!

CONCLUSIONS

It is clear that GRB science is pushing the frontiers of high energy astrophysics and fundamental science, generally. As made clear at this Symposium, *Swift* is opening new vistas into the high redshift and high energy universe. It is now time to push for the Next Generation GRB mission that all of physics and astronomy should expect to make the next quantum leap. The deep questions can be answered, or better constrained, by the very large area and field of view imaging spectrometer that is required to *EXIST* . We collectively await the Beyond Einstein program to enable this next step which is, truly, Beyond Einstein.

ACKNOWLEDGMENTS

I thank D. Band, A. Copete, D. Conte, G. Fishman, N. Gehrels, J. Hong, H. Krawczynski and G. Skinner of the *EXIST* team for contributions. This work was supported in part by NASA grant NNG04GK33G.

REFERENCES

1. G. J. Fishman and C. A. Meegan, *Ann. Rev. Astron. Astrophys.* **33**, 415–458 (1995).
2. E. Costa, F. Frontera, J. Heise et al., *Nature* **387**, 783–785 (1997).
3. J. van Paradijs, P. J. Groot, T. Galama, et al., *Nature* **386**, 686–689 (1997).
4. N. Gehrels, et al., *ApJ* **611**, 1005-1020 (2004).
5. N. Gehrels, et al., *Nature* **437**, 851–854 (2005).
6. P. Jakobsson, et al., these proceedings, and astro-ph/0602071 (2006).
7. A. Burrows, et al., *ApJ* in press, and astro-ph/0510687 (2006).
8. J. S. Villasenor, et al., *Nature* **437**, 855–858 (2005).
9. J. S. Bloom and J. X. Prochaska, these proceedings, and astro-ph/0602058 (2006).
10. K. Hurley, et al., *Nature* **434**, 1098–1103 (2005).
11. A. Crider, these proceedings, and astro-ph/0601019 (2006).
12. N. Shaposhnikov and L. Titarchuk, *ApJ*, in press, and astro-ph/0602091 (2006).
13. J. Grindlay, S. Portegies Zwart, and S. McMillan, *Nature Physics* **2**, 116–119, and astro-ph/0512654 (2006).
14. J. D. Dewi, Ph. Podsiadlowski, and O. R. Pols *MNRAS* **363**, L71–L75 (2005).
15. G. Cusumano, et al., *Nature*, submitted, and astro-ph/0509737 (2005).
16. V. Bromm and A. Loeb *ApJ* in press, and astro-ph/0509303, and these proceedings (2006).
17. G. Ghirlanda, G. Ghisellini, and C. Firmani *MNRAS* **361**, L10–L14 (2005).
18. W. Coburn and S. E. Boggs *Nature* **423**, 415–417 (2003).
19. E. Waxman *ApJ* **606**, 988–993 (2004).
20. M. R. Martinez and T. Piran, astro-ph/0601219 (2006).
21. J. Grindlay, W. Craig, N. Gehrels, F. Harrison and J. Hong *Proc. SPIE* **4851**, 331–344 (2003).
22. J. Grindlay et al., in preparation (2006).
23. J. Grindlay and J. Hong *Proc. SPIE* **5168**, 402–410 (2004).
24. S. Vadawale, J. Hong, J. Grindlay and G. Skinner *Proc. SPIE* **5900**, 338–349 (2005).
25. A. Garson, H. Krawczynski, J. Grindlay, G. Fishman and C. Wilson *A&A Suppl. Ser.*, submitted (2006).
26. V. Bromm and A. Loeb *ApJ* **575**, 111–116 (2002).
27. J. Hong, et al., *Proc. SPIE* **5540**, 63–72 (2004).
28. A. Copete, et al., in preparation (2006).
29. A. M. Read et al., *MPE Report* **288**, 137–139, and astro-ph/0506380 (2005).

GLAST, LAT and GRBs

Nicola Omodei* and GLAST/LAT GRB Science Group[†]

INFN Pisa, Largo B.Pontecorvo, 3 56100 Pisa
[†]*http://glast.gsfc.nasa.gov/science/grbst/members.html*

Abstract. The GLAST Large Area Telescope (LAT) is the next generation satellite experiment for high-energy gamma-ray astronomy. It is a pair conversion telescope built with a plastic anticoincidence shield, a segmented CsI electromagnetic calorimeter, and the largest silicon strip tracker ever built. It will cover the energy range from 30 MeV to 300 GeV, shedding light on many issues left open by its predecessor EGRET. One of the most exciting science topics is the detection and observation of gamma-ray bursts (GRBs). In this paper we present the work done so far by the GRB LAT science group in studying the performance of the LAT detector to observe GRBs. We report on the simulation framework developed by the group as well as on the science tools dedicated to GRBs data analysis. We present the LAT sensitivity to GRBs obtained with such simulations, and, finally, the general scheme of GRBs detection that will be adopted on orbit.

Keywords: Gamma-ray detectors; Gamma-ray telescopes; Gamma-ray bursts;
PACS: 07.85.Fv,29.40.-n; 95.55.Ka; 98.70.Rz;

INTRODUCTION

The Gamma-ray Large Area Space Telescope (GLAST) is an international mission that will study the gamma-rays Universe[1]. GLAST, scheduled for launch in late-2007, is instrumented with a tracker of silicon strip planes with slabs of tungsten converter, followed by an hodoscopic calorimeter; the tracker, is an array of 4×4 identical towers, surrounded by an anticoincidence detector (ACD) which identifies charged cosmic rays. This pair production telescope, called the Large Area Telescope (LAT), is sensitive to gamma rays in the energy range between 30 MeV-300 GeV and above. The LAT's energy range, field-of-view (FoV) and angular resolution are vastly improved in comparison with those its highly-successfull predecessor EGRET (1991-2000), so that the LAT will provide a factor 30 or more advance in sensitivity. This improvement should enable the detection of several thousands of new high-energy sources and allow the study of gamma-ray bursts (GRBs) and other transients, the resolution of the extragalactic diffuse gamma-ray emission, the search for dark matter and the detection of active galactic nuclei (AGNs), pulsars and supernova remnants (SNRs). A detailed description of the scientific goals of GLAST mission and an introduction to the experiment can be found in [1].

The flight hardware production is now complete, the sixteen towers are integrated into the flight grid and the ACD is already placed over the LAT towers. The detector, which represents the biggest silicon strip tracker ever built, is taking data from cosmic rays,

---

[1] For more details, see the GLAST website at: http://glast.gsfc.nasa.gov/

and will be integrated with the spacecraft within the next few months.

The scientific performance of the full LAT detector, i.e., the effective area, the point spread function and the energy dispersion, are obtained from detailed Monte Carlo studies, and are well-parameterized by a series of functions: the Instrument Response Function (IRF)[2].

GAMMA-RAY BURSTS AND THE LARGE AREA TELESCOPE: SIMULATIONS AND DATA ANALYSIS

High energy emission from GRBs is still a puzzling topic and few observations are presently available above 50 MeV. EGRET detected only a few high-energy bursts [2] and no apparent cut-off was detected at these energies. An extraordinary, and still unexplained, discovery was the delayed high energy emission[3], together with the more recent report of the observation of an additional high energy spectral component [4].

For studying the performance of the LAT in observing GRBs, we set up a full simulation chain that starts from a detailed description of the sky, and adopts either a full Monte Carlo simulation of the detector (propagating every single particle through the different materials of the detector), or a fast science simulator which uses a parameterized description of the instrument for processing the incoming fluxes.

We have developed different GRB models within the framework of the LAT software[3]. The Gamma-Ray Burst physical model starts from the well known *fireball* model where the gamma rays are radiated by internal shocks [5]. In this model, shells of matter are emitted with relativistic bulk Lorentz factor; faster shells overtake the slower ones and collide, producing internal shocks. In this scenario electrons and positrons are accelerated and lose energy via synchrotron radiation due in the presence of magnetic fields amplified by the shock compression. In this model the synchrotron cut-off at high energy due to the finite value of the Lorentz factor at which the electrons are accelerated is considered as well as the reprocessing of synchrotron radiation due to inverse Compton emission (i.e., a Synchrotron Self-Compton—SSC—spectrum). In this way we obtain a possible description of the high energy component based on "physical" assumptions. This model is suitable for studying the performance of the LAT detector in spectral analysis. Fig. 1 shows the results of a simulation of a burst with a characteristic Lorentz factor $\Gamma = 180$, which results in a synchrotron cut-off at 4.5 GeV. The large effective area and the good energy resolution of the LAT detector (<15%) allows the study of the detectability of the high energy cut-off. The SSC emission operates in the opposite direction: enhancing the spectrum at high energy. This component has, to first approximation, in the same shape as the synchrotron component but, if γ is the Lorentz factor of the incoming electron, is shifted by a factor of $\gamma^2 \approx 10^4$ to higher energy. Since the spectral energy distribution for the prompt emission of a typical burst has a peak around a few hundred keV, the inverse Compton component, if present, is completely unobservable for detectors operating below tens of MeV (like BATSE); hence the search

---

[2] http://www-glast.slac.stanford.edu/software/IS/glast_lat_performance.htm
[3] http://www-glast.slac.stanford.edu/software/

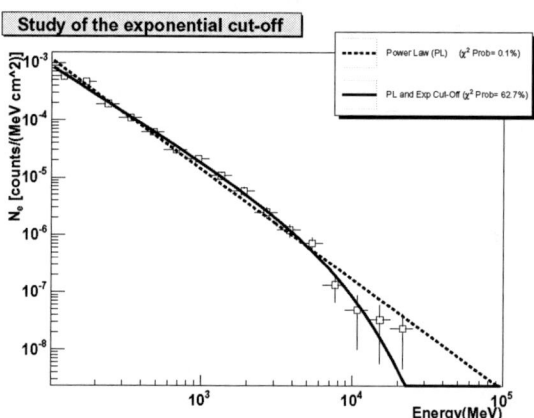

FIGURE 1. Study of the high energy cut-off. The simulated data have been produced using the GRB physical model with a high energy cut-off at 4.5 GeV. The reconstructed data are displayed in the plot and have been fitted with a power law model (dashed line) and with a power law model with an exponential cut-off (solid line). In the legend the values of the χ^2 test are reported which denote the probability of obtaining a greater χ^2 by chance. The estimated cut-off energy is 5.5 ± 1.5 GeV.

for the Compton component represents one of the key goals for the LAT detector onboard the GLAST satellite. In addition, an energy-dependent time lag resulting from Lorentz invariance breaking predicted by some quantum gravity theories[6, 7] has also been included.

Our second approach to simulating GRBs in the LAT energy band, the GRB phenomenological model extrapolates a model of the GRBs spectrum and lightcurve in the ~ 100 keV band to LAT energies. For each simulated GRB the duration is drawn from the observed T_{90} distribution and its fluence is sampled from the BATSE fluence distribution in the 50-300 keV energy range. Each burst has a spectrum described by the Band function[8], where the peak energy E_p and the spectral indices α and β are sampled from the observed distribution[8, 9]. The lightcurve is the sum of temporal pulses described by a general pulse equation with parameters drawn from observed distributions; the pulse width $W(E)$ scales with energy as $W(E) = W_0 \, E^{-0.33}$, as observed by Norris et al.[10]. Short burst and long bursts are treated separately so that the observed hardness-duration correlation is reproduced. Analysis tools are needed for data handling and for science analysis. Analyzing GRBs is different from analyzing other LAT sources. First, the GLAST Burst Monitor (GBM)[11] provides the soft gamma-ray counterparts to the LAT data; the GBM consists of 12 NaI detectors for the 10 keV to 1 MeV range and two BGO detectors for the 150 keV to 30 MeV range. Second, during the burst there are essentially no non-burst photons within the point spread function (PSF) of the LAT detector (3.5° at 100 MeV, 0.15° above 10 GeV). Finally, the Instrument Response Functions that are needed for converting the raw data into astronomical fluxes can be considered constant during the burst, contrary to stationary or long duration sources. Fig.2 shows an intense burst that has been simulated with the full simulation chain. For this burst the LAT and GBM data have been simulated using the response functions for

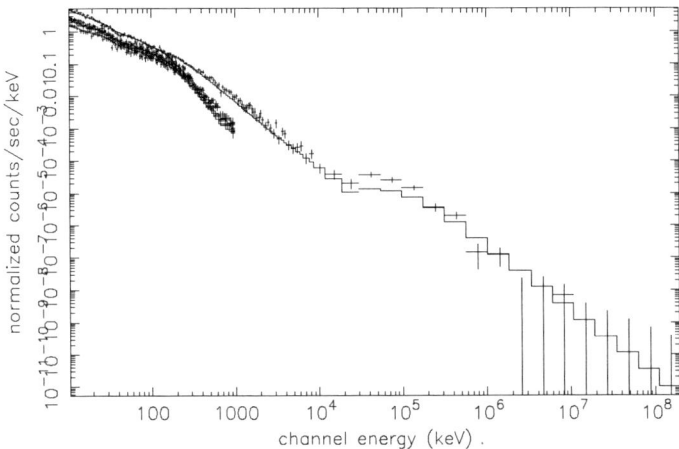

FIGURE 2. GBM (NaI and BGO) and LAT spectral fit. Producing this plot tested the full simulation and analysis chains. The output files are standard HEASARC FITS files, permitting the spectral fit to be performed using XSPEC.

both types of detectors for a given the burst viewing angle. Within a few PSFs from the spatial barycenter we can assume that there are only burst photons for the LAT detector (i.e., we do not need a background model), while a background model has to be assumed for analyzing the GBM signal. The plot shows the result of a joint spectral analysis with one NaI detector (10 keV–1 MeV), one BGO detector (150 keV – 30 MeV) and the LAT detector (> 30MV). GBM and LAT data will be jointly fitted providing spectral information over more than seven energy decades. Here we estimate the LAT GRBs sensitivity, the number of bursts detected per year as a function of the number of photons detected per burst. Of course this quantity depends on the GRB high energy emission where little information is available. In our computations we have adopted the phenomenological GRB description, providing a conservative description of the burst high energy flux.

GRBS LAT SENSITIVITY

In order to compute the performance of the LAT detector to GRBs we simulate one year of observation in scanning mode, assuming a burst rate of 650 bursts per year full sky. Each burst is simulated with the phenomenological model discussed earlier. An observed energy, an observed direction and a detection probability are computed for each simulated burst photon, taking into account the instrument response functions, resulting in an estimate of the number of photons that will be detected by the LAT detector. In addition, the orbit of the GLAST satellite, with SAA passages and Earth occultations, are considered. At high energies (>10 GeV) it is important to consider the attenuation of the flux due to the cosmological absorption given by the interaction of a burst photon with an optical-UV photon of the extragalactic background light (EBL). The uncertain EBL spectral energy distribution resulting from the absence of high redshift data provides a variety of theoretical models for such diffuse radiation. Thus the observation of the high energy cut-off as a function of the GRB distance can, in principle, constrain the infrared background. In our simulation we have included this effect, adopting the EBL model proposed by Primack[12] and assuming the long burst redshift distribution from Porciani and Madau[13] and the short burst distribution from Guetta and Piran[14]. We therefore plot the number of expected bursts per year as a function of the number of photons per burst detected by the LAT (Fig.3). Different colors refer to different energy thresholds (see the legend). The EBL attenuation affects only the high energy curve, as expected from the theory, leaving almost unchanged the sensitivities with thresholds less than 10 GeV. In this calculation, LAT will independently detect more then a hundred photons per burst for a few burst per year; these are the bursts for which a detailed spectral or even time resolved spectral analysis will be possible. Tens of bursts per month will result in more than ten counts in the LAT detector, and, with the assumed high energy emission model, a few bursts per year will show high energy prompt emission, with photons above 100 GeV.

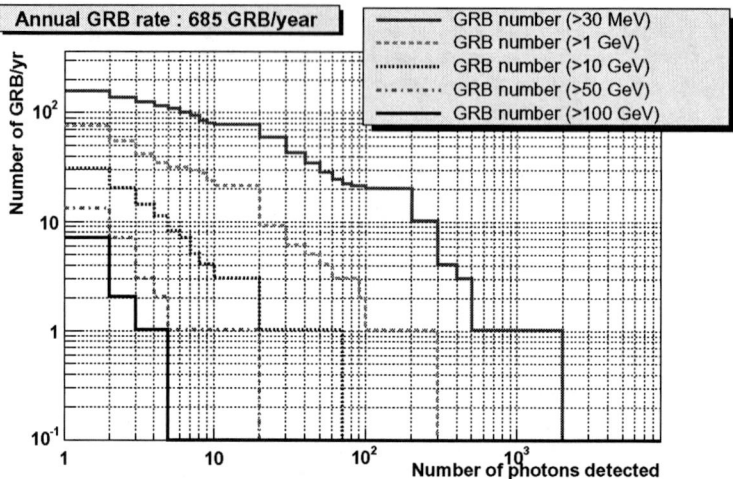

FIGURE 3. Model-dependent LAT GRB sensitivity assuming a burst rate of 650 bursts/yr, including the effect of the EBL absorption. Different curves refer to different energy thresholds.

ALERTS AND COMMUNICATION

GLAST in the first year will operate in the scanning mode, providing uniform full sky coverage every three hours. Starting from the second year of operations, and depending on the Guest Investigator program, GLAST may also be used in pointing mode. The GBM will cover the entire visible sky not occulted by the Earth, and the LAT will cover 20% the sky, with incident inclination up to almost $\sim 70°$ from the normal incident direction. In addition GLAST will also be able to repoint in case of intense burst, to maintain the GRB in the LAT FoV during the prompt emission phase or to search for delayed emission. The GBM and LAT will independently trigger on GRBs: the first on a rapid increase of the count rate, and the second considering spatial and temporal clustering of counts. The GBM will detect ~200 bursts per year, and more than 60 will fall in the LAT FoV, allowing joint observations. In case of a GRB trigger, an alert message will be send to the ground via TDRSS (a communications satellite system) within ~10 seconds. This will provide basic information for follow up observations. The initial on-board GBM localization accuracy is ~15 degrees (within 1.8 s) that can be used by LAT. Updates will come later, reducing the GBM localization error box up to ~5 degrees for a bright burst, while the LAT detector can provide better accuracy, up to tens of arcminutes (depending on the burst intensity). On-board LAT detection of GRBs is under development. A full downlink of all the data will be performed via TDRSS ~ 6 times a day and the scientific data, after a first analysis done by the LAT collaboration, will be delivered to the user community (for more details on GLAST operations in GRB observations, see[15]).

SYNERGY WITH SWIFT

A new mission will soon complement Swift's gamma-ray burst observations. The Gamma-ray Large Area Space Telescope (GLAST) will be launched in late-2007 and will cooperate with Swift in studying bursts. For a few bursts a year the LAT will localize bursts to sufficiently small error boxes that Swift can point to for follow up observations; for these bursts Swift will provide a more precise measurement of the GRB position. On the other hand, GLAST will frequently scan the burst position in the hours after a Swift burst detection, searching for delayed high energy emission. Finally, simple computations show that~20 Swift-detected GRBs per year will also be in the LAT FoV.

REFERENCES

1. P. F. Michelson, "The Gamma-ray Large Area Space Telescope Mission: Science Opportunities," in *AIP Conf. Proc. 587: Gamma 2001: Gamma-Ray Astrophysics*, 2001, pp. 713–+.
2. B. L. Dingus, *Astrophys. & Space Sci.* **231**, 187–190 (1995).
3. K. Hurley et al., *Bulletin of the American Astronomical Society* **26**, 881–+ (1994).
4. M. M. González et al., *Nature* **424**, 749–751 (2003).
5. T. Piran, *Physics Reports* **314**, 575–667 (1999).
6. G. Amelino-Camelia et al., *Nature* **393**, 763–765 (1998).
7. N. Omodei, J. Cohen-Tanugi, and F. Longo, "GLAST and Gamma-Ray Bursts: Probing Photon Propagation over Cosmological Distances," in *AIP Conf. Proc. 727: Gamma-Ray Bursts: 30 Years of Discovery*, 2004, pp. 681–683.
8. D. Band et al., *Ap. J.* **413**, 281–292 (1993).
9. Preece et al., *Ap. J. Supp.* **126**, 19–36 (2000).
10. J. P. Norris et al., *Ap. J.* **459**, 393–+ (1996).
11. C. Kouveliotou et al., *American Astronomical Society Meeting Abstracts* **207**, –+ (2005).
12. J. R. Primack, J. S. Bullock, and R. S. Somerville, "Observational Gamma-ray Cosmology," in *AIP Conf. Proc. 745: High Energy Gamma-Ray Astronomy*, 2005, pp. 23–33.
13. C. Porciani, and P. Madau, *Ap. J.* **548**, 522–531 (2001).
14. D. Guetta, and T. Piran, *Astron. & Astrophys.* **435**, 421–426 (2005).
15. J. McEnery, and S. Ritz, "The Gamma-ray Large Area Space Telescope and Gamma-Ray Bursts," in *These proceedings*, 2006.

AGILE and Gamma-Ray Bursts

Francesco Longo* and M. Tavani, G. Barbiellini, A. Argan, M. Basset,
F. Boffelli, A. Bulgarelli, P. Caraveo, P. Cattaneo, A. Chen, E. Costa,
E. Del Monte, G. Di Cocco, G. Di Persio, I. Donnarumma, M. Feroci,
M. Fiorini, L. Foggetta, T. Froysland, M. Frutti, F. Fuschino, M. Galli,
F. Gianotti, A. Giuliani, C. Labanti, I. Lapshov, F. Lazzarotto, F. Liello,
P. Lipari, M. Marisaldi, M. Mastropietro, E. Mattaini, F. Mauri,
S. Mereghetti, E. Morelli, A. Morselli, L. Pacciani, A. Pellizzoni,
F. Perotti, P. Picozza, C. Pittori, C. Pontoni, G. Porrovecchio, M. Prest,
M. Rapisarda, E. Rossi, A. Rubini, P. Soffitta, A. Traci, M. Trifoglio,
A. Trois, E. Vallazza, S. Vercellone, D. Zanello[†]

Department of Physics, University of Trieste and INFN, section of Trieste
[†]*the AGILE collaboration* http://agile.rm.iasf.cnr.it

Abstract. AGILE is a Scientific Mission dedicated to high-energy astrophysics supported by ASI with scientific participation of INAF and INFN. The AGILE instrument is designed to simultaneously detect and image photons in the 30 MeV - 50 GeV and 15 - 45 keV energy bands with excellent imaging and timing capabilities, and a large field of view covering $\sim 1/5$ of the entire sky at energies above 30 MeV. A CsI calorimeter is capable of GRB triggering in the energy band 0.3-50 MeV. The broadband detection of GRBs and the study of implications for particle acceleration and high energy emission are primary goals of th emission. AGILE can image GRBs with 2-3 arcminutes error boxes in the hard X-ray range, and provide broadband photon-by photon detection in the 15-45 keV, 03-50 MeV, and 30 MeV-30 GeV energy ranges. Microsecond on-board photon tagging and a ~ 100 microsecond gamma-ray detection deadtime will be crucial for fast GRB timing. On-board calculated GRB coordinates and energy fluxes will be quickly transmitted to the ground by an ORBCOMM transceiver. AGILE have recently (December 2005) completed its gamma-ray calibration. It is now (January 2006) undergoing satellite integration and testing. The PLSV launch is planned in early 2006. AGILE is then foreseen to be fully operational during the summer of 2006. It will be the only mission entirely dedicated to high-energy astrophysics above 30 MeV during the period mid-2006/mid-2007.

Keywords: Gamma-ray detectors, X and Gamma-ray telescopes and instrumentation, Gamma-ray Bursts
PACS: 07.85.Fv, 95.55.Ka, 98.70.Rz

INTRODUCTION

AGILE is an ASI Small Scientific Mission dedicated to high-energy astrophysics. The AGILE instrument is designed to detect and image photons in the 30 MeV–50 GeV and 15–45 keV energy bands, with excellent spatial resolution, timing capability, and an unprecedently large field of view covering $\sim 1/5$ of the entire sky at energies above 30 MeV. Primary scientific goals include the study of AGNs, Gamma-Ray Bursts, Galactic sources, unidentified gamma-ray sources, diffuse Galactic gamma-ray emission, high-precision timing studies, and Quantum Gravity testing.

AGILE is planned to be operational during the year 2006, and will be the only Mission entirely dedicated to high-energy astrophysics above 30 MeV to 50 GeV.

Despite its small dimensions ($\sim$0.3 m^3) and weight ($\sim$100 kg) compared to other future instruments such as GLAST, the AGILE performance will be as good as, or better than, that of bigger past instruments such as EGRET, thanks to the new silicon detector technology employed for the AGILE instruments[1].

THE AGILE INSTRUMENT

High-energy photons are converted into e$^+$/e$^-$ pairs in a tracker made of 12 Silicon-Tungsten planes, which allows us to efficiently collect photons with an effective area of $\sim$500 cm^2 and to perform photon direction reconstruction with good angular resolution ($\sim$0.5°, $E\sim$1 GeV) (Figure 1) on a very large field of view, covering about 1/5 of the sky in a single pointing. A mini-calorimeter made of Cesium-Iodide bars supports the tracker for photon energy reconstruction and particle background rejection, together with a set of anticoincidence scintillator panels surrounding 5 sides of the instrument. An additional silicon plane and a coded mask, positioned on top of the tracker, constitute SuperAGILE, an X-ray imager with angular resolution of $\sim$ 6 arcmin and sensitivity of $\sim$ 15 mCrab in the band 15-45 keV. AGILE is characterized by the smallest deadtime ever obtained for gamma-ray detection ($<$200 μs) and absolute time tagging with uncertainty of few μs.

The AGILE scientific instrument is based on an innovative design based on three detecting systems: (1) a Silicon Tracker, (2) a Mini-Calorimeter (MC), and (3) an ultralight coded mask system with Si-detectors (Super-AGILE). AGILE is designed to provide: (1) excellent imaging in the energy bands 30 MeV–50 GeV (5–10 arcmin for intense sources) and 15-45 keV (1–3 arcmin for intense sources); (2) optimal timing capabilities, with independent readout systems and minimal deadtimes for the Silicon tracker, Super-AGILE and Mini-Calorimeter; (3) large fields of view for the gamma-ray imaging detector (GRID) ($\sim$3 sr) and Super-AGILE ($\sim$1 sr)[1].

Despite of its smaller dimensions AGILE will have comparable performances to EGRET on axis and substantially better off axis. The innovative technology will allow AGILE to achieve the smallest deadtime in high-energy astrophysics

Fig. 2 shows the AGILE instrument configuration of total weight of $\sim$ 120 kg including the Si-Tracker, Super-AGILE, Mini-Calorimeter, the Anticoincidence system and electronics.

- The **Silicon-Tracker**, is a gamma-ray pair-converter and imager made of 12 planes, with two Si-layers per plane providing the X and Y coordinates of interacting charged particles. The fundamental Silicon detector unit is a tile of area 9.5 $\times$ 9.5 cm^2. Each Si-Tracker layer is made of 4 ladders (each composed of 4 Si tiles), for a total geometric area of 38 $\times$ 38 cm^2 and 1,536 readout channels. The first 10 planes are made of three elements: a first layer of Tungsten (0.07 X_0) for gamma-ray conversion, and two Si-layers (views) with microstrips orthogonally positioned. Both digital and analog information (charge deposition in Si-microstrip) is read by front-end electronics (FEE). The GRID has an *on-axis* total radiation length

FIGURE 1. The AGILE Satellite during the integration phase at the Carlo Gavazzi Space integration facility in Tortona

near $\sim 0.8\,X_0$. Special algorithms applied off-line to telemetered data will allow optimal background subtraction and reconstruction of the photon incidence angle. Both digital and analog information are crucial for this task.
- **Super-AGILE** is the hard x-ray imager of the mission. It is composed of four indipendent unidimensional coded mask units, two with the same coding direction, the other two with the orthogonal. The detection plane of each unit is made of four square Silicon detectors (9.5 *times* 9.5 cm^2 each). Detection plane and associated FEE is placed on the first GRID tray. The Tungsten coded mask is 14 cm above the detection plane. Super-AGILE characteristics are: *(i)* photon-by-photon detection and imaging of sources in the 15-45 keV energy range; *(ii)* a field of view of 1 sr and a PSF of 6 arcmin; *(iii)* excellent timing (4 μs); *(iv)* quick on-board burst alert and localization; *(v)*GRB trigger for the GRID and MCAL.
- The **Mini-Calorimeter** (MCAL) is made of two planes of Cesium Iodide (CsI) bars, for a total (on-axis) radiation length of 1.5 X_0. The signal from each CsI bar is collected by two photodiodes placed at both ends. The MCAL tasks are: *(i)* obtaining additional information on the energy of particles produced in the Si-Tracker; *(ii)* detecting GRBs and other impulsive events with spectral and intensity information in the energy band $\sim 0.3-50$ MeV. We note that the problem of

FIGURE 2. Lateral view of the AGILE instrument (AC System and spacecraft partially displayed) (from Tavani et al. [2]). The GRID is made of a Silicon Tracker (10 Tungsten and Silicon planes plus 2 additional Silicon-only planes) and a Mini-Calorimeter placed at the bottom of the instrument. Super-AGILE has its detection plane made of 4 Si-detectors placed at the top of the first GRID tray, and an ultra-light coded mask system (CMS) positioned on top (the figure shows the CMS partition configuration). The instrument size is $\sim 63 \times 63 \times 58.5 \, \text{cm}^3$, including Super-AGILE and the AC System for a total weight of ~ 130 kg including the electronics and power supply unit positioned inside the spacecraft.

"particle backsplash" for AGILE is much less severe than in the case of EGRET. AGILE allows a relatively efficient detection of (inclined) photons near 10 GeV and above also because the AC-veto can be disabled for events with more than ~ 100 MeV total energy collected in the MCAL.

- **Data handling system**, for fast processing of the GRID, Mini-Calorimeter and Super-AGILE events. The GRID trigger logic for the acquisition of gamma-ray photon data and background rejection is structured in two main levels: Level-1 and Level-2 trigger stages. The Level-1 trigger is fast ($\lesssim 5\mu$s) and requires a signal in at least three out of four contiguous tracker planes, and a proper combination of fired TA1 chip number signals and AC signals. An intermediate Level-1.5 stage is also envisioned (lasting $\sim 20 \, \mu$s), with the acquisition of the event topology based on the identification of fired TAA1 chips. Both Level-1 and Level-1.5 have a hardware-oriented veto logic providing a first cut of background events. Level 2 data processing includes a GRID readout and pre-processing, "cluster data acquisition" (analog and digital information). The Level-2 processing is asynchronous (estimated duration $\sim$ a few ms) with the actual GRID event processing. The GRID deadtime turns out to be $\lesssim 200 \, \mu$s and is dominated by the Tracker readout.

Appropriate data buffers and burst search algorithms are envisioned to maximize data acquisition for transient gamma-ray events (e.g., GRBs) in the Si-Tracker, Super-AGILE and Mini-Calorimeter, respectively.

The Super-AGILE event acquisition envisions a first "filtering" based on AC-veto signals, and pulse-height discrimination in the dedicated FEE (based on XA chips). The events are then buffered and transmitted to the CPU for burst searching and final data formatting. The four Si-detectors of Super-AGILE are organized in sixteen independent readout units, of $\sim 4\,\mu s$ deadtime each.

AGILE SCIENCE PERFORMANCE

The instrument design of the GRID permits to obtain the following performance:

- **excellent imaging capability in the energy range 100 MeV-50 GeV**, improving the EGRET angular resolution by a factor of 2;
- **a very large field-of-view**, allowing simultaneous coverage of $\sim 1/4$ of the entire sky per each pointing (FOV larger by a factor of ~ 6 than that of EGRET);
- **excellent timing capability**, with absolute time tagging of uncertainty near $1\,\mu s$ and very small deadtimes ($\lesssim 200\,\mu s$ for the Si-Tracker and $\sim 20\,\mu s$ for each of the individual CsI bars);
- **a good sensitivity for point sources**, comparable to that of EGRET for *on-axis* sources, and substantially better for *off-axis* sources;
- **excellent sensitivity to photons in the energy range $\sim$30-100 MeV**, with an effective area above 200 cm^2 at 30 MeV;
- **a very rapid response to gamma-ray transients and gamma-ray bursts**, obtained by a special quicklook analysis program and coordinated ground-based and space observations.

The coded mask imager (Super-AGILE) in addition to the GRID will provide a unique tool for the study of high-energy sources. Super-AGILE can provide important information including:

- **source detection and spectral information in the energy range $\sim$15-45 keV** to be obtained simultaneously with gamma-ray data (15 mCrab sensitivity at 15 keV ($5\,\sigma$) for a 50 ksec integration time);
- **accurate localization ($\sim$2-3 arcmins) of GRBs and other transient events** (for typical transient fluxes above ~ 1 Crab); the expected GRB detection rate is $\sim 1-2$ per month;
- **long-timescale monitoring ($\sim$2 weeks) of hard X-ray sources**;
- **hard X-ray response to gamma-ray transients detected by the GRID**, obtainable by slight repointings of the AGILE spacecraft (if necessary) to include the gamma-ray flaring source in the Super-AGILE FOV.

GRB SCIENCE WITH AGILE

About ten GRBs were detected by the EGRET spark chamber during ~ 7 years of operations [3]. This number was limited by the EGRET FOV and sensitivity and, from what we know today, not by the GRB emission mechanism normally producing gamma-rays above 100 MeV[4].

GRB detection rate by the GRID is expected to be at least a factor of ~ 5 larger than that of EGRET, i.e., ≥ 5–10 events/year). The small GRID deadtime (~ 1000 times smaller than that of EGRET) allows a better study of the initial phase of GRB pulses (for which EGRET response was in many cases inadequate). The remarkable discovery of 'delayed' gamma-ray emission up to ~ 20 GeV from GRB 940217 [5] is of great importance to model burst acceleration processes. AGILE is expected to be highly efficient in detecting photons above 10 GeV because of limited backsplashing. Super-AGILE will be able to locate GRBs within a few arcminutes, and will systematically study the interplay between hard X-ray and gamma-ray emissions. Special emphasis is given to fast timing allowing the detection of sub-millisecond GRB pulses independently detectable by the Si-Tracker, MC and Super-AGILE. AGILE will be able to detect most of the bright long bursts and its two detectors (Super-AGILE and MCAL), with their independent trigger algorithms, compensate each other in detecting soft and high energy GRBs[6].

AGILE is able to detect GRBs and obtain precise localization on-board. The burst coordinates will be transmitted in few minutes to ground with the fast link for the follow-up of GRBs with other telescopes.

CONCLUSIONS

The AGILE scientific instrument is innovative in many ways, and is designed to obtain an optimal gamma-ray detection performance despite its relatively small mass and absorbed power. The combination of hard X-ray and gamma-ray imaging capabilities in a single integrated instrument is unique to AGILE. We anticipate a crucial role of AGILE for studies of GRB in X and gamma ray energy ranges, providing an important step forward in high energy GRB study.

REFERENCES

1. Tavani M. et al., *Proceedings of Gamma2001 Conference*, AIP Conf. Proceedings, eds. Ritz S., Gehrels N. and Shrader C.R., 2001, Vol. 587, p. 729
2. Tavani M. et al: 2005, *Science with AGILE*, AGILE Document AP-27, http://agile.mi.iasf.cnr.it/
3. Schneid E.J. et al., in AIP Conf. Proc., Vol. 384, p.253, 1996
4. Tavani M., PRL, 76, 3478, 1996
5. Hurley K. et al., Nature, 372, 652, 1994
6. Ghirlanda G. et al., AIP Conf. Proc., 727, 704, 2004

Prospects for GRB Polarimetry with GRAPE

M. L. McConnell*, P. F. Bloser*, J. Legere*, J. R. Macri*, T. Narita[†] and J. M. Ryan*

*Space Science Center, University of New Hampshire, Durham, NH 03824
[†]Department of Physics, College of the Holy Cross, Worcester, MA USA

Abstract. This paper discusses the latest progress in the development of GRAPE (Gamma-Ray Polarimeter Experiment), a hard X-ray Compton Polarimeter. The purpose of GRAPE is to measure the polarization of hard X-rays in the 50-300 keV energy range. We are particularly interested in X-rays that are emitted from solar flares and gamma-ray bursts (GRBs). Accurately measuring the polarization of the emitted radiation from these sources will lead to a better understating of both the emission mechanisms and source geometries. The GRAPE design consists of an array of plastic scintillators surrounding a central high-Z crystal scintillator. We can monitor individual Compton scatters that occur in the plastics and determine whether the photon is photo absorbed by the high-Z crystal or not. A Compton scattered photon that is immediately photo absorbed by the high-Z crystal constitutes a valid event. These valid events provide us with the interaction locations of each incident photon and ultimately produces a modulation pattern for the Compton scattering of the polarized radiation. Comparing with Monte Carlo simulations of a 100% polarized beam, the level of polarization of the measured beam can then be determined. The complete array is mounted on a flat-panel multi-anode photomultiplier tube (MAPMT) that can measure the deposited energies resulting from the photon interactions. The design of the detector allows for a large field-of-view ($> \pi$ steradian), at the same time offering the ability to be close-packed with multiple modules in order to reduce deadspace. We present in this paper the latest laboratory results obtained from GRAPE using partially polarized radiation sources along with a brief description of our future plans for the GRAPE design.

Keywords: X-ray instrumentation, gamma-ray instrumentation, polarimetric instrumentation, gamma-ray bursts
PACS: 95.55.Ka, 95.55.Qf, 98.70.Rz

INTRODUCTION

Polarization measurements have become a powerful tool for astronomers throughout the electromagnetic spectrum. It is believed that by accurately measuring hard X-ray polarization levels from solar flares and gamma ray bursts (GRBs) we will be able to better understand both the emission mechanisms and source geometries producing the observed radiation [1]. With this goal in mind, we have been developing a hard X-ray polarimeter design that we call GRAPE (Gamma-Ray Polarimeter Experiment). The purpose of GRAPE is to measure the polarization of hard x-rays in the 50-300 keV energy range. As described here, the GRAPE design is most suitable for studies of either gamma-ray bursts or solar flares, as part of a long-duration balloon platform or as part of a satellite platform.

THE GRAPE CONCEPT

The basic physical process used to measure polarization in the 50–300 keV energy range is Compton scattering, in which photons tend to be scattered at a right angle with respect to the incident electric field vector. In the case of an unpolarized beam of incident photons, there will be no net positive electric field vector and therefore no preferred azimuthal scattering angle (η); the distribution of scattered photons with respect to η will be uniform. However, in the polarized case, the incident photons will exhibit a net positive electric field vector and the distribution in η will be asymmetric. The ultimate goal of a Compton scatter polarimeter is to measure the azimuthal modulation pattern of the scattered photons.

The development and design of the GRAPE detector has evolved through three science models [4–12]. Each one represented a successive improvement, but all three essentially operate under the same underlying principle of operation. A high-Z material, the calorimeter, is surrounded by multiple plastic scintillation detectors that serve as a target for the Compton scattering. The plastic scintillators are made of a low-Z material that maximizes the probability of a Compton interaction. The purpose of the calorimeter is to fully absorb the energy of the scattered photon. Ideally, photons that are incident on the plastic scintillator array will Compton scatter only once, and then be subsequently absorbed by the calorimeter. For such an event we measure the energy of the scattered electron in the plastic and the deposited energy of the scattered photon in the calorimeter. With multiple plastic scintillators surrounding the calorimeter, we can determine the azimuthal scatter angle of each valid event. A histogram of these data represents the azimuthal modulation pattern of the scattered photons, which provides a measurement of the polarization parameters (magnitude of the polarization and polarization angle) of the incident flux.

In order to accurately measure the azimuthal modulation (and hence the polarization parameters), we need to correct for geometric effects specific to the individual detector design. When the azimuthal modulation profile is generated, the distribution not only includes the intrinsic modulation pattern due to the Compton scattering process, but it also includes various geometric effects. One of these effects originates from the specific layout of the detector elements within the polarimeter and the associated quantization of possible scatter angles. Other effects include such things as the nonuniform detection efficiency of the PMT used for detector readout. In order to properly account for these effects, the response of the polarimeter must be properly characterized, for example, by measuring the response of the polarimeter to an unpolarized photon beam.

To determine the polarization level, P, of the incident radiation we need to know the modulation factor for a completely polarized beam (μ_{100}). We have used simulations based on MGEANT (incorporating the GLEPS polarization code) to model the response of the polarimeter to 100% polarized incident radiation. The simulations included all important components of the lab setup.

One significant advantage of the GRAPE design is that it affords a very large FoV, with significant sensitivity for sources as far as $60°$ off axis. This is ideal for GRB studies, where the source comes from random directions on the sky. Previous estimates of the GRB polarization sensitivity of the GRAPE design [9, 10] suggest that a significant number of GRBs could provide useful polarization measurements on a Ultra-Long Du-

 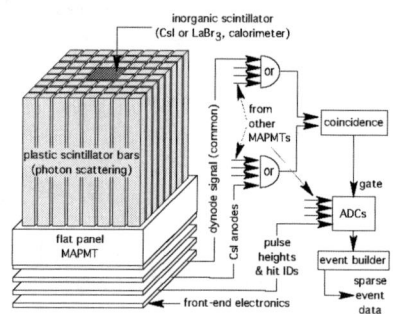

FIGURE 1. GRAPE Science Model 3 (SM3) fully populated with 60 plastic scintillators and a central CsI(Na) calorimeter.

FIGURE 2. GRAPE logic diagram, illustrating the signal processing involved.

ration Balloon (ULDB) platform. Coupled with recent design improvements, we expect that a ULDB version of GRAPE would, depending upon the precise detector configuration, be able to measure GRB polarization at a sensitivity level of about 10% once every 3–4 days. This sensitivity is far below the polarization level of 80% reported for GRB021206 [13].

THE GRAPE DESIGN

The latest version of GRAPE (Science Model 3 – SM3) is a compact design based on the use of a flat-panel MAPMT (Hamamatsu H8500) [11]. The H8500 is a MAPMT with an array of 8×8 independent anodes. The 5 mm anodes are arranged with a pitch of 6 mm, occupying a total area of 52 mm $\times$ 52 mm. The tube depth is only 28 mm. Each GRAPE module includes a complete array of both plastic scattering elements and a CsI(Na) calorimeter element mounted on the front surface of the MAPMT (as seen in Figs. 1 and 2). The MAPMT provides readout of both scintillator types. The entire assembly is housed in a light-tight aluminum conatiner. The square design of the MAPMT also allows for the close packing of multiple modules with minimal dead space.

Experiments were conducted in the laboratory in order to determine the effectiveness of the detector. The polarized radiation used in the tests was produced by Compton scattering radiation from a laboratory gamma-ray source [5]. Radiation from a ^{137}Cs source is collimated with lead shielding and directed toward a plastic scintillation detector, referred to as the polarizer. Photons with an initial energy of 662 keV scatter through an angle of $90°$ before reaching the polarimeter detector. The radiation reaching the polarimeter is highly polarized (at a level of 55-60%) and has a reduced energy of 288 keV. The use of the polarizer also allows us to electronically tag photons. A triple coincidence between the polarizer and the two sets of polarimeter detectors provides an efficient means for recording data from the polarized beam.

Initial laboratory tests of SM3 proved unsuccessful. It was discovered that there were light cross-talk issues between the calorimeter and the twelve plastic elements immedi-

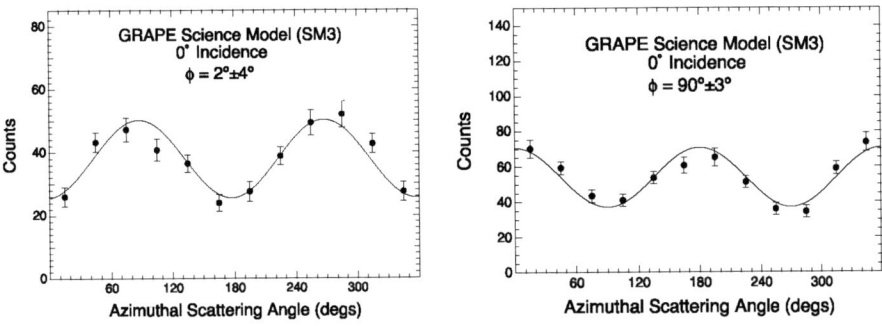

FIGURE 3. Results from SM3 with the inner plastic ring eliminated from the analysis.

ately surrounding the calorimeter (c.f., Fig. 2). We believe that this can be attributed to the close proximity between the calorimeter and adjacent plastics. With a light output of ~ 4 times that of a plastic scintillator, the CsI(NaI) calorimeter was contributing to the energy recorded in this inner plastic ring. We are currently investigating various options for mitigating this cross talk problem. In the meantime, in order to properly analyze the SM3 data, we decided to eliminate the inner twelve plastic scintillator elements from the analysis. This left 48 plastic elements in the array for inclusion in the data analysis. The data shown in Fig. 3 represent the results of an analysis that excluded the innermost plastic elements. Data was collected for $0°$ and $90°$ polarization angle and yielded polarization levels of $56(\pm 9)\%$ and $55(\pm 7)\%$ respectively.

GRAPE FUTURE

The next major step for GRAPE development will be to improve the coincidence timing and energy resolution. We plan to accomplish this by replacing the central CsI calorimeter with one based on Lanthanum Bromide ($LaBr_3$) [14, 15]. This relatively new inorganic scintillator provides an energy resolution that is more than twice as good as NaI at 662 keV (3% vs. 7%). The expected improvement in the GRAPE energy resolution (combining the energy resolution of both the plastic scattering elements and the calorimeter) is shown in Fig. 4. The $LaBr_3$ will also provide better timing characteristics. With decay times of ~ 25 ns, it is comparable to the plastic scintillator and will greatly improve the coincidence timing characteristics of GRAPE.

In order to provide adequate sensitivity, any realistic application of the GRAPE design would involve an array of polarimeter modules, as in Fig. 5 (possibly surrounded by anti-coincidence shielding). Recent simulations have shown that events involving more than one module in a tiled arrangement will provide a significant increase in detection sensitivity and modulation factor. In addition, we have identified a modified arrangement of plastic and calorimeter elements (splitting the calorimeter into four separate pieces and moving them outward from the center of the array) that will also improve the detection efficiency of a single module by nearly a factor of two. We are now in the process of fabricating a 2×2 array of GRAPE modules that would be used for

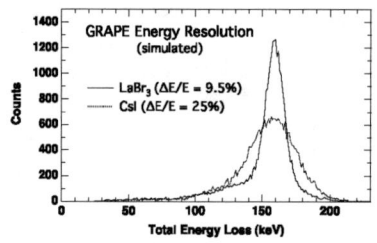 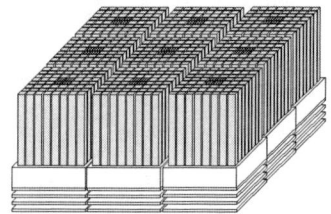

FIGURE 4. Simulated results showing the improved energy resolution that would result from changing the CsI calorimeter to one made from LaBr$_3$.

FIGURE 5. An array of nine GRAPE modules.

laboratory testing and that would also serve as a piggyback balloon package to be flown as an engineering test in late 2006.

One possible deployment option for a GRAPE polarimeter array would be as the primary instrument on an Ultra-Long Duration Balloon (ULDB) payload. The ULDB technology can provide balloon flight durations of up to 100 days. A 1 m^2 array of GRAPE modules would easily fit within the envelope of a ULDB payload. The ideal configuration for GRB studies would be an array that remains pointed in the vertical direction (i.e., towards the zenith) at all times. In this case, there would be no pointing requirements, only a moderate level of aspect information (continuous knowledge of the azimuthal orientation to $\sim 0.5°$). An imaging polarimeter could also be designed to match the payload limitations of a ULDB, although the pointing requirements would be much more severe ($< 1°$ in both azimuth and zenith). A second deployment option would be as part of a spacecraft payload.

ACKNOWLEDGMENT

This work is currently supported by NASA grants NNG04GB83G and NNG04WC16G. We would like to thank Mark Widholm and Paul Vachon for their support with the MAPMT design electronics. We would also like to thank the Laboratory for Advanced Instrumentation Research at Embry Riddle, Sparrow Corp., Drew Weisenberger, and Sergio Brambilla for their support with the VME data acquisition setup.

REFERENCES

1. F. Lei, A.J. Dean and G.L. Hills, *Space Sci. Rev.* **82**, 309–388 (1997).
2. R. Novick, *Space Sci. Rev.* **18**, 389–408 (1975).
3. R. D. Evans, *The Atomic Nucleus*, McGraw-Hill, New York, 1958.
4. M.L. McConnell, D.J. Forrest, J. Macri, J.M. Ryan, and W.T. Vestrand, "Development of a hard X-ray polarimeter for gamma-ray bursts," in AIP Conf. Proc. 428, *Gamma-Ray Bursts*, edited by C.A. Meegan and P. Cushman, AIP, New York, 1998, pp. 889–893.
5. M.L. McConnell, D.J. Forrest, J. Macri, M. McClish, M. Osgood, J.M. Ryan, W.T. Vestrand and C. Zanes, *IEEE Trans. Nucl. Sci.* **45**, 910–914 (1998).

6. M.L. McConnell, J.R. Macri, M. McClish, J. Ryan, D.J. Forrest and W.T. Vestrand, *IEEE Trans. Nucl. Sci* **46**, 890 (1999).
7. M.L. McConnell, J.R. Macri, M. McClish, and J. Ryan, *SPIE Proc.* **3764**, 70–78 (1999).
8. M. L. McConnell, J. R. Macri, and J. M. Ryan, "A modular hard X-ray polarimeter for solar flares," in *High Energy Solar Physics - Anticipating HESSI*, edited by R. Ramaty and N. Mandzhavidze, ASP Conf. Ser. 206, ASP, San Francisco, 2000, pp. 280–283.
9. M. L. McConnell, J. R. Ledoux, J. R. Macri, and J. M. Ryan, *SPIE Proc.* **4851**, 1382-1393 (2002).
10. M. L. McConnell, J. R. Ledoux, J. R. Macri, and J. M. Ryan, "The Development of GRAPE, A Gamma-Ray Polarimeter Experiment," in AIP Conf. Proc. 662, *Gamma-Ray Burst and Afterglow Astronomy*, edited by G. R. Ricker and R. K. Vanderspek, AIP, New York, 2003, pp.503-505.
11. M. L. McConnell, J. R. Ledoux, J. R. Macri, and J. M. Ryan, *SPIE Proc.* **5165**, 334-345 (2004).
12. J. S. Legere, P. Bloser, J. R. Macri, M. L. McConnell, T. Narita and J. M. Ryan, *SPIE Proc.* **5898**, 401-410 (2005).
13. W. Coburn, and S.E. Boggs, *Nature* **423**, 415-417 (2003).
14. van Loef, E. V. D., et al., *Appl. Phys. Lett.* **79**, 1573-1575 (2001).
15. Shah, K. S., et al., *IEEE Trans. Nucl. Sci.* **50**, 2410-2413 (2003).

The Gamma-ray Large Area Space Telescope and Gamma-Ray Bursts

Julie McEnery and Steve Ritz
(for the GLAST Mission Team)

NASA/GSFC, Lab for Astroparticle Physics, MailCode 661, Greenbelt, MD 20771

Abstract.
 The Gamma-ray Large Area Telescope (GLAST) is a satellite-based observatory to study the high energy gamma-ray sky. The main instrument on GLAST, the Large Area Telescope (LAT) is a pair-conversion telescope that will survey the sky from 20 MeV to greater than 300 GeV. With the GLAST launch in 2007, the LAT will open a new and important window on a wide variety of high energy phenomena, including supermassive black holes and active galactic nuclei, gamma-ray bursts, supernova remnants and cosmic ray acceleration and dark matter. A second instrument, the GLAST Burst Monitor (GBM), greatly enhances GLAST's capability to study GRB by providing important spectral and timing information in the 10 keV to 30 MeV range. We describe how the instruments, spacecraft and ground system work together to provide observations of gamma-ray bursts from 8 keV – 300 GeV and to provide rapid notification of bursts to the wider gamma-ray burst community.

HIGH ENERGY OBSERVATIONS OF GRB

EGRET provided a tantalizing glimpse of the properties of gamma-ray bursts above 30 MeV. Only a handful were detected, yet the data are consistent with GRB spectra extending with no cutoffs from the sub-MeV range to the high-energy gamma-ray regime. In a couple of GRB, EGRET also discovered bright, high-energy gamma-ray emission with markedly different temporal and spectral characteristics relative to the sub-MeV component. These observations suggest that, at least for some GRB, the bulk of the luminosity is emitted in the MeV-GeV range by a separate emission mechanism.

 GLAST will address the compelling questions raised by the EGRET observations: How high in energy does the prompt emission extend? Is the hard, independent emission component seen in GRB91017 [1] a common feature of GRB? How common are high energy afterglows such as that seen in GRB940217 [2]?

INSTRUMENTS

GLAST is a general-purpose gamma-ray facility opening a large new window on the high-energy sky. In this section, we focus only on those aspects of the GLAST instruments that are most relevant to GRB science.

CP836, *Gamma-Ray Bursts in the Swift Era*,
edited by S. S. Holt, N. Gehrels, and J. A. Nousek
2006 American Institute of Physics 0-7354-0326-0/06/$23.00

Large Area Telescope

Informed by the experience with the highly-successful EGRET instrument on CGRO, the Large Area Telescope (LAT) will provide a large leap in all key capabilities important for burst observations, including

- very large field of view of about 2.5 sr, which means that approximately 20% of all bursts will start within the LAT FOV;
- large effective area, $> 8,000$ cm^2 on-axis for E greater than a few GeV;
- lage acceptance to $E > 300$ GeV, opening the largely-unexplored region above 10 GeV;
- very small deadtime, approximately 26 μs per event.

Together, these new capabilities will enable the study of high-energy burst emission timing characteristics, and the leap in sensitivity means that more high-energy GRB detections can be anticipated.

More information about LAT performance can be found at `http://www-glast.slac.stanford.edu/software/IS/glast_lat_performance.htm`. The LAT team is a large, international collaboration consisting of more than 120 members from the U.S., Italy, France, Japan, and Sweden. The PI is Peter Michelson (Stanford). The LAT is managed at the Stanford Linear Accelerator Center (SLAC) and is funded by NASA and DOE, along with key international contributions.

GLAST Burst Monitor

The GBM will provide X-ray and gamma-ray observations of GRB from <10 keV to 30 MeV. It has a very large field of view (>8 steradians) so all GRB within the LAT FOV will have high-energy observations spanning over 7 orders of magnitude in energy. In addition, the GBM will detect and characterize bursts outside the LAT FOV. This information can be used onboard to direct the spacecraft to slew autonomously to bring the GRB within the LAT FOV for bursts deemed to be of high interest. The GBM will detect around 200 GRB per year with ~65 of these within the canonical LAT FOV. More details about the GBM instrument can be found at `http://gammaray.msfc.nasa.gov/gbm/`.

The development of the GLAST Burst Monitor and analysis of its observational data is a collaborative effort between institutions in the U.S. and Germany. The Principal Investigator is Dr. Charles Meegan at MSFC. Dr. Giselher Lichti at MPE is co-PI.

TYPICAL GLAST GRB TIMELINE

Upon a burst trigger from either (or both) instrument(s), the observatory will send a sequence of burst alerts to the ground in a similar manner to what is being done for Swift. See Figure 1 for a summary of the processing and messaging timeline. In addition, if the burst is sufficiently interesting, the observatory will repoint autonomously to keep

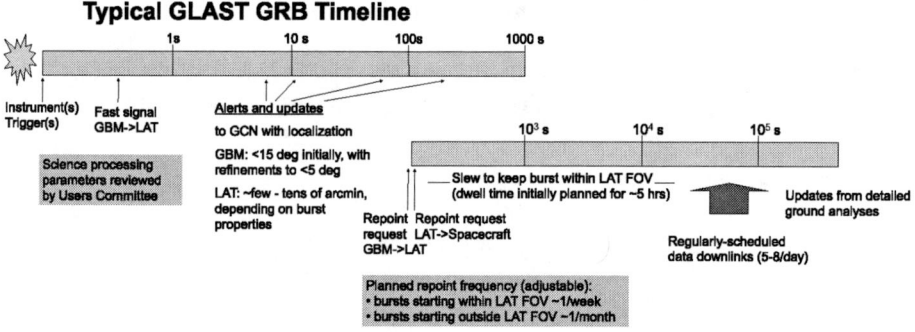

FIGURE 1. GLAST burst processing and action timeline.

the burst within the LAT FOV for a commandable period of time. Based mainly on the remarkable GRB940217[2], from which EGRET observed high-energy emission some 75 minutes after the trigger, the current plan is to set the nominal repoint time to five hours. The autonomous repoint threshold characteristics will be adjusted throughout the mission, in consultation with the community (through the GLAST Users Committee) and the Instrument teams. Bursts that start within the LAT FOV are intrinsically more interesting; remarkable bursts that start outside the LAT FOV will also trigger a repoint request. The thresholds will therefore initially be adjusted to trigger a repoint for LAT-FOV bursts approximately once per week and a repoint for a non-LAT-FOV burst once per month.

SCIENCE OPERATIONS

After inital on-orbit checkout, verification, and calibrations, the first year of science operations will be an all-sky survey. In this mode, the full sky will be covered by the LAT with good uniformity every three hours, with each region of the sky viewed for approximately 30 minutes. The year 1 data will be used for detailed LAT characterization, refinement of the alignment, and key projects including source catalog and diffuse background models, needed by the science community. Thus, during year 1,

- summary data on transients, including GRBs, will be released with caveats;
- autonomous repoints for bright bursts will be enabled;
- burst alerts will be enabled;
- extraordinary targets of opportunity will be supported; and
- workshops on science tools, mission characteristics, and tools for proposal preparation will be held for Guest Investigators.

At the end of year 1, all year 1 data will be released. All subsequent data will be released promptly. The observing plan after year 1 will be driven by Guest Observer proposal selections by peer review, in addition to long periods of sky survey.

MISSION SCIENCE ELEMENTS

In addition to the international instrument teams and the GLAST Project Office at Goddard, there are important science elements including:

- The Science Working Group (SWG), whose membership includes the four GLAST Interdisciplinary Scientists, the Users Committee chair, the two Instrument PIs and instrument team representatives, provides advice on mission science requirements and other aspects of the mission implementation.
- The GLAST Users Committee (GUC) provides external review and feedback on the science tools planning and progress. The GUC is also directly involved in community outreach and Guest Investigator opportuntites planning.
- GLAST data and associated analysis software will be provided to the community by the Science Support Center (GSSC), located at Goddard. The GSSC also supports the Guest Investigator program, provides training workshops for the community, and archives the data to the HEASARC.

More information can be found at `http://glast.gsfc.nasa.gov` and links therein.

SWIFT AND GLAST

Joint observations of bursts by both Swift and GLAST will be extremely valuable as the two missions provide fundamentally different, but complementary, observations. The two instruments on GLAST and the Swift BAT will provide observations of the prompt phase of GRB over a huge energy range. The GLAST LAT and Swift XRT and UVOT will provide afterglow observations at optical, X-ray and high energy gamma-ray wavebands. Assuming a Swift detection rate of 100 GRB per year, if the GLAST and Swift pointing directions are uncorrelated, then around 20 Swift GRB per year will occur within the LAT field of view. The fraction of these actually detected by the LAT, and their characteristics, are important questions for GLAST to answer.

REFERENCES

1. González, M. M., et al, 2003, *Nature*, **424**, 749
2. Hurley, K., et al., 1994, *Nature*, **372**, 652

In-flight calibration of the Swift XRT effective area

G. Cusumano*, S. Campana[†], P. Romano[†], V. Mangano*, A. Moretti[†], A.F. Abbey**, L. Angelini[‡], A.P. Beardmore**, D.N. Burrows[§], M. Capalbi[¶], G. Chincarini[†,∥], O. Citterio[†], P. Giommi[††], M.R. Goad**, O. Godet**, G.D. Hartner[‡‡], J.E. Hill[‡], J.A. Kennea[§], V. La Parola*, T. Mineo*, D. Morris**, J.A. Nousek[§], J.P. Osborne**, K. Page**, C. Pagani[†,§§], M. Perri[¶], G. Tagliaferri[†], F. Tamburelli[¶] and A. Wells**

*INAF-Istituto di Astrofisica Spaziale e Fisica Cosmica di Palermo, Via Ugo La Malfa 153, 90146 Palermo, Italy
[†]INAF – Osservatorio Astronomico di Brera, Via Bianchi 46, 23807 Merate Italy
**Department of Physics and Astronomy, University of Leicester, University Road, Leicester LE1 7RH, UK
[‡]NASA-GSFC, Greenbelt, MD 20771, USA
[§]Department of Astronomy & Astrophysics, Pennsylvania State University, PA 16802, USA
[¶]IASI Science Data Center, via Galileo Galilei, 00044 Frascati, Italy
[∥]Università degli studi di Milano-Bicocca, Dip. di Fisica, Piazza delle Scienze 3, I-20126 Milano, Italy
[††]ASI Science Data Center, via Galileo Galilei, 00044 Frascati, Italy
[‡‡]Max-Planck-Institut für extraterrestrische Physik, Germany
[§§]Pennsylvania State University, 525 Davey Lab, University Park, PA 16802, USA

Abstract. The Swift X-ray Telescope is designed to make astrometric, spectroscopic and photometric observations of the X-ray emission from Gamma-Ray Bursts and their afterglows in the 0.2-10 keV energy band. Here we report some results on the in-flight calibration of the Swift XRT effective area obtained analyzing observations of cosmic sources with different the analysis of cosmic sources intrinsic spectra and using the on-ground calibration and ray-tracing simulations as a starting point. Our analysis includes the study of the effective area for different XRT operating modes.

Keywords: X-ray Telescope, X-ray
PACS: 95.85.Nv, 95.55.Ka

INTRODUCTION

The Swift gamma-ray burst Explorer [1] includes in its payload a wide-field instrument, the gamma-ray (15-350 keV) Burst Alert Telescope (BAT, [2]) which detects the bursts, calculates their position to ~ 1–$4'$ accuracy, and triggers an autonomous slew of the observatory, and two narrow-field instruments: the X-Ray Telescope (XRT, [3]), which operates in the 0.2–10 keV energy range and can provide $\sim 3''$ positions, and the Ultraviolet/Optical Telescope (UVOT, [4]), which operates in the 1700–6000Å wavelength range and provides $\sim 0.5''$ positions.

The XRT is a focusing X-ray telescope that utilizes the third flight mirror module (FM3) developed for the JET–X program [5] and consists of 12 nested, cofocal and coaxial mirror shells arranged in a Wolter I configuration. The mirror diameters range from 191 mm to 300 mm and the focal length is 3.5 m, with a total field of view of ~ 40 arcminutes (50% vignetting level, 1.5 keV). The XRT detector was designed for the EPIC MOS instruments on XMM–Newton and is a MAT-22 CCD consisting of 600 x 602 pixels (40μm $\times 40\mu$m) and being the plate scale of $2.36''$/pixel, it covers an effective field of view of $\sim 24'$ [6]. The XRT Point Spread Function, as measured during the on-ground calibration, is $18''$ and $22''$ (HEW) at 1.5 keV and 8.1 keV, respectively. The XRT effective area is ~ 135 cm^2 at 1.5 keV and ~ 65 cm^2 at 8.1 keV, and depends on XRT read-out modes and grade selection. Four calibration sources are located at the corners of the detector to monitor the spectral resolution as the mission progresses.

XRT supports four different read-out modes to cover the dynamic range and rapid variability expected from GRB afterglows. The switch between modes is performed automatically in order to minimize pile-up and optimize the collected information as the flux of the afterglow diminishes. In Imaging mode the XRT produces an integrated image (no X-ray event recognition takes place) which, for a typical GRB flux, is highly piled up. No spectroscopy is therefore possible, but a very accurate position and a good flux estimate can be obtained. The Photodiode mode (PD) is designed for very bright sources and for high timing resolution. Depending on the source count rate, two telemetry formats are available; at high fluxes (<60 Crab) data are telemetred in Piled-up PD (PuPD) mode in which data are piled-up and spectral information is degraded; at lower fluxes (< 3 Crab) data are telemetred in Low Rate PD (LrPD) and full spectral information is available. The spectra of the four corner sources are superimposed on the spectra of the astronomical sources. High resolution light curves with a time resolution of 0.14 ms are generated. The Windowed Timing (WT) mode is obtained by binning and compressing 10 rows into a single row, and then reading out only the central 200 columns of the CCD. Therefore, it covers the central 8 arcminutes of the field of view and provides one dimensional imaging and full spectral capability with a time resolution of 1.8 ms. This mode is used for fluxes in the range 1–600 mCrab. The Photon Counting mode (PC) allows full spatial and spectral resolution for source fluxes below 1 mCrab with a timing resolution of 2.5 seconds.

X-ray events are classified according to the number and distribution of pixels in which they are detected, and are assigned 'grades' accordingly (XRT uses a library of grades which is derived from the XMM-Newton grading scheme). Default values of grades are 0–5 for PuPD and LrPD, 0–2 for WT and 0–12 for PC mode.

The on-ground calibration of the XRT effective area was carried out in 2002 Sep 23–Oct 4 at the Panter laboratory of the Max-Planck-Institut für extraterrestrische Physik where the integrated system was tested [7]. Here we describe the performance of the present Ancillary Response Files (ARF) generation that were improved, based on the spectroscopic study of several astronomical sources observed by XRT during the in-flight calibration phase, which ended on 2005 Apr 5, and later observations of some cosmic sources performed contemporaneously with XMM and RXTE observatories.

THE IN-FLIGHT CALIBRATION

The XRT effective area is the product of three components: i) the effective area of the mirror, ii) the quantum efficiency (QE) of the CCD and iii) the filter transmission.

In the specific case of XRT, the QE is included in the redistribution matrix (RMF) while the ARFs include the mirror effective area and filter transmission, as well as the vignetting correction and the Point Spread Function (PSF) correction, as a function of the source location and of the size of the extraction region.

We generated ARFs for LrPD, WT and PC XRT modes, and for different grade selections for each mode[1], as an improvement upon the on-ground ones. Based on our knowledge of well-known spectral distribution of some stable astrophysical sources, the on-ground ARFs were modified so that the resulting XRT spectrum becomes consistent with the observed ones through detailed modeling of the residuals. The XRT effective area was calibrated using the Crab nebula (for LrPD and WT modes) and the SNR B0540-69 (for PC mode). Here we report the performance for the default grade selections.

For the LrPD mode the on-ground ARF was modified to reproduce the Crab spectral energy distribution with best fit parameters consistent with those reported in the literature and based on data collected by BeppoSAX [8] and RXTE [9]. Fig. 1 (top panel) shows the residuals obtained by fitting the Crab nebula spectrum with an absorbed power law. The reduced χ^2 is 1.5 (919, dof). The residual systematic uncertainty is lower than 5% level in the 0.6-10 keV energy range with a residual feature at the level of 5% at 5 keV. The flux measurements between XRT and BeppoSAX MECS produce a discrepancy of 2%.

The Crab rate produces, in WT mode, a moderate pile-up that we reduced extracting the off-pulse spectrum (mainly due to the nebula) with a phase-resolved selection. We modified the on-ground WT ancillary files to reproduce the off-pulse spectral model obtained by BeppoSAX-MECS (G. Cusumano, private communication). Fig. 1 (middle panel) shows the residuals obtained by fitting the phase-resolved Crab nebula spectrum with an absorbed power law. The fit yields $\chi^2_{red} = 1.4$ (943 dof). We estimate a residual systematic uncertainty lower than of 5% in the 0.3-10 keV energy range with a residual feature at the level of 10% around 0.7 KeV.

PC mode ARFs were calibrated with the SNR B0540-69 that, thanks to its low count rate ($\sim$ 0.7 c/s) and moderate angular extension ($\sim$ 10-15 arcseconds), produces a negligible pile-up. Fig. 1 (bottom panel) shows the residuals obtained by fitting the SNR spectrum with an absorbed power-law model + a non-equilibrium ionization model (NEI). The reduced χ^2 is 1.4 (943, dof). The statistical uncertainty is lower than 10% level in the 0.7-10 keV energy range, of the same order of the available statistics. Systematics residual are still present below 0.7 keV.

CONCLUSION

The in-flight calibration has allowed to improve the effective area files for all observing modes and grade selections to a level that satisfies the mission requirements. Further

---

[1] The adopted calibration method implies that in the CALDB ARF files we include the residual correction of the CCD quantum efficiency. This explains why the nominal ARF files are different for different grade selections.

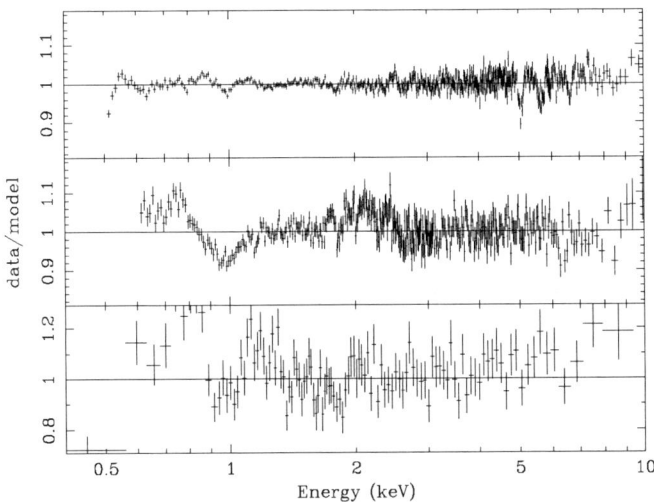

FIGURE 1. Top panel (LrPD mode): residuals (data divided by the folded model) obtained by fitting the Crab nebula spectrum with an absorbed power law. Middle panel (WT mode): residuals obtained by fitting the phase-resolved off-pulse Crab nebula spectrum with an absorbed power law. Bottom panel (PC mode): residuals obtained by fitting the SNR 0540-69 spectrum with an absorbed power-law model + a non-equilibrium ionization model (NEI).

improvement of the ARFs, especially at the low energy end of the spectrum, is coming with the use of more calibration sources and by using XRT and XMM-Newton contemporaneous source observations.

ACKNOWLEDGMENTS

The authors acknowledge support from ASI, NASA and PPARC.

REFERENCES

1. N., Gehrels, G., Chincarini, P., Giommi, et al., *Astrophys. J.* **611**, 1005-1020 (2004)
2. S., Barthelmy, L.M., Barbier, J.R., Cummings, et al., *Space Sci. Rev.*, **120**(2005)
3. D.N., Burrows, J.E., Hill, A., Nousek, et al., *Space Sci. Rev.*, **120**(2005)
4. P.W. Roming, et al., *Space Sci. Rev.*, **120**(2005)
5. O., Citterio, S., Campana, P., Conconi, et al., *Proc. SPIE*, **2805**, 54-65 (1996)
6. D.N., Burrows, J.E., Hill, A., Nousek, et al., *Proc. SPIE*, **4851**, 1320-1325 (1996)
7. G., Tagliaferri, M., A.Moretti, S. Campana, et al., *Proc. SPIE* **5165**, 241-250 (2004)
8. E., Massaro, M., Litterio, G., Cusumano al., *Integral Workshop 4-8 September 2000*, 229-233 (2001)
9. S.H., Pravdo, G., L., Angelini and A.K. Harding, *Astrophys. J.*, **491**, 808-815 (1997)

BOOTES-IR:
The extension of BOOTES towards the near-IR

A. de Ugarte Postigo*, A.J. Castro-Tirado*, M. Jelínek*, P. Kubánek[†],
R. Cunnife*, J. Gorosabel*, S. Vitek*, S. Castillo-Carrión*, S. Guziy*,
S.B. Pandey*, T. de J. Mateo Sanguino**, J.M. Castro Cerón[‡],
F.M. Zerbi[§], P. Conconi[§], S. Covino[§], A. Riva[§], V. de Caprio[§],
P. Amado*, A. Claret*, C. Cardenas*, S. Martín*,
J.M. Trigo-Rodriguez[¶], C. Sánchez-Fernández[‖], M.D. Sabau-Graziati[††],
J. Díaz-Verdejo[‡‡] and F. Vitali[§§]

*Instituto de Astrofísica de Andalucía (IAA-CSIC), P.O. Box 03004, E-18080 Granada, Spain.
[†]Integral Science Data Center, Chemin d'Ecogia 16, CH-1290 Versoix, Switzerland.
**E.P.S., U. de Huelva, E-21819 La Rábida - Palos de la Frontera (Huelva), Spain.
[‡]Niels Bohr Institute, University of Copenhagen, Juliane Maries Vej 30, D-2100 Copenhagen Ø, Denmark.
[§]INAF-Osservatorio Astronomico di Brera Via E. Bianchi, 46, I-23806 Merate (LC), Italy.
[¶]IEEC-CSIC, Campus UAB, Facultat de Ciencies, E-08193 Bellaterra (Barcelona), Spain.
[‖]ESA-VILSPA, E-28080 Villafranca del Castillo, Madrid, Spain.
[††]INTA, Ctra. de Ajalvir, km 4, E-28850 Torrejón de Ardoz, Madrid, Spain.
[‡‡]Area de Ingeniería Telemática, Universidad de Granada, E-18071 Granada, Spain.
[§§]INAF-Osservatorio Astronomico di Roma, I-00040 Monteporzio Catone, Italy.

Abstract. BOOTES-IR is the natural evolution of BOOTES (Burst Observer and Optical Transient Exploring System) towards the near-infrared (NIR). Using the experience gathered during the last 7 years, we are developing a powerful tool for the detection and study of Gamma-Ray Burst (GRB) afterglows. The system includes a custom developed robotic observatory software (RTS2) that manages all the instruments and devices of the observatory to make them work together in an optimal way. Additionally we have created a package of real time reduction and analysis tools (JIBARO) that allow us to automatically identify GRB afterglow candidates. The telescope, a 0.6m Ritchey-Chrétien, now in commissioning phase, is capable of pointing with a typical delay of 5 seconds between arrival of the alert and start of the first exposure. The NIR camera, with 1024 x 1024 pixels and a field of view of 12' x 12' will see first light in Spring 2006 and is expected to be in normal operation during the Summer. An optical camera working in parallel will complement observations.

Keywords: Ground-based ultraviolet, optical and infrared telescopes, Remote observing techniques, Image processing (including source extraction), gamma-ray sources; gamma-ray bursts.
PACS: 95.55.Cs, 95.75.Rs, 95.75.Mn, 98.70.Rz.

INTRODUCTION

BOOTES-IR [1] is the evolution towards the near-infrared (NIR) of BOOTES (Burst Observer and Optical Transient Exploring System, [2]) which is working in southern Spain since 1998. BOOTES is actually made up of three enclosures, two in Huelva (BOOTES-1A & BOOTES-1B) and one in Málaga (BOOTES-2) containing several

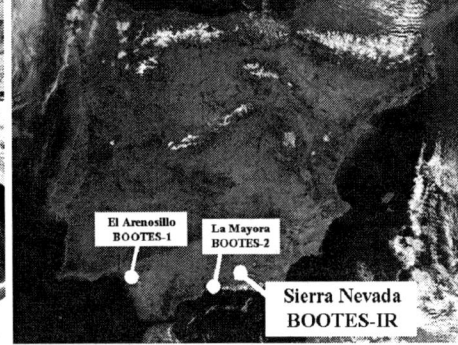

FIGURE 1. BOOTES-IR (left) is located in southern Spain (right), atop Sierra Nevada mountains in Sierra Nevada Observatory (OSN).

robotic telescopes with various apertures of up to 0.3m. During this time our team has developed custom utilities for controlling robotic operations (RTS2, [3]) and for image reduction and analysis (JIBARO, [4]).

In the NIR we will be able to detect high redshift events and afterglows happening in highly extincted regions that would appear invisible in optical wavelengths. It is also a complement to optical BOOTES, as we will have multiband observations that will allow us to study the prompt evolution of the spectral energy distribution of GRB afterglows.

BOOTES-IR was first proposed in 2001. The enclosure was built atop Sierra Nevada in the Summer of 2003. The telescope was installed at the end of 2004 and first light was obtained in February 2005. Since then the telescope is in commissioning phase and operating with an optical camera. The NIR camera will see first light in Spring 2006.

THE SITE

Sierra Nevada Observatory (www.osn.iaa.es), owned by the Instituto de Astrofísica de Andalucía (IAA-CSIC), is located at 2986m above sea level, not far from the city of Granada. Its high altitude and dry climate makes it ideal for NIR observations. It is located in the middle of a sky resort, so access and support is guaranteed all year round (see Fig. 1).

TELESCOPE

The telescope is a 0.6m Ritchey-Chrètien working at f/8. Its direct drives allow it to slew at a speed of 10° per second, being capable of pointing any part of the sky within 20 seconds, with a typical slew time of 5 seconds. It has two Nashmyth, one of which will be occupied by the NIR and optical cameras. Initially the second Nashmyth will be idle, although future instrumentation is foreseen.

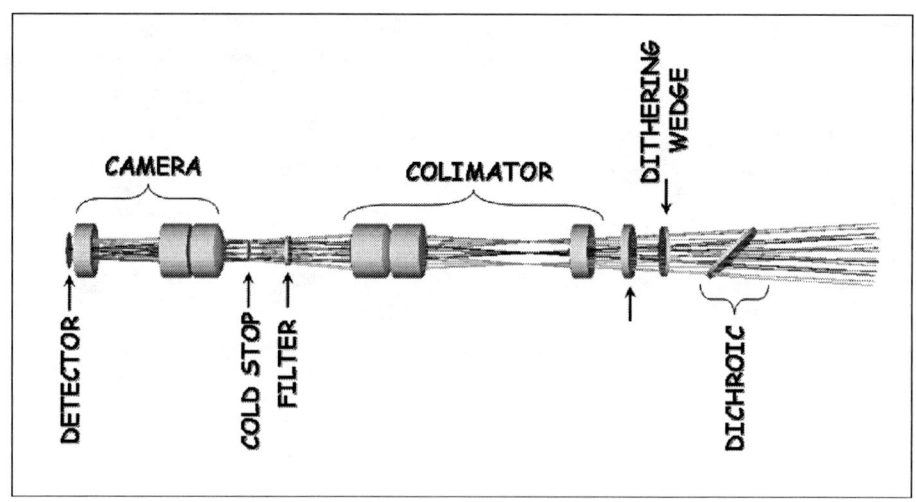

FIGURE 2. Optical layout of the BOOTES NIR camera.

TABLE 1. Estimated limiting magnitudes for BOOTES-IR.

Exposure	J(10σ)	J(5σ)	H(10σ)	H(5σ)	K(10σ)	K(5σ)
5 s	15.5	16.3	14.3	15.1	13.1	13.9
30 s	16.4	17.2	15.2	16.0	14.1	14.8
600 s	17.3	18.0	16.3	17.1	15.4	16.1

CAMERA

The main instrument is a NIR camera with a 1024 x 1024 pixel HgCdTe detector sensible from 0.9 to 2.5 μm. The pixel scale is 0.7", providing a field of view of 12'x12'. Dithering of the images is achieved by using a rotating tilted-window (dithering wedge) that displaces the image in a circular pattern. This allows us to later correct the sky background present in NIR images (see Fig. 2).

A dichroic placed before the entrance of the NIR camera divides the incoming light cone sending the visible light towards an optical camera while transmitting the NIR. This will permit simultaneous observations in two bands. Colour terms will help us to discriminate between GRB afterglows and other objects.

SOFTWARE

The observatory is operated using RTS2, a robotic observatory environment developed by P. Kubánek [3]. The system controls all the devices in the observatory and is able to interrupt observations and immediately start slewing and exposing when a GRB alert arrives.

Image reduction and analysis is achieved using JIBARO, a package of utilities developed in optical BOOTES and now extended for the NIR. It is capable of operating in real-time producing reduction, astrometry, photometry and detection of transient sources (such as GRB afterglows). The position of GRB afterglow candidates can be immediately transmitted to the project scientist, who can then report any dicovery to the scientific comunity and request observations with bigger telescopes to complement the observations by BOOTES-IR.

SCIENCE

The main science of the project is the detection and study of GRBs afterglows. The experience obtained with BOOTES is now extended to the NIR where we can detect high redshift events that would be invisible in optical bands. Furthermore, highly extincted events are more easily detectable in the infrared, as visible bands are strongly dimmed by dust. This makes the NIR camera ideally suited for GRB afterglow studies during the first minutes to hours after the burst onset (see Table 1 for a compilation of the NIR limiting magnitudes). In combination with simultaneous imaging with an optical camera, BOOTES-IR becomes a powerful tool for afterglow detection using colour-colour diagrams.

GRBs will only cover a small fraction of the telescope's observing time. Thus, several parallel programs are foreseen. These "secondary science" projects cover from the study of solar system objects (minor planets, comets), through galactic sources (active stars, brown dwarfs, X-ray binaries and micro quasars) to the extragalactic zoo (AGNs, QSO, Supernovae).

During commissioning phase some scientific programs have began to work with the provisional optical camera and are already producing results, some of which have been published in GCN and IAU circulars ([5], [6]).

ACKNOWLEDGMENTS

This project is supported by Spain's Ministerio de Ciencia y Tecnología under programmes AYA2002-0802 and AYA2004-01515 (including FEDER funds). A. de Ugarte Postigo acknowledges support from an FPU grant from Spain's Ministerio de Educación y Ciencia.

REFERENCES

1. A.J. Castro-Tirado et al., *Il Nuovo Cimento* 28, 715, (2005).
2. A.J. Castro-Tirado et al., *A&AS*, v.138, p.583-585 (1999).
3. P. Kubánek et al., *AIP Conference Proceedings*, Vol. 727 (2004).
4. A. de Ugarte Postigo et al., "JIBARO: Un conjunto de utilidades para la reducción y análisis automatizado de imágenes" in *Astrofísica Robótica en España*, edited by A.J. Castro-Tirado, B.A. de la Morena and J. Torres, Madrid, 2005 pp. 35-50.
5. A. de Ugarte Postigo et al., GCN 3376 (2005).
6. J.M. Pérez-Torres et al., IAUC 8546 (2005).

Real-Time Optical Monitoring of GRBs

René Hudec[1] and Miroslav Křížek[2,1]

[1] *Astronomical Institute of the Academy of Sciences of the Czech Republic, CZ-251 65 Ondřejov, Czech Republic rhudec@asu.cas.cz*
[2] *Astronomical Institute, Faculty of Mathematics and Physics, Charles University, Prague, Czech Republic*

Abstract. Even the fastest alert robotic follow-up telescope is unable to cover the times just after (within first 10 seconds) and before GRB triggers. This time domain is accessible by optical monitors only. We report on analyses of GRB positions on images taken by optical photographic monitors (now operated remotely) within the European meteor network EN. This system is able to provide real-time and pre-burst optical data for GRBs with limiting magnitudes up to 12 in the best cases. The image database is searchable by special software for coincidences with GRBs and the particular images are then scanned and evaluated by computer.

Keywords: Optical All Sky Monitors, Astronomical photography, Gamma-ray bursts
PACS: 95.75 De, 95.75 Mn, 95.75 Pq

INTRODUCTION

The time 0 sec after the GRB onset can be never achieved by alert systems. The alert systems will never achieve coverage for times before the GRB triggers. Both these time coverage's can be easily achieved by sky monitors. There are theoretical expectations that optical flashes may precede GRBs [1].

EN: PHOTOGRAPHIC ALL SKY MONITORING

The European Fireball Network (EN) with 11 stations in the Czech Republic, provides simultaneous optical data for various projects, with almost complete sky monitoring (180 degrees diameter field of view). The optics is Fish-Eye Objective F-Distagon 3.5/30 and the detector is represented by planfilm FOMAPAN 400 ASA or 100 ASA (panchromatic emulsion) 90 x 120 mm, sky diameter 80 mm. Typical exposure time is 3 hrs for guided cameras, and whole night for fixed cameras. Two stations are equipped with guided and fixed cameras and 9 stations are equipped with fixed cameras. The sensitivity for brief 1 sec triggers is 2-3 mag, and for stars up to mag 10... 12. The response is limited to the red light above 400 nm.

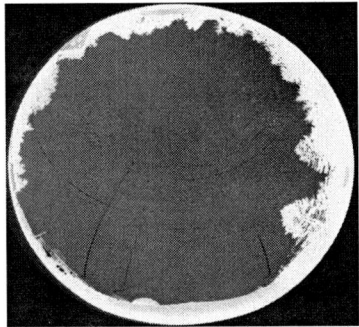
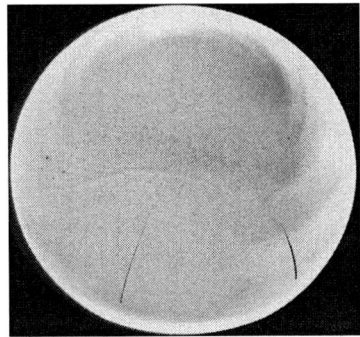

FIGURE 1. Left: Example of digitised trailed all-sky plate. The additional non-stellar trails were caused by aircrafts. Right: Example of digitised guided (pointed) all-sky plate. The lines were created by aircrafts.

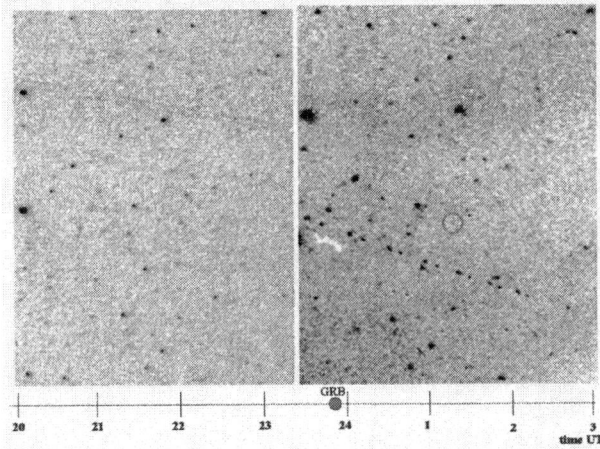

FIGURE 2. The area of the GRB000926 on the plates of the EN network. Left: pre-burst image (end of exposure 29 minutes before the GRB trigger), limiting magnitude 10. Right: simultaneous image, limiting magnitude 8. The position of the GRB is indicated by a circle. Both images contain airplane trails/lights.

RECENT EFFORTS

The recent efforts focus on computer based correlation of plate databases and GRB catalogues to select all plates taken in the time period + - 3 days from the GRB trigger. Novel tools and programs have been developed to make these studies more effective. Recently, for each GRB trigger the plates taken within 3 days from the GRB time are listed including their parameters, the computer-generated sky image analogous to the digitised plate is generated, and the position and error box of GRB is indicated (both for pointed as well as trailed plates). The selected plates relevant to particular GRBs are then scanned by a plate scanner and further evaluated.

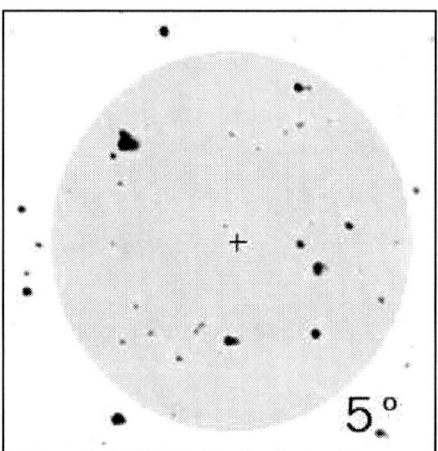

FIGURE 3. The position of the GRB030226 on the digitised sky patrol plate 20A 6300 (taken during the GRB time). The position of the trigger is indicaded by a cross, the size of the error box is equal to the size of the cross.

TABLE 1. Number of guided plates for particular time intervals after GRB.

GRB	Total	[-72;-48)	[-48;-24)	[-24;-12)	[-12;-6)	[-6;-3)	[-3;-1)	[-1;0)	0	(0;1]	(1;3]	(3;6]	(6;12]	(12;24]	(24;48]	(48;72]
JG 051109	11	0	1	2	0	1	1	0	1	1	0	0	0	1	0	3
JG 051006	26	0	3	4	0	0	0	0	2	1	1	1	1	3	5	5
JG 050922C	12	3	2	0	0	0	0	0	1	1	0	0	0	2	1	2
JG 050904	24	5	4	0	0	0	0	0	1	0	0	0	0	4	5	5
JG 050824	12	1	0	0	0	0	0	2	2	0	1	0	0	1	2	3
JG 050714	11	0	2	2	0	0	0	1	2	0	0	0	0	1	1	2
JG 050509	13	2	0	0	0	0	2	1	1	0	0	0	0	2	1	4
JG 050505	12	4	0	0	0	0	0	1	1	2	0	0	0	1	0	3
JG 050215	3	0	0	0	0	1	1	0	1	0	0	0	0	0	0	0
JG 050209	14	4	3	1	0	0	1	0	1	0	0	0	0	4	0	0
JG 041220	15	2	1	2	0	1	1	0	2	0	2	1	0	3	0	0
JG 040916X	20	3	3	0	0	0	0	0	1	1	0	0	0	4	3	5
JG 031111B	37	8	6	4	0	0	0	2	2	0	1	2	0	4	4	4
JG 031109B	8	1	0	1	0	0	0	1	1	0	0	0	0	0	2	2
JG 030323	23	2	6	4	0	0	0	1	1	0	0	0	0	4	1	4
JG 030226	24	3	5	2	0	2	0	1	2	0	0	0	0	5	3	1
JG 020409	13	1	5	3	0	0	0	0	1	0	1	1	0	0	0	1
JG 010115	14	0	1	5	0	0	0	0	1	1	0	1	0	2	2	1
JG 001020	7	0	0	1	0	0	0	0	1	0	0	0	0	2	1	2
JG 001019	21	4	1	0	0	0	1	1	1	0	0	0	0	2	7	4
JG 001017	11	1	3	2	0	0	0	0	1	0	0	0	0	0	2	2
JG 000926	22	3	2	1	0	0	1	1	2	0	1	0	0	2	4	5
JG 990625	5	0	2	0	0	0	0	0	1	0	0	0	0	0	0	2
JG 971214	7	0	0	0	0	1	0	1	1	0	0	0	0	1	2	1
JG 970228	15	0	0	1	0	0	0	0	1	0	0	0	0	3	6	4
Total	380	47	50	35	0	6	8	13	32	7	7	6	1	51	52	65
		0.12	0.13	0.09	0	0.02	0.02	0.03	0.08	0.02	0.02	0.02	0	0.13	0.14	0.17

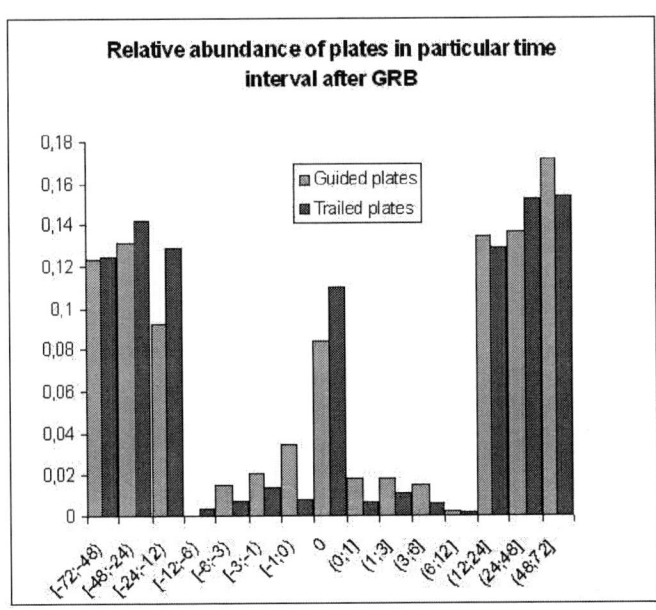

FIGURE 4. Relative abundance of plates in particular time interval before and after GRB. For the results indicated in this figure as well as in the Table 1, the catalog of precisely localized GRBs [2] was used.

ACKNOWLEDGEMENTS

We acknowledge the support provided by the Grant Agency of the Academy of Sciences of the Czech Republic, grant A3003206.

REFERENCES

1. Paczynski, B., 2001, astro-ph/0108522.
2. Greiner, J., 2005, http://www.mpe.mpg.de/~jcg/grbgen.html.

In–flight calibration of the Swift XRT Point Spread Function

A. Moretti*, S. Campana*, M. Capalbi†, G. Chincarini*, S. Covino*, G. Cusumano**, P. Giommi†, V. La Parola**, V. Mangano**, T. Mineo**, M. Perri†, P. Romano* and G. Tagliaferri*

*INAF, Osservatorio Astronomico di Brera, via E. Bianchi 46, I-23807 Merate (LC), Italy
†ASI Science Data Center, via G. Galilei, I-00044 Frascati (Roma), Italy
**INAF, Istituto di Astrofisica Spaziale e Fisica Cosmica di Palermo, via U. La Malfa 153, I-90146 Palermo, Italy

Abstract. The *Swift* X–ray Telescope (XRT) is designed to make astrometric, spectroscopic and photometric observations of the X–ray emission from Gamma–ray bursts and their afterglows, in the energy band 0.2–10 keV. Here we report the results of the analysis of *Swift* XRT Point Spread Function (PSF) as measured in the first four months of the mission during the instrument calibration phase. The analysis includes the study of the PSF of different point–like sources both on–axis and off–axis with different spectral properties. We compare the in–flight data with the expectations from the on–ground calibration. On the basis of the calibration data we built an analytical model to reproduce the PSF as a function of the energy and the source position within the detector which can be applied in the PSF correction calculation for any extraction region geometry. All the results of this study are implemented in the standard public software.

Keywords: *Swift*, XRT, PSF
PACS: 95.55.Ka

The *Swift* satellite was successfully launched on 2004 Nov 20. The XRT first light was on December 12th and the verification and calibration phase ended on April 5th 2005. The XRT is a sensitive, autonomous X-ray CCD imaging spectrometer designed to measure the flux, spectrum, and light curve of GRBs and their afterglows over a wide flux range covering more than seven orders of magnitude in flux in the energy band 0.2–10 keV. XRT utilizes the third flight mirror module (FM3) originally developed for the JET–X program: it consists of 12 nested, confocal and coaxial mirror shells having a Wolter I configuration. The mirror diameters range from 191 mm to 300 mm, the nominal focal length is 3500 mm, the total field of view is about 40 arcminutes (at 50% vignetting level) and the effective area at 1.5 keV is $\sim$ 135 cm^2. The XRT imaging array is a e2v technologies CCD22 consisting of 600 x 600 pixels, each 40μm $\times$40μm, with a nominal plate scale of 2.36 arcseconds per pixel, which makes the effective field of view of the system $\sim$ 24 arcmin (see [1] for a detailed description of the instrument).

Here we present the recent results of the in–flight instrument calibration phase which lasted for the first 4 months of the mission (see [2] for a detailed description of the observations and the analysis). In this period some ad–hoc observations of faint point-like sources with different spectral properties were performed in different positions of the detector in order to observe the surface brightness (SB) profile as function of the energy (E) and the distance from the optical axis of the telescope, usually called off–

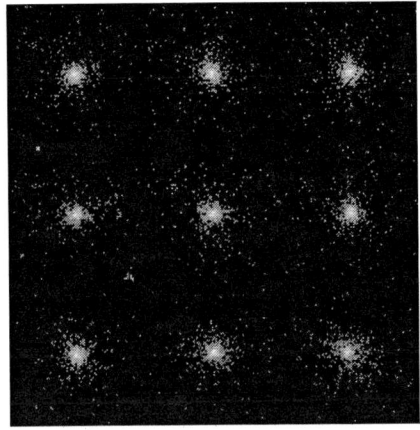 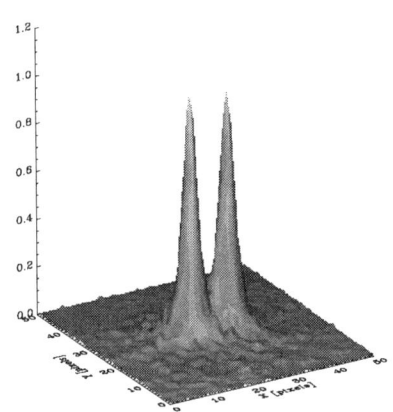

FIGURE 1. Some examples of the images we used in this work. Left panel. Each row corresponds to a different energy bin, 0–1(bottom), 1–2 (middle), 2–10 (top). The columns correspond to different off–axis angles in the range $1'$-$9'$ (increasing from left to right). Right Panel. We show twice the same on–axis observation of RXJ0720.4–3125, where one image has been artificially displaced by $20''$, equivalent to 8.5 pixels (1 pixel=$2.36''$).

axis angle (θ). The in–flight PSF calibration aimed to confirm that the launch and the pre-launch operations did not introduce any distortions of the optics and to measure the PSF shape with a point–like source positioned at infinite distance [3]. The final product of this calibration is an analytical model that reproduces the SB profile of a generic point–like source and can be used for the calculation of the PSF corrections.

The XRT supports different readout modes to enable it to cover the large dynamic range and rapid variability of the GRB afterglows (see [4]). The only readout mode useful for the PSF calibration purposes is the photon counting (PC) mode, which allows full spectral and spatial information for source fluxes below 1 count per second.

For the PSF calibration, low count rate targets are required to completely avoid any PSF distortion due to the pile–up effect. In PC data this effect is significant for count rates > 0.7 counts per second. We used the Isolated Neutron Star RXJ0720.4–3125, which has a count rate of 0.3 counts per second in the 0.2–10 keV band, and a very soft spectrum to study the PSF at low energies (< 2 keV). In order to observe the hard energy PSF we used the observation of the active galactic nucleus Mkn 876 which has a typical count rate of 0.2 counts per seconds in the band 0.2–10keV

We have 12 useful observations, 7 of Mkn876 and 5 of RXJ0720.4–3125, taken in different positions within the field of view, in the $0.1'$ to $9.9'$ range. To study the energy dependence of the PSF, we split the events of all the observations in 3 different energy bins, 0.2–1, 1–2, 2–10 keV (see Fig.1). This binning is a trade–off between having a good energy resolution and a significant number of photons for each bin. Note that the RXJ0720.4-3125 observations have the third energy bin completely empty. This resulted in a list of 31 images ($5 \times 2 + 7 \times 3$). The PSF profile analysis was performed by means of some home–made IDL routines and some DAOPHOT routines within the IDL Astronomy Library(*http://idlastro.gsfc.nasa.gov*).

The Half Energy Width (HEW), defined as the diameter that contains 50% of the total flux, is a very useful parameter to test the performance of our optical system and to study the dependence of the PSF on the different energies and positions in the focal plane. The HEW on–axis (18.0 $''\pm 0.7$ at 1.5 keV) is a local maximum and the HEW decreases for off–axis angles up to 6' (17.2 $''\pm 0.8$ at 1.5 keV). The same feature is expected and was also observed from the ground calibration data. This is due to the fact that the CCD is intentionally slightly offset along the optical axis from the best on–axis focus in order to have a uniform PSF over a large fraction of the field of view. The comparison of the data with the HEWs expected from ray tracing indicated that the offset is about -2 mm [3]. The result is that the optical response of the system is highly uniform over the central $\sim 8'$ radius region of the field of view. Because of the particular operational procedures of the satellite, many XRT observations are performed with the source not perfectly on–axis: the detector position along the optical axis represents a good trade–off between having a good spatial resolution and a larger field of view. We also note that beyond $\sim 8'$ the HEW increases considerably.

The crucial point in our PSF analysis is the construction of an analytical model which describes the PSF as function of (E,θ). The main goal of building this model is the calculation of the PSF correction, which gives for a generic observed source the fraction of the flux contained in the extraction region with a generic shape (square, circle or annulus for example). This is a fundamental ingredient in the photometric measurements and also in the construction of the Ancillary Response File necessary for spectroscopic analysis.

We selected all the observations of the 2 calibration sources up to 9' of off–axis angle and for each observation we split the events in three energy bins (0.2-1,1-2,2-10 keV). For each choice of energy bin and off–axis position, the PSF profile can be well fitted by a King function:

$$PSF(r) = (1+(\frac{r}{r_c})^2)^{-\beta}. \qquad (1)$$

One of the main advantages of this function is that it is analytically integrable in rdr and therefore the integral profile (or Encircled Energy Fraction, EEF) and correspondingly the total flux of a source are also analytically characterized.

$$EEF(r) \equiv \int_0^r PSF(r')2\pi r'dr' = \frac{\pi r_c^2(1-W)}{1-\beta}((1+(\frac{r}{r_c})^2)^{1-\beta} - 1) \qquad (2)$$

$$EEF(\infty) = \pi r_c^2/(\beta-1) \qquad (3)$$

The model has 2 free parameters (plus the normalization), the core radius (r_c) and the slope (β) which are functions of the energy E and position, $r_c = r_c(E,\theta)$, $\beta = \beta(E,\theta)$. To make the model useful for our purposes, i.e. the PSF correction for a generic source, we need to make it predictive and we used the following procedure.

For each of the different sampling points in the energy–position (E,θ) plane, we fitted the best fit PSF parameter values $r_c(E,\theta)$ and $\beta(E,\theta)$ with a plane function:

$$r_c(E,\theta) = a_1 + b_1 \times \theta + c_1 \times E + d_1 \times E \times \theta$$

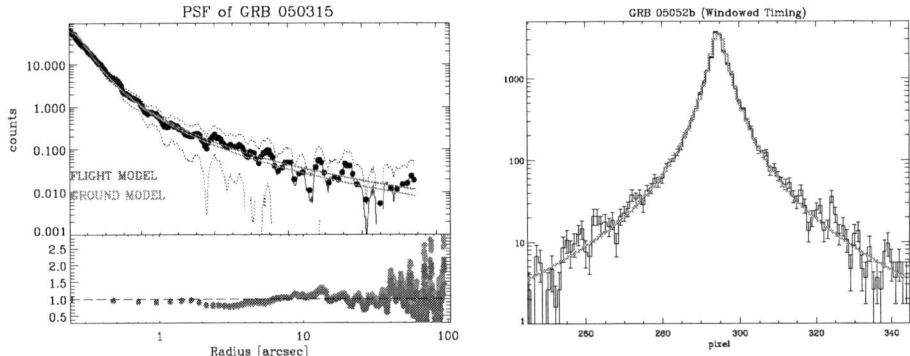

FIGURE 2. Left Panel. As an example we compare the data from the second part of the observation of GRB050315 (without any pile–up) with the analytical model. In the lower panel we plot the ratio between the data and the model. We show also the comparison between the model built from the ground calibration and the updated model built by means of the in–flight calibration. We find good agreement between the model and the data, which demonstrates that our goal is reached with a typical accuracy of 5%. **Right Panel.** The analytical model can also be used for the WT data, where only one spatial dimension is registered by the XRT detector. The observation of GRB05052b is shown. Tthe data (black) are compared to the model (grey). The pixel size is 2.36 arcseconds.

$$\beta(E,\theta) = a_2 + b_2 \times \theta + c_2 \times E + d_2 \times E \times \theta$$

The values of the plane function coefficients are stored in the PSF file within the *Swift* XRT CALDB distribution (http://swift.gsfc.nasa.gov/docs/swift/analysis). In this way, given the position within the field of view and the energy for an hypothetical monochromatic source we can calculate the corresponding values of r_c and β and reproduce the PSF by means of this parametrization. To give an accurate description of the PSF profile of a source with a generic spectrum we have to sum the single monochromatic contributions (Fig. 2). The PSF correction is applied by the task *xrtmkarf*, distributed within the HEADAS software (http://swift.gsfc.nasa.gov/docs/swift/analysis).

REFERENCES

1. D.N. Burrows, et al. 2005, Sp.Sc.Rev , 120 , 165
2. A. Moretti, S. Campana, T. Mineo, et al. 2005, proceedings of SPIE, 5898, 360
3. A. Moretti, S. Campana, G. Tagliaferri, et al. 2004, proceedings of SPIE, 5165, 232
4. J.E. Hill, L. Angelini, D.C. Morris, et al. 2005, proceedings of SPIE, 5898, 313

GRB Astrophysics with Digitised Astronomical Archival Plates

René Hudec[1] and Lukáš Hudec[2,1]

[1] *Astronomical Institute of the Academy of Sciences of the Czech Republic, CZ-251 65 Ondřejov, Czech Republic rhudec@asu.cas.cz*
[2] *Department of Informatics, Faculty of Mathematics and Physics, Charles University, Prague, Czech Republic*

Abstract. There are around 3 millions of astronomical archival plates in the world, representing a very unique and extended data base for various astrophysical projects. Typically each sky position is covered by thousands of plates including those going deep to magnitude 20. The data mining in this database was however very limited so far until the recent wide digitization and evaluation of the plates by powerful (one plate can represent up to 1 GB of data) computers using novel software. We discuss the preferences of use of this type of data in GRB astrophysics such as searches for optical orphans of GRBs. These searches require long-time, wide-field and deep optical coverage and this can indeed be provided by some of plate archives. Some examples of results obtained with archival plates are presented and discussed.

Keywords: Astronomical plate archives, plate digitisation, optical transients, gamma-ray bursts
PACS: 95.75.De, 95.75.Mn, 95.75.Pq

ARCHIVAL SKY PATROL PLATES

There are more than 3 millions archival plates in the world, with limiting magnitude up to 23, and field of view (FOV) > 5 x 5 deg in most cases [1]. These plates are suitable for dense long-term photometry (up to 100 years, up to 2000 points, up to 23 mag). They are at the same time suitable to detect rare events - years of CONTINUOUS monitoring are easily possible [2]. The use of scanners, powerful computers and innovative software allows the effective data evaluation for the first time.

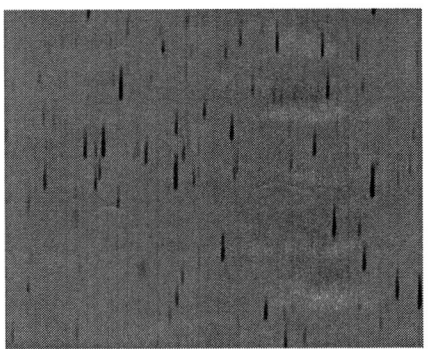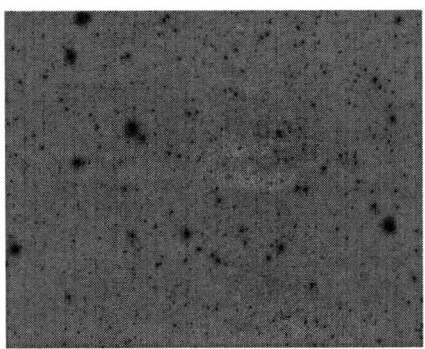

FIGURE 1. Various types of astronomical plates (digitised) suitable for automated analyses with novel algorithms - task mainly for informatics students. Spectral image (top left), direct image (top right), multiple exposure (bottom).

OPTICAL TRANSIENTS OF ASTROPHYSICAL ORIGIN

The plate material is suitable for analyses of optical transients of astrophysical origin, such as optical transients (OTs) and optical afterglows (OAs) of Gamma Ray Bursts (GRB) including orphan afterglows (seen only in optical), flares on blazars and AGNs (up to 10 mag amplitudes revealed by extended plate analyses [3]), stellar OTs such as stellar flares, CVs flares etc. [4].

Orphan afterglows/transients of GRBs

Orphan afterglow is an optical afterglow without detectable gamma-ray emission (due to different beaming). These objects are predicted by theory but not yet confirmed by observation.

The rate of Orphan Optical Afterglows (OOAs) is expected to exceeds the GRB rate, hence the improved GRB statistics is expected with numerous consequences such as improved statistics of host galaxies, redshift distribution, cosmological conclusions, etc.

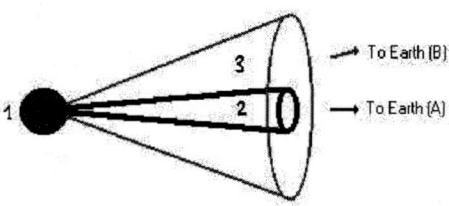

FIGURE 2. The orphan optical afterglow (simplified): 1 – the source, 2 – the angle of gamma-ray emission, 3 – the angle of optical emission. A – GRB burst, B – orphan afterglow.

The digitization of the astronomical plates, the novel software and the use of powerful (one plate can represent up to 1 GB of data) computers allow the automated data mining and scientific evaluation of the plates for the first time. These projects represent an interface between astronomy and informatics.

If we look for optical prompt emission (including orphans) from GRBs, we need to look for short-living transient phenomena lasting minutes. However, on long-exposed deep images and plates, it is very difficult to look for brief transients since the OT image is hidden by typically tens to hundreds of thousands stellar images with similar appearance. The methods of comparing plates and/or comparison with catalogues are still not very effective and reliable. We report on the novel method how to identify brief optical transients (OTs) on sky archival plates.

This method does not require identification of all objects and their comparison with the catalogue, but rely solely on the information included in the particular plate itself.

This is an excellent example how the valuable information obtained in archival astronomical plates can be accessed and scientifically used. Such analyses are possible only based on plate digitisation, novel software and powerful computers.

Analogous procedures can be applied also for trailed star images, spectral plates with objective prisma, etc. - searches for objects with strange spectra, spectral emission lines, rapid light changes etc. (in study), based on collaboration with informatics students.

FIGURE 3. Left: Typical ROB ESO multiple exposure plate, scanned and binned, 1 from 4 pixels read out. Right: Example of OT found: an real object (with 2 images) with duration less than 30 min.

Orphans and archival plates

Independent optical searches for orphan OAs and OTs of GRBs require large FOVs, deep limits, and extended coverage (cumulative exposure times). This can be provided by digitized astronomical archival plates. In some plate collections, the plates taken with filters are available. Hence the OAs of GRBs can be recognized by specific OA colors [5]. The archival astronomical plates represent a valuable data source for searches for optical transients. However the access to the data required special procedures, was laborious and time consuming in the past. The recent novel software development, wide plate digitization, and powerful computers (1 digitised plate can represent up to 1 GB of data) allow the effective data mining for the first time.

ACKNOWLEDGEMENTS

We acknowledge the support provided by the Grant Agency of the Academy of Sciences of the Czech Republic, grant A3003206.

REFERENCES

1. Hudec R., *Acta Historica Astronomiae*, vol. 6, p. 28-40 (1999).
2. Hudec R., *Acta Historica Astronomiae*, vol. 6, p. 127-131 (1999).
3. Hudec, Rene; Vrba, Frederick; Luginbuhl, Chris, in *BL Lac Phenomenon*, Proceedings of the conference held 22-26 June, 1998 in Turku, Finland, p. 115, 1999.
4. Hudec, R. and Wenzel, W., *Astrophysical Letters and Communications*, Vol. 39, 249 (1999).
5. Simon, V.; Hudec, R.; Pizzichini, G.; Masetti, N., *Astronomy and Astrophysics* 377, 450-461 (2003).

BART: real time follow-up of GRBs since 2001

Petr Kubánek[*][†], Martin Jelínek[**], René Hudec[*], Martin Nekola[*] and Jan Štrobl[*]

[*]*Astronomical Institute of Academy of Sciences, Fričova 298, Ondřejov, Czech Republic*
[†]*Integral Science Data Center, Chemin d'Ecógia 16, 1290 Versoix, Switzerland*
[**]*Institute of Astrophysics of Andalusia, Camino Bajo de Huétor 50, 18008 Granada, Spain*

Abstract. BART is a 25 cm aperture, intelligent robotic CCD telescope, devoted for observation of prompt gamma ray burst emission and early afterglow. It's operating since early 2001. Till now, it responded to many GCN GRB alerts.

BART SYSTEMS

The system does not involve only telescope, but forms all-robotic observatory with automatic roof opening, weather sensors, and telescope, running in a software system, with observation scheduling, automatic image analysis and of course robust Virtual Observatory compatible image archiving.

BART was built primarily to perform observations of GRB positions reported by GCN[1], to search for optical transients. Spare time, when there is no GRB to follow, is used for observations of other interesting astronomical objects.

BART OPTICAL SETUP

Over years, BART hardware was vastly modified.

We started with a 25 cm Meade LX200 mount, equipped with a CCD camera. The most important instrument was, however, wide-field camera, for it's importance as an INTEGRAL OMC test device and field of view compatible with GRB error boxes available at that time.

While the narrow field telescope was prone to frequent problems with focusing and collimation, the wide field camera is a source of reliable and stable data over years.

Later we added another wide-field camera to get colour information. Problematic LX200 was replaced by Losmandy Titan mount with Gemini control system. Recently we replaced old parallel port CCD cameras with new USB devices, which have better cooling performance and faster readout times.

Most of the observations in the BART archive was therefore carried with one or two wide-field CCD cameras. The limiting magnitude of this system is typically 15, and can provide reasonable monitoring of objects up to 12th magnitude.

The 25 cm narrow-field telescope, for it's low reliability mostly unused for some time, has received considerable attention lately (digital focuser, new CCD camera, filter wheel) and is supposed to provide reliable results by spring 2006. With current setup, the 3σ limit of 19.0 was reached on a 60 s unfiltered exposure.

RTS2 - OBSERVATION SYSTEM OF BART

Observations on BART are conducted using RTS2[2] software. BART served as reference platform for RTS2, enabling it to grow from a small subset of functions into a full-featured package, which is capable of controlling complicated observatory setups.

RTS2 evolved from RTS1, which was written in Python language. RTS2 was originally written in C and later was completely rewritten in C++. We developed RTS2 to work in two basic layers - abstract, network-based, which handles communication between various RTS2 components, and device-driver layer, which translates network commands to physical hardware.

RTS2 is by our knowledge only currently available system, which is able to control telescope setups with multiple detectors, control full observatory (including dome, weather station etc..), runs on Linux operating system and which source code is available under GNU General Public License (GPL).

RTS2 is designed for automatic observations, so it doesn't have graphical user interface (GUI) for planning and manual observing. As communication protocol is well defined, a graphical user interface (most probably Qt-based) will be available when we will be happy with automatic observations performed.

FOUR YEARS OF OBSERVATION - RESULTS

We started regular observation runs with BART in May 2001. Unfortunately, this time coincided with end of service of CGRO and "dark years" of very few available real-time alerts. Table 1 summarises both real-time and off-line bursts followed by BART.

SYSTEM STABILITY

After extensive bug-fixing, we enjoyed quite stable operation of the software. Thanks to the remote shell access, we have full control over whole software environment, including operating systems. That's critical for RTS2 deployment on remote sites.

OBSERVATION NOTIFICATION

As RTS2 runs mostly in fully automatic mode, it poses mailing capacity. It can send emails to observers when target observations starts, ends, when all images from observation are processed, and at the end of night with list of all observations of given target. Mailing can be widely customised.

TABLE 1. BART GRB OT search observations

GRB	GRB Δ time	GCN Δ time	comments
020124	11h 30m	10h 20m *	R>14 mag
020305	10h 30m	30m	R>14
020317	58m	90s	R>13.5
020331	5h 30m	5h 20m *	R>12.5, bad weather
030824	1h 50m	30m	R>14.3
031111A	1h 10m	30s	overexposed due to moon, GCN 2456
031111B	10m	30s	R>14.5, I>14.2
040403	-93m	-	preburst limit
040606	5.5h	*	I>12 (weather)
050505	9.5m	29s	V>11.5 (poor conditions)
050509A	79s	62s	V>14.9
050520	15m	14m	limit V>14 mag, GCN 3431
050525	10m	16s	**V~15±1, GCN 3481**
050824	10m	31s	V>12: too low

* Daytime burst

TARGET HIERARCHY

In order to solve complex observation tasks, which RTS2 perform, we designed target hierarchy.

Beside Target, we use Observation class for bringing together information about one run of target observation - e.g. one target can have more observations, but one observation is only for one target.

ARCHIVE AND VIRTUAL OBSERVATORY

RTS2 uses PostgreSQL relational database to store all target information (priority, exposures, observing conditions), observation meta-data (observation start, end and list of images taken) and image properties (WCS data, exposure).

We have developed code, which provides Virtual Observatory[3] access to images produced by RTS2. We plan to provide public access to images through Virtual Observatory in near future.

IMAGE ANALYSIS AND IMAGE PROCESSING

We developed an astrometry package JIBARO[4]. This way we can set-up mapping between image (x,y) and sky (α,δ) coordinates (WCS) with plate solution precision typically better than 1/10 of a pixel. This approach further simplifies automatic photometry, object detection and other possible tasks needed in real-time pipeline.

All images obtained and stored in database have proper WCS keywords in their FITS header, are roughly photometrically calibrated (in terms of a limiting magnitude against catalog) and in some cases the relative photometry is performed. Automated transient detection is being actively tested and expected to be routinely used soon.

RTS2 is designed to cooperate with the astrometry package, calls the processing pipeline and depending on result, it stores the image in the archive.

LONG-TERM OPERATION OF A ROBOTIC TELESCOPE

From experiences we gained with robotic follow-ups of GRBs, it's clear to us that building and operating robotic observatory for long period of time isn't as easy as one might think.

It's relatively easy to build and run one purpose instrument setup for few nights, with observer at disposal to solve either technical or software problems which arises during run.

To be able to reproduce results through years, while still preserving ability to react to GCN notice in fastest time possible, and continuously improving and patching control system, so it's able to fulfil tasks it wasn't originally designed to do, is at least a magnitude harder.

ACKNOWLEDGMENTS

We acknowledge the support provided by the Grant Agency of the Academy of Sciences of the Czech Republic, A3003206, and partly by ESA PECS 98023.

REFERENCES

1. S. D. Barthelmy, *American Astronomical Society Meeting Abstracts* **202**, –+ (2003).
2. P. Kubánek, M. Jelínek, M. Nekola, M. Topinka, J. Štrobl, R. Hudec, T. D. J. Mateo Sanguino, A. de Ugarte Postigo, and A. J. Castro-Tirado, "RTS2 – Remote Telescope System, 2nd Version," in *AIP Conf. Proc. 727: Gamma-Ray Bursts: 30 Years of Discovery*, 2004, pp. 753–756.
3. P. J. Quinn, D. G. Barnes, I. Csabai, C. Cui, F. Genova, B. Hanisch, A. Kembhavi, S. C. Kim, A. Lawrence, O. Malkov, M. Ohishi, F. Pasian, D. Schade, and W. Voges, "The International Virtual Observatory Alliance: recent technical developments and the road ahead," in *Ground-based Telescopes. Edited by Oschmann, Jacobus M., Jr. Proceedings of the SPIE, Volume 5493, pp. 137-145 (2004).*, 2004, pp. 137–145.
4. A. de Ugarte Postigo et al., "JIBARO: Un conjunto de utilidades para la reducción y análisis automatizado de imágenes," in *Astrofísica Robótica en España*, edited by J. T. A.J. Castro-Tirado, B.A. de la Morena, Madrid, 2005, pp. 35–50.

GRB follow-up with BOOTES Optical Chapter 5: The *Swift* Era

Martin Jelínek*, Alberto J. Castro-Tirado*, Petr Kubánek[†,**], Stanislav Vítek*, Antonio de Ugarte Postigo* and René Hudec[†]

Instituto de Astrofísica de Andalucía (IAA-CSIC), Camino Bajo de Huétor 50, 18008 Granada, Spain
[†]*Astronomical Institute of the Academy of Sciences, Fričova 298, Ondřejov, Czech Republic*
[**]*Integral Science Data Center, Chemin d'Ecógia 16, 1290 Versoix, Switzerland*

Abstract. BOOTES is a robotic system, whose primary aim is to observe gamma-ray burst prompt emission. Since 1998 BOOTES has provided follow-up observations for more than 70 GRBs; the most important results obtained so far are the detection of an OT in the short/hard GRB 000313 error box, the detection of several optical after-glow for long/soft GRBs and the non-detection of optical emission simultaneous to the high energy emission for several GRBs (both long/soft and short/hard events). During the time of operation we have got triggers from *CGRO/BATSE, BeppoSAX, HETE-2, INTEGRAL* and *Swift*. Here we present our early detections of GRB optical emission using the 30 cm *Bootes-1B* telescope in the R.V and I-bands since the launch of *Swift*.

Keywords: black hole formation, starburts galaxies, gamma ray bursts
PACS: 97.60.-s, 98.54.Ep, 98.62.Py, 98.70.Rz

A STEREOSCOPIC ROBOTIC OBSERVATORY

BOOTES-1, the main BOOTES observing station, is located in Mazagón (Huelva), a very dark sky area in southern Spain. It is hosted by the Estación de Sondeos Atmosféricos (ESAt) at INTA (Instituto Nacional de Tecnica Aeroespacial) in the Centro de Experimentación de El Arenosillo[1, 2]. It has two domes: Bootes-1A and Bootes-1B that hold three schmidt-cassegrain telescopes and several wide-field cameras. Following complementing schemes, instruments carry out systematic exploration of the night sky, responding to GRB alerts at any time via GCN socket connection (see Fig. 1).

The second observatory — BOOTES-2 — is in operation since 2002. It is located at La Mayora (Málaga), a research center under the auspices of the CSIC (Consejo Superior de Investigaciones Científicas), 240 km away from BOOTES-1. On this way, GRB may be observed simultaneously from different locations, discriminating the near earth objects easily.

SOFTWARE

RTS2 is designed as a networked system for driving of robotic telescopes [3]. It is composed of several device servers, central server and various observational clients cooperating over a TCP network. For the communication, there is a private protocol, ensuring speed and reliability. It is intended to be independent on used astronomical

FIGURE 1. The BOOTES-1 telescopes at the Estación de Estudios Atmosféricos (Esat)/INTA in Huelva.

equipments, with access points for controlling of different types of mounts, domes and CCDs. Observation entries, requests and results are kept in database. Positions of GRBs are received from the Internet, and observed either in prompt mode, or added to list of observation targets, depending on weather and other conditions influencing the observation. During the idle time, when there is not any request for GRB observations, the telescope spends monitoring various active galaxies. The database lookup entry point is accessible at http://lascaux.asu.cas.cz/bartdb.

IMAGE ANALYSIS

An astrometric package for BOOTES image analysis, based on JIBARO [4] has been developed. This way, we can set-up mapping between image (x,y) and sky (α,δ) coordinates (WCS) with plate solution precission typically better than 1/10 of a pixel. This approach further simplifies automatization of photometry, object detection and other possible tasks needed in real-time pipeline.

All images obtained and stored in database have proper WCS keywords in their FITS header and are roughly photometrically calibrated (in terms of a limiting magnitude against astronomical catalogues) and in some cases the relative photometry is performed. Automated transient detection is being actively tested and expected to be routinely used soon.

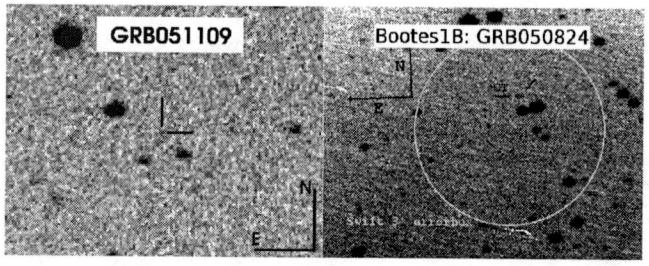

Source: BOOTES

FIGURE 2. An example of optical afterglow detections of *Swift* GRBs by BOOTES.

RTS2 is designed to cooperate with the astrometry package, calls and maintains the pipeline and depending on result, it stores the image in the archive.

MOUNT AND DOME CONTROL

Over the years we have developed BOOTES hardware considerably. The first were Meade LX200 mounts equipped with CCD cameras. In the present time we have Paramount with one 30 cm and two 20 cm telescopes.

All our domes are independent, but controllable over internet. Both stations have weather stations, to provide information necessary for autonomous opening and closing in case of unusable weather conditions. GSM/SMS based control system is used as a back-up for the case the Internet fails.

RAPID GRB OBSERVATIONS BY BOOTES

During 2005, BOOTES has followed many GRB alerts in four cases it has observed an optical transient (see Fig. 2). In some other cases even the magnitude limits were important for constraining early decay rates and so on [5]. In one case, we have detected the OT simultaneously in two colors. Table 1 displays the results of the follow-up observations performed in 2005.

CONCLUSIONS

BOOTES-1B has finally proven it can successfully follow GRB notices. Although we have been on watch for many years, *Swift* gave us the possibility to follow tens of GRB alerts per year, thus allowing the telescope to show its full power. This work has provided in four OT detections and several upper limits which have been obtained using this instrumentation. BOOTES has, however, several intrinsic limitations. Mount speed, telescope diameter and wavelength limitation are being overcome in our project BOOTES-IR [6].

TABLE 1. GRB observations by BOOTES-1B. Only real-time reactions are shown. Δt_{GRB} stands for time between GRB trigger and start of first image, Δt_{GCN} is the time between reception of the GCN coordinates and first image (most of the time is due to telescope slew).

GRB	Δt_{GRB}	Δt_{GCN}	exposure	magnitude	comment
050505	609s	70s	300s	V>15,I>14	GCN 3434
050509B	62s	48s	10s,4min	V>11.5,12.5	Short burst, poor conditions
050525A	383s	28s	25s	V~15.5	V-band data only
050528	71s	28s	60s	V>13.8,I>13.0	light twilight, GCN 3500
050730	233s	172s	300s	R>18.0	during SW update: longer delay
050805A	136s	66s	5x300s	R>18.0	
050805B	62s	7.2s	10s	R>16.0	
050824	636s	55.8s	600s	R=18.2	2 co-added images
050904	124s	43s	10s	R>16.6	GCN 3929, GRB @ z=6.3
050922C	228s	62.3s	2x10s	R~14.5	clouds, three usable windows
051109A	54.8s	27s	10s	R=16.2	GCN 4227, R+I simultaneously

ACKNOWLEDGMENTS

We would like to thank for the support provided to us by Instituto Nacional de Technica Aerospacial (INTA), Consejo Superior de Investigaciones Scientifícas (CSIC) and Grant Agency of Czech Republic (GACR). We also thank Scott Barthelmy for maintaining the Bacodine alert system. This research has been also partially supported by the Ministerio de Ciencia and Tecnología under the programmes AYA2004-01515 and ESP 2002-04124-C3-01 (including FEDER funds).

REFERENCES

1. A. J. Castro-Tirado, J. Soldán, M. Bernas, P. Páta, T. Rezek, R. Hudec, T. J. Mateo Sanguino, B. de La Morena, J. A. Berná, J. Rodríguez, A. Peña, J. Gorosabel, J. M. Más-Hesse, and A. Giménez, *A&A* **138**, 583–585 (1999).
2. A. J. Castro-Tirado, M. Jelínek, T. J. Mateo Sanguino, A. de Ugarte Postigo, and the BOOTES team, *Astronomische Nachrichten* **325**, 679–679 (2004).
3. P. Kubánek, M. Jelínek, M. Nekola, M. Topinka, J. Štrobl, R. Hudec, T. D. J. Mateo Sanguino, A. de Ugarte Postigo, and A. J. Castro-Tirado, "RTS2 – Remote Telescope System, 2nd Version," in *AIP Conf. Proc. 727: Gamma-Ray Bursts: 30 Years of Discovery*, 2004, pp. 753–756 (2004).
4. A. de Ugarte Postigo et al., "JIBARO: Un conjunto de utilidades para la reducción y análisis automatizado de imágenes," in *Astrofísica Robótica en España*, edited by A. J. Castro-Tirado, B. A. de la Morena, and J. Torres, Madrid, 2005, pp. 35–50 (2005).
5. A. J. Castro-Tirado, A. de Ugarte Postigo, J. Gorosabel, T. Fathkullin, V. Sokolov, M. Bremer, I. Márquez, A. J. Marín, S. Guziy, M. Jelínek, P. Kubánek, R. Hudec, S. Vítek, T. J. Mateo Sanguino, A. Eigenbrod, M. D. Pérez-Ramírez, A. Sota, J. Masegosa, F. Prada, and M. Moles, *A&A* **439**, L15–L18 (2005).
6. A. de Ugarte Postigo, A. J. Castro-Tirado, M. Jelínek, P. Kubánek, R. Cunniffe, S. Vítek, S. Castillo-Carrión, S. Guziy, S. B. Pandey, J. M. Castro Cerón, F. M. Zerbi, P. Conconi, A. Riva, V. de Caprio, P. Amado, A. Claret, C. Cárdenas, S. Martín, J. M. Trigo Rodríguez, C. Sánchez-Fernández, M. D. Sabau-Graziati, J. Díez Verdejo, and F. Vitali, "BOOTES-IR: The extension of BOOTES towards the near-IR," in *these proceedings* (2006).

GLAST and GRBs: Probing Photon Propagation over cosmological distances

Francesco Longo*, Nicola Omodei[†], Johann Cohen-Tanugi**, Jeff D. Scargle[‡] and Frederic Piron[§]

Department of Physics, University of Trieste and INFN, section of Trieste
[†]*INFN, section of Pisa*
**SLAC-Stanford University*
[‡]*NASA Ames Research Center*
[§]*Groupe d'Astroparticules de Montpellier*

Abstract. Especially in the framework of Quantum Gravity, it is theoretically possible that photons of different energy propagate at different velocity. Gamma-ray Bursts (GRBs), due to their large distances and rapid variability in a broad energy band, are perhaps the best astronomical sources in which to measure any such dispersion over cosmological distances. GLAST will detect several GRBs per year at GeV energies, where the effect may be detectable. We address problems of optimal sensitivity and discrimination against energy-dependent effects intrinsic to GRB emission, using simulated data and new unbinned lag-detection algorithms.

Keywords: Gamma-ray detectors, X and Gamma-ray telescopes and instrumentation, Gamma-ray Bursts
PACS: 07.85.Fv, 95.55.Ka, 98.70.Rz

INTRODUCTION

Testing new ideas on the fundamental laws of Nature, such as the unification of Quantum Mechanics and General Relativity, requires to go beyond the actual experimental limits of particle accelerators. Recently there has been a growing interest in the field of astroparticle physics, where particles are detected with energies significantly higher than the energies that can be achieved at particle colliders. In particular, the study of the emission from high-energy sources at cosmological distances provides the ground for the so called "quantum gravity phenomenology" (e.g. [1]).

In this theoretical framework, one is looking for small effects whose magnitude is set by the distance of the source and the energy of the particles involved with respect to the Planck energy ($E_p \simeq 10^{28} eV$). In analogy with the particle physics searches for rare phenomena, future astroparticle experiments will test the QG phenomenology through the study of the signals from distant and energetic sources. Among them, Gamma-ray bursts (GRBs) are one of the best candidates due to their high luminosity, variability and distances. In the following we present a brief overview of the QG phenomenology and discuss some predictions in the context of the future space experiment GLAST. Then we discuss the methodology which is needed for QG studies with GRBs, and finally give the example of an analysis which was applied to simulated data coming from a preliminary toy model.

SPACE EXPERIMENTS AND QUANTUM PROPERTIES OF SPACETIME

In several approaches to the quantum-gravity problem one finds some evidence of departures from ordinary Lorentz symmetry. However it was traditionally believed that these effects could not be tested because of their small magnitude, set by the small ratio between the energy of the particles involved and the Planck energy scale. Work on quantum-gravity phenomenology has proven that this old expectation is incorrect[1].

A broken Lorentz symmetry will lead to a a dispersion relation for photon propagation with leading-order in the Planck lenght L_p:

$$E^2 \simeq \vec{p}^2 + m^2 - \eta (L_p E)^n \vec{p}^2 , \qquad (1)$$

without modifying the rules for energy-momentum conservation[1].

In (1) η is a phenomenological parameter of order 1, n, the lowest power of L_p that leads to a nonvanishing contribution, is model dependent.

A deformation term of order $L_p^n E^n p^2$ in the dispersion relation, such as the one in (1), leads to a small energy dependence of the speed of photons $v = dE/dp$ of order $L_p^n E^n$.

Such a dependence is completely negligible in nearly all physical contexts, but, at least for $n = 1$, it can be significant in the analysis of short-duration gamma-ray bursts that reach us from cosmological distances.

The energy dependence of the speed of photons of order $L_p E$ (in the case of $n = 1$) will allow to measure quantum gravity motivated dispersion relations. A space telescope, the GLAST [2] gamma-ray satellite telescope (scheduled to start taking data in 2007), is presently considered to be the best opportunity to look for this effect.

A rough estimate of the experimental sensitivity on this effect could be derived as follows. Supposing the source emits two photons of different energy at the same time, the delay that they accumulate due to the dispersion could be derived as:

$$\Delta t = \eta \frac{L}{c} \frac{\Delta E}{E_{QG}}$$

where L is the source distance, ΔE is the energy difference and E_{QG} is the energy scale for Quantum Gravity effects. Since GLAST has a deadtime below 10^{-3} s, assuming a source at a redshift $z \approx 1$ emitting photons until 100 MeV, we can estimate a rough upper limit on the scale of the Quantum Gravity that could be detected:

$$E_{QG} \approx \frac{10^{28}\text{cm} \times 10^2 \text{MeV}}{10^{-3}\text{s} \times 10^{10}\text{cm/s}} = 10^{20} \text{GeV}$$

GLAST GRB SIMULATION AND QUANTUM GRAVITY

The deformation in the dispersion relation discussed previously leads to a small energy dependence of the speed of photons.

The two instruments onboard the GLAST satellite will cover an energy range of more than seven orders of magnitude : from 10 keV to 30 MeV for the Gamma-ray Burst

Monitor (GBM) and from 30 MeV to more than 100 GeV for the Large Array Telescope (LAT). This huge lever arm will permit to search for QG-induced delay in times of arrival upon comparison of the structure of the signal in different energy channels[3]. Typically, two photons with an energy difference of ~ 100 MeV and which travel over a distance of ~ 1 Gpc, will be detected with a relative time-delay of the order of $t \sim 10^{-3}$s if the QG energy scale is close to the Planck energy.

To investigate such an effect, a detailed simulation of the time structure of GRBs is necessary . In the model we used several features were included[2, 4]. The light curve at LAT energies is derived in the internal shock model where we included the correct high energy absorbtion for sources at high redshift. Moreover the model we used is able to provide the light curve of GRB at different energy ranges. Since the time lags intrinsic to the source could mimic or hide the expected phenomenum due to QG, the final analysis will need to correlate the measured delay between different energy bands with the measured source distance. However, these intrinsic effects have to be studied carefully and somehow included in the model. Indeed, low-energy pulses in the light curves are wider and peak at later times than high-energy pulses[5]. It is also well established that low-luminosity GRBs exhibit larger time lags than high-luminosity GRBs in average[6]. Therefore a GRB sample, where the most distant sources are intrinsically brighter because of the telescope sensitivity, may lead to the observation of lags with no QG origin.

Some preliminary results of this analysis, where the effects of QG are included in the photon propagation, show that the LAT sensitivity permits to reconstruct the time structures of GRB at high energies[7, 8].

ANALYSIS OF SIMULATED DATA AND QUANTUM GRAVITY

To develop new algorithms for detecting such effect, we decided to test them on data from a simple and preliminary toy model with photons in an artificial burst, with a known lag as a function of energy imposed on the times.

The algorithm concept is simple. For a linear quantum gravity model, we bin the events to build an histogram of transformed times

$$t'(i) = t(i) - Ce(i) \qquad (2)$$

compute the entropy of the probability distribution corresponding to this histogram, normalize it and then optimize (in this case, minimize) the entropy as a function of C.

We tried monte carlo results for a case where C = .0005 (in arbitrary units).

Three cases with 20, 30 and 50 photons in the GRB (modeled as a simple step function) and 40 back ground (unshifted) photons. Note that the algorithm works reasonably well, but of course for only 20 or 30 photons, the scatter is pretty large, and one does not get a good value from one measurement of a burst.

An issue still to be addressed is if there is any, even small, coupling between dead-time of the data acquisition system as a whole and the energy of the photon. Even a small

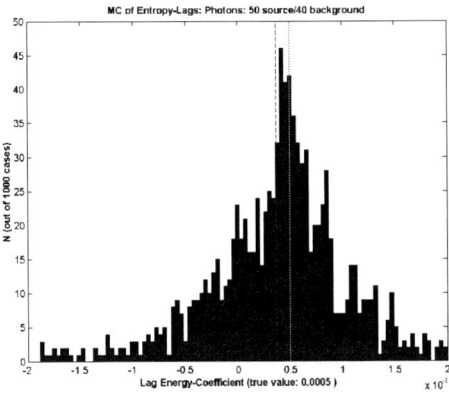

FIGURE 1. Lag entropy detection of energy dependent time lag in GRB

effect might be important at the peak of a bright GRB. This effect will be discussed in future works.

CONCLUSION

Addressing the issues of possible QG and Lorentz symmetry experimental signature has been a long standing question for some years given the new opportunities provided by the new generation of astroparticle experiments.

The space gamma-ray experiment GLAST will have a sensitivity to GRBs which will permit to set a limit on QG phenomenology parameters (e.g. QG energy scale) or to detect such effects. Further developments on GRB modelization are needed and planned in terms of GRB distribution in redshift and GRB intrinsic properties in order to optimize the analysis methods (entropy minimization, wavelet analysis) of light curves.

REFERENCES

1. G. Amelino-Camelia, Gen.Rel.Grav. 36 (2004)
2. N. Omodei et al., *these proceedings*
3. Norris, J.P, et al. (1999) astro-ph/9912136.
4. N. Omodei, *these proceedings*
5. Norris, J.P, et al., ApJ 459, 393 (1996)
6. Norris, J.P, et al., ApJ 534, 340 (2000)
7. Cohen-Tanugi, J., Omodei, N., Longo, F., poster presented at the XXI Symposium on Relativistic Astrophysics (2002).
8. Longo, F., Cohen-Tanugi, J., Omodei, N.,in Gamma Ray Bursts: 30 Years of Discovery: Gamma-Ray Burst Symposium. AIP Conference Proceedings, Vol. 727, Edited by E. E. Fenimore and M. Galassi., 681, (2004)

The CASTER Black Hole Finder Probe

M. L. McConnell*, P. F. Bloser*, G. L. Case[†,**], M. L. Cherry[†], J. Cravens[‡], T. G. Guzik[†], K. Hurley[§], R. M. Kippen[¶], J. R. Macri*, R. S. Miller[∥], W. Paciesas[∥], J. M. Ryan*, B. Schaefer[†], J. G. Stacy[†,**], W. T. Vestrand[¶] and J. P. Wefel[†]

*Space Science Center, University of New Hampshire, Durham, NH 03824
[†]Department of Physics and Astronomy, Louisiana State University, Baton Rouge, LA 70803
[**]Department of Physics, Southern University, Baton Rouge, LA 70813
[‡]Department of Space Science, Southwest Research Institute, San Antonio, TX 78228
[§]Space Sciences Laboratory, University of California, Berkeley, CA 94720
[¶]Los Alamos National Laboratory, Los Alamos, NM 87545
[∥]Department of Physics, University of Alabama, Huntsville, AL 35899

Abstract. The primary scientific mission of the Black Hole Finder Probe (BHFP), part of the NASA Beyond Einstein program, is to survey the local Universe for black holes over a wide range of mass and accretion rate. One approach to such a survey is a hard X-ray coded-aperture imaging mission operating in the 10–600 keV energy band. The development of new inorganic scintillator materials provides improved performance that is well suited to the BHFP science requirements. Detection planes formed with these materials coupled with a new generation of readout devices represent a major advancement in the performance capabilities of scintillator-based gamma cameras. Here, we discuss the Coded Aperture Survey Telescope for Energetic Radiation (CASTER), a concept that represents a BHFP based on the use of the latest scintillator technology.

Keywords: X-ray instrumentation, gamma-ray instrumentation, X-ray sources, gamma-ray sources, black holes
PACS: 95.55.Ka, 98.70.Qy, 98.70.Rz, 97.60.Lf, 97.80.Jp, 98.54.Cm

INTRODUCTION

NASA's Beyond Einstein Program [1] defines a sequence of space missions for exploring the Universe. One aspect of this program is a series of three Einstein Probe missions that would complement the facility-class Einstein Great Observatories (LISA and Con-X). One of the three Einstein Probe missions defined by the Beyond Einstein roadmap is the Black Hole Finder Probe (BHFP). The goal of the BHFP will be to carry out an all-sky census of accreting black holes. It is generally agreed that a hard X-ray coded mask imager covering the 10–600 keV energy band would be an effective tool for achieving this goal. One concept for the BHFP mission, known as EXIST (the Energetic X-ray Imaging Survey Telescope), has been under development for several years [2–4]. Here we offer an alternative concept, one that is similar to EXIST, but based on different detection technologies. We refer to our design concept as the Coded Aperture Survey Telescope for Energetic Radiation (CASTER) [5, 6]. CASTER is designed to employ several new experimental techniques using standard detector technologies, such as inorganic scintillators, wavelength-shifting fibers and photomultiplier tubes (PMTs), all of which have laboratory and space flight heritage. The development of a new inorganic scintillator material, lanthanum bromide (LaBr$_3$), provides improved performance that is well suited to the BHFP science requirements [7, 8]. With LaBr$_3$, we now have the prospect of scintillators with energy resolution and stopping power on par with room-temperature semiconductors (such as CZT), but with far less cost. In addition, scintillator technology offers a practical means to extend the effective energy range beyond 511 keV. We therefore are exploring the implications, benefits and penalties, both practical and scientific, of using inorganic scintillators as the detector technology of a coded aperture imaging BHFP.

THE CASTER CONCEPT

The current concept for CASTER (similar to that of EXIST) envisions an array of 16 separate coded aperture telescope modules with overlapping fields-of-view. Each telescope would have a detection area of 64 cm × 64 cm, providing for a total detection area of 6.6 m^2. The distinguishing feature of the CASTER concept is that the photon detection plane for each coded aperture telescope will be fabricated using traditional scintillator technologies. LaBr$_3$ offers the properties of high stopping efficiency, high light output, good linearity, significantly improved energy resolution, fast response, and (potentially) low cost. The measured energy resolution of 2.6% FWHM at 662 keV is comparable to the energy resolution of 3% at 662 keV for off-the-shelf spectroscopy grade CZT. The fluorescent decay time (~ 25 ns) is much faster than more traditional scintillator materials, thus assuring superior performance in high count-rate situations and in anticoincidence timing. In addition, as is the case with other inorganic scintillators, LaBr$_3$ offers the possibility of fabricating relatively thick detectors to improve the detection efficiency at higher energies.

As part of a NASA-funded concept study, we have identified two detection plane configurations that warrant further study. Both configurations satisfy the basic criteria of the BHFP mission and provide an attractive low cost, low power option.

Anger camera modules, formed with multiple light sensors viewing a layer of scintillation material, are widely used for a variety of applications [9, 10]. The interaction location is determined for each detected photon from the relative signals recorded for each PMT in the array. The spatial resolution of a gamma camera depends on geometrical factors such as the scintillator thickness, its size and the number and density of the light sensors in the array. Statistical and optical factors, such as scintillation yield, surface reflectivity, spectral match and photoelectron yield are also important. Current medical imaging technologies already achieve spatial resolutions of $\sim 1.5 - 2.0$ mm, a level of resolution that may be sufficient for the CASTER design. New gamma cameras employing the combination of higher light yield scintillator material (LaBr$_3$) and a higher density of light sensors, (e.g., multi-anode photomultiplier tubes – MAPMTs) are likely to provide even better spatial resolution capabilities. With proper calibration and analysis of the multiple PMT signals, the achievable energy threshold and spectroscopic performance of these Anger cameras should be similar to that achieved using single PMT spectrometers.

In addition to providing the location in the x- and y-dimensions, the extent of the distribution of scintillation signals within the sensor plane provides a measure of the z-

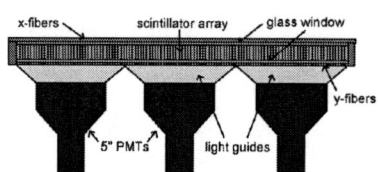

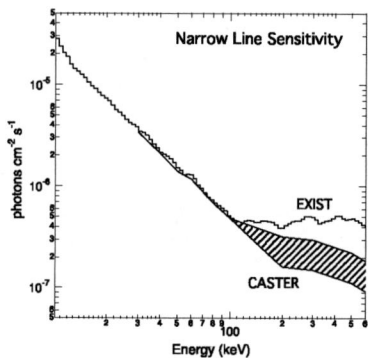

FIGURE 1. One possible imager design consisting of a pixellated scintillator array read out by wavelength shifting fibers. The large area PMTs are used for measuring total energy deposit.

FIGURE 2. The estimated narrow line sensitivity for CASTER as compared to published sensitivity of EXIST (5σ sensitivity, one-year exposure). The hatched area represents a range of possible sensitivity values for CASTER depending on the assumptions for the background scaling.

coordinate or depth of the interaction. Light from scintillations nearer the sensor plane is shared among fewer sensors than light from scintillations farther from the sensor plane. Anger cameras with a higher density of readout sensors will have improved ability to measure the z coordinate. Multi-hit events would be identified and the interaction site locations measured in those cases where spatial resolution is better than the mean free path of the scattered photons.

A second approach, also adapted from medical imaging applications, is analogous to the design of solid state strip detectors [11]. In this case, one layer of wavelength-shifting fibers is laid in the x-direction across the top of a scintillator and a second layer of fibers is laid in the y-direction across the bottom. The light emitted by the fibers is read out at one end of each fiber by a set of MAPMTs. The crossed fiber layers measure the x- and y- position using the center of gravity of the light in the two fiber arrays, and the depth by using the signal distribution across the fiber arrays. Only a small fraction of the light is absorbed, reemitted, and trapped in the fibers, however. The energy measurement, therefore, is performed by a set of large area PMTs viewing the scintillator through the bottom fiber layer. This approach offers the possibility to provide a significant reduction in the power requirements by reducing the number of electronics channels as compared to a pixellated or Anger camera detector geometry. At lower energies, where the light output is reduced, the number of photoelectrons per fiber will become too small for the desired detection efficiency. In order to overcome the limitations of this approach at low energies, segmented scintillator arrays may be used (as in Fig. 1). The segmented nature of such arrays restricts the lateral spreading of the light within the scintillator. Preliminary testing of this concept has recently been conducted at LSU [12]. These tests are continuing with the goal of more fully evaluating the energy resolution and spatial resolution that can be achieved using this approach.

One of the potential advantages of using inorganic scintillation crystals is that it will be much easier to construct thicker detection planes, thus providing improved photon sensitivity at higher energies. The source detection sensitivity will depend not only on the photon detection sensitivity, but also on the mask thickness and, most importantly, on the instrument background. This background is currently being studied using both Monte Carlo simulations and experimental measurements in particle beams. In the meantime, we have estimated the potential improvement in source sensitivity by appropriate scaling of the published sensitivity for EXIST [4]. At lower energies the background is dominated by the cosmic diffuse radiation, so the background will scale as detector area. At higher energies, internal background begins to play a role and so the background will scale roughly as detector volume. The crossover energy for the two different scalings is about 300 keV. Fig. 2 shows the narrow line sensitivity for CASTER based on these assumptions for background scaling. The EXIST sensitivity is based on 5 mm of CZT and a 5 mm thick (tungsten) coded mask. The CASTER sensitivity is based on 2 cm of $LaBr_3$ and a 1 cm thick (tungsten) coded mask. The range of sensitivity values for CASTER at the higher energies corresponds to the range of background values derived from scaling by area (lowest levels) and scaling by volume (highest levels). Regardless of the precise nature of the background scaling, it is clear that the CASTER design has the potential for significant improvement in sensitivity at these highest energies.

The improved high energy sensitivity offered by the CASTER design directly translates into a greater scientific return. Of particular interest to the GRB community will be the more detailed high energy spectra of GRBs that would be measured by CASTER.

ACKNOWLEDGMENTS

The CASTER mission concept study has been supported by NASA grant NNG04GH78G. Some of this work is also supported at UNH by NASA grant NNG05WC26G and at LSU by US Dept. of Energy NNSA Cooperative Agreement DE-FC52-04-NA25683.

REFERENCES

1. White, N., *Adv. Space Res.* **35**, 96 (2005).
2. Grindlay, J., et al., in *GAMMA 2001*, edited by S.Ritz, N. Gehrels & C.R. Shrader, AIP Conf. Proc. 587, AIP, New York, 2001, pp. 899-908.
3. Grindlay, J. E., *Proc. SPIE* **4851**, 331 (2002).
4. Grindlay, J., et al., in *Gamma-Ray Burst and Afterglow Astronomy 2001*, edited by G.R. Ricker & R.K. Vanderspek, AIP Conf. Proc. 662, AIP, New York, 2003, pp. 477-480.
5. McConnell, M. L., et al., *Proc. SPIE* **5488**, 944 (2004)
6. McConnell, M. L., et al., *Proc. SPIE* **5898**, 1 (2005).
7. van Loef, E. V. D., et al., *Appl. Phys. Lett.* **79**, 1573 (2001).
8. Shah, K. S., et al., *IEEE Trans. Nucl. Sci.* **50**, 2410 (2003).
9. Anger, H. O., *Rev. Sci. Instrum.* **29**, 27 (1958).
10. Macovski, A., *Medical Imaging Systems*, Prentice Hall, New York, 1983.
11. Matthews, K. L., et al., *IEEE Trans. Nucl. Sci.* **48**, 1397 (2001).
12. Case, G. L., et al., *Proc. SPIE* **5898**, 144 (2005).

SuperAGILE and Gamma Ray Bursts

Luigi Pacciani[1], Enrico Costa[1], Guido Barbiellini[5], Ettore Del Monte[1],
Immacolata Donnarumma[1], Yuri Evangelista[1], Marco Feroci[1],
Massimo Frutti[1], Francesco Lazzarotto[1], Igor Lapshov[1],
Marcello Mastropietro[4], Ennio Morelli[3], Massimo Rapisarda[2],
Alda Rubini[1], Paolo Soffitta[1], Marco Tavani[1]

[1] *IASF-INAF Rome, Italy,*
[2] *ENEA Frascati, Italy,*
[3] *IASF-INAF-Bologna, Italy,*
[4] *CNR Montelibretti, Italy,*
[5] *INFN Trieste, Italy*

Abstract. The solid-state hard X-ray imager of AGILE gamma-ray mission – SuperAGILE – has a six arcmin on-axis angular resolution in the 15-45 keV range, a field of view in excess of 1 steradian. The instrument is very light: 5 kg only. It is equipped with an on-board self triggering logic, image deconvolution, and it is able to transmit the coordinates of a GRB to the ground in real-time through the ORBCOMM constellation of satellites. Photon by photon Scientific Data are sent to the Malindi ground station at every contact. In this paper we review the performance of the SuperAGILE experiment (scheduled for a launch in the middle of 2006), after its first on-ground calibrations, and show the perspectives for Gamma Ray Bursts.

Keywords: Imager, X-ray, coded mask, Gamma-ray, GRB, Trigger

PACS: 98.70.Rz, 95.75,Tv, 95.55.Ka, 95.75.Mn, 07.05.Pj, 07.05.Hd, 87.59.-e

DESCRIPTION OF SUPERAGILE

Agile[1] is a small satellite mission for gamma-ray, scheduled for launch in the middle of 2006. The SuperAGILE[2] is the GRB monitor of the mission in the hard x-ray energy region. It consists of four detectors with one-dim. coded masks. Two detectors have one coding direction. The other two have the orthogonal coding direction. The GRB localization is performed combining the coordinates found with at least two orthogonal detectors. The collimators define a field of view of $107^{\circ} \times 68^{\circ}$. The mask element is 242 μm wide. The on-axis angular resolution is 6 arcmin.

The detection plane is formed by 6144 silicon μstrip photodiodes with a pitch size of 121 μm, 19 cm long, 410 μm thick. The readout electronic chains are contained in 48 XAA1.2 ASICs produced by the IDE-ASA company. The XAA1.2 are highly configurable, allowing for a fine regulation of shaping time and tuning of individual analog channel threshold.

The on-board hardware and software allow the on-board detection of GRB and localization.

ON-BOARD BURST SEARCH LOGIC

The SuperAGILE on-board *Normal Burst Search* (BS) *logic* operates at three steps. Gamma Ray Bursts produce an increase in counting rate of SuperAGILE detectors. The first step is implemented on ratemeters from each detector. The second step performs the logical combination of ratemeters, the third step makes the burst image.

Since the burst signal is strongly energy and timescale dependent, the first step of BS algorithm operates at 8 different timescales (Search Integration Time or SIT: 1 ms, 16 ms, 64 ms, 256 ms, 1 s, 8.192 s, and other two programmable between 64 μs and 1 ms), and two energy channels (high and low) for each detector (4 detection units x 2 energy channels x 8 SIT = 64 ratemeters). The bound between the high and low energy channels is programmed with TC.

The ratemeter trigger logic can be selected between an *adaptive trigger logic* and a *static trigger logic*. The ratemeter triggers if the following condition is satisfied:

$$R > B + N_\sigma \bullet \sigma_B \quad \text{(adaptive trigger logic)}$$
$$R > \alpha \bullet B + C \quad \text{(static trigger logic)}$$

where R is the count rate of the ratemeter; N_σ is the threshold on the number of standard deviations σ_B of the background counting rate. N_σ, α, C can be programmed with TC. B is the background counting rate. To avoid false trigger from background variation, the estimation of B is performed on-board from the previous ratemeters.

Each ratemeter can be enabled/disabled individually. A fully programmable lookup table (LUT) provides the combinatorial logic from ratemeters (second step) for the on-board burst image accumulation (third step). The background subtracted burst image provides the on-board coordinates and counting rate of the GRB. These informations are sent to the ground through the ORBCOMM network within few minutes and will be distributed to the community for the follow-up with other instruments.

PERSPECTIVES FOR DETECTION OF GAMMA RAY BURSTS

We studied the trigger sensitivity for two peculiar GRBs: the GRB980425 [3] and GRB050709 [4]. The former is associated with the supernova SN1998bw of type Ib/c at a redshift z=0.0085. It is a faint atypical burst with $E_{iso}=10^{49}$ erg and a prompt duration of 40 s. The latter is a short burst detected with HETE 2, with a prompt duration of 200 ms. The optical afterglow was observed and a redshift of z=0.1606 estimated.

The expected counting rate for a burst like GRB980425 on-axis is 104 cts/s in the 15-45 keV energy band. The single detector ratemeters with a time bin of 8.192 s have a statistical significance 7.2. The on-axis energy spectrum of SuperAGILE is shown in figure 1. The expected counting rate for a burst like GRB050709 on-axis is 1505 cts/s (for the duration of the burst) in the 15-45 keV energy band. The single detector ratemeters with a time bin of 256 ms (64 ms) have a statistical significance of 13 (8.5). The on-axis energy spectrum of SuperAGILE is shown in figure 2.

In the calculations we use the adaptive trigger logic option. The sensitivity region for a detector with "X" coding direction is shown in figure 3 and 4 for the two bursts.

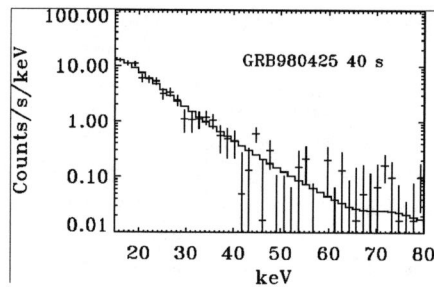

FIGURE 1. Simulation of GRB980425. The continuous line is the model[3] ($N(E) = 3.1E^{-1.41} ph/cm^2/s/keV$), the dots with error bars is the simulation. The integration time is 40 s.

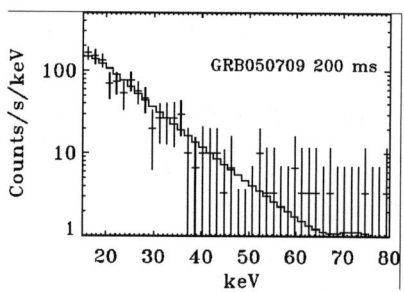

FIGURE 2. Simulation of GRB050709. The continuous line is the model[4] ($N(E) = 3.3E^{-0.53} ph/cm^2/s/keV$), the dots with error bars is the simulation. The integration time is 200 ms.

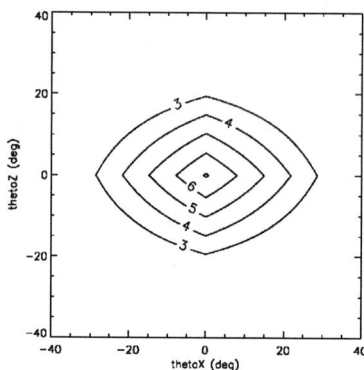

FIGURE 3. Single detector S/NR for GRB980524 for a SIT of 8.192 s.

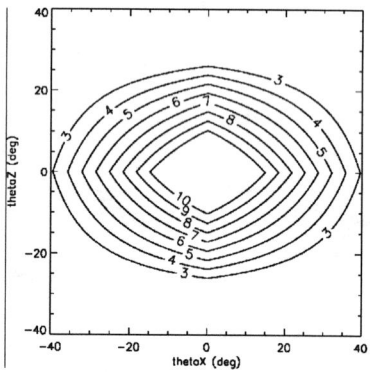

FIGURE 4. Single detector S/NR for GRB050709 for a SIT of 256 ms

The sensitivity region for the detectors with the orthogonal coding direction is similar, but rotated of 90 degrees.

In the following we assume $N_\sigma = 5$ for all the ratemeters, and that the lookup table is configured to select only those triggers coming simultaneously from all detectors, or from 3 of 4 detectors (majority). The final sensitivity level (adopting a majority logic for trigger LUT) is shown in fig. 5 for GRB 980425 and in fig. 6 for GRB 050709. If the simultaneous signals from all four SuperAGILE detectors are required in the LUT, the final trigger sensitivity levels are similar to those reported in fig. 5 and 6, but enhanced of a factor $2/\sqrt{3}$. The BS gives an alert if the burst is inside the dashed octagon. Four sides of the octagonal shape depends on the two detectors with the same coding direction. The other four sides depends on the other two detectors.

The angular size of the octagonal region depends on the N_σ threshold of single detectors. Inside this region, the final sensitivity of the overall BS is drawn. It depends on the chosen combinatorial of the trigger lookup table.

For a burst GRB980425 like, the field of view is 0.10 steradian (0.31 steradian) for $N_\sigma = 5$ ($N_\sigma = 3$). For a burst GRB050709 like, the field of view is 0.45 steradian (0.59 steradian) for $N_\sigma = 5$ ($N_\sigma = 3$).

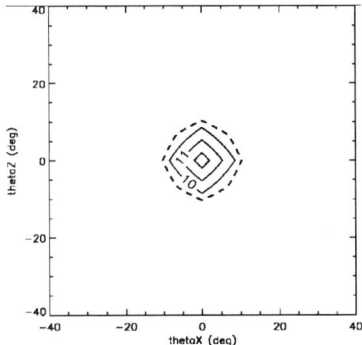

FIGURE 5. GRB980425 trigger sensitivity levels for SIT of 8.192 s and LUT configured to select triggers coming simultaneously from 3 of 4 detectors (majority).

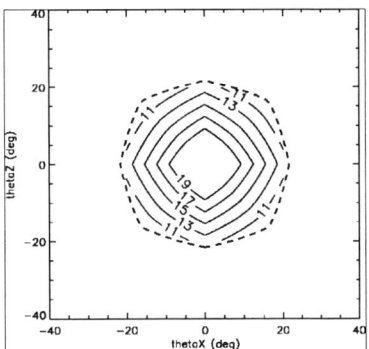

FIGURE 6. GRB050709 trigger sensitivity levels for SIT of 256 ms and LUT configured to select triggers coming simultaneously from 3 of 4 detectors (majority).

The final sensitivity of the BS trigger is better than the chosen N_σ level on single detectors. If the trigger LUT is configured to select triggers from only two detectors simultaneously, the sensitivity of the final BS trigger is at least a factor $\sqrt{2}$ better than the chosen N_σ. If a majority (3 of 4) LUT is used, the final sensitivity is at least a factor $\sqrt{3}$ better. If the simultaneous signals from all four SuperAGILE detectors are required in the LUT, the GRB trigger sensitivity is at least a factor 2 better than the chosen N_σ threshold on a single detector.

The lack of "SIT" higher than 8 s causes the drop in sensitivity of the burst search logic for the longest bursts. Instead the foreseen sensitivity to burst with duration between 16 ms and 8 s (or shorter) is quite high.

REFERENCES

1. M. Tavani et al, "The AGILE Instrument", *X-Ray and Gamma-Ray Telescopes and Instruments for Astronomy*, Edited by Joachim E. Truemper, et al, SPIE Conference Proceedings 4851 (2003), pp. 1151-1162.
2. M. Feroci et al, "SUPERAGILE: The Hard X-ray Imager of AGILE", *Gamma-Ray Bursts: 30 Years of Discovery: Gamma-Ray Burst Symposium*, Edited by E. E. Fenimore et al, AIP Conference Proceedings 727, American Institute of Physics, Melville, NY, 2003, pp. 696-699.
3. E. Pian et al, "BeppoSAX Observations of GRB 980425: Detection of the Prompt Event and Monitoring of the Error Box", *The Astrophysical Journal*, **536**, 2000 June 20, pp. 778-787.
4. J. S. Villasenor et al, "Discovery of the Short Gamma Ray Burst GRB 050709", *Nature*, **437**, 6 October 2005, pp. 855 – 858.

Burst Detector Sensitivity: Past, Present & Future

David L. Band[,†]

Code 661, NASA/Goddard Space Flight Center, Greenbelt, MD 20771
JCA, University of Maryland, Baltimore County, Baltimore, MD 21250

Abstract. I compare the burst detection sensitivity of *CGRO*'s BATSE, *Swift*'s BAT, the *GLAST* Burst Monitor (GBM) and *EXIST* as a function of a burst's spectrum and duration. A detector's overall burst sensitivity depends on its energy sensitivity and set of accumulations times Δt; these two factors shape the detected burst population. For example, relative to BATSE, the BAT's softer energy band decreases the detection rate of short, hard bursts, while the BAT's longer accumulation times increase the detection rate of long, soft bursts. Consequently, *Swift* is detecting long, low fluence bursts (2-3× fainter than BATSE).

Keywords: gamma-ray bursts; detectors
PACS: 98.70.Rz, 95.55.Ka

What is the relative sensitivity of different detectors for detecting gamma-ray bursts, and how should this sensitivity be compared? How do these differences shape the observed burst populations, which must be taken into account in determining the underlying burst distribution? Here I compare BATSE's Large Area Detectors on *CGRO* (the past), the Burst Alert Telescope (BAT)[1] on *Swift* (the present), and the *GLAST* Burst Monitor (GBM) and *EXIST* (the future). BATSE and the GBM are/were sets of NaI(Tl) detectors while the BAT and *EXIST* are/will be CZT coded mask detectors. The energy range of NaI(Tl) detectors is $\sim$20–1000 keV while for CZT it is $\sim$10–150 keV. I apply a semi-analytic methodology using simplified models of the trigger systems of the different detectors.

Most instruments detect bursts using either rate triggers or image triggers. A rate trigger determines whether the increase in the number of counts in a time bin Δt and energy band ΔE over the expected number of background counts is statistically significant. An image trigger determines whether the image formed from the counts in the time bin Δt and energy band ΔE contains a new point source. Usually an image trigger is preceded by a rate trigger that starts the imaging process[2]; the rate trigger is set to permit many false positives that are eliminated by the image trigger. If the number of burst counts is S and the number of non-burst counts is B, then the rate trigger significance σ_r (for BATSE and the GBM) and the image trigger significance σ_i for Δt and ΔE are

$$\sigma_r = \frac{S}{\sqrt{B}} \quad \text{and} \quad \sigma_i = \frac{f_c S}{\sqrt{B+S}} \qquad (1)$$

where f_c accounts for the finite size of the detector pixels. For *Swift* $f_c \sim 0.7$[3], which explains why for a given burst the rate trigger significance is greater than the imaging significance[4]. The BAT uses a more complex rate trigger than shown above. For

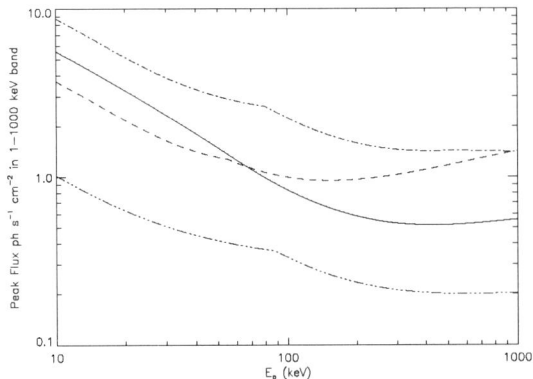

FIGURE 1. Threshold value of F_T (1–1000 keV peak flux) as a function of E_p for BATSE (solid), the BAT (dashed), the GBM (dot-dashed) and one *EXIST* telescope (3 dots-dashed). The spectrum has a low energy spectral index of $\alpha = -0.5$ and a high energy index of $\beta = -2$.

directions other than the burst position the counts S from the burst contribute to the average flux level, and therefore in imaging S is compared to $\sqrt{B+S}$. A trigger occurs when σ_r or σ_i exceeds a threshold value that is sufficiently high for a small probability of false positives. Because of the similarity of σ_r and σ_i, particularly since usually $B \gg S$ near threshold, the methodologies for evaluating the sensitivity of rate and image triggers are essentially the same[5].

Whether a given detector and its trigger system detects a burst depends on the number of counts S the burst produces in time bin Δt and energy band ΔE. If bursts differed only in their intensity, we could use a common measure of burst intensity, but bursts differ in their temporal and spectral properties, and thus a given detector is more sensitive to some burst types and less to others. While it is impossible to characterize a burst completely by only a few parameters, approximate burst types can be described by a few parameters. For burst detection, I characterize bursts by: E_p—the energy at the peak of $E^2 N(E) \propto v f_v$ for the spectrum averaged over Δt; and T_{90}—the time duration for 90% of the flux. I characterize the burst intensity by the peak flux F_T over 1 s in the 1–1000 keV band. A detector does not measure F_T directly; since a spectral fit is necessary to convert from counts to flux, F_T need not be over ΔE[5].

I first evaluate the sensitivity of the four detectors to E_p, disregarding the burst duration. Thus I assume that bursts have constant emission over $\Delta t=1$ s. Then S is proportional to the burst spectrum times the detector efficiency integrated over ΔE. BATSE used only one value of ΔE at any given time, typically $\Delta E=50$–300 keV, but later detectors use a set of ΔE simultaneously. Figure 1 shows the threshold F_T as a function of E_p for BATSE, BAT, GBM and *EXIST*. While I characterize the spectrum primarily by E_p, there remains a dependence on the high and low energy spectral indices, β and α (not shown by Figure 1). In all cases I use the maximum sensitivity over the FOV. The scalloping results from multiple values of ΔE.

As a CZT detector, the BAT's sensitivity is shifted to lower energy than BATSE's. The

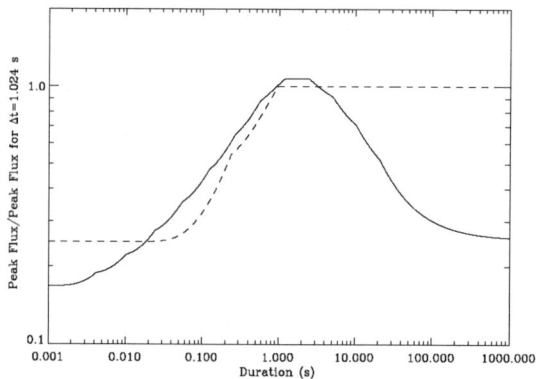

FIGURE 2. The ratio of the detector sensitivity for a trigger system using a set of Δt values to the rate trigger sensitivity for $\Delta t=1.024$ s alone; ratios less than 1 indicate an increase in sensitivity resulting from additional values of Δt. The dashed curve is for BATSE's set of Δt, while the dashed curve is for the BAT.

GBM's detectors will be smaller than BATSE's, while *EXIST* will have larger detectors than the BAT. Note that these sensitivity curves are all for $\Delta t=1$ s; increasing Δt increases the BAT's sensitivity to long bursts.

As Δt increases, the number of burst counts S may increase but the number of background counts B definitely increases. Thus there is a competition between changes in S and B as Δt changes. The affect on the sensitivity of increasing or decreasing Δt depends on the burst lightcurve; the duration T_{90} is a key parameter characterizing the lightcurve. If $T_{90} \ll \Delta t$ then the dilution of the burst counts by the background can be decreased by decreasing Δt, but if $T_{90} \gg \Delta t$ then increasing Δt might increase the number of burst counts relative to the background. Detectors use more than one Δt; the overall sensitivity is the lowest value of F_T for any Δt. Figure 2 shows as a function of T_{90} the ratio of the detector sensitivity for a set of Δt to the rate trigger sensitivity for $\Delta t=1.024$ s alone; ratios less than 1 indicate an increase in sensitivity resulting from the additional values of Δt and the trigger type. The burst is assumed to have an exponential lightcurve. BATSE (dashed curve) used a simple rate trigger with $\Delta t=0.064$, 0.256 and 1.024 s. Thus, for $T_{90} > 1.024$ s the $\Delta t=1.024$ s trigger dominates, and the ratio equals 1, while for $T_{90} < 1.024$ s the smaller Δt values increase the sensitivity (smaller ratios) to short duration bursts. The BAT, GBM and *EXIST* (will) use both smaller and longer Δt values than BATSE did. The solid curve on Figure 2 was calculated for the BAT's image trigger; note the increase in sensitivity over BATSE's set of Δt for both long and short bursts. The increase in sensitivity to short duration bursts is not as dramatic as for long duration bursts because σ_i does not decrease indefinitely as B decreases for fixed S (see eq. 1). The reduced sensitivity of an image trigger relative to a simple rate trigger for very short bursts does NOT mean that a rate trigger is superior to an image trigger: an image trigger localizes the burst, and does not require a model of the background rate.

The spectral and temporal dependencies of the burst sensitivity can be combined to produce the threshold F_T as a function of both E_p and T_{90}. Figure 3 presents the ratio

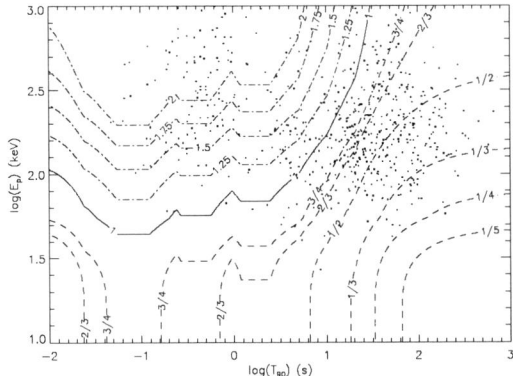

FIGURE 3. Contour plot of the ratio of the sensitivities of the BAT and BATSE as a function of E_p and T_{90}; a ratio less than one indicates that the BAT is more sensitive than BATSE. α=-0.5 and β=-2 are assumed. Also plotted are the E_p and T_{90} for a set of BATSE bursts with enough counts for spectral fits.

of the sensitivities of the BAT and BATSE; a ratio less than one indicates that the BAT is more sensitive than BATSE at that particular set of E_p and T_{90}. Also plotted are the E_p and T_{90} for a set of BATSE bursts. As can be seen, the short, hard bursts are in a region of parameter space where the BAT is less sensitive than BATSE while the BAT is more sensitive to long, soft bursts. The contours' gradient shows that the BAT detects fewer short, hard burst because its energy band is lower than BATSE's was, and the BAT detects more long, soft bursts both because of its lower energy band and its greater sensitivity to long bursts. This explains the shift in the duration distributions. Because the BAT is significantly more sensitive to long bursts, the average fluence detected by the BAT is ~ 2.5 times fainter than BATSE's. As a burst's redshift is increased, its duration is dilated and its spectrum is redshifted. Thus high redshift bursts are shifted towards the parameter space region where *Swift* is particularly sensitive. Note that *Beppo-SAX* and *HETE-II* use(d) imaging triggers with low energy detectors, explaining why these two detectors detect(ed) few short bursts and many X-ray rich bursts and X-ray Flashes.

ACKNOWLEDGMENTS

I thank members of the BAT instrument team for informative discussions; conclusions regarding the BAT's sensitivity are mine, and do not represent the BAT team.

REFERENCES

1. S. Barthelmy, et al., *Space Sci Rev* (2005), in press
2. E. E. Fenimore, et al., *AIPC* **662**, 491
3. G. Skinner, personal communication (2005)
4. D. Palmer, personal communication (2005)
5. D. Band, *Ap. J.* **588**, 945 (2003)

The in-flight spectroscopic calibration of the *Swift* XRT CCD camera

A.P. Beardmore, O. Godet, A.F. Abbey, J.P. Osborne, K.L. Page,
A.A. Wells and M.R. Goad
on behalf of the *Swift* XRT team

Department of Physics and Astronomy, University of Leicester, UK

Abstract.
The *Swift* X-ray Telescope (XRT) focal plane camera houses a front-illuminated MOS CCD, providing a spectral response of 145 eV FWHM at 5.9 keV. We describe the status of the CCD X-ray spectral redistribution matrices, which are made using a Monte-Carlo simulation technique based on physical models of the CCD response. We emphasize how the model has been refined following in-flight experience with celestial and on-board calibration sources. Residuals less than 10% are typically observed for astrophysical sources.

Keywords: gamma-ray bursts
PACS: 98.70.Rz

INTRODUCTION

The X-ray Telescope (XRT; [1]) on-board *Swift* is responsible for localizing the X-ray afterglow emission from gamma-ray bursts (GRBs) and probing their temporal and spectral properties. The XRT utilizes flight spare mirrors from the JET-X program, which provide an effective area of 150 cm^2 at 1.5 keV. These focus incoming X-rays onto an e2v technologies CCD22 detector - the same type of detector is used in the EPIC MOS instruments on the XMM-Newton satellite.

The CCD22 type of detector is a front illuminated device, but is novel in that it was manufactured with an open electrode structure to improve its sensitivity to low energy X-rays. It is also made from high resistivity silicon which provides good high energy efficiency. These enhancements enable it to operate over a useful bandpass of 0.2 – 10 keV.

Due to the dynamic nature of GRBs, and their potentially high initial intensities, the XRT has to be capable of observing over a range of more than 6 orders of magnitude in flux. In order to achieve this, it was designed to operate autonomously and switch between three spectroscopic modes, depending on the source flux, in order to minimize the effects of pile-up. These are (in order of decreasing intensity): Photodiode mode (PD), which has no spatial information; Windowed Timing (WT) mode, which preserves one spatial dimension; Photon Counting (PC) mode, which gives full two dimensional imaging. See Hill et al. [2] for more details.

Two incidents occurred in-orbit which have affected XRT operations. After launch, but just prior to the start of normal operations, the XRT thermo-electric cooler power supply unit became inoperable. This means the CCD can only be cooled passively, down

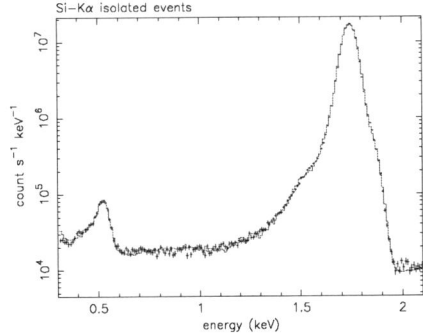

FIGURE 1. Si-K isolated event PC spectrum from the Leicester calibration facility (black). The model (red) includes Si-Kα (1.74 keV), Si-Kβ (1.84 keV), a weak O-Kα (0.525 keV) line and an underlying Bremsstrahlung continuum.

to temperatures between -75C to -45C, rather than the pre-launch expectation of -100C. Then on 2005 May 27, the CCD suffered damage from a micrometeroid impact which has caused the appearance of bright columns and pixels. These are vetoed on-board for WT and PC modes, but this is not possible for PD mode, which is currently disabled for follow-up observations. These events appear to have had no short-term impact on the spectroscopic performance of the XRT.

RESPONSE MATRIX GENERATION

Response Matrix Files (RMFs) are created by a Monte-Carlo simulation code [3, 4]. The code models the following processes: transmission of the incident X-rays through the CCD electrode structure; photo-absorption in the active layers of the device; charge cloud generation, transportation and spreading; silicon fluorescence and its associated escape peak; surface loss effects; mapping of the resultant charge-cloud to the detector pixel array; charge transfer efficiency; addition of electronic read-out noise; event thresholding and classification according to the specific mode of operation.

Ground calibration data taken at the Leicester calibration facility were used to tune the model prior to launch, as illustrated in Fig. 1.

Due to the higher-than-expected in-orbit operating temperatures, the electronic noise has increased compared with the ground calibration expectations; this has been accounted for in the response model.

IN-FLIGHT CALIBRATION

Numerous calibration observations have been made which are being used to refine the CCD response under in-orbit operating conditions. These including the Fe-55 internal calibration door source (prior to the XRT door opening on 2004 Dec 11) and corner source data; the pulsar RXJ0720-31 and neutron star RXJ1856-37 (PC/WT mode effective area and low energy response); the supernova remnants Cas A and 2E0102-72 (PC & WT gain and line kernel); the Crab (PD & WT effective area); the cluster of galaxies PKS0745-19 (PC mode low energy shelf and effective area); the plerion SNR0540-69

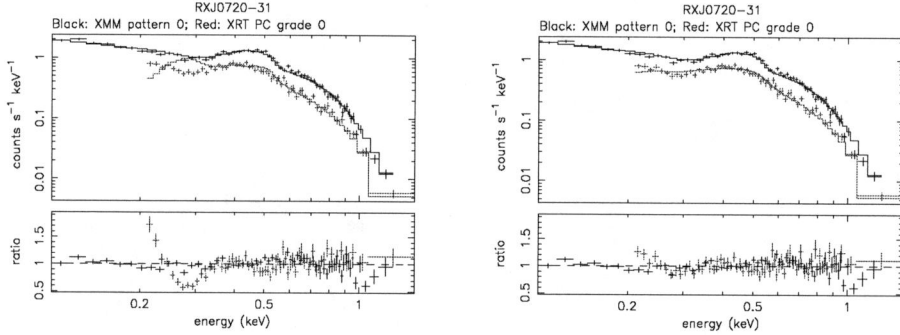

FIGURE 2. XMM MOS and XRT PC grade 0 spectra of the neutron star RXJ0720-31. The left shows the fit with the current RMFs (v007); the right, an improved surface charge loss function (v007b).

(PC mode effective area); the quasar 3C 273 (WT effective area and cross-calibration with XMM/RXTE); the quasar H1426+428 (PC effective area and cross-calibration with XMM). Some results from these observations are given below :

- Using observations of the soft sources mentioned above, the surface loss function has been modified to better match the spectral redistribution at low energies (see Fig. 2).
- The line-rich supernova remnant 2E0102-72 (Fig. 3) shows residuals below 20% above 0.7 keV in WT and PC modes.
- For highly absorbed sources the response model showed an under-estimation of the redistribution down to low energies; the parameterisation of this loss-shelf has been improved (Fig. 4) but needs further refinement.
- A multi-mission cross-calibration campaign was performed on 3C 273. Fig. 5 shows the results of fitting the XRT WT, XMM PN/MOS and RXTE PCA spectra with a power-law and blackbody components for the soft excess. The XRT photon index and flux compare well to the other instruments, with residuals less than 10%.

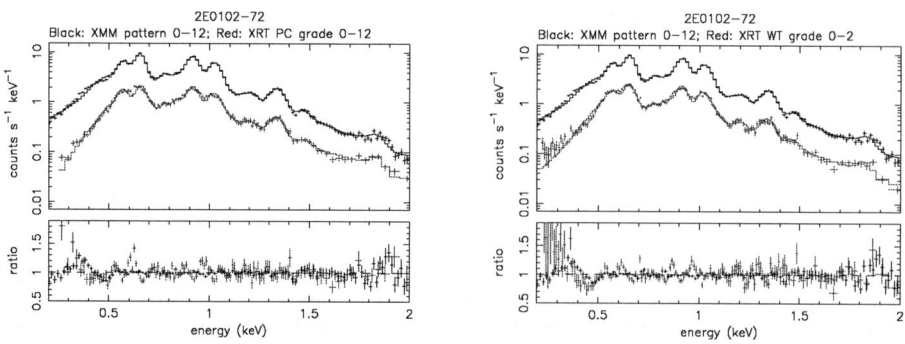

FIGURE 3. Left: XMM MOS and XRT PC grade 0-12 spectra of the line-rich SNR 2E0102-72. Right: as left, but for XRT WT grade 0-2

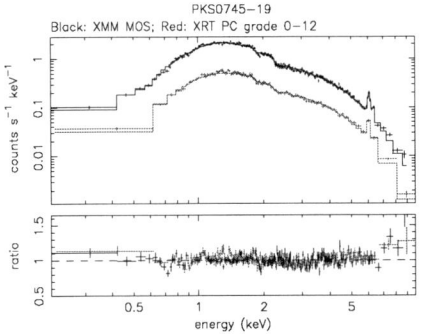
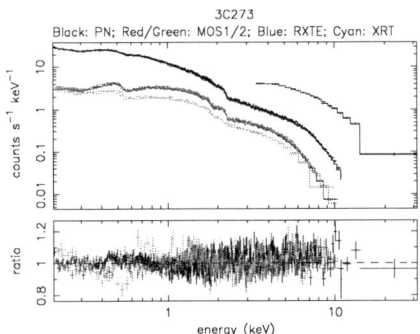

FIGURE 4. XMM MOS and XRT PC spectra of the galaxy cluster PKS0745-19 fit with the improved loss-shelf response (v007b).

FIGURE 5. 3C 273 cross-calibration observation, comparing XMM PN, MOS1/2, XRT WT and RXTE PCA spectra.

FUTURE DEVELOPMENTS

The response model continues to be refined. In particular, the modelling of large event sizes is being improved with the use of a revised field-free region charge-cloud shape description. This better accounts for the sub-threshold losses seen in the door-source calibration data around the line shoulder (Fig. 6).

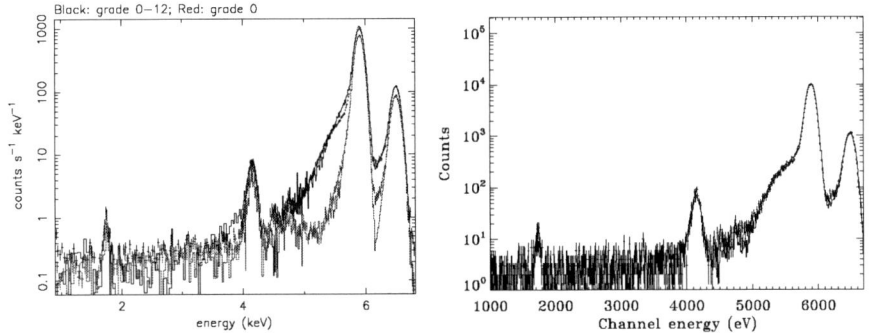

FIGURE 6. Fe-55 PC mode door-source data. The left plot shows a fit with the current released RMF (v007); the right shows the fit from a development RMF using a new charge-cloud model.

REFERENCES

1. D.N. Burrows et al., 2005, Space Science Reviews, 120, 95
2. J.E. Hill et al., 2005, SPIE, 5898, 313
3. A.P. Beardmore et al., 2005, XRT-LUX-CAL-108, http://heasarc.gsfc.nasa.gov/docs/heasarc/caldb/swift/docs/xrt/XRT-LUX-CAL-108_RMF_v7.ps
4. J.P. Osborne et al., 2005, SPIE, 5898, 340

GRB Astrophysics with LOBSTER

R. Hudec[1], L. Pína[2,3], L. Švéda[2], A. Inneman[3]

1 Astronomical Institute, Academy of Sciences of the Czech Republic, CZ-251 65 Ondrejov, Czech Republic rhudec@asu.cas.cz
2 Faculty of Nuclear Science, Czech Technical University, Prague, Czech Republic
3 Center of Advanced X-Ray Technologies, Reflex s.r.o., Prague, Czech Republic

Abstract. We refer on the recent developments of LOBSTER project suggesting novel wide-field Lobster-Eye type of X-ray All Sky Monitor to detect and to analyze GRBs including XRF and X-ray rich GRBs. The triggers can be detected and localized by their X-ray emission in the 0.1 – 8 keV energy range. The system exhibits fine detecting sensitivities of order of 10-12 ergcm^{-2}s^{-1} and the localization accuracy is of order of a few arcmin. The LOBSTER is expected to contribute significantly to analyses of GRBs and especially the XRFs.

Keywords: Gamma-ras bursts, X-ray telescopes, X-ray optics, Lobster-Eye
PACS: 95.55.Ka, 95.85.Nv, 95.85.Pw

INTRODUCTION - LOBSTER-EYE (LE)

Novel Wide Field X-ray Telescopes of LE type allow fields of view (FOV) of 100 sq. deg. and more easily possible (classical X-ray optics only 1 deg or less). They are based on analogy with lobster eyes and have been designed for astronomy, but laboratory applications are also possible [1], [2], [3], [4].

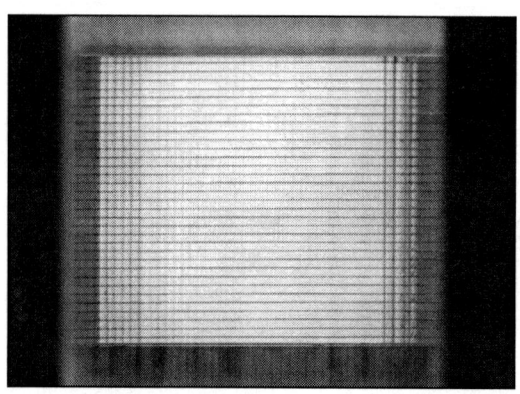

FIGURE 1. The front view of the mini - lobster module, Schmidt arrangement, based on 100 micron thick plates spaced by 300 microns, 23 x 23 mm each.

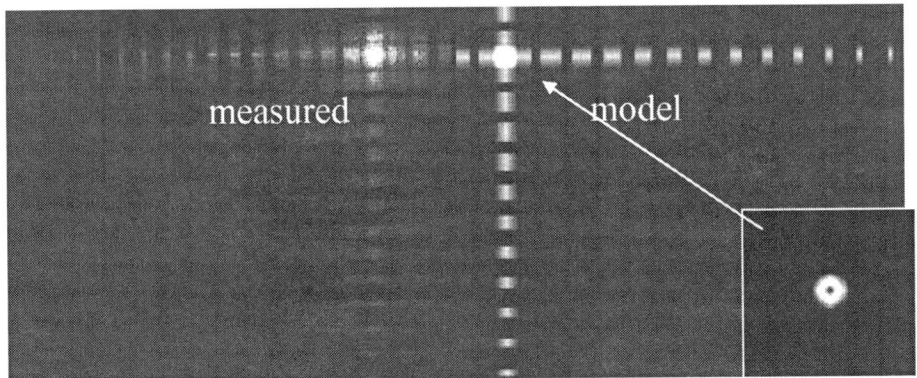

FIGURE 2. X-ray experiment vs. simulation. Point-to-point focusing system, LE Schmidt mini. Source-detector distance 1.2 m, 8 keV photons. Image width: 2x512 pixels, 24 mm pixel. Gain: ~570 (measured) vs. ~584 (model).

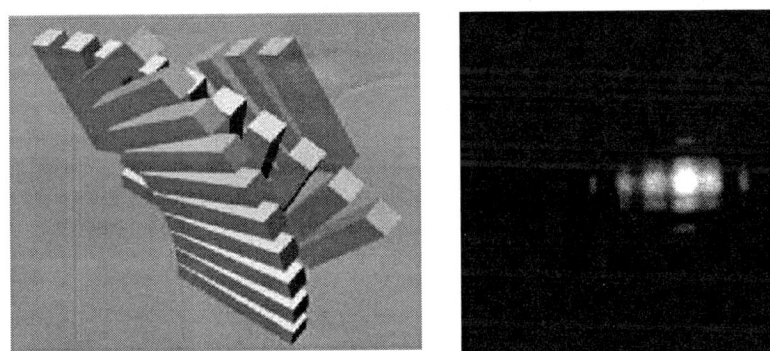

FIGURE 3. Left: LE All Sky Monitor – Proposal. Modular concept - design for ISS. Easy modification for EXIST, HXMT or other satellites is possible. Right: Focal image of the micro LE test module (3 x 3 mm, 30 microns thick gold coated glass foils) at 8 keV.

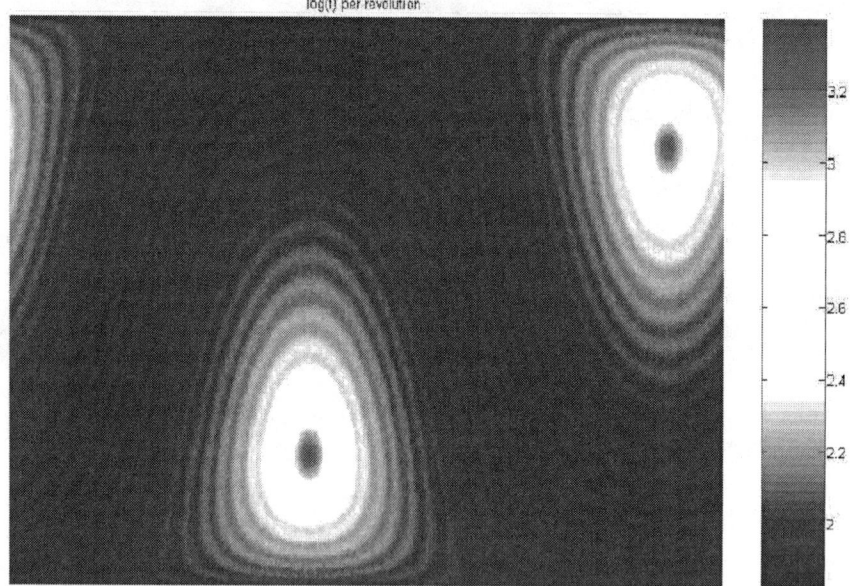

FIGURE 4. The LE Sky coverage per revolution (arrangement according to the Fig. 3 and 90 min revolution)

The estimated daily limiting flux for the arrangement illustrated in Fig.3 is 10^{-12} erg/s/cm^2. One module is represented by 2 x 195 plates 78 x 11.5 x 0.1 mm, with 0.3 mm spacing. The expected detector pixel size is 150 microns. The total front area represents 1825 cm^2 and the energy range covers 0.1 - 10 keV. The FOV represents 180 x 6 degrees (30 modules 6 x 6 degrees) with angular resolution 3 - 4 arcmin and a total mass < 200 kg.

SCIENTIFIC GOALS

1. Alert System for X-ray transients

The LE X-ray All Sky Monitor (ASM) is able to provide fast recognition of new X-ray sources and/or to detect sudden changes in X-ray flux of known sources, with related prompt emission study, precise positioning, and alert system for narrow-field instruments. The LE ASM is expected to detect the GRB prompt and afterglow X-ray emission (20-60 triggers/year), the X-ray flashes (> 8 triggers/year), orphan GRBs (detectable in X-rays but not in gamma), SNe prompt emission (thermal flash) 10-20 triggers/year, X-ray binary & CVs flux changes, and stellar events in the Sun's vinicity.

In the study of GRBs, the recent results indicate that a vast majority (e.g. about 90% for the BeppoSAX GRBs [5]) of GRBs has a detectable X-ray emission well above the detection limit of the LE ASM. This means that not only the GRBs can be independently detected by their X-ray emission with relatively good localization accuracy, but that also analyses of their fading light curves over hours will be possible in numerous cases. While the prompt X-ray emission of GRBs detected by BeppoSAX could be easily detected by a LE ASM in all cases, even after 11 hours there still the fading X-ray afterglow emission remains to be detectable in some cases [5].

2. Long-term X-ray source monitoring

The LE ASM will also enable the long-term monitoring of large number of X-ray sources with sampling of hours to days (depending on the source flux). Light curves for all the sources together with rough spectra (continuum monitoring, strong lines, iron detectable) are expected to be obtained. In the following list we assume the limiting flux of 10^{-12} erg/s/cm^2 (but we can go deeper): X-ray binaries ~ 700 triggers, Cataclysmic Variables ~ 200 triggers, stars ~ 600 triggers, AGN ~ 4 000 triggers, galaxy clusters ~ 400 triggers, SN remnants.

CONCLUSION

The LE ASM will contribute to various regions of recent astrophysics. The necessary technical background is already available, making proposals for space project based on Lobster Eye optics possible.

ACKNOWLEDGEMENTS

We acknowledge the support (in considering independent detections of GRBs by their lower energy emission) by the Grant Agency of the Academy of Sciences of the Czech Republic, grant A3003206. The optics technology is partly related to the project FD-K3/052 by the Ministry of Industry and Trade of the Czech Republic.

REFERENCES

1. Sveda, Libor; Hudec, Rene; Pina, Ladislav; Inneman, Adolf, *X-Ray Sources and Optics*. Edited by MacDonald, Carolyn A.; Macrander, Albert T.; Ishikawa, Tetsuya; Morawe, Christian; Wood, James L. Proceedings of the SPIE, Volume 5539, pp. 116-125 (2004).
2. Hudec, Rene; Sveda, Libor; Inneman, Adolf; Pina, Ladislav, *Proceedings of the SPIE*, Volume 5488, pp. 449-459 (2004).
3. Hudec, Rene; Inneman, Adolf V.; Pina, Ladislav; Hudcova, V.; Sveda, L.; Ticha, Hana, *X-Ray and Gamma-Ray Telescopes and Instruments for Astronomy*. Edited by Joachim E. Truemper, Harvey D. Tananbaum. Proceedings of the SPIE, Volume 4851, pp. 578-586 (2003).
4. Hudec, Rene; Inneman, Adolf V.; Pina, Ladislav Proc. SPIE Vol. 4012, p. 432-441, *X-Ray Optics, Instruments, and Missions III*, Joachim E. Truemper; Bernd Aschenbach; Eds. (2000).
5. DePasquale M. et al. astro-ph/0507708, 2005.

High Redshift GRBs observed by GLAST

Nicola Omodei for the GLAST/LAT GRB Science Group

INFN Pisa, Largo B.Pontecorvo, 3 56100 Pisa

Abstract. The Gamma-ray Large Area Space Telescope (GLAST) is the next generation satellite for high energy astronomy. The flight hardware production is now complete, the sixteen towers have been integrated into the flight grid and the ACD is already placed onsite. The detector will now be integrated with the spacecraft in the next few months. GLAST will be launched in 2007 and it will cover the energy range from 10 keV to more than 300 GeV. Inspiring ourself to the huge explosion observed the 4th September of 2005 by the Swift mission [1], we use the full simulation chain developed by the GLAST collaboration to simulate an high-redshift Gamma-Ray Burst, combining all the information available in literature on GRB 050904 with some assumptions, especially for the high energy emission. Our simulation takes care both of the effect of the cosmological expansion on the spectra and on the light curve, as well as the absorption of radiation by photon-photon interaction with the Extragalactic Background Light (EBL).

Keywords: Gamma-ray detectors; Gamma-ray telescopes; Gamma-ray bursts;
PACS: 07.85.Fv,29.40.-n ; 95.55.Ka; 98.70.Rz;

THE GAMMA-RAY LARGE AREA SPACE TELESCOPE

The heart of GLAST is the Large Area telescope (LAT), a pair production telescope, made by an array of 16 identical towers each one composed by a high precision tracker, based on silicon strip detector, and a hodoscopic CsI calorimeter of 8.5 radiation length. The array of towers is surrounded by a segmented anticoincidence detector (ACD) for shielding the tracker by the incoming cosmic rays. The large effective area (~1 square meter at 10 GeV), the wide field of view (2.5 sr) the good angular resolution (~0.1° at 10 GeV), the good energy resolution (<15%), and the dynamical range (from 30 MeV to more than 300 GeV), provide a factor of 30 or more advance in sensitivity with compare to those of its highly-successfull predecessor EGRET (1991-2000). GLAST is also instrumented with a burst hunter, the GLAST Burst Monitor (GBM), which cover the typical range of the Gamma Ray Burst prompt emission. An array of 12 Sodium Iodide (NaI) detector covers an energy range from 10 KeV to 1 MeV, providing low energy coverage and a wide FoV. Two Bismunth Germanade detector (BGO), operating between 150 keV and 30 MeV, guarantee the spectral overlap with the LAT over a large field of view.

HIGH REDSHIFT GAMMA-RAY BURSTS

Gamma-Ray Bursts are one of the most exciting topics in high-enerry astrophysics. They are extragalactic sources now observed, thanks to the Swift mission, up to the most remote region of the Universe ($z\sim7$). At these very high redshifts the transparency of the Universe is strictly related to the Extragalactic Background Light: high-energy photons interact with the EBL producing pairs. The high-energy spectrum of distant sources presents a cut-off, whose position is related to the density of the background photons. Given the uncertainties of the theoretical predictions due to the lack of measurements, it can not be determined a priori, but it is reasonable that, at GLAST (LAT) energies, the Universe remains transparent up redshift > 10 [2]. Nevertheless, the measurement of the cut-off of a large sample of distant sources (AGN, GRB) has been proposed as method for scanning the EBL spectral energy distribution [3]. GLAST will naturally play an important role in this scenario: by combining observations from the GBM and the LAT covering a wide spectral range, and providing strong constrain on the high energy emission; i.e the high energy spectral index obtained by GLAST observations, can be use as *prior* for ground-based experiments. In this scenario a fast alert is provided by GLAST via GCN circulars; fast repositioning telescopes (like MAGIC) can point the GRB localization by looking for high-energy photons. Telescopes involved in this game are the ones with the lowest energy threshold (or the largest collective area), hopefully overlapping the LAT energy range.

SIMULATING HIGH-REDSHIFT GRBS

In the GRB simulator adopted here, the GRB spectrum is described with a Band function [4], and the temporal behavior is reproduced by adopting a universal pulse shape [5]. We then combine several pulses reproducing the typical observed variability. To extrapolate the spectrum at high energy we adopt the "pulse paradigm" observed by Norris et al.[5], for which the pulse width W scales with the energy as $W \propto E^{-0.33}$, as well as the peak positions at different energies (high energy pulses peak before). Finally, we have introduced the redshift dependence in our simulator: time dilation effects make the observed time-scales longer than the ones in the source rest frame by a factor $(1+z)$, and the observed energies lower than those emitted by the same factor. We have also introduced the EBL absorption, by adopting the EBL model proposed by Primack et al. [6]. We constrain our simulations in order to obtain a reasonable spectrum and light curve at Swift energies and we extrapolate the spectrum at LAT energies. In the observation made by Cusumano et al. [7] the duration of the prompt emission of GRB050904 at BAT energy range (15-150 keV) in the GRB rest frame is ~60 s, and the low energy spectral index is ~ -1.2. Unfortunately, we don't know the peak energy for the prompt emission, as well as the high-energy spectral index, so we have to assumed those two values: E_p=1 MeV (the upper-bound for the BAT detector, shifted by the factor 1+z) and $\beta = -2.25$ (typical value for GRBs [8]). The BAT fluence (15 keV-150 keV) is 5.4×10^{-6} erg cm^{-2}. In our simulation we normalize the spectrum with the value of the fluence observed in the BAT energy range, and compute the peak flux in the same range, obtaining 0.83 cm^{-2}s^{-1} (the observed measured 0.8±0.2 cm^{-2}s^{-1}, GCN 3910). In the GBM trigger energy range (50 keV-300 keV), the simulated peak flux is 0.42 cm^{-2}s^{-1}, which is close to the GBM sensitivity (currently 0.35 cm^{-2}s^{-1} on ground [9]). To compute the Swift/BAT response (Fig. 1) we use a simplified effective area of the Swift's instrument just considering a constant value of A_{BAT}=2000 cm^2 between 15 keV and 150 keV, and zero outside. Poisson fluctuation are considered in each bin. For the LAT detector, a different approach is used. We first sample the number of photons that *illuminate* the LAT

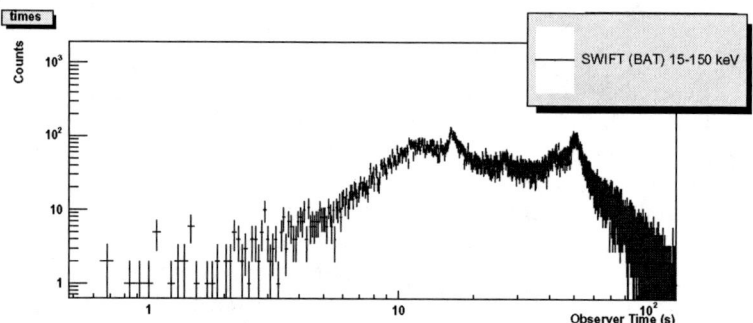

FIGURE 1. Simulated BAT lightcurve. Poissonian noise has been added to the light curve in order to have an estimation of the observed counts. A simplified effective area of the BAT detector is adopted. The burst fluence is 5.4×10^{-6} erg cm^{-2}. Error bars are statistical fluctuations.

detector, and than we process these photons with the fast simulator (*observationSim*) which uses parameterized IRF to fast compute the response of the detector. Finally, the GLAST Burst Monitor simulator, provided by the GBM collaboration, can be combined with the LAT simulator to obtain the *same burst* described for the both GBM and LAT detector.

We simulate a on-axis burst at redshift $z = 6.3$: LAT and GBM simulators provide data files that are equivalent, in terms of data format and data handling, to the real data (standard *HEASARC*): tools already developed by the LAT collaboration allow LAT and GBM files to be combined. In Fig. 2 we combined 6 NaI low energy spectra (all the illuminated detectors), one BGO high energy spectrum, and the LAT spectrum to obtain a coverage over more than seven decades in energy. Counts in the LAT detector have been obtained selecting a region of 20° around the burst localization. Given the low background photon contamination, no background rejection in needed. GBM detectors are, on the other hand, background dominated and the subtraction of the background is needed (∼10% of the total counts are GRB counts). A spectral analysis with *xspec* has been done. The best results have been obtained firstly using the only LAT data and a power law as fitting function. A spectral index of -2.25 ± 0.2 with a reduced χ^2 of 0.4 for 13 degree of freedom (96% of c.l.) has been estimated. Than this value has been used to constrain the Band function high energy spectral index, used to fit the combined GBM and LAT data. The best model parameters are $\alpha = -1.25 \pm 0.02, \beta - 2.25 \pm 0.1$ and $E_p = 191 \pm 10$ with a reduced χ^2 (with 764 dof) of 1.1. The folded model and the data are shown in Fig.2. Different lines are referred to different detectors.

Different fitting functions have also been used, in particular the exponential cut-off at high energy. In this case, the best result has been obtained using the GBM detectors alone, for constraining the high energy spectral index, and fitting the LAT spectrum with a constrained power-law with an exponential cut-off. The low counts in the high-energy bin of the LAT detector make anyway low the significance of this fit. For the simulated burst, the possibility of directly detect the high energy cut-off is low. Only for very bright bursts (one order of magnitude more in fluence) this could be possible: future works will investigate on this topic.

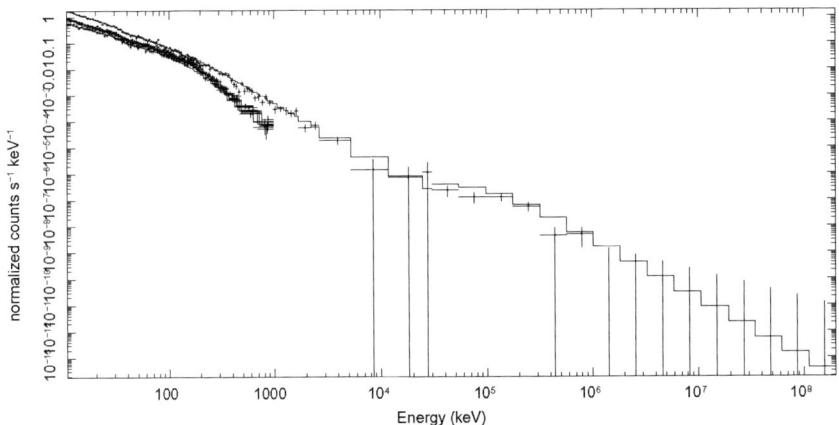

FIGURE 2. Joint spectral analysis: 6 NaI crystals, 1 BGO and the LAT detector. Different lines are the same model folded with different instrument response functions for different detectors.

DISCUSSION AND CONCLUSIONS

The combination with GBM and LAT detectors allows to constrain the spectral parameters over a wide energy range (several order of magnitudes), and a measurement of these parameters is possible, even for week bursts (i.e. considering the GBM burst sensitivity threshold). Moreover, the peak energy, required for empirical relations like the Amati [10] or the Ghirlanda [11] relations, is well covered by the GBM detector and spectral analysis provides good estimation of this parameter. The Large Area Telescope, on the other hand, provides the spectral coverage between the typical prompt GRB emission and Cherencov telescopes: joint observations of GRBs from space and ground will hence allow the study of the cosmological high-energy cut-off and its implication on EBL attenuation.

REFERENCES

1. N. Gehrels et al., *Ap. J.* **611**, 1005–1020 (2004).
2. T. M. Kneiske, T. Bretz, K. Mannheim, and D. H. Hartmann, *Astron. & Astrophys.* **413**, 807–815 (2004).
3. D. H. Hartmann, T. M. Kneiske, K. Mannheim, and K. Watanabe, "Gamma-Ray Bursts and Cosmic Radiation Backgrounds," in *AIP Conf. Proc. 662: Gamma-Ray Burst and Afterglow Astronomy 2001: A Workshop Celebrating the First Year of the HETE Mission*, 2003, pp. 442–445.
4. D. Band et al., *Ap. J.* **413**, 281–292 (1993).
5. J. P. Norris et al., *Ap. J.* **459**, 393–+ (1996).
6. J. R. Primack, J. S. Bullock, and R. S. Somerville, "Observational Gamma-ray Cosmology," in *AIP Conf. Proc. 745: High Energy Gamma-Ray Astronomy*, 2005, pp. 23–33.
7. G. Cusumano et al., *ArXiv Astrophysics e-prints* (2005), arXiv:astro-ph/0509737.
8. R. D. Preece et al., *Ap. J.* **581**, 1248–1255 (2002).
9. C. Kouveliotou et al., *American Astronomical Society Meeting Abstracts* **207**, –+ (2005).
10. L. Amati et al., *Astron. & Astrophys.* **390**, 81–89 (2002).
11. G. Ghirlanda, G. Ghisellini, and D. Lazzati, *Ap. J.* **616**, 331–338 (2004).

LIST OF ATTENDEES

Abe, Keiichi	Saitama University	abe@heal.phy.saitama-u.ac.jp
Adams, Vera	Vera Initiative for Peace	vera.adams@btopenworld.com
Akerlof, Carl	University of Michigan	akerlof@umich.edu
Alatalo, Katherine	University of California	Berkeley, kalatalo@berkeley.edu
Almohammad, Abdalla,	Irbid-Huson Univ – Jordan	ASEEL_00@YAHOO.COM
Angelini, Lorella	NASA GSFC	angelini@milkyway.gsfc.nasa.gov
Aoki, Kentaro	Subaru Telescope	kaoki@subaru.naoj.org
Atteia, Jean-Luc	Observatoire Midi-Pyrenees	atteia@ast.obs-mip.fr
Bagchi, Manjari	Presidency College	mnj2003@vsnl.com
Band, David	NASA GSFC UMBC	dband@lheapop.gsfc.nasa.gov
Bandstra, Mark	UC Berkeley / SSL	bandstra@ssl.berkeley.edu
Barbiellini, Guido	Univ of trieste and INFN	guido.barbiellini@ts.infn.it
Barbier, Louis,	NASA GSFC	louis.m.barbier@nasa.gov
Barnes, Sandy	NASA GSFC USRA	sbarnes@milkyway.gsfc.nasa.gov
Barthelmy, Scott	NASA GSFC	scott@milkyway.gsfc.nasa.gov
Beardmore, Andrew	University of Leicester	apb@star.le.ac.uk
Beckmann, Volker	NASA GSFC	beckmann@milkyway.gsfc.nasa.gov
Beloborodov, Andrei	Columbia University	amb@phys.columbia.edu
Berger, Edo	Carnegie Observatories	eberger@ociw.edu
Bernardini, Maria Grazia	ICRANet and Dipartimento di Fisica	maria.bernardini@icra.it
Bernas, Martin	Czech Technical University	bernas@fel.cvut.cz
Bersier, David	Liverpool John Moores Univ	dfb@astro.livjm.ac.uk
Bhat, Narayana	Univ of Alabama, Huntsville	bhatn@uah.edu
Bianco, Carlo Luciano	ICRANet and Dipartimento di Fisica	bianco@icra.it
Bloom, Joshua		jbloom@astro.berkeley.edu
Boer, Michel	OHP	Michel.Boer@oamp.fr
Bonnell, Jerry	NASA GSFC USRA	bonnell@grossc.gsfc.nasa.gov
Boyd, Patricia	NASA GSFC	padi@milkway.gsfc.nasa.gov
Braga, Joao	INPE	braga@das.inpe.br
Braito, Valentina	Johns Hopkins University	vale@milkyway.gsfc.nasa.gov
Brandt, Soren	Danish National Space Center	sb@spacecenter.dk
Briggs, Michael	UAH	michael.briggs@cspar.uah.edu
Bulik, Tomasz	CAMK	bulik@camk.edu.pl
Burrows, David	Penn State	burrows@astro.psu.edu
Butler, Nathaniel	UC Berkeley	nat@astro.berkeley.edu
Campana, Sergio	Osservatorio astronomico di Brera	campana@merate.mi.astro.it
Cannizzo, John	NASA GSFC UMBC	cannizzo@milkyway.gsfc.nasa.gov
Capalbi, Milvia	ASI Science Data Center	capalbi@asdc.asi.it
Casanova, Sabrina	Los Alamos National Laboratory	casanova@lanl.gov
Castro-Tirado, Alberto	IAA-CSIC	ajct@iaa.es
Cenko, Brad	Caltech	cenko@srl.caltech.edu
Chen, Hsiao-Wen	University of Chicago	hchen@oddjob.uchicago.edu
Chen, Wenxin	Columbia University	wxchen@phys.columbia.edu
Chincarini, Guido	INAF / OAB	guido@merate.mi.astro.it

Claret, Arnaud	CEA-SACLAY	aclaret@cea.fr
Cleave, Mary	NASA Headquarters	mary.cleave@nasa.gov
Cline, Thomas	NASA GSFC	Thomas.L.Cline@nasa.gov
Cominsky, Lynn	Sonoma State University	lynnc@charmian.sonoma.edu
Cordier, Bertrand	CEA-SACLAY	bcordier@cea.fr
Covino, Stefano	INAF / OAB	covino@merate.mi.astro.it
Crawford, Hank	UCB / SSL	hjcrawford@lbl.gov
Crider, Anthony	Elon University	acrider@elon.edu
Csatorday, Peter	MIT Kavli Institute	peterc@space.mit.edu
Cummings, Jay	NASA GSFC NRC	jayc@milkyway.gsfc.nasa.gov
Cusumano, Giancarlo	IASFPa-INAF	cusumano@pa.iasf.cnr.it
Daigne, Frederic	Institut d'Astrophysique de Paris	daigne@iap.fr
Dantzler, Andrew	NASA Headquarters	andrew.dantzler@nasa.gov
Della Valle, Massimo	INAF-Arcetri (Florence)	massimo@arcetri.astro.it
Dermer, Charles	Naval Research Laboratory	dermer@ssd5.narl.navy.mil
Desai, Upendra	Retired/Emeritus-GSFC	desai@stars.gsfc.nasa.gov
Di Pippo, Simonetta	Italian Space Agency	simonetta.dipippo@asi.it
Dingus, Brenda	Los Alamos National Laboratory	dingus@lanl.gov
Donaghy, Timothy	University of Chicago	quinn@oddjob.uchicago.edu
Dwek, Eli	NASA GSFC	eli.dwek@nasa.gov
Fahey, Dick	NASA GSFC	Dick.Fahey@NASA.gov
Falcone, Abe	Pennsylvania State University	afalcone@astro.psu.edu
Fenimore, Ed	Los Alamos National Laboratory	efenimore@lanl.gov
Fisher, Richard	NASA Headquarters	richard.r.fisher@nasa.gov
Fox, Derek	Penn State Astronomy	dfox@astro.psu.edu
Frail, Dale	NRAO	dfrail@nrao.edu
Frederic, Malacrino	LATT OMP	fmalacri@ast.obs-mip.fr
Friedman, Andrew	Harvard University	afriedman@cfa.harvard.edu
Fruchter, Andrew	Space Telescope Science Inst.	fruchter@stsci.edu
Galassi, Mark	Los Alamos National Laboratory	mark_grb2005@galassi.org
Garimella, Kiran	Clemson University	kgarime@clemson.edu
Geithner, Paul	NASA Headquarters	paul.geithner@nasa.gov
Gendre, Bruce	IASF-Rome/INAF	gendre@rm.iasf.cnr.it
Genet, Franck	Institut d'Astrophysique de Paris	genet@iap.fr
Giblin, Timothy	The College of Charleston	giblint@cofc.edu
Goad, Michael	University of Leicester	mrg@star.le.ac.uk
Godet, Olivier	University of Leicester	og19@star.le.ac.uk
Goldoni, Paolo	DSM / DAPMIA / SAP – APC	pgoldomi@cea.fr
Gonzalez, Maria-Magdalena	Instituto de Astronomia, UNAM	magda@astroscu.unam.mx
Gou, Lijun	Penn State University	lijun@astro.psu.edu
Graber, James	LC	jgrd@loc.gov
Graham, John	JHU & STScI	graham@stsci.edu
Granot, Jonathan	KIPAC, Stanford	granot@slac.stanford.edu
Graziani, Carlo	University of Chicago	carlo@oddjob.uchicago.edu
Greiner, Jochen	MPE Garching	jcg@mpe.mpg.de
Grindlay, Josh	Harvard University	josh@cfa.harvard.edu

Grupe, Dirk	Pennsylvania State University	grupe@astro.psu.edu
Guido, Di Cocco	INAF, IASF-Bo	dicocco@bo.iasf.cnr.it
Guidorzi, Cristiano	Liverpool John Moores Univ.	crg@atros.livjm.ac.uk
Gull, Theodore	NASA GSFC	
Gursky, Herbert	Naval Research Laboratory	hgursky@ssd5.nrl.navy.mil
Hakkila, Jon	College of Charleston	hakkilaj@cofc.edu
Harding, Alice	NASA GSFC	harding@twinkie.gsfc.nasa.gov
Hartman, Colleen	NASA Headquarters	colleen.hartman@nasa.gov
Hartman, Robert	NASA GSFC	Rober t.C.Har tm an@nas a.gov
Hartmann, Dieter	Clemson University	hdieter@clemson.edu
Hasan, Hashima	NASA Headquarters	
Hellings, Ron	NASA Headquarters	hellings@hq.nasa.gov
Hill, Joanne	NASA GSFC	jhill@milkyway.gsfc.nasa.gov
Holland, Stephen	NASA GSFC USRA	sholland@milkyway.gsfc.nasa.gov
Holt, Steve	Olin College	steve.holt@olin.edu
Homewood, Autumn	Clemson University	ahomewo@clemson.edu
Howard, Richard	NASA Headquarters	richard.j.howard@nasa.gov
Hudec, Rene	Astronomical Institute	rhudec@asu.cas.cz
Immler, Stefan	NASA USRA GSFC	immler@milkyway.gsfc.nasa.gov
Ioka, Kunihito	Kyoto University	ioka@tap.scphys.kyoto-u.ac.jp
Iping, Rosina	Catholic University of America	rosina@taotaomona.gsfc.nasa.gov
Jacques, Paul	APC-CEA	jpaul@cea.fr
Jahoda, Keith	NASA GSFC	keith.m.jahoda@nasa.gov
Jakobsson, Pall	Dark Cosmology Centre	pallja@astro.ku.dk
Jones, Frank	NASA GSFC	frank.c.jones@gsfc.nasa.gov
Kallman, Tim	NASA GSFC	timothy.r.kallman@nasa.gov
Kaluzienski, Louis	NASA GSFC	louis.j.kaluzienski@nasa.gov
Kaneko, Yuki	NASA GSFC USRA	yuki.kaneko@nsstc.nasa.gov
Kawai, Nobuyuki	Tokyo Tech	nkawai@phys.titech.ac.jp
Kaye, Jack	NASA Headquarters	jack.a.kaye@nasa.gov
Kaye, Tom		tom@tomkaye.com
Kazanas, Demosthenes	NASA GSFC	Demos.Kazanas-1@nasa.gov
Kelley, Richard	NASA GSFC	kelley@milkyway.gsfc.nasa.gov
Kennea, Jamie	Penn State University	kms52@psu.edu
Klose, Sylvio	Landessternwarte Tautenburg	klose@tls-tautenburg.de
Klotz, Alain	O.H.P.	alain.klotz@free.fr
Kniffen, Donald	NASA GSFC USRA	dak@milkyway.gsfc.nasa.gov
Kocevski, Daniel	University of California	kocevski@berkeley.edu
Kouveliotou, Chryssa	NASA MSFC	chryssa.kouveliotu-2@n
Krimm, Hans	NASA GSFC USRA	krimm@milkyway.gsfc.nasa.gov
Kuehn, Kyler	University of California	kuehn@HEP.ps.uci.edu
Kupcu Yoldas, Aybuke	MPE Garching	ayoldas@mpe.mpg.de
Kurfess, James	Naval Research Laboratory	kurfess@nrl.navy.mil
Kwak, Kyujin	Stony Brook University	kkwak@grad.physics.sunysb.edu
LaCluyze, Aaron	UNC - Chapel Hill	lacluyze@physics.unc.edu
Laming, Martin	Naval Research Laboratory	laming@nrl.navy.mil

LaPiana, Lia	NASA Headquarters	lia.s.lapiana@nasa.gov
Lazzati, Davide	JILA - University of Colorado	lazzati@colorado.edu
Le, Truong	NRL / NRC	truong.le@nrl.navy.mil
Le Floc'h, Emeric	University of Arizona	elefloch@as.arizona.edu
Li, Chao	Caltech Tapir	lichao@caltech.edu
Liang, Edison	Rice University	liang@spacibm.rice.edu
Liang, Enwei	University of Nevada, Las Vegas	lew@physics.unlv.edu
Loeb, Abraham	Harvard University	aloeb@cfa.harvard.edu
Longo, Francesco	UofTrieste INFN sezione di Trieste	francesco.longo@ts.infn.it
Lyutikov, Maxim	UBC	lyutikov@physics.ubc.ca
MacFadyen, Andrew	Institute for Advanced Study	aim@ias.edu
Maran, Steven	Press	
Marcum, Pam	NASA Headquarters	
Marka, Szabolcs	Columbia University	smarka@phys.columbia.edu
Markwardt, Craig	NASA GSFC	craigm@milkyway.gsfc.nasa.gov
Marshall, Francis	NASA GSFC	frank.marshall@gsfc.nasa.gov
Mason, Keith	PPARC	keith.mason@pparc.ac.uk
McBreen, Sheila	ESA / ESTEC	smcbreen@rssd.esa.int
McConnell, Mark	University of New Hampshire	mark.mcconnell@unh.edu
McEnery, Jule	NASA GSFC	mcenery@milkyway.gsfc.nasa.gov
McGlynn, Sinead	University College Dublin	smcglynn@bermuda.ucd.ie
McGrath, Melissa	NASA Headquarters	melissa.a.mcgrath@nasa.gov
McMahon, Erin	Univ. of Texas, Austin / GRBlog	emcmahon@astro.as.utexas.edu
Medvedev, Mikhail	University of Kansas	medvedev@ku.edu
Meegan, Charles	NASA MSFC	charles.a.meegan@nasa.gov
Meszaros, Peter	Pennsylvania State University	pmeszaros@astro.psu.edu
Metzger, Brian	University of California, Berkeley	bmetzger@berkeley.edu
Mitchell, John	NASA GSFC	john.w.mitchell@nasa.gov
Mizuno, Yosuke	NASA MSFC NRC	Yosuke.Mizuno@msfc.nasa.gov
Mochkovitch, Robert	Institut d'Astrophysique de Paris	mochko@iap.fr
Modjaz, Maryam	CfA	mmodjaz@cfa.harvard.edu
Moore, Michael	NASA Headquarters	michael.r.moore@nasa.gov
Moretti, Alberto	INAF / OABrera	moretti@merate.mi.astro.it
Morris, David	Penn State University	kms52@psu.edu
Moseley, Harvey	NASA GSFC	samuel.h.moseley@nasa.gov
Mushotzky, Richard	NASA GSFC	
Nakagawa, Yujin	Aoyama Gakuin University	yujin@phys.aoyama.ac.jp
Nakamura, Takashi	Kyoto University	takashi@tap.scphys.kyoto-u.ac.jp
Neff, Susan	NASA GSFC EUD	susan.g.neff@nasa.gov
Nicastro, Fabrizio	SAO / UNAM	fnicastro@cfa.harvard.edu
Nishikawa, Ken	UAH	ken-ichi.nishikawa@nsstc.nasa.gov
Noguchi, Koichi	Rice University	noguchi2@llnl.gov
Norris, Jay	NASA GSFC	jnorris@lheapop.gsfc.nasa.gov
Nousek, John	Penn State University	kms52@psu.edu
Noyes, David	University of Maryland	dnoyes@umdgrb.umd.edu
Nysewander, Melissa	UNC - Chapel Hill	mnysewan@astro.unc.edu

O'Brien, Paul	University of Leicester	pto@star.le.ac.uk
Oegerle, William	NASA GSFC	William.R.Oegerle@nasa.gov
Omodei, Nicola	INFN Pisa	nicola.omodei@pi.infn.it
Onda, Kaori	Saitama University	onda@heal.phy.saitama-u.ac.jp
Osborne, Julian	University of Leicester	julo@star.le.ac.uk
Pacciani, Luigi	IASF-INAF Rome, Italy	luigi.pacciani@rm.iasf.cnr.it
Paciesas, William	UAH	william.paciesas@msfc.nasa.gov
Pagani, Claudio	Penn State University	kms52@psu.edu
Page, Kim	University of Leicester	kpa@star.le.ac.uk
Palazzi, Eliana	INAF-IASF di Bologna	palazzi@bo.iasf.cnr.it
Palmer, David	Los Alamos National Laboratory	palmer@lanl.gov
Panaitescu, Alin	Los Alamos National Laboratory	adp@astro.as.utexas.edu
Parsons, Ann	NASA GSFC	parsons@milkyway.gsfc.nasa.gov
Patel, Sandy	NASA GSFC	sandeep.k.patel@nasa.gov
Pe'er, Asaf	University of Amsterdam	apeer@science.uva.nl
Pelangeon, Alexandre	LaB d'Astrophy. de Toulouse	apelange@ast.obs-mip.fr
Perley, Daniel	University of California, Berkeley	dperley@astro.berkeley.edu
Perri, Matteo	ASI Science Data Center	perri@asdc.asi.it
Petro, Larry	STScI	petro@stsci.edu
Petry, Dirk	UMBC	petry@milkyway.gsfc.nasa.gov
Piran, Tsvi	Hebrew University	tsvi@phys.huji.ac.il
Piro, Luigi	IASF/INAF-ROMA	PIRO@RM.IASF.CNR.IT
Pizzichini, Graziella	INAF/IASF Bologna	pizzichini@bo.iasf.cnr.it
Polcar, Jiri		polcar@physics.muni.cz
Prochaska, Xavier	UC Santa Cruz	xavier@ucolick.org
Proga, Daniel	Princeton University	dproga@astro.princeton.edu
Racusin, Judy	Penn State University	kms52@psu.edu
Reeves, James	NASA GSFC JHU	jnr@milkyway.gsfc.nasa.gov
Retter, Alon	Penn State University	retter@astro.psu.edu
Reyes, Luis	UMD GSFC	lreyes@milkyway.gsfc.nasa.gov
Ricker, George	MIT Kavli Institute	grr@space.mit.edu
Ritz, Steven	NASA GSFC	steven.m.ritz@nasa.gov
Rol, Evert	University Of Leicester	er45@star.le.ac.uk
Romano, Patrizia	INAF / OAB	romano@merate.mi.astro.it
Roming, Pete	Penn State University	roming@astro.psu.edu
Rosenbaum, Doris		drteplitz@aol.com
Rossi, Elena Maria	JILA - University of Colorado	emr@mpa-garching.mpg.de
Ruffini, Remo	ICRANet and Dipartimento di Fisica	ruffini@icra.it
Ryde, Felix	Stockholm University	felix@astro.su.se
Rykoff, Eli	University of Michigan	erykoff@umich.edu
Sakamoto, Takanori	NASA GSFC	takanori@milkyway.gsfc.nasa.gov
Salgado, Carlos	Norfolk State University	salgado@jlab.org
Sarira, Sarira	Instituto de Ciencias Nucleares	sarira@nucleares.unam.mx
Sato, Goro	ISAS JAXA	gsato@astro.isas.jaxa.jp
Savaglio, Sandra	Johns Hopkins University	savaglio@pha.jhu.edu
Saz Parkinson, Pablo	UC Santa Cruz	pablo@scipp.ucsc.edu

Schady, Patricia	MSSL/PSU	ps@mssl.ucl.ac.uk
Schanne, Stephane	CEA-SACLAY	schanne@hep.saclay.cea.fr
Schwamb, Megan	University of Pennsylvania	mschwamb@astro.upenn.edu
Serlemitsos, Peter	NASA GSFC	pjs@astron.gsfc.nasa.gov
Shen, Rongfeng	University of Texas at Austin	rfshen@astro.as.utexas.edu
Shrader, Chris	NASA GSFC	Chris.R.Shrader@gsfc.nasa.gov
Silverberg, Robert	NASA GSFC	Robert.Silverberg@nasa.gov
Smale, Alan	NASA Headquarters	alan.smale@nasa.gov
Smith, Eric	NASA Headquarters	
Soderberg, Alicia	Caltech	ams@astro.caltech.edu
Sonneborn, George	NASA GSFC	George.Sonneborn-1@nasa.gov
Stamatikos, Michael	University of Wisconsin-Madison	michael.stamatikos@icecube.wisc.edu
Stecker, Floyd	NASA GSFC	Floyd.W.Stecker@nasa.gov
Stratta, Giulia	OMP/LAT	gstratta@ast.obs-mip.fr
Streitmatter, Robert	NASA GSFC	Robert.E.Streitmatter@nasa.gov
Sugita, Satoshi	Aoyama Gakuin University	sugita@phys.aoyama.ac.jp
Sugiyama, Shinya	Rice University	shinya@rice.edu
Swank, Jean	NASA GSFC	swank@milkyway.gsfc.nasa.gov
Tagliaferri, Gianpiero	INAF / OAB	tagliaferri@merate.mi.astro.it
Tanvir, Nial	University of Hertfordshire	nrt@star.herts.ac.uk
Teplitz, Vic	NASA GSFC	Vigdor.L.Teplitz@nasa.gov
Toma, Kenji	Kyoto University	toma@tap.scphys.kyoto-u.ac.jp
Trasco, John	Universtiy of Maryland	jtrasco@astro.umd.edu
Trimble, Virginia	Univ of CA, Irvine & Las Cumbres Obs	vtrimble@astro.umd.edu
Uhm, Zuhngwhi	Columbia University	zu2001@columbia.edu
Vanderspek, Roland	MIT	roland@space.mit.edu
Vestrand, Tom	LANL	vestrand@lanl.gov
Vetere, Loredana	IASF-INAF (Rome) / Univ. "La Sapienza"	loredana.vetere@rm.iasf.cnr.it
Villasenor, Jesus Noel	MIT KIASR	jsvilla@space.mit.edu
von Kienlin, Andreas	Max-Planck-Institut fuer extraterrestrische Physik	azk@mpe.mpg.de
Wang, XiangYu	Nanjing University	xywang@nju.edu.cn
Waxman, Eli		waxman@wicc.weizmann.ac.il
Weiler, Kurt	Naval Research Laboratory	Kurt.Weiler@nrl.navy.mil
Williams, Christopher	CPI / Naval Research Lab	cwilliams@cpi.com
Wilson, Robert	Retired Industrial Chemist (Kodak)	wilson99@ix.netcom.com
Wilson, Robert	NASA MSFC	robert.b.wilson@nasa.gov
Woosley, Stan	UCSC	woosley@ucolick.org
Wozniak, Przemek	Los Alamos National Laboratory	wozniak@lanl.gov
Yamaoka, Kazutaka	Aoyama Gakuin University	yamaoka@phys.aoyama.ac.jp
Yamazaki, Ryo	Hiroshima University	ryo@theo.phys.sci.hiroshima-u.ac.jp
Yoshida, Atsumasa	Aoyama Gakuin University	ayoshida@phys.aoyama.ac.jp
Yost, Sarah	University of Michigan	sayost@umich.edu
Zhang, Bing	Univ. Las Vegas	bzhang@physics.unlv.edu
Zhang, William	NASA GSFC	William.W.Zhang@nasa.gov

Author Index

A

Abbey, A. F., 289, 664, 708
Abe, K., 201
Aharonian, F., 349
Akerlof, C. W., 349
Alatalo, K., 277
Amado, P., 668
Anderson, M. I., 357
Angelini, L., 289, 297, 317, 408, 564, 664
Antonelli, L. A., 54, 408, 467
Aptekar, R., 43
Argan, A., 648
Armus, L., 528
Ashley, M. C. B., 349
Atteia, J.-L., 43, 145, 149, 173, 177, 271, 333, 448

B

Bagoly, Z., 125
Balázs, L. G., 125
Banat, P. L., 570
Band, D. L., 133, 599, 704
Bandstra, M. E., 109
Barbiellini, G., 91, 648, 700
Barbier, L., 43, 578
Barthelmy, S., 43, 141, 145, 289, 317, 321, 349, 564, 578
Basset, M., 648
Basta, M., 440
Beardmore, A. P., 85, 244, 281, 289, 297, 321, 408, 564, 708
Begelman, M. C., 513
Belczyński, K., 68
Bellm, E., 109
Beloborodov, A. M., 189, 193
Berger, E., 33
Bernardini, M. G., 103
Bersier, D., 420
Bianco, C. L., 103
Bihain, G., 79
Biryukov, V. V., 289
Bloom, J. S., 277, 473, 534
Bloser, P. F., 654, 696

Blustin, A. J., 436
Bo, M., 271, 558
Boer, M., 305, 333
Boffelli, F., 648
Boggs, S. E., 109
Boyd, P. T., 564
Bremer, M., 79
Briggs, M. S., 133
Bromm, V., 503
Brown, P. J., 432
Buchholtz, G., 305
Bulgarelli, A., 648
Bulik, T., 68
Burrows, D. N., 85, 141, 215, 289, 297, 317, 321, 386, 408, 432, 564, 570, 574, 664
Butler, N., 259, 277

C

Caballero, J. A., 79
Campana, S., 54, 285, 289, 297, 317, 321, 386, 408, 467, 564, 570, 574, 664, 676
Capalbi, M., 285, 289, 297, 408, 570, 574, 664, 676
Caraveo, P., 648
Cardenas, C., 668
Carrillo-Barragán, M., 612
Casanova, S., 113
Case, G. L., 696
Castillo-Carrión, S., 668
Castro Cerón, J. M., 552, 582, 668
Castro-Tirado, A. J., 79, 668, 688
Cattaneo, P., 648
Celotti, A., 91
Chardonnet, P., 103
Charmandaris, V., 528
Chen, A., 648
Chen, H.-W., 534
Chen, W., 193
Cherry, M. L., 696
Chincarini, G., 54, 215, 285, 289, 297, 317, 321, 386, 408, 467, 564, 570, 574, 664, 676
Citterio, O., 664

Claret, A., 668
Cline, T., 43
Cohen-Tanugi, J., 692
Cominsky, L. R., 564
Conciatore, M. L., 54
Conconi, P., 668
Corsi, A., 428
Costa, E., 341, 648, 700
Covino, S., 54, 285, 408, 467, 570, 668, 676
Cravens, J., 696
Crider, A., 64
Cucchiara, A., 467
Cummings, J., 43, 321, 432, 578
Cunnife, R., 668
Cusumano, G., 285, 289, 321, 386, 408, 564, 570, 574, 664, 676

D

Dai, Z. G., 72
Daigne, F., 209, 313, 353, 452, 546
Damerdji, Y., 271, 305
D'Avanzo, P., 54, 467
de Caprio, V., 668
de J. Mateo Sanguino, T., 668
de Jong, J., 79
Della Valle, M., 54, 367, 467
Del Monte, E., 648, 700
De Luca, A., 285, 408, 467
De Pasquale, M., 428
Dermer, C. D., 97
de Ugarte Postigo, A., 79, 668, 688
Devost, D., 528
Díaz-Verdejo, J., 668
Di Cocco, G., 648
Dingus, B. L., 113, 612
Di Persio, G., 648
Donaghy, T. Q., 117, 145, 253
Donnarumma, I., 648, 700

E

Endo, Y., 201
Enoto, T., 201
Evangelista, Y., 700
Evans, S., 345
Eysseric, J., 305

F

Falcone, A. D., 317, 386, 570
Fan, Y. Z., 72
Fenimore, E. E., 43, 173, 177, 564, 578
Feroci, M., 648, 700
Ferrero, P., 79
Filgas, R., 440
Filippenko, P., 277
Filliatre, P., 467
Fiore, F., 54
Fiorini, M., 648
Firmani, C., 513
Fishman, G. J., 265
Foggetta, L., 648
Foley, R., 277
Foley, S., 165, 467, 582
Forrest, B., 528
Fraschetti, F., 103
French, J., 165
Froysland, T., 648
Frutti, M., 648, 700
Fugazza, D., 54, 467
Fukazawa, Y., 201
Fuller, S., 293
Fuschino, F., 648
Fynbo, J. P. U., 552, 582

G

Galassi, M., 173, 177
Galli, A., 424
Galli, M., 648
Garimella, K. V., 293
Garnavich, P. M., 420
Gebet, F., 353
Gehrels, N., 14, 43, 85, 141, 145, 289, 297, 317, 321, 349, 386, 408, 467, 564, 578
Gendre, B., 271, 428, 558
Genet, F., 313, 452
Ghirlanda, G., 513
Ghisellini, G., 513
Gianotti, F., 648
Giblin, T. W., 121
Giommi, P., 215, 285, 289, 321, 386, 408, 564, 570, 574, 664, 676
Giuliani, A., 648
Glazebrook, K., 540

Goad, M. R., 85, 244, 281, 321, 386, 408, 564, 664, 708
Godet, O., 85, 244, 281, 321, 386, 408, 574, 664, 708
Göğüs, E., 349
Goldoni, ., 467
Golenetskii, S., 43, 141, 145
González, M. M., 612
Gorosabel, J., 79, 552, 668
Götz, D., 467
Graham, J., 321
Graziani, C., 117
Greiner, J., 414, 522
Grindlay, J. E., 631
Gronwall, C., 564
Grupe, D., 317, 432
Guetta, D., 54, 58
Guida, R., 103
Güver, T., 349
Guzik, T. G., 696
Guziy, S., 79, 668

H

Hajdas, W., 129
Hakkila, J., 121
Hanlon, L., 161, 165, 467
Hardee, P., 265
Hartmann, D. H., 293, 464
Hartner, G. D., 664
Hededal, C., 265
Heger, A., 398
Henden, A. A., 357
Henson, G., 293
Hill, J. E., 289, 297, 317, 386, 408, 664
Hjorth, J., 552, 582
Holland, S. T., 436
Homewood, A. L., 293
Hong, S., 201
Horns, D., 349
Horváth, I., 125
Hroch, F., 460
Hudcová, V., 460
Hudec, L., 680
Hudec, R., 129, 329, 361, 440, 460, 672, 680, 684, 688, 712
Hullinger, D., 43, 145, 578
Hurely, K., 696
Hurkett, C., 141

Hurley, K., 25, 54, 173, 564

I

Ibrahimov, M. A., 289
Inneman, A., 712
Ioka, K., 205, 289, 301, 337
Ishikawa, N., 177
Israel, G. L., 54, 467

J

Jakobsson, P., 552, 582
Jelínek, M., 79, 357, 668, 684, 688
Jensen, B. L., 552

K

Kaneko, Y., 133, 321, 612
Kann, D. A., 464
Kawai, N., 43, 145, 173, 177
Kazanas, D., 137
Kenji, T., 301
Kennea, J. A., 215, 289, 297, 317, 321, 386, 408, 564, 574, 664
Kippen, R. M., 696
Kiziloğlu, Ü., 349
Klose, S., 464
Klotz, A., 271, 305, 333
Kobayashi, S., 289, 301, 317, 386
Kocevski, D., 277
Kokubun, M., 201
Kouveliotou, C., 265
Krimm, H. A., 43, 141, 145, 349, 578
Křížek, M., 672
Kubánek, P., 668, 684, 688
Kwak, K., 153

L

Labanti, C., 648
Lamb, D. Q., 43, 117, 145, 149
La Parola, V., 285, 408, 564, 570, 574, 664, 676
Lapshov, I., 648, 700
Lazzarotto, F., 648, 700

Lazzati, D., 54, 386, 408, 513
Le Borgne, D., 540
Le Floc'h, E., 528
Legere, J., 654
Levan, A., 552
Li, W., 277
Liang, E., 157, 197, 444, 456
Liello, F., 648
Lipari, P., 648
Loeb, A., 503
Longo, F., 91, 648, 692
Lyutikov, M., 483

M

MacFadyen, A. I., 48
Macri, J. R., 654, 696
Maetou, M., 173
Maiorano, E., 357
Makishima, K., 201
Malacrino, F., 448
Malesani, D., 54, 285, 467, 570
Mangano, V., 285, 321, 408, 564, 570, 574, 664, 676
Manning, A., 293
Marisaldi, M., 648
Márka, S., 605
Markwardt, C., 43, 289, 321, 578
Marshall, F. E., 309, 564
Martín, S., 668
Masetti, N., 357, 460
Mason, K. O., 224, 564
Massaro, E., 341
Mastichiadis, A., 137
Mastropietro, M., 648, 700
Matheson, T., 420
Matone, L., 605
Matsuoka, M., 173, 177
Mattaini, E., 648
Mauri, F., 648
Mazets, E., 43
Mazzali, P., 420
McBreen, B., 161, 165, 467
McBreen, S., 79, 161, 165, 467, 582
McConnell, M. L., 654, 696
McEnery, J., 660
McGlynn, S., 165, 467
McIntyre, T., 293
McKay, T. A., 349

Medvedev, M. V., 169
Meegan, C. A., 133
Melady, G., 165
Mereghetti, S., 467, 648
Mészáros, A., 125, 129
Mészáros, P., 181, 234, 289, 386, 432, 564
Miller, R. S., 696
Mineo, T., 285, 408, 564, 574, 664, 676
Mirabal, N., 349
Mirabel, F., 528
Misra, K., 79
Miyawaki, R., 201
Mizuno, Y., 265
Mochkovitch, R., 209, 313, 353, 452, 546
Moran, L., 467
Morelli, E., 648, 700
Moretti, A., 285, 289, 297, 317, 408, 570, 574, 664, 676
Morris, D. C., 289, 297, 317, 386, 408, 664
Morselli, A., 648
Morsony, B., 513

N

Nakagawa, Y. E., 173, 177
Nakamura, T., 205, 301, 337
Nakazawa, K., 201
Narita, T., 654
Natarajan, P., 552
Nava, L., 513
Ne, D., 460
Nekola, M., 684
Nishikawa, K.-I., 265
Noguchi, K., 157, 197, 456
Nolan, C. M., 121
Norris, J. P., 141
Nousek, J. A., 215, 289, 297, 317, 386, 408, 432, 467, 564, 664
Noyes, D., 620

O

O'Brien, P. T., 85, 215, 281, 289, 321, 386, 408
Ohno, M., 201

Omodei, N., 91, 642, 692, 716
Onda, M, K., 201
Osborne, J. P., 85, 141, 215, 244, 281, 289, 297, 321, 386, 408, 564, 664, 708
Özell, M., 349

P

Pacciani, L., 648, 700
Paciesas, W. S., 133, 696
Pagani, C., 289, 297, 317, 386, 408, 570, 664
Page, K., 244, 317, 386, 574, 664
Page, K. L., 85, 281, 321, 408
Palazzi, E., 357, 460
Palmer, D., 43, 564, 578
Pal'shin, V., 43, 141, 145
Panaitescu, A., 564
Pandey, S. B., 79, 357, 668
Parsons, A., 43, 141, 578
Pedersen, K., 552, 582
Pe'er, A., 181
Pélangeon, A., 149
Pellizzoni, A., 642
Perley, D., 277
Perna, R., 467
Perotti, F., 648
Perri, M., 141, 285, 289, 408, 570, 574, 664, 676
Peters, C. D., 121
Peterson, B. A., 357
Petry, D., 325
Phillips, A., 349
Pian, E., 357
Picozza, P., 648
Pína, L., 712
Piran, T., 58
Piro, L., 424, 428
Piron, F., 692
Pittori, C., 648
Pizzichini, G., 329, 361, 460
Polcar, J., 460
Pollas, C., 305
Pontoni, C., 648
Pooley, D., 277
Porrovecchio, G., 648
Pozanenko, A. S., 289
Preece, R. D., 133, 165, 612
Prest, M., 648

Price, P. A., 357
Priddey, R., 552
Prochaska, J. X., 473, 534

Q

Quimby, R. M., 349

R

Racusin, J. L., 297, 317, 408, 574
Ramirez-Ruiz, E., 48, 493
Rapisarda, M., 648, 700
Rau, A., 414, 522
Rees, M. J., 181
Reeves, J. N., 285, 582
Reichart, D. E., 321
Retter, A., 432
Richaud, Y., 305
Ricker, G. R., 43, 145, 149, 173, 177
Riddle, C., 293, 464
Řípa, J., 129
Ritz, S., 660
Riva, A., 668
Rol, E., 321, 357
Romano, P., 285, 317, 386, 408, 570, 574, 664, 676
Roming, P. W. A., 224, 564
Rossi, E., 546, 648
Rowell, G., 349
Rubini, A., 648, 700
Rudak, B., 68
Ruffini, R., 103
Rujopakarn, W., 349
Rumyantsev, V. V., 289
Ryan, J. M., 654, 696
Ryde, F., 125
Rykoff, E. S., 349

S

Sabau-Graziati, M. D., 668
Sahu, D. K., 79
Sakamoto, T., 43, 145, 173, 289, 317, 321, 564, 578
Sánchez-Fernández, C., 668
Sato, G., 43, 145, 201, 289, 578

Savaglio, S., 357, 540
Saz Parkinson, P. M., 624
Scargle, J. D., 692
Schady, P., 289
Schaefer, B. E., 349, 696
Schwarz, R., 414
Sharapov, D. A., 289
Shirasaki, Y., 173, 177
Shrader, C. R., 185
Simon, V., 329, 361, 440
Smith, D. A., 349
Soderberg, A. M., 380
Soffitta, P., 341, 648, 700
Sollerman, J., 552, 582
Sonnett, S. M., 121
Stacy, J. G., 696
Stamatikos, M., 599
Stanek, K. Z., 420
Stella, L., 54, 467
Stratta, G., 271, 333
Strobl, J., 684
Sugita, S., 201
Sugiyama, S., 157, 197
Suzuki, M., 173, 177
Svéda, L., 712
Swan, H. F., 349
Swesty, F. D., 153

T

Tagliaferri, G., 54, 215, 285, 289, 297, 317, 321, 386, 408, 467, 564, 570, 574, 664, 676
Takabe, H., 197
Takahashi, T., 201
Tamagawa, T., 173, 177
Tamburelli, F., 664
Tanaka, K., 173, 177
Tanvir, N. R., 321, 552
Tashiro, M., 201
Tavani, M., 91, 648, 700
Terada, Y., 201
Tiengo, A., 285
Toma, K., 205, 337
Topinka, M., 440, 460
Traci, A., 648
Trifoglio, M., 648
Trigo-Rodríguez, J. M., 668
Trimble, V., 3

Trois, A., 648
Tueller, J., 43, 578

U

Uhm, Z., 189

V

Vallazza, E., 648
Vanderspek, R., 177
Vercellone, S., 648
Vestrand, W. T., 345, 349, 696
Vetere, L., 341
Vietri, M., 54
Vitali, F., 668
Vítek, S., 668, 688
von Kienlin, A., 165
Vreeswijk, P. M., 357

W

Wang, X. Y., 72
Watson, D., 552, 582
Waxman, E., 589
Wefel, J. P., 696
Wells, A., 289, 297, 564, 664
Wells, A. A., 321, 408, 708
Wheeler, J. C., 349
White, N. E., 564
White, R., 345
Willingale, R., 85
Winston, E., 161
Woosley, S. E., 398
Wozniak, P. R., 345
Wren, J., 345, 349
Wu, X. F., 72

X

Xue, S.-S., 103

Y

Yamaoka, K., 201
Yamazaki, R., 205, 301, 337
Yoldaş, A. K., 522
Yoshida, A., 173, 177
Yost, S. A., 349
Yuan, F., 349

Z

Zanello, D., 648
Zeh, A., 464
Zerbi, F. M., 668
Zhang, B., 72, 141, 289, 301, 321, 386, 392, 432, 444, 564
Zhang, W., 48
Zitouni, H., 209